Geotechnical, Geological and Earthquake Engineering

Volume 52

Series Editor

Atilla Ansal, School of Engineering, Özyegin University, Istanbul, Turkey

Editorial Board Members

Julian Bommer, Imperial College, London, UK

Jonathan D. Bray, University of California, Berkeley, Walnut Creek, USA

Kyriazis Pitilakis, Aristotle University of Thessaloniki, Thessaloniki, Greece

Susumu Yasuda, Tokyo Denki University, Hatoyama, Japan

The book series entitled *Geotechnical, Geological and Earthquake Engineering* has been initiated to provide carefully selected and reviewed information from the most recent findings and observations in these engineering fields. Researchers as well as practitioners in these interdisciplinary fields will find valuable information in these book volumes, contributing to advancing the state-of-the-art and state-of-the-practice. This book series comprises monographs, edited volumes, handbooks as well as occasionally symposia and workshop proceedings volumes on the broad topics of geotechnical, geological and earthquake engineering. The topics covered are theoretical and applied soil mechanics, foundation engineering, geotechnical earthquake engineering, earthquake engineering, rock mechanics, engineering geology, engineering seismology, earthquake hazard, etc.

Prospective authors and/or editors should consult the **Series Editor Atilla Ansal** for more details. Any comments or suggestions for future volumes are welcomed.

EDITORS AND EDITORIAL BOARD MEMBERS IN THE GEOTECHNICAL, GEOLOGICAL AND EARTHQUAKE ENGINEERING BOOK SERIES:

Series Editor:

Atilla Ansal
Ozyegin University
School of Engineering
Alemdag Çekmeköy 34794, Istanbul, Turkey
Email: atilla.ansal@ozyegin.edu.tr

Advisory Board

Julian J. Bommer
Imperial College
Department of Civil & Environmental Engineering
Imperial College Road
London SW7 2AZ, United Kingdom
Email: j.bommer@imperial.ac.uk

Jonathan D. Bray
University of California, Berkeley
Department of Civil and Environmental Engineering
453 David Hall, Berkeley, CA 94720, USA
Email: jonbray@berkeley.edu

Kyriazis Pitilakis
Aristotle University Thessaloniki
Department of Civil Engineering
Laboratory of Soil Mechanics and Foundations
University Campus
54124 Thessaloniki, Greece
Email: pitilakis@civil.auth.gr

Susumu Yasuda
Tokyo Denki University
Department of Civil and Environmental Engineering
Hatoyama-Cho
Hiki-gun
Saitama 350-0394, Tokyo, Japan
Email: yasuda@g.dendai.ac.jp

More information about this series at https://link.springer.com/bookseries/6011

Lanmin Wang · Jian-Min Zhang ·
Rui Wang
Editors

Proceedings of the 4th International Conference on Performance Based Design in Earthquake Geotechnical Engineering (Beijing 2022)

Set 2

Springer

Editors
Lanmin Wang
Lanzhou Institute of Seismology
China Earthquake Administration
Lanzhou, China

Jian-Min Zhang
Department of Hydraulic Engineering
Tsinghua University
Beijing, China

Rui Wang
Department of Hydraulic Engineering
Tsinghua University
Beijing, China

ISSN 1573-6059   ISSN 1872-4671 (electronic)
Geotechnical, Geological and Earthquake Engineering
ISBN 978-3-031-11897-5   ISBN 978-3-031-11898-2 (eBook)
https://doi.org/10.1007/978-3-031-11898-2

© The Editor(s) (if applicable) and The Author(s), under exclusive license to Springer Nature Switzerland AG 2022
This work is subject to copyright. All rights are solely and exclusively licensed by the Publisher, whether the whole or part of the material is concerned, specifically the rights of translation, reprinting, reuse of illustrations, recitation, broadcasting, reproduction on microfilms or in any other physical way, and transmission or information storage and retrieval, electronic adaptation, computer software, or by similar or dissimilar methodology now known or hereafter developed.
The use of general descriptive names, registered names, trademarks, service marks, etc. in this publication does not imply, even in the absence of a specific statement, that such names are exempt from the relevant protective laws and regulations and therefore free for general use.
The publisher, the authors, and the editors are safe to assume that the advice and information in this book are believed to be true and accurate at the date of publication. Neither the publisher nor the authors or the editors give a warranty, expressed or implied, with respect to the material contained herein or for any errors or omissions that may have been made. The publisher remains neutral with regard to jurisdictional claims in published maps and institutional affiliations.

This Springer imprint is published by the registered company Springer Nature Switzerland AG
The registered company address is: Gewerbestrasse 11, 6330 Cham, Switzerland

Preface

The 4th International Conference on Performance-based Design in Earthquake Geotechnical Engineering (PBD-IV) will be held on July 15–17, 2022, in Beijing, China. The PBD-IV conference is organized under the auspices of the International Society of Soil Mechanics and Geotechnical Engineering—Technical Committee on Earthquake Geotechnical Engineering and Associated Problems (ISSMGE-TC203). The PBD-I, PBD-II, and PBD-III events in Japan (2009), Italy (2012), and Canada (2017), respectively, were highly successful events for the international earthquake geotechnical engineering community. The PBD events have been excellent companions to the International Conference on Earthquake Geotechnical Engineering (ICEGE) series that TC203 has held in Japan (1995), Portugal (1999), USA (2004), Greece (2007), Chile (2011), New Zealand (2015), and Italy (2019). The goal of PBD-IV is to provide an open forum for delegates to interact with their international colleagues and advance performance-based design research and practices for earthquake geotechnical engineering.

The proceedings of PBD-IV is the outcome of more than two years of concerted efforts by the conference organizing committee, scientific committee, and steering committee. The proceedings include 14 keynote lecture papers, 25 invited theme lecture papers, one TC203 Young Researcher Award lecture paper, and 187 accepted technical papers from 27 countries and regions. Each accepted paper in the conference proceedings was subject to review by credited peers. The final accepted technical papers are organized into five themes and six special sessions: (1) Performance Design and Seismic Hazard Assessment; (2) Ground Motions and Site Effects; (3) Foundations and Soil–Structure Interaction; (4) Slope Stability and Reinforcement; (5) Liquefaction and Testing; (6) S1: Liquefaction experiment and analysis projects (LEAP); (7) S2: Liquefaction Database; (8) S3: Embankment Dams; (9) S4: Earthquake Disaster Risk of Special Soil Sites and Engineering Seismic Design; (10) S5: Special Session on Soil Dynamic Properties at Micro-scale: From Small Strain Wave Propagation to Large Strain Liquefaction; and (11) S6: Underground Structures.

The Chinese Institution of Soil Mechanics and Geotechnical Engineering - China Civil Engineering Society and China Earthquake Administration provided tremendous support to the organization of the conference. The publication of the proceedings was financially supported by the National Natural Science Foundation of China (No. 52022046 and No. 52038005).

We would like to acknowledge the contributions of all authors and reviewers, as well as members of both Steering Committee and Scientific Committee for contributing their wisdom and influence to confer a successful PBD-IV conference in Beijing and attract the participation of many international colleagues in the hard time of COVID pandemic.

March 2022

Misko Cubrinovski
ISSMGE-TC203 Chair

Jian-Min Zhang
Conference Honorary Chair

Lanmin Wang
Conference Chair

Rui Wang
Conference Secretary

Organization

Honorary Chair

Jian-Min Zhang

Conference Chair

Lanmin Wang

Conference Secretary

Rui Wang

Steering Committee

Atilla Ansal
Charles W. W. Ng
George Gazetas
Izzat Idriss
Jian Xu
Jian-Min Zhang
Jonathan Bray
Kenji Ishihara
Kyriazis Pitilakis

Liam Finn
Lili Xie
Misko Cubrinovski
Ramon Verdugo
Ross Boulanger
Takaji Kokusho
Xianjing Kong
Xiaonan Gong
Yunmin Chen

Scientific Committee

Anastasios Anastasiadis
A. Murali Krishna
Alain Pecker
Amir Kaynia
Andrew Lees
António Araújo Correia
Antonio Morales Esteban

Arnaldo Mario Barchiesi
Carl Wesäll
Carolina Sigarán-Loría
Christos Vrettos
Deepankar Choudhury
Dharma Wijewickreme
Diego Alberto Cordero Carballo

Duhee Park
Eleni Stathopoulou
Ellen Rathje
Farzin Shahrokhi
Francesco Silvestri
Gang Wang
George Athanasopoulos
George Bouckovalas
Ioannis Anastasopoulos
Ivan Gratchev
Jan Laue
Jay Lee
Jean-François Semblat
Jin Man Kim
Jorgen Johansson
Juan Manuel Mayoral Villa
Jun Yang
Kemal Onder Cetin

Louis Ge
M. A. Klyachko
Masyhur Irsyam
Mitsu Okamura
Nicolas Lambert
Paulo Coelho
Sjoerd Van Ballegooy
Rubén Galindo Aires
Ryosuke Uzuoka
Sebastiano Foti
Seyed Mohsen Haeri
Siau Chen Chian
Stavroula Kontoe
Valérie Whenham
Waldemar Świdziński
Wei F. Lee
Zbigniew Bednarczyk

Organizing Committee

Ailan Che
Baitao Sun
Degao Zou
Gang Wang
Gang Zheng
Guoxing Chen
Hanlong Liu
Hongbo Xin
Hongru Zhang
Jie Cui
Jilin Qi
Maosong Huang
Ping Wang
Shengjun Shao

Wei. F. Lee
Wensheng Gao
Xianzhang Ling
Xiaojun Li
Xiaoming Yuan
Xiuli Du
Yanfeng Wen
Yangping Yao
Yanguo Zhou
Yong Zhou
Yufeng Gao
Yunsheng Yao
Zhijian Wu

Acknowledgements

Manuscript Reviewers
The editors are grateful to the following people who helped to review the manuscripts and hence assisted in improving the overall technical standard and presentation of the papers in these proceedings:

Amir Kaynia
Atilla Ansal
Brady cox
Dongyoup Kwak
Ellen Rathje
Emilio Bilotta
Haizuo Zhou
Jiangtao Wei
Jon Stewart
Kyriazis Pitilakis
Lanmin Wang
Mahdi Taiebat
Majid Manzari
Mark Stringer
Ming Yang

Rui Wang
Scott Brandenberg
Tetsuo Tobita
Ueda Kyohei
Wenjun Lu
Wu Yongxin
Xiao Wei
Xiaoqiang Gu
Xin Huang
Yang Zhao
Yanguo Zhou
Yifei Cui
Yu Yao
Yumeng Tao
Zitao Zhang

Contents

Keynote Lectures

Performance-Based Seismic Assessment of Slope Systems 3
Jonathan D. Bray and Jorge Macedo

Recent Advances in Helical Piles for Dynamic and Seismic
Applications 24
M. Hesham El Naggar

Seismic Response of Offshore Wind Turbine Supported by Monopile
and Caisson Foundations in Undrained Clay 50
Maosong Huang, He Cui, Zhenhao Shi, and Lei Liu

Development of the Earthquake Geotechnical Engineering (EGE) in
ISSMGE. 67
Kenji Ishihara

Transient Loading Effects on Pore Pressure Generation and the
Response of Liquefiable Soils 74
Steven L. Kramer and Samuel S. Sideras

Performance-Based Design for Earthquake-Induced Liquefaction:
Application to Offshore Energy Structures. 100
Haoyuan Liu and Amir M. Kaynia

Factors Affecting Liquefaction Resistance and Assessment by Pore
Pressure Model. 120
Mitsu Okamura

How Important is Site Conditions Detailing and Vulnerability
Modeling in Seismic Hazard and Risk Assessment at Urban Scale?. . . . 140
Kyriazis Pitilakis, Evi Riga, and Stefania Apostolaki

The Influence of Soil-Foundation-Structure Interaction on the Seismic Performance of Masonry Buildings 162
Francesco Silvestri, Filomena de Silva, Fulvio Parisi, and Annachiara Piro

Coseismic and Post-seismic Slope Instability Along Existing Faults 195
Ikuo Towhata

Static Liquefaction in the Context of Steady State/Critical State and Its Application in the Stability of Tailings Dams 214
Ramon Verdugo

Seismic Behaviour of Retaining Structures: From Fundamentals to Performance-Based Design 236
Giulia M. B. Viggiani and Riccardo Conti

Study and Practice on Performance-Based Seismic Design of Loess Engineering in China 258
Lanmin Wang, Jinchang Chen, Ping Wang, Zhijian Wu, Ailan Che, and Kun Xia

Large-Scale Seismic Seafloor Stability Evaluation in the South China Sea Incorporating Soil Degradation Effects 288
Yuxi Wang, Rui Wang, and Jian-Min Zhang

Seismic Inspection of Existing Structures Based on the Amount of Their Deformation Due to Liquefaction 296
Susumu Yasuda

Invited Theme Lectures

Site Characterization for Site Response Analysis in Performance Based Approach 319
Atilla Ansal and Gökçe Tönük

Seismic Landslide Susceptibility Assessment Based on Seismic Ground Motion and Earthquake Disaster Analysis 327
Ailan Che, Hanxu Zhou, Jinchang Chen, Yuchen Wu, and Ziyao Xu

SEM-Newmark Sliding Mass Analysis for Regional Coseismic Landslide Hazard Evaluation: A Case Study of the 2016 Kumamoto Earthquake 342
Zhengwei Chen and Gang Wang

Understanding Excess Pore Water Dissipation in Soil Liquefaction Mitigation 353
Siau Chen Chian and Saizhao Du

Buried Pipeline Subjected to Ground Deformation and Seismic Landslide: A State-of-the-Art Review 363
Deepankar Choudhury and Chaidul H. Chaudhuri

Performance-Based Assessment and Design of Structures on Liquefiable Soils: From Triggering to Consequence and Mitigation 376
Shideh Dashti, Zachary Bullock, and Yu-Wei Hwang

Calibration and Prediction of Seismic Behaviour of CFRD Dam for Performance Based Design 397
Barnali Ghosh, Vipul Kumar, and Sergio Solera

Uncertainties in Performance Based Design Methodologies for Seismic Microzonation of Ground Motion and Site Effects: State of Development and Applications for Italy 412
Salvatore Grasso and Maria Stella Vanessa Sammito

Mechanisms of Earthquake-Induced Landslides: Insights in Field and Laboratory Investigations 428
Ivan Gratchev

Regionalization of Liquefaction Triggering Models 437
Russell A. Green

Different Aspects of the Effects of Liquefaction-Induced Lateral Spreading on Piles, Physical Modelling 452
S. Mohsen Haeri

Hybrid Type Reinforcement of Highway Embankment Against Earthquake Induced Damage 472
Hemanta Hazarika, Chengjiong Qin, Yoshifumi Kochi, Hideo Furuichi, Nanase Ogawa, and Masanori Murai

Multiple Liquefaction of Granular Soils: A New Stacked Ring Torsional Shear Apparatus and Discrete Element Modeling 500
Duruo Huang, Zhengxin Yuan, Siyuan Yang, Pedram Fardad Amini, Gang Wang, and Feng Jin

Liquefaction-Induced Underground Flow Failures in Gently-Inclined Fills Looser Than Critical 514
Takaji Kokusho, Hazarika Hemanta, Tomohiro Ishizawa, and Shin-ichiro Ishibashi

Prediction of Site Amplification of Shallow Bedrock Sites Using Deep Neural Network Model 527
Duhee Park, Yonggook Lee, Hyundong Roh, and Jieun Kang

Response of Pumice-Rich Soils to Cyclic Loading 530
Mark Stringer

In-Situ Liquefaction Testing of a Medium Dense Sand Deposit and Comparison to Case History- and Laboratory-Based Cyclic Stress and Strain Evaluations 545
Armin W. Stuedlein and Amalesh Jana

Flowslide Due to Liquefaction in Petobo During the 2018 Palu Earthquake 565
Mahdi Ibrahim Tanjung, Masyhur Irsyam, Andhika Sahadewa, Susumu Iai, Tetsuo Tobita, and Hasbullah Nawir

Challenge to Geotechnical Earthquake Engineering Frontier: Consideration of Buildings Overturned by the 2011 Tohoku Earthquake Tsunami 580
Kohji Tokimatsu, Michitaka Ishida, and Shusaku Inoue

Multipoint Measurement of Microtremor and Seismic Motion of Slope Using Small Accelerometers 597
Lin Wang, Takemine Yamada, Kentaro Kasamatsu, Kazuyoshi Hashimoto, and Shangning Tao

Cyclic Failure Characteristics of Silty Sands with the Presence of Initial Shear Stress 609
Xiao Wei, Zhongxuan Yang, and Jun Yang

Particle Fabric Imaging for Understanding the Monotonic and Cyclic Shear Response of Silts 617
Dharma Wijewickreme and Ana Maria Valverde Sancho

Technical Framework of Performance-Based Design of Liquefaction Resistance 631
Jinyuan Yuan, Lanmin Wang, and Xiaoming Yuan

Deformation Mechanisms of Stone Column-Improved Liquefiable Sloping Ground Under Earthquake Loadings 643
Yan-Guo Zhou and Kai Liu

Liquefaction-Induced Downdrag on Piles: Insights from a Centrifuge and Numerical Modeling Program 660
Katerina Ziotopoulou, Sumeet K. Sinha, and Bruce L. Kutter

Performance Design and Seismic Hazard Assessment

A Simplified Method to Evaluate the 2D Amplification of the Seismic Motion in Presence of Trapezoidal Alluvial Valley 685
Giorgio Andrea Alleanza, Anna d'Onofrio, and Francesco Silvestri

Critical Acceleration of Shallow Strip Foundations 693
Orazio Casablanca, Giovanni Biondi, and Ernesto Cascone

Seismic Vulnerability Analysis of UHV Flat Wave Reactor Based on Probabilistic Seismic Demand Model 702
Jiawei Cui and Ailan Che

Seismic Performance of the "Mediterranean Motorway" Piers Founded on Soft Soil 711
Filomena de Silva, Michele Boccardi, Anna d'Onofrio, Valeria Licata, Enrico Mittiga, and Francesco Silvestri

Performance-Based Estimation of Lateral Spread Displacement in the State of California: A Case Study for the Implementation of Performance-Based Design in Geotechnical Practice 720
Kevin W. Franke, Clay Fullmer, Delila Lasson, Dallin Smith, Sarah McClellan, Ivy Stout, and Riley Hales

On a Novel Seismic Design Approach for Integral Abutment Bridges Based on Nonlinear Static Analysis 730
Domenico Gallese, Davide Noè Gorini, and Luigi Callisto

Influence of Seismic Displacement Models on Landslide Prediction at the Egkremnoi Coastline During the 2015 Lefkada Earthquake 739
Weibing Gong, Dimitrios Zekkos, and Marin Clark

Assessment of the Seismic Performance of Large Mass Ratio Tuned Mass Dampers in a Soil-Structure System 747
Davide Noè Gorini, Guglielmo Clarizia, Elide Nastri, Pasquale Roberto Marrazzo, and Rosario Montuori

Required Strength of Geosynthetics in Seismic Reinforced Soil Retaining Wall in Multi-tiered Configuration 755
Shilin Jia, Fei Zhang, Xiaoyi Lu, and Yuming Zhu

Study on Design Method of Vertical Bearing Capacity of Rock-Socketed Piles Based on Reliability 762
Zhongwei Li, Hanxuan Wang, Guoliang Dai, and Fan Huang

Performance-Based Probabilistic Assessment of Liquefaction Induced Building Settlement in Earthquake Engineering 773
Chenying Liu, Jorge Macedo, and Gabriel Candia

Seismic Hazard Analysis, Geotechnical, and Structural Evaluations for a Research Reactor Building in the Philippines 783
Roy Anthony C. Luna, Patrick Adrian Y. Selda, Rodgie Ello B. Cabungcal, Luis Ariel B. Morillo, Stanley Brian R. Sayson, and Alvie J. Asuncion-Astronomo

Integrating Local Site Response Evaluations in Seismic Hazard Assessments 792
Roy Anthony C. Luna, Ramon D. Quebral, Patrick Adrian Y. Selda, Francis Jenner T. Bernales, and Stanley Brian R. Sayson

Performance-Based Design Review of a Reinforced Earth Retaining Wall for a Road Embankment Project in the Philippines 801
Roy Anthony C. Luna, Jenna Carmela C. Pallarca, Patrick Adian Y. Selda, Rodgie Ello B. Cabungcal, Marvin Renzo B. Malonzo, and Helli-Mar T. Trilles

Geotechnical and Seismic Design Considerations for Coastal Protection and Retaining Structures in Reclaimed Lands in Manila Bay 809
Gian Paulo D. Reyes, Roy Anthony C. Luna, John Michael I. Tanap, Marvin Renzo B. Malonzo, and Helli-mar T. Trilles

A Framework for Real-Time Seismic Performance Assessment of Pile-Supported Wharf Structures Incorporating a Long Short-Term Memory Neural Network 818
Liang Tang, Yi Zhang, Zheng Zhang, Wanting Zhang, and Xianzhang Ling

Evaluation of the Liquefaction Hazard for Sites and Embankments Improved with Dense Granular Columns 826
Juan Carlos Tiznado, Shideh Dashti, and Christian Ledezma

History of Liquefaction Hazard Map Development and a New Method for Creating Hazard Maps for Low-Rise Houses 834
Susumu Yasuda

Dynamic Response Analysis of Slope Based on 3D Mesh Model Reconstruction and Electrical Resistivity Topography 845
Hanxu Zhou and Ailan Che

Ground Motions and Site Effects

In-Situ Characterization of the Near-Surface Small Strain Damping Ratio at the Garner Valley Downhole Array Through Surface Waves Analysis 855
Mauro Aimar, Mauro Francavilla, Brady R. Cox, and Sebastiano Foti

A Seismic Microzonation Study for Some Areas Around the Mt. Etna Volcano on the East Coast of Sicily, Italy 863
Antonio Cavallaro, Salvatore Grasso, and Maria Stella Vanessa Sammito

Influence of Local Soil Conditions on the Damage Distribution in Izmir Bay During the October 30, 2020, Samos Earthquake 871
Anna Chiaradonna, Eyyub Karakan, Giuseppe Lanzo, Paola Monaco, Alper Sezer, and Mourad Karray

Topographic Amplification of Seismic Ground Motion in Hills 879
Sukanta Das and B. K. Maheshwari

Input Ground Motion Selection for Site Response Analysis at the Port of Wellington (New Zealand) 888
Riwaj Dhakal, Misko Cubrinovski, and Jonathan Bray

Responses of a Cantilever Retaining Wall Subjected to Asymmetric Near-Fault Ground Motions.................................. 896
Seong Jun Ha, Hwanwoo Seo, Hyungseob Kim, and Byungmin Kim

Modeling Two-Dimensional Site Effects at the Treasure Island Downhole Array... 904
Mohamad M. Hallal, Peyman Ayoubi, Domniki Asimaki, and Brady R. Cox

Variability in Kappa (κ_r) Estimated with Coda Waves for California 912
Chunyang Ji, Ashly Cabas, Marco Pilz, and Albert Kottke

Structural Response to High Frequency Motion Released in Nonlinear Soil 920
Piotr Kowalczyk

Linear Combination of Ground Motion Models with Optimized Weights Using Quadratic Programming......................... 928
Dongyoup Kwak, Dongil Jang, and Jae-Kwang Ahn

An Evaluation of V_{SZ} Estimates from the P-wave Seismogram Method for Sites in California...................................... 936
Meibai Li and Ellen M. Rathje

The Uncertainty of In-situ S and P Wave Velocity Test at Xichang Experimental Field of CSES.................................. 944
Yongbo Liu, Zhuoshi Chen, Xiaoming Yuan, and Longwei Chen

Assessment of Spatial Variability of Site Response in Japan 952
Cristina Lorenzo-Velazquez and Ashly Cabas

Ground Deformation Evaluation Using Numerical Analyses and Physics-Based Mega-thrust Subduction Zone Motions 961
Diane Moug, Arash Khosravifar, and Peter Dusicka

V_{S30} Correlations from Shear Wave Velocity Profiles in the UAE 971
Deepa Kunhiraman Nambiar, Tadahiro Kishida, Tareq Fahmy Abdallatif, and Mohammad H. AlHamaydeh

Insight into the Relationship Between Dynamic Shear Strain and Vibration Velocity of Horizontally Layered Sites 980
Qi Xia, Rui-shan Li, and Xiao-ming Yuan

Investigating the In-Situ Properties of Poisson's Ratio Based on KiK-Net Data 991
Yang Shi, Hao Zhang, and Yu Miao

Effectiveness of Distributed Acoustic Sensing for Acquiring Surface Wave Dispersion Data Using Multichannel Analysis of Surface Waves 1000
Joseph P. Vantassel, Brady R. Cox, Peter G. Hubbard, Michael Yust, Farnyuh Menq, Kyle Spikes, and Dante Fratta

Relationships Between Ground-Motion Intensity Measures and Earthquake-Induced Permanent Slope Displacement Based on Numerical Analysis 1009
Mao-Xin Wang, Dian-Qing Li, and Wenqi Du

Accuracy of Complex Moduli in Seismic Response Analysis of Ground 1019
Nozomu Yoshida and Kenji Adachi

Scattering of Incident Plane Waves by Underground Cylindrical Cavity in Unsaturated Poroelastic Medium 1027
Aichen Zhang, Weihua Li, and Fengcui Feng

Dynamic Soil-Structure Interaction for a SDOF with Rigid Foundation Embedded in a Radially Inhomogeneous Bearing Stratum Under SH Waves 1041
Ning Zhang, Xinyu Sun, Haijun Lu, and Denghui Dai

Ground Motion Amplification by a Rectangular Tunnel in a Saturated Poroelastic Half-Space 1049
Jun Zhu, Xiaojun Li, Jianwen Liang, and Mianshui Rong

Foundations and Soil-Structure Interaction

The Effect of Soil Damping on the Soil-Pile-Structure Interaction Analyses in Liquefiable and Non-liquefiable Soils 1059
Ozan Alver and E. Ece Eseller-Bayat

Bearing Capacity of Shallow Strip Foundations Adjacent to Slopes 1067
Orazio Casablanca, Giovanni Biondi, Giuseppe Di Filippo, and Ernesto Cascone

Seismic Response of Anchored Steel Sheet Pile Walls in Dry and Saturated Sand 1075
Alessandro Fusco, Giulia M. B. Viggiani, Gopal S. P. Madabhushi, Riccardo Conti, and Cécile Prüm

Response of Suction Bucket Foundation Subjected to Wind and Seismic Loading During Soil Liquefaction 1084
Bin Gao, Guanlin Ye, Qi Zhang, and Wenxuan Zhu

A Class of Thermodynamic Inertial Macroelements for Soil-Structure Interaction 1095
Davide Noè Gorini and Luigi Callisto

Simulation Analyses of Centrifuge Model Tests on Piled Raft Foundation with Deep Mixing Walls 1103
Junji Hamada, Takehiro Okumura, Yoshimasa Shigeno, and Yoshihiro Fujiwara

Seismic Interactions Among Multiple Structures on Liquefiable Soils Improved with Ground Densification 1111
Yu-Wei Hwang, Shideh Dashti, and Juan Carlos Tiznado

Numerical Simulation of Real-Scale Vibration Experiments of a Steel Frame Structure on a Shallow Foundation 1119
Marios Koronides, Stavroula Kontoe, Lidija Zdravković, Athanasios Vratsikidis, Dimitris Pitilakis, Anastasios Anastasiadis, and David M. Potts

Experimental Behavior of Single Pile with Large Mass in Dry Medium Sand Under Centrifuge Shaking Table Test 1128
Longyu Lu, Chunhui Liu, Mengzhi Zhang, and Tiqiang Wang

Numerical Investigation on Dynamic Response of Liquefiable Soils Around Permeable Pile Under Seismic Loading 1136
Chi Ma, Guo-Xiong Mei, and Jian-Gu Qian

Rotation of a Cantilevered Sheet-Pile Wall with Different Embedment Ratios and Retaining a Liquefiable Backfill of Various Relative Densities 1144
Satish Manandhar, Seung-Rae Lee, and Gye-Chun Cho

Foundation Alternatives for Bridges in Liquefiable Soils 1152
Juan Manuel Mayoral, Daniel De La Rosa, Mauricio Alcaraz, Nohemi Olivera, and Mauricio Anaya

A Case Study of Seismic Design of Pile Foundation Subject to Liquefaction, Cyclic Softening, and Lateral Spreading 1161
Yasin Mirjafari and Malcolm Stapleton

Dynamic Centrifuge Model Tests on Plate-Shaped Building Supported by Pile Foundation on Thin Load-Bearing Stratum Overlying Soft Clay Layer 1168
Takehiro Okumura and Junji Hamada

The Behaviour of Low Confinement Spun Pile to Pile Cap Connection 1176
Mulia Orientilize, Widjojo Adi Prakoso, Yuskar Lase, and Carolina Kalmei Nando

Rocking Pilegroups Under Seismic Loading: Exploring a Simplified Method 1185
Antonia Psychari, Saskia Hausherr, and Ioannis Anastasopoulos

Remediation of Structure-Soil-Structure Interaction on Liquefiable
Soil Using Densification 1193
Shengwenjun Qi and Jonathan Adam Knappett

Seismic Responses Analysis on Basements of High-Rise Buildings
Considering Dynamic Soil-Structure Interaction 1201
Yan-Jia Qiu, Hong-Ru Zhang, and Zhong-Yang Yu

Centrifuge and Numerical Simulation of Offshore Wind Turbine
Suction Bucket Foundation Seismic Response in Inclined
Liquefiable Ground 1215
Xue-Qian Qu, Rui Wang, Jian-Min Zhang, and Ben He

Unconventional Retrofit Design of Bridge Pile Groups: Benefits
and Limitations 1222
L. Sakellariadis, S. Alber, and I. Anastasopoulos

Shaking Table Tests on Level Ground Model Simulating
Construction of Sand Compaction Piles 1231
Hiroshi Yabe, Junichi Koseki, Kenji Harada, and Keiichi Tanaka

3D Numerical Lateral Pushover Analysis of Multiple Pile Group
Systems 1240
Amelia Yuwono, Widjojo A. Prakoso, and Yuskar Lase

Responses of Adjacent Building Pile to Foundation Pit
Dewatering 1248
Chao-Feng Zeng, Hai-Yu Sun, Hong-Bo Chen, Xiu-Li Xue,
Yun-Si Liu, and Wei-Wei Song

Dynamic Interaction Between Adjacent Shallow Footings in
Homogeneous or Layered Soils 1257
Enza Zeolla, Filomena de Silva, and Stefania Sica

Effects of Nonliquefiable Crust on the Seismic Behavior of Pile
Foundations in Liquefiable Soils............................... 1265
Gang Zheng, Wenbin Zhang, and Haizuo Zhou

Numerical Implementation of Ground Behaviors Beneath
Super-Tall Building Foundations During Construction 1277
Youhao Zhou and Takatoshi Kiriyama

Slope Stability and Reinforcement

Failure Mechanism Analysis of Loess Slope Under the Coupling
Effect of Rainfall and Earthquake Using Shaking Table Test 1289
Jinchang Chen, Lanmin Wang, and Ailan Che

Seismic Stability Analysis of Earth Slopes Using Graphical
Chart Solution 1297
Hong-zhi Cui and Jian Ji

Influence of Cyclic Undrained Shear Strength Degradation on the Seismic Performance of Natural Slopes 1308
Giuseppe Di Filippo, Orazio Casablanca, Giovanni Biondi, and Ernesto Cascone

Dynamic Analysis of Geosynthetic-Reinforced Soil (GRS) Slope Under Bidirectional Earthquake Loading 1316
Cheng Fan, Kui Cai, and Huabei Liu

Seismic Performance of Slopes at Territorial Scale: The Case of Ischia Island 1324
Francesco Gargiulo, Giovanni Forte, Anna d'Onofrio, Antonio Santo, and Francesco Silvestri

Numerical Simulation of Seismic Performance of Road Embankment Improved with Hybrid Type Steel Pile Reinforcement 1332
Chengjiong Qin, Hemanta Hazarika, Divyesh Rohit, Nanase Ogawa, Yoshifumi Kochi, and Guojun Liu

Distribution of Deformations and Strains Within a Slope Supported on a Liquefiable Stratum 1340
Zhijian Qiu and Ahmed Elgamal

Probabilistic Seismic Hazard Curves and Maps for Italian Slopes 1348
Fabio Rollo and Sebastiano Rampello

New Soil-Pile Spring Accounting for a Tree-Root System in the Evaluation of Seismic Slope Stability 1356
Yoshikazu Tanaka, Kyohei Ueda, and Ryosuke Uzuoka

Numerical Study on Delayed Failure of Gentle Sloping Ground 1362
Tetsuo Tobita, Hitomi Onishi, Susumu Iai, and Masyhur Irsyam

Seismic Fragility Assessment for Cohesionless Earth Slopes in South Korea 1377
Dung Thi Phuong Tran, Hwanwoo Seo, Youngkyu Cho, and Byungmin Kim

Liquefaction-Induced Lateral Displacement Analysis for Sloping Grounds Using Long-Duration Ground Motions 1386
Qiang Wu, Dian-Qing Li, and Wenqi Du

Performance of Slopes During Earthquake and the Following Rainfall 1395
Jiawei Xu, Kyohei Ueda, and Ryosuke Uzuoka

Seismic Stability Analysis of Anti-dip Bedding Rock Slope Based on Tensile Strength Cut-Off 1404
Qiangshan Yu, Dejian Li, and Yingbin Zhang

Seismic Stability Analysis of High Steep Slopes Considering Spatial Variability of Geo-Materials Based on Pseudo Static Method 1413
Wengang Zhang, Fansheng Meng, Jianxin Li, and Changjie He

Liquefaction and Testing

Urban Scale Fragility Assessment of Structures Considering Soil-Structure-Interaction 1425
C. Amendola and D. Pitilakis

Seismic Performance Assessment of Port Reclaimed Land Incorporating Liquefaction and Cyclic Softening 1434
Ioannis Antonopoulos, Alex Park, and Grant Maxwell

An Attempt to Evaluate In Situ Dynamic Soil Property by Cyclic Loading Pressuremeter Test 1446
Keigo Azuno, Tatsumi Ishii, Youngcheul Kwon, Akiyoshi Kamura, and Motoki Kazama

Effect of Refinements to CPT-Based Liquefaction Triggering Analysis on Liquefaction Severity Indices at the Avondale Playground Site, Christchurch, NZ 1454
John R. Cary, Armin W. Stuedlein, Christopher R. McGann, Brendon A. Bradley, and Brett W. Maurer

Effect of Membrane Penetration on the Undrained Cyclic Behavior of Gravelly Sands in Torsional Shear Tests 1467
Matthew Gapuz Chua, Takashi Kiyota, Masataka Shiga, Muhammad Umar, and Toshihiko Katagiri

Implementation and Verification of an Advanced Bounding Surface Constitutive Model 1475
Tony Fierro, Stefano Ercolessi, Massimina Castiglia, Filippo Santucci de Magistris, and Giovanni Fabbrocino

Performance of Advanced Constitutive Models in Site Response Analyses of Liquefiable Soils 1483
Tony Fierro, Massimina Castiglia, and Filippo Santucci de Magistris

A Study on Liquefaction Characteristics of Sandy Soil in Large Strain Levels to Improve the Accuracy of Large Deformation Analysis 1491
Noriyuki Fujii, Takashi Kiyota, Muhammad Umar, and Kyohei Ueda

A New Biaxial Laminar Shear Box for 1g Shaking Table Tests on Liquefiable Soils 1499
Salvatore Grasso, Valentina Lentini, and Maria Stella Vanessa Sammito

**Assessing the Limitations of Liquefaction Manifestation Severity
Index Prediction Models** 1508
Russell A. Green, Sneha Upadhyaya, Brett W. Maurer,
and Adrian Rodriguez-Marek

**Assessment of Stone Column Technique as a Mitigation Method
Against Liquefaction-Induced Lateral Spreading Effects on
2×2 Pile Groups** 1516
S. Mohsen Haeri, Morteza Rajabigol, Milad Zangeneh,
and Mohammad Moradi

**Influence of Lateral Stress Ratio on N-value and Cyclic Strength of
Sands Containing Fines** 1524
Kenji Harada, Kenji Ishihara, and Hiroshi Yabe

**Effect of Water Flow Rate and Insertion Velocity on Soil Disturbance
Due to Insertion of Small-Scale Self-boring Tube**................ 1534
Pei-Chen Hsieh, Takashi Kiyota, Toshihiko Katagiri, Masataka Shiga,
and Manabu Takemasa

**Liquefaction Countermeasure for Existing Structures Using
Sustainable Materials** 1543
Yutao Hu, Hemanta Hazarika, Gopal Santana Phani Madabhushi,
and Stuart Kenneth Haigh

**Undrained Monotonic Compression, Cyclic Triaxial and Cyclic
Simple Shear Response of Natural Soils: Strength and Excess Pore
Water Pressure Response** 1552
Majid Hussain and Ajanta Sachan

**Simple Countermeasure Method to Mitigate the Settlement and
Tilting of Existing Detached Houses Owing to Liquefaction**.......... 1571
Keisuke Ishikawa, Susumu Yasuda, Motomu Matsuhashi,
and Toshifumi Fukaya

**Effect of Artesian Pressure on Liquefaction-Induced Flow-Slide:
A Case Study of the 2018 Sulawesi Earthquake, Indonesia** 1579
Takashi Kiyota, Masataka Shiga, Toshihiko Katagiri, Hisashi Furuichi,
and Hasbullah Nawir

**Physical Modeling and Reliability Assessment of Effectiveness of
Granular Columns in the Nonuniform Liquefiable Ground to Mitigate
the Liquefaction-Induced Ground Deformation** 1587
Ritesh Kumar and Akihiro Takahashi

**Experimental Study on the Effect of Coexistence of Clay and Silt on
the Dynamic Liquefaction of Sand** 1607
Tao Li and Xiao-Wei Tang

The Prediction of Pore Pressure Build-Up by an Energy-Based Model Calibrated from the Results of In-Situ Tests 1622
Lucia Mele, Stefania Lirer, Alessandro Flora, Alfredo Ponzo, and Antonio Cammarota

CDSS Tests for Evaluation of Vibration Frequency in Liquefaction Resistance of Silica Sand 1630
Zhen-Zhen Nong, Sung-Sik Park, and Peng-Ming Jiang

Some Important Limitations of Simplified Liquefaction Assessment Procedures 1638
Nikolaos Ntritsos and Misko Cubrinovski

Dynamic Behavior of Pipe Bend Subjected to Thrust Force Buried in Liquefiable Sand 1647
Kohei Ono and Mitsu Okamura

Liquefaction Resistance of Solani Sand Under Normal and Sequential Shaking Events 1656
Gowtham Padmanabhan and B. K. Maheshwari

Numerical Simulation of Caisson Supported Offshore Wind Turbines Involving Uniform Liquefiable Sand Layer 1664
Alfonso Estepa Palacios, Manh Duy Nguyen, Vladimir Markovic, Sina Farahani, Amin Barari, and Lars Bo Ibsen

Pore-Pressure Generation of Sands Subjected to Cyclic Simple Shear Loading: An Energy Approach 1674
Daniela Dominica Porcino, Giuseppe Tomasello, and Roohollah Farzalizadeh

Constitutive Modeling of Undrained Cyclic Shearing of Sands Under Non-zero Mean Shear Stress 1683
Andrés Reyes, Mahdi Taiebat, and Yannis F. Dafalias

Investigation of Lateral Displacement Mechanism in Layered and Uniform Soil Models Subjected to Liquefaction-Induced Lateral Spreading 1692
Anurag Sahare, Kyohei Ueda, and Ryosuke Uzuoka

Probabilistic Calibration and Prediction of Seismic Soil Liquefaction Using quoFEM 1700
Aakash Bangalore Satish, Sang-ri Yi, Adithya Salil Nair, and Pedro Arduino

Fluid-Solid Fully Coupled Seismic Response Analysis of Layered Liquefiable Site with Consideration of Soil Dynamic Nonlinearity 1708
Yiyao Shen, Zilan Zhong, Liyun Li, and Xiuli Du

Variation in Hydraulic Conductivity with Increase in Excess Pore Water Pressure Due to Undrained Cyclic Shear Focusing on Relative Density 1717
Toshiyasu Unno, Akiyoshi Kamura, and Yui Watanabe

Framework and Demonstration of Constitutive Model Calibration for Liquefaction Simulation of Densified Sand 1725
Hao Wang, Armin W. Stuedlein, and Arijit Sinha

Three-Dimensional Numerical Simulations of Granular Column Improved Layered Liquefiable Soil Deposit 1737
Zhao Wang, Rui Wang, and Jian-Min Zhang

Fundamental Study on Laboratory Test Method for Setting Parameters of Effective Stress Analysis 1745
Masanori Yamamoto, Ryuichi Ibuki, Yasutomo Yamauchi, Taku Kanzawa, and Jun Izawa

Experimental and Numerical Study of Rate Effect in Cone Penetration Tests 1754
Jian-Hong Zhang and Hao Wang

S1: Special Session on Liquefaction Experiment and Analysis Projects (LEAP)

Lessons Learned from LEAP-RPI-2020 Simulation Practice 1763
Long Chen, Alborz Ghofrani, and Pedro Arduino

Numerical Simulations of the LEAP 2020 Centrifuge Experiments Using PM4Sand 1772
Renzo Cornejo and Jorge Macedo

LEAP-2021 Cambridge Experiments on Cantilever Retaining Walls in Saturated Soils 1785
Xiaoyu Guan, Alessandro Fusco, Stuart Kenneth Haigh, and Gopal Santana Phani Madabhushi

Repeatability Potential and Challenges in Centrifuge Physical Modeling in the Presence of Soil-Structure Interaction for LEAP-2020 1794
Evangelia Korre, Tarek Abdoun, and Mourad Zeghal

Numerical Modeling of the LEAP-RPI-2020 Centrifuge Tests Using the SANISAND-MSf Model in FLAC3D 1802
Keith Perez, Andrés Reyes, and Mahdi Taiebat

Numerical Simulations of LEAP Centrifuge Experiments Using a Multi-surface Cyclic Plasticity Sand Model 1812
Zhijian Qiu and Ahmed Elgamal

Validation of Numerical Predictions of Lateral Spreading Based on Hollow-Cylinder Torsional Shear Tests and a Large Centrifuge-Models Database 1821
R. Vargas, Z. Tang, K. Ueda, and R. Uzuoka

Centrifuge Modeling on the Behavior of Sheet Pile Wall Subjected Different Frequency Content Shaking 1829
Yi-Hsiu Wang, Jun-Xue Huang, Yen-Hung Lin, and Wen-Yi Hung

S2: Special Session on Liquefaction Database

Liquefaction Cases and SPT-Based Liquefaction Triggering Assessment in China 1839
Longwei Chen, Gan Liu, Weiming Wang, Xiaoming Yuan, Jinyuan Yuan, Zhaoyan Li, and Zhenzhong Cao

Hammer Energy Measurement of Standard Penetration Test in China 1848
Longwei Chen, Tingting Guo, Tong Chen, Gan Liu, and Yunlong Wang

Empirical Magnitude-Upper Bound Distance Curves of Earthquake Triggered Liquefaction Occurrence in Europe 1857
Mauro De Marco, Francesca Bozzoni, and Carlo G. Lai

Laboratory Component of Next-Generation Liquefaction Project Database 1865
Kenneth S. Hudson, Paolo Zimmaro, Kristin Ulmer, Brian Carlton, Armin Stuedlein, Amalesh Jana, Ali Dadashiserej, Scott J. Brandenberg, John Stamatakos, Steven L. Kramer, and Jonathan P. Stewart

Liquefaction Characteristics of 2011 New Zealand Earthquake by Cone Penetration Test 1875
Zhao-yan Li, Si-yu Zhang, and Xiao-Ming Yuan

Insights from Liquefaction Ejecta Case Histories from Christchurch, New Zealand 1884
Z. Mijic, J. D. Bray, and S. van Ballegooy

Chilean Liquefaction Surface Manifestation and Site Characterization Database 1893
Gonzalo A. Montalva, Francisco Ruz, Daniella Escribano, Felipe Paredes, Nicolás Bastías, and Daniela Espinoza

S3: Special Session on Embankment Dams

Deterministic Seismic Hazard Analysis of Grand Ethiopia Renaissance Dam (GERD) 1903
Mohammed Al-Ajamee, Abhishek Baboo, and Sreevalsa Kolathayar

Parametric Study of Seismic Slope Stability of Tailings Dam 1914
T. S. Aswathi and Ravi S. Jakka

Seismic Performance of a Zoned Earth Dam 1929
Orazio Casablanca, Andrea Nardo, Giovanni Biondi, Giuseppe Di Filippo, and Ernesto Cascone

Evaluation of the Seismic Performance of Small Earth Dams 1937
Andrea Ciancimino, Renato Maria Cosentini, Francesco Figura, and Sebastiano Foti

Liquefaction Potential for In-Situ Material Dams Subjected to Strong Earthquakes 1946
Hong Nam Nguyen

Seismic Assessment of a Dam on a Clayey Foundation 1959
Franklin R. Olaya and Luis M. Cañabi

Seismic Fragility Analysis of Two Earth Dams in Southern Italy Using Simplified and Advanced Constitutive Models 1968
Gianluca Regina, Paolo Zimmaro, Katerina Ziotopoulou, and Roberto Cairo

Comparison of the Monotonic and Cyclic Response of Tailings Sands with a Reference Natural Sand 1976
David Solans, Stavroula Kontoe, and Lidija Zdravković

Seismic Performance of a Bituminous-Faced Rockfill Dam 1984
M. Tretola, E. Zeolla, and S. Sica

Liquefaction-Induced River Levee Failure at Miwa of Inba Along Tone River During 2011 Tohoku Earthquake 1995
Yoshimichi Tsukamoto, Naoki Kurosaka, Shohei Noda, and Hiroaki Katayama

S4: Special Session on Earthquake Disaster Risk of Special Soil Sites and Engineering Seismic Design

Effect of Rice Husk Ash and Sisal Fibre on Strength Behaviour of Soil 2007
Benard Obbo, Gideon Okurut, Pragnesh J. Patel, Nisha. P. Soni, and Darshil V. Shah

Cyclic Behaviour of Brumadinho-Like Reconstituted Samples of Iron Ore Tailings 2024
Paulo A. L. F. Coelho, David D. Camacho, Felipe Gobbi, and Luís M. Araújo Santos

Dynamic Properties of Organic Soils 2033
Vincenzo d'Oriano and Stavroula Kontoe

Experimental Study on the Correlation Between Dynamic Properties and Microstructure of Lacustrine Soft Clay 2041
Yurun Li, Zhongchen Yang, Jingjuan Zhang, Yingtao Zhao, and Chuang Du

Small-Strain Shear Modulus of Coral Sand with Various Particle Size Distribution Curves 2054
Ke Liang, Guoxing Chen, and Qing Dong

Flowability of Saturated Calcareous Sand 2073
Lu Liu, Jun Guo, Armin W. Stuedlein, Xin-Lei Zhang, Hong-Mei Gao, Zhi-Hua Wang, and Zhi-Fu Shen

Liquefaction Characteristics of Saturated Coral Sand Under Anisotropic Consolidations 2080
Weijia Ma and Guoxing Chen

Microscopic Test Analysis of Liquefaction Flow of Saturated Loess ... 2088
Xingyu Ma, Lanmin Wang, Qian Wang, Ping Wang, Xiumei Zhong, Xiaowu Pu, and Fuqiang Liu

Geotechnical Behavior of the Valley Bottom Plain with Highly Organic Soil During an Earthquake 2098
Motomu Matsuhashi, Keisuke Ishikawa, and Susumu Yasuda

Development of Constitutive Model Describing Unsaturated Liquefaction Characteristics of Volcanic Ash Soil 2106
Takaki Matsumaru and Toshiyasu Unno

Field Observations and Direct Shear Tests on the Volcanic Soils Responsible for Shallow Landslides During 2018 Hokkaido Eastern Iburi Earthquake 2114
Shohei Noda, Hiroya Tanaka, and Yoshimichi Tsukamoto

Investigation on Saturation State of Loess Using P-wave Velocity 2123
Zehua Qin, Yuchuan Wang, and Xin Liu

Relationship Between Shear Wave Velocity and Liquefaction Resistance in Silty Sand and Volcanic Sand 2131
Masataka Shiga and Takashi Kiyota

Small to Medium Strain Dynamic Properties of Lanzhou Loess 2141
Binghui Song, Angelos Tsinaris, Anastasios Anastasiadis, Kyriazis Pitilakis, and Wenwu Chen

Cyclic Simple Shear Tests of Calcareous Sand 2151
Kai-Feng Zeng and Hua-Bei Liu

S5: Special Session on Soil Dynamic Properties at Micro-scale: From Small Strain Wave Propagation to Large Strain Liquefaction

Effect of Fabric Anisotropy on Reliquefaction Resistance of Toyoura Sand: An Experimental Study 2161
Pedram Fardad Amini and Gang Wang

Variation of Elastic Stiffness of Saturated Sand Under Cyclic Torsional Shear 2171
Yutang Chen and Jun Yang

Liquefaction Resistance and Small Strain Shear Modulus of Saturated Silty Sand with Low Plastic Fines 2180
Xiaoqiang Gu, Kangle Zuo, Chao Hu, and Jing Hu

Volumetric Strains After Undrained Cyclic Shear Governed by Residual Mean Effective Stress: Numerical Studies Based on 3D DEM 2188
Mingjin Jiang, Akiyoshi Kamura, and Motoki Kazama

One-Dimensional Wave Propagation and Liquefaction in a Soil Column with a Multi-scale Finite-Difference/DEM Method 2196
Matthew R. Kuhn

Microscopic Insight into the Soil Fabric During Load-Unload Correlated with Stress Waves 2204
Yang Li, Masahide Otsubo, and Reiko Kuwano

Shear Work and Liquefaction Resistance of Crushable Pumice Sand 2212
Rolando P. Orense, Jenny Ha, Arushi Shetty, and Baqer Asadi

Digital Particle Size Distribution for Fabric Quantification Using X-ray μ-CT Imaging 2220
Ana Maria Valverde Sancho and Dharma Wijewickreme

The Relationship Between Particle-Void Fabric and Pre-liquefaction Behaviors of Granular Soils 2229
Jiangtao Wei, Minxuan Jiang, and Yingbin Zhang

Changes in Sand Mesostructure During Sand Reliquefaction Using Centrifuge Tests 2237
Xiaoli Xie and Bin Ye

Examining the Seismic Behavior of Rock-Fill Dams Using DEM Simulations 2245
Zitao Zhang, Rui Wang, Jing Hu, Xuedong Zhang, and Jianzheng Song

DEM Simulation of Undrained Cyclic Behavior of Saturated Dense Sand Without Stress Reversals 2253
Xin-Hui Zhou and Yan-Guo Zhou

S6: Special Session on Underground Structures

Seismic Behaviour of Urban Underground Structures in Liquefiable Soil 2265
Emilio Bilotta

Parametric Analyses of Urban Metro Tunnels Subject to Bedrock Dislocation of a Strike-Slip Fault 2277
Zhanpeng Gan, Junbo Xia, Jun Du, and Yin Cheng

Resilience Assessment Framework for Tunnels Exposed to Earthquake Loading 2285
Z. K. Huang, D. M. Zhang, and Y. T. Zhou

Stability Analysis of Tunnel in Yangtze Estuary Under Dynamic Load of High-Speed Railway 2295
Liqun Li, Qingyu Meng, Leming Wang, and Zhiyi Chen

Seismic Performance of an Integrated Underground-Aboveground Structure System 2304
Wen-Ting Li, Rui Wang, and Jian-Min Zhang

A Simplified Seismic Analysis Method for Underground Structures Considering the Effect of Adjacent Aboveground Structures 2313
Jianqiang Liu, Tong Zhu, Rui Wang, and Jian-Min Zhang

Seismic Response Analysis on the Tunnel with Different Second-Lining Construction Time 2322
Weigong Ma, Lanmin Wang, and Yuhua Jiang

Coupled Seismic Performance of Underground and On-Ground Structures 2331
Juan Manuel Mayoral, Daniel De La Rosa, Mauricio Alcaraz, and Enrique Barragan

Simplified Numerical Simulation of Large Tunnel Systems Under Seismic Loading, CERN Infrastructures as a Case Study 2339
A. Mubarak and J. A. Knappett

Shaking Table Test of the Seismic Performance of Prefabricated Subway Station Structure 2348
Lianjin Tao, Cheng Shi, Peng Ding, Linkun Huang, and Qiankun Cao

Numerical Study on Seismic Behavior of Shield Tunnel Crossing Saturated Sand Strata with Different Densities 2362
Hong-Wu, Zhi-Ye, Hua-Bei Liu, and Yu-Ting Zhang

Seismic Response of a Tunnel-Embedded Saturated Sand Ground Subject to Stepwise Increasing PGA 2371
Mingze Xu, Zixin Zhang, and Xin Huang

Groundwater Response to Pumping Considering Barrier Effect of Existing Underground Structure 2381
Xiu-Li Xue, Long Zhu, Shuo Wang, Hong-Bo Chen, and Chao-Feng Zeng

Ranking Method for Strong Ground Motion Based on Dynamic Response of Underground Structures 2391
Wei Yu and Zhiyi Chen

Author Index 2399

Ground Motions and Site Effects

In-Situ Characterization of the Near-Surface Small Strain Damping Ratio at the Garner Valley Downhole Array Through Surface Waves Analysis

Mauro Aimar[1]($\boxtimes$), Mauro Francavilla[1], Brady R. Cox[2], and Sebastiano Foti[1]

[1] Politecnico di Torino, Corso Duca degli Abruzzi 24, 10129 Torino, Italy
mauro.aimar@polito.it
[2] Utah State University, Logan 84322, USA

Abstract. The quantification of attenuation properties of soils has great relevance in geotechnical earthquake engineering. The small strain damping ratio is generally obtained from either direct laboratory tests on small samples or generic empirical relationships. Alternatively, some promising techniques for extracting in-situ small strain damping ratio rely on the analysis of surface wave data. This paper presents a subset of results from a massive dynamic site characterization study at the Garner Valley Downhole Array, wherein in-situ damping ratio profiles have been extracted from several multichannel analysis of surface waves (MASW) datasets. Waveforms generated from both a sledgehammer and a dynamic shaker were recorded, allowing for comparisons between the damping estimates obtained from both types of sources. Dispersion data and attenuation curves were derived from the waveforms using several approaches presented in the literature, and one new approach developed by the authors. This paper documents the inter-method differences and similarities across approaches in terms of uncertainties in the wavefield attenuation, together with the impacts on the amplification of the soil deposit. This study contributes towards better, in-situ characterization of the attenuation properties of soil deposits, enhancing the accuracy of ground models used in dynamic analyses. It is expected that progress in this area will lead to greater reliability of predicted ground motion amplification.

Keywords: Small strain damping ratio · Surface waves · MASW · Uncertainties

1 Introduction

An accurate evaluation of site effects is crucial to define the expected ground motion at the surface. Among the various parameters affecting the stratigraphic amplification, the soil small-strain damping ratio D_0 and the related uncertainties are gaining attention. This parameter, in fact, is believed to control variations in motion amplitude and frequency content, especially when a low-intensity shaking is involved (e.g., [1]).

The D_0 is typically estimated from laboratory tests. However, experimental evidence from back-analysis of Down-Hole seismic arrays showed D_0 values in the field larger than the ones obtained through laboratory tests. In fact, at the site scale, complex wave propagation phenomena (e.g., wave scattering [2]) induce energy dissipation mechanisms that cannot be captured by laboratory tests. Therefore, an in-situ estimate of D_0 should be adopted in ground response simulations.

A potentially effective way for obtaining soil dissipative parameters relies on geophysical tests, especially on Multi-Channel Analysis of Surface Waves (MASW). The MASW-based estimate of the shear-wave velocity V_S and D_0 refers to the measurement of variations of phase and amplitude of surface waves along linear arrays with active sources, from which the dispersion and the attenuation curves are obtained. These curves describe variations of the Rayleigh wave phase velocity and the phase attenuation with the frequency. This dependence is the combined effect of both geometrical dispersion, i.e. the variation in mechanical properties with depth, and intrinsic dispersion, linked to the frequency-dependence of material parameters in linear, viscoelastic media (e.g., [3]). Then, the V_S and D_0 profiles are estimated through an inversion scheme, where a theoretical soil model is calibrated to match the experimental dispersion and attenuation data.

This note shows some relevant results of a MASW survey performed at the Garner Valley Downhole Array (GVDA) site, utilizing a sledgehammer and a vibroseis truck as active sources. Experimental dispersion and attenuation curves were obtained according to different methods, and results were compared to highlight strengths and limitations of each one. Furthermore, the influence of source characteristics on the quality of estimated data was investigated. The second part of the paper focuses on the inversion problem for the coupled estimation of the V_S and D_0 profiles, that are compared with results from other geophysical surveys and empirical estimates from the literature, respectively. The corresponding amplification is finally computed and compared with the one obtained from observed data, to check the reliability of the derived soil models in capturing the site response.

2 Site Description and Data Acquisition

The Garner Valley Down-Hole Array is a site located in Southern California. The site stratigraphy is characterized by alluvial soil with a shallow water table, that overlies a layer of decomposed granite transitioning to competent granite bedrock. This site is instrumented with a seismic monitoring system. The equipment includes an instrumented borehole, with accelerometers on the surface and at various depths, to capture variations in the ground motion in the soil deposit. For this reason, the site represents an effective benchmark for testing the validity of ground motion amplification models.

The investigation of the small-strain parameters of the soil deposit at the GVDA site was carried out by means of a MASW survey. The testing involved a two-dimensional array (Fig. 1), made by a regular grid of geophones, to develop a three-dimensional model of the soil deposit. This study, however, focuses on the South-East line, which is a linear array composed by 14 geophones, with inter-receiver distance equal to 5 m. The receivers are Magseis Fairfield Nodal ZLand 3C sensors, that are three-component geophones suitable for both active and passive surveys. Waveform were generated through the NHERI@UTexas Thumper vibroseis truck [4] and an instrumented sledgehammer. The shaker generated a chirp signal, with frequency content ranging between 5 and 30 Hz, at two shot points with reversal. The source-offsets of road-side shot points are 4.5 m and 35.5 m respectively, whereas the ones on the parking side are 2.5 m and 33.5 m far from the closest sensor. As for the sledgehammer, two source-offsets of 5 m and 15 m were used off both ends.

Fig. 1. MASW array setup. The larger circles represent the receivers belonging to the array analyzed in this study. The blue area identifies GVDA, where the instrumented boreholes are located.

3 Data Processing

The estimation of the Rayleigh phase velocity V and phase attenuation α was carried out by using different methods. For simplicity, this section compares results from shaker data, and the influence of source characteristics will be reported later.

The considered methods are the Transfer Function Method (TFM) [3], the Generalized Half-Power Bandwidth Method (GHPB) [5] and the Wavefield Decomposition Method (WD) [6]. In addition, this study considers a new technique, i.e. the Frequency-Domain BeamForming – Attenuation (FDBFa) [7]. This approach is a generalization of the Frequency-Domain BeamForming (FDBF) method [8] for the attenuation estimate and it is based on the following wavefield transformation:

$$v(r, \omega) = [u(r, \omega)]^j \quad (1)$$

where $u(r, \omega)$ is the particle displacement (as a function of the distance r and the circular frequency ω) and j is the imaginary unit. In the case of a single dominant mode, it can be demonstrated that the wavenumber of the transformed wave $v(r,\omega)$ corresponds to the attenuation of the original one [7]. Hence, α can be estimated through the dispersion analysis of $v(r,\omega)$ – in this case, by using the FDBF approach. In the estimation, the geometric spreading effect on the wave amplitude is removed by multiplying recorded data by $\sqrt{r}$ (e.g., [3]).

Figure 2 compares the dispersion and attenuation curves obtained for each method. Results are described in terms of mean μ and coefficient of variation (CoV), i.e. the ratio between the standard deviation and μ. Data statistics were estimated through the multiple source-offset technique [9]. In all the cases, only the fundamental mode was identified. The dispersion curves are rather close to each other and affected by low variability, for almost each approach. As for α, all the approaches agree quite well at short wavelengths, although the WD scheme is not able to estimate wave parameters at wavelengths greater than 30 m, for this site. On the other side, the TFM matches the average values of all the other methods, even with less variability. However, it tends to overestimate α at greater wavelengths, probably because of near-field effects, that are not modeled in this case. Finally, the GHPB and the FDBFa methods provide similar results, though the former is affected by rather large variability both on V and α.

As for the source effect, Fig. 3 compares the normal statistics of the estimated V and α from shaker data and sledgehammer data. For simplicity, only results derived through the FDBFa method are considered, but similar considerations are valid for the other methods. The mean estimates are close to each other, independently from the investigated wavelength. However, shaker data extend to longer wavelengths. Furthermore, sledgehammer-based α is affected by larger variability, differently from V. A potential reason of the different scatter might be noise, as the sledgehammer is not a high-energy source and the signal-to-noise ratio of recorded traces might be low, especially at high frequencies. However, the mean trend in the attenuation curve can still be captured.

4 Data Inversion

The estimated V and α data were used to estimate the V_S and D_0 profiles, through a joint inversion of the experimental curves. For simplicity, the analysis focused on results from shaker data, interpreted according to the FDBFa method.

The inversion was carried out by using a Monte-Carlo-based global search algorithm. The algorithm is based on a smart sampling technique of the model parameter space, by exploiting the scaling properties of the modal curves [7; 10]. Thus, a good quality result can be achieved with a moderately small number of generated ground models. The randomization of earth models investigated an adequate range of layer thicknesses, S-wave velocities and damping ratios, whereas P-wave velocities and mass densities were inferred from borehole logs [11]. Instead, the P-wave damping ratio was kept equal to the corresponding one for S-waves. Forward dispersion and attenuation modeling was carried out through the ElastoDynamics Toolbox [12].

The degree of matching between synthetic and experimental data is measured through the following misfit function:

$$M = \frac{1}{2n}\left[\sum_{i=1}^{n}\frac{\left(V_{t,i} - V_{e,i}\right)^2}{\sigma_{V,i}^2} + \sum_{i=1}^{n}\frac{\left(\alpha_{t,i} - \alpha_{e,i}\right)^2}{\sigma_{\alpha,i}^2}\right] \quad (2)$$

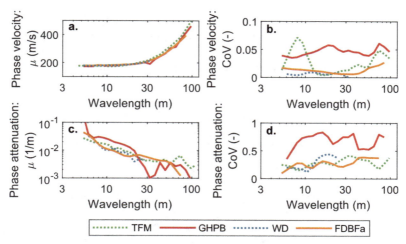

Fig. 2. Inter-method comparison: a) mean and b) coefficient of variation of the estimated phase velocity; c) mean and d) coefficient of variation of the estimated phase attenuation.

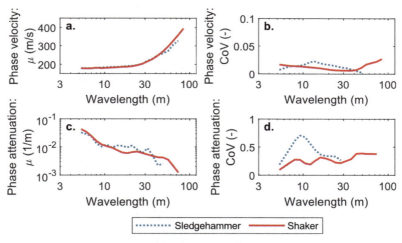

Fig. 3. Influence of source characteristics: a) mean and b) coefficient of variation of the estimated phase velocity; c) mean and d) coefficient of variation of the estimated phase attenuation

where $V_{t,i}$ and $V_{e,i}$ are the theoretical and experimental velocity values and $\sigma_{V,i}$ is the corresponding standard deviation, $\alpha_{t,i}, \alpha_{e,i}$ and $\sigma_{\alpha,i}$ are the corresponding ones for phase attenuation and n is the number of sample points in the experimental data.

Figure 4 shows results for the best fitting 10 models out of 10,000 trial profiles. Inverted V_S profiles are poorly scattered, whereas D_0 profiles are affected by large variability. This is an effect of the high CoV in the experimental attenuation data, that does not allow an effective constraint of D_0. For comparison purposes, Fig. 4 includes profiles obtained from P-S suspension logging [11] and a Down-Hole test [13], where the corresponding D_0 was estimated through Darendeli's empirical relationship [14]. Resulting profiles are quite compatible with each other, although MASW-based D_0 rises up to 5% on the surface layer. This increase might be an effect of heterogeneities on the top of the soil deposit, resulting in wave scattering phenomena.

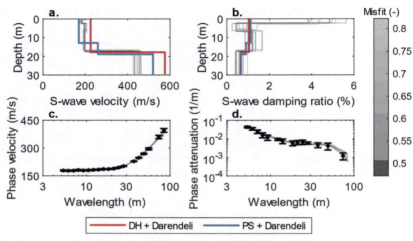

Fig. 4. Inverted S-wave velocity (a) and damping ratio profiles (b); c) Theoretical vs. experimental dispersion curves; d) Theoretical vs. experimental attenuation curves. Theoretical data correspond to the best fitting 10 models. Profiles obtained from a Down-Hole (DH) test [13] and PS suspension logging (PS) [11] are included, where D_0 was estimated through Darendeli's relationship [14].

Finally, implications of the inverted soil models into the site response were addressed, by comparing the estimated stratigraphic amplification with the one observed at GVDA. In this study, the amplification is described as acceleration transfer function (TF), i.e. the ratio of the Fourier spectra between acceleration time histories at different depths. Specifically, empirical TFs between the surface sensor and the ones at 6 m and 15 m depth were considered, the values of which were taken from Vantassel and Cox [15]. The corresponding theoretical TFs were computed through linear visco-elastic simulations, assuming "within" conditions at each reference depth, for compatibility with empirical data. TFs were computed from MASW data and the invasive surveys mentioned above. Different testing procedures result in TFs with nonidentical location and amplitude of peaks, due to different V_S and D_0 values especially in the shallow layer. However, MASW data compare moderately well with the empirical TF when considering the shallower sensor (Fig. 5a), especially for the fundamental peak. On the contrary, the fitting quality for the fundamental mode is poor when the deeper sensor is adopted, although the compatibility slightly improves at high frequencies (Fig. 5b). However, this discrepancy may be an effect of inaccuracies in the low-frequency data recorded in the sensor at 15 m depth [15].

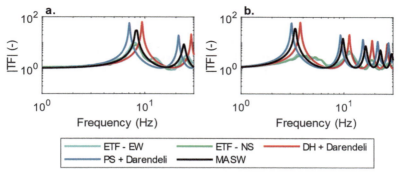

Fig. 5. Comparison between experimental transfer function (ETF, extracted from Vantassel and Cox [15]) and median theoretical transfer function for the best fitting 10 models in the inversion (MASW), for the reference depths of 6 m (a) and 15 m (b). Theoretical transfer functions obtained from results of a Down-Hole (DH) test [13] and PS suspension logging (PS) [11] are included, where D_0 was estimated through Darendeli's relationship [14]. Experimental data are labelled as North–South (NS) and East–West (EW), corresponding to the components of seismometer records from which they were derived.

5 Conclusions

The present paper investigated the efficiency of surface wave analysis in estimating the small-strain damping ratio from a MASW survey carried out at the Garner Valley Downhole Array. First, uncertainties in the estimation of the Rayleigh wave phase velocity and attenuation were assessed, focusing on inter-method differences and on the effect of source characteristics. On the one side, all the considered approaches provide similar dispersion estimates, whereas some divergence is observed in attenuation data. Furthermore, the novel FDBFa method returns results compatible with other approaches, both in terms of mean and variability. As for the source effect, data from vibroseis and sledgehammer compare quite well, though the latter is characterized by larger variability. However, this result positively contributes to the capability of the sledgehammer for the attenuation estimate. This is helpful for ordinary applications, where high-energy sources are not typically available.

Then, profiles of S-wave velocity and damping ratio were extracted from experimental data through an inversion process. The resulting models exhibit well-constrained velocity profiles, whereas damping ratios are more scattered. Such variability may be the combined effect of the large CoV in attenuation data and the limited wavelength range investigated. Nonetheless, the estimated amplification is compatible with empirical data and with results obtained from other surveys.

Future work will focus on an extended characterization of dissipation properties on the whole array. Furthermore, the possibility of extracting attenuation from ambient noise data will be addressed, to better constrain estimated damping ratios at greater depths.

Acknowledgements. The study was partially supported by the ReLUIS 3 project, funded by the Italian Civil Protection Agency. Data were extracted from the NSF Project "Collaborative

Research: 3D Ambient Noise Tomography for Natural Hazards Engineering" grant CMMI-1931162. However, any opinions, findings, and conclusions or recommendations expressed in this material are those of the authors and do not necessarily reflect the views of the NSF. The seismic instruments were provided by IRIS through the PASSCAL Instrument Center at New Mexico Tech. Data collected will be available through the IRIS Data Management Center. The facilities of the IRIS Consortium are supported by the NSF SAGE Award under Cooperative Support Agreement EAR-1851048.

References

1. Foti, S., Aimar, M., Ciancimino, A.: Uncertainties in small-strain damping ratio evaluation and their influence on seismic ground response analyses. In: Latest Developments in Geotechnical Earthquake Engineering and Soil Dynamics, pp. 175–213 Springer (2021)
2. Thompson, E.M., Baise, L.G., Kayen, R.E., Guzina, B.B.: Impediments to predicting site response: seismic property estimation and modeling simplifications. Bull. Seismol. Soc. Am. **99**(5), 2927–2949 (2009)
3. Lai, C.G., Rix, G.J., Foti, S., Roma, V.: Simultaneous measurement and inversion of surface wave dispersion and attenuation curves. Soil Dyn. Earthq. Eng. **22**(9–12), 923–930 (2002)
4. UT Austin NHERI Experimental Facility https://utexas.designsafe-ci.org/equipment-portfolio/ Accessed 22 Oct 2021
5. Badsar, S.A., Schevenels, M., Haegeman, W., Degrande, G.: Determination of the material damping ratio in the soil from SASW tests using the half-power bandwidth method. Geophys. J. Int. **182**(3), 1493–1508 (2010)
6. Bergamo, P., Maranò, S., Imperatori, W., Hobiger, M., Fäh, D.: Wavefield decomposition technique applied to active surface wave surveys: towards joint estimation of shear modulus and dissipative properties of the near-surface. In: EPOS@ SERA "Strong Motion Site Characterization" workshop (2019)
7. Aimar, M.: Uncertainties in the estimation of the shear-wave velocity and the small-strain damping ratio from surface wave analysis. Ph.D. Dissertation, Politecnico di Torino, Italy (2022) (In preparation)
8. Zywicki, D. J.: Advanced signal processing methods applied to engineering analysis of seismic surface waves, Ph.D. Dissertation, Georgia Institute of Technology, Georgia (1999).
9. Cox, B.R., Wood, C.M., Teague, D.P.: Synthesis of the UTexas1 surface wave dataset blind-analysis study: inter-analyst dispersion and shear wave velocity uncertainty. In: Geo-Congress 2014 Technical Papers, Atlanta, GA, pp. 850–859 (2014)
10. Socco, L.V., Boiero, D.: Improved Monte Carlo inversion of surface wave data. Geophys. Prospect. **56**(3), 357–371 (2008)
11. Steller, R.: New borehole geophysical results at GVDA. UCSB Internal report (1996)
12. Schevenels, M., Degrande, G., François, S.: EDT: an elastodynamics toolbox for MATLAB. Comput. Geosci. **35**(8), 1752–1754 (2009)
13. Gibbs, J.F.: Near-surface P- and S-wave velocities from borehole measurements near Lake Hemet, California. U.S. Geological Survey Open File Report (1989)
14. Darendeli, M.B.: Development of a new family of normalized modulus reduction and material damping curves. Ph.D. Dissertation, The University of Texas, Austin (2001)
15. Vantassel, J.P., Cox, B.R.: Multi-reference-depth site response at the Garner Valley Downhole Array. In: Proceedings of the VII ICEGE, p. 8 (2019)

A Seismic Microzonation Study for Some Areas Around the Mt. Etna Volcano on the East Coast of Sicily, Italy

Antonio Cavallaro[1], Salvatore Grasso[2(✉)], and Maria Stella Vanessa Sammito[2]

[1] CNR-ISPC National Research Council, Catania, Italy
[2] University of Catania, Catania, Italy
sgrasso@dica.unict.it

Abstract. On the night of December 26th, 2018, a strong earthquake, with a magnitude of 4.9 on the Richter scale, hit the southeastern side of the Mt. Etna Volcano (Sicily), with the epicenter between the Municipalities of Viagrande and Trecastagni. The hypocenter of the strong earthquake was located just 1 km deep and, for this reason, the effects of the shock were greatly amplified on the ground surface. They have been counted not only material damages to churches and buildings, but also 10 injured around the epicenter area. Therefore, an investigation campaign was carried out with the aim of planning the reconstruction of the damaged areas. In situ soil investigations were carried out in order to determine the soil profile and the geotechnical parameters for the area under consideration. Among in situ tests, borings, Down Hole Tests (D-H), Multichannel Analysis Surface Wave Tests (MASW), Seismic Refraction Tomography and Horizontal to Vertical Spectral Ratio Tests (HVSR) were carried out, with the aim to evaluate the soil profile of shear wave velocity (V_S). Moreover, laboratory tests were carried out on undisturbed samples in the static field: Shear Tests (ST). The results have been grouped and allowed the characterisation of the following soil categories: clay, sandy clay, silica sand, volcanic sand, volcanic rock. The seismic behaviour of these soil categories has been used for the site response analysis and for the seismic microzonation of the studied area.

Keywords: Shear modulus · Damping ratio · Geotechnical characterisation · Seismic microzonation · Shear wave velocity

1 Introduction

The paper deals with the seismic microzonation studies for five areas around the Mt. Etna Volcano on the East Coast of Sicily, Italy (Fig. 1), after the effects of the 2018 seismic sequence. It is considered as one of the zones of Italy with greater high seismic risk, basing on seismic history and on the huge patrimony of historical and industrial buildings [1–3]. The earthquake of 26 December, 2018, is the most energetic event that has occurred on Etna in the last 70 years. It caused serious damage in the epicentral area, between Fleri and Pennisi.

Fig. 1. Overview of the areas under investigation

The Seismic Microzonation can be defined as a process aimed at identifying and mapping the subsoil local response in an area [4]. The detailed evaluation of the spatial variability in seismic responses is the first and most important step towards a seismic risk analysis and mitigation strategy in densely populated regions, as the areas under consideration [5–7].

2 Seismicity of Area

On 26 December 2018, at 03:19 Italian time, an earthquake of magnitude Mw 4.9 occurred on the eastern flank of Etna Vulcano at a depth of less than 1 km. This event is the main one among those located during the intense Etna seismic activity that began on 23 December 2018 and characterized by about seventy events with a magnitude M >2.5, concomitant with the eruptive activity.

The earthquake of 26 December occurred on the Fiandaca fault, producing superficial faulting. The Fiandaca fault is located in the southernmost sector of a complex system of active faults, known as the Timpe, which characterizes the eastern sector of the continuously deforming volcano. This fault has variable directions between NW and NNO and predominantly right-transcurrent kinematics with vertical [8]. The Timpe system participates in a general movement of the side of the volcano towards the east which is constantly monitored by the sensors of the INGV Etneo Observatory. This moving area is clearly bounded to the north by the Pernicana fault with E-W direction and left transtensive kinematics, while to the south there are more sub-parallel fault systems in which the movement gradually attenuates. Most of the damage following the 2018 earthquake appears to be localized in correspondence with the coseismic superficial faulting and in the areas adjacent to it.

3 Investigation Program and Basic Soil Properties

To evaluate the local seismic response in the areas under study, it is essential to have a wide and in-depth knowledge of the geological, geotechnical and geophysical characteristics of the site. The following in situ and laboratory tests have been carried out in the areas: Boreholes, Down Hole Tests (DH), Multichannel Analysis of Surface Wave Tests (MASW), Seismic Refraction Tomography and Horizontal to Vertical Spectral Ratio Tests (HVSR) and Direct Shear Tests (DST).

The study reported in this paper takes into consideration the data collected near the City Hall of Milo, a building of strategic importance for Civil Protection interventions. Milo is a small town near the Etna Volcano that has an altitude of 720 m above sea level. Given the nature of the soils found, no ground water table was identified. The value of the natural moisture content w_n prevalently ranges from between 13% at depth of 20.43 m and 22% at depth of 8.15 m while characteristic values for e (void ratio) ranged between 0.56 and 0.64 for Milo test site, as an example. As regards strength parameters of the deposits mainly encountered in this area c' from DST ranged between 7 and 38 kPa while ϕ' from DST ranged between 32° and 40°. In Fig. 2, the shear and compression wave velocities are shown against depth, for Milo. The shear V_s and compression V_p wave velocities initially increase with depth. At the depth of about 5 and 10 m there is a peak in velocities due to the presence of volcanic pebbles and basaltic lava in a sandy matrix.

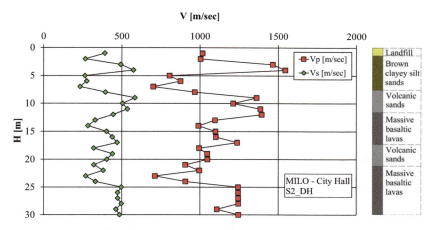

Fig. 2. Vs and V_p from Down Hole tests in Milo area

The subsequent degrading trend of the shear V_s and compression V_p wave velocities depends on the lower mechanical characteristics of the soils encountered. At a depth of about 25 m, the velocity values re-stabilize in the presence of massive basaltic lavas with sub-vertical fractures.

The reliability of the shear wave velocity data acquired at the Milo site is supported by comparison with similar data available in adjacent areas with comparable conditions and stratigraphic succession of the soil. Given the nature of the soils present, no undisturbed samples were taken suitable for performing dynamic laboratory tests. Therefore, to describe the dependence of the shear modulus and damping on deformation, it was necessary to refer to decay curves from the literature [9–13]. For the purpose of describing the decay curves of the shear modulus and damping as a function of the level of deformation, the equations proposed by Yokota et al. [14] were used suitably calibrated to the soil under study.

$$\frac{G(\gamma)}{G_o} = \frac{1}{1 + \alpha\gamma(\%)^\beta} \quad (1)$$

where: $G(\gamma)$ = strain dependent shear modulus; γ = shear strain; α, β = soil constant.

$$D(\gamma)(\%) = \eta \cdot \exp\left[-\lambda \cdot \frac{G(\gamma)}{G_o}\right] \quad (2)$$

where: $D(\gamma)$ = strain dependent damping; γ = shear strain; η, λ = soil constant. Table 1 shows the obtained soil constants.

Table 1. Soil constants for milo soil

Decay curves	α	β	η	λ
1	7.5	0.897	90	4.5
2	6.9	1	23	2.21
3	16	1.2	33	2.4

4 Ground Response Analyses

4.1 Input Motion Selection

In order to estimate local site effects, one-dimensional ground response analyses have been performed using the equivalent-linear site response program STRATA [15].

One of the most important aspects is the definition of the input motions based on deterministic or probabilistic studies. It is recommended to use the deterministic seismic hazard analysis (DSHA) in area where severe earthquakes are observed [16–18]. However, the actual Italian seismic code NTC 2018 [18] considers a probabilistic seismic hazard analysis (PSHA) for the calculation of seismic design actions on structures. In this study, ESM_REXELweb [20], available at https://esm-db.eu/#/rexel, has been used to search for a set of seven accelerograms, from the European Strong-motion Database, compatible to the reference spectrum in a defined range of periods according to the new Italian seismic code [19, 21].

The latitudine and longitudine (37.735; 15.139) have been defined considering the central point of the five municipalities showed in Fig. 1. The disaggregation data have been obtained considering a probability of excess of 10% in 50 years (from http://esse1-gis.mi.ingv.it). The preliminary and matching parameters used in ESM_REXELweb are summarized in Table 2. Finally, Figs. 3 and Table 3 report the suites of waveforms compatible to the reference spectra.

Table 2. Input parameters used in ESM_REXELweb (https://esm-db.eu)

Preliminary		Matching	
Inferred soil class	yes	Lower tol. [%]	10
Late trigger events	no	Upper tol. [%]	30
Soil Type	A	T1[s]	0.1
T1[s]	0.1	T2[s]	1.0
T2[s]	1.0	Set size	7
Criterion	Magnitude-distance	Number of combinations	3
Minimum event magnitude	3.5	Adimensional flag	no
Maximum event magnitude	6.2	Check on PGA	yes
Minimum epicentral distance [km]	0	-	-
Maximum epicentral distance [km]	20	-	-

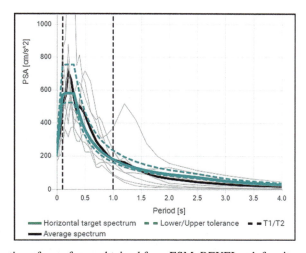

Fig. 3. Combination of waveforms obtained from ESM_REXELweb for site response analyses

Table 3. Suites of waveforms compatible to the reference spectra

n.	Waveform ID	Orientation	Mw	D [km]
1	E.SRC0.00.HN.IT-1976-0030	E	6.0	15.8
2	IT.ACC.00.HG.EMSC-20161030_0000029	E	6.5	18.6
3	IT.SVN.00.HG.EMSC-20181226_0000014	N	4.9	4.5
4	IV.EVRN.00.HN.EMSC-20181226_0000014	E	4.9	5.3
5	IV.T1212.00.HN.EMSC-20161026_0000077	E	5.4	15.2
6	IV.ILLI.00.HN.IT-2010-0032	E	4.7	11.4
7	IV.T1212.00.HN.EMSC-20161026_0000077	E	5.4	15.2

4.2 Analysis of the Results

In Fig. 4, results are presented in term of response spectra at the surface, obtained by setting a structural damping of 5%. For comparison, the elastic response spectrum provided by the Italian seismic code [19] and the mean spectrum are also shown. Figure 5 reports the amplification functions A(f) evaluated as the ratio between the Fourier spectrum at the surface level and the Fourier spectrum of the input motions applied at the base of the model. Finally, according to the guidelines for Seismic Microzonation studies [22] FA factor has been determined (Table 4).

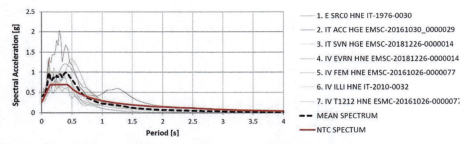

Fig. 4. Comparison between the elastic response spectra obtained by numerical analyses and the same provided by NTC 2018 [18]

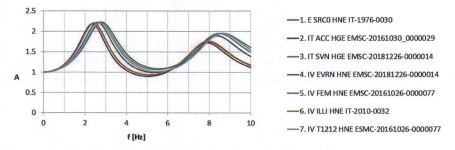

Fig. 5. Amplification functions obtained by numerical modelling

Table 4. evaluation of the amplification factor FA [21]

Period corresponding to the max. spectral acc. of the input spectrum (TAi)	0.20 s
Period corresponding to the max. spectral acc. of the output spectrum (TAo)	0.40 s
0.5 TAi	0.10 s
1.5 TA$_i$	0.30 s
0.5 TA$_o$	0.20 s
1.5 TA$_o$	0.60 s
Mean Value of the input spectrum in 0.5 TAi and 1.5 TAi, SA$_{m,i}$	0.62
Mean Value of the output spectrum in 0.5 TAo and 1.5 TAo, SA$_{m,o}$	0.87
FA = SA$_{m,o}$/SA$_{m,i}$	1.41

5 Concluding Remarks

To evaluate the local seismic response, laboratory and in situ tests have been performed. Results of the site response analyses show high values in soil amplification effects. Moreover, only for periods greater than 0.7 s, the spectrum provided by NTC 2018 [19] is more conservative than that obtained from numerical analyses. Finally, the main resulting frequency is equal to 2.59 Hz and corresponding to the natural frequency of the soil deposit.

References

1. Castelli, F., Cavallaro, A., Grasso, S., Lentini, V.: Seismic microzoning from synthetic ground motion earthquake scenarios parameters: the case study of the city of Catania (Italy). Soil Dyn. Earthq. Eng. **88**(2016), 307–327 (2016)
2. Castelli, F., Grasso, S., Lentini, V., Massimino, M.R.: In situ measuremnts for evaluating liquefaction potential under cyclic loading. In: Proceedings of the 1st IMEKO TC4 International Workshop on Metrology for Geotechnics, Benevento, Italy, 17–18 March 2016, pp. 79–84 (2016)
3. Cavallaro, A., Castelli, F., Ferraro, A., Grasso, S., Lentini, V.: Site response analysis for the seismic improvement of a historical and monumental building: the case study of augusta hangar. Bull. Eng. Geol. Environ. **77**(3), 1217–1248 (2017). https://doi.org/10.1007/s10064-017-1170-9
4. Pergalani, F., Pagliaroli, A., Bourdeau, C., et al.: Seismic microzoning map: approaches, results and applications after the 2016–2017 Central Italy seismic sequence. Bull. Earthq. Eng. **18**, 5595–5629 (2020)
5. Grasso, S., Maugeri, M.: The seismic microzonation of the city of Catania (Italy) for the maximum expected scenario earthquake of January 11, 1693. Soil Dyn. Earthq. Eng. **29**(6), 953–962 (2009). https://doi.org/10.1016/j.soildyn.2008.11.006
6. Grasso, S., Maugeri, M.: The seismic microzonation of the city of Catania (Italy) for the Etna scenario earthquake (M=6.2) of February 20, 1818. Earthq. Spectra **28**(2), 573–594 (2002). https://doi.org/10.1193/1.4000013
7. Grasso, S., Maugeri, M.: Seismic microzonation studies for the city of Ragusa (Italy). Soil Dynamics Earthquake Eng. **56**, 86–97 (2014). ISSN: 0267-7261

8. Gresta, S.: Zone di Fratturazione al Suolo. Rilievo e Perimetrazione. In: Azzaro, R., Carocci, C.F., Maugeri, M.: (eds.) Torrisi, Microzonazione Sismica del Versante Orientale dell'Etna. Studi di Primo Livello, Regione Siciliana, Dipartimento della Protezione Civile. Le Nove Muse editrice, pp. 81–98 (2010)
9. Cavallaro, A., Maugeri, M.: Non linear behaviour of sandy soil for the city of Catania. Seismic Prevention of Damage: A Case Study in a Mediterranean City, Wit Press Publishers, Editor by Maugeri, M.: Advanced in in Earthquake Engineering, vol. 14, pp. 115–132 (2005)
10. Cavallaro, A., Grasso, S., Maugeri, M.: Site Characterisation and Site Response for a Cohesive Soil in the City of Catania. Proceedings of the Satellite Conference on Recent Developments in Earthquake Geotechnical Engineering, Osaka, 10 September 2005, pp. 167–174 (2005)
11. Cavallaro, A., Grasso, S., Maugeri, M.: A dynamic geotechnical characterization of soil at Saint Nicola alla Rena Church damaged by the South Eastern Sicily Earthquake of 13 December 1990. In: Proceedings of the 15th International Conference on Soil Mechanics and Geotechnical Engineering, Satellite Conference "Lessons Learned from Recent Strong Earthquakes", Istanbul, 25 August 2001, pp. 243–248 (2001)
12. Castelli, F., Cavallaro, A., Ferraro, A., Grasso, S., Lentini, V., Massimino, M.R.: Static and Dynamic Properties of Soils in Catania City (Italy). Annals of Geophysics **61**(2), SE221 (2018)
13. Castelli, F., et al.: Caratterizzazione Geotecnica e Risposta Sismica Locale dell'Area del Monastero dei Benedettini a Catania. XXVI Convegno Nazionale di Geotecnica – L'Ingegneria Geotecnica nella Conservazione e Tutela del Patrimonio Costruito, Roma, 20 - 22 Giugno 2017, pp. 507–515 (2017)
14. Yokota, K., Imai, T., Konno, M.: Dynamic deformation characteristics of soils determined by laboratory tests. OYO Tec. Rep. **3**, 13–37 (1981)
15. Kottke, A.R., Rathje, E.M.: Technical Manual for STRATA. PEER Report 2008/10; Univ. of California: Berkeley, CA, USA (2008)
16. Grasso, S., Massimino, M.R., Sammito, M.S.V.: New stress reduction factor for evaluating soil liquefaction in the Coastal Area of Catania (Italy). Geosciences **11**, 12 (2021). https://doi.org/10.3390/geosciences11010012
17. Castelli, F., Grasso, S., Lentini, V., Sammito, M.S.V.: Effects of Soil-Foundation-Interaction on the Seismic Response of a Cooling Tower by 3D-FEM Analysis. Geosciences **11**, 200 (2021) https://doi.org/10.3390/geosciences11050200
18. Ferraro, A., Grasso, S., Massimino, M.R.: Site effects evaluation in Catania (Italy) by means of 1-D numerical analysis. Ann. Geophys. **61**(2), SE224 (2018)
19. NTC D.M. New Technical Standards for Buildings (2018). https://www.gazzettaufficiale.it/eli/gu/2018/02/20/42/so/8/sg/pdf
20. Iervolino, I., Galasso, C., Cosenza, E.: REXEL: computer aided record selection for code-based seismic structural analysis. Bull. Earthquake Eng. **8**, 339–362 (2010). https://doi.org/10.1007/s10518-009-9146-1
21. Iervolino, I., Chioccarelli, E. Convertito, V.: Engineering design earthquakes from multimodal hazard disaggregation. Soil Dynamics Earthquake Eng. **31**(9), 1212–1231 (2011)
22. Dipartimento della Protezione Civile e Conferenza delle Regioni e delle Province Autonome: Indirizzi e criteri per la microzonazione sismica https://www.protezionecivile.gov.it

Influence of Local Soil Conditions on the Damage Distribution in Izmir Bay During the October 30, 2020, Samos Earthquake

Anna Chiaradonna[1]($\boxtimes$), Eyyub Karakan[2], Giuseppe Lanzo[3], Paola Monaco[1], Alper Sezer[4], and Mourad Karray[5]

[1] University of L'Aquila, 67040 L'Aquila, Italy
anna.chiaradonna1@univaq.it
[2] Kilis 7 Aralik University, 79000 Kilis, Turkey
[3] Sapienza University of Rome, 00197 Rome, Italy
[4] Ege University, 3500 Izmir, Turkey
[5] Universitè de Sherbrooke, Sherbrooke, QC J1K 2R1, Canada

Abstract. On October 30, 2020, a damaging earthquake of moment magnitude 6.6 struck about 14 km northeast of the island of Samos, Greece, and about 70 km from the center of the city of Izmir in Turkey. Even though the epicenter was relatively far away, the effects of the seismic event in the highly populated city center of Izmir were destructive causing over 100 fatalities and significant structural damage. Multiple failures of high buildings constituted the major source of the fatalities.

This paper aims to understand the link between the localized damage distribution and the nature of amplification effects that have been observed in Izmir Bay, starting from collection and data analysis interpretation of seismic records and targeted damage assessment of the built environment, as well as geological and morphological characteristics of the area and the geotechnical properties of soils. Critical analysis of the numerous recorded signals shows the key role of the young alluvium and shallow marine deposits of the basin on which Izmir Bay was growing. The coupling mechanism between the frequency content of the shaking and the fundamental frequencies of the damaged buildings contributed to exacerbating the inertial forces acting on the collapsed buildings.

Keywords: Site effects · Double resonance · Recording station · Building damage · 2020 Samos earthquake

1 Introduction

On October 30, 2020, a damaging earthquake of moment magnitude 6.6 struck approximately 14 km northeast of the island of Samos, Greece, and approximately 70 km from the center of the city of Izmir in Turkey (Fig. 1). The rupture occurred on a previously mapped normal fault north of Samos Island, referred to as the North Samos Fault or the Kaystrios Fault, in the Aegean Sea.

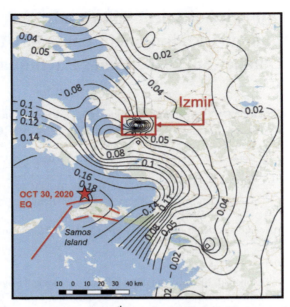

Fig. 1. Map showing the epicentre of 30th October 2020, fault lines around Samos region and contours of peak ground acceleration (PGA) in western Turkey.

The city of Izmir is the third largest city in the country according to the number of inhabitants (about 4 million), and the second largest port in Turkey, after Istanbul. It is located by the shore of the Aegean Sea -Izmir Gulf. Izmir is composed of 30 districts, but the population is concentrated around the Inner Bay, where the majority of the damage was observed.

Although the epicenter was relatively far away from Izmir (Fig. 1), the effects of the seismic event in the city center were destructive, resulting in 117 fatalities, over 1030 injuries, and significant structural damage [1]. Approximately 20 buildings collapsed in the city after the earthquake. The earthquake left more than 5000 people homeless, who were temporarily housed in tents, and a hotel that was partially managed by Izmir metropolitan municipality. In contrast, limited damage was observed on the Greek island, Samos, consisting of only two fatalities and 19 minor injuries, despite the short distance from the epicenter.

In this paper, a critical analysis of the numerous recorded signals was performed to highlight the key role of the young alluvium and shallow marine deposits of the basin on which Izmir was growing, in the amplification and elongation of the incoming motions. In addition, a specific damage assessment of the built environment, as well as geological and morphological characteristics of the area were carried out. The cross checking of the available data was used to preliminary identify possible coupling mechanisms between the frequency content of the shaking and the fundamental frequencies of the damaged buildings which could have been contributed to exacerbate the inertial forces acting on the collapsed buildings.

2 Analysis of the Recorded Motion

The mainshock of the 30th October 2020 event was recorded at several seismic stations in the Turkish strong-motion network. More than 4700 aftershocks were recorded [2], 48 of which had magnitude greater than 4. Figure 1 presents the peak ground acceleration (largest of the two horizontal components) contours based on 160 records obtained from the Turkish seismic network [3], showing that the recorded motion between 60 and 70 km is generally in the range 0.02–0.08 g.

Conversely, Fig. 2 focused on the area around the Izmir, by providing the measured peak ground acceleration (PGA) recorded by the stations located around the Izmir Bay, along with the station number and the subsoil class according to Eurocode 8 code [4] superimposed to the geological map.

Geologically speaking, the city center of Izmir is an actively growing shallow marine basin controlled by E-W trending normal faults active in the stretching Aegean Province. Young alluvium (Holocene) and the fan delta with shallow marine deposits are confined and controlled by the Izmir fault to the south, and the Karşıyaka-Bornova fault to the north constitute the highest sediments in the basin. The basin fill exceeds 300 m in thickness in the middle part of the Bornova Plain [5].

As shown in Fig. 2, the maximum acceleration on the rock subsoil (class A) around the Izmir city center is generally less than 0.06 g. Accelerations reaching 0.14 g around station 3519 and 0.11 g near station 3521 were recorded: these two stations are located on thick alluvial deposits in the center of the valley.

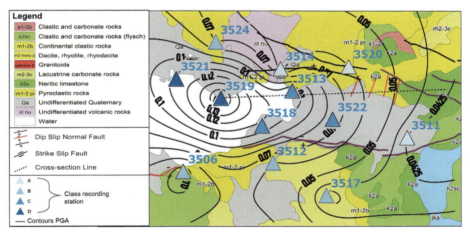

Fig. 2. Recording stations in and around Izmir on different subsoil class sites superimposed to the geological map of the Izmir Bay.

The acceleration response spectra of the recorded signals by stations installed on the rock (class A) were compared with those of class B, C and D in the region of Izmir in Fig. 3. The latter clearly shows that the records from the alluvial basin are richer in terms of energy at lower frequencies (larger periods). The spectra in Fig. 3c show very little variation in spectral acceleration along with a wide period band (plateau between 0.25 s and 1.5 s).

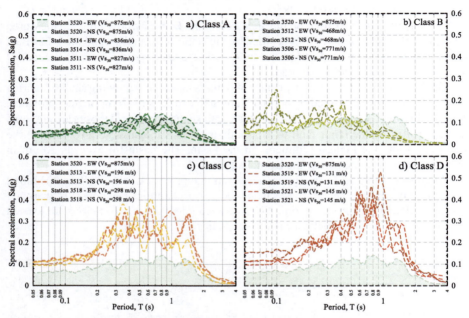

Fig. 3. Acceleration response spectra of the two horizontal component motions (east–west and north–south) registered by different stations installed around Izmir Bay on different classes of soil: a) Class A, b) Class B, c) Class C, d) Class D.

Figure 4 reports the trend of the Arias intensity, calculated for the recorded signals in the range of the Joyner-Boore distance, R_{jb}, between 62.5–79 km, as a function of the equivalent shear wave velocity, $V_{S,30}$ of the correspondent recording stations. Moving from class A to class D (decreasing of $V_{S,30}$), the calculated Arias intensity shows a marked increase, clearly due to the valley effect (subsoil class C and D). Additional details about the interpretation of the recorded motions can be found in [6].

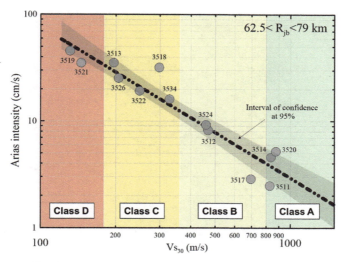

Fig. 4. Arias intensity vs. $V_{S,30}$ in the region of Izmir Bay.

3 Observed Damage Distribution

After the event, teams of engineers employed by the Ministry of Environment and Urbanization (including two authors of this study) conducted reconnaissance studies in Izmir city. Moreover, an independent on-site reconnaissance of the damage was carried out by a team guided by some of the authors at the time of the earthquake. Data merging of all the collected information was then performed, and the classification of the damage was finally carried out on the entire dataset.

Typical damages observed during the Samos earthquake includes inappropriate superstructural design and field application, low quality of concrete and reinforcement detailing, under designed beam-column joints, formation of soft stories, and short columns were more frequently observed (Fig. 5).

Fig. 5. Observed damage to buildings: (a) partially collapsed building (b) improperly designed beam column joints (c) formation of cracks due to torsional irregularity.

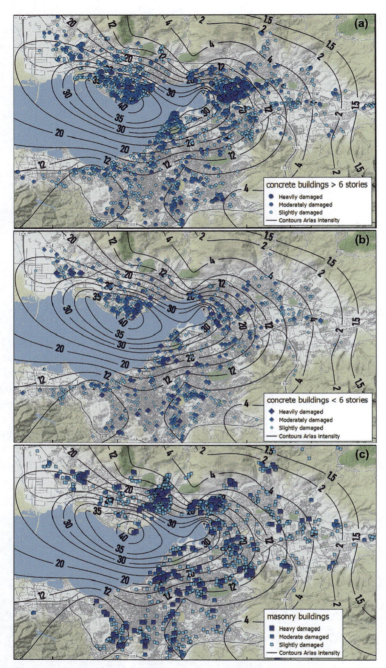

Fig. 6. Distribution of slightly, moderately, and heavily damaged (a) concrete buildings with more than 6 stories, (b) concrete buildings with less than 6 stories, (c) masonry structures correlated with the contours of Arias Intensity in cm/s.

Building stock in Izmir is composed of several types of structures that can be classified into three categories: reinforced concrete buildings with six floors or more, reinforced concrete buildings with less than six floors, and masonry buildings with one to two floors. These structures were built during different periods and were designed according to different seismic design codes ranging from those of 1975, 1998, 2007, and 2018.

During the post-earthquake reconnaissance, buildings with low, moderate, and high damage were determined by observations (Fig. 6). Slightly damaged buildings were distinguished by slender cracks and fissures in the building plaster, walls, and paint caused by ground shaking. These observations lead to the fact that there are no objections to the use of buildings. Next, moderately damaged buildings were defined by discontinuities in frames and slender cracks on load-bearing elements. Building use is not permitted until the damaged elements are reinforced or repaired. Highly damaged buildings showed wide shear cracks and cleavages in the load-bearing elements of the structure, which cannot be reverted to their initial performance by reinforcement and maintenance.

Considering the structural features, separate maps based on building type and height can be made. In this regard, Arias Intensity (AI) contours of the recorded motion was compared to the distribution of slightly, moderately, and heavily damaged concrete buildings with number of stories greater than or equal to 6 (Fig. 6a), less than 6 (Fig. 6b) and masonry structures (Fig. 6c).

Regarding the concrete buildings with number of stories greater than or equal to 6, it is very clear that the heavily damaged buildings are concentrated in a limited area, where the natural period of the structure (0.6–1 s) tends to be strongly excited by the amplified shaking in the same range of periods (0.3–1 s) (see also Fig. 3).

For the concrete buildings with number of stories less than 6, the basin effects are not loudly pronounced as for higher concrete buildings, and the damage distribution is widespread. AI values as low as 6 and 3 can be associated with high and moderate damages, respectively.

Figure 6c shows the distribution of damage in the masonry buildings in Izmir. In fact, after 1970, owing to high seismic threat, almost all buildings are constructed with reinforced concrete, while older structures remain to be masonry. Because there is a great need for housing in Izmir, many of the old structures are demolished and these parcels are used to construct higher reinforced concrete structures. In most houses, these structures can still be encountered throughout Izmir. The masonry structures with a maximum of two stories throughout the city center and southern districts were heavily damaged.

4 Concluding Remarks

The particular features of the Samos earthquake and its effects on the Izmir city center were highlighted along with analyses of the recorded motions in and around Izmir. Site effects due to the local basin overlap with the main characteristics of this far-field event. The amplitude of the shaking is amplified in the basin because of the presence of soft sediments, that is, stations on subsoil classes C and D, exhibit a PGA approximately three times higher than the ground motion recorded on the rock (Fig. 3). The duration of the shaking was increased by multiple reflections in the basin, as revealed by the Arias intensity distribution (Fig. 4). The shaking coming from the south direction, already rich in low frequencies owing to the long distance from the epicenter, is further modified by the deformability of the basin, because the natural period of the basin is estimated to be approximately 1 s (Fig. 3). The resonance between the frequency content of the incoming shaking and the natural period of the basin induced a large plateau in the acceleration response spectra (Fig. 3c) between 0.6 and 1.5 s.

A comparison between the observed damage and mapped AI of the recorded shaking shows the concentration of moderately and highly damaged concrete buildings with more than six stories in the basin area (Fig. 5a). An approximate estimation of the fundamental periods of these high structures is 0.6–0.9 s (the number of stories divided by 10), which is further increased by the soil-structure interaction effects under dynamic loading. It is included in the 0.6–1.5 s range of periods where the higher spectral accelerations were observed (Fig. 3c). This highlights a double resonance effect among ground motion, soil deposits, and high concrete buildings, which reasonably explains the severe damage suffered by such structures.

References

1. Çetin, K.Ö., Mylonakis, G., Sextos, A., Stewart, J.P., et al.: Seismological and Engineering Effects of the M 7.0 Samos Island (Aegean Sea) Earthquake, Geotechnical Extreme Events Reconnaissance Association: Report GEER-069 (2021). https://doi.org/10.18118/G6H088
2. Utku, M.: The aftershock activity of the Samos earthquake (Mw = 7.0) of October 30, 2020: Aftershock regime and a new method for estimating aftershock duration. Arabian J. Geosci. **15**(19), 1–20 (2022)
3. AFAD Homepage. www.afad.gov.tr. Accessed 14 Oct 2021
4. CEN: European Committee for Standardisation. Eurocode 8: Design of Structures for Earthquake Resistance. Part 1: General Rules, Seismic Actions and Rules for Buildings, European Standard EN 1998–1: 2004 (stage 51) (2004)
5. Pamuk, E., Akgün, M., Özdağ, Ö.C., Gönenç, T.: 2D soil and engineering-seismic bedrock modeling of eastern part of Izmir inner bay/Turkey. J. Appl. Geophys. **137**, 104–117 (2017)
6. Karakan, E., Chiaradonna, A., Monaco, P., Lanzo, G., Sezer, A., Karray, M.: Identification of the site effects induced by the 2020 Samos earthquake in Izmir. In: Proceedings of the Mediterranean Geosciences Union Annual Meeting (MedGU-21), Istanbul, Turkey, pp. 25–28, November 2021. Paper number 819 (2021)

Topographic Amplification of Seismic Ground Motion in Hills

Sukanta Das and B. K. Maheshwari

Department of Earthquake Engineering, Indian Institute of Technology Roorkee, Roorkee, India
bk.maheshwari@eq.iitr.ac.in

Abstract. Effect of slopes on topographic amplification has been rarely studied. This is an important issue for performance based design of foundations in hilly areas. The presented numerical study examines the seismic response of slope topography for different site conditions under vertically propagating SV waves. A single face slope, of constant base width, is considered. The influence of slope angle (β) and frequency of excitation on amplification of seismic ground motion is investigated. The two-dimensional Finite Element Analysis (2D-FEA) is adopted for the present seismic analysis of slope-topography. The behavior of soil is assumed as linear elastic. The side boundary is considered as a free field boundary to avoid reflection of ground motion. Complaint base condition is assigned at the bottom of the model. The incident ground motion is applied at the base of the FEA model and a far-field point from toe and crest is defined. The Seismic-Slope Topographic Amplification Factor (S-STAF) is expressed as a ratio of seismic response at the near field (along and surrounding the slope) and response at free field condition (horizontal ground surface). It was observed that the amplification increases as the slope angle and frequency of excitation increases. Further, the distance of point of the maximum peak ground acceleration from the crest increases as the slope angle increases.

Keywords: Seismic-slope topographic amplification factor · Finite element analysis · Vertically propagating SV waves

1 Introduction

For performance based design (PBD) of foundations for earthquake loads, it is important to consider the amplification of ground motion. This amplification has been much studied for horizontal ground surface, howevwer, for slopes, this study has been rarely carried out.

Many monuments and historical sites are situated in hilly areas worldwide. The safety of these structures and sites during earthquakes is an important issue. In many past earthquakes, it was observed that the amplification of seismic ground motion was responsible for damage to structures. This amplification depends on the source, traveling path of the earthquake and site condition. In hilly areas, the topographic amplification is a crucial issue for earthquake-induced disasters. The 1991 Uttarkashi earthquake, 1999 Chamoli earthquake, 1999 Athens earthquake, 2011 Sikkim earthquake and 2015 Nepal earthquake are some of the examples where massive damage occurred due to topographic amplification in hills. The amplification factor depends on several parameters such as type of topography, soil materials properties and frequency contents of the earthquake.

Field reports also indicate that the topography has a significant role in ground motion amplification. However, field measurement needs an adequate recording station and detailed plan, which is problematic. Therefore, numerical techniques are widely used to predict Seismic-Slope Topographic Amplification Factor (S-STAF) quantitatively. Many researchers have studied the effect of slope topography on seismic amplification using the analytical and numerical methods [1–5]. During 1994 Northridge Earthquake in California, a Peak Ground Acceleration (PGA) of 1.58 g was measured at a ridge near the Pacoima Dam, while the PGA at the surrounding horizontal area was calculated as 0.5 g [6]. Similarly, significant topographic amplification was observed during Wenchun Earthquake [7]. Some literature reports that the amplification on the side of the hills ranges between 1 to 3.5 times [8–10], while other studies reported 30 times [11]. After experimental research, Brennan and Madabhushi [12] reported that the amplification of the slope influenced up to approximately 5 m from the crest and 2.5 m deep, away from the slope. However, only few pieces of literature are available on slope topographic amplification using finite element analysis, whereas some of the literature is based on the finite difference method.

In the present study, two-Dimensional Finite Element Analysis (2D-FEA) has been carried out to investigate the seismic-slope topographic amplification factor (S-STAF). Further, a parametric study is performed with varying slope angle and frequency of excitation.

2 Finite Element Modeling

A series of FEA has been carried out in PLAXIS2D to investigate the ground motion amplification due to slope topography. In the present study, homogeneous soil slopes of width w = 50 m with varying slope angles as $\beta = 10°, 20°, 30°$ and $40°$ have been considered. The height of the slope towards toe (H) is considered as 200 m as shown in Fig. 1. In the FEA model, the domain is extended to 700 m from the crest as weill as from the toe. The toe of the slope has been marked as 0 (zero) point i.e. the origin of the coordinates. Two far-field points, a_{ffc} and a_{fft} significantly away from the crest and from the toe, respectively have been considered where the effect of slope topography is insignificant. For soil, the linear elastic model has been considered and details of material properties are listed in Table 1. The material properties of the soil slope considered are the same as reported by Zhang et al. [2].

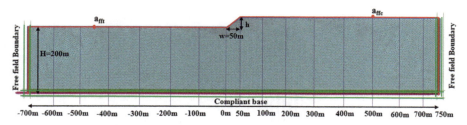

Fig. 1. 2D-FE model used to study the seismic-slope topographic amplification (S-STA)

Table 1. Material properties assumed for the present study (after Zhang et al. [2])

Properties	Value
Unit Weight, γ (kN/m^3)	20
Poisson's Ratio, ν	0.30
Youngs modulus, E (MPa)	500
Shear Wave Velocity, V_s (m/s)	310
Damping Ratio, ζ (%)	5

The maximum size of finite element is determined based on the highest frequency considered (smallest wavelength). Thus the size considered for all the elements is less than 1/8 times of the minimum excitation wavelength [13]. The free-field boundary condition has been assigned at the two sides of the FEA model while the base is subjected to the compliant boundary condition. The type of artificial boundary conditions affect results of the problem. Here for the simplicity the free-field boundary condition is considered at the edges. However, based on the sensitivity analysis, the size of domain considered is large enough so that it does not affect the accuracy of the results.

For the input seismic excitation, the Gabor wavelet or Chang's time history with peak ground acceleration (PGA) of 0.5 m.s^{-2} of vertically propagating shear waves (SV) has been used at the base of the FEA model. The Gabor wavelet is generated using

$$a(t) = \sqrt{\beta e^{-\alpha t} t^\gamma} \sin(2\pi f t) \tag{1}$$

where $a(t)$ is acceleration in m.s^{-2} varying with time t, f is the central frequency in Hz. Further, α, β & γ are shape controlling parameters with values as 6.5, 3.7 and 5, respectivelly as reported by Zhang et al. [2]. Many other researchers [14, 15] have also used the Gabor wavelet as seismic excitation for seismic slope topographic amplification. A typical Gabor wavelet time history with central frequency 6 Hz is shown in Fig. 2. The central frequency is varied to check the influence of frequency on S-STAF.

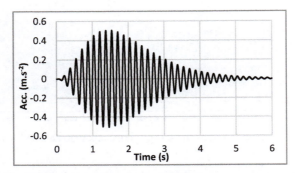

Fig. 2. A typical Gabor wavelet input time history of central frequency 6 Hz

2.1 Validation and Comparison of 1D and 2D Seismic Response

The analytical solution is available to estimate the amplification of horizontal surface ground. Several one-dimensional software (e.g. DEEPSOIL, SHAKE and PEER etc.) are available to examine the amplification factor. In the present study, the seismic response for the horizontal surface ground was computed and compared those obatyined using DEEPSOIL and PLAXIS 1D. The comparison is also made between those obtained using both PLAXIS 2D and PLAXIS 1D. Here PLAXIS 1D means to carry out 1D analysis using the PLAXIS 2D software putting the restrains to convert a 2D problem into a 1D problem. The Gabor wavelet time history at the ground surface has been presented in Fig. 3a and b for central frequency 6 Hz. The results obtained from DEEPSOIL and PLAXIS 1D are in very good agreement. However, the PLAXIS 2D results follow a similar trend and have slightly lesser value than those obtained using PLAXIS 1D due to more release of energy. This validates, the results of PLAXIS 2D and further analysis is carried out using PLAXIS 2D.

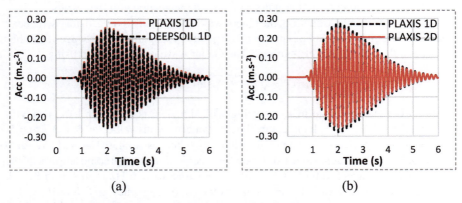

Fig. 3. Gabor wavelet time history at horizontal ground surface for central frequency 6 Hz (a) Comparison between PLAXIS1D and DEEPSOIL (b) Comparison between PLAXIS1D and PLAXIS2D

3 FEA Results and Discussions for S-STAF

The seismic-slope topographic amplification factor (S-STAF) has been investigated using PLAXIS 2D and obtained results have been presented in this section. The S-STAF has been defined as the ratio of PGA along slope surface ($a\beta$) and PGA of free field site (horizontal ground surface) with the same height (a_{ff}), as given in the equation 2.

$$S - STAF = \frac{a_\beta}{a_{ff}} \qquad (2)$$

The effect of slope angle and frequency of the excitation on S-STAF has been examined in following subsections.

3.1 Effect of Variation in Slope Angle

To examine the effect of slope angle, a slope of constant width w = 50 m has been considered with varying slope height according to slope angle. The variation of slope topographic amplification along the slope surface has been presented in Fig. 4. The negative values represent the distance from the toe away from the slope, while the positive value represent the disctance on the crest side, as shown in Fig. 1. The values of coordinate marked in Fig. 4 represents the location of the maximum PGA and value of S-STAF. From Fig. 4(a-d), it can be observed that the seismic-slope topographic amplification factor (S-STAF) is maximum near the crest of the slope. Also as the slope angle increases, the distance of the point of the maximum amplification factor from the crest increases. Further, the value of S-STAF increases with the slope angle.

For all slope angles, maximum ground motion amplification is observed near the crest or bit away from the crest. As distance increases, de-amplification and re-amplification occurs. However, after a certain distance (called as a setback from the crest), the effect of slope topography, gets minimized means the response is similar to that of the free field condition of the horizontal surface ground. In the present study this has been achieved by satisfying the criteria (0.95≤ S-STF ≤1.05). The setback distance (d_s) from crest also depends on the slope angle and it is found that d_s = 170 m, 225 m, 300 m and 325 m for slope angle β = 10°, 20°, 30° and 40°, respectively. Similarly, de-amplification and re-amplification take place as distance increases from the toe of the slope. But after a setback distance (d_s') from the toe, the influence of slope has been found insignificant. For the model shown in Fig. 1, ds' = 30 m, 50 m, 100 and 150 m for slope angle β = 10°, 20°, 30° and 40°, respectively.

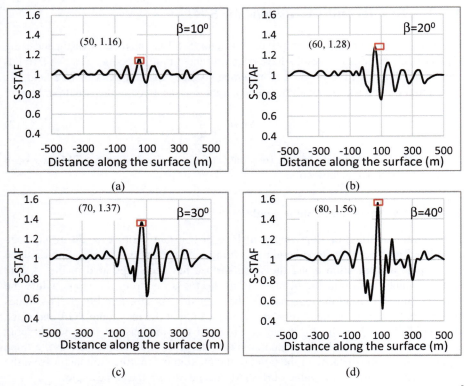

Fig. 4. The variation of S-STF with distance along the surface of slope for $f = 6$ Hz (a) $\beta = 10°$ (b) $\beta = 20°$ (c) $\beta = 30°$ and (d) $\beta = 40°$.

3.2 Influence of Ground Motion Excitation

To study the effect of frequency, the central frequency of the wavelet (f_c) is varied from 1 to 6 Hz. The PGA at the crest has been recorded and values of S-STF has been presented in Fig. 5 for all the slope angles. From Fig. 5a, for $(f_c = 1$ Hz$)$, a small increase in S-STAF with the increase of slope angle can be observed. A moderate change has been found at the central frequency of 2 Hz. However, the S-STAF increases significantly as the frequency increases beyond 2 Hz. For all values of (f_c), the maximum change S-STAF has been observed at slope angle $40°$ and maximum values are for $(f_c = 6$ Hz$)$.

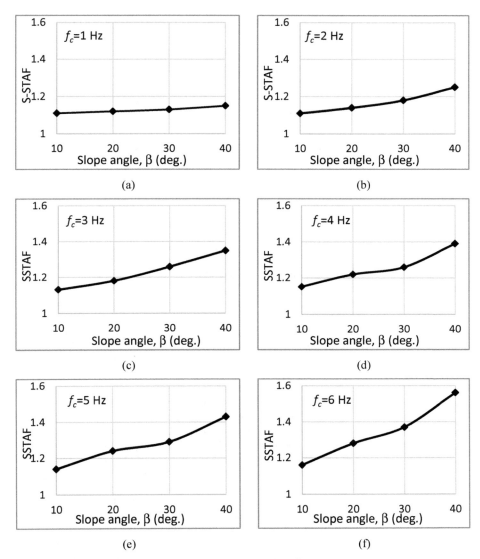

Fig. 5. Influence of ground motion frequency on S-STAF (a) $f_c = 1$ Hz (b) $f_c = 2$ Hz (c) $f_c = 3$ Hz (d) $f_c = 4$ Hz (e) $f_c = 5$ Hz and (f) $f_c = 6$ Hz

4 Summary

Finite element analysis for seismic-slope topographic amplification factor (S-STAF) has been performed. The results indicated that the S-STAF is strongly influenced by slope angle and frequency of the excitation. Based on the present study, the following major conclusions may be drawn:

1. The maximum seismic-slope topographic amplification factor (S-STAF) has been observed near the crest of the slope. As the slope angle increases, its value as well as the distance of the point where the maximum S-STAF occurs also increases. Thus S-STAF increases with the increase of slope angle for the given frequency of the excitation and soil properties.
2. De-amplification occurs at the toe. After a certain distance from the crest and the toe, the effect of slope topography gets minimized. This distance, called as a setback, increases as the slope angle increases.
3. For constant soil properties, the influence of seismic amplification has been found significant at high frequency and at large slope angle.

References

1. Bouckovalas, G.D., Papadimitriou, A.G.: Aggravation of seismic ground motion due to slope topography. In: Proceedings of the First European Conference on Earthquake Engineering and Seismology, p. 1171. Geneva, Switzerland (2006)
2. Zhang, Z., Fleurisson, J.-A., Pellet, F.: The effects of slope topography on acceleration amplification and interaction between slope topography and seismic input motion. Soil Dyn. Earthq. Eng. **113**, 420–431 (2018)
3. Shabani, M.J., Ghanbari, A.: Design curves for estimation of amplification factor in the slope topography considering nonlinear behavior of soil. Indian Geotech. J. **50**(6), 907–924 (2020)
4. Zhang, N., Pan, J., Gao, Y., Zhang, Y., Dai, D., Chen, X.: Analytical approach to scattering of SH waves by an arbitrary number of semicircular canyons in an elastic half space. Soil Dyn. Earthq. Eng. **146**, 106762 (2021)
5. Dai, D., Zhang, N., Lee, V.W., Gao, Y., Chen, X.: Scattering and amplification of SV waves by a semi-cylindrical hill in a half-space by a wavefunction-based meshless method using mapping and point-matching strategies. Eng. Anal. Boundary Elem. **106**, 252–263 (2019)
6. Sepulveda, S.A., Murphy, W., Jibson, R.W., Petley, D.N.: Seismically induced rock slope failures resulting from topographic amplification of strong ground motions: The case of Pacoima Canyon California. Eng. Geol. **80**(3–4), 336–348 (2005)
7. Qi, S., Xu, Q., Lan, H., Zhang, B., Liu, J.: Spatial distribution analysis of landslides triggered by 2008.5.12 Wenchuan Earthquake China. Eng. Geol. **116**, 95–108 (2010)
8. Pedersen, H., Le Brun, B., Hatzfeld, D., Campillo, M., Bard, P.Y.: Ground-motionamplitude across ridges. Bull. Seismol. Soc. Am. **84**, 1786–1800 (1994)
9. Assimaki, D., Kausel, E.: Modified topographic amplification factors for a single-faced slope due to kinematic soil-structure interaction. J. Geotech. Geoenviron. Eng. **133**(11), 1414–1431 (2007)
10. Assimaki, D., Gazetas, G., Kausel, E.: Effects of local soil conditions on the topographic aggravation of seismic motion: Parametric investigation and recorded field evidence from the 1999 Athens earthquake. Bull. Seismol. Soc. Am. **953**, 1059–1089 (2005)
11. Geli, L., Bard, P.Y., Jullien, B.: The effect of topography on earthquake ground motion: A review and new results. Bull. Seismol. Soc. Am. **78**(1), 42–63 (1988)
12. Brennan, A.J., Madabhushi, S.P.G.: Amplification of seismic accelerations at slope crests. Can. Geotech. J. **45**(5), 585–594 (2009)
13. Kuhlemeyer, R.L., Lysmer, J.: Finite element method accuracy for wave propagation problems. J Soil Mech. Found. Div. **99**, 421–427 (1973)

14. Ashford, S.A., Sitar, N., Lysmer, J., Deng, N.: Topographic effects on the seismic response of steep slopes. Bull. Seismol. Soc. Am. **87**(3), 701–709 (1997)
15. Tripe, R., Kontoe, S., Wong, T.K.C.: Slope topography effects on ground motion in the presence of deep soil layers. Soil Dyn. Earthq. Eng. **50**, 72–84 (2013)

Input Ground Motion Selection for Site Response Analysis at the Port of Wellington (New Zealand)

Riwaj Dhakal[1(✉)], Misko Cubrinovski[1], and Jonathan Bray[2]

[1] University of Canterbury, Christchurch 8041, New Zealand
riwaj.dhakal@canterbury.ac.nz
[2] University of California, Berkeley, CA 94720, USA

Abstract. Semi-empirical liquefaction evaluation procedures have several limitations and challenges when applied to nonstandard soils such as those found in reclaimed land. Hence, there is a growing need and benefit of additionally performing advanced numerical analyses. Such techniques can provide unique insights on the dynamic response of soils which are beyond the scope of simplified procedures. A key issue in applying dynamic analyses for back-analysis of case history records is the determination of appropriate input earthquake motions. In this study, recorded ground motions of three earthquakes at sites around Wellington city are deconvolved and used to derive input motions for 1D site response analysis of the reclaimed land at the port of Wellington, New Zealand (CentrePort). Three of the deconvolved sites sit on native soil deposits 200–600 m away from the port, and one nominal rock site is approximately 1 km away. The deconvolved motions are then used for forward analysis at neighboring sites and rigorously compared to recorded motions to scrutinize the appropriateness of the input motions. Surface ground motions of the three sites within the basin incorporate the effects of the complex basin geometry (i.e., basin-edge effects) which are not present at the nominal rock site. Since 1D site response analysis does not simulate these features, the study finds that the three basin sites provide reasonable estimates of the recorded surface motions for 1D analysis compared to the nominal rock site. The recommendations in this study will inform the definition of input motions for future analyses of case-history sites in CentrePort.

Keywords: Site response analysis · Reclaimed land · Earthquake ground motion · Deconvolution

1 Introduction

Reclaimed land is highly vulnerable to liquefaction as recently shown by the major liquefaction-induced damage caused by relatively moderate levels of shaking (recorded peak horizontal ground accelerations 0.2–0.3 g) during the 2016 Kaikōura earthquake (moment magnitude M_w7.8) at the port of Wellington, New Zealand (CentrePort) [1]. Several challenges and limitations in the engineering evaluation of soil liquefaction for

CentrePort reclamations have been identified when utilizing the simplified approach [2, 4–6]. Therefore, insights are needed from more rigorous dynamic site response analyses to further advance our understanding of soil liquefaction in reclaimed land. These analyses utilize advanced numerical techniques able to accurately simulate dynamic soil response observed in experiments, providing important details on the timing of onset of liquefaction, dynamic interactions between soil layers, and consequent severity of liquefaction. However, there are limited studies formalizing the requirements for such analysis procedures, particularly for complex case history settings [11].

A key requirement for performing numerical simulations of the response observed (recorded) in past earthquakes is the definition of a realistic input ground motion, for the site and earthquake of interest [10]. The input motion needs to be defined for a selected reference layer (at a specific depth) which provides the link between the recording site and the analysis site. In this study, recorded motions from several neighboring strong motion station (SMS) sites in the central area of Wellington city are deconvolved and used as candidate reference ground motions for forward 1D analysis at other nearby SMS sites in and around CentrePort. The quality of the input ground motion is then assessed by comparing the computed surface ground motions from the forward analysis with those recorded at the SMS sites. The implicit assumption is that if the surface ground motion is well-estimated, then the input motion is appropriate for a 1D analysis at the site of interest. Findings from this study will inform the selection of input motions for future dynamic analysis of case-history sites at CentrePort including seismic effective stress analyses.

2 Background

The central area of Wellington city is affected by a fault-angle depression on the southeast side of the Wellington Fault, which runs along the northern boundary of CentrePort (Fig. 1). Much of the city sits on naturally deposited weathered alluvium and colluvium, beach, estuarine, and swamp deposits [8]. The original coastline is approximately 200 m to 500 m inland from the current revetment line delineating a belt of reclaimed land developed over the past 170 years, of which CentrePort makes up approximately 0.5 km^2. Reclaimed materials comprise locally sourced end-tipped quarried gravel-sand-silt mixtures (up to 22 m thick at Thorndon Reclamation) and hydraulic fill, consisting of sand and mud from the nearby seabed (up to 10 m thick). Underlying these reclamations are Pleistocene deposits consisting of weathered alluvium, colluvium, and marine deposits. The greywacke basement rock under these deposits is 100–150 m deep at CentrePort, but it outcrops in the hills surrounding central Wellington, at a short distance of 250 m northwest of CentrePort. An aerial view of central Wellington is shown in Fig. 1, including the location of CentrePort, existing coastline, approximate boundary delineating reclaimed land from native soil deposits, and approximate depth to a deep basement rock layer [8].

Local site effects of earthquake ground motions are largely a result of interactions with two major basin structures which characterize central Wellington. Figure 1 identifies the location of the Te Aro and Thorndon basins in the southern and northern areas of central Wellington, respectively. CentrePort reclamations sit on top of the Thorndon

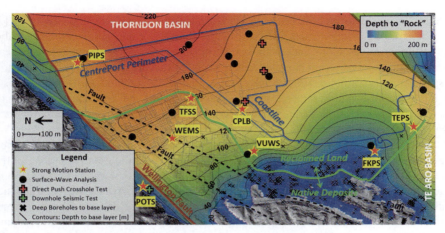

Fig. 1. Aerial view of central Wellington, present coastline, and the CentrePort area with contours of the depth to base layer, major faults (red and dashed black lines), approximate reclamation boundary (green line), locations of strong motion stations, locations of deep boreholes, and location of V_S profiles (base map from [8]).

basin, which is bounded on its western side by the Wellington Fault resulting in a valley structure due to the 80° dip of the fault [8].

3 Defining 1D Subsurface Stratigraphy

When performing site response analyses, stratification of principal subsurface layers is required [11]. These layers must identify characteristics and differences in the elastic properties up until a depth where there is a significant change in the layer's stiffness (i.e., a large impendence contrast), which indicates the depth to the base layer from which earthquake motions are propagated upwards. Elastic properties are characterized by measured shear-wave velocity (V_S). In central Wellington, a reasonably large impedance contrast (and therefore appropriate choice for a base layer) exists at the boundary of deeper alluvial deposits and Greywacke basement rock (characterized by $V_S \approx 1200$ m/s). Multichannel analysis of surface-waves (MASW) have been performed at several regions around central Wellington following the Kaikōura earthquake [12], which are used to develop V_S profile at the locations shown in Fig. 1. Note that for some locations, such as near POTS and CPLB inside CentrePort, downhole and crosshole seismic tests supplement MASW to refine the shallow (top 25 m) V_S profiles.

A key challenge in the use of MASW is that the inversion process to obtain V_S profiles is nonunique. Therefore, information from other in-situ tests are commonly used to constrain the inversion to obtain physically appropriate V_S profiles. At CentrePort, over 100 Cone Penetration Tests (CPTs) and over 30 borehole logs are available to constrain the V_S profile [5], and the profiles can be corroborated to direct push crosshole tests performed at three locations (Fig. 1). The V_S profiles are used to then find a layer boundary where a significant impedance contrast is observed, which is assumed as the depth to the base layer. The basin model developed by [8], which utilizes deep boreholes (available throughout central Wellington as shown in Fig. 1), are also used to constrain the V_s-based estimated depth to the base layer. Depth to the base layer is observed at around 60–90 m around the city (at a distance of 200–600 m from the port), and as deep as 120 m at CentrePort. POTS is located on the uphill side of the Wellington Fault (1 km from the port) and only has 22 m depth to the base layer. POTS is therefore a nominal rock site. Representative V_s profiles for six SMS sites in the Thorndon basin closest to CentrePort are shown in Fig. 2a and 2b. Figure 2c and 2d show V_s profiles defined for different areas of CentrePort based on the constraints from surrounding in-situ and MASW data. TFSS and WEMS are associated with the same V_s profile since only one set of MASW has been performed in the area of the two SMS sites.

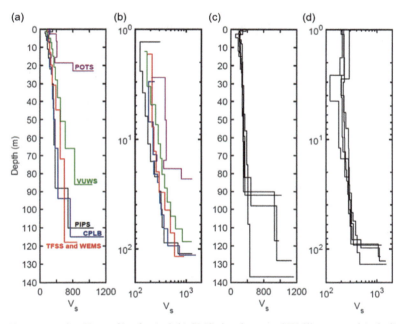

Fig. 2. Representative V_s profiles for (a & b) SMS sites in central Wellington and (c & d) several regions within CentrePort, both shown in (a & c) linear and (b & d) log scale.

4 Selection of Input Ground Motions

4.1 Deconvolution Analysis

Once V_s profiles and reference base layers are identified, the motion at the base layer at any SMS site, for a given earthquake, is derived through deconvolution analysis [10]. This study performs deconvolution using equivalent-linear analysis (ELA) which can relate the response at a given layer to that of another layer through a transfer function. While a key advantage of ELA is its ability to approximate the nonlinear soil response while still performing linear analysis in the frequency domain, it is limited to shear strains of up to 0.5%–1% (e.g., [11]).

While there is evidence of severe soil liquefaction at parts of CentrePort during the Kaikōura earthquake [1], the two SMS sites at CentrePort (CPLB and PIPS) are not located in areas of severe liquefaction. The absence of reduced frequency content and shaking amplitude, typically observed in liquefied sites [10], suggests liquefaction may not have occurred, or may have occurred in a limited capacity at the SMS sites. However, given the possibility of liquefaction occurring, and since liquefaction is associated with large shear strains in the excess of 1%, CPLB and PIPS are not considered appropriate for deconvolution analysis.

Of the remaining six SMS sites in central Wellington, two (FKPS and TFSS) are located in the Te Aro basin, and hence are considered inappropriate for analyses of target sites at CentrePort, in the Thorndon basin. This leaves only four candidate SMS sites in the Thorndon basin for selection of a reference input motion, i.e., VUWS, WEMS, TEPS and POTS.

Deconvolution analyses are performed for two orthogonal horizontal components of recorded ground motions for the Kaikōura earthquake (M_w7.8) and two M_w6.6 earthquakes from 2013, using the 1D ground response analysis program STRATA [9] which approximates soil nonlinearity using ELA with generic modulus reduction and damping curves recommended by [3]. Results from deconvolution of the Kaikōura earthquake

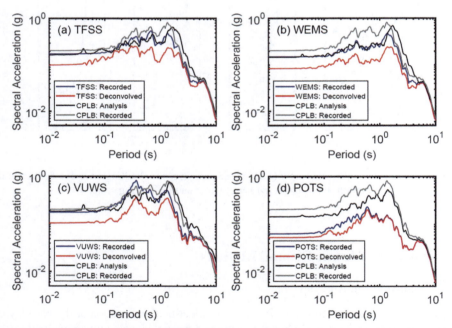

Fig. 3. 5%-damped spectral accelerations of the recorded motion at the surface of a SMS (blue lines), deconvolved equivalent outcrop motion (red lines), calculated motion at the surface of CPLB using the deconvolved motion as input (black lines), and recorded motion at the ground surface of CPLB (grey lines), for SMS sites: (a) TFSS, (b) WEMS, (c) VUWS and (d) POTS.

motion (north-south direction) for the four SMS sites are shown in Fig. 3 as spectral-acceleration plots (red lines). The deconvolution analyses all result in maximum shear strains below 0.1%, except for the TFSS deconvolution of the Kaikōura earthquake north-south component, which reaches strains of 0.15%.

4.2 Forward Analysis for Ground Motion Selection

The deconvolved motions are then used as input motions in a total stress nonlinear analysis at five other SMS sites within the Thorndon basin using the program DEEPSOIL [7]. DEEPSOIL is chosen because it provides a more rigorous modelling of the nonlinear stress-strain behaviour of soils including a more realistic hysteretic damping response and frequency-independent damping formulation.

The resulting spectral accelerations at the ground surface of CPLB are shown as black lines for the four deconvolved motions in Fig. 3. The ratio between computed and observed intensity measures (IMs) and spectral accelerations (i.e., black versus grey lines in Fig. 3) for the Kaikōura earthquake (north-south direction) are shown for CPLB and PIPS in Figs. 4 and 5, respectively. The IMs include amplitude-based measures (i.e., PGA, PGV, and PGD), several energy-based measures expressing cumulative intensity (i.e., AI, SED, SAV, ASI, and VSI), a measure of the mean ground motion frequency (i.e., T_m), and two measures of strong motion duration (i.e., D_{5-75} and D_{5-95}). The average of the spectral acceleration residuals (μ_R) are also shown.

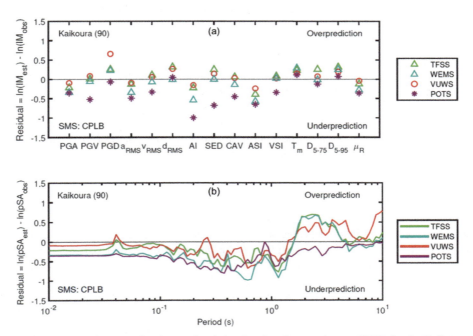

Fig. 4. Residuals between the observed and simulated surface motions at CPLB for the Kaikōura earthquake (east-west direction) using deconvolved motions from four SMS as input: (a) Intensity measures; (b) spectral accelerations.

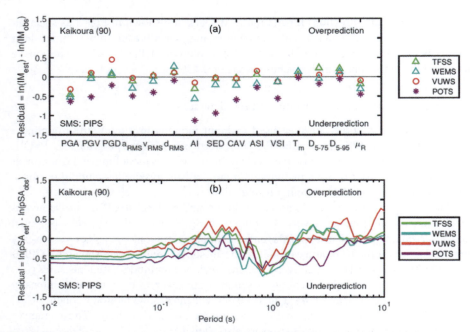

Fig. 5. Residuals between the observed and simulated surface motions at PIPS for the Kaikōura earthquake (east-west direction) using deconvolved motions from four SMS as input: (a) Intensity measures; (b) spectral accelerations.

The results at CPLB and PIPS show that analyses with input motions obtained from VUWS, TFSS and WEMS have the lowest residuals despite the larger depth to the base layer (up to 120 m), while POTS with the shallowest depth to the base layer (<25 m) had the largest residuals. Overall, VUWS generally performed better (albeit marginally) compared to TFSS and WEMS. Therefore, the deconvolved VUWS motions are chosen as the reference input motion for most of CentrePort, with TFSS or WEMS as reasonable alternative options for choice of reference input ground motion.

5 Conclusions

This paper identifies appropriate input motions for 1D site response analysis of sites within the reclaimed land at the port of Wellington, which are affected by complex basin structure. V_s profiles and depth to a base (rock) layer are obtained for the sites within the port reclamations targeted for 1D analysis, the three SMS sites for deconvolution (200–600 m from the port), and the nominal rock SMS site also considered for deconvolution (1 km from the port), all within the same basin structure. The V_s profiles of the four deconvolution sites are used for deconvolving recorded motions from three recent earthquakes, which are assigned as input motions for nonlinear total stress site response analysis at neighboring SMS sites.

Scrutiny of the quality of the forward analysis results through rigorous comparisons with recorded ground motions show that the three SMS sites on native deposits are more appropriate as they result in the smallest residuals of several intensity measures and spectral acceleration, compared to the nominal rock site. The surface ground motions of the three SMS sites within the basin structure contain effects of the complex geometry (e.g., basin-edge effects) which are not present at the nominal rock site. Therefore, deconvolution of such motions for subsequent 1D forward analysis at a neighboring basin site preserves the basin-edge features present in the original recorded motion, which leads to the observed improvement in the quality of the forward simulation and hence compensates for the fact that 1D analyses cannot generate such basin-edge effects. The results demonstrate the ability of the recommended motions to capture key aspects of the ground response, which can be used for future site response studies of CentrePort sites.

References

1. Cubrinovski, M., Bray, J.D., de la Torre, C., Olsen, M., Bradley, B.A., Chiaro, G., Stocks, E., Wotherspoon, L.: Liquefaction effects and associated damages observed at the Wellington CentrePort from the 2016 Kaikoura Earthquake. Bull. N. Z. Soc. Earthq. Eng. **50**(2), 152–173 (2017). https://doi.org/10.5459/bnzsee.50.2.152-173
2. Cubrinovski, M., et al.: Liquefaction-induced damage and CPT characterization of the reclamation at CentrePort Wellington. Bull. Seismol. Soc. Am. **108**(3), 1695–1708 (2018). https://doi.org/10.1785/0120170246
3. Darendeli, M.B.: Development of a new family of normalized modulus reduction curves and material damping curves. PhD thesis, Austin: University of Texas (2001)
4. Dhakal, R., Cubrinovski, M., Bray, J.D., de la Torre, C.: Liquefaction assessment of reclaimed land at CentrePort, Wellington. Bull. N. Z. Soc. Earthq. Eng. **53**(1), 1–12 (2020). https://doi.org/10.5459/bnzsee.53.1.1-12
5. Dhakal, R., Cubrinovski, M., Bray, J.D.: Geotechnical characterization and liquefaction evaluation of gravelly reclamations and hydraulic fills (port of Wellington, New Zealand). Soils Found. **60**(6), 1507–1531 (2020). https://doi.org/10.1016/j.sandf.2020.10.001
6. Dhakal, R., Cubrinovski, M., Bray, J.D.: Evaluating the applicability of conventional CPT-based liquefaction assessment procedures to reclaimed gravelly soils. Soil Dyn. Earthq. Eng. **155**, p. 107176 (2022). https://doi.org/10.1016/j.soildyn.2022.107176
7. Hashash, Y.M.A., Musgrove, M.I., Harmon, J.A., Groholski, D.R., Phillips, C.A., Park, D.: DEEPSOIL 6.1, User manual (2016)
8. Kaiser, A.E., et al.: Updated 3D basin model and NZS 1170.5 subsoil class and site period maps for the Wellington. GNS Science Consultancy Report 2019/01, Project 2017-GNS-03-NHRP (2019)
9. Kottke, A.R., Rathje, E.M.: Technical manual for strata. PEER Report No. 2008/10, University of California Berkeley, Berkeley, California (2009)
10. Ntritsos, N., Cubrinovski, M., Bradley, B.A.: Challenges in the definition of input motions for forensic ground-response analysis in the near-source region. Earthq. Spectra (2021). https://doi.org/10.1177/87552930211001376
11. Stewart, J.P., Afshari, K., Hashash, Y.M.A.: Guidelines for performing hazard-consistent one-dimensional ground response analysis for ground motion prediction. PEER Report No. 2014/16, University of California, Berkeley (2014)
12. Vantassel, J., Cox, B., Wotherspoon, L., Stolte, A.: Deep shearwave velocity profiling and fundamental site period measurements at CentrePort, Wellington and implications for locate site amplification. Bull. Seismol. Soc. Am. **108**(3) (2018). https://doi.org/10.1785/0120170287

Responses of a Cantilever Retaining Wall Subjected to Asymmetric Near-Fault Ground Motions

Seong Jun Ha, Hwanwoo Seo, Hyungseob Kim, and Byungmin Kim[✉]

Ulsan National Institute of Science and Technology, 50 UNIST-gil, Ulju-gun,
Ulsan 44919, Republic of Korea
byungmin.kim@unist.ac.kr

Abstract. Fault rupture produces pulse-like ground motions in the near-fault regions, which are significantly different from ordinary ground motions in the far-field regions with respect to both intensity and frequency contents. The pulse-like ground motions are characterized as high amplitude and long period pulses that could cause severe damage to structures. In this study, we investigated how the pulse-like ground motions have effects on the responses of a cantilever retaining wall by performing a series of numerical simulations. The seismic behaviors of the retaining wall were numerically modeled by adopting a finite difference scheme. High amplitude ground motions scaled to a fixed PGA were first collected as an input dataset and classified into pulse-like, non-pulse-like, and ambiguous motions by a pulse indicator. Then, differences in the development of displacements at the wall were quantitatively compared between the types of the motions. Additional simulations were carried out with the original and inverted input ground motions to investigate the effect of asymmetry of ground motion on the wall responses. It turned out that the asymmetrical ground motions with larger velocity amplitudes in the direction coincide with the relative wall movement could generate significant wall displacements.

Keywords: Retaining walls · Near-fault motions · Pulse-like motions · Fling-step motions · Wave asymmetry

1 Introduction

The fault rupture process could generate pulse-like ground motions near the rupture front in the near-fault regions. The pulse-like ground motions typically have high amplitude and long-period pulses in a velocity history, which are different motion characteristics with far-field ground motions [1]. Strong earthquakes in Northridge (the USA, 1994), Kocaeli, (Turkey, 1999), and ChiChi (Taiwan, 1999) had enormously increased the PGV of the pulse-like motions by more than 100 cm/s, which had caused severe damages to infrastructures and private properties [2]. As such instances were frequently reported in the 1990s, the destructive nature of the pulse-like ground motions began to draw attention in related industries and academia.

This study investigated the response characteristics of the retaining wall subjected to the pulse-like ground motions using numerical modeling. The difference from the previous studies was that we selected the near-field ground motions as an input database for numerical simulation, not the far-field ground motions. The near-field ground motions not involving velocity pulses (i.e., non-pulse-like) may still give as much damage to the structures as the pulse-like ground motions. The high-amplitude near-field ground motions were collected from PEER NGA-west2 database and then, classified into pulse-like, non-pulse-like, and ambiguous ground motions. Numerical results using the collected ground motions were quantitatively analyzed by comparing the wall responses between the types of motions.

We additionally investigated the wave asymmetry effect on the response characteristics of the retaining walls. We inferred that this wave asymmetry could have a significant effect in the case of retaining walls that behave differently depending on the direction of lateral movement. Particularly, the responses of retaining wall might be unexpectedly amplified when it is subjected to the pulse-like ground motion that inherently features high asymmetrical characteristics owing to one-sided pulse. Then, we additionally performed numerical simulations with the inverted versions of the collected ground motions and compared the wall responses between original and inverted motion cases.

2 Near-Fault Ground Motions

The input dataset for numerical simulations included high amplitude ground motions with PGV above 30 cm/s of PEER NGA-west2 database. The ground motions in the dataset were classified into three types of pulse-like, non-pulse-like, and ambiguous motions using the wavelet analysis given by Baker [3]. The wavelet analysis includes the extraction of velocity pulses from a ground motion using the wavelet decomposition and then the classification of the motion according to a pulse indicator that depends on both PGV and energy of the extracted pulses. A ground motion is a pulse-like motion if a pulse indicator is above 0.85 and a non-pulse-like motion if it is below 0.15. It is considered an ambiguous motion if a motion does not belong to pulse-like or non-pulse-like motions. Figure 1a shows the velocity histories of pulse-like and non-pulse-like motions that were classified. The pulse-like motion possesses a noticeable velocity pulse with a high PGV value while the non-pulse-like motion just looks like a typical far-field motion.

We representatively selected 24 pulse-like motions, 24 non-pulse-like motions, and 10 ambiguous motions from the dataset as input ground motions for numerical simulations. Figure 1b shows the distribution of the selected motions against their epicentral distance. It is confirmed that most of the motions were previously recorded within 60 km of the epicenter. That is, the selected motions are considered near-fault ground motions. In addition, the motions occurred from strong earthquakes whose moment magnitude scales (M) were 5 or higher. Finally, the PGA values of the selected input motions were scaled to 0.4 g to compare numerical results between the pulse-like and the non-pulse-like motion cases under the fixed PGA value.

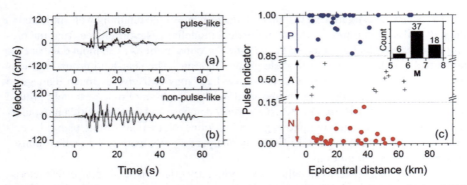

Fig. 1. (a) Velocity-time histories of pulse-like and non-pulse like motions (RSN = 316 and 786 in PEER NGA-west2 database), and (b) characteristics of input ground motions: pulse indicator against epicentral distance (The letters "P", "A", and "N" represent pulse-like, ambiguous, and non-pulse-like motions, respectively).

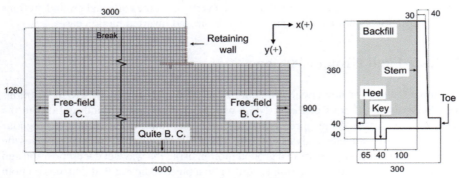

Fig. 2. Drawing of a modeled cantilever retaining wall with its dimensions (left) and numerical domain including boundary conditions used in the analyses (right).

3 Numerical Modeling

Seismic responses of a retaining wall were numerically studied by performing explicit finite difference analyses, given by FLAC2D (Itasca Consulting Group). The retaining wall modeled was inverted T-shape geometry with 4-m height, as shown in Fig. 2. The cantilever retaining wall was modeled with elastic beam elements with unit weight of 24.5 kN/m^3 and design strength of 24 MPa based on JSCE report. Sand was modeled as backfill and foundation soils, which behave as elastoplastic with the Mohr-Coulomb yield criterion. The soils also were modeled with hysteretic and Rayleigh damping ratios, using the Default model given by FLAC2D. Interface elements with interface properties (normal and shear stiffness, interface friction angle, and adhesion, listed in Table 1) were used for modeling interaction between the wall and the backfill soil. Properties of modeled materials were summarized in Table 1. Free-field and quite boundary conditions, respectively, were assigned at the sides and the bottom of the numerical domain.

The ground motions were finally input at the bottom of the domain as shear-stress time series. Details on the numerical mode are available in Seo et al. [4].

Table 1. Properties of modeled retaining wall, soil, and interface

Materials	Properties	Values
Retaining wall	Unit weight, γ_{rw} (kN/m^3)	24.5
	Poisson's ratio, ν_{rw}	0.2
	Design strength, f_{ck} (MPa)	24
	Elastic modulus, E_{rw} (MPa)	$2.3 \cdot 10^4$
	Second moment of area, I_{rw} ($10^{-3} \cdot$m^4)	2.25 to 5.33
Soil	Unit weight, γ_s (kN/m^3)	18
	Poisson's ratio, ν_s	0.3
	Elastic modulus, E_s (kPa)	$1.91 \cdot 10^5$
	Cohesion, c (kPa)	0
	Internal friction angle, ϕ (°)	32
	Dilation angle, ψ (°)	0
Interface	Normal stiffness, k_n (kPa)	$8.6 \cdot 10^6$
	Shear stiffness, k_s (kPa)	$8.6 \cdot 10^6$
	Friction angle, δ (°)	21.3
	Adhesion, a (kPa)	0

4 Responses of a Retaining Wall: Results and Discussion

4.1 Effect of Pulse-Like Motions

Figure 3 shows velocity-time histories of pulse-like and non-pulse-like motions and corresponding evolutions of horizontal displacement at the top of the wall. The distinct difference between the two velocity-time histories was visually observed such that the pulse-like motion includes a high amplitude and long period pulse at 9 s, unlike the non-pulse-like motion. Besides, long periods were noticeably more pronounced overall in the pule-like motion.

It was also observed that total displacement profiles in both cases were clearly different with respect to amplitude and period of oscillations. In the pulse-like motion case, total displacement largely oscillated with a long period over the whole duration. This oscillation occurred largest up to the range of 80 cm near the velocity pulse. The total displacement in the non-pulse-like motion case generally developed in a positive direction (right side of a wall) with high-frequency oscillations, and then shortly converged to an asymptotic value. In both motion cases, the total displacement profiles seemed to be

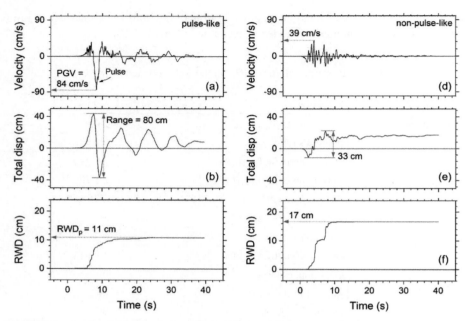

Fig. 3. Velocity-time histories and corresponding simulation results of total displacement and RWD histories at the top of a retaining wall for (a-c) pulse-like and (d-f) non-pulse-like motions (RSN = 178 and 753 in PEER NGA-west2 database)

greatly influenced by a peak amplitude and a major period component of each velocity history.

We also calculated relative wall displacement (i.e., RWD) by subtracting free-field displacement from total displacement. The RWD value has been considered as a primary measure to evaluate the stability of a wall during design procedures in geotechnical works [5, 6]. The free-field displacement was obtained at the same height of the top of the wall in the left-side free-field zone. In Fig. 3c and 3f, the RWD value continued to increase in a positive direction and eventually converged to a maximum value. This maximum value is herein called permanent RWD (i.e., RWD_p) because it still holds after earthquake events. The RWD_p was comparably larger in the non-pule-like motion case (17 cm) than in the pulse-like motion case (11 cm). Unlike the total displacement, the velocity pulse in the pulse-like motion did not contribute to the increase of the RWD_p and even the gradient of the RWD.

The RWD_p values were obtained from the numerical results of all the input motions and compared between the types of the motions, as shown in Fig. 4a. The values of RWD_p for each type of motion were distributed in a similar range and were not particularly large in the pulse-like motion cases. This implied that the velocity pulse itself does not greatly increase the RWD_p value even with its high amplitude. Malhotra [7] also reported that the responses of buildings subjected to a pule-like motion are affected more by peak motion parameters (e.g., PGA, PGV, PGD) than by velocity pulse. Then, Fig. 4b-d shows how the RWD_p value was related to PGV, PGD, and I_{arias} values in the case of the retaining wall. In Fig. 4b, the RWD_p value roughly increased with the PGV value, but its increasing extent was different for each type of motion. The pulse-like motion should have a larger PGV value than the non-pulse-like motion to induce the same RWD_p. In Fig. 4c, it was difficult for the PGD values to find a correlation with the RWDp values. Finally, Fig. 4d shows that the RWD_p values were determined by the I_{arias} of the motion regardless of the type of the motion. Among the ground motion parameters, the I_{arias} was judged to be most reliable in predicting the permanent displacement of the retaining wall subjected to earthquake loading.

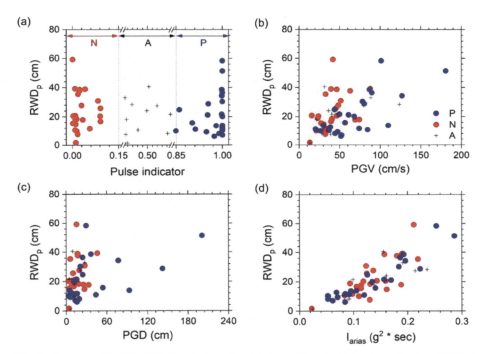

Fig. 4. Permanent RWD (RWD_p) at the top of a retaining wall against (a) pulse indicator, (b) PGV, (c) PGD, and (d) Arias intensity (I_{arias}) of pulse-like, non-pulse-like, and ambiguous motions

4.2 Effect of Wave Asymmetry

The effect of wave asymmetry on the wall responses was investigated with the numerical simulations using original and inverted input motions of the dataset. Figure 5 shows the evolutions of RWD values for original and inverted motions, in representative pulse-like and non-pulse-like motion cases. In the pulse-like motion case (Fig. 5a and b), the RWD values for the original and inverted motions (i.e., RWD_o and RWD_i) differently evolved even though a pair of motions shares the same intensity and frequency contents. The RWD_o value increased to the large extent for the original motion in which the pulse pointed in a positive direction (right side of a wall). As shown in Fig. 5c, the difference between RWD_o and RWD_i values suddenly widened from the vicinity of the velocity pulse and then converged to 17.5 cm. Thus, the direction of the velocity pulse appeared to play a vital role in causing this difference for the pulse-like motion case. The final value of 17.5 cm was also nonnegligible such that it was almost half of permanent RWD_o. The same direction (right side of a wall) of the positive pulse and the wall movement probably provided favorable conditions to further develop the RWD value for the original motion. Besides, the evolutions of the RWD values for the original and inverted motions closely resembled each other over an interval other than the vicinity of the velocity pulse. This similarity between a pair of motions would probably be attributed to the low and symmetrical amplitude of the motions during the corresponding interval.

In the case of the non-pulse-like motion, there was also a clear difference in RWD histories between original and inverted motions. Unlike the pulse-like motion case, the RWD_i value for the inverted motion developed more rapidly and recorded a larger permanent RWD_i. The difference between RWD_o and RWD_i values gradually decreased over a longer duration, compared to the pulse-like motion case. These results confirmed that the responses of the retaining wall could be significantly different depending on the direction of the ground motions against the wall even if the motions have the same amplitude and frequency contents.

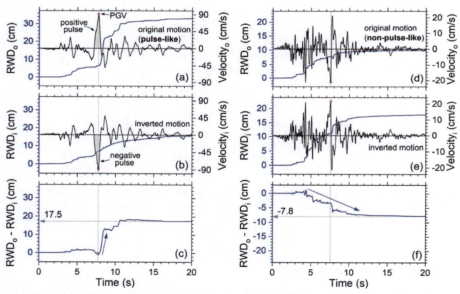

Fig. 5. Time series of (a) RWD using the original ground motion (RWD_o); (b) RWD using the inverted ground motion (RWD_i); and (c) the difference between RWD_o and RWD_i for the pulse-like ground motion (RSN = 159 in PEER NGA-west2 database). (d-f) RWD_o; RWD_i; and RWD_o-RWD_i for the non-pulse-like ground motion (RSN = 952)

5 Conclusions

The pulse-like motions generated a larger oscillation of the total horizontal displacement at the top of the wall with its high amplitude PGV, compared to the non-pulse-like motions. Unlike the total wall displacement, the development of the relative wall displacement (i.e., RWD) was not distinctly different between the pulse-like and non-pulse-like motion cases. The velocity itself did not greatly contribute to the increase of the permanent RWD value and the gradient of the RWD. It was rather revealed that Arias intensity of the ground motions strongly controls the development of the RWD than other motion parameters (e.g., PGV and PGD). Therefore, distinguishing whether the ground motion is pulse-like or not might be meaningless in predicting the relative wall movement against backfill soil in the near-fault regions. In addition, the pulse-like motions are highly asymmetrical owing the one-sided velocity pulse, which could result in significant RWD if the direction of the pulse is aligned with that of the relative wall movement. This wave asymmetry effect would probably act on all structures that exhibit asymmetric behavior according to the moving direction.

References

1. Somerville, P.G.: Engineering characterization of near fault ground motions. In: Proceedings of the Conference NZSEE, pp. 1–8 (2005)
2. Moustafa, A., Takewaki, I.: Deterministic and probabilistic representation of near-field pulse-like ground motion. Soil Dyn. Earthq. Eng. **30**, 412–422 (2010)
3. Baker, J.W.: Quantitative classification of near-fault ground motions using wavelet analysis. Bull. Seismol. Soc. Am. **97**, 1486–1501 (2007)
4. Seo, H., Kim, B., Park, D.: Seismic Fragility evaluation of inverted t-type wall with a backfill slope considering site conditions. KSCE J. Civ. Environ. Eng. Res. **41**, 533–541 (2021)
5. Richards, R., Jr., Elms, D.G.: Seismic behavior of gravity retaining walls. J. Geotech. Eng. Div. **105**, 449–464 (1979)
6. Anderson, D.G.: Seismic analysis and design of retaining walls, buried structures, slopes, and embankments, vol. 611. Transportation Research Board (2008)
7. Malhotra, P.K.: Response of buildings to near-field pulse-like ground motions. Earthq. Eng. Struct. Dyn. **28**, 1309–1326 (1999)

Modeling Two-Dimensional Site Effects at the Treasure Island Downhole Array

Mohamad M. Hallal[1](✉) , Peyman Ayoubi[2] , Domniki Asimaki[2] , and Brady R. Cox[3]

[1] The University of Texas at Austin, Austin, TX 78712, USA
mhallal@utexas.edu
[2] California Institute of Technology, Pasadena, CA 91125, USA
[3] Utah State University, Logan, UT 84322, USA

Abstract. Numerous studies have challenged the wide applicability of one-dimensional (1D) ground response analyses (GRAs), seemingly due to the variable subsurface geologic conditions present at most sites. However, while two- and three-dimensional (2D and 3D, respectively) GRAs could be used to overcome these limitations, the lack of site-specific 3D subsurface models and the daunting computational costs of current software are impediments to improved modeling of site effects in engineering practice. In this paper, we aim to address both of these challenges by utilizing: (1) a practical framework called the 'H/V geostatistical approach' to develop large-scale, site-specific, 3D shear wave velocity models, and (2) an optimized open-source, finite element software called 'SeismoVLAB' to perform large-scale 2D/3D finite element analyses. The investigations are performed at the Treasure Island Downhole Array (TIDA). By incorporating cross-sections across different lateral extents and azimuths and validating the site response predictions relative to recorded earthquake motions in the downhole array, we show that the site-specific 3D Vs model from the H/V geostatistical approach is capable of replicating wave scattering and more complex wave propagation phenomena observed in the field.

Keywords: Site response · Spatial variability · Ground motions · Numerical simulation · Treasure Island downhole array

1 Introduction

Over the past decade, numerous studies have examined ground motions recorded at borehole array sites and found that, on average, more than 50% of sites are poorly modeled using one-dimensional (1D) ground response analyses (GRAs) [1]. These discrepancies have been generally attributed to limitations of conventional 1D site assumptions, which disregard complex wave propagation effects resulting from laterally variable subsurface conditions. Given the sizable percentage of poorly modeled sites based on 1D GRAs, and that subsurface spatial variability has been found to significantly influence site response, many studies have attempted to perform two- and three-dimensional (2D and 3D, respectively) GRAs. While these studies have provided useful insights on how complex subsurface conditions influence site response, the vast majority have modeled spatial variability using stochastic methods, such as spatially correlated random fields [2–4]. These methods require assuming generic input parameters, which may not account for the site-specific variability in a meaningful way, introducing significant uncertainties. In addition, many existing studies have been limited to idealized single-layer profiles, have performed theoretical assessments without validation against observations, and/or have been scaled down due to the poor computational performance of existing finite element software.

The lack of site-specific 2D/3D subsurface models needed for multi-dimensional GRAs and their daunting computational costs are fundamental limitations that hinder improved modeling of site effects in engineering practice. In this paper, we aim to address both of these challenges by utilizing a framework called the 'H/V geostatistical approach' to develop large-scale, site-specific, 3D shear wave velocity (Vs) models, and (2) a new, open-source, finite element software called 'Seismo-VLAB' (SVL) to optimize large-scale finite element analyses. Although this study is working towards the goal of modeling 3D site response in SVL, in this paper we present results from 2D GRAs as a steppingstone to 3D. The investigations are performed at the Treasure Island Downhole Array (TIDA) and involve 2D GRAs using cross-sections across different azimuths and lateral extents. The influence of the incorporated cross-section is assessed relative to recorded earthquake motions.

2 Methods

2.1 Modeling Spatial Variability Using the H/V Geostatistical Approach

Hallal and Cox [5, 6] developed a framework called the H/V geostatistical approach to build site-specific, pseudo-3D Vs models over large areas (hundreds of square meters). The approach integrates simple, horizontal-to-vertical spectral ratio (H/V) noise measurements with geostatistical modeling tools and a single borehole Vs profile. The framework and its application at TIDA are thoroughly discussed in Refs. [5, 6]. First, spatially distributed fundamental site frequency (f_0) estimates from H/V noise measurements ($f_{0,H/V}$) were collected across the site and then kriged to obtain a uniformly estimated $f_{0,H/V}$ map, as shown in Fig. 1a. Then, an invasively measured Vs profile from PS suspension logging [7] was repeatedly modified by scaling all of its layer thicknesses such that the modified Vs profiles match the $f_{0,H/V}$ of each cell in the grid (i.e., Fig. 1a) and form a pseudo-3D Vs model, as shown in Fig. 1b. As further discussed in Refs. [5, 6], the choice of scaling only layer thicknesses without changing Vs of each layer is a simplifying assumption, but which has been shown to yield representative subsurface models while maintaining a reasonable degree of simplicity.

The H/V geostatistical approach has been implemented to develop 3D Vs models covering large areas at two sites, including the TIDA site investigated in this paper. By performing 1D GRAs for each grid-point Vs profile in the pseudo-3D Vs model and statistically combining the results as a means to roughly account for spatial variability, Hallal et al. and Hallal and Cox [1, 6] argued that this approach yields site response predictions that reasonably match recorded earthquake ground motions in the downhole

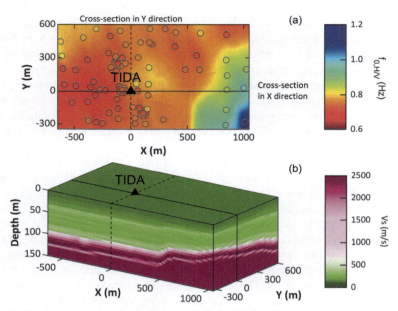

Fig. 1. (a) Plan view of the measured (circle symbols) and uniformly estimated $f_{0,H/V}$ at TIDA and (b) the resulting pseudo-3D Vs model developed using the H/V geostatistical approach.

array. In addition, the authors hypothesized that the site response at TIDA is influenced by subsurface conditions as much as 1 km from the downhole array. The current study expands on the previous work by performing 2D GRAs to: (1) further examine the ability of the pseudo-3D Vs model in capturing spatial variability of interest to site response, and (2) test the hypothesis regarding the large lateral extent of the spatial area influencing site response.

2.2 Description of Numerical Model

We perform 2D viscoelastic site response using SVL, which is designed to optimize mesoscale (i.e., order of kilometers) simulations, and has been shown to be computationally efficient compared to other dynamic finite element software (e.g., OpenSees) [8]. Figure 2 illustrates many of the features of the 2D site response model for the full cross-section in the X direction (refer to Fig. 1). The model is 150-m thick and 4000-m wide and is discretized using 5-m quad elements. This allows resolving frequencies up to 2 Hz, which includes the fundamental (0.75 Hz) and first higher mode (1.9 Hz). Free-field (FF) boundary conditions were prescribed at sufficient distances to minimize boundary effects. This comprises shear- and compression-wave Lysmer dashpots [9], and FF equivalent forces corresponding to 1D wave propagation conditions, as shown in Fig. 2. A vertically propagating plane SV Ricker wavelet is used as the input excitation and is applied as a shear force at the base. Rayleigh damping was used with a target damping ratio of 0.75% at the fundamental and first higher mode frequencies to approximately match laboratory-based material damping in that frequency range [10].

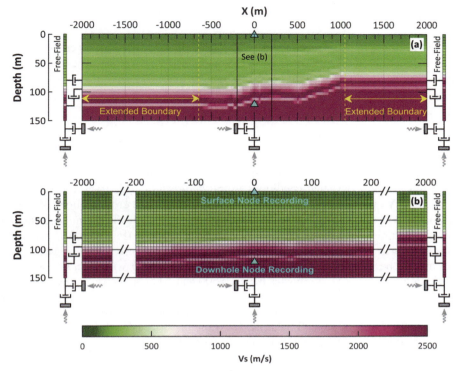

Fig. 2. (a) Schematic of the finite element model, boundary conditions, surface and downhole acceleration recorder node locations, and extended boundaries for the full 2D cross-section in the X direction (Fig. 1a). The vertical scale is stretched by a factor of 7. (b) Zoomed in view of (a) illustrating the finite element mesh, which is discretized using 5 m × 5 m quad elements.

Several 2D GRAs were performed by incorporating cross-sections along different azimuths/directions and different lateral extents. Cross-sections along both X and Y directions from the pseudo-3D Vs models were analyzed (refer to Fig. 1). In addition, different lateral extents from each cross-section, such as the full cross-sections (e.g., between X of [−650, 1050] m) or narrower portions (e.g., between X of [−250, 250] m and X of [−600, 600] m), were investigated to assess the influence of lateral extent on the predicted site response at TIDA. Regardless of the incorporated lateral extent, the material properties at either boundary were extended such that the full model width is 4000 m (Fig. 2a) for all 2D GRAs. Therefore, all analyses had a total of 24000 elements. To capture the full wavefield and enable accurate Fourier Spectra computations, the total duration of the analyses was 15 s with 1501 time steps. Analyses were performed on the Texas Advanced Computing Center's (TACCs) cluster Stampede2. Each analysis took around 13 min using a single Skylake (SKX) node.

3 Results

We first illustrate the influence of subsurface variability from the full Vs cross-sections in each direction on the simulated ground motions and then we compare site response predictions from different lateral extents to observations at TIDA.

3.1 Illustrative Examples for Two Cross-Sections

Figure 3 compares normalized acceleration time series at ground surface nodes for the full cross-sections in the X and Y directions (refer to Fig. 1). From Fig. 1a, it is evident that there is relatively higher variability in $f_{0,H/V}$ in the X direction compared to the much more consistent conditions in the Y direction. This translates to a more pronounced Vs variability in the X direction, as evident in the cross-sections of Fig. 3, which consequently influences the simulated surface wavefield. The simulated surface acceleration time series for the cross-section in the X direction (Fig. 3a) show scattering and surface waves can be seen travelling from right to left. These surface waves appear to primarily be generated from the refraction of non-vertical wave incidences above the sloping bedrock at x-coordinates between 500 and 1000 m. However, the simulated wavefield for the cross-section in the Y direction (Fig. 3b) is very consistent, and similar to what would be observed for vertical wave propagation in 1D conditions. To further scrutinize these effects, we will focus next only on the surface and downhole recordings at the location of the TIDA accelerometers (refer to Fig. 2).

3.2 Influence of Incorporated 2D Cross-Sections on Transfer Functions

Figure 4 compares 2D GRA results from cross-sections with different lateral extents in the X and Y directions. Also shown are simulation results from 1D conditions (no variability) using the measured Vs profile at TIDA. To validate the results against earthquake motions, theoretical transfer functions (TTFs) were calculated as the ratio of the Fourier Amplitude Spectrum (FAS) of the simulated surface recording to that of the simulated downhole recording for each cross-section. Next, suites of more than 30 recorded, low-amplitude, earthquake motions were used to compute an empirical transfer function (ETF), which is then represented using the lognormal median and the lognormal standard deviation ($\sigma_{\ln ETF}$) of all recorded earthquake ground motions. For more details on the empirical site response calculations, please refer to Ref. [1].

Figure 4 highlights the significance of the azimuth and incorporated lateral extent of the 2D cross-sections on the site response predictions. Similar to Fig. 3, results from cross-sections in the Y direction with different lateral extents are very comparable to those from 1D conditions and overpredict the peak amplitudes relative to the ETF (Fig. 4g). On the contrary, results from cross-sections in the X direction show increased wave scattering with increasing lateral extent. This is evident in the reduced TTF peak amplitudes, which are in good agreement with the ETF for cross-sections wider than 500 m (Fig. 4c). These results support those from Hallal and Cox [6], who suggested that 400-m cross-sections might be considered as a minimum over which to appropriately account for spatial variability in GRAs.

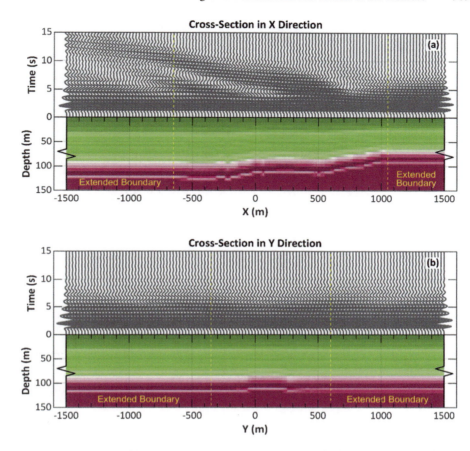

Fig. 3. Comparison of simulated surface node acceleration time series for cross-sections in (a) the X direction and (b) the Y direction (Fig. 1). The acceleration time series are normalized by the peak acceleration and are extracted at 1 in every 5 surface nodes (i.e., every 25 m). The cross-sections show the respective Vs models with the depth being stretched by a factor of 4.

Interestingly, Fig. 4c shows that the 3D Vs model also captures more complex wave propagation phenomena, even using 2D cross-sections. As previously discussed, it appears that the shallower bedrock causes the generation of surface waves, which are evident in the surface acceleration time series beyond 5 s for the X [−650, 1050] m cross-section (Fig. 4b). These reverberations result in a secondary TTF peak at approximately 1 Hz, similar to the ETF (Fig. 4c). However, this secondary peak is not observed for 1D conditions or when incorporating smaller cross-sections. These results corroborate the hypothesis of Hallal and Cox [6] that subsurface conditions at even 1 km away from the TIDA site are influencing its recorded earthquake motions.

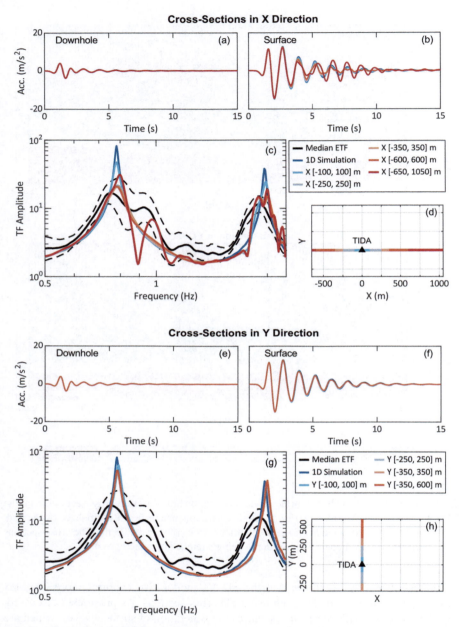

Fig. 4. Comparison of 2D GRAs results obtained using different cross-sections from the 3D Vs model. Shown are: (a, e) simulated acceleration time series at the downhole recording node, (b, f) simulated acceleration time series at the surface recording node, (c, g) comparison of theoretical transfer functions from simulations with the empirical transfer function (ETF) from actual recorded ground motions, and (d, h) plan view of the different incorporated cross-sections in the X and Y directions, respectively.

4 Conclusions

A new 3D site characterization approach and finite element analysis tool were utilized to perform 2D GRAs at the TIDA site. By incorporating 2D cross-sections along different azimuths and lateral extents, we show that the site-specific 3D Vs model is capable of replicating wave scattering and more complex wave propagation phenomena observed in the field. Our results indicate that the azimuth and lateral extent of the incorporated cross-section have a significant influence on the site response predictions. In addition, the 2D simulations presented in this paper corroborate previous observations regarding the significant lateral area influencing site response, which could extend to distances as much as 1 km for certain sites. Future work will extend these analyses to 3D GRAs and to other downhole array sites.

References

1. Hallal, M.M., Cox, B.R., Vantasel, J.P.: Comparison of state-of-the-art approaches usedto account for spatial variability in 1D ground response analyses. J. Geotech. Geoenviron. Eng. **148**(5), p. 04022019 (2022). https://doi.org/10.1061/(ASCE)GT.1943-5606.0002774
2. El Haber, E., Cornou, C., Jongmans, D., et al.: Influence of 2D heterogeneous elastic soil properties on surface ground motion spatial variability. Soil Dyn. Earthq. Eng. **123**, 75–90 (2019). https://doi.org/10.1016/j.soildyn.2019.04.014
3. de la Torre, C., McGann, C., Bradely, B., Pletzer, A.: 3D seismic site response with soil heterogeneity and wave scattering in OpenSees. In: Proceedings of the 1st Eurasian Conference on Opensees: Opensees Days Eurasia. Hong Kong SAR, China, 20–21 June (2019)
4. Huang, D., Wang, G., Du, C., Jin, F.: Seismic amplification of soil ground with spatially varying shear wave velocity using 2D spectral element method. J. Earthq. Eng. (2019). https://doi.org/10.1080/13632469.2019.1654946
5. Hallal, M.M., Cox, B.R.: An H/V geostatistical approach for building pseudo-3D Vs models to account for spatial variability in ground response analyses Part I: Model development. Earthq. Spectra **37**, 2013–2040 (2021). https://doi.org/10.1177/8755293020981989
6. Hallal, M.M., Cox, B.R.: An H/V geostatistical approach for building pseudo-3D Vs models to account for spatial variability in ground response analyses Part II: Application to 1D analyses at two downhole array sites. Earthq. Spectra **37**, 1931–1954 (2021). https://doi.org/10.1177/8755293020981982
7. Graizer, V., Shakal, A.: Analysis of CSMIP strong-motion geotechnical array recordings. In: Proceedings of the International Workshop for Site Selection, Installation, and Operation of Geotechnical Strong-Motion Arrays. Sacramento, CA (2004)
8. Kusanovic, D., Seylabi, E., Kottke, A., Asimaki, D.: Seismo-vlab: A parallel object-oriented platform for reliable nonlinear seismic wave propagation and soilstructure interaction simulation. Submitt. Comput. Methods Appl. Mech. Eng. (2020)
9. Lysmer, J., Kuhlemeyer, R.L.: Finite dynamic model for infinite media. J. Eng. Mech. Div. **95**, 859–877 (1969). https://doi.org/10.1061/jmcea3.0001144
10. Darendeli, M.B.: Development of a New Family of Normalized Modulus Reduction and Material Damping Curves. PhD Thesis University of Texas at Austin (2001)

Variability in Kappa (κ_r) Estimated with Coda Waves for California

Chunyang Ji[1]($\boxtimes$), Ashly Cabas[1], Marco Pilz[2], and Albert Kottke[3]

[1] North Carolina State University, Raleigh, NC 27695, USA
cji3@ncsu.edu
[2] Deutsches GeoForschungsZentrum GFZ, Telegrafenberg, 14473 Potsdam, Germany
[3] Pacific Gas and Electric Company, San Francisco, CA 94105, USA

Abstract. Characterizing and quantifying the effects of local soil conditions are essential for site-specific seismic hazard assessment and site response analysis. The high-frequency spectral decay parameter κ_r and its site-specific component, κ_0, have gained popularity due to their abilities to characterize near-surface attenuation in situ. Values of κ_0 for rock conditions are of particular interest for site-specific seismic hazard analysis for critical facilities. However, ground motions (GM) recorded at sites underlain by stiff soils or rocks are scarce, which limits the computation of κ_0 values via the classic acceleration spectrum method. Recent research has found that κ values computed using the coda wave of a GM (i.e., the multiple-scattered wave that is less sensitive to the earthquake source and local site effects) can capture regional variations of the attenuation of hard rock materials regardless of the subsurface conditions near the surface. However, there are still large uncertainties in κ estimates based on the coda wave per GM, κ_{r_coda}, associated with the absence of consistent guidelines for the computation procedure and a user-orientated GM processing protocol. This work uses California GMs to examine the variability associated with the computation process of κ_{r_coda}, including the choice of onset of the coda wave and its duration. The objective of this paper is to understand and quantify the variabilities in κ values based on coda waves, which has potentially large implications in its applicability in future ground motion models.

Keywords: Kappa · Coda wave · Variabilities and uncertainties · Computation procedure

1 Introduction

Site-specific seismic assessment and site response analysis require the quantification of the effects of local soil columns on ground motions (GMs), which are a complex function of multiple factors (e.g., the depth to bedrock, the stiffness of the materials, and damping) [1]. Commonly used site parameters such as the time-averaged shear wave velocity for the top 30 m, V_{s30}, are not able to capture comprehensive effects from shallow and deep soil layers. High-frequency attenuation parameter kappa (κ), which was proposed to describe the linear spectral decay of Shear-wave (S-wave) Fourier Amplitude Spectrum (FAS) in log-linear scale [2], and its site-specific component κ_0 have been shown to capture near-surface attenuation (although there is no agreement about the soil depth captured by κ_0) [3–5]. However, the classic acceleration spectrum (AS) approach for κ estimations has limitations for hard-rock sites because of the scarcity of records and potential bias from site amplification in high frequencies. The appropriate characterization of site effects at hard-rock sites is critical for the definition of reference stations, which can enable host-to-target evaluations among seismically active and inactive regions and allow the applicability of ground motion models (GMM) from soft to stiff and hard rock sites [6].

The coda wave, which results from the scattering of S-waves by the heterogeneous lithosphere [7–9], is affected by the Earth's crust properties and it can capture regional attenuation characteristics associated with hard-rock sites [10–11]. For example, a regional pattern of the site-specific κ_0 computed with coda waves, κ_{0_coda}, for rock conditions has been observed in Europe [12]. However, there are still uncertainties and significant variability in the κ_r (i.e., κ value measured directly from FAS per record) and κ_0 [13–14] estimations. Additionally, κ_r computed with coda waves, which is denoted as κ_{r_coda}, is also affected by the definition of the coda wave (i.e., its onset and duration). The objective of this work is to investigate and quantify the influence of the selection of the coda wave onset selection and duration on κ_{r_coda} estimates using GMs from a recording station in California, USA.

2 Methods

A Garner Valley Downhole Array (GVDA) station operated by UC Santa Barbara Engineering Seismology network (SB) is selected to explore the variability in κ_{r_coda} associated with the coda wave window pickup. The V_{s30} at the station is 281.5 m/s. Twelve GMs with sampling frequency of 200 Hz are included in this work, which are observed from the events with magnitude larger than 4.0, epicentral distance less than 150 km, and focal depth less than 35 km. The earthquakes involved in this work are queried from the USGS Comprehensive Catalog (ComCat, [15]), and the event IDs are available in Table 1. Although the seismic moment magnitude (M_w) is preferred, other available types of magnitude are used in ComCat query (e.g., local magnitude M_l). Figure 1 shows the location of station SB.GVDA, locations of the epicenters of selected earthquakes, and the shear-wave velocity profile at SB.GVDA [16].

Processing of the GMs was performed with USGS gmprocess [17], which is a Python toolkit for GM retrieving and processing. The gmprocess toolkit processes GMs with an automatic protocol and a quality assurance check [18]. The trend and instrument response are corrected per record, and a bandpass filter is applied to remove the noise. The processed GMs are trimmed 3 s prior to the P-wave arrival. The usable frequency band (with signal-to-noise ratio (SNR) larger than 3.0) is at least 10 Hz (for each horizontal component). All records have peak ground acceleration (PGA) less than 0.5 m/s^2 to minimize soil nonlinear behavior effects.

The onset of the coda wave should be selected after the end of the S-wave window and the acceleration time series amplitude decay becomes regular. The lapse time, which is twice the S-wave travel time, has been commonly used as the onset of coda wave in previous studies [8, 19]. However, the origin time of earthquake is not available or not reliable for the trimmed GMs in this work. Hence, an equivalent model for the onset of the coda wave proposed [12, 20] is used in this work:

$$T_c = 2.3(T_s - T_p) + T_s \tag{1}$$

where T_c is the onset of the coda wave, T_s and T_p are the onsets of the S-wave and P-wave, respectively. T_s is set to be 1 s before the maximum horizontal amplitude [12]. Visual inspection is necessary to ensure that wave phases are identified appropriately.

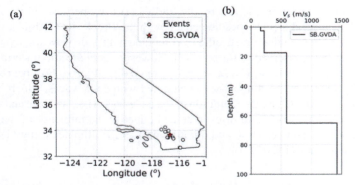

Fig. 1. (a) Location of SB.GVDA and observed earthquakes, and (b) shear-wave velocity profile at SB.GVDA

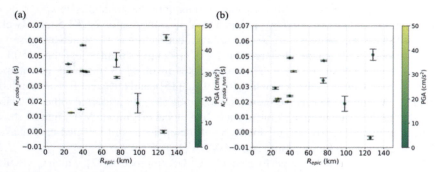

Fig. 2. Variations of κ_{r_coda} with coda wave durations at SB.GVDA for (a) NS and (b) EW component. The dots show the mean estimate of κ_{r_coda} estimated from coda windows with potential durations, the error bars represent the corresponding σ_κ, and marker colors show the PGA per horizontal component.

To study the influence of coda wave durations on κ_{r_coda} estimates, the onset of coda (i.e., T_c) is fixed and the duration is increased with increments of 1 s. To investigate the effect of coda wave onset, the width of the coda window is fixed to be 20 s and the onset of the coda wave is increased with the interval of 1 s. κ_{r_coda} estimates per horizontal component are computed with the classic AS approach [2]. The same frequency band for the κ_{r_coda} computation is used for coda waves with all possible durations or coda onsets per record to minimize the bias associated with frequency band selections [20]. The fixed frequency band ($[f_1, f_2]$) is selected with the following criteria: (1) SNR is larger than 3.0, (2) the higher frequency limit is less than 35 Hz with support of visual inspections, and (3) the lower limit of the frequency band is larger than 10 Hz, which is the earthquake corner frequency for magnitude of 3.5 and stress drop of 10 MPa [21]. The ordinary least square regression is applied to the FAS over the selected frequency band to compute κ_{r_coda}.

3 Results and Discussions

The variation in κ_{r_coda} estimates as a function of coda wave durations is shown in Fig. 2 and Table 1. There are at least five coda windows with various widths analyzed per horizontal component. The influence of the coda wave duration on κ_{r_coda} estimates is insignificant for nine out of twelve GMs at the site of interest. The other three records are associated with higher standard deviations (σ_κ) for at least one horizontal component. The marker colors represent the corresponding PGA values, and there is no pattern observed between the κ_{r_coda} variations and PGA. To further evaluate the observed scatter in the data, two earthquake GMs, one that is significantly affected by the coda duration, and another one that is not, are selected to inspect their acceleration time series and FAS shapes.

Table 1. Variability in κ_{r_coda} as a function of the coda wave durations and coda wave onset for the selected GMs at SB.GVDA

Event ID	Magnitude[a]	R_{epic} (km)	Influence of coda duration				Influence of coda onset			
			EW component		NS component		EW component		NS component	
			μ_κ[b](s)	σ_κ[c](s)	μ_κ(s)	σ_κ(s)	μ_κ(s)	σ_κ(s)	μ_κ(s)	σ_κ(s)
ci15296281	M_h 4.7	27	0.0124	0.0001	0.0222	0.0002	0.0222	0.0026	0.0234	0.0036
ci37510616	M_w 4.4	38	0.0147	0.0001	0.0202	0.0001	0.0349	0.0069	0.0235	0.0026
ci37701544	M_w 4.3	98	0.0190	0.0065	0.0190	0.0050	0.0151	0.0156	0.0184	0.0100
ci14746172	M_l 4.3	126	0.0002	0.0009	−0.0036	0.0011	0.0056	0.0034	0.0111	0.0017
ci38167848	M_w 4.5	40	0.0397	0.0004	0.0240	0.0005	0.0124	0.0023	0.0152	0.0012
ci15520985	M_w 4.6	75	0.0473	0.0048	0.0340	0.0017	0.0328	0.0113	0.0335	0.0016
ci10701405	M_w 4.5	40	0.0570	0.0003	0.0490	0.0003	0.0487	0.0142	0.0476	0.0074
ci10370141	M_w 4.5	76	0.0357	0.0006	0.0472	0.0004	0.0380	0.0035	0.0482	0.0076
ci10530013	M_l 4.3	44	0.0393	0.0004	0.0400	0.0004	0.0223	0.0055	0.0284	0.0050
ci38245496	M_w 4.4	24	0.0444	0.0002	0.0291	0.0007	0.0518	0.0028	0.0489	0.0112
ci14403732	M_l 4.1	26	0.0393	0.0005	0.0207	0.0003	0.0385	0.0039	0.0314	0.0066
ci14745580	M_w 5.7	128	0.0622	0.0019	0.0511	0.0037	0.0278	0.0044	0.0312	0.0051

[a] M_w: seismic moment magnitude; M_h: Non-standard magnitude method; M_l: local magnitude.
[b] μ_κ: the mean of κ_{r_coda} estimates
[c] σ_κ: the standard deviation of κ_{r_coda} estimates

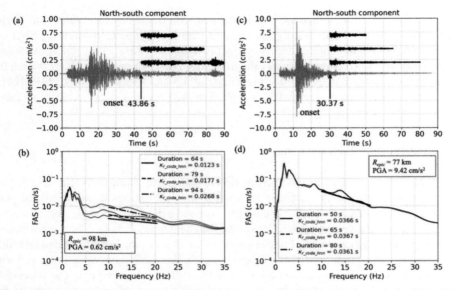

Fig. 3. Illustrations of coda duration influence on κ_{r_coda} estimates with example GMs (NS component). Three coda windows with different durations are determined with the same onset, and the corresponding FAS and κ_{r_coda} values are plotted (bottom). (a) and (b) are observed from event ci37701544, (c) and (d) are from event ci10370141.

Figure 3 shows the acceleration time series and FAS of two example earthquake GMs, where one of them corresponds to event with R_{epic} of 98 km and M_w of 4.3 and the other one to the event with R_{epic} of 77 km and M_w of 4.5. Significant influence of the coda wave window duration (Fig. 3a) is observed in Fig. 3b. The σ_κ of κ_{r_coda} estimates based on different coda wave window widths is 0.0050 s for the north-south (NS) component shown in Fig. 3a. In contrast, the σ_κ of κ_{r_coda} estimates for the GM shown in Fig. 3c (event with R_{epic} of 77 km and M_w of 4.5) is 0.0004 s, and little variation of its FAS is observed in Fig. 3d. Variations in spectral shape observed in Fig. 3b may be caused by the presence of larger amplitudes of waves at the end of time series (see Fig. 3a, which could be an artifact due to the noise or disturbance. Figure 3c and d provide evidence that the duration of the coda wave window does not affect κ_{r_coda} estimates significantly (e.g., the FAS shapes are almost the same in high frequencies). Hence, if appropriate GM quality controls are put in place during the processing protocol (including visual inspection to discard GMs with uncommon waveforms), the effect of the selected duration of the coda wave window should have a negligible effect on κ_{r_coda} calculations.

The influence of the onset of the coda wave on κ_{r_coda} estimates is discussed in this section. Table 1 and Fig. 4 show the variations in κ_{r_coda} estimates with the change of three selected onset of the coda wave (i.e., the coda wave onset estimated with Eq. 1 is increased with increments of 1 s). The duration of the coda wave window has been fixed to 20 s for these analyses to isolate the effects of the selected onset. Compared with Fig. 2, the selection of the onset of the coda wave plays an important role on κ_{r_coda} values. All the selected GMs show a larger σ_κ of κ_{r_coda} values resulting from the assumption of different coda windows for at least one horizontal component. No relationship between computed σ_κ and PGA or Repic is observed in Fig. 4. Figure 5 shows the time series and FAS for the same GMs displayed in Fig. 4. Comparing spectral shapes in Fig. 3 (d) and Fig. 5 (d) shows that the FAS displays significant changes in high frequencies as a function of the selected onset of the coda wave window.

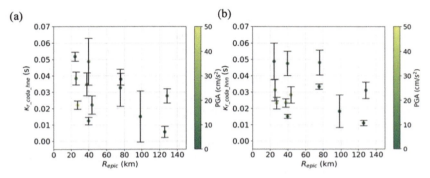

Fig. 4. Variations of κ_{r_coda} with coda wave onset selections at SB.GVDA for (a) NS and (b) EW component. The dots show the mean estimate of κ_{r_coda} estimates with all moving coda window, the error bars represent the corresponding standard deviation, and marker colors show the PGA per horizontal component.

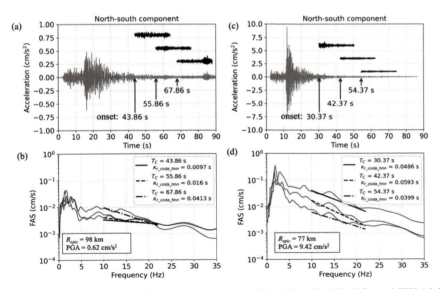

Fig. 5. Variations of κ_{r_coda} with coda wave onset at SB.GVDA for NS (left) and EW (right) component. The dots show the mean estimate of κ_{r_coda} estimated from coda windows with different start times, the error bars represent the corresponding standard deviation, and marker colors show the PGA per horizontal component.

Earlier onset times, such as $T_c = 30$ s in Fig. 5b may capture more energetic portions of the time series (e.g., some of S-waves and surface waves), which results in larger amplitude FAS. Further investigation is required to develop recommendations for selections of coda wave onset window to minimize potential bias in κ_{r_coda} values.

4 Conclusions

This paper studied the influence of the computation procedure on κ_{r_coda} values, particularly the selection of the coda wave window onset and duration. Using a recording station in California (i.e., the Garner Valley Downhole Array) and twelve GMs recordings, the effects of the onset of the coda wave window and its duration were analyzed. We found that, generally, the effect of the coda wave onset selection on κ_{r_coda} estimates were prominent. In contrast, the influence of coda wave durations was negligible. The early stage of coda wave was governed by the angular dependence of scattering and is sensitive to the source radiation pattern, but the latter coda waves are dominated by the multiple scattering of S-waves [9, 19]. Thus, the selection of the coda wave onset implies the explicit consideration (or lack thereof) of the influence of multiple scattering waves of S-wave and disturbance by surface waves.

Thus, we conclude that identifying the appropriate coda wave is critical for a robust κ_{r_coda} computation. Picking later onsets of the coda wave than the lapse time (i.e., twice the S-wave travel time) could help avoid the potential influence of surface waves. However, the lower amplitude of late portions of the seismogram usually has lower SNR and narrower usable frequency ranges, which introduces additional difficulties for analyses in the Fourier domain (e.g., for κ studies). Since there is no consensus on wave phase identifications yet, the resulting uncertainties in κ_{r_coda} caused by coda wave selection remain unavoidable and require further investigation.

References

1. Kaklamanos, J., Cabas, A., Parolai, S., Guéguen, P.: Introduction to the special section on advances in site response estimation. Bull. Seismol. Soc. Am. **111**(4), 1665–1676 (2021)
2. Anderson, J.G., Hough, S.E.: A model for the shape of the Fourier amplitude spectrum of acceleration at high frequencies. Bull. Seismol. Soc. Am. **74**(5), 1969–1993 (1984)
3. Cabas, A., Rodriguez-Marek, A., Bonilla, L.F.: Estimation of site-specific kappa ($\kappa 0$) - consistent damping values at KiK-net sites to assess the discrepancy between laboratory-based damping models and observed attenuation (of seismic waves) in the field. Bull. Seismol. Soc. Am. **107**(5), 2258–2271 (2017)
4. Ktenidou, O.J., Gélis, C., Bonilla, L.F.: A study on the variability of kappa (κ) in a borehole: implications of the computation process. Bull. Seismol. Soc. Am. **103**(2A), 1048–1068 (2017)
5. Ktenidou, O.J., Abrahamson, N.A., Drouet, S., Cotton, F.: Understanding the physics of kappa (κ): insights from a downhole array. Geophys. J. Int. **203**(1), 678–691 (2015)
6. Hashash, Y.M., et al.: Reference rock site condition for central and eastern North America. Bull. Seismol. Soc. Am. **104**, 684–701 (2014)
7. Aki, K., Chouet, B.: Origin of coda waves: source, attenuation, and scattering effects. J. Geophys. Res. **80**, 3322–3342 (1975)

8. Biswas, N.N., Aki, K.: Characteristics of coda waves: central and southcentral Alaska. Bull. Seismol. Soc. Am. **74**(2), 493–507 (1984)
9. Herraiz, M., Espinosa, A.F.: Coda waves: a review. Pure Appl. Geophys. **125**(4), 499–577 (1987)
10. Rautian, T.G., Khalturin, V.I.: The use of the coda for determination of the earthquake source spectrum. Bull. Seismol. Soc. Am. **68**, 923–948 (1978)
11. Mayor, J., Traversa, P., Calvet, M., Margerin, L.: Tomography of crustal seismic attenuation in Metropolitan France: implications for seismicity analysis. Bull. Earthq. Eng. **16**(6), 2195–2210 (2018)
12. Pilz, M., Cotton, F., Zaccarelli, R., Bindi, D.: Capturing regional variations of hard-rock attenuation in Europe. Bull. Seismol. Soc. Am. **109**(4), 1401–1418 (2019)
13. Ktenidou, O.J., Gélis, C., Bonilla, L.F.: A study on the variability of kappa (κ) in a borehole: implications of the computation process. Bull. Seismol. Soc. Am. **103**(2A), 1048–1068 (2013)
14. Ji, C., Cabas, A., Cotton, F., Pilz, M., Bindi, D.: Within-station variability in kappa: evidence of directionality effects. Bull. Seismol. Soc. Am. **110**(3), 1247–1259 (2020)
15. Guy, M.R., et al.: National Earthquake Information Center Systems Overview and Integration. US Department of the Interior, US Geological Survey (2015)
16. Kwak, D.Y., Ahdi, S.K., Wang, P., Zimmaro, P., Brandenberg, S.J., Stewart, J.P.: Web portal for shear wave velocity and HVSR databases in support of site response research and applications. UCLA Geotechnical Engineering Group. https://doi.org/10.21222/C27H0V. Accessed 02 Oct 2021
17. Hearne, M., Thompson, E.M., Schovanec, H., Rekoske, J., Aagaard, B.T., Worden, C.B.: USGS automated ground motion processing software. USGS Software Release (2019). https://doi.org/10.5066/P9ANQXN3
18. Rekoske, J.M., Thompson, E.M., Moschetti, M.P., Hearne, M.G., Aagaard, B.T., Parker, G.A.: The 2019 Ridgecrest, California, earthquake sequence ground motions: processed records and derived intensity metrics. Seismol. Res. Lett. **91**(4), 2010–2023 (2020)
19. Sato, H., Fehler, M.: Scattering and Attenuation of Seismic Waves in Heterogeneous Earth. Springer, New York (1998)
20. Perron, V., et al.: Selecting time windows of seismic phases and noise for engineering seismology applications: a versatile methodology and algorithm. Bull. Earthq. Eng. **16**(6), 2211–2225 (2018)
21. Hanks, T.C., Kanamori, H.: A moment magnitude scale. J. Geophys. Res.: Solid Earth **84**(B5), 2348–2350 (1979)

Structural Response to High Frequency Motion Released in Nonlinear Soil

Piotr Kowalczyk[✉]

University of Trento, Trento, Italy
pk.piotrkowalczyk.pk@gmail.com

Abstract. High frequency motion is often registered in experimental measurements in laminar boxes placed on shaking tables in centrifuge or 1-g laboratory tests when investigating soil free field response under sinusoidal input motions. The source of the high frequency motion is often attributed to inaccuracies of experimental setups. On the other hand, some numerical studies suggested physical explanations, due to soil complex mechanical behavior, such as general soil nonlinearity or dilation. The most recent numerical studies suggest initially another potential source of the high frequency motion in tested soil specimens, i.e. the hypothetical release of unloading elastic waves in the steady state response. Moreover, these studies show that soil-released high frequency motion can potentially impact structural response.

This paper presents a brief example of a numerical study on dynamic soil-structure interaction analyzed under harmonic input motion of a single sinusoidal driving frequency. The soil is modelled using an advanced soil constitutive model to accurately represent soil cyclic behavior. The analyzed structure is shown to experience the motion of the driving frequency and the motion of high frequency in the steady state response. Importantly, the computed high frequency motion is not present at the specimen base and is apparently introduced into the dynamic system by soil nonlinearity.

Keywords: Soil-structure interaction · Finite element modelling · High frequency

1 Background

Soil-structure interaction under seismic loading conditions is of great interest to earthquake engineering scientists. Two methodologies are typically employed to advance the knowledge on seismic soil-structure interaction. First of all, small scale laboratory works are carried out on soil containers placed on shaking tables in 1-g and multiple g (centrifuge) loading environment. Numerous experimental research studies on seismic soil-structure interaction were carried out, including investigation of the behavior of flexible retaining walls [1], tunnels [2] and piles [3]. Secondly, finite element numerical studies using advanced soil constitutive models is another way of reliable investigation

---

P. Kowalczyk—Independent Researcher.

© The Author(s), under exclusive license to Springer Nature Switzerland AG 2022
L. Wang et al. (Eds.): PBD-IV 2022, GGEE 52, pp. 920–927, 2022.
https://doi.org/10.1007/978-3-031-11898-2_68

of soil-structure interaction in earthquake engineering problems, for example to investigate the behavior of a pile in a layered soil profile [4], seismic response of a retaining wall [5], or to determine the impact of pore pressure on temporary natural frequency wandering of structures [6].

One of the intriguing aspects in the experimental and numerical modelling of soil dynamic behavior is the presence of high frequency components of motion registered in spectral response when specimens are subjected to harmonic excitation. The presence of the high frequency components has been studied and attempted to be explained in past research (e.g. [7–11]). Many ideas on the origin of high frequency motion have been suggested ranging from experimental [7] and numerical [8] inaccuracies to the propagation of unloading waves in soil [11]. In fact, the most recent explanation is the novel idea of the potential release of unloading elastic waves in the steady state response of nonlinear hysteretic materials [12]. The authors [12] present in their numerical studies compared with benchmark experimental work from the past how unloading elastic waves can potentially be released in the steady state response of free field analyzed under harmonic excitation.

Nevertheless, although some research studies on the presence of high frequency components in free field were conducted in the past, the impact of soil high frequencies on structural response appears to be the research subject only in the author's previous works (e.g. [11–13]). In particular, the dynamic response of kinematic piles [11, 12] and a simple structure [13] was shown and revealed how high frequency motion, apparently generated in soil, affects the response of the piles and the structure.

This paper presents a short finite element numerical study of a boundary value problem representative of typical experimental setups carried out in flexible soil containers. In detail, a single pile with a mass on its top (to simulate a single degree of freedom structure) and embedded in homogenous dry sand is analyzed under harmonic excitation. The results show how the modelled structure is excited by the soil-released high frequency motion.

2 Methodology

The finite element numerical model has been carried out in Abaqus [14]. A boundary value problem representative of the soil geometry used in typical experimental works with flexible container (e.g. [15, 16]) has been modelled as shown in Fig. 1. The soil volume in the flexible container has the size of 1.2 m length, 0.55 m width and 0.8 m height. Note that only half of the flexible container has been modelled to reduce the computational time costs.

The size of the finite elements for soil modelling has been chosen in a way to ensure accurate wave propagation in soil according to the standard 'rule of thumb'. Quadratic brick elements of varying size of 0.02 m (at the top) to 0.06 m (at the bottom) have been specified in the mesh in order to account for slower wave propagating in the upper part of the soil mass. The soil container has not been explicitly modelled, instead the nodes of the short sides were linked together in order to ensure the same horizontal displacements on both sides.

The applied input motion has been defined as horizontal acceleration of a perfect sinusoidal shape of 5 Hz frequency and the amplitude of 0.06 g. The input motion has been introduced in a smooth manner with steadily increasing amplitude in order to minimize introduction of spurious waves. The natural frequency of the modelled soil is approximately 25 Hz which is representative of a typical soil profile in a flexible container of 0.8 m height (e.g. [16]).

The soil has been modelled with an advanced soil constitutive model, i.e. hypoplastic sand constitutive model [17] including its further developments of the intergranular strain concept to account for small strain stiffness [18] and an addition of one more model parameter to optimize the accumulation of cyclic strain [19]. This sand constitutive model represents soil stress-strain nonlinearity and irreversibility in a form of a hysteretic loop, thus it complies with the necessary constitutive features indicated in [12] in order to observe the release of unloading elastic waves in the steady state response under harmonic excitation. The model implementation used in this work is as per the available 'umat' file on the *soilmodels.info* webpage [20]. Figure 2 shows the prediction of G/G_0 stiffness reduction simulated with the chosen calibration of the constitutive model being close to the recommended fit [15] and well within the recommended limits [21] for the shear strain levels of interest (i.e. typically less than 10^{-3} for seismic applications). Table 1 presents the chosen calibration of the model parameters, the same as per one of the previous works of the author [13].

The pile of an external diameter of 0.02 m has been modelled as a tube (wall thickness 0.7 mm) with a linear elastic model with material properties of aluminum, i.e. with stiffness E = 70 GPa and Poisson's ratio v = 0.3. To simulate a structure, a mass representative of a typical value in experiments from the past [16] has been modelled on a 0.1m high aluminum column fixed to the top of the pile.

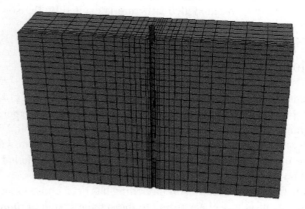

Fig. 1. Geometry and mesh of the analyzed boundary value problem.

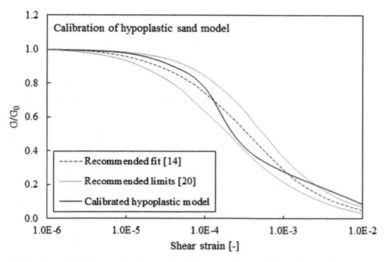

Fig. 2. Calibration of the hypoplastic model in terms of the stiffness degradation curve G/G_0.

Table 1. Input parameters for the hypoplastic sand model.

	Parameter	Description	Value
Basic hypoplastictiy	φ_c	Critical friction angle	33.0
	h_s	Granular hardness [MPa]	2500
	n	Stiffness exponent ruling pressure-sensitivity	0.42
	e_{d0}	Limiting minimum void ratio at $p' = 0$ kPa	0.613
	e_{c0}	Limiting void ratio at $p' = 0$ kPa	1.01
	e_{i0}	Limiting maximum void ratio at $p' = 0$ kPa	1.21
	α	Exponent linking peak stress with critical stress	0.13
	β	Stiffness exponent scaling barotropy factor	0.8
Intergranular strain concept	R	Elastic range	0.00004
	m_R	Stiffness multiplier	4.0
	m_T	Stiffness multiplier after 90° change in strain path	2.0
	β_R	Control of rate of evolution of intergranular strain	0.8
	χ	Control on interpolation between elastic and hypoplastic response	0.5
	ϑ	Control on strain accumulation	5.0

3 Results

The results in this paper are shown in terms of computed horizontal accelerations on the structure and axial strains on the pile. Figure 3 shows that, although perfect sinusoidal input motion with gradually increasing amplitude of motion up to 0.06 g has been introduced at soil base, the response on the structure reveals presence of high frequency motion, firstly irregular in the transient response and subsequently repetitive in the steady state response. In detail, the pattern of high frequencies emerges from the very beginning and reaches its repetitive form (i.e. steady state response) after the approximate time of 0.4 s, where the transient response does not seem to affect the results any longer.

Figure 4 shows spectral evaluation of the computed horizontal accelerations for the steady state part of the response (i.e. taking into consideration sine cycles after 0.4 s). It can be observed that the high frequency motion apparently generated in the soil and registered at the top of the structure has the dominant frequency of 25 Hz, thus apparently being representative of soil elastic waves (to remind the natural frequency of the modelled soil is around 25 Hz). In general, one may expect high frequency motion representative of soil elastic waves in the transient response. However, the presence of soil elastic waves in the steady state response may be found unexpected. The explanation to this observation could be therefore the novel concept in stress wave propagation, i.e. unloading elastic waves described in more details in [12].

Further to the computed horizontal accelerations, the computed pile axial strains at shallow depth are shown on Fig. 5. The computed strain values confirm the presence of high frequency motion on the pile in the steady state response. Figure 6 shows that the frequency of 25 Hz is dominant also in the case of the computed axial strains.

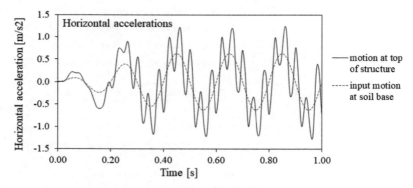

Fig. 3. Horizontal accelerations computed at the top of the structure.

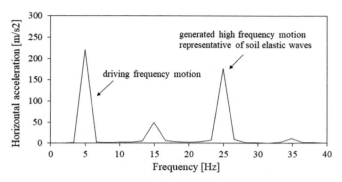

Fig. 4. Spectral response of the computed horizontal accelerations.

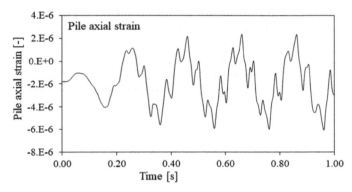

Fig. 5. Pile axial strains computed in the top part of the pile (approximate depth of 30 mm).

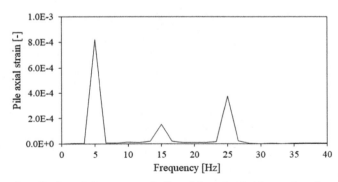

Fig. 6. Spectral evaluation of the computed pile axial strains (with very low frequencies <1Hz filtered out).

4 Summary

This brief numerical study has shown how high frequency motion, apparently representative of soil elastic waves, can be generated by nonlinear soil and how it can influence the dynamic response of the analyzed structure.

Note that the geometry of the analyzed case is somehow representative of a perfect experimental setup not available previously to the author. In the past, the author compared his results ([11–13]) with the experimental work on a group of five piles placed in a bi-layered soil profile ([3, 16]), where the measured high frequency motion could somehow be influenced also by wave reflection on the soil layer interface or due to the interaction between the structural elements. Therefore, it appears interesting to analyze such simplified experimental setup as the one presented in this paper in the future research.

In conclusion, although the presented numerical study showed some proof supporting the idea of the hypothetical presence of soil elastic waves in the steady state response of a soil-structure system, more research is needed. Firstly, to explicitly confirm the existence of soil unloading elastic waves released in nonlinear hysteretic soil, secondly, if such existence is confirmed, to quantify the effects of such waves on the dynamic response of structures when subjected to simplified input motions and real earthquake motions.

Acknowledgments. The author would like to acknowledge University of Trento for the provision of computational resources.

References

1. Conti, R., Madabhushi, G.S.P., Viggiani, G.M.B.: On the behavior of flexible retaining walls under seismic actions. Géotechnique **62**(12), 1081–1094 (2012)
2. Lanzano, G., Bilotta, E., Russo, G., Silvestri, F., Madabhushi, S.P.G.: Centrifuge modelling of seismic loading on tunnels in sand. Geotech. Test. J. **35**(6), 854–869 (2012)
3. Durante, M.G., Di Sarno, L., Mylonakis, G., Taylor, C.A., Simonelli, A.L.: Soil-pile-structure interaction: experimental outcomes from shaking table tests. Earthq. Eng. Struct. Dyn. **45**(7), 1041–1061 (2016)
4. Martinelli, M., Burghignoli, A., Callisto, L.: Dynamic response of a pile embedded into a layered soil. Soil Dyn. Earthq. Eng. **87**, 16–28 (2016)
5. Miriano, C., Cattoni, E., Tamagnini, C.: Advanced numerical modelling of seismic response of a propped R.C. diaphragm wall. Acta Geotecnica **11**(1), 161–175 (2016)
6. Kowalczyk, P., Gajo, A.: Influence of pore pressure on natural frequency wandering of structures under earthquake conditions. Soil Dyn. Earthq. Eng. **142** (2021)
7. Brennan, A.J., Thusyanthan, N.I., Madabhushi, S.P.G.: Evaluation of shear modulus and damping in dynamic centrifuge tests. J. Geotech. Geoenviron. Eng. **131**(12), 1488–1497 (2005)
8. Tsiapas, Y.Z., Bouckovalas, G.D.: Selective filtering of numerical noise in liquefiable site response analyses. In: Proceedings of Geotechnical Earthquake Engineering and Soil Dynamics Conference, Austin, Texas, 10–13 June 2018 (2018)

9. Kutter, B.L., Wilson, D.W.: De-liquefaction shock waves. In: Proceedings of 7th U.S. - Japan Workshop on Earthquake Resistant Design for Lifeline Facilities and Countermeasures Against Soil Liquefaction, Technical Report MCEER-99-0019, pp. 295–310 (1999)
10. Mercado, V., El-Sekelly, W., Abdoun, T., Pajaro, C.: A study on the effect of material non-linearity on the generation of frequency harmonics in the response of excited soil deposits. Soil Dyn. Earthq. Eng. **115**, 787–798 (2018)
11. Kowalczyk, P.: Validation and application of advanced soil constitutive models in numerical modelling of soil and soil-structure interaction under seismic loading. Ph.D. thesis. University of Trento (2020). https://doi.org/10.15168/11572_275675
12. Kowalczyk, P., Gajo, A.: Introductory consideration supporting the idea of the potential presence of unloading elastic waves in seismic response of irreversible soil. Soil Dyn. Earthq. Eng. (2022, under completion)
13. Kowalczyk, P.: Resonance of a structure with soil elastic waves released in nonlinear hysteretic soil upon unloading. Studia Geotechnica et Mechanica (2022 Accepted)
14. Dassault Systèmes: Abaqus Standard Software Package (2019)
15. Dietz, M., Muir Wood, D.: Shaking table evaluation of dynamic soil properties. In: Proceedings of 4th International Conference of Earthquake Geotechnical Engineering, Thessaloniki, Greece, 25–28 June 2007 (2007)
16. Durante, M.G.: Experimental and numerical assessment of dynamic soil-pile-structure interaction. Ph.D. Thesis. Università degli Studi di Napoli Federico II (2015)
17. Von Wolffersdorff, P.A.: A hypoplastic relation for granular materials with a predefined limit state surface. Mech. Cohes. Frict. Mater. **1**(3), 251–271 (1996)
18. Niemunis, A., Herle, I.: Hypoplastic model for cohesionless soils with elastic strain range. Mech. Cohes.-Frict. Mater. **2**, 279–299 (1997)
19. Wegener, D.: Numerical investigation of permanent displacements due to dynamic loading. Ph.D. thesis. TU Dresden (2013)
20. Gudehus, G., Amorosi, A., Gens, A., et al.: The soilmodels.info project. Int. J. Numer. Anal. Methods Geomech. **32**(12), 1571–1572 (2008)
21. Seed, H.B., Idriss, I.M.: Soil moduli and damping factors for dynamic response analysis. EERC report 70-10. University of California, Berkeley (1970)

Linear Combination of Ground Motion Models with Optimized Weights Using Quadratic Programming

Dongyoup Kwak[1](✉), Dongil Jang[1], and Jae-Kwang Ahn[2]

[1] Hanyang University, Ansan, Gyeonggi-do 15588, Republic of Korea
dkwak@hanyang.ac.kr
[2] Korea Meteorological Administration, Seoul 07062, Republic of Korea

Abstract. The epistemic uncertainty of a prediction model can be reduced by linearly combining independent models with appropriate weights. Linear combination indicates that weights are assigned at each model, where their sum is unity, and weighted models are summed for the final combined model. This paper suggests a framework selecting optimized weights minimizing standard deviation of combined models using the quadratic programming technique. Estimating optimized weights for the combination of two models is straightforward and a mathematical equation can be easily derived. However, finding optimized weights for multiple models is not trivial and a numerical approach such as a grid search technique was often used. The quadratic programming, optimizing a quadratic objective function, can be used to find the optimized weights for multiple models effectively. We applied the quadratic programming to the combination of ground motion models to evaluation the effectiveness of the proposed method. The spectral accelerations from seismic records observed at the downhole sensors in South Korea were used as the dataset. We found that the quadratic programming successfully suggested optimized weights minimizing the uncertainty of the combined model.

Keywords: Linear combination · Optimized weight · Quadratic programming · Ground motion model

1 Introduction

A predictive model contains aleatory and epistemic uncertainties, where the epistemic uncertainty can be reduced if modelers have more information regarding the input parameters (e.g., more data points, new parameters constraining output, or new modeling strategy). However, the level of information and the way to develop a model is different for each modeler, so different models predict different outcomes even if they used the same dataset for model development (e.g., GMPEs by the NGA West2 project [1]).

One way reducing the epistemic uncertainty is to combine models. The linear combination of multiple predictive models with appropriate weights can reduce the epistemic uncertainty [2]. If the objective of the linear combination is to minimize the standard deviation of residuals from the combined model, various weight schemes can be selected depending on the availability of information. If only mean predictions from models are available, the equivalent weight scheme can be used; if standard deviations of model predictions are available additionally, the inverse-variance weight scheme can be used; if a covariance matrix between models is available, the optimized weight scheme can be selected which results in the minimum standard deviation of the combined model [2]. Previously the optimized weight was used for the linear combination of V_{S30} and ground motion model (GMM) predictions [2, 3].

This study suggests a way to select optimized weights using the quadratic programming (QP) optimization technique. In the previous study, the optimized weights for multiple models were found using the Monte-Carlo (MC) simulation [2], but the MC simulation is not efficient in terms of the time and repeatability. The QP technique is much faster than the MC simulation and has the better prediction performance. This paper illustrates the usage of QP technique in the linear combination problem which finds optimized weights and the application for the GMM prediction.

2 Quadratic Programming on Linear Combination

2.1 Optimized Weights

Linear combination of multiple models with weights can be represented as:

$$\hat{y} = w_1\hat{y}_1 + w_2\hat{y}_2 + \ldots + w_n\hat{y}_n \text{ where } \sum_{i=1}^{n} w_i = 1 \qquad (1)$$

where $\hat{y}$ is the combined prediction, $\hat{y}_i$ is the prediction from model i, and w_i is the weight assigned to model i. Summation of w_i equal to unity and w_i is within the range of 0–1. Optimized weights would be the set of w_i minimizing the mean square error (MSE) of $\hat{y}$.

If the covariance matrix between models is available, the MSE of $\hat{y}$ can be calculated as below [2, 3]:

$$\sigma^2 = \mathbf{w}^T \boldsymbol{\sigma}^2 \mathbf{w} \qquad (2)$$

where σ^2 is the variance of residuals of the combined model (i.e., MSE), $\mathbf{w}$ is the weight vector, and $\boldsymbol{\sigma}^2$ is the covariance matrix of residuals from each model. If we find $\mathbf{w}$ minimizing σ^2, that will become optimized $\mathbf{w}$.

If the number of models is two, $\mathbf{w}$ can be easily derived and suggested as [2]:

$$\mathbf{w} = \begin{bmatrix} w_1 \\ w_2 \end{bmatrix} = \begin{bmatrix} \frac{\sigma_2^2 - \sigma_{12}}{\sigma_1^2 + \sigma_2^2 - 2\sigma_{12}} \\ \frac{\sigma_1^2 - \sigma_{12}}{\sigma_1^2 + \sigma_2^2 - 2\sigma_{12}} \end{bmatrix} \qquad (3)$$

where σ_i^2 is the variance for model i and σ_{ij} is the covariance between model i and j. For multiple models, it is not trivial deriving $\mathbf{w}$ analytically. Therefore, in the previous study, as an alternative, the Monte-Carlo (MC) simulation was suggested [2]. However, the MC simulation was time consuming and $\mathbf{w}$ for multiple models was not fixed unless a large number of iterations was used.

2.2 Quadratic Programming

The QP numerically solves an optimization problem involving a quadratic function [4], which can be expressed as:

$$\min_{x} : \tfrac{1}{2}\mathbf{x}^T\mathbf{G}\mathbf{x} + \mathbf{a}^T\mathbf{x}$$
$$s.t. : \mathbf{C}^T\mathbf{x} \geq \mathbf{b} \qquad (4)$$

where $\mathbf{x}$ and $\mathbf{a}$ are n vectors, $\mathbf{G}$ is an $n \times n$ symmetric positive definite matrix, $\mathbf{C}$ is an $n \times m$ matrix, and $\mathbf{b}$ is an m vector. Equation (4) contains the quadratic part ($\tfrac{1}{2}\mathbf{x}^T\mathbf{G}\mathbf{x}$) and the linear part ($\mathbf{a}^T\mathbf{x}$) of the optimization problem by variable vector $\mathbf{x}$, and the inequality constraints ($\mathbf{C}^T\mathbf{x} \geq \mathbf{b}$) of $\mathbf{x}$ at the second line. The equation of optimized weight problem (Eq. 2) is also a quadratic function, so we can use this QP technique to find optimized weights.

To convert notations used in Eq. (4) to the optimized weight problem in Eq. (2), we can simply replace $\mathbf{x}$ to $\mathbf{w}$, $\tfrac{1}{2}\mathbf{G}$ to σ^2, and set the $\mathbf{a}$ vector as zero. Next, to be compatible with weight conditions (i.e., summation of $\mathbf{w}$ is unity and the range is 0–1), $\mathbf{C}^T$ and $\mathbf{b}$ can be set as below (for the case that the number of models = 3):

$$\mathbf{C}^T = \begin{bmatrix} 1 & 1 & 1 \\ 1 & 0 & 0 \\ 0 & 1 & 0 \\ 0 & 0 & 1 \end{bmatrix} \qquad (5)$$

$$\mathbf{b} = \begin{bmatrix} 1 \\ 0 \\ 0 \\ 0 \end{bmatrix} \qquad (6)$$

By setting $\mathbf{C}^T$ and $\mathbf{b}$ as Eqs. (5) and (6), the summation condition and the range constraint of $\mathbf{w}$ are met. With these constraints, the QP technique can output $\mathbf{w}$ minimizing the MSE of the combined model.

3 Application to Ground Motion Models

3.1 Ground Motion Data

We used ground motion records from national seismic network in South Korea operated by the Korea Meteorological Administration (KMA) for the examination of linear combination model using the QP technique. Total of 614 ground motions recorded during 2000 to 2018 for the local magnitude (M_L) range of 3.5–5.8 and epicentral distance range of 0–210 km from 61 seismic stations were collected. The sensor condition for the records is the within-rock condition for which the sensor depth is 20–70 m and the adjacent layer is soft rock or harder. We selected this condition because we can reduce the site effect uncertainty on the ground motion predictions. Note that the major portion of the data are located at the range of $R_{epi} > 100$ km and $M_L < 4$.

3.2 Ground Motion Models

Seven local ground motion models (GMMs) developed for South Korea region were used as predictive models ([5] denoted Pea01; [6] denoted Jea02; [7] denoted JB03; [8] denoted Yea08; [9] denoted R13a and R13b; [10] denoted Eea15). We selected these models because they were the most representative models used in the national PSHA development [11]. Among seven GMMs, six GMMs (Pea01, Jea02, JB03, Yea08, R13a, and R13b) were developed using the stochastic simulation method (i.e., random vibration theory [12]) adjusting parameters based on recorded ground motions and one GMM (Eea15) was developed empirically using local records. Figure 1 shows PGA trends versus moment magnitude (M_W) and hypocentral distance (R_{hypo}), respectively. Even though the target region is the same, there is a prediction difference between models. This difference implies that linear combination will improve the prediction and reduce the uncertainty.

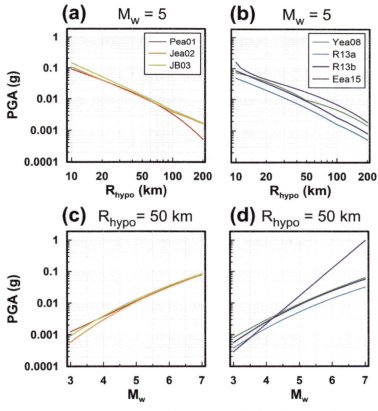

Fig. 1. PGA versus R_{hypo} with $M_W = 5$ for (a) Pea01, Jea02, and JB03 and (b) Yea08, R13a, R13b, and Eea15, and PGA versus M_W with $R_{hypo} = 50$ km for (c) Pea01, Jea02, and JB03 and (d) Yea08, R13a, R13b, and Eea15.

3.3 Performance of Combined Ground Motion Models

For the sake of the performance evaluation of the combined GMM, we first calculated residuals of intensity measures (IMs) for seven GMMs selected. The residual is the difference between the prediction and the observation in logarithm, which is calculated as:

$$res_{ij,k} = \ln(IM_{ij}) - \ln(\widehat{IM}_{ij,k}) \tag{7}$$

where $\widehat{IM}_{ij}$ is the median prediction for event i and site j for a GMM k. For res_{ij} in Eq. (7), we calculated its mean ($\mu_{ij,k}$) and standard deviation ($\sigma_{ln,k}$). The $\mu_{ij,k}$ is not necessarily zero because the ground motion data set is not consistent with the original data set that each GMM modeler used. Since the purpose of this study is to reduce uncertainty of the model prediction by linear combination, we shifted $\ln(\widehat{IM}_{ij,k})$ by $\mu_{ij,k}$ so we computed modified residual ($res'_{ij,k}$) which has zero mean and the standard deviation as $\sigma_{ln,k}$ as follows:

$$res'_{ij,k} = \ln(IM_{ij}) - \left(\ln(\widehat{IM}_{ij,k}) + \mu_{ij,k}\right) = \ln(IM_{ij}) - \ln(\widehat{IM}'_{ij,k}) \tag{8}$$

where $\ln(\widehat{IM}'_{ij,k})$ indicates the shifted $\ln(\widehat{IM}_{ij,k})$ by $\mu_{ij,k}$. Figure 2 shows σ_{ln} of spectral accelerations ranging from T = 0.01 s–1 s. We limit the T up to 1 s because a ground motion model has period limitation up to 1 s. Except Pea01, σ_{ln} ranges 0.6–0.85. Pea01 has high σ_{ln} over 1.0 for short and long periods.

After evaluation of σ_{ln}, we calculated optimized weights using the QP technique introduced in the Sect. 2.2. We created the covariance matrix, σ^2 in Eq. (2) using σ_{ln} and residuals for each GMM, and calculated the weight vector, **w**, following the procedure shown in the Sect. 2.2. The weights assigned for seven GMMs for selected IMs are listed in Table 1. JB03 and Eea15 were evaluated as the major models having the most weights, while other models have very limited weights for the entire periods. The zero weight indicates that the model prediction does not contribute to the reduction of uncertainty.

Using the optimized **w** from Table 1, we calculated the residual using the median prediction of the combined GMM ($res'_{ij,comb}$) as:

$$res'_{ij,comb} = \ln(IM_{ij}) - \sum_{k=1}^{n} w_k \ln(\widehat{IM}'_{ij,k}) = \ln(IM_{ij}) - \ln(\widehat{IM}'_{ij,comb}). \tag{9}$$

where w_k is the weight for the k GMM and $\ln(\widehat{IM}'_{ij,comb})$ is the logarithm of the combined GMM prediction. The mean of $res'_{ij,comb}$ is equal to zero, and the standard deviation ($\sigma_{ln,comb}$) is shown in Fig. 2. The $\sigma_{ln,comb}$ ranges from 0.58 – 0.71. Comparing to σ_{ln} of individual GMMs, $\sigma_{ln,comb}$ was reduced as much as 10% at 0.3 s, and the reduction was negligible at 1 s. The combination effect was greater at mid-period (0.15–0.4 s) range.

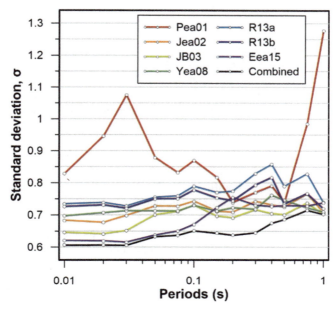

Fig. 2. Standard deviation of residuals for seven GMMs and the combined GMM using QP technique.

Table 1. Optimized weights for seven GMMs for the selected spectral accelerations.

Periods	Weights						
	Pea01	Jea02	JB03	Yea08	R13a	R13b	Eea15
PGA	0	0	0.3872	0	0	0	0.6128
PGV	0	0	0.7511	0	0	0	0.2489
0.01	0	0	0.3709	0	0	0	0.6291
0.02	0	0	0.3730	0	0	0	0.6270
0.03	0	0	0.3080	0	0	0	0.6920
0.05	0	0	0.0962	0.0979	0	0	0.8059
0.075	0	0	0	0.2849	0	0	0.7151
0.1	0	0.0699	0	0.2545	0	0	0.6756
0.15	0	0.5202	0	0	0	0	0.4798
0.2	0	0	0.6024	0	0	0	0.3976
0.3	0	0	0.5237	0	0	0	0.4763
0.4	0	0	0.5717	0	0	0	0.4283
0.5	0	0	0.6342	0	0	0	0.3658
0.75	0	0	0.4332	0	0	0	0.5668
1	0	0	0.8252	0.040	0	0.1348	0

4 Conclusion

We suggested the QP technique for the optimization problem of linear combination. The modification of the constraints in the QP technique was successfully matched the weight condition of linear combination. The performance of linear combination was tested using seven ground motion models and the ground motion data set collected for South Korea. We followed the below process for the estimation of the combined GMM:

1) Take the natural logarithm on the prediction of each GMM;
2) Modify the predicted IM in logarithm by adding the mean misfit of residual;
3) Calculate optimized weights using QP technique;
4) Multiply weights at each modified IM in logarithm;
5) Sum the weighted IMs in logarithm and take exponential.

The standard deviation of residuals from the combined model was reduced up to 10% comparing to the standard deviation from an individual model. The combination was effective especially at the mid-period range (0.15–0.4 s). The effectiveness of the linear combination for the GMMs selected in this study was proved, and the analysis on the limitation of uncertainty improvement will be performed in future study.

Acknowledgement. This study was supported by Korea Meteorological Administration (grant number: KMI2021-01910). Authors greatly appreciate their help.

References

1. Gregor, N., et al.: Comparison of NGA-West2 GMPEs. Earthq. Spectra **30**(3), 1179–1197 (2014)
2. Kwak, D.Y., Seyhan, E., Kishida, T.: A method of linear combination of multiple models for epistemic uncertainty minimization. In: Proceedings of the 11th National Conference on Earthquake Engineering, Los Angeles, CA, pp. 25–29 (2018)
3. Kwok, O.L.A., Stewart, J.P., Kwak, D.Y., Sun, P.-L.: Taiwan-specific model for VS30 prediction considering between-proxy correlations. Earthq. Spectra **34**(4), 1973–1993 (2018)
4. Goldfarb, D., Idnani, A.: A numerically stable dual method for solving strictly convex quadratic programs. Math. Program. **27**, 1–33 (1983)
5. Park, D.H., Lee, J.M., Baag, C.-E., Kim, J.K.: Stochastic prediction of strong ground motions and attenuation equations in the southeastern Korean Peninsula. J. Geol. Soc. Korea **37**, 21–30 (2001). (in Korean)
6. Junn, J.-G., Jo, N.-D., Baag, C.-E.: Stochastic prediction of ground motions in southern Korea. Geosci. J. **6**, 203–214 (2002)
7. Jo, N., Baag, C.: Estimation of spectrum decay parameter k and stochastic prediction of strong ground motions in southeastern Korea. J. Earthq. Eng. Soc. Korea **7**, 59–70 (2003). (in Korean)
8. Yun, K.-H., Park, D.-H., Chang, C.-J., Sim, T.-M.: Estimation of aleatory uncertainty of the ground-motion attenuation relation based on the observed data. In: Proceedings of EESK Conference, Seoul, Korea, pp. 116–123 (2008). (in Korean)

9. Hong, T.K.: Establishment of standardized procedure and development of national seismic hazard maps. Report R&D 2017-MPSS31-107. Ministry of Interior and Safety, Sejong, Korea (2019). (in Korean)
10. Emolo, A., et al.: Ground-motion prediction equations for South Korea Peninsula. Bull. Seismol. Soc. Am. **105**(5), 2625–2640 (2015)
11. Hong, T.K.: Guideline for national seismic hazard map generation procedure. National Disaster Management Research Institute, No. 2017-MPSS31-007
12. Boore, D.M.: Stochastic simulation of high-frequency ground motions based on seismological models of the radiated spectra. Bull. Seismol. Soc. Am. **73**(6A), 1865–1894 (1983)

An Evaluation of V_{SZ} Estimates from the P-wave Seismogram Method for Sites in California

Meibai Li[✉] [iD] and Ellen M. Rathje[iD]

University of Texas at Austin, Austin, TX 78712, USA
mli@utexas.edu

Abstract. The P-wave seismogram method estimates the average shear wave velocity over a representative depth (V_{SZ}) from earthquake recordings at a site. The V_{SZ} is computed from the amplitudes of the radial and vertical P-wave arrivals on the earthquake recordings and an estimate of the seismological ray parameter (p). The ray parameter is estimated from the depth of the event, the epicentral distance, and the regional crustal velocity model. We evaluated the P-wave seismogram approach to estimating V_{SZ} at 153 seismic recording stations in California for which shear wave velocity profiles are available and tested the effect of different crustal models on the estimated ray parameter and the resulting V_{SZ}. Across all the sites, the estimated V_{SZ} values were, on average, about 24% larger than the measured V_{SZ}, although the difference was negligible for softer sites and as large as 45% at stiffer sites (>1000 m/s). Two crustal velocity models for California were considered: a simplified four-layer crustal model for the entire state and a set of more detailed crustal models used for different parts of the state. The effect of the assumed crustal velocity profile was not significant for earthquake events with focal depths greater than about 3–5 km, but for shallower events the detailed crustal velocity model produced V_{SZ} values significantly smaller than the V_{SZ} from the simplified crustal velocity model due to the effect of the shallow low velocity layers and large gradient on the ray parameter.

Keywords: V_{SZ} · Shear wave velocity · P-wave seismogram

1 Introduction

The influence of geologic and soil conditions on the intensity of ground shaking has been widely recognized in earthquake resistance design. People have been using shear wave velocity to characterize site conditions and model wave propagation through geologic materials. For example, the parameter V_{S30}, which represents the time-averaged shear wave velocity over the upper 30 m, has become a standard site parameter that is widely used in a number of predictive ground motion models and building codes.

There are various approaches to acquire information about near surface shear wave velocity, and the most common way is to measure shear wave velocity profiles with in-situ geophysical field tests, such as downhole testing, cross-hole testing, suspension logging, the seismic cone penetration test (SCPT), Multichannel Analysis of Surface Waves (MASW), Spectral Analysis of Surface Waves (SASW), and Microtremor array measurements (MAM). Since it is economically not practical to measure the shear wave velocity profile at all ground motion recording stations, people have been using proxy methods to estimate shear wave velocities (e.g., V_{S30}) from relevant parameters such as topographic slope (e.g., [1]), terrain (e.g., [2]), surficial geology and geotechnical indices (e.g., [3]), or combinations of multiple parameters (e.g., [4]). Alternatively, Ni et al. [5] proposed the P-wave seismogram method which estimates the near surface shear wave velocity using seismograms recorded at ground motion recording stations.

The P-wave seismogram method has been used to estimate V_{S30} in central and eastern North America (CENA) [6, 7] and Japan [8, 9] and has shown promising performance. Compared to proxy-based methods, the P-wave seismogram method has the advantage of being site-specific as it uses seismograms recorded by the seismic stations and regional crustal models.

In this study, we evaluate the estimates of shear wave velocity from the P-wave seismogram method using data in California and test the influence of using a more detailed crustal model on the estimated shear wave velocity. We analyzed 1558 recordings from 153 seismic recording stations in California, and compare the results that are analyzed by using two different sets of crustal models.

2 Application of the P-wave Seismogram Method

The P-wave seismogram method infers the shear wave velocity from the amplitudes of the radial and vertical P-wave arrivals recorded at the ground surface. The physical basis of the method can be qualitatively explained by Snell's Law, which states that a body wave that travels from a layer with higher velocity to a layer with lower velocity will be refracted to a more vertical position. Aki and Richards [10] have derived theoretical expression of the particle motion caused by an incident P-wave at the ground surface, and from their derivation Ni et al. [5] showed that the near surface shear wave velocity can be directly related to the ratio between the radial and vertical components of the first P-wave arrival and the ray parameter p, which is associated with the ray path of the arriving P-wave. Thus, it is possible to estimate the near surface shear wave velocity from the initial P-wave arrival of recordings. Ni et al. [5] also conducted simulations to show that the estimated shear wave velocity by the P-wave seismogram method is actually an average shear wave velocity from the ground surface to a resolvable depth z. Therefore, this shear wave velocity is denoted as V_{SZ}, where z can be estimated using $z = \tau_p \cdot V_{SZ}$, with τ_p representing the earthquake source time function [5]. Kim et al. [6] recommended 0.1 second to be used as τ_p, which is appropriate for earthquake events with magnitude between 3 and 4.

This study used 1558 recordings from 712 earthquake events to estimate V_{SZ} for 153 ground motion recording stations. The earthquake magnitude is limited to be below 5.5 so that the assumption of $\tau_p = 0.1$ s is valid, and the epicentral distance is controlled to be below 200 km to avoid noisy recordings. We expect to analyze ~10 recordings for each seismic recording station considered, but for some stations less than 10 recordings are available. For all stations considered, at least 3 recordings are used for analysis. The earthquake recordings are processed by instrument response removal, baseline correction and bandpass filtering before used in the P-wave seismogram method. Subsequently, the ratio between radial and vertical components ($\dot{U}_R/\dot{U}_Z$) at the time of first peak in the vertical component is extracted from the velocity time series.

To obtain the ray parameter p, we utilized the crustal model for southern California as utilized in [11], which is a simplified four-layer model. Based on Snell's Law, the ray parameter p associated with each recording is estimated from the depth of the event, the epicentral distance, and the regional crustal velocity model, and then used in combination with $\dot{U}_R/\dot{U}_Z$ to compute V_{SZ}. More thorough discussion of the theoretical basis and the details in procedure of applying the P-wave seismogram method is provided in [12].

3 Comparison of V_{SZ} Estimates with V_{SZ} Measurements

To evaluate the accuracy of the estimated V_{SZ}, we compared all estimates with measured V_{SZ} for the 153 sites. The measured V_{SZ} was computed from the shear wave velocity profiles that were obtained from in-situ geophysical tests at each site. The shear wave velocity profiles are documented in the shear-wave velocity profile database (VSPDB) developed by Ahdi et al. [13] and available for cloud-based analysis in the DesignSafe cyberinfrastructure [14]. For each station analyzed by the P-wave seismogram method, the resolvable depth z associated with the estimated V_{SZ} is taken as 0.1 s times the mean estimated V_{SZ} from all recordings for the station, and the measured V_{SZ} is then computed from the V_S profile measured the site over this same depth. In conditions where the depth of the measured V_S profile is smaller than the resolvable depth z, we extrapolate the shear wave velocity profile by assigning the value at the bottom of the V_S profile to the remaining depths and then compute the measured V_{SZ} from the extrapolated profile. This extrapolation will introduce uncertainty in the comparisons of the measured and estimated V_{SZ}.

In Fig. 1, we compare the measured V_{SZ} with the estimated V_{SZ} from the individual 1558 recordings (Fig. 1a) and with the average estimated V_{SZ} computed in log scale for the 153 stations (Fig. 1b). In Fig. 1b, the data points are binned based on the average estimated V_{SZ}. An overall overestimation is observed for the estimated V_{SZ} across all the data, but the comparison between the measured and estimated V_{SZ} varies with V_{SZ}, as demonstrated by the binned means in Fig. 1b. Specifically, for sites with average estimated V_{SZ} less than 500 m/s, the P-wave seismogram method is almost unbiased, with the measured and estimated V_{SZ} within 4%. However, for sites with average estimated V_{SZ} between 500 and 700 m/s, the overestimation is about 20%, and for sites with average estimated V_{SZ} over 1000 m/s the overestimation is about 45%, on average.

The apparent overestimation by the P-wave seismogram method likely is influenced by the procedure of extrapolating the measured V_S profile to the resolvable depth z. Extrapolation using the V_S at the base of the profile is certainly a lower-bound estimate of the deeper V_S structure, which will likely underestimate V_{SZ} since the shear wave velocity typically increases with depth. Therefore, the true V_{SZ} should be greater than the values used here. As shown in Fig. 1b, the sites for which the V_S profiles are not extrapolated show little bias of about 3%. We considered other extrapolation methods, but they introduced additional uncertainty into the comparison. A final note is that the bias is expected to increase with increasing estimated V_{SZ} because of the dependence of the resolvable depth z on estimated V_{SZ}. As z increases with estimated V_{SZ}, the bottom layer V_S needs to be extended to a greater depth, leading to more underestimation of measured V_{SZ}. As a result, the increase in bias as shown by the binned means in Fig. 1b is reasonable.

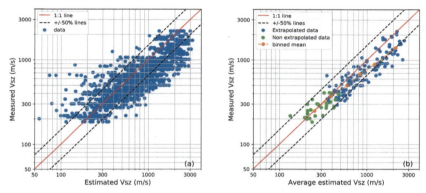

Fig. 1. (a) Measured V_{SZ} and the V_{SZ} estimated from 1558 records; (b) Measured V_{SZ} and the average estimated V_{SZ} for the 153 stations with V_S profile available on VSPDB.

Although the measured V_{SZ} is not the true V_{SZ} for some of the analyzed sites, the comparison still provides us with essential information about the accuracy and variation of the estimated V_{SZ}. To statistically evaluate the performance of the P-wave seismogram method, we computed the total residual ($y_{i,j}$) between the measured and estimated V_{SZ} using Eq. 1, where j represents each record and i represents each site. Using the same mixed-effect model as [6], the total residual $y_{i,j}$ is partitioned into three components: the overall mean residual a, the between-site residual η_{Si}, and the remaining within-site residual $\varepsilon_{i,j}$. Equations 2–4 explain this partitioning, with $\overline{y_i}$ representing the average residual for site i.

$$y_{i,j} = \ln\left(V^i_{SZ,mea}/V^{i,j}_{SZ,est}\right) \qquad (1)$$

$$y_{i,j} = a + \eta_{Si} + \varepsilon_{i,j} \qquad (2)$$

$$\eta_{Si} = \overline{y_i} - a \qquad (3)$$

$$\varepsilon_{i,j} = y_{i,j} - \eta_{Si} - a \qquad (4)$$

The mean residual a for the dataset is computed to be −0.216, which indicates an average overestimation of about 24%. However, the actual bias is expected to be less than 24%, due to the potential underestimation of the measured V_{SZ}, as discussed earlier. The standard deviation of the total residual, i.e., σ is 0.429, is decomposed into the standard deviation of the between-site residual ($\tau = 0.261$) and the standard deviation of the within-site residual ($\phi = 0.341$), as shown in Eq. 5:

$$\sigma = \sqrt{\tau^2 + \phi^2} \qquad (5)$$

We compared the mean a and standard deviation σ of total residuals obtained from this study with values in [6] and [8]. The mean total residual $a = -0.216$ of this study is greater than the mean total residual of -0.13 (about 14% overestimation) from [6] in absolute value, indicating more overestimation of V_{SZ} in California than in CENA. However, it should be noted that most of the base V_S of the profiles used by [6] in computing the measured V_{SZ} is greater than 900 m/s, while the base V_S of only 35% of the profiles used in this study is greater than 900 m/s. Therefore, the influence of extrapolation is expected to be smaller for [6], which can lead to less overestimation. Compared to the mean total residual of 0.039 from [8], the absolute value of a obtained from this study is again greater. The standard deviations of the total residuals are 0.38 and 0.39 from [6] and [8], respectively, which are comparable to the standard deviation of total residuals from this study, i.e., 0.429. The reported values of τ and ϕ from [6] and [8] are similar to those from the current study.

In general, the comparison between estimated and measured V_{SZ} suggests about 24% overestimation of V_{SZ} in California, and one reason of the overestimation may be the potential bias that is caused by the procedure used to extrapolate the V_S profiles to compute the measured V_{SZ}. The mean bias from this study is larger than the mean bias obtained from [6] and [8] for CENA and Japan, but the variation of total residuals from this study is comparable to the values from [6] and [8].

4 Effect of Using a More Detailed Crustal Model for Estimating Ray Parameter p

As a crustal model with more detailed layers at shallow depth is available for California, from the Broadband Platform of the Southern California Earthquake Center (BBP, https://github.com/SCECcode/bbp/wiki/File-Format-Guide), we evaluate the influence of using this more detailed crustal model on the estimated V_{SZ}.

As shown in Fig. 2a, the BBP crustal models divide California into four regions: Southern California (SoCal), Mojave, Northern California (NoCal), and Central Coast, and different crustal models are assigned to different regions (Fig. 2b). The original crustal model used in this study (Simplified four-layer model) is plotted in Fig. 2b as well. Compared to the BBP crustal models, the simplified four-layer model has only one layer in the top 5.5 km, while the BBP models use thinner layers with smaller P-wave velocities and a steeper gradient to characterize shallow depths. The four crustal models differ at depths greater than 6 km, but the difference is not significant.

When applying the P-wave seismogram method, using a different crustal model only affects the estimation of the ray parameter and does not affect the ($\dot{U}_R/\dot{U}_Z$) of the P-wave arrival. Thus, we estimate the ray parameter separately using the BBP crustal models. Based on the location of a ground motion recording station, the BBP crustal model for the specific region is assigned and the ray parameter is estimated. This value of p is used to estimate V_{SZ} in combination with the $\dot{U}_R/\dot{U}_Z$, which remains unchanged from the original analysis. The effect of the more detailed BBP crustal models on the estimated V_{SZ} is shown in Fig. 3a where we plot the total V_{SZ} residual as a function of earthquake focal depth for the different crustal models. Significantly large positive residuals are observed for focal depths shallower than 3 km when the BBP crustal models are used, indicating that V_{SZ} is severely underestimated. However, this trend is not present in the V_{SZ} total residuals obtained when the simplified four-layer model is used. The total V_{SZ} residuals computed using the BBP crustal models and the simplified four-layer crustal model show little difference for events with focal depth greater than about 5 km. The ray parameters that are obtained using the two sets of crustal models are plotted against focal depth in Fig. 3b. The ray parameter estimated when the BBP models are used is much greater (by as much as a factor of 2) than the ray parameter estimated from the simplified four-layer crustal model for events with focal depth less than about 3 km. As demonstrated by the equation to estimate V_{SZ} [12], an increase in the ray parameter leads to a decrease in the estimated V_{SZ} for the same $\dot{U}_R/\dot{U}_Z$, which explains the underestimation of V_{SZ} when the BBP crustal models are used.

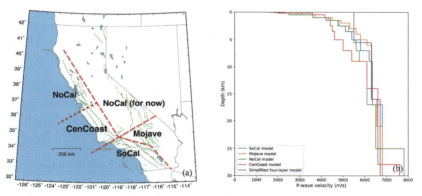

Fig. 2. (a) Four regions that are divided from California where different crustal models are used; (b) P-wave velocity models for the four different regions and the simplified P-wave velocity crustal model used by this study.

The unreasonably large value of ray parameter that is estimated from the more detailed BBP crustal models is caused mainly by the small P-wave velocity and significant change in the P-wave velocity at shallow depth in the BBP models. Based on Snell's Law, which is used to estimate the ray parameter in this study, the steeper gradient at depths shallower than 3 km in the BBP models cause the estimated take-off angle to be greater than the take-off angle estimated from the simplified four-layer crustal model, because the refraction at shallow layers of BBP models are more significant.

Due to the observed underestimation of V_{SZ} when the more detailed crustal models are used, we recommend using the simplified four-layer crustal model to estimate ray parameter.

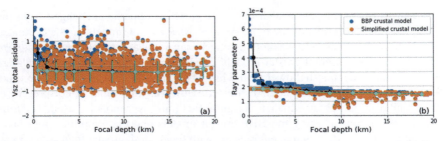

Fig. 3. (a) Total V_{SZ} residuals and (b) ray parameters computed assuming the more detailed BBP crustal models and the simplified southern California crustal model.

5 Conclusions

This study applied the P-wave seismogram method to evaluate the near surface shear wave velocity of 153 stations in California using 1558 recordings from 712 earthquake events. V_{SZ}, which represents the time averaged shear wave velocity over the upper z meters, is estimated based on the ratio between the radial and vertical components of the first P-wave arrival, along with the ray parameter. The resolvable depth z of V_{SZ} can be approximated using the equation $z = 0.1 \text{ s} \cdot V_{SZ}$, where 0.1 s is suggested by [6] for magnitude 3–4 events and is considered appropriate for the range of magnitudes used in this study, which is limited to be below 5.5.

The estimated V_{SZ} was compared with the measured V_{SZ} that was computed from in-situ measured V_S profiles. For V_S profiles with bottom depths shallower than the required depth z, the V_S at the bottom of the profile is extrapolated to z for computation of the measured V_{SZ}. The comparison of estimated and measured V_{SZ} suggests that V_{SZ} may be overestimated, on average, by the P-wave seismogram method, but the degree of disagreement varies with estimated V_{SZ}, with almost no bias when estimated $V_{SZ} < 500$ m/s and 45% overestimation when estimated $V_{SZ} > 1000$ m/s. However, this overestimation may partially be due to the fact that the measured V_{SZ} tends to be underestimated due to the V_S extrapolation procedure. The total residual between measured and estimated V_{SZ} was computed and a mixed-effect model was applied to decompose the total residual $y_{i,j}$ into the mean overall residual, the between-site residual, and the within-site residual. The mean total residual from this study was -0.216, indicating on average 24% of overestimation, which is greater in absolute value than the mean total residual from other studies [6, 8]. The standard deviation of total residual σ is 0.429, which is comparable to those from CENA (0.38) and Japan (0.39).

Finally, the effect of using a more detailed crustal model at shallow depth is evaluated using the BBP crustal models. The result indicates that V_{SZ} is severely underestimated for events with focal depths less than about 3 km when the BBP crustal models are used. This bias is caused by the relatively small P-wave velocities and steep gradient of the BBP models at shallower depths, which lead to unreasonably large ray parameters. For events with focal depth greater than 5 km, the influence of using the BBP models become negligible.

References

1. Wald, D.J., Allen, T.I.: Topographic slope as a proxy for seismic site conditions and amplification. Bull. Seisol. Soc. Am. **97**, 1379–1395 (2007)
2. Yong, A.: Comparison of measured and proxy-based values in California. Earthq. Spectra **32**(1), 171–192 (2016)
3. Wills, C.J., Clahan, K.B.: Developing a map of geologically defined site-condition categories for California. Bull. Seismol. Soc. Am. **96**, 1483–1501 (2006)
4. Parker, G.A., et al.: Proxy-based estimation in central, and eastern North America. Bull. Seisol. Soc. Am. **107**(1), 117–131 (2017). https://doi.org/10.1785/012016010
5. Ni, S., Li, Z., Somerville, P.: Estimating subsurface shear velocity with radial to vertical ratio of local P waves. Seismol. Res. Lett. **85**(1), 82–89 (2014)
6. Kim, B., et al.: Subsurface shear-wave velocity characterization using P-wave seismograms in Central, and Eastern North America. Earthq. Spectra **32**(1), 143–169 (2016)
7. Zalachoris, G., Rathje, E.M., Paine, J.G.: VS30 characterization of Texas, Oklahoma, and Kansas using the P-wave seismogram method. Earthq. Spectra **33**(3), 943–961 (2017)
8. Miao, Y., Shi, Y., Wang, S.-Y.: Estimating near-surface shear wave velocity using the P-wave seismograms method in Japan. Earthq. Spectra **34**(4), 1955–1971 (2018). https://doi.org/10.1193/011818EQS015M
9. Kang, S., Kim, B., Park, H., Lee, J.: Automated procedure for estimating VS30 utilizing P-wave seismograms and its application to Japan. Eng. Geol. **264**, 1–16 (2020). https://doi.org/10.1016/j.enggeo.2019.105388
10. Aki, K., Richards, P.G., Ellis, J. (ed.): Quantitative Seismology. University Science Books, Sausalito (2002). 700 p.
11. Wald, L.A., Hutton, L.K., Given, D.D.: The southern California network bulletin: 1990–1993 summary. Seismol. Res. Lett. **66**(1), 9–19 (1995)
12. Rathje, E.M., Li, M.: Evaluation of the P-wave seismogram approach to estimate. U. S. Geological Survey Final Technical Report (2021). https://earthquake.usgs.gov/cfusion/external_grants/reports/G19AP00104.pdf
13. Ahdi, S.K., Stewart, J.P., Ancheta, T.D., Kwak, D.Y., Mitra, D.: Development of Vs profile database and proxy-based models for Vs30 prediction in the Pacific Northwest region of North America. Bull. Seis. Soc. Am. **107**(4), 1781–1801 (2017)
14. Rathje, E., et al.: DesignSafe: a new cyberinfrastructure for natural hazards engineering. ASCE Nat. Hazards Rev. (2017). https://doi.org/10.1061/(ASCE)NH.1527-6996.0000246

The Uncertainty of In-situ S and P Wave Velocity Test at Xichang Experimental Field of CSES

Yongbo Liu, Zhuoshi Chen[✉], Xiaoming Yuan, and Longwei Chen

Institute of Engineering Mechanics, China Earthquake Administration, Key Laboratory of Earthquake Engineering and Engineering Vibration of China Earthquake Administration, Harbin, China
various@163.com

Abstract. In this study, a blind in-situ wave-velocity experiment was carried out in the Xichang Experimental Field of the CSES (China Earthquake Science Experiment Site). Relying on 286 times of boreholes velocity results, a statistics model was established to describe the test uncertainty of Vs and Vp. Based on the blind test, the uncertainty of the Vs and Vp test does not relay on the objective factors of the field. The relative deviation from the mean of the velocity can be described by Standard Normal Distribution $N(0, 0.08^2)$ at Xichang. Furthermore, when consider the impact of the uncertainty on the site rigidity judgement, it is only need to consider the situation when the overburden depth is less than 10 m and the site is stiff. In conclusion, all these points can be used for the construction of CSES.

Keywords: Uncertainty · Velocity test · Site rigidity · CSES

1 Introduction

The CSES (China Earthquake Science Experimental Site) plan was officially proposed on the tenth anniversary of the Wenchuan earthquake at 2018 (Wu et al. 2020). The area of CSES includes the Sichuan-Yunnan rhombic block near the eastern tectonic junction of the Qinghai-Tibet Plateau and its surrounding areas. It aims to make a deep understanding of the law of occurrence of earthquakes and the mechanism of seismic disasters. Specific to Xichang Experimental Field of CSES, it is expected to carry out a long-term site and typical structural seismic response under the resilience considerations of cities. For the construction of CSES, the accuracy and reliability of basic parameters become an important problem and it is very important to know "How well we doing in the engineering". The first-going work is to figure out the uncertainty of important parameters (Terzaghi 1943, Casagrande 1965).

How to objectively and scientifically understand the uncertainty of the parameters becomes an important issue in engineering (Morgan and Henrion 1990, Chen 2018). At 2003, JT Christian was invited to be the keynote speaker at the 39th Terzaghi Lecture. He gave a keynote report which was entitled "Geotechnical Engineering Reliability:

How Well Do We Know What We Are Doing?", pointed out that the uncertainty of mechanical parameters system analysis is an indispensable part of engineering risk and reliability analysis (Christian 2004). The NEHRP, the US Earthquake Preparedness and Disaster Mitigation Plan, particularly emphasizes that accurate understanding of system uncertainty is an essential analysis process for reducing engineering risks.

Focused on the parameters of the site, compression wave velocity (Vp) and shear wave velocity (Vs) are two critical parameters, which are commonly used to define the overall rigidity. Their uncertainty may cause obvious influence to related parameters. In order to have a clear understanding of the uncertainty of the S-wave test, a blind experiment was carried out at the Parkfield Earthquake Science Experimental Site in California (Real 1988, Annie et al. 2008). Later, the InterPACIFIC project in Europe (Garofalo et al. 2016a, 2016b) showed some meaningful findings about the uncertainty of Vs results which gave out by different methods at same sites. These two experiments showed that the in-situ velocity test results have obvious deviation. Limited by the database, it is very hard to describe the uncertainty.

Xichang Experiment Field is expected to become a multi-function experiment site. For the goal of a high-quality seismic science experiment field, we designed a blind parallel in-situ test to describe the uncertainty of Vs and Vp. Furthermore, its impact on the overall rigidity judgement of the site was given to identify the influence to the site class classification. This paper provides a reasonable statistic model of the uncertainty of in-situ Vs and Vp at Xichang. It can be used for the design and construction of CSES.

2 In-situ Test

The blind in-situ test was settled in the Anning River Valley and Qionghai Basin. These two basins are covered by thick sedimentary soil layers and selected for the potential of CSES seismic observation arrays. Experimental locations are shown in Fig. 1.

Fig. 1. Locations of in-situ Vs and Vp test

Vs and Vp tests were taken in five boreholes with depth 100 m (No. ZK1 to No. ZK5) and eight boreholes with depth 40 m (No. SK1 to No. SK8). For each borehole, 11 times Vs and Vp test were separately finished by six research institutes and companies. All these team use the same P-S log velocity instrument made by the OYO company. As a blind test, all information about the site, such as soil layer, stiffness, groundwater level and so on, is totally anonymous for all test teams. All teams show off the results of Vs and Vp. Figure 2, 3, 4, 5 and 6 show all test results and the average of Vs and Vp of each site.

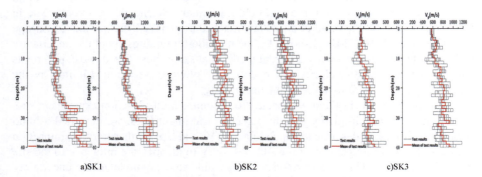

Fig. 2. Blind test results of SK1, SK2 and SK3

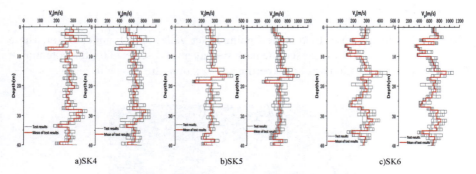

Fig. 3. Blind test results of SK4, SK5 and SK6

Even all teams used the same equipment, there are obvious deviation of the Vs and Vp of the sites. This phenomenon proves that the uncertainty of in-situ velocity test exists objectively once again. For further consideration of the uncertainty and corresponding influence, authors tried to build a statistics model to describe the uncertainty.

The Uncertainty of In-situ S and P Wave Velocity Test   947

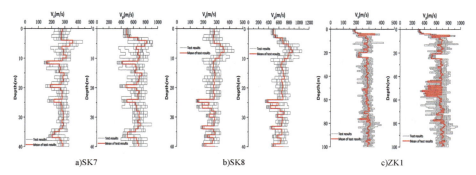

Fig. 4. Blind test results of SK7, SK8 and ZK1

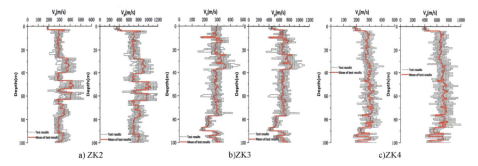

Fig. 5. Blind test results of ZK2, ZK3 and Zk4

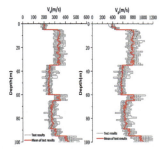

Fig. 6. Blind test results of ZK5

3 Statistics Model of the Uncertainty

Based on all data showed above, the test uncertainty exists in all tests. Because all sites are quite different, it is too hard to describe the uncertainty by using one unified parameter. For the consideration of its impact, this research used relative deviation of Vs and Vp of each site as the parameter to build the statistics model. As shown in Eq. 1, the deviation of each result from the mean value is calculated. By doing this transition, the uncertainty of Vs turns into a dimensionless value for different kinds of site.

$$\text{BiasV}_s = (V_{S\ Data} - V_{S\ Mean})/V_{S\ Mean} \times 100\% \quad (1)$$

In Eq. 1, BiasV$_s$ is the deviation of the V$_s$ test value from the average value of each borehole, which is dimensionless. V$_{S\ Data}$ is the V$_s$ value which obtained from 11 tests of each borehole with unite m/s. V$_{s\ Mean}$ is the average V$_s$ obtained from 11 tests of each borehole with unite m/s. In this experiment, 13 sites have a total of 143 V$_s$ curves calculated by Eq. 1. Among them, there are 143 per-meter BiasV$_s$ data at the depth from surface to 40 m and 55 per meter BiasVs data at the depth from 41 m to100 m. Based on the descriptive statistics theory, all depth data quantity satisfys the minimal need to build a statistical model (William M. and Terry L.S. 2015). Similarly, the relative deviation of P wave-velocity from the mean value of each site can be calculated by Eq. 2.

$$\text{BiasV}_p = (V_{P\ Data} - V_{P\ Mean})/V_{P\ Mean} \times 100\% \quad (2)$$

By using Eq. 1 and 2, the BiasV$_s$ and BiasV$_p$ of all 13 boreholes are shown in Fig. 7.

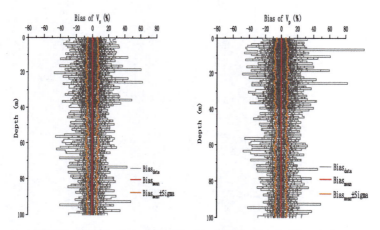

Fig. 7. Statistical results of BiasVs and BiasVp of all boreholes

Based on Fig. 7, BiasVs and BiasVp are all fluctuate in a small range between 5% to 10%. We used SPSS software to carry out descriptive statistics results and normal hypothesis test of P-P graphs for BiasVs and BiasVp. The distribution of BiasVs and BiasVp obey Normal Distribution. We can use the average value 8% to define the uncertainty model of the relative deviation as $N(0, 0.08^2)$.

Twenty depth fitting results are randomly selected of all 100-m data. The normal distribution histogram fitting sampling results and normal distribution hypothesis results of BiasV$_s$ are shown in Fig. 8. Similar statistics results of BiasV$_p$ are shown in Fig. 9.

By examining all fitting model, we found out that the uncertainty of each site is similar. It indicates that the site object factors, such as site category, layer condition, soil property and groundwater level, have little influence on the test uncertainty. We can use

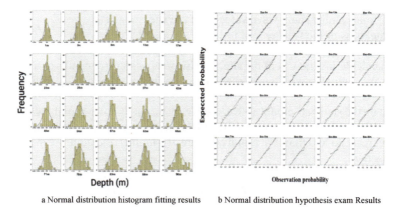

a Normal distribution histogram fitting results   b Normal distribution hypothesis exam Results

Fig. 8. P-P fitting and hypothesis exam results of BiasV_s at twenty random depths

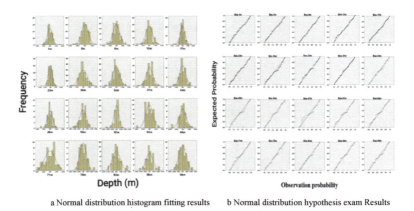

a Normal distribution histogram fitting results   b Normal distribution hypothesis exam Results

Fig. 9. P-P fitting and hypothesis exam results of BiasV_p at twenty random depths

this relationship to make a preliminary estimate of the reliability of the test results of Xichang Experimental Filed.

4 Influence of vs Uncertainty on the Site Overall Rigidity Judgement

The overall rigidity of the site is judged by the thickness of layer and the S wave-velocity of the layer. Time average shear wave velocity $V_{s,z}$ is used to describe the rigidity of the site, which is shown in Eq. 3.

$$V_{s,z} = \frac{\sum_{i=1}^{n} d_i}{\sum_{i=1}^{n} \frac{d_i}{V_{s,i}}} \tag{3}$$

$V_{s,z}$: Time average shear wave velocity of the site
d_i: Thickness of the i-th layer of the site
$V_{s,i}$: Shear wave-velocity of the i-th layer of the site

When calculating the $V_{s,z}$ of the site, we always consider $V_{s,i}$ results as a fixed value. Based on the data all above, when the uncertainty is considered, $V_{s,z}$ shifted with the $V_{s,i}$.

Based on the statistics model above, the CV of $V_{s,z}$ goes down rapidly with the depth and we got the peak at the site surface. When the depth more than 10 m from the surface, the CV decrease to less than 5% rapidly. Figure 10 summarized all thirteen CV average curves together.

A fitting curve is used to describe the regular of CV in Xichang Experimental Field which can be used to estimate the uncertainty of $V_{s,z}$. It is shown as Eq. 4.

$$CV = 7.034 * Z^{-0.38} \tag{4}$$

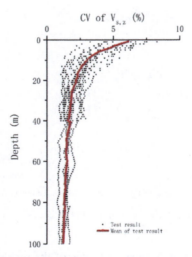

Fig. 10. The average fitting curve of CV of $V_{s,z}$

When the sublayer is thicker than 10 m, the uncertainty of Vs have little influence on $V_{s,z}$. Only when the site is stiff and the sublayer is thin, it is necessary to do further consideration of the influence on site rigidity judgement.

5 Conclusions

The blind in-situ wave-velocity experiment was carried out in the Xichang Experimental Field of the CSES (China Earthquake Science Experiment Site). It aims at figure out how well we finished Vs and Vp tests. By finishing 143 times parallel experiment tests of Vs and Vp in 13 boreholes, the statistics model was built to describe the uncertainty. The uncertainty of Vs and Vp can be described by Standard Normal Distribution $N(0, 0.08^2)$.

When considering the impact on the site rigidity judgement, it only need to analysis the misfit when the overburden depth is less than 10 m and the site is stiff. All these points can be used for the construction of CSES.

Acknowledgement. This study is financially supported by the Scientific Research Fund of Institute of Engineering Mechanics, China Earthquake Administration (Grant No. 2019B04).

References

Wu, Z.L., et al.: China seismic experimental site: retrospective and prospective. Rev. Geophys. Planet. Phys. **52**, 234–238 (2021). https://doi.org/10.16738/j.dqyxx.2020-023

Terzaghi, K.: Theoretical Soil Mechanics. Wiley, New York (1943)

Casagrande, A.: Role of the "calculated risk" in earthwork and foundation engineering. J. Soil Mech. Found. Div. **91**(4), 1–40 (1965)

Morgan, M.G., Henrion, M.: Uncertainty: A Guide to Dealing with Uncertainty in Quantitative Risk and Policy Analysis. Cambridge University Press, New York (1990)

Chen, Z.Y.: Reliability analysis and safety criterion in geotechnical engineering based on the index of safety margin. Chin. J. Rock Mech. Eng. **37**(3), 521–544 (2018)

Christian, J.T.: Geotechnical Engineering Reliability: how well do we know what we are doing? J. Geotech. Geoenviron. Eng. **130**(10), 985–1003 (2004)

Real, C.R.: Turkey Flat, USA site effects test area: report 2, site characterization (1998)

Kwok, A.O.L., Stewart, J.P., Hashash, Y.M.A.: Nonlinear ground-response analysis of Turkey flat shallow stiff-soil site to strong ground motion. Bull. Seismol. Soc. Am. **98**, 331–343 (2008)

Garofalo, F., et al.: InterPACIFIC project: comparison of invasive and non-invasive methods for seismic site characterization. Part I: intra-comparison of surface wave methods. Soil Dyn. Earthq. Eng. **82**, 222–240 (2016a)

Garofalo, F., et al.: InterPACIFIC project: comparison of invasive and non-invasive methods for seismic site characterization. Part II: inter-comparison between surface-wave and borehole methods. Soil Dyn. Earthq. Eng. **82**, 241–254 (2016b)

Roeloffs, E.: The Parkfield California earthquake experiment: an update in 2000. Curr. Sci. **79**, 1226–1236 (2000)

William, M., Terry, L.S.: Statistics for Engineering and the Sciences. CRC Press (2015)

Code for Seismic Design of Buildings: GB 50011-2010 (2010)

European Committee for Standardization: Eurocode 8: Design of Structures for Earthquake Resistance (2004)

International Code Council: International Building Code, INC (2009)

Author, F.: Article title. Journal **2**(5), 99–110 (2016)

Author, F., Author, S.: Title of a proceedings paper. In: Editor, F., Editor, S. (eds.) CONFERENCE 2016. LNCS, vol. 9999, pp. 1–13. Springer, Heidelberg (2016)

Author, F., Author, S., Author, T.: Book title, 2nd edn. Publisher, Location (1999)

Author, F.: Contribution title. In: 9th International Proceedings on Proceedings, pp. 1–2. Publisher, Location (2010)

LNCS Homepage. http://www.springer.com/lncs. Accessed 21 Nov 2016

Assessment of Spatial Variability of Site Response in Japan

Cristina Lorenzo-Velazquez(✉) and Ashly Cabas

North Carolina State University, Raleigh, NC 27695, USA
clorenz@ncsu.edu

Abstract. Local soil conditions influence earthquake-induced ground shaking and deformation; a phenomenon known as site effects. The latter correlate with the concentration of damages to the built environment in areas prone to ground motion (GM) amplification. However, the characterization of geotechnical parameters affecting spatially varying earthquake GMs is often oversimplified. Spatial variability of GMs stems from source heterogeneities, varying ray propagation paths, varying local soil conditions, wave scattering, and directivity effects. Regionalization of site effects may also be necessary when specific geologic structures found within a given region control systematic amplifications in the area (e.g., deep sedimentary columns in the US Atlantic and Gulf coastal plains). Evaluating spatially variable GMs is also essential when investigating large, distributed infrastructure, such as water distribution systems. Because soil properties can be spatially correlated at nearby locations, the expected site response will also be spatially correlated. This study focuses on quantifying spatial correlations in site parameters from the Japanese databases, Kyoshin Network (K-Net) and Kiban-Kyoshin Network (KiK-net). Current spatial correlation models for intensity measures are based on correlation length, which neglects for instance, the effects of depositional processes in the spatial distribution of local soil conditions. In this work, we use Kriging to evaluate the significance of the spatial correlation for different site parameters and evaluate potential causes for the observed differences.

Keywords: Spatial variability · Seismic hazard · Regional assessment · Site response

1 Introduction

Seismic activity induces widespread ground shaking that potentially leads to infrastructure damage in vulnerable areas. In particular, damage induced to geographically distributed systems, known as lifelines, can be critical in a community's path to recovery after an extreme event. Hence, assessing the reduction of seismic risk is crucial yet complex due to the geographic distribution of such infrastructure which are exposed to varying ground responses and geotechnical failure mechanisms [8].

The spatial variability of ground motions is relevant for regions with high seismic activity and influenced by source heterogeneities, varying ray propagation paths, and local soil conditions, as well as wave scattering, and directivity effects. Site effects need to be taken into account within a greater spatial resolution with reasonable accuracy, as they can lead to damage concentration in areas prone to ground motion amplification.

Site conditions have been characterized through empirical approaches commonly using time-average seismic shear-wave velocity from the surface to a depth of 30 m (V_{S30}) as a proxy. The latter is an often used site parameter to estimate site responses single or combined with other parameters, as a continuous or discrete predictor variable [1, 16]. However, at regional scale analyses, site specific V_{S30} measurements are needed although not always available and even when available, they fail to capture the effects of deeper geologic structures. Therefore, site fundamental frequency, f_0, has been proposed as an alternative site parameter or to complement the V_{S30} parameter in recent studies [3, 4, 9, 15].

Because soil properties can be spatially correlated at nearby locations, we investigate the significance of the spatial correlation of site terms, namely V_{S30} and f_0 using kriging. Kriging is a geostatistical method based on statistical models that include autocorrelation. This interpolation method relies on semivariogram models of spatial autocorrelation, fitted to an experimental semivariogram model. The semivariogram indicates the degree of variability of observations or measured values as a function of their separation distances. It indicates that closer observations tend to be more similar to each other than distant observations. That theoretical semivariogram model is used in kriging to obtain the predicted spatial distribution of the site terms.

2 Database

K-NET and KiK-net stations are high sensitivity seismograph networks located in Japan, operated by the National Research Institute for Earth Science and Disaster Resilience (NIED). KiK-net stations have seismographs installed at the surface and at depth at each station; boreholes are between 100 m and a few thousand meters deep, often located at sites with stiff soils and/or shallow rock. K-net stations record data only on the free surface and are uniformly distributed every 20 km.

Zhu et al. [16], developed an open-source site database including 1,742 sites from both, K-NET (1045) and KiK-net (697) networks. The database includes site characterization parameters from available 1D velocity profiles, inferred from earthquake horizontal to vertical spectral ratio (HVSR), and inferred from regional models (e.g., topography- and geology-based V_{S30} proxies). This database also includes average wave velocities and depths to bedrock, and peak frequencies.

To evaluate the significance of the spatial correlation among site terms in Japan, we extracted the data from the Zhu et al. [16] database. However, all the stations within the database were not used. The spatial distribution of KiK-net [6] stations, having only a few sites within 10 km of each other [5] presented a challenge for this work. However, NIED in Japan maintains another large seismic network, the Kyoshin Network [7]. Considering stations from both seismic networks, 758 pairs of stations were found within 10-km of each other. From those pairs, a total of 1033 stations are being used in this study (483 KiK-net stations and 550 K-NET stations). Pairs of stations may include combinations of two K-net, two KiK-net, or one K-net and one KiK-net station. Figure 1 presents the distribution of the 1033 stations used in this study. The following parameters were compiled for all the study sites: measured V_{S30} from 1D velocity profiles, proxy-based V_{S30} from topographic slopes [12] and topographic proxies [11], and first peak and predominant peak frequencies from HVSRs. The peak frequency of amplification is related to depth to bedrock. Regional variations of soft and stiff sediments overlaying bedrock mean varying impedance contrasts, which result in potential variations of the amplification of ground motions [2].

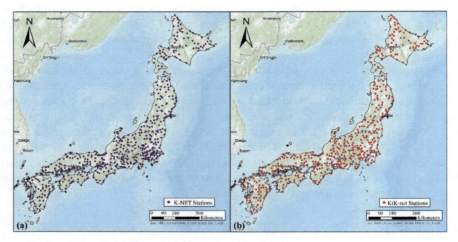

Fig. 1. Spatial distribution of (a) K-NET and (b) KiK-net stations used for this study.

3 Methodology

The spatial analyses in this study were performed using the kriging algorithm within ArcGIS. Specifically, we use ordinary kriging, the most widely used kriging method. It uses a weighted average of neighboring points to produce an estimated value for an unobserved point [10]. This technique involves quantifying the spatial correlation structure of the data and determining the predicted values. First, it is important to explore the distribution of the data to determine the transformation that should be used in case it is not normally distributed. Histograms allow visualization of the distribution, from which the skewness value (i.e., a measure of the asymmetry of the distribution) could be used to determine the transformation that should be used. According to Webster and Oliver [13], root transformation is a special case of Box-Cox transformation used if the skewness is between 0.5 and 1, and if the skewness exceeds 1 the logarithmic transformation.

Then, to measure the spatial autocorrelation, kriging uses the semivariance, γ, computed as follows:

$$\gamma(h) = \frac{1}{2}[z(x_i) - z(x_j)]^2; \tag{1}$$

where γ is the semivariance between known points, x_i and x_j separated by a distance h, and z is the attribute value. Subsequently, to average the semivariance data by distance, the algorithm groups sample points into lag classes and pairs of sample points by direction. The average semivariance is computed as:

$$\hat{\gamma}(h) = \frac{1}{2n} \sum\nolimits_{i=0}^{n} \left[z(x_i) - z(x_j + h)\right]^2; \qquad (2)$$

where $\hat{\gamma}$ is the average semivariance (or experimental semivariogram) between known points, separated by a distance h, n is the number of pairs of sample points in the bin, and z is the attributed value. After the experimental semivariogram is obtained, it is fitted to a mathematical model that describes the variability of the site terms with distance, which is used in kriging. The model influences the prediction of the unknown values; thus the selected model type should provide the best fit to the experimental semivariogram. Particularly, the influence of the closest neighbors on the prediction will be more significant the steeper the model's curve is near the origin.

The kriging algorithm within ArcGIS Pro, uses the semivariogram model and the search neighborhood to determine the weights for the measured locations. Then, the prediction for the locations with no observations or measurements within the study area can be made using the measured values and weights to obtain the spatially distributed surface. In this study, a standard neighborhood type was used as it assigns weights based on distance from the target location.

4 Spatial Dependence Results and Discussion

The spatial autocorrelation of selected site terms in the study are examined with the semivariograms obtained from ArcGIS. The semivariogram indicates, based on spatial dependence, that pairs of data points that are closer have similar values. First, ordinary kriging was chosen to obtain the predicted distribution of the site terms under study. The data distribution was analyzed and based on the skewness, a box-cox or logarithmic transformation was chosen. Then the experimental semivariogram generated was fitted with an exponential model for all the site terms. The fitted exponential model (theoretical semivariogram) was chosen because it was found to provide the best fit to the experimental semivariogram. This model exhibits an exponential decrease of the spatial dependence with increasing distance until it levels off at some distance. It levels off at the variance value (known as sill), where the range (i.e., a distance beyond which there is little or no autocorrelation) is reached (see Fig. 2); the values of the sill and range for the selected site parameters are presented in Table 1. Figure 2 presents the binned values showing the local variation of the semivariogram values for proxy-based V_{S30}. Whereas the average values presented on Fig. 2 are generated from the binned values, showing a smooth semivariogram. The model is fitted to the latter as it offers a clearer view of the spatial autocorrelation in the data.

The observed trend in semivariograms presented in Figs. 2, 3, and 4 suggest that the variability from the distances at which the model levels off (i.e., distances exceeding the range) may be of interest as there is little or no autocorrelation beyond that point. Comparing the results from proxy based V_{S30} from Wakamatsu and Matsuoka [11]

Table 1. Site parameters used from Zhu et al. [16] database and semivariogram characteristics.

Site parameter	Units	Available	Sill	Range (km)
V_{S30} (measured)	m/s	445	1.91E−1	18.82
V_{S30} (Wald and Allen [12])	m/s	1033	4.10E+4	116.55
V_{S30} (Wakamatsu and Matsuoka [11])	m/s	1033	2.74E+4	115.49
f_0	Hz	856	1.11E+0	51.31
f_p	Hz	856	8.90E−1	49.65

and Wald and Allen [12] in Fig. 3, the Wald and Allen [12] model presents a greater variance, but the range of their spatial autocorrelation is nearly the same (i.e., around 116 km). Figure 4 presents the semivariance from measured V_{S30} showing that the spatial autocorrelation is around 100 km less compared to the observed in proxy based V_{S30} values. This observation may result from the lack of measured V_{S30} values, as only 445 out of 1033 stations considered in this study had measured V_{S30} values, while both proxy-based V_{S30} estimates are available for the 1033 stations.

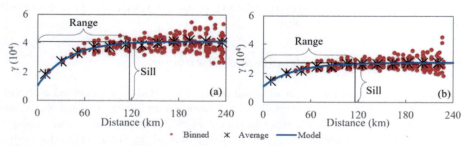

Fig. 2. Empirical (binned and average) and model semivariograms for proxy based V_{S30} from (a) Wald and Allen [12] and (b) Wakamatsu and Matsuoka [11].

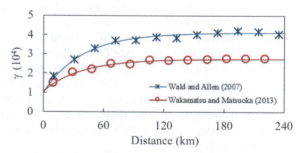

Fig. 3. Empirical semivariogram (symbols) and semivariogram model (curve) for the two proxy-based V_{S30} values.

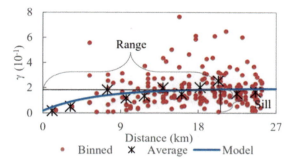

Fig. 4. Semivariograms of measured V_{S30} values from 1D velocity profiles.

The spatial autocorrelation of the site resonant frequencies from HVSR of earthquake recordings obtained from the Zhu et al. [16] database was also analyzed (Fig. 5). Comparing the resulting spatial autocorrelation range (Table 1) of the analyzed frequencies, f_0 and f_p, the correlation is similar, being both available for the same 856 stations. Thus, the spatial distribution of f_p according to Zhu et al. [14] is preferable for regionalization of site response, as f_0 underestimates the predominant frequency.

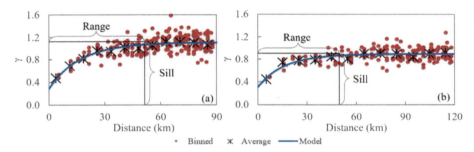

Fig. 5. Semivariograms of the (a) f_0 and (b) f_p values from Zhu et al. [16] database.

After performing kriging to map the predicted spatial distribution of the site terms, spatial distributions could be compared as shown in Figs. 6 and 7. Even though, the lack of measured V_{S30} values for about 57% of the selected stations the predicted spatial distribution maps generated from the kriging analyses performed for proxy-based V_{S30} distributions and measured V_{S30} (Fig. 6) have similar patterns. The similarity of both proxy-based V_{S30} spatial distribution patterns is expected as the range obtained from empirical semivariograms is almost the same. For the site frequency distribution of both f_0 and f_p, the spatial distribution is based on the stations with frequencies values available which are 83% of the selected stations.

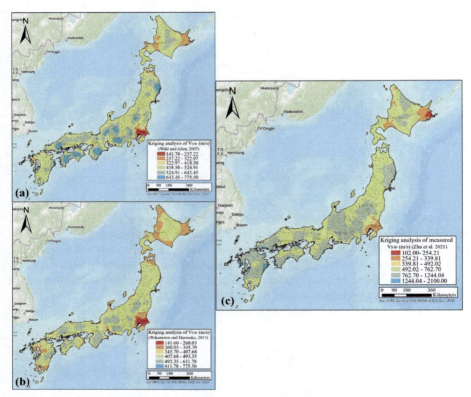

Fig. 6. Spatial distributions for proxy based V_{S30} from (a) Wald and Allen [12] and (b) Wakamatsu and Matsuoka [11], and V_{S30} (c) from measured values.

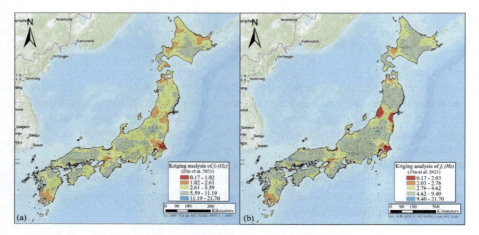

Fig. 7. Spatial distribution of the (a) first (f_0) and the (b) highest peak (f_p) frequencies.

According to Zhu et al. [14], f_p can help predict linear site amplification in a more robust manner, as it is associated with large velocity contrasts at shallower depths. From Fig. 7, can be observed that the highest peak frequencies take place at higher frequencies. Comparing the maps depicting the site frequency distributions (Fig. 7) with those of measured and proxy-based V_{S30} (Fig. 6), the areas with the lowest frequencies are observed throughout the areas with the lowest V_{S30} values.

5 Conclusions

Accounting for the spatial variability and the correlation of site terms is relevant to the next generation of ground motion models (GMMs) and the regionalization of site effects. This study recognized the need to understand geostatistical approaches to have a representation of the spatial distribution of subsurface conditions.

Our findings suggest that the spatial correlation of different site parameters varies significantly, even between proxy-based and measured V_{S30} values, where the correlation length varied from around 115 km to 18 km, respectively. The lack of data may influence the extent of the spatial autocorrelation; particularly, when comparing the distances at which the measured V_{S30} and the proxy-based V_{S30} are not spatially correlated according to experimental semivariograms. The correlation length of proxy-based V_{S30} is greater than that corresponding to measured V_{S30}, possibly because not as many values of the latter are available for all the station locations used in this study, and because of inherent limitations of proxy-based approaches in capturing real variations of subsurface conditions in the field. Analogously, the correlation length for fundamental and predominant frequencies (i.e., approximately 50 km for our database) was also different from their counterparts corresponding to V_{S30} values. Further research is necessary to provide recommendations on appropriate spatial correlations to use for site terms in GMMs. Furthermore, as subsurface conditions are spatially correlated rather than random, accounting for reliable site terms spatial variability using geostatistical approaches such as kriging should support the development of more robust GMMs in the future.

Acknowledgment. This material is based upon work supported by the National Science Foundation Graduate Research Fellowship under Grant No. DGE-2137100. Any opinion, findings, and conclusions or recommendations expressed in this material are those of the authors and do not necessarily reflect the views of the National Science Foundation.

References

1. Borcherdt, R.D.: Estimates of site-dependent response spectra for design (methodology and justification). Earthq. Spectra **10**, 617–654 (1994)
2. Braganza, S., Atkinson, G.M., Molnar, S.: Assessment of the spatial variability of site response in southern Ontario. Seismol. Res. Lett. **88**(5), 1415–1426 (2017)
3. Hashash, Y.M., Ilhan, A.O., Hassani, B., Atkinson, G.M., Harmon, J., Shao, H.: Significance of site natural period effects for linear site amplification in central and eastern North America: empirical and simulation-based models. Earthq. Spectra **36**, 87–110 (2020)

4. Hassani, B., Atkinson, G.M.: Application of a site-effects model based on peak frequency and average shear-wave velocity to California. Bull. Seismol. Soc. Am. **108**, 351–357 (2018)
5. Jayaram, N., Baker, J.W.: Correlation model for spatially distributed ground-motion intensities. Earthq. Eng. Struct. Dyn. **38**(15), 1687–1708 (2009)
6. National Research Institute for Earth Science and Disaster Prevention (NIED): Kiban Kyosin Network (KiK-net) (2011a). http://www.kik-net.bosai.go.jp/. Accessed 30 July 2019
7. National Research Institute for Earth Science and Disaster Prevention (NIED): Kyosin Network (K-net) (2011b). http://www.k-net.bosai.go.jp/. Accessed 30 July 2019
8. O'Rourke, T.D.: Geohazards and large, geographically distributed systems. Géotechnique **60**(7), 505–543 (2010)
9. Pitilakis, K., Riga, E., Anastasiadis, A.: New code site classification, amplification factors and normalized response spectra based on a worldwide ground-motion database. Bull. Earthq. Eng. **11**, 925–966 (2013)
10. Trauth, M.H.: MATLAB® Recipes for Earth Sciences, 4th edn. Springer, Heidelberg (2015). https://doi.org/10.1007/978-3-662-46244-7
11. Wakamatsu, K., Matsuoka, M.: Developing a 7.5-sec site-condition map for Japan based on geomorphology classification. WIT Trans. Built Environ. **120**, 101–112 (2013)
12. Wald, D.J., Allen, T.I.: Topographic slope as a proxy for seismic site conditions and amplification. Bull. Seismol. Soc. Am. **97**, 1379–1395 (2007)
13. Webster, R., Oliver, M.A.: Geostatistics for Environmental Scientists. John Wiley & Sons, New York (2001)
14. Zhu, C., Cotton, F., Pilz, M.: Detecting site resonant frequency using HVSR: Fourier versus response spectrum and the first versus the highest peak frequency. Bull. Seismol. Soc. Am. **110**(2), 427–440 (2020)
15. Zhu, C., Pilz, M., Cotton, F.: Which is a better proxy, site period or depth to bedrock, in modeling linear site response in addition to the average shear-wave velocity? Bull. Seismol. Soc. Am. **18**, 797–820 (2020)
16. Zhu, C., Pilz, M., Cotton, F.: An open-source site database of strong-motion stations in Japan: K-NET and KiK-net (v1.0.0). Earthq. Spectra **37**(3), 2126–2149 (2021)

Ground Deformation Evaluation Using Numerical Analyses and Physics-Based Mega-thrust Subduction Zone Motions

Diane Moug[✉], Arash Khosravifar, and Peter Dusicka

Portland State University, Portland, OR 97201, USA
dmoug@pdx.edu

Abstract. A large earthquake is anticipated for the Pacific Northwest of the United States and western Canada from the Cascadia Subduction Zone (CSZ). The earthquake is expected to cause significant damage to infrastructure, including at Oregon's Critical Energy Infrastructure (CEI) hub which extends for over 9 km and handles 90% of Oregon's liquid fuel. There is particular concern about damage to fuel storage tanks from earthquake-induced ground deformations. CSZ earthquake events are infrequent and there do not exist any recorded ground motions. Recently, a suite of broadband synthetic physics-based broadband ground motions was developed by modeling full rupture of the CSZ as part of The M9 Project. The present study estimates likely lateral displacements in the CEI hub using numerical simulations with the M9 ground motions. Two alluvial soil types are modeled: coarse-grained soils with the PM4Sand model and fine-grained soils with the PM4Silt model. Lateral displacements are estimated with one-dimensional (1D) site response analysis combined with strain-potential and empirical procedures, and two-dimensional (2D) nonlinear dynamic simulations. The sensitivity of estimated lateral displacements to ground motion characteristics are presented to evaluate the use of M9 motions for geotechnical seismic studies, and to evaluate the use of 1D and 2D methods with fine-grained soils.

Keywords: Numerical analyses · Liquefaction · Cyclic softening in clays and plastic silts · Lifelines

1 Introduction

A Cascadia Subduction Zone (CSZ) earthquake is anticipated to cause widespread damage throughout the Pacific Northwest (PNW) of the United States and Western Canada. Anticipating the magnitude and extent of ground deformations from a CSZ event is critical for seismic hazard mitigation throughout the PNW. However, a significant challenge is the lack of historic strong earthquake events, due to the long return period of CSZ rupture. Therefore, there is large uncertainty regarding anticipated ground motion characteristics, for example peak ground acceleration (PGA) or cumulative absolute velocity (CAV_5), and anticipated ground deformations. To fill these knowledge gaps, The M9

Project developed a suite of broadband synthetic physics-based ground motions for locations throughout the PNW based on a full rupture of the CSZ [1]. These ground motions have been used to evaluate likely structural damage from a CSZ event [2–5], but their use for geotechnical studies is limited. This study uses the set of M9 ground motions to examine how earthquake-induced lateral ground deformations relate to ground motion characteristics for a site of sand-like and clay-like alluvial soils.

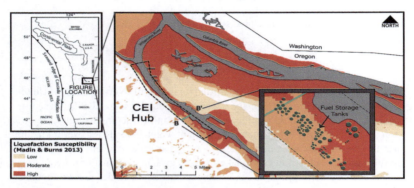

Fig. 1. Location of the CEI hub and liquefaction susceptibility in the Portland, Oregon area.

A case study was selected using the Critical Energy Infrastructure (CEI) hub in Portland, Oregon. The CEI hub extends for 9 km along the Willamette River in Northwest Portland, as shown in Fig. 1. This industrial area handles 90% of Oregon's liquid fuel and 100% of the jet fuel for the Portland International Airport. The site is located on young alluvial soils which are considered highly susceptible to liquefaction [6]. There is a concern that a CSZ event will cause large earthquake-induced ground deformations throughout the CEI hub that critically damage fuel storage tanks [7]. The present study examines lateral deformations at a soil profile that was characterized by publicly-available subsurface data [8]. Lateral ground deformations are estimated with one-dimensional (1D) seismic site response analysis with empirical and strain-potential methods; and two-dimensional (2D) non-linear deformations analysis (NDA) using PM4Silt [9] and PM4Sand [10] soil models. The results of 1D and 2D analyses are compared against each other and against ground motion characteristics for the input ground motions. The objectives of this study are: (i) to compare 1D and 2D methods for estimating lateral deformations with a liquefiable soil and a cyclic-softening soil, (ii) evaluate the ground motion characteristics that strongly affect ground deformations, and (iii) use physics-based ground motions to characterize geotechnical earthquake hazards in the CEI hub.

2 Subsurface Conditions

Four primary geologic units were identified throughout the CEI hub [8, 11]. Engineered fill (FILL) is the shallowest soil layer and is made of dredged soils from the Willamette River. Quaternary-aged alluvium (QAL) lies below the FILL unit, was deposited by the Willamette River, and generally ranges from low plasticity silts to fine to medium sand. The presence of fine-grained QAL (FG-QAL) and coarse-grained QAL (CG-QAL) varies throughout the CEI hub, reflecting historic river deposition patterns. The bedrock is Columbia River Basalt (CRB), which is overlain by approximately 1.5 m of the weathered Columbia River Basalt (WCRB) unit.

A representative profile, referred to as B-B' was analyzed; its location is illustrated in Fig. 1 and stratigraphy is detailed in Fig. 2. Some nearby geotechnical boreholes are plotted along B-B' in Fig. 2, showing examples of the data used to interpret the stratigraphic profile. The borehole data indicate heterogeneous distribution of FG-QAL and CG-QAL. To bookend the possible range of lateral ground deformations, this study performs analysis on uniform profiles of both FG-QAL and CG-QAL.

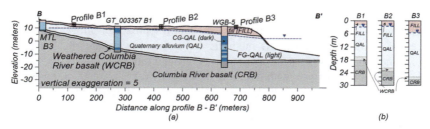

Fig. 2. (a) Stratigraphic profile for B-B', and (b) profiles for 1D analysis at B1, B2, and B3.

3 Ground Motions

The M9 project developed a suite of 30 scenarios for a potential Mw 9.0 earthquake on the CSZ, each considering a different hypocenter location and rupture parameters [1]. The input acceleration time histories were developed for outcrop motions from shear wave velocity (Vs) = 800 m/s, which is the approximate Vs of the CRB unit. The synthetic M9 motions for the CEI hub were available via the CSZ@PDX online tool (m9csz.cee.pdx.edu) for coordinates within the CEI hub, therefore, effects from the Portland basin are inherent in the ground motions. Only the x-component of the ground motion acceleration time histories were used in this study.

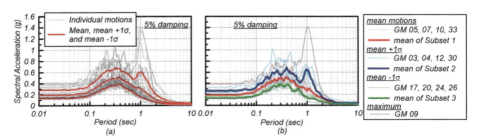

Fig. 3. Acceleration spectra of M9 ground motions used in this study: (a) all motions, and (b) motion subsets used for 2D analyses.

All 30 motions were used for 1D site response and are shown in Fig. 3a along with the mean (determined by the geometric mean), mean +1 standard deviation (σ), and mean −1σ. 13 selected motions were used for 2D modeling to reasonably approximate the entire suite of 30 motions. Of the 13 motions, four subsets were selected to represent the mean of all 30 ground motions (05, 07, 10, 33), mean +1σ (03, 04, 12, and 30), mean −1σ (17, 20, 24, and 26), and the maximum ground motion (09). The spectral accelerations of the ground motions within each subset are depicted in Fig. 3b.

4 Numerical Models

4.1 1D and 2D Models

Earthquake-induced lateral deformations were estimated with 1D seismic site response analysis and 2D NDA of profile B-B'. Using both approaches allowed comparison between simplified and advanced methods for estimating ground deformations of clay-like and sand-like soils with physics-based ground motions.

1D site response analysis with the finite difference program FLAC 8.0 was used to estimate deformations at locations B1, B2, and B3, as illustrated in Fig. 2. Total stress analysis was performed to estimate profiles of maximum shear strain (γ_{max}) and cyclic shear stress ratio (CSR) for each ground motion. Established methods [12–16] were used to estimate lateral ground deformations for each ground motion based on γ_{max} and/or factor of safety against liquefaction triggering following the guidance in Kramer [12]. It is worth noting that a M_w 9 CSZ event is outside the range of data used to develop Youd et al. [12], however, as will be shown in the results, the estimates are approximately in-line with the other methods and therefore were included in this study. These methods were developed for sand-like soils, however, since few simple methods are available for clay-like soils, they were also used to estimate lateral deformations of FG-QAL profiles. Figure 4 shows the 2D FLAC mesh developed for profile B-B'. Free-field boundary conditions were applied to the lateral boundaries of the 2D model. The models were subjected to 1D horizontal dynamic loading at the base of the model using acceleration time histories. The ground motions were applied as outcrop motions using the compliant-base procedure in combination with the FLAC quiet boundary [17]. Lateral ground deformations were estimated along B-B' from model geometry at the end of simulated earthquake shaking.

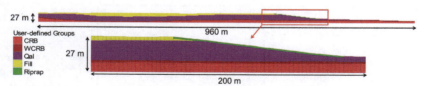

Fig. 4. 2D FLAC model geometries for B-B' profile.

4.2 Selection of Soil Model Parameters

This section describes the model parameters for the CG-QAL and FG-QAL units since they were the primary concern for ground failure. CRB and WCRB were modeled with the elastic model; FILL and Riprap units were modeled with the Mohr-Coulomb model.

CG-QAL. The PM4Sand model was used to capture the cyclic response of CG-QAL soil for 2D NDA. The model parameters are summarized in Table 1. The parameter G_o was calibrated to approximate the relationship of G_{max} versus depth based on Vs measurements in QAL [8]. The shear modulus degradation (G/G_{max}) relationship to shear strain (γ) was calibrated with the h_o parameter to approximate standard relationships for sand [18, 19]. Calibration of cyclic strength behavior of CG-QAL was performed with 9 CPT profiles from the CEI hub [20, 21] the sand-like soils of CG-QAL had a representative clean sand normalized cone tip resistance ($q_{c1n,cs}$) of 90, that related to a relative density (Dr) of 53% and a CRR at M_w 7.5 of 0.13 based on commonly used relationships [22, 23]. $H_{po} = 0.3$ was selected to achieve CRR = 0.13 in cyclic undrained direct simple shear (UDSS) single element simulations, where CRR is considered as the CSR that results in 3% single amplitude shear strain after 15 uniform cycles (N = 15).

Table 1. Model parameters for 2D NDA

Model parameter[a]	CG-QAL calibration (PM4Sand)	FG-QAL calibration (PM4Silt)
G_o	650	650
G_{exp}	-	0.75
h_o	Default	0.6
h_{po}	0.30	40
ϕ'_{cs}	35°	30°
Dr	53%	-
s_u/σ'_v	-	0.35
$n_{b,wet}$	-	1.0

[a]Other model parameters use default values [9, 10]

FG-QAL. The PM4Silt model was used to capture the cyclic response of FG-QAL through clay-like cyclic softening behavior. Since there was no differentiation between coarse-grained and fine-grained QAL in the Vs data of Roe and Madin [8], the G_{max} versus depth relationship was calibrated with parameters G_o and G_{exp} to be consistent with the relationship for CG-QAL. The $G/G_{max} - \gamma$ relationship was calibrated with the parameter h_o to agree with published relationships for soil with plasticity index of 15 at 1 atmosphere of overburden stress [24, 25]. PM4Silt model parameters that control monotonic and cyclic shear strength were calibrated to laboratory test results from two nearby sites on the Willamette River that are underlain by similar soils as the CEI hub

[26]. FG-QAL CPT data in the CEI hub indicated an OCR of 1.5. Monotonic UDSS tests indicated a relationship between normalized undrained shear strength (s_u/σ'_{vc}) and OCR as: $s_u/\sigma'_{vc} = 0.25 OCR^{0.87}$. Therefore, the s_u/σ'_{vc} is approximately 0.35. ϕ'_{cs} was estimated as 30° from monotonic UDSS tests at high strain. The final calibration priority was the CRR based on results of cyclic UDSS tests. At OCR = 1.5, the approximate CRR is 0.28, which was approximated with the parameter $h_{po} = 40$.

5 Results of Numerical Analyses

5.1 1D Analysis Results

CSR profiles were obtained from 1D seismic site response analysis at B1, B2, and B3 using standard methodologies [12]. The CSR profiles were then compared to CRR to calculate factor of safety against liquefaction. Estimated lateral displacements from 1D analyses are summarized in Fig. 5. The lateral deformations estimated at profile B1 with Zhang et al. [13] are notably larger than those estimated by the other methods. This may reflect that profile B1 is located on sloping ground, which is accounted for in some methods [12, 13] but not others. Estimated CG-QAL and FG-QAL lateral deformations at B2 show relatively consistent variation between the different estimation methods. The lateral deformation estimates at B3 are more variable between ground motions than estimates at B1 and B2. This may be due to higher ground motion sensitivity near the free face. Overall, lateral deformations are estimated to be notably lower for FG-QAL profiles than CG-QAL.

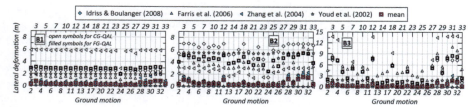

Fig. 5. Lateral deformations estimated with 1D site response analysis for FG-QAL and CG-QAL

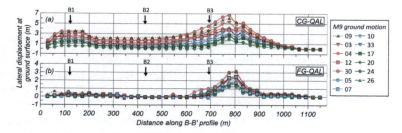

Fig. 6. Lateral deformations at B-B' from 2D NDA for (a) CG-QAL, and (b) FG-QAL

5.2 2D Analysis Results

The lateral displacements from 2D NDA are shown in Fig. 6. The 2D NDA results indicate that lateral deformations will be significant for sites dominated by CG-QAL compared to sites composed of FG-QAL, which is consistent with estimates from 1D analyses. Lateral deformations of the CG-QAL are largest close to the river free face and near the 2% ground slope at the location of Profile B1. The lateral deformations at the CG-QAL profile appear to be sensitive to the intensity of ground motions, where the maximum ground motion and the mean $+1\sigma$ ground motions resulted in the largest deformations. Lateral displacements of FG-QAL were concentrated near the free face and had some deformation along the ground slope, however displacements were near zero over the rest of the profile. The FG-QAL results did not show notable grouping between the ground motion subsets.

The differences in lateral displacement between FG-QAL and CG-QAL profiles are likely attributable to: (i) lower CRR for CG-QAL soils than FG-QAL soil, as described in the model properties section, and (ii) CG-QAL being susceptible to excess porewater pressure generation and loss of stiffness and strength once liquefaction is triggered, whereas FG-QAL was subject to cyclic softening.

6 Discussion

The analyses performed allow examination of how the estimated ground deformations vary with use of 1D methods or 2D NDA, and ground motion characteristics PGA and CAV_5. These are compared in Fig. 7. The estimates from 1D analysis represent the mean values presented in Fig. 5. For CG-QAL, the 1D estimates appear to be more conservative than estimates from 2D NDA, as is expected. In contrast, for FG-QAL, lateral deformation estimates do not show that either 1D or 2D NDA estimates are consistently more conservative: at B1, representing a 2% slope far from the free-face, 1D estimates are slightly higher than 2D NDA estimates; at B2, representing flat ground near the free-face, 1D estimates are much higher than 2D NDA estimates; and at B3, representing flat ground very close to the free-face, the 2D NDA estimates are larger than 1D estimates. These inconsistencies for FG-QAL may reflect that the 1D methods were developed for liquefaction triggering of sand-like soils. Reliable estimation of lateral deformation for clay-like soils with 1D methods likely require investigation into how lateral deformations relate to CRR, CSR, ground slope, and distance from the free face for cyclic-softening soils.

Lateral deformation of both CG-QAL and FG-QAL appear to be similarly sensitive to PGA and CAV_5, based on 2D NDA results. From Fig. 7a, as PGA and CAV_5 increase, the estimated lateral deformation increases for CG-QAL. However, 1D estimates for CG-QAL are not as sensitive to the ground motion parameters. This is partly attributed to estimated deformations reaching maximum potential shear strains for most ground motions with B1 and B2, and some ground motions at B3. The FG-QAL results in Fig. 7b show that 1D estimates of deformation increase with PGA and CAV_5. 2D NDA estimates of deformation are also sensitive to PGA and CAV_5 at B1 and B3, however, B2 does not appear to be unlikely to undergo lateral deformations in a CSZ event.

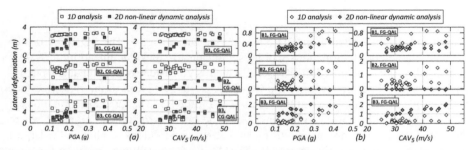

Fig. 7. Lateral deformations from 1D and 2D analysis with M9 ground motions related to peak ground acceleration and CAV_5 for: (a) CG-QAL, and (b) FG-QAL.

Although the FG-QAL and CG-QAL analyses are meant to book-end lateral deformations, these cases assume that the site is composed of a single alluvial soil type. However, alluvial deposition patterns typically result in heterogeneous soil deposits, as shown in the geotechnical borehole WGB-5 in Fig. 2b. Heterogeneous distribution of CG-QAL may reduce lateral displacements, especially when the liquefiable layers are not continuous [27].

7 Summary and Conclusions

This study examined a range of likely ground deformations for a profile of alluvial soils in Portland, Oregon's CEI hub. The study focused on earthquake-induced lateral deformations for soils with sand-like liquefaction triggering cyclic behavior or soils with clay-like cyclic softening behavior. A set of physics-based ground motions from The M9 Project were used. Lateral deformations were estimated with 1D seismic response analysis, and 2D NDA analyses of a site profile. The results indicated that (i) lateral deformation estimates are conservative with 1D methods compared to 2D NDA for CG-QAL, however, the 1D estimates for FG-QAL appear to be conservative or non-conservative which may indicate that 1D methods for clay-like soils should be used cautiously; (ii) lateral displacement for both CG-QAL and FG-QAL positively relate to the ground motion PGA and CAV_5, however deformations will likely be most severe for sites where CG-QAL is largely present, and (iii) the M9 project physics-based ground motions are useful to overcome the case-history gap of subduction zone earthquake events in the PNW and western Canada.

Estimates of earthquake-induced lateral deformations are valuable to assess potential damage to CEI hub fuel storage tanks. Future work will benefit from addressing several knowledge gaps related to this study. First, fragility of above ground liquid storage tanks are primarily based on PGA and do not account for ground deformations [28]. Second, further characterization of soil behavior and stratigraphy will reduce the uncertainty of soil model calibration parameters. Third, development of simple methods to relate 1D seismic site response to lateral ground deformation of clay-like soils will allow efficient hazard assessment.

Acknowledgements. The authors acknowledge the support from the Oregon Department of Defense and Portland State University. The authors also thank Ericka Koss at the City of Portland for help gathering geotechnical reports from the CEI hub. The findings, opinions, and conclusions expressed in this paper are solely those of the authors.

References

1. Frankel, A., Wirth, E., Marafi, N., Vidale, J., Stephenson, W.: Broadband synthetic seismograms for magnitude 9 earthquakes on the Cascadia megathrust based on 3D simulations and stochastic synthetics, part 1: methodology and overall results. Seismol. Soc. Am. Bull. **108**(5A), 2347–2369 (2018)
2. Mirafi, N.A., Makdisi, A.J., Eberhard, M.O., Berman, J.W.: Impacts of an M9 Cascadia subduction zone earthquake and Seattle basin on performance of RC core wall buildings. J. Struct. Eng. **146**(2), 04019201 (2019)
3. Mirafi, N.A., Makdisi, A.J., Berman, J.W., Eberhard, M.O.: Design strategies to achieve target collapse risks for reinforced concrete wall buildings in sedimentary basins. Earthq. Spectra **36**(3), 1038–1073 (2020)
4. Eksir Monfared, A., Molina Hutt, C., Kakoty, P., Kourehpaz, P., Centeno, J.: Effects of the georgia sedimentary basin on the response of modern tall RC shear-wall buildings to M9 Cascadia subduction zone earthquakes. J. Struct. Eng. **147**(8), 05021003 (2021)
5. Kourehpaz, P., Hutt, C.M., Marafi, N.A., Berman, J.W., Eberhard, M.O.: Estimating economic losses of midrise reinforced concrete shear wall buildings in sedimentary basins by combining empirical and simulated seismic hazard characterizations. Earthq. Eng. Struct. Dyn. **50**(1), 26–42 (2021)
6. Madin, I.P., Burns, W.J.: Ground motion, ground deformations, tsunami inundation, coseismic subsidence, and damage potential maps for the 2012 Oregon Resilience Plan for Cascadia Subduction Zone Earthquakes. Open-file report O-13-06 for DOGAMI (2013)
7. Wang, Y., Bartlett, S.F., Miles, S.B.: Earthquake risk study for Oregon's critical energy infrastructure hub: final report to Oregon Department of Energy and Oregon Public Utility Commission. DOGAMI (2013)
8. Roe, W.P., Madin, I.: 3D geology and shear-wave velocity models of the Portland, Oregon, Metropolitan Area. Oregon Department of Geology and Mineral Industries (DOGAMI), Open-File Report O-13-12, p. 48 (2013)
9. Boulanger, R.W., Ziotopoulou, K.: PM4Silt (Version 1): A Silt Plasticity Model for Earthquake Engineering Applications. Report No. UCD/CGM-18/01, Center for Geotechnical Modeling, Department of Civil and Environmental Engineering, University of California, Davis, CA (2018)
10. Boulanger, R.W., Ziotopoulou, K.: PM4Sand (Version 3.1): A Sand Plasticity Model for Earthquake Engineering Applications. Report No. UCD/CGM-17/01, Center for Geotechnical Modeling, Department of Civil and Environmental Engineering, University of California, Davis, CA (2017)
11. Madin, I.P., Ma, L., Niewendorp, C.A.: Preliminary geologic map of the Linnton 7.5' quadrangle, Multnomah and Washington Counties, Oregon. Oregon Department of Geology and Mineral Industries Open-File Report O-08-06 (2008)
12. Youd, T.L., et al.: Liquefaction resistance of soils: summary report from the 1996 NCEER and 1998 NCEER/NSF workshops on evaluation of liquefaction resistance of soils. J. Geotech. Geoenviron. **127**(4), 297–313 (2001)

13. Zhang, G., Robertson, P.K., Brachman, R.W.I.: Estimating liquefaction-induced lateral displacements using the standard penetration test or cone penetration test. J. Geotech. Geoenviron. **130**(8), 861–871 (2004)
14. Faris, A.T., Seed, R.B., Kayen, R.E., Wu, J.: A semi-empirical model for estimation of maximum horizontal displacement due to liquefaction-induced lateral spreading. In: Proceedings of the 8th U.S. National Conference on Earthquake Engineering (2006)
15. Idriss, I.M., Boulanger, R.W.: Soil liquefaction during earthquakes. Earthquake Engineering Research Institute (2008)
16. Yoshimine, M., Nishizaki, H., Amano, K., Hosono, Y.: Flow deformation of liquefied sand under constant shear load and its application to analysis of flow slide in infinite slope. Soil Dyn. Earthq. **26**, 253–264 (2006)
17. Mejia, L.H., Dawson, E.M.: Earthquake deconvolution for FLAC. In: Proceedings of Fourth International FLAC Symposium on Numerical Modeling in Geomechanics, Madrid (2006)
18. Seed, H.B., Idriss, I.M.: Soil moduli and damping factors for dynamic response analysis. Report No. UCB/EERC-70/10, Earthquake Engineering Research Center, University of California, Berkeley, California (1970)
19. EPRI: Guidelines for determining design basis ground motions, early site permit demonstration program, vol. 1, RP3302, Electric Power Research Institute, Palo Alto, California (1993)
20. GeoDesign: Report of Geotechnical Engineering Services for Inter-Fluve, Inc. Report no. InterFluve-2-01 (2016)
21. Moug, D.M., et al.: Field evaluation of microbially induced desaturation for liquefaction mitigation of silty soils. In: 17th World Conference of Earthquake Engineering, Sendai, Japan (2020)
22. Boulanger, R.W., Idriss, I.M.: CPT and SPT based liquefaction triggering procedures. Report No. UCD/CGM.-14/1 (2014)
23. Boulanger, R.W., Idriss, I.M.: CPT-based liquefaction triggering procedure. J. Geotech. Geoenviron. **142**(2), 04015065 (2016)
24. Vucetic, M., Dobry, R.: Effect of soil plasticity on cyclic response. J. Geotech. Eng. **117**(1), 89–107 (1991)
25. Darendelli, M.B.: Development of a new family of normalized modulus reduction and material damping curves. The University of Texas at Austin (2001)
26. Dickenson, S.E., et al.: Cyclic and Post-Cyclic Behavior of Silt-Rich, Transitional Soils of the Pacific Northwest; A Database for Geo-professionals in Practice and Research, Data report prepared for the Oregon Department of Transportation, Bridge Engineering Section, Salem, Oregon, by New Albion Geotechnical, Inc., Reno, NV (2021)
27. Boulanger, R.W., Moug, D.M., Munter, S.K., Price, A.B., DeJong, J.T.: Evaluating liquefaction and lateral spreading in interbedded sand, silt, and clay deposits using the cone penetrometer. Aust. Geomech. **51**(4), 109–128 (2016)
28. FEMA: Hazus Earthquake Model Technical Manual, Hazus 4.2 SP3, Federal Emergency Management Agency (2020)

V_{S30} Correlations from Shear Wave Velocity Profiles in the UAE

Deepa Kunhiraman Nambiar[1], Tadahiro Kishida[1(✉)], Tareq Fahmy Abdallatif[2,3], and Mohammad H. AlHamaydeh[4]

[1] Khalifa University of Science and Technology, Abu Dhabi, United Arab Emirates
tadahiro.kishida@ku.ac.ae
[2] National Research Institute of Astronomy and Geophysics, Helwan, Egypt
[3] GSC Geophysical Consulting, Abu Dhabi, United Arab Emirates
[4] American University of Sharjah, Sharjah, United Arab Emirates

Abstract. The time averaged shear wave velocity to 30-m depth (V_{S30}) is used for seismic hazard assessment for National Earthquake Hazard Reduction Program (NEHRP) site classification. There are several models to predict V_{S30} from average shear wave velocity to a depth (z) less than 30 m (V_{Sz}). This study evaluates the prediction capability of the existing models in the United Arab Emirates (UAE). The analyses show that there is a significant prediction biases in the existing models developed in different regions. By reviewing borehole data from the selected sites, it reveals that these biases are mainly because of the characteristic geology of the UAE where weak bedrock is encountered at shallow depths. This bedrock mostly shows a considerable variation in both clay content and the porosity which also have a direct effect on the prediction uncertainty.

Keywords: V_{S30} · V_{Sz} · Regression model · UAE

1 Introduction

The prediction of time-averaged shear-wave velocity in the upper 30 m of sites (V_{S30}) is important to access seismic hazards. Many studies investigated the prediction of V_{S30} from various variables. Boore (2004) collected 135 shear wave velocity profiles in California and proposed regression models of V_{S30} using averaged V_S to a shallow depth (z) (V_{Sz}). Boore et al. (2011) similarly developed the models using data from Japan, Turkey, and Europe and concluded that these models are regionally dependent. Kwak et al. (2017) summarized the recent studies for various regions. However, a regional model to predict V_{S30} from V_{Sz} does not exist in the UAE.

Wald and Allen (2007) and Allen and Wald (2009) collected V_{S30} measurements from active tectonic regions (California, Italy, and Taiwan) and stable continental regions (Australia and Tennessee). They proposed the prediction models of V_{S30} from topographic slope gradient at 30 arcsec resolution using Digital Elevation Model (DEM). Iwahashi and Pike (2007) used the same DEM to define geomorphic terrain category by combining slope gradient and metrics of convexity and texture. The terrain category has been used to predict region-specific V_{S30} predictions such as California (Yong et al. 2012; Yong 2016), Greece (Stewart et al. 2014), Taiwan (Kwok et al. 2018), Pacific Northwest (Kwak et al. 2017). Ahdi et al. (2017) developed two proxy-based V_{S30} estimation procedures, one is hybrid geology-slope approach and the other is geomorphic terrain-based approach, in Pacific Northwest region. In the literatures, however, we did not find any specific study for DEM proxy based V_{S30} prediction in the UAE nor Arabian Plate. For instance, Kiuchi et al. (2019) estimated V_{S30} values at Saudi Arabia based on H/V response spectral ratio because proxy-based methods did not exist.

Wills and Clahan (2006) divided the state of California in to simplified geologic units and predicted V_{S30} for each unit group. Wills and Gutierrez (2011) studied the effect of different groupings of geologic units and introduced a system of subdividing younger alluvium based on surface slopes that have distinct V_{S30} ranges. Wills et al. (2015) developed a new geologically defined V_{S30} site condition map of California based on the works of Wills and Clahan (2006) and Wills and Gutierrez (2011). Kwok et al. (2018) developed a prediction model of V_{S30} in Taiwan. They developed a hybrid geology-slope-elevation proxy model. However, based on our knowledge, there is no V_{S30} prediction model in the UAE using surface geology.

This paper mainly addresses the limitations in the V_{S30} prediction models in the UAE. A database of shear wave velocity (V_S) profiles is developed by various geophysical methods in the UAE. Mean and standard deviations of V_{S30} in natural log scale are predicted from V_{Sz}. Site-specific proxies such as surface geology, topographic slope, and terrain category are extracted at the locations where V_s profiles are available. The prediction models of V_{S30} are developed by using surface geology and digital elevation models (DEMs) which are available across the UAE.

2 V_S Profile Database in UAE

2.1 Data Sources

A V_S profile database was compiled for UAE that consists of a collection of 89 sites. The profiles are obtained using MASW and collected from various studies. Among which 14 sites were explored by MASW and have the Vs profiles along the line with a separation distance of 2 m. If the V_S profiles at MASW inversion locations are counted independently, it constitutes 428 V_S profiles in the database. The contents of UAE profile database include site identification number, data source, geodetic coordinates, V_S profile information, depth to bedrock, geology, and DEM parameters. The parameters from geology and DEM are described in the following sections.

2.2 Computation of V_{Sz}

General formula to calculate V_{Sz} is presented in Eq. (1):

$$V_{Sz} = \frac{z}{\sum_{i=1}^{n}\left(\frac{z_i}{v_i}\right)} \qquad (1)$$

where z_i is the thickness of ith soil layer, v_i is the shear wave velocity of the ith layer and n is the number of layers. The V_{Sz} is defined as wave travel distance divided with wave travel time. This time averaged method is based on the travel time of shear waves as they travel through 30 m of soil beneath the ground surface explicitly. This formula preserves the behavior of slow layers, which tend to be the most likely to exhibit amplification during an earthquake.

3 V_{S30} Prediction from Surface Geology and Terrain Category in UAE

The country is relatively low lying, with near surface geology dominated by Quaternary to late Pleistocene age, whereas older deposits also exist such as Tertiary, Jurassic to Cretaceous and Permo-Tertiary. Quaternary deposits mainly consist of mobile Aeolian sand dunes and evaporates deposits of sabkha. The surface deposits typically comprise fine grained silty calcareous sand, which is commonly dense and variably cemented. Although variable, the degree of cementation generally increases with depth, such that the variably cemented sand grades to predominantly calcareous sandstone. Very silty gypsiferous sabkha and evaporate layers occur occasionally within the Aeolian sand deposits. The bedrock is typically made of limestone or calcareous sandstone interbedded with gypsum. The gypsum is the weathered form of limestone (Mitchell and Soga 2005) and it has a relatively low V_S compared to the limestone. These rocks are relatively weak with average V_S of 600–800 m/s. Figure 1 shows the relationship between V_{S30} and bedrock depth. The $Z_{0.7}$ and $Z_{1.0}$ represent the bedrock depth when V_S reaches 0.7 and 1.0 km/s, respectively. As V_{S30} increases, the bedrock depth decreases. The $Z_{0.7}$ exposes around the depth of 0–30 m. It is approximately 10 m when V_{S30} is 600 m/s. The number of data points for $Z_{1.0}$ is much smaller than that of $Z_{0.7}$ because many V_S profiles does not encounter the bedrock with $V_S = 1.0$ km/s. The figure shows that the

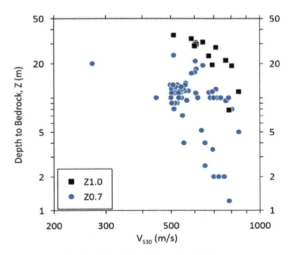

Fig. 1. Depth to bedrock with V_{S30}

$Z_{1.0}$ is 40 m when V_{S30} is 600 m/s. Chiou and Youngs (2014) reported that the $Z_{1.0}$ is about 300 m in California when V_{S30} is 600 m/s.

Figure 2 shows histogram of V_{S30} for geologic categories. The V_S profiles are located on Aeolian sand dunes, Coastal Sabkha, Delta or Shoal deposits and Fluviatile deposits of Quaternary ones. The V_{S30} ranges from 600–800 m/s for Aeolian sand dunes, where it ranges from 500–700 m/s for Coastal Sabkha, Delta or Shoal deposits and Fluvial deposits. Table 1 shows the summary of mean and standard deviations of V_{S30} based on surface geology category. The average V_{S30} for each geology is relatively high (>500 m/s) compared to the past studies. Ahdi et al. (2017) reported that V_{S30} is 339 m/s in the Pacific Northwest region when the surface geology is beach, bar, dune deposits, indicating that the V_{S30} in the UAE is much higher. These observations are related to soil stratigraphy in UAE (Fig. 1) where the bedrocks expose at relatively shallow depths.

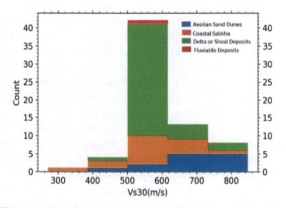

Fig. 2. Histogram describing measured V_{S30} value in each geologic category

Table 1. Geologic scheme for estimating V_{S30}

Geologic category	Mean V_{S30} (m/s)	Standard deviation (σ_{lnV})
Aeolian sand dunes	662	0.22
Coastal sabkha	568	0.29
Delta or shoal deposits	557	0.14
Fluviatile deposits	511	NA

Terrain category is also studied in the UAE using the terrain type (TT) classification map by Iwahashi and Pike (2007) (hereinafter IP07). Figure 3 shows a percent distribution of terrain category in the UAE, where TT5, TT9, TT14, and TT16 are the major types. A small percentage are associated with TT2. Using the locations of available V_{S30} values, we determined the corresponding terrain category. Table 2 summarizes the mean and standard deviation of V_{S30} for each terrain category. The V_{S30} ranges from 550 to

700 m/s for TT9, TT13, TT15, TT16, which is much larger than the data in California (200–500 m/s, Yong et al. 2012). From Tables 1 and 2, the standard deviation using the terrain type seems to be slightly smaller than those by surface geology.

Table 2. Terrain category-based scheme for estimating V_{S30}.

Terrain category	Description	Mean V_{S30} (m/s)	Standard deviation ($\sigma_{\ln V}$)
9	Well eroded plain of weak rocks	689	0.22
13	Incised terrace	623	0.24
15	Dune, Incised terrace	656	0.19
16	Fluvial plain, alluvial fan, low lying flat plains	561	0.15

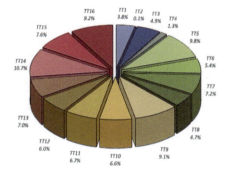

Fig. 3. 3D pie chart describing terrain types (TT) and percentage of occurrence found in UAE

4 V_{S30} Prediction from Shallow V_S Profiles

We computed V_{Sz} in 1-m increments from 428 profiles. These data are used in the regression analyses to evaluate their performance. Figure 4 shows the V_{Sz} vs. V_{S30} for four depths (z = 5, 10, 15, and 20 m). The figure shows data from the UAE as well as California and Japan for comparison purpose. The data from California and Japan are obtained from Boore (2004) and KiK-net (https://www.kyoshin.bosai.go.jp/pub/sitedat/), respectively. The figure shows that data distribution of UAE overlaps with California, whereas its variation is relatively narrow compared to other regions. Most UAE sites have V_{S30} > 500 m/s except one site with V_{S30} of 280 m/s that locates at a reclaimed island. Correlation coefficients of V_{Sz} with V_{S30} are presented in Table 3. The value increases with depth. The correlations among the regions are very similar with z > 20 m even though the geological conditions are very different. However, the correlation significantly decreases in UAE when z < 20 m compared to the other regions. Figure 4 also presents that the V_{S30} in UAE becomes around 800 m/s when $V_{Sz} > 600$ m/s. These observations are related to the fact that the weak bedrock exposes in UAE with the depth shallower than 30 m. These characteristics are unique in the UAE compared to the regions in the past studies.

Table 3. Correlation coefficient (r) between V_{Sz} and V_{S30} for four depths (z)

Dataset	No. of profiles	r (z = 5 m)	r (z = 10 m)	r (z = 15 m)	r (z = 20 m)
Japan (Kik-net)	624	0.80	0.90	0.96	0.98
UAE	428	0.72	0.80	0.84	0.94
California	263	0.75	0.92	0.97	0.99

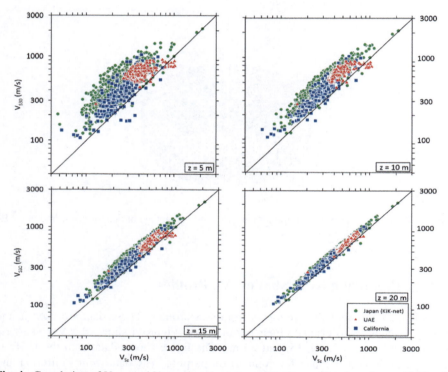

Fig. 4. Correlation of V_{S30} and V_{SZ} using shear wave velocity profiles from Japan, California, and UAE

4.1 Regression Models

For comparison purpose, we regressed V_{S30} against V_{Sz} by using the formula presented in the past studies (Boore 2004; Boore et al. 2011; Midorikawa and Nogi 2015). Boore (2004) used the following:

$$\log(\overline{V_{S30}}) = c_0 + c_1 \log(\overline{V_{Sz}}) \qquad (2)$$

where c_0 and c_1 are model coefficients. Boore et al. (2011) updated Eq. (2) by including a quadratic term in Eq. (3). It can model the saturation of V_{S30} with V_{Sz} as bedrock approaches for KiK-net in Japan:

$$\log V_{S30} = c_0 + c_1 \log V_{sz} + c_2 (\log V_{Sz})^2 \qquad (3)$$

where c_0, c_1 and c_2 are model coefficients. Midorikawa and Nogi (2015) used the following formula in which V_{S30} is modeled by V_{Sz} and V_S at the depth of z ($V_S(z)$):

$$\log V_{S30} = c_0 + c_1 \log V_{sz} + c_2 (\log V_s(z)) \qquad (4)$$

where c_0, c_1 and c_2 are model coefficients depending on z. Figure 5 shows the variation of σ_{res} for different regression formulas. When z < 16 m, the variation in the σ_{res} is very small between models. The model by Midorikawa and Nogi (2015) resulted in the smaller σ_{res} when z > 16 m. This is because most sites encounter the bedrock at the depth of 16 m.

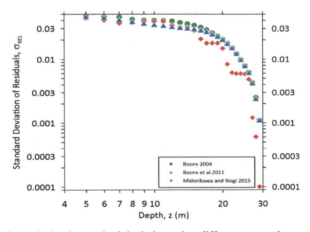

Fig. 5. Variation in standard deviation using different regression models

5 Summary and Conclusions

In this paper we focused on the prediction of V_{S30} from V_{Sz} for the sites in UAE. The main challenge was the lack of widely variable data availability across the region. We compiled a V_S profile database in UAE along with information of surface geology and terrain class. The bedrock depths of $Z_{0.7}$ and $Z_{1.0}$ are plotted against V_{S30}. Weak bedrock exposes at shallow depth in the UAE. Correlations between V_{Sz} and V_{S30} are computed in the UAE and compared with different regions. The correlations are very similar among the regions when z > 20 m even though these geological conditions are very different. However, the correlation substantially decreases in the UAE when z < 20 m compared to the other regions. Three V_{S30} prediction models by V_{Sz} (Boore 2004; Boore et al. 2011; Midorikawa and Nogi 2015) have been evaluated using the UAE database. The results show that the standard deviation of residuals using the method by Midorikawa and Nogi (2015) is lowest among the models from z = 16 to 30 m.

References

1. Ahdi, S.K., Stewart, J.P., Ancheta, T.D., Kwak, D.Y., Mitra, D.: Development of VS profile database and proxy-based models for V_{S30} prediction in the Pacific Northwest region of North America. Bull. Seismol. Soc. Am. (2017). https://doi.org/10.1785/0120160335
2. Allen, T.I., Wald, D.J.: On the use of high-resolution topographic data as a proxy for seismic site conditions (Vs30). Bull. Seismol. Soc. Am. **99**, 935–943 (2009)
3. Boore, D.M.: Estimating $\overline{V_{S30}}$ (or NEHRP site classes) from shallow velocity models (depths < 30 m). Bull. Seismol. Soc. Am. **94**, 591–597 (2004)
4. Boore, D.M., Thompson, E.M., Cadet, H.: Regional correlations of V_{S30} and velocities averaged over depths less than and greater than 30 meters. Bull. Seismol. Soc. Am. **101**(6), 3046–3059 (2011)
5. Chiou, B.S.-J., Youngs, R.R.: Update of the Chiou and Youngs NGA model for the average horizontal component of peak ground motion and response spectra. Earthq. Spectra **30**, 1117–1153 (2014)
6. Iwahashi, J., Pike, R.J.: Automated classifications of topography from DEMs by an unsupervised nested-means algorithm and a three-part geometric signature. Geomorphology **86**(3/4), 409–440 (2007)
7. Kiuchi, R., Mooney, W.D., Zahran, H.M.: Ground-Motion prediction equations for western Saudi Arabia. Bull. Seismol. Soc. Am. **109**, 2722–2737 (2019)
8. Kwak, D.Y., et al.: Performance evaluation of V_{SZ}-to-V_{S30} correlation methods using global V_S profile database. In: Proceedings Third International Conference on Performance-Based Design in Earthquake Geotechnical Engineering, 16–19 July 2017, Vancouver, Canada (2017)
9. Kwok, O.L.A., Stewart, J.P., Kwak, D.Y., Sun, P.L.: Taiwan-specific model for V_{S30} prediction considering between-proxy correlations. Earthq. Spectra **34**, 1973–1993 (2018)
10. Mitchell, J.K., Soga, K.: Fundamentals of Soil Behavior. John Wiley & Sons Inc., Hoboken (2005)
11. Midorikawa, S., Nogi, Y.: Estimation of V_{S30} from shallow velocity profile. J. Japan Assoc. Earthq. Eng. **15**(2), 91–96 (2015)
12. Seyhan, E., Stewart, J.P., Ancheta, T.D., Darragh, R.B., Graves, R.W.: NGA-West2 site database. Earthq. Spectra **31**(3), 1007–1024 (2014)
13. Stewart, J.P., et al.: Compilation of a local VS profile database and its application for inference of V_{S30} from geologic-and terrain-based proxies. Bull. Seismol. Soc. Am. **104**, 2827–2841 (2014)
14. Wald. D.J., Allen, T.I.: Topographic slope as a proxy for seismic site conditions and amplification. Bull. Seismol. Soc. Am. **97**(5), 1379–1395 (2007)
15. Wills, C.J., Clahan, K.B.: Developing a map of geological defined site-condition categories for California. Bull. Seismol. Soc. Am. **96**(4A), 1483–1501 (2006)

16. Wills, C.J., Gutierrez, C.I.: Investigation of geographic rules for improving site-conditions mapping. In: Proceedings of the 4th IASPEI/IAEE International Symposium: Effects of Surface Geology on Seismic Motion, 23–26 August 2011, University of California, Santa Barbara (2011)
17. Wills, C.J., Gutierrez, C.I., Perez, F.G., Branum, D.M.: A next generation V_{S30} map for California based on geology and topography. Bull. Seismol. Soc. Am. **105**, 3083–3091 (2015)
18. Yong, A., Hough, S.E., Iwahashi, J., Braverman, A.: Terrain-based site conditions map of California with implications for the contiguous United States. Bull. Seismol. Soc. Am. **102**(1), 114–128 (2012)
19. Yong, A.: Comparison of measured and proxy-based V_{S30} values in California. Earthq. Spectra **32**, 171–192 (2016)

Insight into the Relationship Between Dynamic Shear Strain and Vibration Velocity of Horizontally Layered Sites

Qi Xia, Rui-shan Li($^{\boxtimes}$), and Xiao-ming Yuan

Key Laboratory of Earthquake Engineering and Engineering Vibration of China Earthquake Administration, Institute of Engineering Mechanics, China Earthquake Administration, Harbin 150080, China
lrshan22@hotmail.com

Abstract. Equivalent linear method in frequency domain, represented by SHAKE2000, is the mainstream approach for seismic response analysis of soil layers. Due to the seriously unreasonable results in soft soil sites, its improvement becomes a research hotspot, which mainly adopts frequency-dependent method (FDM), but there has been no substantial effect. In this paper, the assumption of dynamic shear strain and vibration velocity behave a constant proportional relationship in FDM was studied. Based on wave motion equation, exact solutions for the relationship of the two variables were derived on one-dimensional horizontally layered sites. The reasonability and deviation degree of the assumption were studied through numerical experiments. Results show that the constant proportional hypothesis only holds in the case of unbounded homogeneous medium with one-way traveling wave. For actual layered soil sites, the relationship between dynamic shear strain and vibration velocity strongly depends on the frequency of the wave and the location of the observation point. If we ignore the reflection wave in ground seismic response analysis, the use of constant proportional assumption under one-way traveling wave will make significant deviation to the results. The deviation caused by the constant proportional relationship assumption may span four order ranges. Even for a single uniform deposit, the deviation is also very significant. For practical ground motion calculation, this assumption clearly exists qualitative error in theory, and the quantitative deviation can not be acceptable.

Keywords: Dynamic shear strain · Vibration velocity · Seismic response analysis · Equivalent linear method

1 Introduction

The seismic fortification of various engineering structures is the most direct and fundamental means to reduce earthquake disasters. One of the key issues is to determine reasonable design ground motion parameters. For major projects, such as large-span, high-rise, nuclear power, and important water conservancy facilities, the work is particularly important [1]. Previous seismic survey data and strong motion observation data show that local site conditions have a significant impact on ground motion and engineering damage. Therefore, the estimation of soil ground motion has become a key link in engineering seismic design [2].

At present, the most important method of soil ground motion estimation is the method of calculation and analysis. One-dimensional soil seismic response analysis is most widely used, and it is also the basis for studying two-dimensional and three-dimensional soil seismic response analysis methods. One-dimensional soil seismic response analysis methods are mainly composed of two types of solutions: nonlinear solutions in time domain and equivalent linear in frequency domain. Although time-domain nonlinear solution is more advanced in theory, they are not effectual in fact. At present, the typical one-dimensional frequency-domain equivalent linear method is SHAKE2000, and the typical one-dimensional time-domain nonlinear method is DEEPSOIL. The former is developed by the University of California, Berkeley (UC Berkeley), and the latter is developed by the University of Illinois at Urbana-Champaign (UIUC). The comparison results of the two programs indicate that [3], although DEEPSOIL performs better than SHAKE2000 in calculating the high frequency part of the response spectra, both the peak ground acceleration (PGA) and the overall performance of the response spectra are slightly deficient to the latter, therefore the frequency-domain equivalent linear analysis is still the current international mainstream method for site seismic response calculation. However, in the calculation and analysis of ground motions for soft soil sites, many cases did not reflect the actual magnification effect of the near-surface soil mass, and the calculation results were seriously unreasonable [4]. The development of classical equivalent linear method is one of the main research objects of scholars in geotechnical earthquake engineering.

Recently, some scholars tried to improve the equivalent linear method in frequency domain [5–8]. The basic concept is to change the calculation method of the effective shear strain from the maximum shear strain × 0.65 approach in the traditional equivalent linear method (including SHAKE2000) to the frequency-dependent approach, which is called the frequency-dependent method (FDM). The fundamental concept of these studies is correct, because seismic loads have strong randomness and irregularity, ground motions usually show outstandingly various spectral characteristics, and different frequency components contribute differently to the ultimate vibration response. The traditional constant effective shear strain calculation approach is of obvious unreasonableness. However, in these improved methods, although the effective shear strain varies with frequency is introduced, in order to correlate the shear strain with the frequency characteristics of ground motions, one basic assumption is adopted: the dynamic shear strain is in constant proportion to the vibration velocity. Current comparison results indicate that these improved methods have no substantial effects. They did not overcome the deficiency that traditional calculation methods cannot reflect the real amplification effect of soft soil sites. Therefore, it is necessary to explore these improved methods thoroughly to examine the basic assumptions of FDM.

Based on the fundamental solution of wave equation, this paper derives analytical solutions for the relationship between dynamic shear strain and vibration velocity on horizontally layered soil site. The necessary conditions to establish the basic assumptions of FDM method are given. The rationality of using velocity spectra to represent the shear strain spectra is evaluated. The deviation degree under the proportion assumptions is examined.

2 Wave Equation and Basic Solutions

Assuming that the actual site is horizontally layered and extends infinitely. The input seismic load is a shear wave perpendicular to soil layer, incident upward from the bedrock. Each soil layer is characterized by its layer thickness h, gravity ρ, shear wave velocity v_s, and initial damping ratio ξ. The layer model is shown in Fig. 1. The local coordinate origin of each layer is at the top of the layer, and the direction is downward. Assuming that the soil is *Kelvin-Voigt* viscoelastic material, the one-dimensional wave propagating vertically in the medium can be expressed as:

$$\rho \frac{\partial^2 u}{\partial t^2} = G \frac{\partial^2 u}{\partial z^2} + \eta \frac{\partial^3 u}{\partial z^2 \partial t} \tag{1}$$

Under the incidence of harmonic wave with circular frequency ω, the displacement of the n-th layer of soil can be expressed by its spatial local coordinate z_n and time coordinate t as:

$$u_n(z_n, t) = \left[E_n \exp(ik_n z_n) + F_n \exp(-ik_n z_n)\right] e^{i\omega t} \tag{2}$$

In the formula, E and F are the amplitudes of the upward and downward waves propagating in the soil layer, and the formula $k = w/v_s$, representing the wave number, and i is an imaginary unit.

According to the displacement expression, the vibration velocity and shear strain formulas can be derived as:

$$\begin{aligned}\dot{u}_n(z_n, t) &= \frac{\partial u_n(z_n, t)}{\partial t} = i\omega \left[E_n \exp(ik_n z_n) + F_n \exp(-ik_n z_n)\right] e^{i\omega t} \\ \gamma_n(z_n, t) &= \frac{\partial u_n(z_n, t)}{\partial z_n} = ik_n \left[E_n \exp(ik_n z_n) - F_n \exp(-ik_n z_n)\right] e^{i\omega t}\end{aligned} \tag{3}$$

On the layered interface, its displacement continuity and shear stress compatibility conditions can be expressed as:

$$\begin{aligned} u_n(h_n, t) &= u_{n+1}(0, t) \\ G_n \gamma_n(h_n, t) &= G_{n+1} \gamma_{n+1}(0, t) \end{aligned} \tag{4}$$

Substituting Eqs. (2) and (3) into (4), the relationship between the wave amplitude vectors of the two adjacent layers can be derived as [9]:

$$\begin{bmatrix} E_{n+1} \\ F_{n+1} \end{bmatrix} = \begin{bmatrix} \frac{1+\alpha_n}{2} \exp(ik_n h_n), & \frac{1-\alpha_n}{2} \exp(-ik_n h_n) \\ \frac{1-\alpha_n}{2} \exp(ik_n h_n), & \frac{1+\alpha_n}{2} \exp(-ik_n h_n) \end{bmatrix} \begin{bmatrix} E_n \\ F_n \end{bmatrix} \tag{5}$$

In the formula, $\alpha_n = (\rho_n v_{sn})/(\rho_{n+1} v_{s(n+1)})$ is the wave impedance between two layers of media.

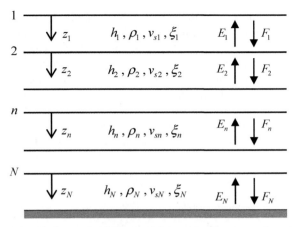

Fig. 1. Horizontally layered soil site model (used for frequency domain solution)

Based on the above derivation, this paper will further elaborate the relationship be-tween dynamic shear strain and vibration velocity on horizontally layered soil site.

3 Analytical Solution on Layered Site

3.1 Theoretical Solution

We adopt the simplified horizontal layered model to represent the actual engineering site. Seismic waves are incident perpendicularly. There are wave transmission and reflection phenomena on the ground surface and soil layer interface. The upward wave and downward wave in the soil layer coexist, which together determine the soil layer response. The relationship between the upward and downward wave amplitude vector of each layer is shown in formula (5).

In an unbounded homogeneous medium, the incident wave will not be reflected, and only unidirectional traveling waves exist in the medium. In this condition, the displacement of soil layer, expression (2), degenerates into:

$$u(z,t) = Ee^{i(kz+\omega t)} \tag{6}$$

Meanwhile, the specific form of the vibration velocity and dynamic shear strain, expression (3), in an unbounded medium is:

$$\dot{u}(z,t) = \frac{\partial u(z,t)}{\partial t} = i\omega E e^{i(kz+\omega t)}$$
$$\gamma(z,t) = \frac{\partial u(z,t)}{\partial z} = ikE e^{i(kz+\omega t)} \tag{7}$$

Thus, the relationship between the dynamic shear strain and the vibration velocity caused by the unidirectional traveling wave in an unbounded homogeneous medium can be obtained as:

$$\frac{\gamma(z,t)}{\dot{u}(z,t)} = \frac{ikEe^{i(kz+\omega t)}}{i\omega Ee^{i(kz+\omega t)}} = \frac{k}{\omega} = \frac{1}{v_s}\omega \tag{8}$$

It can be clearly seen from formula (8) that the dynamic shear strain exhibits a constant proportional relationship with the vibration velocity. The proportional coefficient is the reciprocal of the shear wave velocity of soil medium, which has nothing to do with the input wave frequency characteristic. Taking into account the linear superposition property of the Fourier transform, this conclusion is valid not only for harmonics, but also for irregular seismic waves. It means that the normalized dynamic shear strain spectra are exactly the same as the velocity spectra in the frequency domain, in the case of unidirectional traveling wave within uniform medium.

For uniform semi-infinite space situation, the free surface boundary conditions can be expressed as:

$$\tau(z,t)|_{z=0} = G \cdot \gamma(z,t)|_{z=0} = 0 \qquad (9)$$

Substituting the expression $y(z, y)$ in formula (3) into the above formula, $E = F$ can be obtained, indicating that the steady-state downward wave amplitude and upward wave amplitude are equal. This conclusion is also applicable to the surface layer of a stratified site. In this situation, the specific expression form of the soil displacement can be rewritten as:

$$u(z,t) = 2E\cos(kz)e^{i\omega t} \qquad (10)$$

From this, the soil vibration velocity and dynamic shear strain in the uniform half space can be expressed as:

$$\begin{aligned}\dot{u}(z,t) &= \frac{\partial u(z,t)}{\partial t} = 2i\omega E \cos(kz)e^{i\omega t} \\ \gamma(z,t) &= \frac{\partial u(z,t)}{\partial z} = -2kE\sin(kz)e^{i\omega t}\end{aligned} \qquad (11)$$

Hence, the relationship between the dynamic shear strain and the vibration velocity of the uniform half space can be derived as:

$$\frac{\gamma(z,t)}{\dot{u}(z,t)} = \frac{-2kE\sin(kz)e^{i\omega t}}{2i\omega E\cos(kz)e^{i\omega t}} = \frac{i}{v_s}\cdot\tan(kz) \qquad (12)$$

Formula (12) is obviously different from formula (8). In the case of single uniform semi-infinite space, the constant proportional relationship between the dynamic shear strain and vibration velocity no longer exists. The ratio of them is related to frequency. Whether it is time history or frequency spectra, they are not only different in value, but also in shape. In other words, even for the simple half space, the primary assumptions of FDM are also unreasonable.

As for two-layer model, according to the free boundary conditions, $E_1 = F_1$ can be obtained for the surface layer. The specific derivation process is nearly the same as that of the single uniform semi-infinite site above, so it will not repeat here. The ratio of the shear wave velocity to the vibration velocity of the top soil layer can also be expressed by formula (12), but the coordinates z in the formula should be replaced by the first local coordinates z_1.

For the second layer, substituting $E_1 = F_1$ into formula (5), the upward and downward wave amplitudes can be obtained as:

$$E_2 = \frac{1}{2}E_1\left[(1+\alpha_1)e^{ik_1h_1} + (1-\alpha_1)e^{-ik_1h_1}\right] = E_1[\cos(k_1h_1) + i\alpha_1 \sin(k_1h_1)]$$
$$E_2 = \frac{1}{2}E_1\left[(1-\alpha_1)e^{ik_1h_1} + (1+\alpha_1)e^{-ik_1h_1}\right] = E_1[\cos(k_1h_1) - i\alpha_1 \sin(k_1h_1)] \quad (13)$$

In this formula, α_1 is the wave impedance between the second and the first soil layer. According to formula (3), the relationship between the dynamic shear strain of the second layer and its vibration velocity can be obtained as:

$$\frac{\gamma_2(z_2,t)}{\dot{u}_2(z_2,t)} = \frac{ik_n\left[E_n \exp(ik_n z_n) - F_n \exp(-ik_n z_n)\right]e^{i\omega t}}{i\omega\left[E_n \exp(ik_n z_n) + F_n \exp(-ik_n z_n)\right]e^{i\omega t}}$$
$$= \frac{1}{v_{s2}} \cdot \frac{E_2 \exp(ik_2 z_2) - F_2 \exp(-ik_2 z_2)}{E_2 \exp(ik_2 z_2) + F_2 \exp(-ik_2 z_2)} \quad (14)$$

Substituting the expressions E_2 and F_2 in formula (13) into formula (14), we can get:

$$\frac{\gamma_2(z_2,t)}{\dot{u}_2(z_2,t)} = \frac{i}{v_{s2}} \cdot \frac{\cos(k_1h_1)\sin(k_2z_2) + \alpha_1 \sin(k_1h_1)\cos(k_2z_2)}{\cos(k_1h_1)\cos(k_2z_2) - \alpha_1 \sin(k_1h_1)\sin(k_2z_2)} \quad (15)$$

It can be degenerated to a single uniform half-space solution, when $\alpha_1 = 1$ and $k_1 = k_2$.

In terms of multi-layer soil site, the derivation process is similar to the two-layer soil model. First, the upward and downward wave amplitude vectors of each layer are obtained through recursion, and then the displacement, velocity and strain expressions (2) and (3) can be obtained. The accurate solution expressions can be derived, which will not be described in detail here. In most situations, the numerical result is generally used by recursion through programming, so the specific expressions of the multi-layer soil model are not listed here.

3.2 Analytical Result on a Two-Layer Site

A typical calculation example of two-layer site is used to illustrate the variation of the ratio between the dynamic shear strain and the vibration velocity. In this example, the parameters of the first layer are $H_1 = 10$ m, $v_{s1} = 150$ m/s, and the parameters of the second layer are $H_2 = 20$ m, $v_{s2} = 250$ m/s. Assuming that the soil layers have the same unit weight, the wave impedance between the two layers is $\alpha = 0.6$, and the soil damping ratio is $\xi = 5\%$. Figure 2 shows the curve of the ratio between soil dynamic shear strain and vibration velocity versus frequency at depths of 5 m and 25 m separate. The dotted line in the figure is the result of unidirectional traveling wave in uniform media, that is, $1/v_{s1} = 1/150$ and $1/v_{s2} = 1/250$. Figure 2 reveals that when seismic waves propagate from the bedrock to the surface, there are strong reflection phenomena at the ground surfaces and soil layer interfaces. The steady-state response of the soil layer is the interaction results of upward and downward waves. Figure 3 shows the ratio curves between shear strain and vibration velocity along soil depths when the harmonic waves with different frequencies are input. When the frequency is 0.1 Hz, 0.3 Hz, 0.5 Hz, 0.8 Hz and 1.2 Hz separately, comparing the result at $h_1 = 8$ m with $1/v_{s1} = 6.67 \times 10^{-3}$, the relative values are 3%, 10%, 17%, 28%, and 43%.

The results at $h_2 = 20$ m are 5%, 15%, 26%, 43%, and 72%, when compared to $1/v_{s2} = 4 \times 10^{-3}$.

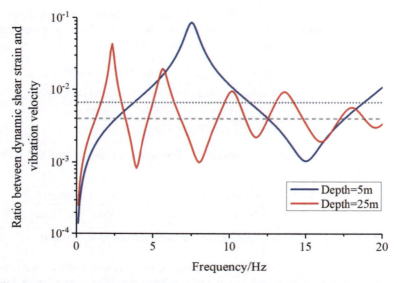

Fig. 2. Ratio between dynamic shear strain and vibration velocity at different depth

3.3 Numerical Result on Actual Layered Sites

Figure 4 shows the shear wave velocity profiles of two actual sites, which are Kobe Port Island [10] in Japan and Treasure Island in San Francisco Bay [11]. They are frequently used by scholars to analyze local site effects. According to site category classification standards in Chinese code for seismic design of buildings, soil thickness of Port Island site is $H = 85$ m (limited to the drilling depth, not reaching the bedrock), and the average shear wave velocity is $v_{sm} = 257$ m/s, estimated site period is $T = 4H/v_{sm} = 1.32$ s, equivalent shear wave velocity within 20 m is $v_{s20} = 197$ m/s, it should be a typical Class III site. Soil thickness of Treasure Island site is $H = 88$ m, cover layer average shear wave velocity is $v_{sm} = 240$ m/s, site period is $T = 4H/v_{sm} = 1.47$ s, equivalent shear wave velocity within 20 m is $v_{s20} = 169$ m/s, which should be a Class III site, but it is softer than Port Island, which means it is closer to Class IV.

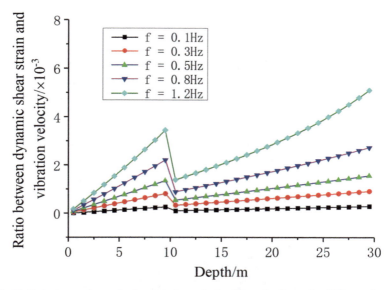

Fig. 3. Ratio between dynamic shear strain and vibration velocity under different frequency

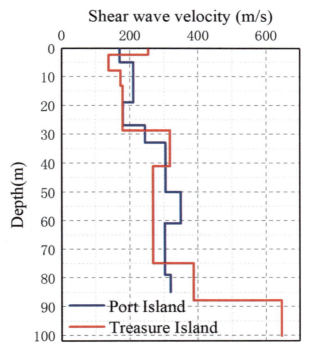

Fig. 4. Shear wave velocity profiles of Port Island and Treasure Island

Figures 5 and 6 show the ratio of the dynamic shear strain to the vibration velocity at different depths and frequencies on the two actual thickness soil sites, Port Island and Treasure Island, respectively. It can be seen from the figure that the ratio on the two sites display remarkable variations. In the frequency range below 20 Hz, the ratio spans nearly four orders of magnitude from maximum to minimum. Especially in the case of low-frequency wave input, the ratio fluctuation in shallow layers becomes more obvious. This result indicates significant differences to the theoretical solution of unidirectional traveling waves. It can be inferred from this that it is seriously unreasonable to use the assumption of a constant proportion ratio to improve the traditional equivalent linear method for seismic response analysis.

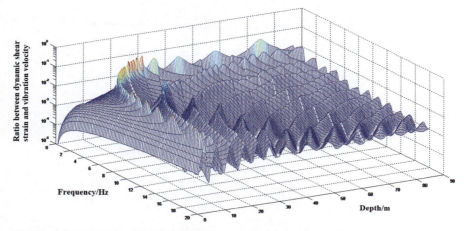

Fig. 5. Relationship between dynamic shear strain and vibration velocity on Port Island site

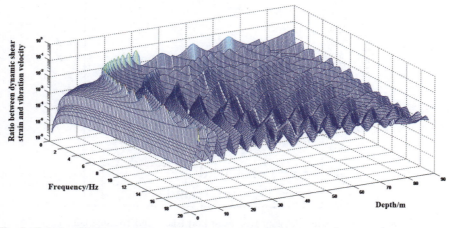

Fig. 6. Relationship between dynamic shear strain and vibration velocity on Treasure Island site

4 Conclusion

In view of the importance and development state of ground motion calculation methods in earthquake resistant engineering, this paper discusses the rationality and feasibility of the basic assumption that the dynamic shear strain behaves constant proportional relationship with the vibration velocity in current mainstream improved methods. The analytical solution for the relationship between the dynamic shear strain and the vibration velocity on horizontally layered sites was derived based on one-dimensional wave equation. Numerical experiments were conducted to study the relationship between the velocity spectra and the shear strain spectra. The rationality of the available frequency-dependent methods (FDM) and deviation degree of the basic assumptions were examined. The results show that:

(1) In the case of unidirectional traveling wave in uniform media, there exists a constant proportion relationship between dynamic shear strain and the vibration velocity. The ratio is the reciprocal of the media shear wave velocity and has nothing to do with frequency. Therefore, the time history and frequency spectra for the normalized dynamic shear strain and vibration velocity are exactly the same and can be substituted for each other.
(2) As for the uniform half-space and layered soil sites, there exists wave reflection at the free ground surface and the soil layer interface. The steady-state vibration response is the result of the interaction of the incident and the reflected wave. The relationship between the dynamic shear strain and the vibration velocity significantly depends on the wave frequency and the observation point location.
(3) Only when it is in the situation of unidirectional traveling wave propagation in uniform media does the relationship between dynamic shear strain and vibration velocity have a constant proportion. However, in terms of near surface soil layers, the variation of the ratio is close to four orders of magnitude. The deviation caused by constant proportion assumption within the frequency band of engineering interest can reach more than 95%.

Acknowledgement. This work was supported by the Scientific Research Fund of Institute of Engineering Mechanics, China Earthquake Administration (No. 2019C04) and Natural Science Foundation of Heilongjiang Province (No. LH2019E093).

References

1. National Standard of the People's Republic of China: GB 17741-2005 Evaluation of Seismic Safety for Engineering Sites China Building Industry Press, Beijing (2005)
2. Hu, Y., Sun, P., Zhang, Z.: Effect of Site Conditions on Earthquake Damage and Ground Motion Seismic Microzonation - Theory and Practice. Seismological Press, Beijing (1989)
3. Yang, Y., Sun, R., Yang, H., et al.: Contrasting study between two international typical soil layers seismic response analysis programs. World Earthq. Eng. **33**(03), 17–23 (2017)
4. Li, R.: Research on a new generation technique for ground seismic response analysis. Institute of Engineering Mechanics. China Earthquake Administration, Harbin (2016)

5. Yoshida, N., Kobayashi, S.: SUETOMI I: equivalent linear method considering frequency dependent characteristics of stiffness and damping. Soil Dyn. Earthq. Eng. **22**(03), 205–222 (2002)
6. Assimaki, D., Kausel, E.: An equivalent linear algorithm with frequency-and pressure-dependent moduli and damping for the seismic analysis of deep sites. Soil Dyn. Earthq. Eng. **22**(9–12), 959–965 (2002)
7. Kausel, E., Assimaki, D.: Seismic simulation of inelastic soils via frequency-dependent moduli and damping. J. Eng. Mech. **128**(01), 34–47 (2002)
8. Jiang, T., Xing, H.: An equivalent linear method considering frequency-dependent soil properties for seismic response analysis. Chin. J. Geotech. Eng. **29**(02), 218–224 (2007)
9. Liao, Z.: Introduction to Wave Motion Theories for Engineering, 2nd edn. Science Press, Beijing (2002)
10. Iwasaki, Y., Tai, M.: Strong motion records at Kobe Port Island. Soils Found. **36**(Special), 29–40 (1996)
11. Gibbs, J.F., Fumal, T.E., Boore, D.M.: Seismic velocities and geologic logs from borehole measurements at seven strong-motion stations that recorded the Loma Prieta earthquake. U.S. Geological Survey Open-File Report 92-287 (1992)

Investigating the In-Situ Properties of Poisson's Ratio Based on KiK-Net Data

Yang Shi[✉], Hao Zhang, and Yu Miao

Huazhong University of Science and Technology, Wuhan, China
shiyang1@hust.edu.cn

Abstract. Poisson's ratio is an important parameter in the engineering field. However, most related studies have been performed using laboratory test or in-situ test, the studies based on seismic observations are limited. In this study, the methods for estimating the site Poisson's ratio and its variation with shear strain based on HVSR are proposed combined with the theories of wave propagation and one-dimensional site response analysis. Then, the in-situ properties of Poisson's ratio are investigated by applying the proposed method to KiK-net seismic database. It is observed that the shear strain change threshold of the site Poisson's ratio is around 10^{-5}, and the maximum Poisson's ratio is smaller than 0.5, without the effects of groundwater. The results of this study are verified by the comparison with the laboratory test results, and they can be utilized as the reference to the engineering practice and related studies.

Keywords: Site Poisson's ratio · In-situ properties · HVSR · KiK-net seismic data

1 Introduction

Poisson's ratio plays an important role in the analysis of a geotechnical structure under both static and dynamic loads in pratice. Nowadays, the effect of Poisson's ratio has been emphasized by many researchers and its properties have been studied from different aspects [1–6].

However, almost all studies of Poisson's ratio have been performed by using the laboratory test [1, 2, 4] or the in-situ test [7–9], the studies performed based on seismic observations are rare [10]. The main advantages of using the seismic observation records can be summarized in the following two points [11]. The first is that the results from the seismic observation records reflect the in-situ properties compared with the laboratory test results which are usually limited by scaling effect and sampling disturbance [12, 13]. The second is that the seismic records document the ground motions caused by actual earthquakes, while for the in-situ test, it is hard to simulate an actual earthquake particularly the strong earthquake hence its use is only limited in the study of the small-strain or elastic properties, i.e., shear strain level less than 10^{-5} [7, 8]. All in all, the study of the Poisson's ratio behaviour based on seismic observations can provide the results which properly reflect the in-situ characteristics, especially the nonlinearity [14–17].

Horizontal-to-vertical spectral ratio (HVSR) method has been widely used since its publication, it can be utilized to extract the site fundamental frequency based on the seismic observation records logged in the surface seismograph [18, 19]. For the site which can be regarded as the uniform soil on rigid rock, the site fundamental frequency can be transformed into the site dynamic modulus [11] combined with one-dimensional site response analysis [9], which makes it possible to investigate the in-situ properties of Poisson's ratio based on seismic observations [1, 2].

In this study, the methods for estimating the site Poisson's ratio and its variation with shear strain based on HVSR are firstly proposed. Then, the in-situ properties of the site Poisson's ratio are investigated by applying the proposed methods to KiK-net seismic observations. The estimated results are verified by the comparison with the laboratory test results, including the small-strain value, the change threshold and the maximum value.

2 Research Data

KiK-net is a nation-wide strong-motion seismograph network in Japan [20]. It consists of about 700 seismic stations, and each station includes a borehole and a pair of three-component seismometers. The depth of the borehole mostly ranges from 100 m to 200 m. One three-component seismometer is installed at the bottom of the borehole, and the other is installed on the ground surface. The detailed geological information and borehole profiles of the seismic stations are provided on the NIED website: https://doi.org/10.17598/NIED.0004. As mentioned in the NIED website, the sampling frequency of seismometer was changed from 200 Hz into 100 Hz around 2007 and there are instrumental relocations were operated around 2014 for some seismic stations. In this

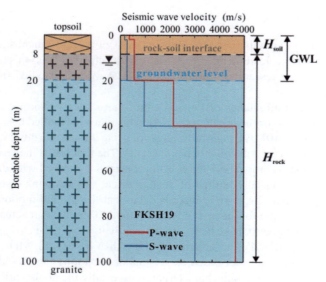

Fig. 1. Borehole profile of FKSH19 (H_{soil} and H_{rock} represent the thicknesses of soil layer and rock layer, respectively. GWL represents the groundwater level)

study, we only used the records which were documented between the date of above two operations to eliminate their possible interferences.

Table 1. Information of the selected stations

Station	Num$_{total}$	Depth$_B$/m	Num$_{100gal}$	H_{soil}/m	GWL/m	Latitude	Longitude
AKTH04	1222	100	11	24	20	39.17	140.71
FKSH19	1959	100	28	8	20	37.47	140.72
IWTH03	1257	100	8	5	5	39.80	141.65
IWTH04	1779	106	10	15	15	39.18	141.39
IWTH27	1955	100	23	4	4	39.03	141.53
TCGH07	632	100	17	27	32	36.88	139.45

Num$_{total}$ is the total number of the seismic records. Depth$_B$ is the borehole depth. Num$_{100gal}$ is the number of the seismic records with PHA larger than 100 cm/s^2.

In this study, seismic station selection consisted of 5 steps as follows: (1) The selected stations must have at least 400 seismic records. (2) The selected stations must have at least 2 records with horizontal components of peak ground acceleration (PHA) larger than 100 cm/s^2, to ensure the nonlinearity can be observed. (3) The borehole depth must range from 100 m to 120 m. (4) The site can be basically regarded as the uniform soil on rigid rock to meet the requirements of one-dimensional site response analysis. (5) The groundwater level (GWL) should be greater than or near the thickness of soil layer (H_{soil}) to eliminate the effects of groundwater. H_{soil} was determined manually based on the borehole profiles of the stations (Fig. 1). GWL was determined based on the borehole profiles with the result of Pavlenko and Irikura [21] which is that, P-wave velocity, when exceeding about 1000 m/s, indicates saturation of soil with water. Finally, 6 seismic stations were selected and some information of these stations are provided in Table 1.

3 Method

Firstly, 1–13 Hz bandpass filters were applied to the records [22]. Then, moving windows and short time Fourier transform (STFT) [23] were applied to the records with PHA larger than 100 cm/s^2 to increase the amount of the strong motion records. As shown in Fig. 2a, the window width and step size of the moving window were 10.24 s and 1.00 s, respectively. A Hanning taper of 2.5% was applied to each window before further computation. Moreover, the signal-to-noise ratio (SNR) of the short time signal was required to be not less than 50 and the noise was defined as the short time signal with the minimum energy.

Based on the theory of Nakamura [19]. HVSR of the seismic record can be calculated as

$$SR_{HV} = \frac{AF_{SH}}{AF_{SV}} \quad (1)$$

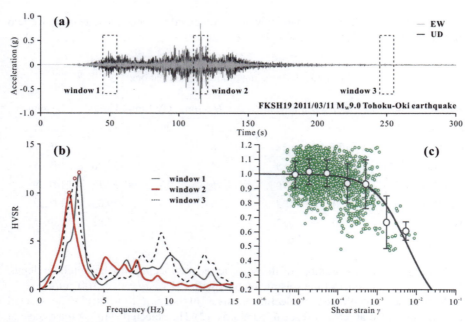

Fig. 2. Application of HVSR combined with STFT to a seismic record (main shock of 2011 M_W 9.0 Tohoku-Oki earthquake) in FKSH19 (a) Acceleration time histories for the EW and UD components of the example record (b) HVSR of the short time signals for three example windows (The hollow circles represent the site fundamental frequency) (c) Estimates of the shear modulus degradation curve for FKSH19

in which AF_{SH} and AF_{SV} represent the Fourier amplitude spectra of the horizontal and vertical components of the ground surface record. In this study, the Fourier amplitude spectrum of the horizontal component was defined as the vector sum of the results of the EW and NS components.

Then, as shown in Fig. 2b, the site fundamental frequency was estimated as the location of the first distinct crest in HVSR [18]. Based on one-dimensional site response analysis [9], for the site which can be regarded as the uniform soil on rigid rock, the site fundamental frequency can be transformed into the shear modulus [11] as

$$\frac{G}{G_0} = (\frac{f_H}{f_{H0}})^2 \tag{2}$$

in which G and f_H represent the shear modulus and the site fundamental frequency, respectively. G_0 and f_{H0} represent the small-strain values. F_{H0} was defined as the mean value of the f_H calculated from the records with PHA less than 10 cm/s².

Based on theory of wave propagation [24], the effective shear strain γ and the effective normal strain ε can be estimated by the following equation [25].

$$\begin{cases} \gamma = \dfrac{du}{dz} = \dfrac{du/dt}{dz/dt} = \dfrac{PHV}{V_{S30}} \\ \varepsilon = \dfrac{dw}{dz} = \dfrac{dw/dt}{dz/dt} = \dfrac{PVV}{V_{P30}} \end{cases} \quad (3)$$

where z, u and w are vertical depth, horizontal and vertical displacement, respectively. PHV is the peak value of the horizontal component of particle velocity which was calculated by Eq. (4) in this study [26]. PVV is the peak value of the vertical component of particle velocity. V_{S30} and V_{P30} are the 30 m average S-wave velocity and the 30 m average P-wave velocity derived from borehole profile.

$$PHV = \dfrac{\sum_{\theta=10°,10°}^{180°} \max\left[|V_{surface,\theta}(t)|\right]}{18} \quad (4)$$

in which $V_{surface,\theta}(t)$ is the velocity time history of θ azimuth recorded at the surface seismometer.

The shear modulus degradation curve was obtained by applying a one-parameter hyperbolic model Eq. (5) [14] to the estimated results of G/G_0 and γ (Fig. 2c).

$$\dfrac{G}{G_0} = \dfrac{1}{1+k\gamma} \quad (5)$$

in which k is a fitting parameter.

Based on one-dimensional site response analysis [9] and the results of Tsai and Liu [27], the site constrained modulus degradation curve (constrained modulus ratio M/M_0 varies with normal strain) can be regarded as being consistent with the site shear modulus degradation curve (shear modulus ratio G/G_0 varies with shear strain) when the effects of groundwater can be neglected. Moreover, Shi et al. [16] found that there is a log-linear relationship between the effective shear strain and the effective normal strain. Hence, the site constrained modulus degradation curve which varies with shear strain can be indirectly estimated by completing the following steps: (1) Estimate f_H/f_{H0} from HVSR using Eq. (1). (2) Calculate G/G_0 using Eq. (2). (3) Calculate γ and ε using Eqs. (3) and (4). (4) Calculate the empirical relationship between γ and ε using Eq. (6). (5) Regard G/G_0 and γ as M/M_0 and ε, respectively. Convert ε into γ using the empirical relationship. (6) Calculate the site constrained modulus degradation curve based on M/M_0 (G/G_0) and γ using Eq. (5).

$$\ln\gamma = k\ln\varepsilon + b \quad (6)$$

in which b and k are fitting parameters.

Finally, the site Poisson's ratio ν and its variation with shear strain γ can be estimated by the following formula [27]:

$$\nu = \dfrac{1}{2}\left(1 - \dfrac{1}{\frac{M}{G}-1}\right) = \dfrac{1}{2}\left(1 - \dfrac{1}{\frac{M}{M_0}\frac{G_0}{G}\left(\frac{V_{P0}}{V_{S0}}\right)^2 - 1}\right) \quad (7)$$

where V_{S0} and V_{P0} are the H_{soil} average S-wave and P-wave velocities derived from borehole profile.

In reality, soil deposits are never purely homogeneous, and due to the limitation of the method proposed in this paper. We mainly focused on the qualitative properties of the site Poisson's ratio, and the comparison of the results observed based on KiK-net database in this study and obtained using laboratory tests in previous studies.

4 Results and Discussion

For the selected stations, the fitting parameters of the empirical relationships between shear strain and normal strain are provided in Table 2, it was found that the empirical relationships agreed well with the estimated results. Moreover, the variations of the site Poisson's ratio with shear strain are depicted in Fig. 3, and the corresponding fitting parameters are also provided in Table 2. It was found from Fig. 3 that the attenuations of constrained modulus and shear modulus were out of sync and shear modulus attenuated faster, which are consistent with the laboratory test results in Chen et al. [1].

Table 2. Fitting parameters of the selected stations

Station	$G/G_0 - \gamma$			$M/M_0 - \gamma$			$\gamma - \varepsilon$			
	k	R	σ	k	R	σ	k	b	R	σ/($\times 10^{-5}$)
AKTH04	357	0.9341	0.0803	83	0.9012	0.0888	0.9792	0.5887	0.9804	6.8945
FKSH19	167	0.9436	0.0540	42	0.9727	0.0336	1.0193	0.8057	0.9484	10.4936
IWTH03	2712	0.9891	0.0445	1107	0.9874	0.0301	0.9192	0.1566	0.8550	9.5981
IWTH04	644	0.9838	0.0328	173	0.9882	0.0267	0.9625	0.4703	0.9383	8.0532
IWTH27	1252	0.9847	0.0378	281	0.9905	0.0272	0.9662	0.5653	0.9209	10.2231
TCGH07	565	0.9350	0.0756	89	0.9532	0.0740	1.0308	0.8860	0.9564	9.9925

R and σ represent the Pearson correlation coefficient and the standard error of the fitting formula, respectively.

In this study, the in-situ properties of Poisson's ratio were investigated by three indicators, which are the small-strain value v_0, the change threshold and the maximum value v_{max}. To make the analysis more intuitive, the estimated results of the selected stations are gathered in Fig. 4, the small-strain values and the maximum values are provided in Table 3. As shown in Fig. 4 and Table 3, it was found that the small-strain Poisson's ratio of IWTH27 which consists almost exclusively of topsoil layer (the part with low V_S of granite layer could affect the result of FKSH19) was larger than those of other sites. The reason might be that, topsoil is soft, unstable and wet, which make the site Poisson's ratio larger [28]. Moreover, It was observed that the shear strain change threshold of the site Poisson's ratio was around 10^{-5} (Fig. 4b), and the maximum values of the selected stations were all smaller than 0.5. These results are consistent with the laboratory test results as reported in many previous articles [1, 2].

Table 3. Estimated results of the selected stations

Station	Soil type	v_0	v_{max}
AKTH04	Gravel	0.34	0.467
FKSH19	Topsoil	0.35	0.473
IWTH03	Gravel	0.33	0.434
IWTH04	Clay and sand	0.33	0.458
IWTH27	Topsoil	0.39	0.478
TCGH07	Topsoil and gravel	0.35	0.481

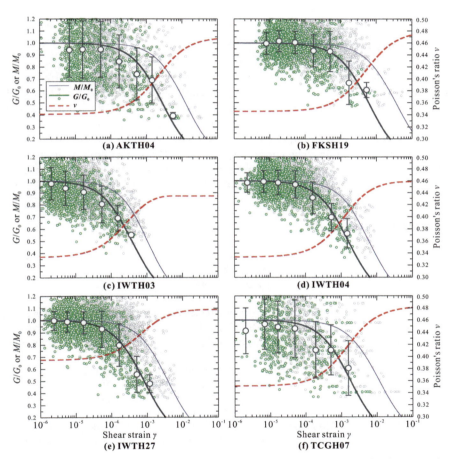

Fig. 3. Estimates of the variations of the site Poisson's ratio with shear strain, the site constrained and shear modulus degradation curves with the estimated values of the selected stations

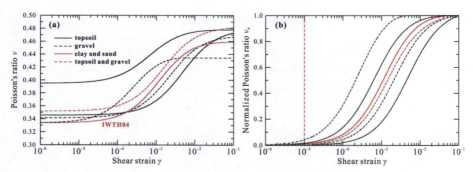

Fig. 4. Estimates of the variations of the site Poisson's ratio (a) and the normalized site Poisson's ratio (b) with shear strain of the selected stations

5 Conclusions

In this study, the methods for estimating the site Poisson's ratio and its variation with shear strain based on HVSR were proposed. By applying the proposed methods to KiK-net data, the in-situ properties of the site Poisson's ratio were investigated. It was found that the small-strain Poisson's ratio of topsoil was larger compared with gravel, sand and clay, which might be interpreted by the fact that topsoil is soft, unstable and wet. Moreover, the shear strain change threshold of the site Poisson's ratio was around 10^{-5}, and the maximum Poisson's ratio was lower than 0.5, without the effects of groundwater. These results are verified by their consistency with the laboratory test results.

References

1. Chen, S., Kong, L., Xu, G.: An effective way to estimate the poisson's ratio of silty clay in seasonal frozen regions. Cold Reg. Sci. Technol. **154**, 74–84 (2018)
2. Dutta, T.T., Saride, S.: Influence of shear strain on the poisson's ratio of clean sands. Geotech. Geol. Eng. **34**(5), 1359–1373 (2016)
3. Gu, X., Yang, J., Huang, M.: Laboratory measurements of small strain properties of dry sands by bender element. Soils Found. **53**(5), 735–745 (2013)
4. Kokusho, T.: Cyclic triaxial test of dynamic soil properties for wide strain rate. Soils Found. **20**(2), 45–60 (1980)
5. Kumar, J., Madhusudhan, B.N.: Effect of relative density and confining pressure on Poisson's ratio from bender and extender elements tests. Géotechnique **60**(7), 561–567 (2010)
6. Nakagawa, K., Soga, K., Mitchell, J.K.: Observation of Biot compressional wave of the second kind in granular soils. Géotechnique **47**(1), 133–147 (1997)
7. Ayres, A., Theilen, F.: Relationship between P- and S-wave velocities and geological properties of near-surface sediments of the continental slope of the Barents Sea. Geophys. Prospect. **47**(4), 431–441 (1999)
8. Cosentini, R.M., Foti, F.: Evaluation of porosity and degree of saturation from seismic and electrical data. Géotechnique **64**(4), 278 (2014)
9. Kramer, S.L.: Geotechnical Earthquake Engineering. Prentice Hall, Upper Saddle River (1996)

10. White, D.J., Milkereit, B., Salisbury, M.H., Percival, J.A.: Crystalline lithology across the Kapuskasing Uplift determined using in situ Poisson's ratio from seismic tomography. J. Geophys. Res. Solid Earth **97**(B13), 19993–20006 (1992)
11. Yang, Z., Yuan, J., Liu, J., Han, B.: Shear modulus degradation curves of gravelly and clayey soils based on KiK-Net in situ seismic observations. J. Geotech. Geoenviron. Eng. **143**(9), 06017008 (2017)
12. Liao, T., Massoudi, N., McHood, M., Stokoe, K.H., Jung, M.J., Menq, F.Y. Normalized shear modulus of compacted gravel. In: Proceedings of the 18th International Conference on Soil Mechanics and Geotechnical Engineering. Elsevier, Amsterdam (2013)
13. Tsai, C.C.: Seismic site response and interpretation of dynamic soil behavior from downhole array measurements. Department of Civil and Environmental Engineering, University of Illinois at Urbana-Champaign, Urbana, IL (2007)
14. Chandra, J., Guéguen, P., Bonilla, L.F.: PGA-PGV/V_S considered as a stress-strain proxy for predicting nonlinear soil response. Soil Dyn. Earthq. Eng. **85**, 146–160 (2016)
15. Guéguen, P., Bonilla, L.F., Douglas, J.: Comparison of soil nonlinearity (in situ stress-strain relation and G/Gmax reduction) observed in strong-motion databases and modeled in ground-motion prediction equations. Bull. Seismol. Soc. Am. **109**(1), 178–186 (2019)
16. Shi, Y., Wang, S.-Y., Cheng, K., Miao, Y.: In situ characterization of nonlinear soil behavior of vertical ground motion using KiK-net data. Bull. Earthq. Eng. **18**(10), 4605–4627 (2020)
17. Wang, H., Jiang, W., Wang, S., Miao, Y.: In situ assessment of soil dynamic parameters for characterizing nonlinear seismic site response using KiK-net vertical array data. Bull. Earthq. Eng. **17**(5), 2331–2360 (2019)
18. Gallipoli, M.R., Mucciarelli, M.: Comparison of site classification from V_{S30}, V_{S10}, and HVSR in Italy. Bull. Seismol. Soc. Am. **99**(1), 340–351 (2009)
19. Nakamura, Y.: A method for dynamic characteristics estimation of subsurface using microtremor on the ground surface. Railw. Tech. Res. Inst. Q. Rep. **30**(1) (1989)
20. National Research Institute for Earth Science and Disaster Resilience. NIED K-NET, KiK-net, National Research Institute for Earth Science and Disaster Resilience (2019)
21. Pavlenko, O.V., Irikura, K.: Estimation of nonlinear time-dependent soil behavior in strong ground motion based on vertical array data. Pure Appl. Geophys. **160**(12), 2365–2379 (2003)
22. Nakata, N., Snieder, R.: Estimating near-surface shear wave velocities in Japan by applying seismic interferometry to KiK-net data. J. Geophys. Res. Solid Earth **117**(1), B01308 (2012)
23. Bonilla, L.F., Guéguen, P., Ben-Zion, Y.: Monitoring coseismic temporal changes of shallow material during strong ground motion with interferometry and autocorrelation. Bull. Seismol. Soc. Am. **109**(1), 187–198 (2019)
24. Rathje, E.M., Chang, W.J., Stokoe, K.H., Cox, B.R.: Evaluation of ground strain from in situ dynamic response. In: Proceeding of the 13th World Conference on Earthquake Engineering, 1–6 August 2004, Vancouver, British Columbia, Canada (2004)
25. Idriss, I.M.: Use of V_{S30} to represent local site condition. In: 4th IASPEI/IAEE International Symposium: Effects of Source Geology on Seismic Motion, 23–26 August 2011, University of Santa Barbara California (2011)
26. Chandra, J., Guéguen, P., Steidl, J.H., Bonilla, L.F.: In situ assessment of the G-γ curve for characterizing the nonlinear response of soil: application to the Garner Valley downhole array and the wildlife liquefaction array. Bull. Seismol. Soc. Am. **105**(2A), 993–1010 (2015)
27. Tsai, C.C., Liu, H.W.: Site response analysis of vertical ground motion in consideration of soil nonlinearity. Soil Dyn. Earthq. Eng. **102**, 124–136 (2017)
28. Kumar Thota, S., Duc Cao, T., Vahedifard, F.: Poisson's ratio characteristic curve of unsaturated soils. J. Geotech. Geoenviron. Eng. **147**(1), 04020149 (2021)

Effectiveness of Distributed Acoustic Sensing for Acquiring Surface Wave Dispersion Data Using Multichannel Analysis of Surface Waves

Joseph P. Vantassel[1(✉)], Brady R. Cox[2], Peter G. Hubbard[3], Michael Yust[1], Farnyuh Menq[1], Kyle Spikes[1], and Dante Fratta[4]

[1] The University of Texas at Austin, Austin, TX 78712, USA
jvantassel@utexas.edu
[2] Utah State University, Logan, UT 84322, USA
[3] University of California Berkeley, Berkeley, CA 94720, USA
[4] University of Wisconsin-Madison, Madison, WI 53706, USA

Abstract. Distributed acoustic sensing (DAS) is a rapidly expanding tool to sense vibrations and system deformations in many engineering applications. In terms of site characterization, DAS presents the ability to make static and dynamic strain measurements on a scale (e.g., kilometers), density (e.g., meter-scale), and fidelity (e.g., microstrain) that was previously unattainable with traditional measurement technologies. In this study, we assess the effectiveness of using DAS to extract surface wave dispersion data using the multichannel analysis of surface waves (MASW) technique. We utilized both highly-controlled, broadband vibroseis shaker truck and more-variable, narrow-band sledgehammer sources to excite the near surface and compared the DAS-derived dispersion data directly with concurrently acquired traditional geophone-derived dispersion data. We report that the differences between the two sensing approaches are minimal and well within the uncertainty bounds associated with each individual measurement for the following DAS testing conditions: (a) a tight-buffered or strain-sensing fiber optic cable is used, (b) the cable is buried in a shallow trench to enhance coupling, and (c) short gauge lengths and small channel separations are used. Our deployed conditions are more promising than previous attempts documented in the literature, thereby demonstrating that DAS can provide accurate measurements of surface wave dispersion data of the same quality as geophones. We show that frequency-dependent normalization of the dispersion image removes the effects of scaling, integration, and differentiation of the measured data, thereby removing the need to post-process the geophone-derived and DAS-derived waveforms into equivalent units before performing dispersion processing. Finally, we summarize the important effect of gauge length on the dispersion data for future reference. This study demonstrates that DAS, when appropriate considerations are made, can be used in-lieu of traditional sensors (i.e., geophones) for making high-quality measurements of surface wave dispersion data using the MASW technique.

Keywords: High density arrays · DAS · Geophones · Dispersion · Surface waves

© The Author(s), under exclusive license to Springer Nature Switzerland AG 2022
L. Wang et al. (Eds.): PBD-IV 2022, GGEE 52, pp. 1000–1008, 2022.
https://doi.org/10.1007/978-3-031-11898-2_77

1 Introduction

Distributed acoustic sensing (DAS) is an emerging technology with broad applications in infrastructure health monitoring and site characterization (Hubbard et al. 2022; Hubbard et al. 2021; Lindsey et al. 2020; Spikes et al. 2019; Wang et al. 2018). DAS permits the acquisition of static and dynamic signals at scales of tens of kilometers and meter spatial resolutions previously unattainable with traditional sensing technologies (Soga and Luo 2018). DAS, and for that matter, the larger area of distributed fiber optic sensing (DFOS), requires three main system components: a fiber-optic cable, an interrogator unit (IU), and a dedicated storage, computation, and visualization resource. The fiber-optic cable is the sensing instrument whose elongation or compression (i.e., strain) is measured by the DAS system. Fiber-optic cables are specially designed to propagate the light emitted by the interrogator unit with minimal loss, allowing light to travel (and therefore strain measurements to be made) over large distances (i.e., tens of kilometers) (Lindsey and Martin 2021). Note that the fiber must be selected and installed carefully to ensure acceptable results; however, as even specially designed fiber-optic cables for strain-sensing applications are relatively inexpensive (between $3 and $7 per meter) they are typically not retrieved for re-use after testing concludes. The second component, the IU, is connected to one end of the fiber-optic cable to send pulses of light down the length of the fiber and measure the returned Rayleigh backscatter events. The IU senses the Rayleigh backscattering caused by imperfections in the silica within the fiber and interprets the phase change as measurements proportional to the local strain along the fiber (Karrenbach et al. 2019). The IU uses precise timing and fast sampling rates (i.e., up to 100s of kHz) to interpret the scattering events as one-dimensional measurements of local strain at various physical distances along the fiber. The dedicated storage, computational, and visualization resource is typically a high-end computer, with large amounts of dedicated storage (at least multiple terabytes), and a real-time data acquisition software interface to facilitate the visualization and interpretation of DAS measurements.

The multichannel analysis of surface waves (MASW) is an active-source surface-wave testing technique for measuring a site's surface wave dispersion from recordings of dynamic signals with strong surface wave content (Park et al. 1999). The MASW technique most commonly involves a linear array of receivers, typically geophones (i.e., velocity transducers), and a surface wave source located collinear with the array and operated by the experimenters. Geophones are most commonly oriented vertically, but they can be oriented horizontally in the in-line or cross-line direction (i.e., sensing particle motion collinear with or perpendicular to the array, respectively). Just as with the geophones, the seismic source may be oriented vertically, horizontally in-line, or horizontally cross-line, depending on the types of surface waves one desires to utilize (Vantassel and Cox 2022).

In this paper, we examine the fitness of DAS to acquire dynamic strains to extract surface wave dispersion data using MASW. This study compares a 94-m section of a 200-m long strain-sensing fiber-optic cable with an adjacently deployed 94-m long geophone array (48 receivers at a 2-m spacing). The array recorded dynamic signals rich in surface wave energy generated by off-end vibroseis and impulse sources. The dispersion extracted from both measurement systems shows excellent agreement. This study demonstrates that DAS can be used as a replacement for traditional geophone deployments for extracting surface wave dispersion data.

2 Experimental Setup

This experiment was conducted at the NHERI@UTexas (Stokoe et al. 2020) Hornsby Bend test site in Austin, Texas, USA. Figure 1 shows a plan view of the experimental setup: a 94-m geophone array deployed alongside 200 m of fiber-optic cable. The DAS array consisted of two fiber-optic cables buried alongside one another, one from NanZee Sensing Technology (NZS-DSS-C02) and the other from AFL (X3004955180H-RD). However, only results from the NanZee cable will be discussed here for brevity. Both cables have tight buffered optical fibers and are constructed such that strain is transferred from the exterior to the core. The AFL cable was designed for tactical telecommunications applications while the NanZee cable was specifically designed for strain sensing (Zhang et al. 2021). Note that the fiber-optic cables are sensitive to strain along their length (i.e., in the horizontal in-line direction). The signals on the two cables were recorded simultaneously by splicing the NanZee and AFL cables together at the far end of the array. On the near-side of the array, the fiber was connected to an OptaSense ODH4 IU, and the other end was appropriately terminated to reduce end reflections. This IU allows for a variable gauge length, we selected the minimum available gauge length of 2.04 m. For those readers who may not be familiar, the gauge length represents the length of fiber that the elongation (or strain) is measured/averaged over at each sensing position. The channel separation, as configured in this study, is the distance between each gauge lengths' centers, and it was set at 1.02 m. The effect of gauge length and channel separation on surface wave dispersion will be discussed later in this work. The IU's sampling frequency (or ping rate) was set at 100 kHz. After acquisition, the purposely over-sampled raw measurements (for the purposes of increasing the acquisition's signal-to-noise and dynamic range) were appropriately downsample to 1 kHz and high-pass filtered at 3 Hz. Immediately adjacent to the fiber-optic cables, two geophone arrays (one vertical and one horizontal in-line) were deployed. However, for brevity, only the results from the horizontal in-line geophones (i.e., those sensing in the same direction as the DAS array) will be discussed in this study. The 48 vertical and 48 horizontal 4.5-Hz Geospace Technologies (GS-11D) geophones were deployed at a constant 2-m spacing. The geophones were mounted in PC21 land cases and coupled to the ground surface with 7.6-cm aluminum spikes. Note that geophone arrays with equal length and channel spacing to the DAS could not be deployed due to equipment constraints. Signals from the geophone arrays were recorded simultaneously at 1 kHz using four interconnected 24-channel Geometric Geode seismographs.

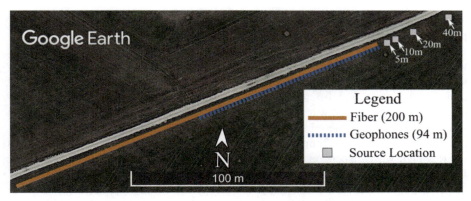

Fig. 1. Plan view of the experimental setup at the Hornsby Bend test site in Austin, Texas, USA, where 200 m of fiber-optic cable was deployed alongside a 94-m long geophone array (48 receivers at a 2 m spacing) to assess the effectiveness of using distributed acoustic sensing (DAS) for recording surface wave propagation. Surface wave energy was produced from four distinct shot locations denoted as 5 m, 10 m, 20 m, and 40 m.

3 Data Acquisition

The DAS and geophone arrays were used to simultaneously record actively generated surface waves using highly controlled vibroseis shaker trucks and more variable sledgehammer impact sources. The vibroseis sources include the specialized three-dimensional shaker T-Rex and the highly-mobile one-dimensional shaker Thumper from the NHERI@UTexas experimental facility (Stokoe et al. 2020). T-Rex was used to shake the ground in all three directions (i.e., vertically, horizontally in-line, and horizontally crossline). However, only the vertical and horizontal in-line shakes are discussed here. T-Rex was used to produce a 12-s chirp with frequencies swept linearly from 3 to 80 Hz. The other vibroseis source Thumper was used to shake vertically following a 12-s chirp with frequencies swept linearly from 5 to 200 Hz. The impact source used for this study was an instrumented 5.4 kg sledgehammer from PCB Piezotronics. The frequency content produced by the sledgehammer is highly variable and depends on the operator and the tested material. All four sources (i.e., T-Rex shaking vertically, T-Rex shaking horizontally in-line, Thumper shaking vertically, and the sledge hammer striking vertically) were deployed in-line at 5, 10, 20, and 40 m away from the start of the DAS fiber-optic cable and geophone array. The source positions relative to the array are shown in Fig. 1. Three vibroseis chirps and five sledge hammer impacts were stacked in the time-domain to produce a single vibroseis or sledge hammer waveform at each sensing location with higher signal-to-noise ratio.

4 MASW Processing

Only the first 94 m of the 200-m long DAS array was used during MASW processing to ensure a fair comparison between the geophone-derived and DAS-derived dispersion. All MASW processing was performed using the frequency-domain beamformer with cylindrical-steering vector and square-root weighting (Zywicki and Rix 2005). A comparison of dispersion images from the geophone array and NanZee cable DAS array is made in Fig. 2 for a vertical sledgehammer impact at the 5 m source location. We utilize frequency-dependent normalization of the dispersion image, where we normalize the surface wave energy at each frequency by its maximum, to remove the effects of scaling, integration, and differentiation and thereby allow us to process the geophone-derived and DAS-derived waveforms in their raw units. The small white circles shown in each panel indicate the relative phase velocity maximums selected via an algorithmic relative peak search. From examining Fig. 2, we first observe the complexity of the Rayleigh-wave dispersion at the Hornsby Bend site, with the presence of multiple Rayleigh-wave modes. Second, when comparing Fig. 2a and b, it is clear that the geophone array acquires shorter wavelengths than the DAS array. Specifically, the lower right-hand corner of Fig. 2b indicates that the DAS dispersion data for this offset cannot be extracted for wavelengths less than about 2 m indicated by a light-colored dashed line. This observation, which is confirmed by other data acquired at the site (see Fig. 3), shows a relationship between the minimum wavelength of surface wave dispersion acquired and the DAS system's gauge length. In particular, there is very limited coherent surface wave energy at wavelengths less than approximately 2 m for the DAS array for any source-offset combination. In contrast, we are able to clearly resolve such wavelengths using the geophone array. This is especially notable given that the DAS data is acquired with a 1.02 m channel separation whereas the geophone data is acquired at a 2 m receiver spacing. This effect shows that when sufficiently small channel separations are used it is the gauge length that controls the shortest wavelength surface waves DAS can measure. While unfortunately there is insufficient room in the current work to fully discuss the physics behind the gauge length's limiting effect, we do note that it is the result of a 180 degree phase shift that occurs for wavelengths between 0.5 and 1 gauge lengths and more generally every $(n+1)^{-1}$ to n^{-1} gauge lengths where n is an odd integer. While this limitation of DAS is not particularly troublesome for the present study, because a short gauge length was able to be used, this observation is critically important for those using DAS systems with much larger minimum gauge lengths. Therefore, when using DAS systems for surface wave acquisition for engineering applications, where short wavelengths are critical for correctly resolving the stiffness of near-surface layers, two factors must be considered. Those two factors are the trace separation, and especially, the gauge length, which must be selected to be sufficiently short to permit good near-surface resolution. The reader will note that the gauge length limitation discussed here for MASW is less restrictive than those proposed by others for other imaging techniques (e.g., Dean et al. 2017) as MASW is primarily sensitive to changes in phase and is not greatly affected by amplitude-related effects provided they are consistent across the array as they are in this case.

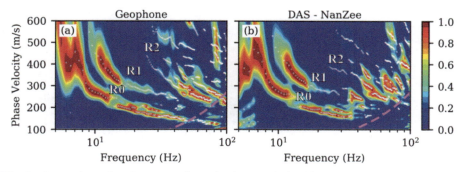

Fig. 2. Comparison of surface wave dispersion images derived from the stacked vertical sledge hammer impacts at a distance of 5 m away from the array as derived from (a) geophone array and (b) NanZee cable of the DAS array. The small white circles shown in each panel indicate the relative phase velocity maximums selected from the dispersion images and later used to identify multiple Rayleigh modes of surface wave propagation. The apparent fundamental, first-higher, and second-higher Rayleigh wave modes are denoted as R0, R1, and R2, respectively. The dashed line in each panel's lower right denotes a wavelength equal to the IU's gauge length (i.e., 2.04 m).

5 Extraction of Experimental Dispersion Data

The frequency-phase velocity peaks from all 16 source-position and source-type combinations were extracted from their respective dispersion images. These peaks were interactively trimmed to isolate three clear Rayleigh modes and summarized into dispersion statistics following the workflow developed by Vantassel and Cox (2022). The interactively-trimmed peaks and dispersion statistics (mean ± one standard deviation) are shown for the geophone array and NanZee cable of the DAS array in Fig. 3a and b, respectively. To facilitate a more direct comparison, Fig. 4 plots the dispersion statistics from the geophone-derived and DAS-derived experimental dispersion data directly on top of one another. Excellent agreement is observed between the two sensing systems. We note that at high frequencies the consistency between the geophone-derived and DAS-derived dispersion data decreases slightly for the R0 and R1 modes. This slight decrease in consistency is due in part to less clear dispersion trends in these regions (recall Fig. 2) that make consistent interactive trimming difficult. Nonetheless, the geophone-derived and DAS-derived experimental dispersion data is in excellent agreement, thereby demonstrating that when appropriate considerations are made (i.e., proper cable selection, good cable-soil coupling, and sufficiently short gauge length and trace separation) DAS can be used to measure surface wave dispersion data that is of equal quality to that acquired using geophones.

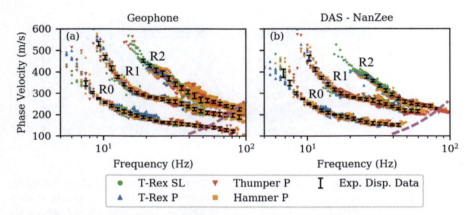

Fig. 3. Experimental dispersion data after performing interactive trimming to isolate the first three Rayleigh modes (i.e., R0, R1, and R2) for the: (a) geophone array, and (b) NanZee cable of the DAS array. The statistical representation of each mode is denoted by error bars that delineate the ± one standard deviation range. The dashed line in each panel's lower right denotes a wavelength equal to the IU's gauge length (i.e., 2.04 m).

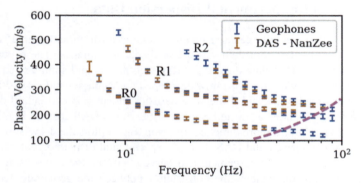

Fig. 4. Comparison between the geophone-derived and DAS-derived experimental dispersion data at the Hornsby Bend test site. The vertical error bars at each frequency represent the mean ± one standard deviation of the experimental dispersion data for the fundamental, first-higher, and second-higher Rayleigh modes (R0, R1, and R2, respectively). The dashed line to the lower right denotes a wavelength equal to the IU's gauge length (i.e., 2.04 m).

6 Conclusions

We assess the effectiveness of using DAS for extracting dispersion data using the MASW technique. The DAS data from a tightly-buffered, strain-sensing fiber-optic cable buried in a shallow trench were compared with dispersion data extracted from horizontal geophones coupled to the ground surface using an aluminum spike. Wavefields with strong Rayleigh-type surface wave content were generated by using highly-controlled vibroseis sources and more-variable impact sources at four distinct source positions. The use of frequency-dependent normalization allowed the raw geophone-derived waveforms, proportional to velocity, and the raw DAS-derived waveforms, proportional to strain/displacement, to be used for dispersion processing in their raw units and eliminate the need to convert the measurements into consistent units. Using a relative-peak search three Rayleigh wave modes of propagation were able to be extracted over a relatively broad frequency range for active-source studies (~6 to 70 Hz). A limiting gauge length effect was observed for the DAS-derived dispersion data, where wavelengths shorter than approximately the gauge length could not be resolved despite using a 1.02-m trace separation, thereby making gauge length selection an important factor to consider in future near-surface studies using DAS. The limitation of DAS at short wavelengths aside, the experimental dispersion data recovered from the geophone and DAS systems show excellent agreement for all three recovered Rayleigh modes. Therefore, when appropriate considerations are made to ensure proper cable selection, good cable-soil coupling, and sufficiently short gauge lengths and trace separations, DAS can be an effective alternative to geophones for the purpose of acquiring dynamic signals for the intent of extracting surface wave dispersion data using MASW.

Acknowledgements. This work was supported in part by the U.S. National Science Foundation (NSF) grants CMMI-2037900, CMMI-1520808, and CMMI-1931162. However, any opinions, findings, and conclusions or recommendations expressed in this material are those of the authors and do not necessarily reflect the views of NSF. Special thanks to Dr. Kevin Anderson at Austin Water - Center for Environmental Research for the access to the Hornsby Bend Biosolids Management Plant test site. Special thanks to Dr. Kenichi Soga for the contribution of the NanZee cable used in this study. Special thanks to Todd Bown and the OptaSense team for their assistance in configuring the ODH4, extracting and filtering the DAS-derived seismic waveforms, and permitting us to publish these results. Active-source surface wave processing and calculation of dispersion statistics were performed using the Python package *swprocess* v0.1.0b0 (Vantassel 2021a). Subsequent operations with the dispersion statistics were performed using the Python package *swprepost* v1.0.0 (Vantassel 2021b).

References

Dean, T., Cuny, T., Hartog, A.H.: The effect of gauge length on axially incident P-waves measured using fibre optic distributed vibration sensing: Gauge length effect on incident P-waves. Geophys. Prospect. **65**, 184–193 (2017). https://doi.org/10.1111/1365-2478.12419

Hubbard, P.G., et al.: Road Deformation Monitoring and Event Detection using Asphalt-embedded Distributed Acoustic Sensing (DAS) (2022). https://doi.org/10.31224/osf.io/mer43

Hubbard, P.G., et al.: Dynamic structural health monitoring of a model wind turbine tower using distributed acoustic sensing (DAS). J. Civ. Struct. Heal. Monit. **11**(3), 833–849 (2021). https://doi.org/10.1007/s13349-021-00483-y

Karrenbach, M., et al.: Fiber-optic distributed acoustic sensing of microseismicity, strain and temperature during hydraulic fracturing. Geophysics **84**, D11–D23 (2019). https://doi.org/10.1190/geo2017-0396.1

Lindsey, N.J., Martin, E.R.: Fiber-Optic Seismology. Annu. Rev. Earth Planet. Sci. **49**, 309–336 (2021). https://doi.org/10.1146/annurev-earth-072420-065213

Lindsey, N.J., Yuan, S., Lellouch, A., Gualtieri, L., Lecocq, T., Biondi, B.: City-scale dark fiber das measurements of infrastructure use during the COVID-19 pandemic. Geophys. Res. Lett. **47** (2020). https://doi.org/10.1029/2020GL089931

Park, C.B., Miller, R.D., Xia, J.: Multichannel analysis of surface waves. Geophysics **64**, 800–808 (1999). https://doi.org/10.1190/1.1444590

Soga, K., Luo, L.: Distributed fiber optics sensors for civil engineering infrastructure sensing. Journal of Structural Integrity and Maintenance **3**, 1–21 (2018). https://doi.org/10.1080/24705314.2018.1426138

Spikes, K.T., Tisato, N., Hess, T.E., Holt, J.W.: Comparison of geophone and surface-deployed distributed acoustic sensing seismic data. Geophysics **84**, A25–A29 (2019). https://doi.org/10.1190/geo2018-0528.1

Stokoe, K.H., Cox, B.R., Clayton, P.M., Menq, F.: NHERI@UTexas experimental facility with large-scale mobile shakers for field studies. Front. Built Environ. **6**, 575973 (2020). https://doi.org/10.3389/fbuil.2020.575973

Vantassel, J.: jpvantassel/swprocess: v0.1.0b0. Zenodo (2021a). https://doi.org/10.5281/zenodo.4584129

Vantassel, J.: jpvantassel/swprepost: v1.0.0. Zenodo (2021b). https://doi.org/10.5281/zenodo.5646771

Vantassel, J.P., Cox, B.R.: SWprocess: a workflow for developing robust estimates of surface wave dispersion uncertainty. Journal of Seismology Accepted (2022)

Wang, H.F., et al.: Ground motion response to an ML 4.3 earthquake using co-located distributed acoustic sensing and seismometer arrays. Geophys. J. Int. **213**, 2020–2036 (2018). https://doi.org/10.1093/gji/ggy102

Zhang, C.-C., et al.: Microanchored borehole fiber optics allows strain profiling of the shallow subsurface. Sci Rep **11**, 9173 (2021). https://doi.org/10.1038/s41598-021-88526-8

Zywicki, D.J., Rix, G.J.: Mitigation of near-field effects for seismic surface wave velocity estimation with cylindrical beamformers. J. Geotech. Geoenviron. Eng. **131**, 970–977 (2005). https://doi.org/10.1061/(ASCE)1090-0241(2005)131:8(970)

Relationships Between Ground-Motion Intensity Measures and Earthquake-Induced Permanent Slope Displacement Based on Numerical Analysis

Mao-Xin Wang, Dian-Qing Li, and Wenqi Du(✉)

State Key Laboratory of Water Resources and Hydropower Engineering Science,
Institute of Engineering Risk and Disaster Prevention, Wuhan University,
Wuhan 430072, People's Republic of China
wqdu309@whu.edu.cn

Abstract. In engineering applications, the permanent displacement (D) commonly serves as a useful indicator of the seismic performance of slopes. When developing empirical displacement models as a function of ground-motion intensity measures (IMs), the IMs that are best correlated to D are preferred. On the other hand, the predictability of IMs, in terms of the standard deviations using ground motion models, is also of concern in developing D models. This study aims to: (1) investigate the efficiency of IMs in developing D models for a cohesive-frictional slope based on numerical analysis; and (2) compare the means and standard deviations of randomized D by considering uncertainties in predicting both the IMs and D via Monte Carlo simulation (MCS). A total of 10 scalar IMs and 38 vector-IMs, are employed to develop D models. The results indicate that the spectral acceleration at a degraded period of the soil layer ($SA(1.5T_{s,layer})$) and Arias intensity (IA) are the two most efficient scalar IMs. Additionally, the vector-IMs consisting of [IA, spectrum intensity] and [IA, mean period] are the two most efficient vectors. The MCS results illustrate that the rankings for standard deviations of D models and total standard deviations (i.e., including ground motion variability) may be considerably different. The results are also found to be dependent on earthquake magnitudes and site conditions. This study could provide guidance on the development of numerical-based D models especially within a probabilistic seismic slope displacement analysis framework.

Keywords: Seismic slope performance · Numerical analysis · Intensity measure · Displacement prediction · Model variability

1 Introduction

Evaluating the seismic stability of slopes is an important task of geotechnical engineers. The performance-based permanent displacement analysis has attracted increasing attention in assessing the seismic safety of slopes, as usually conducted by the Newmark-type sliding block procedures [1–3]. Such procedures that approximately estimate the earthquake-induced permanent slope displacement (D) are particularly useful for preliminary screening-level analyses and regional landslide hazard mapping. For assessing the seismic performance of important slopes such as those involved in critical projects, the numerical stress-deformation analysis is necessary to provide a more accurate estimate of the slope performance [4].

The probabilistic displacement hazard approach has been well developed to estimate the hazard-consistent D [5–7]. Recently, this approach has been improved with the numerical slope displacement analysis [8]. As a key component, the relationship between one or multiple ground motion (GM) intensity measure (IMs) and D should be explicitly described by a regression model, in which the IMs better correlated to D (i.e., higher efficiency) are recommended as predictor variables. Therefore, it is of interest to investigate the efficiency of different IMs. However, most of the existing studies were based on the Newmark-type procedures (e.g., [9]), so the findings and conclusions may not hold for the numerical cases [8, 10]. Also, the efficiency rankings of various IMs may be dependent on slope materials, slope geometry, etc. Hence, more research efforts should be devoted to this topic.

This study aims to: (1) investigate the performance of IMs in developing D models for a cohesive-frictional slope based on numerical analysis; and (2) compare the means and standard deviations of D for these models considering uncertainties in predicting both the IMs and D via Monte Carlo simulation (MCS). The remaining part of this paper starts with a description of the procedure employed, followed by the slope model development and comparative results.

2 Procedure for Seismic Slope Displacement Analysis

The seismic slope displacement analysis procedure includes the following steps.

(i) Select GM acceleration-time series, and compute their IMs of interest. The GMs selected should cover a wide range of the earthquake magnitude (M), rupture distance (R), and shaking intensity, etc.
(ii) Perform dynamic analysis for the slope model using each of the GMs to obtain D.
(iii) Develop D models based on the obtained data of D. The trend of ln$D$ versus lnIM (i.e., in natural logarithmic scale) is fitted via one of the following formulas:

$$\ln D = \mu_{\ln D} + \varepsilon_{\ln D} \cdot \sigma_{\ln D} = a_1 \ln IM_1 + a_0 + \varepsilon_{\ln D} \cdot \sigma_{\ln D} \quad (1)$$

$$\ln D = \mu_{\ln D} + \varepsilon_{\ln D} \cdot \sigma_{\ln D} = a_1 \ln IM_1 + a_2 \ln IM_2 + a_0 + \varepsilon_{\ln D} \cdot \sigma_{\ln D} \quad (2)$$

where D is in units of cm; a_0, a_1, and a_2 are regression coefficients; $\varepsilon_{\ln IM}$ denotes a standard normal variable; $\sigma_{\ln D}$ is the model-specific standard deviation, so ln$D$ follows the normal distribution with mean of $\mu_{\ln D}$ and standard deviation of $\sigma_{\ln D}$.

(iv) Perform MCS to generate N_{IM} samples of the correlated IMs. The probability distribution of lnIM is described by the associated ground motion model (GMM) with the following general expression [11]:

$$\ln(IM) = \mu_{\ln IM}(M, R, \text{etc.}) + \varepsilon_{\ln IM} \cdot \sigma_{\ln IM} \quad (3)$$

where $\mu_{\ln IM}(M, R,$ etc.) represents the mean of lnIM estimated by using M, R, etc.; $\sigma_{\ln IM}$ denotes the standard deviation of lnIM given by GMM; $\varepsilon_{\ln IM}$ is a standard normal variable. Based on the joint normal distribution of lnIMs, the N_{IM} samples can be readily generated by specifying the correlation coefficients among IMs (e.g., [12]).

(v) Substitute each sample of IM (or IMs) into Eq. (1) (or Eq. (2)) to derive $\mu_{\ln D}$, and then perform MCS to generate N_D samples of D based on the distribution of lnD. This process is repeated for N_{IM} times, resulting in $N_{IM} \times N_D$ data points of D.

3 Slope Model Establishment

Figure 1 shows the slope model established in the finite difference code *FLAC* [13], where the slope height (H) and slope angle are 20 m and 30°, respectively. The cyclic soil behavior is described by the hysteretic damping model Sig4, in which the model parameters are calibrated according to the Darendeli [14] modulus reduction and damping ratio curves with plasticity index of 0 and the effective vertical stress at the depth of H. Sig4 is combined with the Mohr-Coulomb plasticity criterion for modelling the plastic behavior of soils. To remove high-frequency noises, a small amount of stiffness-proportional Rayleigh damping (0.2%) is specified. The bedrock layer is considered linear-elastic with 0.5% mass- and stiffness-proportional Rayleigh damping. Table 1 summarizes the parameters assigned to the slope model. To minimize wave reflection effects, the quiet boundary is applied along the bedrock base, and the free-field boundary that simulates a quiet boundary is implemented at both lateral sides [13]. The sum of L_1 and L_2 is equal to $12H$, as illustrated in Fig. 1.

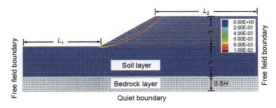

Fig. 1. Numerical model of slope and the associated maximum shear strain increment contours obtained from pseudostatic strength reduction technique. The CSS from Bishop's simplified method is shown using a red line.

Table 1. Summary of the geotechnical parameters assigned to the slope model.

Soil layer				Bedrock layer	
ρ (kg/m^3)	c (kPa)	ϕ (°)	G_{max} (MPa)	ρ (kg/m^3)	G_{max} (MPa)
2000	10	35	320	2300	1472

Note: ρ, c, ϕ, and G_{max} denote the density, cohesion, friction angle, and initial shear modulus of soil, respectively.

Before conducting dynamic analysis, pseudostatic slope stability analysis is conducted using both the strength reduction technique (SRT) and Bishop's simplified limit equilibrium method. As also shown in Fig. 1, the resulting shear band from SRT agrees well with the critical slip surface (CSS) from the Bishop method. The fundamental period of the failure mass (above CSS) is estimated as $T_{s,mass} = 4h/V_s = 0.077$ s, where V_s is the soil shear-wave velocity (i.e., $V_s = \sqrt{G_{max}/\rho}$), and h denotes the maximum thickness of the failure mass. The fundamental period of the soil layer ($T_{s,layer}$) is similarly calculated using $h = 1.5H$ (average of the downhill and uphill soil layer thicknesses). In the next section, two degraded periods determined as $T_{d1} = 1.5T_{s,mass} = 0.116$ s and $T_{d2} = 1.5T_{s,layer} = 0.45$ s will be considered to construct the spectral acceleration (SA) for predicting earthquake-induced slope displacements (e.g., [2]).

4 Development of Slope-Specific Displacement Models

4.1 Construction of Displacement Models

The NGA-West2 database (https://ngawest2.berkeley.edu/) is used to select 83 GM records, which cover a wide range of peak ground acceleration (PGA) from 0.05 g to 1.41 g, allowing for a proper consideration of the GM record-to-record variability. Following Step (ii) in Sect. 2, the D for each GM is calculated as the maximum horizontal displacement along the slope surface.

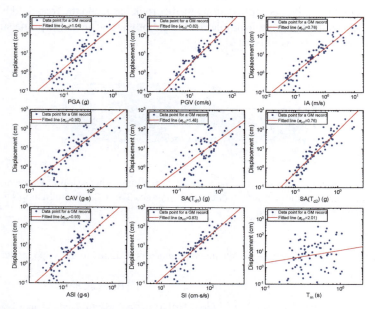

Fig. 2. Distribution of D versus different IMs. Also shown are trends fitted by Eq. (1) or (2).

For a thorough comparison, this study incorporates 10 representative IMs, including PGA, peak ground velocity (PGV), Arias intensity (IA), cumulative absolute velocity

(CAV), SA(T_{d1}), SA(T_{d2}), acceleration spectrum intensity (ASI), spectrum intensity (SI), mean period (T_m), and significant duration Ds_{5-75}. Among them, SA(T_{d1}) and SA(T_{d2}) represent the 5%-damped spectral accelerations at the periods T_{d1} and T_{d2}, respectively. Note that most of the existing Newmark-type models correlate SA(T_{d1}) to D. Step (iii) is conducted for the 10 scalar IMs and 38 combinations of IMs, resulting in 48 D models.

Figure 2 shows the D versus IM distributions for 9 scalar IMs and the fitted linear trends (and the associated $\sigma_{\ln D}$). It is found that the ln$D$ versus lnIM relationship generally follows a linear pattern. The SA(T_{d2}) model fits the data better than the others; yet, the more usually used SA(T_{d1}) leads to much larger scatter (i.e., $\sigma_{\ln D}$ is almost double). Also, the scatter for T_m and Ds_{5-75} is significantly larger, indicating the low efficiency of the two scalar IMs in predicting D. Such wide ranges of D and IMs imply the capabilities of the models for estimating D in various shaking levels.

4.2 Standard Deviations of the Displacement Models

Figure 3a and b further compare $\sigma_{\ln D}$ for the scalar-IM and vector-IM models, respectively. The order of SA(T_{d2}) > IA > PGV > SI > CAV > ASI > PGA > SA(T_{d1}) > T_m > Ds_{5-75} is observed for the efficiency of scalar IMs. The smallest $\sigma_{\ln D}$ of 0.76 for SA(T_{d2}) may be attributed to that $T_{d2} = 0.45$ s (for the soil layer's fundamental period) is comparable to the value of T_m (e.g., see Fig. 2) and is more related to the dynamic response of the slope. This indicates that the degraded period for the failure mass (i.e., T_{d1}) should be used cautiously with the consideration of GMs' period range. Besides, IA and PGV, which were identified as the two most efficient IMs by Cho and Rathje [8], yield similar $\sigma_{\ln D}$ (0.78 and 0.82) to that of SA(T_{d2}). Hence, the three IMs may be preferred for deriving D models. Although it is not a common IM, SI also results in relatively small $\sigma_{\ln D}$ (≈ 0.8). On the other hand, T_m and Ds_{5-75} lead to significantly large $\sigma_{\ln D}$ (≈ 2).

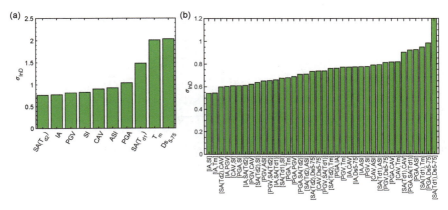

Fig. 3. Model-specific standard deviations for different (a) scalar-IM and (b) vector-IM models.

It is observed from Fig. 3b that the vector-IM models yield much smaller $\sigma_{\ln D}$ than the scalar-IM models as a result of more complementary information carried by two IMs. The order of [IA,SI] > [IA,T_m] > [SA(T_{d2}),CAV] > [IA,PGV] > [CAV,SI] is observed for the five most efficient vectors (i.e., $\sigma_{\ln D} = 0.54$–0.61). Specifically, the $\sigma_{\ln D}$ values (0.54 and 0.55) for [IA,SI] and [IA,T_m] are noticeably smaller, indicating the attractiveness of the two vectors. The [SA(T_{d2}),CAV], [IA,PGV], [CAV,SI], [PGA,SI], [IA,SA(T_{d2})] and [PGV,CAV] are the subsequently efficient vectors that achieve comparative $\sigma_{\ln D}$ within the range of 0.60–0.61. These vectors include either IA or CAV, which capture multiple characteristics (amplitude, duration, etc.) of GMs.

5 Scenario-Based Comparison of the Models Using MCS

Following Step (iv), the mean and variability of the D prediction for different models are compared under some representative scenarios. Multiple GMMs are adopted for individual IMs [11, 15–18] following the logic tree method. The correlation coefficient matrix for modeling the joint distribution of the 10 IMs is derived according to the references [12, 19, 20]. Both N_{IM} and N_D are specified as 200, yielding 40000 displacement values. The geometric mean of these values (termed as D_{mean} hereafter) and the standard deviation of $\ln D$ (St.d. of predicted $\ln D$) are investigated as follows.

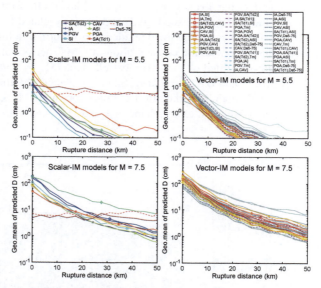

Fig. 4. Mean prediction trends associated with different displacement models for $V_{S30} = 760$ m/s

5.1 Mean of the Displacement Prediction

Figure 4 shows the D_{mean} versus R for $V_{S30} = 760$ m/s. Both $M = 5.5$ and 7.5, and the scalar- and vector-IM models are included. Note that the legend is shown in the way of illustrating the ranking of $\sigma_{\ln D}$ for different displacement models. The results for $V_{S30} = 360$ m/s are not shown due to the limited space, yet are also discussed as follows. First, D_{mean} decreases with increasing R, while the decreasing trend becomes slower for larger R. Second, the trends for the scalar- and vector-IM models are generally comparable, and the difference of the results for most vector-IM models (especially for the 10 most efficient ones) is smaller in comparison with the scalar-IM models. Third, the model-to-model difference is slightly dependent on V_{S30} and M. Regarding the soil site condition ($V_{S30} = 360$ m/s), $M = 5.5$ generally corresponds to smaller difference of D_{mean} than $M = 7.5$; yet, this observation appears to reverse for the rock site condition. Fourth, the SA(T_{d1}) model generally produces the upper and lower bounds among the scalar-IM models for $M = 5.5$ and 7.5, respectively. The T_m and Ds_{5-75} models are two exceptional cases with almost unchanged D_{mean}, indicating that the two IMs should not be used solely to predict D.

5.2 Variability in the Displacement Prediction

Figure 5 displays the St.d. of predicted $\ln D$ versus R for $V_{S30} = 760$ m/s. The $V_{S30} = 360$ m/s case produces slightly smaller St.d., and is not shown here for brevity. No evident trend of St.d. versus R is observed. The St.d. values for the scalar-IM models are generally comparable to those for the vector-IM models, although the 10 most efficient vector-IM models tend to result in smaller St.d. values. As a superposition of $\sigma_{\ln IM}$ and $\sigma_{\ln D}$, St.d. is generally within the range from 1.4 to 1.8.

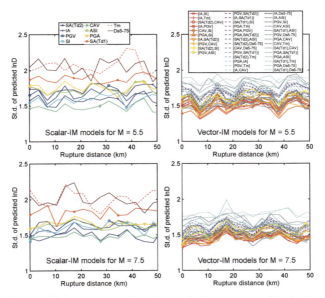

Fig. 5. Variability trends associated with different displacement models for $V_{S30} = 760$ m/s.

For comparison, Table 2 lists the different types of standard deviations for the scalar-IM models, where μ_{Std} represents the average of St.d. for different scenarios (various

V_{S30} and R). It is seen that the IM for the smallest μ_{Std} may not produce the smallest $\sigma_{\ln D}$, due to the additionally included uncertainty in the IM prediction. In general, CAV tends to yield smaller μ_{Std}, and μ_{Std} for IA is still relatively small. The μ_{Std} values for the 38 vector-IM models are derived and are shown in Fig. 6. A similar observation of different rankings of μ_{Std} and $\sigma_{\ln D}$ can be made. Specifically, the low uncertainty in estimation of CAV is illustrated again; that is, some IM vectors including CAV yield relatively small total variability (μ_{Std}). The larger μ_{Std} is, the more uncertain the displacement prediction. Though $\sigma_{\ln D}$ can generally be smaller than 0.75, the total uncertainty is still large, and in most cases μ_{Std} has exceeded 1.4. Take the predicted D of 40 cm given an earthquake scenario as an example; the expected range of D considering the total uncertainty ($\mu_{Std} = 1.4$) can be estimated as $[\exp(\ln(40)-\mu_{Std}), \exp(\ln(40) + \mu_{Std})] =$ [10 cm, 162 cm] [2]. Such a large interval should be shrunk for a more robust estimation of D. Thus, more powerful GMMs with smaller aleatory variability should be developed and applied to the seismic slope displacement analysis.

Table 2. Standard deviation parameters for the scalar-IM models.

IM_1	$\sigma_{\ln IM}$ ($M = 5.5$)	$\sigma_{\ln IM}$ ($M = 7.5$)	$\sigma_{\ln D}$	μ_{Std} ($M = 5.5$)	μ_{Std} ($M = 7.5$)
PGA	0.58–0.65	0.52–0.58	1.04	1.73	1.63
PGV	0.63–0.64	0.59–0.60	0.82	1.67	1.58
IA	1.01–1.08	0.94–0.99	0.78	1.57	1.46
SA (T_{d1})	0.59–0.71	0.54–0.64	1.48	1.86	1.79
SA (T_{d2})	0.66–0.69	0.61–0.64	0.76	1.72	1.62
CAV	0.39–0.49	0.39–0.43	0.90	1.46	1.42
SI	0.60–0.62	0.60–0.61	0.83	1.59	1.60
ASI	0.56–0.63	0.52–0.57	0.93	1.71	1.60

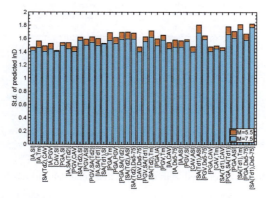

Fig. 6. Comparison of μ_{Std} for different vector-IM displacement models.

6 Summary and Conclusions

This study investigated the relationship between ground motion intensity measures (IMs) and earthquake-induced permanent slope displacement (D) based on numerical stress-deformation analyses. Ten scalar IMs and 38 vector-IM combinations, were used to develop the D models, and the efficiencies of various scalar and vector-IMs were compared in terms of the model standard deviation ($\sigma_{\ln D}$). Furthermore, a Monte Carlo simulation-based procedure was utilized to compare these D models under representative earthquake scenarios, in which the uncertainties in predicting both IMs and D are considered. The comparative results lead to the following conclusions:

1. The SA($T_{d2} = 1.5T_{s,layer}$)- and IA-based displacement models were identified as the most efficient scalar-IM models with the smallest $\sigma_{\ln D}$ of 0.77, while the more commonly used SA($T_{d1} = 1.5T_{s,mass}$) resulted in much larger $\sigma_{\ln D}$. In constrast, the scalar IMs of T_m and Ds_{5-75} exhibit the lowest efficiency in regressing D.
2. The [IA,SI] and [IA,T_m] resulted in the smallest $\sigma_{\ln D}$ of about 0.55 for the vector-IM models. The subsequent six most efficient models generally included ether IA or CAV. Only 6 among the 38 models yielded $\sigma_{\ln D}$ greater than 0.77, indicating the advantage of vector-IM models for improving the efficiency of regressing D.
3. The total standard deviation contributed by both the uncertainties in IM and D predictions is considerably larger than $\sigma_{\ln D}$, and the models with the smallest $\sigma_{\ln D}$ do not necessarily yield the smallest total standard deviation. This is more evident for models including CAV in which a smaller variability is involved in predicting this IM. Recent studies on developing non-ergodic ground motion models shed light in the direction of reducing variability in the IM prediction (e.g., [21]).

The results of the mean and standard deviation of D obtained in this study could be useful to select IMs in developing predictive models for seismically-induced slope displacements. Such models can be implemented within the fully-probabilistic framework [7, 8], allowing practitioners to estimate the hazard-compatible D for seismic design of slopes.

References

1. Newmark, N.M.: Effects of earthquakes on dams and embankments. Geotechnique **15**(2), 139–160 (1965)
2. Bray, J.D., Travasarou, T.: Simplified procedure for estimating earthquake-induced deviatoric slope displacements. Journal of Geotechnical and Geoenvironmental Engineering **133**(4), 381–392 (2007)
3. Du, W., Huang, D., Wang, G.: Quantification of model uncertainty and variability in Newmark displacement analysis. Soil Dyn. Earthq. Eng. **109**, 286–298 (2018)
4. Jibson, R.W.: Methods for assessing the stability of slopes during earthquakes—A retrospective. Eng. Geol. **122**(1–2), 43–50 (2011)
5. Rathje, E.M., Saygili, G.: Probabilistic seismic hazard analysis for the sliding displacement of slopes: scalar and vector approaches. Journal of Geotechnical and Geoenvironmental Engineering **134**(6), 804–814 (2008)

6. Li, D.Q., Wang, M.X., Du, W.: Influence of spatial variability of soil strength parameters on probabilistic seismic slope displacement hazard analysis. Engineering Geology 105744 (2020)
7. Wang, M.X., Li, D.Q., Du, W.: Probabilistic seismic displacement hazard assessment of earth slopes incorporating spatially random soil parameters. J. Geotechn. Geoenvironmen. Eng. **147**(11), 04021119 (2021)
8. Cho, Y., Rathje, E.M.: Displacement hazard curves derived from slope-specific predictive models of earthquake-induced displacement. Soil Dyn. Earthq. Eng. **138**, 106367 (2020)
9. Wang, G.: Efficiency of scalar and vector intensity measures for seismic slope displacements. Front. Struct. Civ. Eng. **6**(1), 44–52 (2012)
10. Fotopoulou, S.D., Pitilakis, K.D.: Predictive relationships for seismically induced slope displacements using numerical analysis results. Bull. Earthq. Eng. **13**(11), 3207–3238 (2015). https://doi.org/10.1007/s10518-015-9768-4
11. Gregor, N., et al.: Comparison of NGA-West2 GMPEs. Earthq. Spectra **30**(3), 1179–1197 (2014)
12. Bradley, B.A.: A ground motion selection algorithm based on the generalized conditional intensity measure approach. Soil Dyn. Earthq. Eng. **40**, 48–61 (2012)
13. Itasca Consulting Group: FLAC-Fast Lagrangian analysis of continua. Version 8.0. User's manual (2016)
14. Darendeli, M.B.: Development of a new family of normalized modulus reduction and material damping curves. Ph.D. thesis. Dept. of Civil, Architectural and Environmental Engineering, Univ. of Texas at Austin (2001)
15. Campbell, K.W., Bozorgnia, Y.: Ground motion models for the horizontal components of Arias intensity (AI) and cumulative absolute velocity (CAV) using the NGA-West2 database. Earthq. Spectra **35**(3), 1289–1310 (2019)
16. Du, W., Wang, G.: A simple ground-motion prediction model for cumulative absolute velocity and model validation. Earthquake Eng. Struct. Dynam. **42**(8), 1189–1202 (2013)
17. Afshari, K., Stewart, J.P.: Physically parameterized prediction equations for significant duration in active crustal regions. Earthq. Spectra **32**(4), 2057–2081 (2016)
18. Rathje, E.M., Faraj, F., Russell, S., Bray, J.D.: Empirical relationships for frequency content parameters of earthquake ground motions. Earthq. Spectra **20**(1), 119–144 (2004)
19. Bradley, B.A.: Correlation of Arias intensity with amplitude, duration and cumulative intensity measures. Soil Dyn. Earthq. Eng. **78**, 89–98 (2015)
20. Du, W.: Empirical correlations of frequency-content parameters of ground motions with other intensity measures. J. Earthquake Eng. **23**(7), 1073–1091 (2019)
21. Abrahamson, N.A., Kuehn, N.M., Walling, M., Landwehr, N.: Probabilistic seismic hazard analysis in California using nonergodic ground-motion models. Bull. Seismol. Soc. Am. **109**(4), 1235–1249 (2019)

Accuracy of Complex Moduli in Seismic Response Analysis of Ground

Nozomu Yoshida[1](✉) and Kenji Adachi[2]

[1] Kanto Gakuin University, Yokohama 243-0031, Japan
nyoshida@kanto-gakuin.ac.jp
[2] Jiban Soft Factory, Tokyo 171-0033, Japan

Abstract. The complex moduli used for the seismic response analysis of ground, the Sorokin model used in original SHAKE, the Lysmer model proposed to improve SHAKE, and the YAS model proposed by the authors are discussed. The Sorokin model overestimates maximum stress. The Lysmer model shows maximum stress the same as the test but underestimates the damping ratio. The YAS model gives the maximum stress and the damping ratio the same as the test result. The case study using 269 ground and 10 observed strong earthquake motions in Japan shows that the Sorokin model overestimates PGA about 15% at maximum compared with the YAS model. Overestimation of PGA by the Lysmer model is as small as 3% or less, although it is theoretically incorrect.

Keywords: Seismic response analysis · Equivalent linear · Complex modulus

1 Introduction

The seismic response analysis based on the equivalent linear method, represented by SHAKE [1] and FLUSH [2], has been used in engineering practice. The name SHAKE has been used as if it is a common noun. However, the complex moduli or method of analysis had several disadvantages and has been improved by several researchers. The first improvement is made by Lysmer [3] by proposing a new complex modulus. As far as the authors understand, almost all computer codes use this new complex modulus, and they are called SHAKE. One of the authors pointed out that it overestimates maximum stress and maximum acceleration [4]. This was proved by comparing the truly nonlinear earthquake response [5]. The overestimation of PGAs comes from two reasons. The one is to use equivalent strain in the equivalent linear analysis [4] and the other is the overestimation of the maximum stress which is discussed in this paper. The authors checked the accuracy of the conventional complex moduli and found incorrect use of the technical term and error of the theory, and proposed a new complex modulus named YAS model [6]. This paper reviews past studies on complex moduli and discusses the accuracy through numerical analyses.

2 Complex Modulus and Material Damping Constant

2.1 Material Damping Constant

The idea to use complex modulus for the nonlinear system is firstly proposed by Sorokin [7], although Schnabel et al. [1] did not cite it. The hysteretic behavior is expressed by the strain-dependent modulus G and a damping ratio h. The modulus is the secant modulus, and the damping ratio is defined as

$$h = \frac{1}{4\pi} \frac{\Delta W}{W}, \quad W = \frac{1}{2} G \gamma_0^2 \qquad (1)$$

The stress–strain relationship of the Voigt model is expressed as

$$\tau = G\gamma + C\dot{\gamma} \qquad (2)$$

where τ and γ denote the stress and the strain, respectively, and dot denotes time derivative with time. Under the action of harmonic wave $\gamma = \gamma_0 \cos\omega t$ (ω: circular frequency), the stress–strain relationship yields

$$\tau = G\gamma_0 \cos\omega t - C\gamma_0\omega \sin\omega t = G\gamma \pm C\omega\sqrt{\gamma_0^2 - \gamma^2} \qquad (3)$$

The stress–strain relationship is supposed to be frequency independent [e.g. 8], but Eq. (3) shows frequency dependence. Then, the damping coefficient is set to be frequency-dependent by introducing a new parameter β as

$$C = 2\beta G/\omega \qquad (4)$$

Thus, the stress–strain relationship yields,

$$\tau = G\gamma \pm 2\beta G\sqrt{\gamma_0^2 - \gamma^2} \qquad (5)$$

This stress–strain relationship is frequency independent. By using this stress–strain relationship in Eq. (1), the damping ratio yields

$$h = \beta \qquad (6)$$

Schnabel et al. [1] called β critical damping ratio, and Lysmer [3] called it a fraction of critical damping. However, these names are incorrect usage of the technical term. These terms come from the vibration of a one-degree-of-freedom system. Here, the critical damping ratio β_1 with the mass m and the spring constant k is defined as

$$2\beta_1 = C/\sqrt{mk} \qquad (7)$$

By comparing Eqs. (4) and (7), it is clear that the definitions are different; β in Eq. (4) is a constant depending only on the material, whereas β_1 in Eq. (7) is a constant depending on the system. Since there is no relevant name on β, we named it a material damping constant.

It is also noted that since we start from the Voigt model, the viscous coefficient C appears in Eq. (4). However, since we considers anergy absorption caused by the nonlinear behavior, we use the stiffness and the damping ratio as parameters.

2.2 Sorokin Model

The idea to use complex modulus to express the nonlinear behavior is firstly proposed by Sorokin [8] although Schnabel et al. [1] did not cite it. When considering the harmonic strain $\overline{\gamma} = \gamma_0 e^{i\omega t}$ and substituting it into the stress–strain relationship (complex expression of Eq. (2)), we obtain

$$\overline{\tau} = G(1 + 2i\beta)\gamma_0 e^{i\omega t} = \overline{G_S^*}\overline{\gamma} \tag{8}$$

where

$$\overline{G_S^*} = G(1 + 2i\beta) \tag{9}$$

is a complex modulus. This kind of frequency-independent complex modulus is called Sorokin's damping (hypothesis). Thus, it is called the Sorokin model in this paper.

The stress–strain curve in Eq. (5) is an elliptical shape, and the area or the energy absorption of the stress-strain model, ΔW in Eq. (1), is calculated as

$$\Delta W_V = \int_{-\gamma_0}^{\gamma_0} \left\{ \left(G\gamma + 2\beta G\sqrt{\gamma_0^2 - \gamma^2}\right) - \left(G\gamma - 2\beta G\sqrt{\gamma_0^2 - \gamma^2}\right) \right\} d\gamma = 2\beta G\pi \gamma_0^2 \tag{10}$$

This means that the absorbed energy depends only on the modulus G and the imaginary part of the complex modulus. This holds for other models shown below. From Eqs. (1) and (10), we obtain $h = \beta$ (same as Eq. (6)). Therefore, the damping ratio h and the material damping constant β are not distinguished in engineering practice. However, this relation does not always hold as shown in the next section.

2.3 Lysmer Model

Lysmer focused on the response of a one-degree-of-freedom system subjected to harmonic loading with a Voigt model and with a complex modulus in Eq. (9). He showed that the displacement amplitudes of the two models are different, and proposed a new complex modulus [3]

$$G_L^* = G(1 - 2\beta^2 + 2i\beta\sqrt{1 - \beta^2}) \tag{11}$$

Unfortunately, there is no theoretical background of this equation. We showed that there are infinite number complex moduli that result in the same displacement amplitude [6]. When the complex modulus $\overline{G}^* = G(G_r + iG_i)$ satisfies

$$1 - 2\alpha^2 + 4\alpha^2\beta^2 = G_r^2 - 2G_r\alpha^2 + G_i^2 \tag{12}$$

displacement amplitudes of the two systems become the same, where $\alpha = \omega/\omega_0$ (ω and ω_0 are circular frequencies of the excitation and the system, respectively) is a tuning ratio. Among Eq. (12), possibly Eq. (11) is only the one that does not include the tuning ratio.

There are two problems in Lysmer's proposal. The one is that, as discussed above, the material damping constant and critical damping ratio are to be distinguished. He discussed only the displacement amplitude and did not discuss the phase. The displacements on one-degree-of-freedom systems by using the Voigt model, Lysmer model, and YAS model which will be shown in the next section becomes as follows.

$$u = \frac{\cos(\omega t - \phi)}{k\sqrt{(1-\alpha^2)^2 + (2\alpha\beta_1)^2}}, \quad \tan\phi = \frac{2\alpha\beta_1}{1-\alpha^2} \quad (13)$$

$$u = \frac{\cos(\omega t - \phi)}{k\sqrt{(1-\alpha^2)^2 + (2\alpha\beta_1)^2}}, \quad \tan\phi = \frac{2\beta_1\sqrt{1-\beta_1^2}}{1-\alpha^2 - 2\beta_1^2} \quad (14)$$

$$u = \frac{\cos(\omega t - \phi)}{k\left(\sqrt{(\sqrt{1-4\beta_1^2}-\alpha^2)^2 + (2\beta_1)^2}\right)}, \quad \tan\phi = \frac{2\beta_1}{\sqrt{1-4\beta_1^2}-\alpha^2} \quad (15)$$

It is clear that displacements obtained by the Voigt and the Lysmer models show the same displacement amplitude, but the phases are different. Therefore, as shown in Fig. 1, the hysteretic behaviors are different. This means that discussion by Lysmer does not seem important.

The second problem is the damping ratio. The damping ratio h calculated from the Lysmer model based on Eq. (1) results in

$$h = \frac{1}{4\pi} \frac{2\beta G \gamma_0^2 \pi \sqrt{1-\beta^2}}{G\gamma_0^2/2} = \beta\sqrt{1-\beta^2} \quad (16a)$$

or

$$\beta = \sqrt{\frac{1-\sqrt{1-4h^2}}{2}} \quad (16b)$$

This means the values of the damping ratio and the material damping constant are different. Since both the damping ratio and the material damping constant have the same value

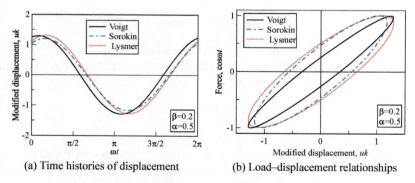

(a) Time histories of displacement   (b) Load–displacement relationships

Fig. 1. Comparison of behavior of one-degree-of-freedom system

in the Sorokin model (used in original SHAKE) and the YAS model (explained later), we need not notice the difference. However, in the Lysmer model, they are different. Therefore, when the Lysmer model is used, the material damping constant β is to be calculated from the damping ratio h by using Eq. (16b).

Christian et al. [9] pointed that the difference between the Sorokin and the Lysmer model is very small when $\beta < 0.3$ and suggested using the Sorokin model. Kramer [10] also made a similar discussion. They seem to consider that the Sorokin model is sufficient. However, their discussions lack another important feature.

The maximum stress τ_{max} by the Sorokin model is calculated from the real part of Eq. (8) as

$$\tau_{max} = G\gamma_0\sqrt{1+4\beta^2} = \tau_0\sqrt{1+4\beta^2} \tag{17}$$

It is noted that the maximum stress of the test is $\tau_{max} = \tau_0 = G\gamma_0$. This means that the maximum stress is $(1+4h^2)^{0.5}$ times overestimated, which will result in the overestimation of the peak acceleration. On the other hand, the maximum stress by the Lysmer model becomes $G\gamma_0$, the same as the test result. This is an important feature because the maximum stress is strongly related to PGA.

2.4 YAS Model

The YAS (Yoshida–Adachi–Sorokin) model is proposed to give both the maximum stress and the damping ratio same as those of the test. The model is expressed as

$$\overline{G}_Y^* = G\left(\sqrt{1-4\beta^2} + 2i\beta\right) \tag{18}$$

It is easily understood that $\tau_{max} = \tau_0 = G\gamma_0$. The damping ratio h is calculated from the stress–strain curve and Eq. (1), which results in $\beta = h$ (as discussed in the previous

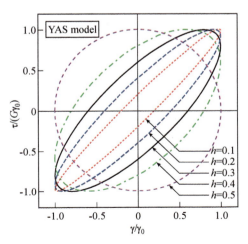

Fig. 2. Damping ratio dependency of stress-strain curve of YAS model

section, the imaginary part is the same as that of the Sorokin model). It is noted that the same complex modulus is obtained by substituting Eq. (16b) into the Lysmer model, Eq. (11).

Figure 2 shows stress–strain curves with $\beta = h$ as parameters. Since the real part of the complex modulus becomes zero at $\beta = h = 0.5$, the stress–strain curve becomes a complete circle. This means that applicability of the YAS model is up to the material constant 0.5. Since maximum damping ratio (or material damping constant) is 0.3 or a little larger, this does not become disadvantage.

3 Comparison of Three Models

3.1 Stress–Strain Curve

Since energy absorption of the hysteretic loop depends on the imaginary part as shown in Eq. (10), the imaginary parts are compared in Fig. 3(a). Both the Sorokin model and the YAS model show linear relationships indicating that $h = \beta$. On the other hand, the damping ratio h begins to decrease as β increases in the Lysmer model. However, the difference between other models is small when $\beta < 0.3$, an important damping ratio in practice. The damping ratios are 0.05%, 0.4%, 1.4% smaller than the material damping constants at $\beta = 0.1, 0.2, 0.3$, respectively. Figure 3(b) compares the stress-strain loops of three models at $\beta = 0.3$. It is clearly seen that the Sorokin model shows larger maximum stress than that of other models. The difference between the Lysmer model and the YAS model is small because their maximum stresses are the same and the damping ratio differs only 1.4%.

3.2 Earthquake Response

In order to see the effect of the complex moduli on the seismic response of ground, seismic response analyses are conducted. In total 269 grounds are analyzed under 10

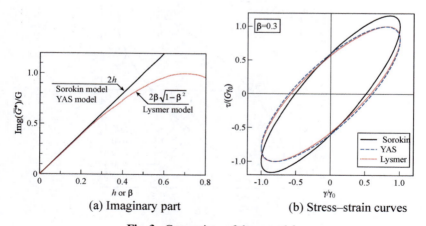

(a) Imaginary part (b) Stress–strain curves

Fig. 3. Comparison of three models

observed strong earthquake motions in Japan. Among the 10 earthquake motions, 5 were obtained under the inland earthquakes and the rest 5 were obtained under the ocean trench type earthquakes. Details of the ground and the earthquake motions are shown in Ref. [5]. Since the purpose is the effect on PGA, cases where the effect appears large are chosen. Thus, the shear moduli are evaluated from the SPT-N value and kept constant. The material damping constant is set to be 0.3.

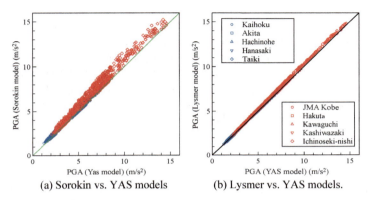

Fig. 4. Comparison of PGA under 11 earthquake motions

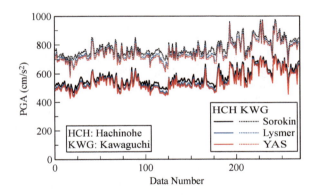

Fig. 5. Comparison of PGA under ocean type and inland type earthquakes

Figure 4 compares PGA on 2690 cases (269 grounds and 10 earthquake motions). It is seen that the Sorokin model shows a larger PGA than the YAS model. The maximum difference is about 15% and 3% between Sorokin and YAS models and Lysmer and YAS models. Figure 5 compares PGAs for two earthquake motions, one is the Hachinohe wave obtained during the 1968 Tokachi-oki earthquake (Ocean trench type eq.) and the other is the Kawaguchi wave obtained during the 2004 Niigata-ken-Chuetsu earthquake (Inland earthquake). In all cases, PGAs are the largest in the Sorokin mode, and they are the smallest in the YAS model. The Lymser model is between them.

4 Conclusion

A new technical term "material damping constant" is proposed to express the complex modulus because the conventionally used term such as the critical damping ratio is not relevant; it is the term for the vibration of a one-degree-of-freedom system and depends on the system such as the mass and the spring constant whereas the material damping constant depends on material property such as stiffness and hysteretic energy absorption.

Three complex moduli, the Sorokin model, the Lysmer model, and the YAS model are discussed and compared from the point of view of the hysteretic behavior. The Sorokin model overestimates the maximum stress. The Lysmer model underestimates the damping ratio although the maximum stress is the same as the test. The maximum stress and the damping ratio same as the test are obtained by the YAS model.

The use of the material damping constant instead of the damping ratio is theoretically incorrect in the Lysmer model. Equation (16b) is to be used to evaluate β from the damping ratio h to make the model theoretically correct.

References

1. Schnabel, P.B., Lysmer, J., Seed, H.B.: SHAKE A Computer program for earthquake response analysis of horizontally layered sites. In: EERC72–12. University of California (1972)
2. Lysmer, J., Udaka, T., Tsai, C.-F., Seed, H.B.: FLUSU a computer program for approximate 3-D analysis of soil-structure interaction problems. In: EERC, pp. 75–30. University of California (1975)
3. Lysmer, J.: Modal damping and complex stiffness. University of California Lecture Note. University of California, Berkeley (1973)
4. Yoshida, N.: Applicability of conventional computer code SHAKE to nonlinear problem. In: Proceedings of the Symposium on Amplification of Ground Shaking in Soft Ground, pp. 14–31. JSSMFE, Tokyo (1994). (in Japanese)
5. Yoshida, N.: Comparison of seismic ground response analyses under large earthquakes. Indian Geotechnical Journal **44**(2), 119–131 (2014). https://doi.org/10.1007/s40098-014-0104-8
6. Yoshida, N., Adachi, K.: Complex moduli for seismic response analysis of ground. Journal of JAEE **21**(1), 65–81 (2021). (in Japanese); English version is in print
7. Hara, A.: Kinematic property of ground and its application. In: Proc. The 2nd Symposium of Earthquake Ground Motion, AIJ, pp. 33–39 (1973). (in Japanese)
8. Sorokin, F.: Internal and external friction by vibrations of rigid bodies, CNIISK 162, Moscow (1957). In: Tseitlin, A.I., Kusalnov, A.A. (eds.) Role of internal friction in dynamic analysis of structures, 215 pp. CRC Press, London (1999)
9. Christian, J.T., Roesset, J.M., Desai, C.S.: Two- and three-dimensional dynamic analyses (chapter 20). In: Numerical Methods in Geotechnical Engineering, 683–718. McGraw Hill (1977)
10. Kramer, S.L.: Geotechnical earthquake engineering, 653 pp. Prentice Hall (1996)

Scattering of Incident Plane Waves by Underground Cylindrical Cavity in Unsaturated Poroelastic Medium

Aichen Zhang, Weihua Li(✉), and Fengcui Feng

School of Civil Engineering and Architecture, Beijing Jiaotong University,
Beijing 100044, China
whli@bjtu.edu.cn

Abstract. At present, the study of the dynamic characteristics of unsaturated porous media is still in its infancy. Therefore, it is particularly important to study the dynamic characteristics of cavity in unsaturated media, which is similar to the actual situation of cylindrical underground structures such as tunnels are encountered with seismic wave. Based on the dynamic theory of unsaturated porous media considering the mixture theory, analytical solutions are presented for two-dimensional of scattering of plane P-wave and SV-wave by cylindrical cavity by using wave function expansion method. The validity of the solutions is confirmed by comparing with the one in which the infinite medium is assumed as a pure elastic medium. A detailed parametric study is presented to illustrate the effect of saturation, Poisson's ration, and porosity of the unsaturated poroelastic medium on the dynamic stress concentration factor around the cavity is analyzed.

Keywords: Unsaturated porous media · Cylindrical cavity · Plane waves · Scattering and diffraction · Analytic solutions

1 Introduction

With the optimization of engineering technology and the acceleration of urbanization, underground space is being developed and utilized as a resource. In the event of an earthquake, the seismic action can cause great damage to underground structures such as tunnels and pipelines in the underground space. The existence of underground cavities will produce inhomogeneity in soil medium, which causes the wave scattering of cavities, changes the original wave field, and causes stress concentration around cavities. Therefore, it is necessary to study the dynamic characteristics of cavity in unsaturated media, which is similar to the actual situation of cylindrical underground structures such as tunnels are encountered with seismic wave.

In previous studies, scholars have made a lot of results in studying the dynamic response characteristics of the soil around underground cavities. Initially, when studying the fluctuation of elastic medium, the soil was generally assumed to be a pure elastic medium, and the influence of pores in the soil was not considered. The publication of Biot's [1, 2] classic theory established the basic dynamic theory of saturated porous media, laying the foundation for future research. T. Senjuntichai et al. [3] studied the dynamic response of a cavity in an infinitely saturated medium with transient loads by using the Laplace transform method. Based on the revised Biot wave equation, Hu [4] obtained analytical solutions to the plane wave scattering and refraction problems of a cylinder in a saturated porous medium. They also gave the solution when the cylinder is a cylindrical cavity, a rigid column, a pure elastic soil medium and a pure elastic fluid medium. Liu et al. [5–8] studied the dynamic response of a series of cylindrical tunnels in viscoelastic saturated soil. Based on Biot's dynamic theory of saturated porous media, Li and Zhao [9, 10] established a wave function expansion method to solve the problem of plane wave scattering from cylindrical cavities in the half-space of saturated soil for the first time, and obtained analytical solutions of plane P waves and SV waves by cylindrical cavities in the half-space of saturated soil. Zhou [11] used the complex variable function method to obtain an analytical solution for the dynamic response of a cylindrical-lined cavity in saturated porous media. Li [12] gave the solution to the transient plane wave scattering by a deep-buried cylindrical-lined cavern in saturated porous media. Song [13] used the wave function expansion method to obtain an analytical solution to the dynamic stress concentration problem of deep-buried composite linings under the action of SV waves, and degenerate it into the problem of SV wave scattering by a single-layer lining in an elastic medium, which verify the reliability of the results.

It can be seen from the above introductions that these studies either regard the surrounding soil as a pure elastic medium or as a saturated porous medium to analyze the dynamic characteristics of the soil. However, in actual engineering, most of the soil is in an unsaturated state. Xu [14] used the Hankel transformation method to study the steady-state dynamic response of the layered unsaturated soil foundation, and pointed out that the saturation has a significant effect on the vibration of the foundation plate. Based on the effective stress principle of Bishop unsaturated soil, Liu [15] used analytical methods to study the dynamic characteristics of the unsaturated soil-lining structure of deep-buried cylindrical tunnels in the frequency domain, and indicated the influence of saturation on the dynamic response of the tunnel. Based on the dynamic model of unsaturated porous media proposed by Wei [17, 18], Li [16, 19] obtained the answer to the seismic response of unsaturated soil free site under plane wave incidence, and systematically analyzed the soil saturation influence of parameters such as degree on seismic response of unsaturated soil free site. Later, Li [20] used the same model to obtain the answer to the fluctuation problem of unsaturated soil stratified site. Xu [21], Li and Feng [22] considered the transient response of a long cylindrical cavity under pressure in an infinite unsaturated porous medium for the first time. Analytical solution of the transient response of a cylindrical cavity in unsaturated porous media under pressure. Yu et al. [23] gave the dynamic response of a lined tunnel with an incomplete interface embedded in an unsaturated porous elastic medium under the incidence of P waves, and studied the effects of spring stiffness, frequency and saturation on the dynamic stress concentration factor and displacement. At present, there are few researches on the dynamic response of cavities in unsaturated media, especially the scattering of waves by cavities in unsaturated media, and they mainly focus on the dynamic response of layered foundations and the dynamic stress concentration of cavities. With the gradual improvement of the wave theory of unsaturated media and the success of the research on the dynamic characteristics of the cavity in saturated porous media, it is natural to extend it to the more practical unsaturated porous media.

In this paper, the dynamic equation of unsaturated porous media established by Wei is used as the governing equation, and the total scattered wave field of P wave and SV wave is obtained by using the method of separation of variables and the method of wave function expansion. By degrading unsaturated porous media to pure elastic state, and comparing with the existing analytical solutions, the validity of the results in this paper is verified. Select appropriate unsaturated porous media parameters to analyze the calculation example, and study the influence of saturation on the dynamic response of the cavity.

2 Model and Solution

2.1 Mathematical Models and Basic Equations

Consider an infinitely long cavity in full space of the infinite unsaturated porous medium (see Fig. 1.). The radius of the cavity is a, and the plane wave is incident horizontally along the positive x-axis from the left side of the cavity. The filling medium outside the cavity is unsaturated porous medium. The research is based on the dynamic theory of unsaturated porous medium established by Wei. The governing equation is of the following form:

$$n_0^S \rho_0^S \ddot{u}^S = \left(M_{SS} + n_0^S \mu_S\right) \nabla \nabla \cdot u^S + n_0^S \mu_S \nabla \cdot \nabla u^S + M_{SW} \nabla \nabla \cdot u^W \\ + M_{SN} \nabla \nabla \cdot u^N + \hat{\mu}^W \left(\dot{u}^W - \dot{u}^S\right) + + \hat{\mu}^N \left(\dot{u}^N - \dot{u}^S\right) \quad (1a)$$

$$n_0^W \rho_0^W \ddot{u}^W = M_{SW} \nabla \nabla \cdot u^S + M_{WW} \nabla \nabla \cdot u^W + M_{WN} \nabla \nabla \cdot u^N - \hat{\mu}^W \left(\dot{u}^W - \dot{u}^S\right) \quad (1b)$$

$$n_0^N \rho_0^N \ddot{u}^N = M_{SN} \nabla \nabla \cdot u^S + M_{WN} \nabla \nabla \cdot u^W + M_{NN} \nabla \nabla \cdot u^N - \hat{\mu}^N \left(\dot{u}^N - \dot{u}^S\right) \quad (1c)$$

where the superscript S denotes the solid component; W and N denote the wetting and non-wetting fluids, respectively; μ_S represents the shear modulus of the unsaturated porous medium; and n_0^α, ρ_0^α and $u^\alpha (\alpha = S, W, N)$ are the initial volume fraction, initial density, and displacements of individual components, respectively; M_{SS}、M_{WW}、M_{NN}、M_{SW}、M_{SN} and M_{WN} respectively represent the bulk modulus K_α ($\alpha = S, W, N$) related to the unsaturated porous medium of each phase, the bulk modulus of the solid framework $\hat{K}$, the shear modulus G, the effective stress coefficientα_B, and the parameters Θ_f describing the capillary equilibrium conditions Parameter. $\hat{\mu}^f$ (f = W, N) is the coefficient related to permeability, and the calculation method of each parameter can be found in the literature [27].

The V-G model of unsaturated soil using soil-water characteristic curve can be expressed as

$$k_r^W = \sqrt{S_e}\left[1 - \left(1 - S_e^{1/m}\right)^m\right]^2 \quad (2a)$$

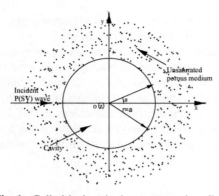

Fig. 1. Cylindrical cavity in unsaturated medium

$$k_r^N = \sqrt{1-S_e}\left(1-S_e^{1/m}\right)^{2m} \quad (2b)$$

where m is the model parameter in the soil-water characteristic curve.

$$S_e = \begin{cases} 0 & S_r \leq S_{rW} \\ \frac{S_r-S_{rW}}{S_{rN}-S_{rW}} & S_{rW} \leq S_r \leq S_{rN} \\ 1 & S_r \geq S_{rN} \end{cases} \quad (3)$$

In which S_r is the saturation of the unsaturated porous medium, S_{rW} is the irreducible degree of saturation, and S_{rN} is the air-entry degree of saturation.

Based on the poroelastic model proposed by Wei, the stress tensor in an unsaturated porous medium can be expressed as

$$\begin{aligned} \overline{\sigma}^S &= \left(M_{SS}\nabla\cdot\boldsymbol{u}^S + M_{SW}\nabla\cdot\boldsymbol{u}^W + M_{SN}\nabla\cdot\boldsymbol{u}^N\right)I + 2n_0^S\mu_S\left[\nabla\boldsymbol{u}^S + (\nabla\boldsymbol{u}^S)^T\right] \\ \overline{\sigma}^W &= \left(M_{SW}\nabla\cdot\boldsymbol{u}^S + M_{WW}\nabla\cdot\boldsymbol{u}^W + M_{WN}\nabla\cdot\boldsymbol{u}^N\right)I \\ \overline{\sigma}^N &= \left(M_{SN}\nabla\cdot\boldsymbol{u}^S + M_{WN}\nabla\cdot\boldsymbol{u}^W + M_{NN}\nabla\cdot\boldsymbol{u}^N\right)I \end{aligned} \quad (4)$$

where $\overline{\sigma}^\alpha$ ($\alpha = S, W, N$) indicates the stress tensors of each phase, and I is the unit tensor matrix.

2.2 Wave Field Analysis

Based on the Helmholtz theorem, the displacement vector $\boldsymbol{u}$ may be written as

$$\boldsymbol{u} = \nabla\varphi + \nabla\times\boldsymbol{\Psi} \quad (5)$$

where φ, $\boldsymbol{\Psi}$ are the scalar and vector displacement potentials, respectively. Substituting Eq. (5) into Eq. (1), there are three compression waves P1, P2, P3 and one shear wave S wave in unsaturated porous media. For details, please refer to the literature [14].

Consider the incident P1(SV) wave propagates in the positive direction of the x-axis, the potential function of the incident wave can be expressed as

$$\varphi^i = \varphi_0 e^{i(k_0 x - \omega t)} \quad \text{(for an incident P wave)} \quad (6)$$

$$\Psi^i = \Psi_0 e^{i(k_0 x - \omega t)} \quad \text{(for an incident SV wave)} \tag{7}$$

where $k_0 = \omega/c_p$ (or $k_0 = \omega/c_s$) is the wave number of the compression wave (or shear wave), $\varphi_0(\Psi_0)$ is the incident wave amplitude, ω is the angular frequency and $i = \sqrt{-1}$. It may be expanded in terms of the Bessel-Fourier series

$$\varphi^i = \varphi_0 \sum_{n=0}^{\infty} \varepsilon_n i^n J_n(k_0 r) \cos n\theta e^{-i\omega t} \tag{8}$$

$$\Psi^i = \Psi_0 \sum_{n=0}^{\infty} \varepsilon_n i^n J_n(k_0 r) \cos n\theta e^{-i\omega t} \tag{9}$$

When the plane wave hits the surface of the cavern, the wave form conversion occurs, scattered waves φ_1^S、φ_2^S、φ_3^S and Ψ^S are generated on the surface of the cavity. The scattering wave potential function of the solid phase can be expressed as.

The scattered wave generated by the incident P wave:

$$\varphi_j^S(r, \theta) = \sum_{n=0}^{\infty} A_{n,j} K_n(k_j r) \cos n\theta \, (j = 1, 2, 3) \tag{10}$$

$$\Psi^S(r, \theta) = \sum_{n=0}^{\infty} B_n K_n(k_4 r) \sin n\theta \tag{11}$$

When the SV wave is incident, due to the change of the incident wave, the parity of displacement and stress has changed. In order to satisfy the parity and boundary conditions of the displacement and stress, the circumferential function in the general solution of the scalar potential function and vector potential function of the generated scattered wave has changed.

$$\varphi_j^S(r, \theta) = \sum_{n=0}^{\infty} A_{n,j} K_n(k_j r) \sin n\theta \, (j = 1, 2, 3) \tag{12}$$

$$\Psi^S(r, \theta) = \sum_{n=0}^{\infty} B_n K_n(k_4 r) \cos n\theta \tag{13}$$

where $\varphi_j^S (j = 1, 2, 3)$ is the scalar potential function of the solid skeleton, Ψ^S is the solid skeleton vector potential function, (j = 1, 2, 3); $A_{n,j}(j = 1, 2, 3)$, B_n refer to the amplitudes of a P wave and S wave; $J_n(.)$ represents the Bessel function of the first kind; $\varepsilon_n = 1$ if $n = 0$ and $\varepsilon_n = 2$ if $n \geq 1$.

The total wave field in the unsaturated porous medium studied is composed of the incident wave field and the scattered wave field. The potential function of the total wave field in the unsaturated porous medium is expressed as

Total scalar potential functions of each phase

$$\begin{cases} \varphi^S = \varphi_1^S + \varphi_2^S + \varphi_3^S + \varphi^i \\ \varphi^W = \delta_1^W (\varphi_1^S + \varphi^i) + \delta_2^W \varphi_2^S + \delta_3^W \varphi_3^S \quad \text{(Incident P wave)} \\ \varphi^N = \delta_1^N (\varphi_1^S + \varphi^i) + \delta_2^N \varphi_2^S + \delta_3^N \varphi_3^S \end{cases} \tag{14a}$$

$$\begin{cases} \varphi^S = \varphi_1^S + \varphi_2^S + \varphi_3^S \\ \varphi^W = \delta_1^W \varphi_1^S + \delta_2^W \varphi_2^S + \delta_3^W \varphi_3^S \quad \text{(Incident SV wave)} \\ \varphi^N = \delta_1^N \varphi_1^S + \delta_2^N \varphi_2^S + \delta_3^N \varphi_3^S \end{cases} \quad (14b)$$

Vector potential functions of each phase

$$\begin{cases} \Psi^S = \Psi^S \\ \Psi^W = \zeta^W \Psi^S \quad \text{(Incident P wave)} \\ \Psi^W = \zeta^W \Psi^S \end{cases} \quad (15a)$$

$$\begin{cases} \Psi^S = \Psi^S + \Psi^i \\ \Psi^W = \zeta^W (\Psi^S + \Psi^i) \quad \text{(Incident SV wave)} \\ \Psi^N = \zeta^N (\Psi^S + \Psi^i) \end{cases} \quad (15b)$$

where δ_j^W, δ_j^N, ζ^W and ζ^N can refer to Ref [14].

Next, by solving the total wave field in the unsaturated porous medium, and then using the boundary conditions to determine the unknown coefficients in the general solution, the complete displacement and stress analytical expressions will be obtained, that is, the dynamic response of the cavity under the action of the incident wave will be obtained.

2.3 Boundary Condition

Consider the closed and impermeable boundary condition of the cavity and the continuity condition of the displacement between the fluid and the solid skeleton:

If $r = a$,

$$\begin{cases} u_r^S = u_r^W = u_r^N \\ \sigma_{rr}^S + \sigma^W + \sigma^N = 0 \\ \sigma_{r\theta} = 0 \end{cases} \quad (16)$$

After calculating the undetermined coefficients of the general solutions of the various functions from Eq. (16), the expressions of the potential functions are substituted into the constitutive relations in the cylindrical coordinate system to obtain the expressions of the stress and displacement in the soil.

$$\bar{\sigma}_{rr}^S = \left(M_{SS} \nabla^2 \varphi^S + M_{SW} \nabla^2 \varphi^W + M_{SN} \nabla^2 \varphi^N \right) + 2n_0^S \mu_S \left[\frac{\partial^2 \varphi^S}{\partial r^2} + \frac{\partial}{\partial r} \left(\frac{1}{r} \frac{\partial \Psi^S}{\partial \theta} \right) \right] \tag{17}$$

$$\bar{\sigma}_{\theta\theta}^S = \left(M_{SS} \nabla^2 \varphi^S + M_{SW} \nabla^2 \varphi^W + M_{SN} \nabla^2 \varphi^N \right) + \frac{2n_0^S \mu_S}{r} \left(\frac{\partial \varphi^S}{\partial r} + \frac{1}{r} \frac{\partial^2 \varphi^S}{\partial \theta^2} \right) - 2n_0^S \mu_S \left[\frac{\partial}{\partial r} \left(\frac{1}{r} \frac{\partial \Psi^S}{\partial \theta} \right) \right] \tag{18}$$

$$\bar{\sigma}_{r\theta}^S = 2n_0^S \mu_S \left(\frac{1}{r} \frac{\partial^2 \varphi^S}{\partial r \partial \theta} - \frac{1}{r^2} \frac{\partial \varphi^S}{\partial \theta} \right) + n_0^S \mu_S \left[\frac{1}{r^2} \frac{\partial^2 \Psi^S}{\partial \theta^2} - r \frac{\partial}{\partial r} \left(\frac{1}{r} \frac{\partial \Psi^S}{\partial r} \right) \right] \tag{19}$$

$$\bar{\sigma}^W = M_{SW} \nabla^2 \varphi^S + M_{WW} \nabla^2 \varphi^W + M_{WN} \nabla^2 \varphi^N \tag{20}$$

$$\overline{\sigma}^N = M_{SN}\nabla^2\varphi^S + M_{WN}\nabla^2\varphi^W + M_{NN}\nabla^2\varphi^N \tag{21}$$

$$u_r^\alpha = \frac{\partial \varphi^\alpha}{\partial r} + \frac{1}{r}\left(\frac{\partial \Psi^\alpha}{\partial \theta}\right) \tag{22}$$

$$u_\theta^\alpha = \frac{1}{r}\left(\frac{\partial \varphi^\alpha}{\partial \theta}\right) - \frac{\partial \Psi^\alpha}{\partial r} \tag{23}$$

where σ_{rr}^S represents the normal stress component of the solid in unsaturated porous media, $\sigma_{\theta\theta}^S$ represents the hoop stress component of the solid in unsaturated porous media, $\sigma_{r\theta}^S$ represents the tangential stress component of the solid framework in unsaturated porous media, $\overline{\sigma}^W$ represents the tangential stress component of the solid framework in unsaturated porous media, $\overline{\sigma}^N$ represents the pore gas pressure in unsaturated porous media; u_r^S represents the normal phase displacement component of the solid framework in unsaturated porous media, u_θ^S represents the tangent phase displacement component of the solid framework in unsaturated porous media.

Substituting the above displacement and stress expressions into the boundary condition Eq. (16), we get.

Incident P wave

$$\begin{bmatrix} T_1 & T_2 & T_3 & T_5 \\ U_1 & U_2 & U_3 & U_5 \\ V_1 & V_2 & V_3 & V_5 \\ Y_1 & Y_2 & Y_3 & Y_5 \end{bmatrix}\Bigg|_{r=a} \begin{bmatrix} A_{n,1} \\ A_{n,2} \\ A_{n,3} \\ B_n \end{bmatrix} = \begin{bmatrix} -T_4 \\ -U_4 \\ -V_4 \\ -Y_4 \end{bmatrix}\Bigg|_{r=a} \tag{24}$$

Incident SV wave

$$\begin{bmatrix} T_1 & T_2 & T_3 & T_4 \\ U_1 & U_2 & U_3 & U_4 \\ V_1 & V_2 & V_3 & V_4 \\ Y_1 & Y_2 & Y_3 & Y_4 \end{bmatrix}\Bigg|_{r=a} \begin{bmatrix} A_{n,1} \\ A_{n,2} \\ A_{n,3} \\ B_n \end{bmatrix} = \begin{bmatrix} -T_5 \\ -U_5 \\ -V_5 \\ -Y_5 \end{bmatrix}\Bigg|_{r=a} \tag{25}$$

The expressions of coefficients T_j, U_j, V_j and Y_j can be found in Appendix 1. By solving the above equations, the undetermined wave amplitude coefficients in Eqs. (24) and (25), that is, the undetermined coefficients in the general solution of the potential function, can be obtained to determine the scattered wave field.

3 Verification and Examples

3.1 Verification

Based on the above analysis, the stress field and displacement field scattered by the elastic wave on the surface of the cavity in the unsaturated porous medium can be obtained. From the perspective of engineering seismic resistance, the main goal of the above research is to study the dynamic response of the cavity. Since there is no calculation about the plane wave scattering by the cavity in the whole space of saturated soil, in order to discuss the validity of the method in this paper, the solution of the incident SV wave scattering by the cylindrical cavity in the unsaturated poroelastic medium obtained in this paper is degenerated to the solution of the incident SV wave scattering by a cylindrical cavity in a pure elastic medium. Set the parameter related to the permeability coefficient k in the unsaturated medium dynamic Eq. (1) $\mu^W = 0$, $\mu^N = 0$, and porosity $n_0 = 0$, and degenerate to pure elastic medium. The calculation result is obtained and compared with the reference [24] for comparison. Saturated medium parameters are as follows: Poisson's ratio $v = 0.25$, $k_4 a = 0.1$. Calculate the dynamic stress concentration factor of hoop pressure $\sigma_{\theta\theta}^S/\sigma_0$, and compared with reference. The result is shown in Fig. 2, it can be seen from the figure that the degradation result is consistent with the present solution, which can verify that the result of this paper is effective.

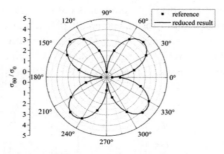

Fig. 2. Comparison of results of degenerate solution by the present method and Ref [24]

3.2 Case Analysis

Through the above research, it has been deduced that when the P(SV) wave is incident, the complete analytical expression of the displacement stress around the cylindrical cavity in the unsaturated poroelastic medium has been derived. In order to analyze the influence of saturation S_r on the dynamic response of the cavity, the parameters of unsaturated porous medium are shown in Table 1. Take the saturation $S_r = 0.2$, $S_r = 0.4$, $S_r = 0.6$, $S_r = 0.8$, $S_r = 1.0$ and, $k_4 a = 0.2$, $k_4 a = 1.0$, $k_4 a = 3.0$ respectively to analyze the influence of saturation on the circumferential stress $\sigma_{\theta\theta}^S$ and displacement u_r^S of the cavity around the cavity. When $k_4 a = 0.2, 1.0$ and 3.0, respectively indicate the incidence of low-frequency, intermediate-frequency, and high-frequency incident waves. The dynamic response around the cavity focuses on the displacement amplification factor and dynamic stress concentration factor around the cavity (DSCF, i. e. Dynamic stress concentration factor). The calculation formulas are u_r^S/u_0, $\sigma_{\theta\theta}^S/\sigma_0$, where u_0 is the displacement amplitude of the incident wave, σ_0 is the amplitude of the shear stress caused by the incident wave.

Figure 3 and Fig. 4 shows the distribution curves of dynamic stress concentration factor and displacement amplification factor around the cavity under different frequencies and different saturations when the surface of the cavity is undrained and airtight when plane P waves are incident. It can be seen from the figure that the displacement and stress

Table 1. Material parameters of unsaturated poroelastic medium

Material parameters	n_0	ρ_0^S (kg/m^3)	ρ_0^W (kg/m^3)	ρ_0^N (kg/m^3)	K_S (kPa)	K_W (kPa)	K_N (kPa)	υ^W (Pas)	υ^N (Pas)
Numerical value	0.34	2650	1000	1.1	3.6×10^7	2×10^6	110	1.0×10^{-3}	1.8×10^{-5}
Material parameters	κ (m/s)	G (kPa)	α_B	$\hat{K}$ (kPa)	υ	α	d	S_{rW}	S_{rN}
Numerical value	2.5×10^{-12}	1.01×10^6	1	2.189×10^6	0.3	2.0×10^{-5}	4.15	0.05	1.0

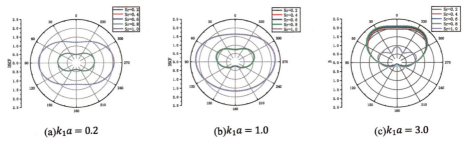

(a)$k_1 a = 0.2$ (b)$k_1 a = 1.0$ (c)$k_1 a = 3.0$

Fig. 3. The curve of DSCF around cavity in different saturation with P wave incident

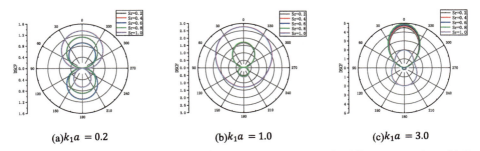

(a)$k_1 a = 0.2$ (b)$k_1 a = 1.0$ (c)$k_1 a = 3.0$

Fig. 4. The curve of amplification of displacement around cavity in different saturation with P wave incident

distribution around the cavity are symmetrically distributed on both sides of the incident wave. When the medium saturation $S_r = 1.0$, the response of the stress and displacement around the cavity is significantly different from that when the medium saturation $S_r < 1.0$, the saturation change has little effect on the distribution of stress and displacement around the cavity. It can be seen that the existence of the gas phase has a significant effect on the dynamic response of the cavity in the unsaturated medium, because the existence of the gas phase will significantly affect the wave velocity of the P wave in the unsaturated medium [25]. The distribution of stress and displacement around the cavity changes significantly with the frequency of the incident wave. At low frequency incidence ($k_1 a = 0.2$), the hoop stress appears at the minimum in the direction of the normal of the incident wave (0° and 180°), and the maximum appears in the direction perpendicular to the normal of the incident wave (90° and 270°). The displacement has a maximum value in the direction of 0°, and a minimum value in the direction perpendicular to the normal of the incident wave (90° and 270°), and increases with an increase in saturation in the directions of 0°–90° and 270°–360°. In the direction of 90°-270°, it decreases with an increase in saturation. In the case of intermediate frequency incidence ($k_1 a = 1.0$), the hoop stress has a maximum value in the direction perpendicular to the normal of the incident wave (90° and 270°), and a minimum value when $\theta = 0°$. The displacement peaks in the direction of 0°, and the value is the smallest in the incident direction (180°). Under medium frequency incidence, when the saturation $S_r < 1.0$, the stress and displacement distribution around the cavity do not change significantly with the saturation. At high frequency incidence ($k_1 a = 3.0$), the distribution of stress and displacement around the cavity at saturation $S_r = 1.0$ is significantly different from that at saturation $S_r < 1.0$. When the saturation $S_r < 1.0$, the hoop stress peaks at 30° and 330°, and the minimum value appears at 180°, and as the saturation increases, the hoop stress tends to increase. T The displacement has a minimum value in the incident direction (180°) and a maximum value away from the incident wave direction. When saturation $S_r = 1.0$, the hoop stress has a maximum value near the normal of the incident wave, and a minimum value in the incident direction of 180°. The displacement value is the smallest in the directions of 60° and 300°, and the maximum is at 180°.

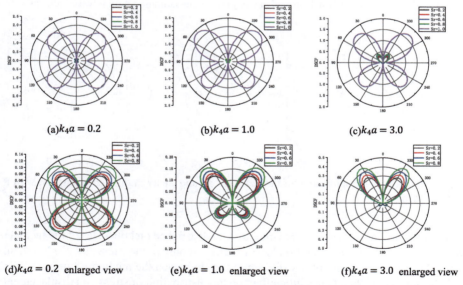

Fig. 5. The curve of DSCF around cavity in different saturation with SV wave incident

Figure 5 and Fig. 6 respectively show the distribution curves of the displacement and dynamic stress concentration factor around the cavity when the cavity surface is

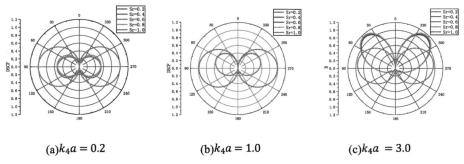

(a) $k_4a = 0.2$ (b) $k_4a = 1.0$ (c) $k_4a = 3.0$

Fig. 6. The curve of amplification of displacement around cavity in different saturation with SV wave incident

impervious to water and air under the action of the plane SV incident wave. As can be seen, some results are similar to those for incident P wave. When the saturation $S_r = 1.0$, the response of stress and displacement around the cavity is significantly different from that when the saturation $S_r < 1.0$. When the saturation $S_r < 1.0$, the saturation change has little effect on the distribution of stress and displacement around the cavity. This is because the existence of the gas phase will also have a significant impact on the velocity of the SV wave in the unsaturated medium [25]. The distribution of stress and displacement around the cavity also changes with the frequency of the incident wave. When low-frequency incident ($k_4a = 0.2$), the hoop stress has a minimum value in the directions of 0° and 180°, its value is 0, and its maximum value is in the directions of 45°, 135°, 225° and 315°. The minimum displacement occurs in the direction close to the normal of the incident wave (0° and 180°), and the maximum displacement occurs in the direction perpendicular to the normal of the incident wave (90° and 270°). In the case of intermediate frequency incidence ($k_4a = 1.0$), the hoop stress has a minimum value at 0° and 180°, its value is 0, and a peak at 45° and 315°. When the saturation $S_r < 1.0$, the displacement does not change significantly with the saturation, and there is a peak in the direction 60° and 300°. When saturation $S_r = 1.0$, the minimum displacement appears in the direction close to the normal line of the incident wave (0° and 180°), and the maximum displacement appears in the direction away from the normal line of the incident wave (90° and 270°). When high frequency is incident ($k_4a = 3.0$), the hoop stress appears minimum at 0° and 180°, its value is 0, and peaks appear at 45° and 315°. The change of displacement with saturation is not obvious. When the saturation $S_r < 1.0$, the displacement peaks in the directions of 45° and 315°, and when the saturation $S_r = 1.0$, the displacement peaks in the directions of 75° and 285°.

4 Conclusion

In this paper, based on the dynamic theory of unsaturated poroelastic media established by Wei, we use the wave function expansion method and the variable separation method to solve the analytical solution of the dynamic response of the cylindrical cavity when the plane wave is incident. By degrading the unsaturated poroelastic medium to the saturated state, and verifying with the reference, the validity of the results of this paper is proved. The results are used to solve the dynamic stress concentration factor around the cavity under undrained boundary conditions, and the influence of different saturation on the dynamic stress concentration factor is analyzed, and the following conclusions are obtained.

(1) The dynamic response of the cavity will change with the change of saturation, especially when saturation $S_r = 1.0$, the dynamic stress concentration factor of each stress and displacement will change greatly, so the influence of saturation on the dynamic response cannot be ignored. But in the unsaturated state, the saturation change has little effect on the distribution of stress and displacement around the cavity.
(2) The distribution of stress and displacement around the cavity also changes with the frequency of the incident wave. At different frequencies, the dynamic stress concentration factors of stress and displacement have different changing laws.

Appendix 1

The expressions of coefficients T_j, U_j, V_j and Y_j in Eqs. (24) and (25) are as follows:

$S_1 = M_{SS}L_n^{21} + M_{SW}\delta_1^W L_n^{21} + M_{SN}\delta_1^N L_n^{21} + 2n_0^S \mu_S N_n^{21}$; $S_2 = M_{SS}L_n^{22} + M_{SW}\delta_2^W L_n^{22} + M_{SN}\delta_2^N L_n^{22} + 2n_0^S \mu_S N_n^{22}$;

$S_3 = M_{SS}L_n^{23} + M_{SW}\delta_3^W L_n^{23} + M_{SN}\delta_3^N L_n^{23} + 2n_0^S \mu_S N_n^{23}$;

$S_4 = \varphi_0 \varepsilon_n i^n \left(M_{SS}L_n^{10} + M_{SW}\delta_1^W L_n^{10} + M_{SN}\delta_1^N L_n^{10} + 2n_0^S \mu_S N_n^{10} \right)$ (Incident P wave);

$S_4 = \frac{2n_0^S \mu_S}{r^2} nE_n^{24} - \frac{2n_0^S \mu_S}{r} nM_n^{24}$ (Incident SV wave);

$S_5 = \frac{2n_0^S \mu_S}{r} nM_n^{24} - \frac{2n_0^S \mu_S}{r^2} nE_n^{24}$ (Incident P wave); $S_5 = \Psi_0 \varepsilon_n i^n n \left(\frac{2n_0^S \mu_S}{r^2} E_n^{10} - \frac{2n_0^S \mu_S}{r} M_n^{10} \right)$ (Incident SV wave).

$W_1 = M_{SW}L_n^{21} + M_{WW}\delta_1^W L_n^{21} + M_{WN}\delta_1^N L_n^{21}$; $W_2 = M_{SW}L_n^{22} + M_{WW}\delta_2^W L_n^{22} + M_{WN}\delta_2^N L_n^{22}$;

$W_3 = M_{SW}L_n^{23} + M_{WW}\delta_3^W L_n^{23} + M_{WN}\delta_3^N L_n^{23}$; $W_4 = \varphi_0 \varepsilon_n i^n \left(M_{SW}L_n^{10} + M_{WW}\delta_1^W L_n^{10} + M_{WN}\delta_1^N L_n^{10} \right)$.

$X_1 = M_{SN}L_n^{21} + M_{WN}\delta_1^W L_n^{21} + M_{NN}\delta_1^N L_n^{21}$; $X_2 = M_{SN}L_n^{22} + M_{WN}\delta_2^W L_n^{22} + M_{NN}\delta_2^N L_n^{22}$;

$X_3 = M_{SN}L_n^{23} + M_{WN}\delta_3^W L_n^{23} + M_{NN}\delta_3^N L_n^{23}$; $X_4 = \varphi_0 \varepsilon_n i^n L_n^{10} \left(M_{SN} + M_{WN}\delta_1^W + M_{NN}\delta_1^N \right)$.

$Y_1 = \frac{2n_0^S \mu_S}{r^2} nE_n^{21} - \frac{2n_0^S \mu_S}{r} nM_n^{21}$ (Incident P wave); $Y_1 = \frac{2n_0^S \mu_S}{r} nM_n^{21} - \frac{2n_0^S \mu_S}{r^2} nE_n^{21}$ (Incident SV wave);

$Y_2 = \frac{2n_0^S \mu_S}{r^2} nE_n^{22} - \frac{2n_0^S \mu_S}{r} nM_n^{22}$ (Incident P wave); $Y_2 = \frac{2n_0^S \mu_S}{r} nM_n^{22} - \frac{2n_0^S \mu_S}{r^2} nE_n^{22}$ (Incident SV wave);

$Y_3 = \frac{2n_0^S \mu_S}{r^2} nE_n^{23} - \frac{2n_0^S \mu_S}{r} nM_n^{23}$ (Incident P wave); $Y_3 = \frac{2n_0^S \mu_S}{r} nM_n^{23} - \frac{2n_0^S \mu_S}{r^2} nE_n^{23}$ (Incident SV wave)

$Y_4 = \varphi_0 \varepsilon_n i^n \left(\frac{2n_0^S \mu_S}{r^2} n E_n^{10} - \frac{2n_0^S \mu_S}{r} n M_n^{10} \right)$ (Incident P wave); $Y_4 = -\frac{n_0^S \mu_S}{r^2} n^2 E_n^{24} - n_0^S \mu_S N_n^{24} + \frac{n_0^S \mu_S}{r} M_n^{24}$ (Incident SV wave);

$Y_5 = \frac{n_0^S \mu_S}{r} M_n^{24} - \frac{n_0^S \mu_S}{r^2} n^2 E_n^{24} - n_0^S \mu_S N_n^{24}$ (Incident P wave); $Y_5 = \Psi_0 \varepsilon_n i^n \left(-\frac{n_0^S \mu_S}{r^2} n^2 E_n^{10} - n_0^S \mu_S N_n^{10} + \frac{n_0^S \mu_S}{r} M_n^{10} \right)$ (Incident SV wave).

$Z_1 = M_{SS} L_n^{21} + M_{SW} \delta_1^W L_n^{21} + M_{SN} \delta_1^N L_n^{21} + \frac{2n_0^S \mu_S}{r} M_n^{21} - \frac{2n_0^S \mu_S}{r^2} n^2 E_n^{21}$; $Z_2 = M_{SS} L_n^{22} + M_{SW} \delta_2^W L_n^{22} + M_{SN} \delta_2^N L_n^{22} + \frac{2n_0^S \mu_S}{r} M_n^{22} - \frac{2n_0^S \mu_S}{r^2} n^2 E_n^{22}$;

$Z_3 = M_{SS} L_n^{23} + M_{SW} \delta_3^W L_n^{23} + M_{SN} \delta_3^N L_n^{23} + \frac{2n_0^S \mu_S}{r} M_n^{23} - \frac{2n_0^S \mu_S}{r^2} n^2 E_n^{23}$;

$Z_4 = \varphi_0 \varepsilon_n i^n \left(M_{SS} L_n^{10} + M_{SW} \delta_1^W L_n^{10} + M_{SN} \delta_1^N L_n^{10} + \frac{2n_0^S \mu_S}{r} M_n^{10} - \frac{2n_0^S \mu_S}{r^2} n^2 E_n^{10} \right)$
(Incident P wave);

$Z_4 = -\frac{2n_0^S \mu_S}{r^2} n E_n^{24} + \frac{2n_0^S \mu_S}{r} n M_n^{24}$ (Incident SV wave);

$Z_5 = \frac{2n_0^S \mu_S}{r^2} n E_n^{24} - \frac{2n_0^S \mu_S}{r} n M_n^{24}$ (Incident P wave); $Z_5 = \Psi_0 \varepsilon_n i^n \left(-\frac{2n_0^S \mu_S}{r^2} n E_n^{10} + \frac{2n_0^S \mu_S}{r} n M_n^{10} \right)$ (Incident SV wave).

$T_1 = (1 - \delta_1^W) M_n^{21}$, $T_2 = (1 - \delta_2^W) M_n^{22}$, $T_3 = (1 - \delta_3^W) M_n^{23}$,
$T_4 = (1 - \delta_1^W) \varphi_0 \varepsilon_n i^n M_n^{10}$ (Incident P wave), $T_4 = -\frac{1}{r} n (1 - \zeta^W) E_n^{24}$ (Incident SV wave),
$T_5 = \frac{1}{r} n (1 - \zeta^W) E_n^{24}$ (Incident P wave), $T_5 = -\frac{1}{r} n (1 - \zeta^W) \Psi_0 \varepsilon_n i^n E_n^{10}$ (Incident SV wave).

$U_1 = (1 - \delta_1^N) M_n^{21}$, $U_2 = (1 - \delta_2^N) M_n^{22}$, $U_3 = (1 - \delta_3^N) M_n^{23}$,
$U_4 = (1 - \delta_1^N) \varphi_0 \varepsilon_n i^n M_n^{10}$ (Incident P wave), $U_4 = -\frac{1}{r} n (1 - \zeta^N) E_n^{24}$ (Incident SV wave),
$U_5 = \frac{1}{r} n (1 - \zeta^N) E_n^{24}$ (Incident P wave), $U_5 = -\frac{1}{r} n (1 - \zeta^N) \Psi_0 \varepsilon_n i^n E_n^{10}$ (Incident SV wave).

$V_1 = S_1 + W_1 + X_1$, $V_2 = S_2 + W_2 + X_2$, $V_3 = S_3 + W_3 + X_3$, $V_4 = S_4 + W_4 + X_4$ (Incident P wave), $V_4 = S_4$ (Incident SV wave), $V_5 = S_5$.

With $E_n^{10} = J_n(k_0 r)$; $E_n^{21} = K_n(k_1 r)$; $E_n^{22} = K_n(k_2 r)$; $E_n^{23} = K_n(k_3 r)$; $E_n^{24} = K_n(k_4 r)$; $M_n^{ij} = \frac{dE_n^{ij}}{dr}$; $N_n^{ij} = \frac{d^2 E_n^{ij}}{dr^2}$; L_n^{ij} Represents the Laplacian ∇^2.

References

1. Biot, M.A.: Theory of propagation of elastic wave in fluid-saturated porous soil (I.lower-frequency range). J. Acoust. Soc. Am. **28**(2), 168–178 (1956)
2. Biot, M.A.: Theory of propagation of elastic wave in fluid-saturated porous soil (II.higher-frequency range). J. Acoust. Soc. Am. **28**(2), 179–191 (1956)
3. Senjuntichai, T., Rajapakse, R.K.N.D.: Transient response of a circular cavity in a poroelastic medium. Int. J. Numeri. Analyt. Geomechan. **17**, 357–383 (1993)
4. Yayuan, H., Lizhong, W., Chen Yunmin, W., Shiming, Z.Z.: Scattering and refraction of plane strain waves on a cylinder in saturated soil. Act a Seism Ologica Sin Ica. **20**(3), 300–307 (1998). (In Chinese)

5. Ganbin, X.K., Shi, Z.: Dynamic response of partially sealed circular tunnel in viscoelastic saturated soil. Acta Mechanica Solida Sinica. **25**(3):287–290 (2004). (In Chinese)
6. Liu, G., Dun, Z., Xie, K., Shi, Z.: Steady state response of a partially sealed circular tunnel in viscoelastic saturated soil. China Railway science **25**(5): 78–83 (2004). (In Chinese)
7. Liu, G., Xie, K., Shi, Z.: Frequency response of a cylindrical cavity in poro-viscoelastic saturated medium. Acta Mechanica Sinica. **36**(5), 557–563 (2004). (In Chinese)
8. Ganbin, L., Kanghe, X., Zuyuan, S.: Dynamic interacion of circular tunnel in viscoelastic saturated soil. J. Zhejiang University (Engineering Science). **39**(10), 194–201 (2005). (In Chinese)
9. Li, W., Zhao, C.: An analytical solution for diffraction of plane P-waves by cylindrical cavity in a fluid-saturated porous media semi-space. Rock and Soil Mechanics. (12), 1867–1872 (2004). (In Chinese)
10. Li, W., Zhao, C.: An analytical solution for the scattering of plane SV-waves around cylindrical cavity in a fluid-saturated porous media half space **28**(6), 1–7 (2008). (In Chinese)
11. Zhou, X., Zhou, G., Wang, J.: Scattering of elastic wave by circular cavity with lining in saturated soil. Chin. J. Rock Mechan. Eng. **24**(9), 1572–1576 (2005). (In Chinese)
12. Li, W.H., Zhang, Z.: Scattering of transient plane waves by deep buried cylindrical lining cavity in saturated soil. Chinese J. Geophys. **56**(1), 325–334 (2013). (In Chinese)
13. Song, J., Song, R.: Dynamic response of deep circula composite-lined tunnel in saturated soil subjected to incident SV waves. Journal of Vibaration Engineering. **33**(02), 383–390 (2020)
14. Xu, M.: Investigation on Dynamic Response of Unsaturated Soils and Foundation. South China University of Technology (2010). (In Chinese)
15. Liu, H., Liu, J., Wen, M.: Dynamic characteristics of an unsaturated soil and lining structure system with a deeply embedded circular tunnel. J. Vibr. Shock **35**(13), 36–41 (2016). (In Chinese)
16. Li, W., Zheng, J.: Effects of saturation on free-field response of site due to plane P-wave incidence. Chinese J. Geotechinal Eng. **39**(3):427–435 (2017). (In Chinese)
17. Wei, C., Muraleetharan, K.K.: A continuum theory of porous media saturated by multiple immiscible fluids. I. Linearporoelasticity. Int. J. Eng. Sci. **40**, 1807–1833 (2002)
18. Wei, C., Muraleetharan, K.K.: Acoustical waves in unsaturated porous media. School of Civil Engineering and Environmental Science, 73019–1024. University of Oklahoma, 202 W. Boyd St., Room 334, Norman, Oklahoma (2002)
19. Li, W., Hu, Y., Zhao, C., Zheng, J.: Analytic solution for wave propagations in layered unsaturated soil and its application. Chinese J. Geotechn. Eng. **40**(10), 1790–1798 (2017). (In Chinese)
20. Li, W.-H., Zheng, J., Trifunac, M.D.: Saturation effects on ground motion of unsaturated soil layer - bedrock system excited by plane P and SV waves. Soil Dyn. Earthq. Eng. **37**(3), 482–492 (2018)
21. Zhao, X.: Transient Dynamic Response of Deep Buried Cylindrical Cavity in Unsaturated Porous Media. Beijing Jiao-tong University (2016)
22. Li, W.-H., Feng, F.-C., Xu, Z.: Analytical solution for the transient response of cylindrical cavity in unsaturated porous medium. In: Proceedings of China-Europe Conference on Geotechnical Engineering. Springer Series in Geomechanics and Geoengineering. Springer, Cham (2018). https://doi.org/10.1007/978-3-319-97112-4
23. Tan, Y., Yang, M., Li, X.: Dynamic response of a circular lined tunnel with an imperfect interface embedded in the unsaturated poroelastic medium under P wave. Computers and Geotechnics **122** (2020)
24. Pao, Y.H., Mow, C.C.: Diffraction of Elastic Waves and Dynamic Stress Concent- ration [M], pp. 239–274. Crane, Russak and Company, New York (1973)
25. Xu, C., Shi, Y.: Characteristics of wave propagation in unsaturated soils. Rock and Soil Mechanics. **03**, 354–358 (2004). (In Chinese)

Dynamic Soil-Structure Interaction for a SDOF with Rigid Foundation Embedded in a Radially Inhomogeneous Bearing Stratum Under SH Waves

Ning Zhang[1(✉)], Xinyu Sun[2], Haijun Lu[1], and Denghui Dai[1]

[1] Hohai University, Nanjing, Jiangsu, China
`240401zn@163.com`
[2] China Design Group Co., Ltd., Nanjing, Jiangsu, China

Abstract. Dynamic soil-structure interaction will occur when the seismic wave passes through a soil-structure system. It involves the scattering of incident wave by the foundation of superstructure, the transfer of incident wave energy to the superstructure and the radiation of structural vibration energy back into the soil. In this kind of wave process, the local soil stratum condition of the bearing stratum has an important role, especially for those engineering sites with either sedimentation- or load-induced inhomogeneous shear modulus profile. In this study, an analytical model of dynamic interaction between inhomogeneous bearing stratum and superstructure in bedrock half space under plane SH wave is established. The shear modulus of the inhomogeneous bearing stratum is assumed to vary with radius in a power-law manner, and the superstructure adopts the classical single degree of freedom (SDOF) oscillator model. By using a wave function expansion method, an analytical series solution for the foundation displacement and relative displacement of superstructure is derived. The influence of the inhomogeneous bearing stratum on the system response is studied. When the inhomogeneous bearing stratum is more flexible than the bedrock half space, the peak value of foundation displacement is obviously larger than that without considering the inhomogeneous bearing stratum. With the decrease of the stiffness of the superstructure, the peak frequency of the system response shifts to the low frequency.

Keywords: Dynamic soil-structure interaction · Wave propagation and scattering · Inhomogeneous bearing stratum · Oscillator · SH waves

1 Introduction

The dynamic soil-structure interaction (SSI) has always been a key research topic in the field of earthquake engineering. Luco (1969) and Trifunac (1972) proposed classical analytical models of SSI by using a semi-circular rigid foundation to connect the superstructure and bedrock half space. Afterwards, many scholars extended it to more general situations. For example, Liang et al. (2016), Jin and Liang (2018) and Jin & Liang (2021) focused on the effect of foundation flexibility. Recently, some scholars pointed out that there is a disturbed zone around the foundation, which has a certain impact on the dynamic SSI (Pak and Ashlock 2011; Jiang and Ashlock 2020). Considering the effect of disturbed zone near foundation, this study focuses on the dynamic SSI for a SDOF oscillator with rigid foundation embedded in a radially inhomogeneous bearing stratum under SH waves.

2 Methodology

2.1 Model

The two-dimensional model in this paper is shown in Fig. 1. The SDOF oscillator is supported by an infinite semi cylindrical rigid foundation (radius a), which is embedded in an infinite semi-circular annular inhomogeneous bearing stratum (maximum radius b). The contact interfaces among each part are assumed to be perfectly bonded. The undamped natural frequency of the SDOF oscillator is ω_s, and its mass per unit length in the z direction is M_s. The mass per unit length of the rigid foundation and the bearing stratum in the z direction are M_f and M_a, respectively with their mass densities being ρ_f and ρ_a, respectively. The half space medium is assumed to be elastic, isotropic and homogeneous with constant mass density ρ_b, shear modulus μ_b, and shear wave velocity β_b. The center point O of the rigid foundation is set as the origin of both Cartesian coordinate system (x, y) and polar coordinate system (r, θ). The angle θ is measured from the horizontal x-axis clockwise towards the y-axis. The excitation is plane SH waves with circular frequency ω and incident angle γ.

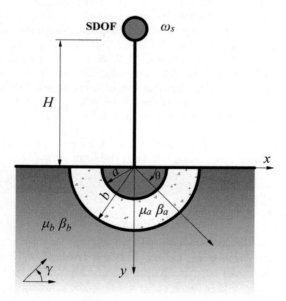

Fig. 1. Model of inhomogeneous bearing stratum-rigid foundation-SDOF oscillator system in half space

Assuming that the shear wave velocity of the inhomogeneous bearing stratum has a power-law variation with the radial depth r, it can be expressed as

$$\beta_a(r) = \beta_b \left(\frac{r}{b}\right)^\xi, \quad a \leq r \leq b \tag{1}$$

where ξ indicates the degree of inhomogeneity of the bearing stratum, and β_b is the shear wave velocity of the half space. For brevity, the shear wave velocity on the lower surface of the bearing stratum is assumed to equal to that in the underlying half space, i.e. $\beta_a(b) = \beta_b$.

2.2 Boundary Conditions

The wave fields in inhomogeneous bearing stratum and half space are represented by v^a and v^b, respectively. v^a and v^b must satisfy the zero stress condition on the free surface, and the continuity of displacements and stresses on their contact interface ($r = b$)

$$\sigma_{\theta z}^b = \frac{\mu_b}{r} \frac{\partial v^b(r, \theta)}{\partial \theta} = 0, \quad \theta = 0, \pi, \quad r \geq b \tag{2}$$

$$\sigma_{\theta z}^a = \frac{\mu_a(r)}{r} \frac{\partial v^a(r, \theta)}{\partial \theta} = 0, \quad \theta = 0, \pi, \quad a \leq r \leq b \tag{3}$$

$$v^b(r, \theta) = v^a(r, \theta), \quad r = b, \ 0 \leq \theta \leq \pi \tag{4}$$

$$\mu_b \frac{\partial v^b(r, \theta)}{\partial r} = \mu_a(b) \frac{\partial v^a(r, \theta)}{\partial r}, \quad r = b, \ 0 \leq \theta \leq \pi \tag{5}$$

where $\mu_a(r) = \rho_a \beta_a^2(r)$ is the varying shear modulus in the inhomogeneous bearing stratum.

In addition, the contact surface ($r = a$) between inhomogeneous bearing stratum and rigid foundation must satisfy the displacement continuity condition:

$$v^a(r, \theta) = \Delta, \quad r = a, \ 0 \leq \theta \leq \pi \tag{6}$$

where Δ represents the displacement amplitude of the rigid foundation to be determined. For the anti-plane steady-state problem studied herein, the time factor $e^{-i\omega t}$ is omitted here and after.

2.3 Wave Field in the Half Space

The total wave field v^b in the half space is composed of the free wave field v^{i+r} and the scattered wave field v^s due to the existence of the soil-structure system. It is noted as

$$v^b(r, \theta) = v^{i+r}(r, \theta) + v^s(r, \theta), \quad r \geq b, \ 0 \leq \theta \leq \pi \tag{7}$$

where v^{i+r} includes the incident wave v^i and its reflected wave v^r by the horizontal surface. The free field motion, v^{i+r}, expanded in cylindrical wave function, has the form

$$v^{i+r}(r, \theta) = \sum_{n=0}^{\infty} p_n J_n(k_b r) \cos(n\theta) \tag{8}$$

where $k_b = \omega/\beta_b$, and $J_n(x)$ is the Bessel function of the first kind, with argument x and order n. Here, $p_n = 2\varepsilon_n i^n \cos(n\gamma)$ is the free field coefficient, and ε_n is *Neumann* factor ($\varepsilon_0 = 1; \varepsilon_n = 2, n \geq 1$).

The scattered wave field v^s can be expanded as

$$v^s(r,\theta) = \sum_{n=0}^{\infty} A_n H_n^{(1)}(k_b r) \cos(n\theta) \qquad (9)$$

where A_n is the unknown complex coefficient of the scattered wave field. $H_n^{(1)}(x)$ is the Hankel function of the first kind with order n, which represents the outward traveling wave towards infinity and satisfies the *Sommerfeld* radiation condition at infinity.

In this way, the half space wave field automatically satisfies the *Helmholtz* equation and the surface stress free condition.

2.4 Wave Field in the Inhomogeneous Bearing Stratum

Following Zhang et al. (2017, 2019), the wave field of the inhomogeneous bearing stratum satisfying the zero stress condition (3) for this study is constructed as

$$v^a(r,\theta) = \sum_{n=0}^{\infty} [B_n U_n(r) + C_n V_n(r)] \cos(n\theta), \quad a \leq r \leq b, 0 \leq \theta \leq \pi \qquad (10)$$

where B_n and C_n are the unknown complex coefficients. The radial wave functions $U_n(r)$ and $V_n(r)$ are defined as follows

$$U_n(r) = \begin{cases} r^{-\xi} J_{\frac{\sqrt{\xi^2+n^2}}{1-\xi}}(\frac{k_0 a^\xi}{1-\xi} r^{1-\xi}), & \xi \neq 1 \\ r^{-1} r^{\sqrt{1-(k_0^2 a^2 - n^2)}}, & \xi = 1, k_0^2 a^2 - n^2 < 1 \\ r^{-1}, & \xi = 1, k_0^2 a^2 - n^2 = 1 \\ r^{-1} \cos(\sqrt{k_0^2 a^2 - n^2 - 1} \ln r), & \xi = 1, k_0^2 a^2 - n^2 > 1 \end{cases} \qquad (11)$$

$$V_n(r) = \begin{cases} r^{-\xi} Y_{\frac{\sqrt{\xi^2+n^2}}{1-\xi}}(\frac{k_0 a^\xi}{1-\xi} r^{1-\xi}), & \xi \neq 1 \\ r^{-1} r^{-\sqrt{1-(k_0^2 a^2 - n^2)}}, & \xi = 1, k_0^2 a^2 - n^2 < 1 \\ r^{-1} \ln r, & \xi = 1, k_0^2 a^2 - n^2 = 1 \\ r^{-1} \sin(\sqrt{k_0^2 a^2 - n^2 - 1} \ln r), & \xi = 1, k_0^2 a^2 - n^2 > 1 \end{cases} \qquad (12)$$

where Y_n is Bessel function of the second kind with order n.

The force f_z^a acting on the base of rigid foundation per unit length by the inhomogeneous bearing stratum can be expressed as

$$f_z^a = -\mu_0 \pi a [B_0 U_0'(a) + C_0 V_0'(a)] \qquad (13)$$

where the superscript $'$ represents the differential of the corresponding function, and μ_0 is the shear modulus at the upper surface of the bearing stratum ($r = a$).

2.5 Force from the SDOF Oscillator Acting on the Foundation

Assuming that Δ_s is the relative displacement of the oscillator with respect to the foundation with natural frequency ω_s. According to the dynamic equilibrium equation of the oscillator, the relative displacement of the oscillator can be deduced as

$$\Delta_s = \frac{\omega^2}{\omega_s^2 - \omega^2} \Delta \tag{14}$$

Furthermore, the force exerted on the foundation by the upper undamped SDOF oscillator system can be expressed as

$$f_z^s = -\omega^2 M_s (\Delta_s + \Delta) \tag{15}$$

2.6 Determination of Unknown Coefficients of Wave Fields

According to Luco (1969) and Trifunac (1972), the steady-state motion equation of the foundation displacement Δ is

$$-\omega^2 M_f \Delta = -(f_z^a + f_z^s) \tag{16}$$

By substituting the forces f_z^a and f_z^s into Eq. (16) and satisfying the boundary conditions (4), (5) and (6), all unknown coefficients of the above wave fields can be expressed explicitly as follows:

$$A_n = \left\{ \frac{2i/(\pi k_b b)}{H_n^{(1)'}(k_b b) H_n^{(1)}(k_b b) - [H_n^{(1)}(k_b b)]^2 \mu_a(b) F_n/(\mu_b k_b E_n)} - \frac{J_n(k_b b)}{H_n^{(1)}(k_b b)} \right\} p_n, \quad n = 0, 1, 2\ldots \tag{17}$$

$$B_n = \frac{2i[\delta_{0n}\Delta_0 - V_n(a)]/\pi k_b b}{E_n H_n^{(1)'}(k_b b) - F_n H_n^{(1)}(k_b b) \mu_a(b)/(\mu_b k_b)} p_n, \quad n = 0, 1, 2, \ldots \tag{18}$$

$$C_n = \frac{2i U_n(a)/\pi k_b b}{E_n H_n^{(1)'}(k_b b) - F_n H_n^{(1)}(k_b b) \mu_a(b)/(\mu_b k_b)} p_n, \quad n = 0, 1, 2, \ldots \tag{19}$$

where $\delta_{0n} = 1$ for $n = 0$ and $\delta_{0n} = 0$ for $n \geq 1$. E_n, F_n and Δ_0 are defined as

$$E_n = U_n(a) V_n(b) - U_n(b) V_n(a) + \delta_{0n} \Delta_0 U_n(b) \tag{20}$$

$$F_n = U_n(a) V_n'(b) - U_n'(b) V_n(a) + \delta_{0n} \Delta_0 U_n'(b) \tag{21}$$

$$\Delta_0 = \frac{U_0'(a) V_0(a) - U_0(a) V_0'(a)}{\frac{k_0^2 a}{2} U_0(a)[\frac{M_s}{M_a} T(\omega) + \frac{M_f}{M_a}] + U_0'(a)} \tag{22}$$

Then, the displacement response of rigid foundation Δ can be written as

$$\Delta = \frac{2i U_0(a)/\pi k_b b}{E_0 H_0^{(1)'}(k_b b) - F_0 H_0^{(1)}(k_b b) \mu_a(b)/(\mu_b k_b)} p_0 \Delta_0 \tag{23}$$

At this point, all wave field coefficients are obtained, and therefore the displacement response of any position in the entire system can be calculated.

2.7 Dimensionless Parameters

Similar to previous studies, the dimensionless frequency $\eta = \omega a/\beta_b$, the mass ratios M_b/M_a for the half space, M_f/M_a for the rigid foundation and M_s/M_a for the oscillator are adopted. Note that M_a is the mass of the bearing stratum replaced by the foundation and M_b is the mass of the soil in the half space replaced by the foundation. Since the effects of the mass ratios have been studied previously, all three mass ratios are set to be 1 herein. It is necessary to further define the dimensionless parameter $\eta_s = \omega_s a/\pi \beta_b$ to describe the stiffness of the oscillator. Three values of η_s are adopted. $\eta_s = 1/6$ corresponds to a SDOF oscillator system with lower stiffness, $\eta_s = 1/3$ corresponds to a stiffer system, and $\eta_s = \infty$ (100 in calculation) is for a rigid one.

3 Numerical Results and Analysis

To show the influence of the degree of inhomogeneity of the bearing stratum, the foundation displacement and the relative displacement of the superstructure are both normalized by the surface displacement of the free field as $|\overline{\Delta}| = |\Delta|/|v^{i+r}(0,0)|$ and $|\overline{\Delta}_s| = |\Delta_s|/|v^{i+r}(0,0)|$, respectively. The inhomogeneity coefficient ξ takes the values $-1/2$, 0, 1/2 and 1, respectively. As shown in Fig. 2, the wave velocity of the bearing stratum varies with depth as $\beta_a(r) = \beta_b(\frac{r}{b})^\xi$. Note that $\xi = 0$ corresponds to the case of a homogeneous bearing stratum, $\xi < 0$ corresponds to the case of a stiffer inhomogeneous bearing stratum relative to the homogeneous one and $\xi > 0$ is for a softer one. The size of the bearing stratum is set to be $b/a = 2$.

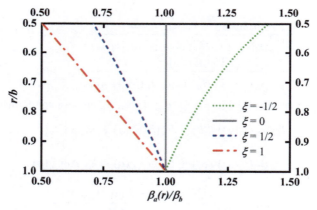

Fig. 2. Variations of the wave velocity of the bearing stratum with depth under different ξ

Figure 3 shows the normalized foundation displacements varying with the dimensionless frequency η under different η_s. Different curves in each figure correspond to different inhomogeneity coefficients ξ. Among them, the solid line corresponds to $\xi = 0$, that is, the case where the bearing stratum is homogeneous. For $\xi < 0$, the foundation displacement is not obviously changed in a low frequency range, but decreases significantly as $\eta > 0.4$. For $\xi > 0$, there is an obvious amplification effect ($|\overline{\Delta}| > 1$)

in the low frequency range of $\eta < 0.6$. This effect becomes stronger with the increase of ξ from 1/2 to 1. When $\eta > 2$, the foundation displacements for different ξ tend to decrease and be close. Take Fig. 3(a) as an example, when ξ is taken as 0, 1/2, and 1, the peak values of the foundation displacement are 1.032, 1.126, and 1.418, respectively, showing a gradually increasing manner. In addition, as the inhomogeneity degree of the bearing stratum increases, the dimensionless frequency corresponding to the peak of the foundation displacement tends to be lower. This implies that the less stiffness of the bearing stratum makes the flexibility of the whole system increases and therefore the resonance frequency decreases. Another phenomenon is that the peak of foundation displacement is less affected by the stiffness of the superstructure. For example, when $\xi = 1$ and η_s takes 1/6, 1/3 and ∞, respectively, the peak of foundation displacement is 1.558, 1.451 and 1.417, respectively. These values are not much different.

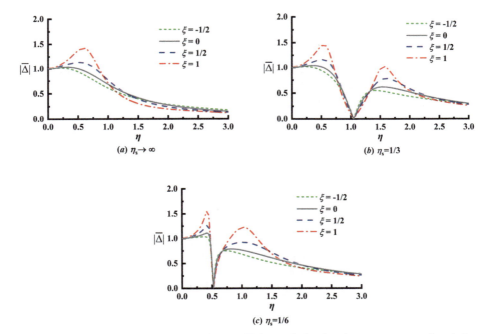

Fig. 3. The influence of inhomogeneity coefficient of the bearing stratum on foundation displacement

Figure 4 shows the normalized relative displacement of the superstructure with the dimensionless frequency for $\eta_s = 1/3$ and 1/6 (when η_s approaches ∞, the relative displacement of the superstructure is 0). Different curves in each figure correspond to different inhomogeneity coefficients ξ. When the stiffness of the superstructure is relatively large ($\eta_s = 1/3$), the peak values of relative displacements of the superstructure are between 1.47 and 1.90, while they increase to values between 3.65 and 5.22 for $\eta_s = 1/6$. This shows that the stiffness of the superstructure can affect its peak relative displacement.

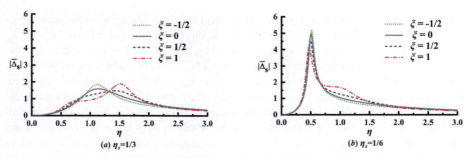

Fig. 4. The influence of the inhomogeneity coefficient of the bearing stratum on the relative displacement of the superstructure

4 Conclusions

In this paper, an analytical solution to dynamic interaction between inhomogeneous bearing stratum and a SDOF oscillator through a semi-circular rigid foundation under plane SH wave is proposed. It is found that an inhomogeneous bearing stratum may amplify the dynamic response of the foundation and the superstructure, especially for the bearing stratum with a low shear velocity.

References

Luco, J.E.: Dynamic interaction of a shear wall with the soil. J. Eng. Mech. Div. Am. Soc. Civil. Eng. **95**, 333–346 (1969)

Trifunac, M.D.: Interaction of a shear wall with the soil for Incident Plane SH-waves. Bull. Seismol. Soc. Am. **62**(1), 63–68 (1972)

Liang, J., Jin, L., Todorovska, M.I., Trifunac, M.D.: Soil-structure interaction for a SDOF oscillator supported by a flexible foundation embedded in a half-space: closed-form solution for incident plane SH-waves. Soil Dyn. Earthq. Eng. **90**, 287–298 (2016)

Jin, L., Liang, J.: The effect of foundation flexibility variation on system response of dynamic soil-structure interaction: an analytical solution. Bull. Earthq. Eng. **16**, 113–127 (2018)

Jin, L., Liang, J.: Dynamic soil-structure interaction with a flexible foundation embedded in a half-space: closed-form analytical solution for incident plane SH waves. J. Earthquake Eng. **25**(8), 1565–1589 (2021)

Pak, R., Ashlock, J.C.: A fundamental dual-zone continuum theory for dynamic soil-structure interaction. Earthquake Eng. Struct. Dynam. **40**(9), 1011–1025 (2011)

Jiang, Z., Ashlock, J.C.: Computational simulation of three-dimensional dynamic soil-pile group interaction in stratumed soils using disturbed-zone model. Soil Dyn. Earthq. Eng. **130**, 105928 (2020)

Zhang, N., Gao, Y., Pak, R.: Soil and topographic effects on ground motion of a surficially inhomogeneous semi-cylindrical canyon under oblique incident SH waves. Soil Dyn. Earthq. Eng. **95**, 17–28 (2017)

Zhang, N., Zhang, Y., Gao, Y., Pak, R., Wu, Y., Zhang, F.: An exact solution for SH-wave scattering by a radially multistratumed inhomogeneous semicylindrical canyon. Geophys. J. Int. **217**, 1232–1260 (2019)

Ground Motion Amplification by a Rectangular Tunnel in a Saturated Poroelastic Half-Space

Jun Zhu[1], Xiaojun Li[1(✉)], Jianwen Liang[2], and Mianshui Rong[1]

[1] Beijing Key Laboratory of Earthquake Engineering and Structural Retrofit, Beijing University of Technology, Beijing 100124, China
beerli@vip.sina.com
[2] Department of Civil Engineering, Tianjin University, Tianjin 300354, China

Abstract. The ground motion around underground tunnels can vary significantly from the free-field site response due to wave scattering, thus influencing the seismic demand for the design of aboveground structures in the vicinity. This paper numerically investigates the effects of a rectangular tunnel on the ground motion of a saturated poroelastic half-space for obliquely incident seismic waves. The results show that the ground surface accelerations can be significantly modified by the rectangular tunnel, while the modification is a function of the soil property. It is shown that the differences in the ground motion amplification between the saturate and dry poroelastic half-spaces are noticeable, both in terms of the magnitude and period contents of the ground surface accelerations, and moreover, the saturation degree is a critical factor that can influence the ground motion amplification to a great degree.

Keywords: Ground surface accelerations · Underground tunnels · Saturated soil · Obliquely incident seismic waves

1 Introduction

The presence of underground tunnels may modify the input motion for the seismic design of aboveground structures [1, 2]. The ground motion amplification/de-amplification induced by the wave scattering of the underground tunnels has been recognized and extensively studied by theoretical analysis and experimental measurements [e.g., 3–8], but this issue is not thoroughly addressed given the large number of factors involved. One of the critical factors is the soil property, while the majority of the previous studies consider tunnels embedded in dry soils (single-phase media). However, the soil is naturally porous and usually saturated under the groundwater table, which can be regarded as a two-phase medium consisting of the solid skeleton and pore water. Due to the presence of the pore water, the dynamic characteristic of the saturated soil can significantly differ from that of the dry soil [9–11], thus further influencing the ground motion above the underground tunnels.

This paper presents the preliminary results of our study on the possible effects of underground tunnels on the ground motion. We employ a hypothetical model of a rectangular tunnel embedded in a saturated poroelastic half-space. The seismic waves are obliquely incident from the elastic bedrock and not perpendicular to the longitudinal axis of the tunnel, and therefore, the wave scattering of the tunnel structure has a three-dimensional feature. The ground surface accelerations of the saturated poroelastic half-space are compared with those of a dry poroelastic half-space to investigate the influence of the pore water. Moreover, the results for different saturation degree values are presented to study the effects of the partial saturation on the ground motion amplification. The findings can be helpful for the improvement of the seismic design of aboveground structures which relies on proper input motions that account for the below-ground influences.

2 Methodology

The ground motion above an underground tunnel in a poroelastic half-space for obliquely incident seismic waves is obtained by a 2.5D finite element and boundary element coupling method in the frequency domain. The fast Fourier transform technique is used for the transform between the time and frequency domains. The viscoelastic material behavior of the soil and tunnel structure is taken into account by replacing the Lame constants with their complex counterparts. The longitudinal axis of the tunnel is parallel to the y-axis, and it is assumed that the soil-tunnel system is infinite and invariant in the y-direction. By the Fourier transform of the y coordinate to the wavenumber, the original 3D dynamic response of the soil-tunnel system can be obtained via a 2D mesh with different wavenumbers.

The tunnel structure is modelled by 2.5D finite elements which have four nodes with three degrees of freedom per node, i.e., the translational displacements in the x-, y- and z-directions. The saturated poroelastic half-space is modelled as a two-phase medium based on the Biot's theory [12, 13]. Along the soil-tunnel interface, two-node boundary elements apply, and the wave scattering in the saturated poroelastic half-space is solved by a boundary element method based on the Green's functions for moving line loads, and hence the radiation condition at infinity can be exactly satisfied. The finite element method and boundary element method are coupled through the displacement compatibility and traction equilibrium conditions along the soil-tunnel interface. Further details about this 2.5D finite element and boundary element coupling method can be found in [14, 15].

3 Results and Analysis

Consider a rectangular tunnel embedded in a saturated poroelastic half-space. The rectangular tunnel is 26.5 m (wide) × 8 m (high), with an overburden $H = 8$ m. The thickness of the top slab and center wall of the rectangular tunnel is 0.8 m, the thickness of the bottom slab and side wall is 1.0 and 1.1 m, respectively. The tunnel structure has the elastic modulus $E_t = 34.5$ GPa, mass density $\rho_t = 2500$ kg/m^3, Poisson's ratio $\upsilon_t = 0.2$ and damping ratio $\zeta_t = 0.02$. The soil-tunnel interface is tied and impermeable, and thus the separation and slippage along the interface are not allowed. The saturated poroelastic half-space is assumed to be single-layered, and the saturated soil layer overlaying the bedrock has a thickness of 29 m, which is characterized by the shear modulus $\mu = 111$ MPa, mass densities of the soil grain and pore water, $\rho_s = 2600$ kg/m^3 and $\rho_f = 1000$ kg/m^3, porosity $\varphi = 0.35$, Poisson's ratio $\upsilon_s = 0.25$, damping ratio $\zeta_s = 0.02$, and the coefficients of the Biot's model, $m = 5510$ kg/m^3, $b = 1.0 \times 10^7$ N·s/m^4, $M = 5650$ MPa, and $\alpha = 0.995$. The bedrock is a single-phase medium which has the shear modulus equal to 1303 MPa, mass density equal to 2200 kg/m^3, Poisson's ratio equal to 0.25, and damping ratio equal to 0.02.

It is assumed that the soil-tunnel interaction system behaves in a linear elastic manner. Only the seismic loading is considered in the analysis, while the static loads including the self-weight of the surrounding soil and tunnel structure, as well as the excavation process, are not considered. The scaled El Centro record is employed as the incident SV-waves from the elastic bedrock, as presented in Fig. 1. The seismic waves are obliquely incident with the horizontal incident angle $\theta_h = 45°$ and vertical incident angle $\theta_v = 45°$; the horizontal incident angle θ_h corresponds to the angle between the y-axis and the projection of the propagation direction of the incident waves in the x-o-y plane, and the vertical incident angle θ_v corresponds to the angle between the z-axis and the propagation direction of the incident waves (see Fig. 2).

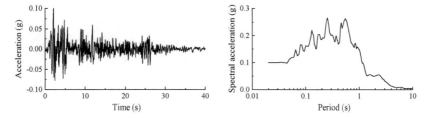

Fig. 1. The input motion from the bedrock.

3.1 General Patterns of the Ground Motion Amplification

Figure 2 shows the effects of the rectangular tunnel on the ground surface accelerations, in which $\ddot{u}_x/\ddot{u}_x^f$ and $\ddot{u}_y/\ddot{u}_y^f$ denote the ratios of the two horizontal peak ground accelerations above the tunnel in the x- and y-directions, respectively, to those of the free field, $\ddot{u}_z/\ddot{u}_z^f$ denotes the ratio of the vertical peak ground acceleration above the tunnel to that of the free field. For this example, the tunnel tends to significantly de-amplify the two horizontal accelerations right above it, with the minimum values of $\ddot{u}_x/\ddot{u}_x^f$ and $\ddot{u}_y/\ddot{u}_y^f$ being 0.8 and 0.5 which appear at $x/a = 2$ and 1, respectively. Slight amplification of the two horizontal accelerations appears at a distance from the tunnel, and the maximum values of $\ddot{u}_x/\ddot{u}_x^f$ and $\ddot{u}_y/\ddot{u}_y^f$ are 1.1 and 1.2, respectively, at $x/a = 4$ and 2. As for the vertical acceleration ratio $\ddot{u}_z/\ddot{u}_z^f$, it is close to unity in the range of $-2 < x/a < 2$, and its maximum and minimum values are 1.1 and 0.95, respectively, appearing around $x/a = 7$ and 6.

The ratios of the peak ground accelerations for a dry poroelastic half-space are also presented in Fig. 2 for comparison. The soil parameters of the dry poroelastic half-space are the same as those of the saturated poroelastic half-space, while the coefficients that related to the pore water (ρ_f, α, M, m and b) are set to be zero. Pronounced differences in the ratios of the peak ground accelerations between the saturated and dry poroelastic half-space cases can be found, highlighting the important role of the pore water. Moreover, the maximum amplification ratios for the saturated poroelastic half-space case are smaller than those for the dry poroelastic half-space case, and this may be due to the low compressibility of the pore water which results in the smaller ground motion amplification.

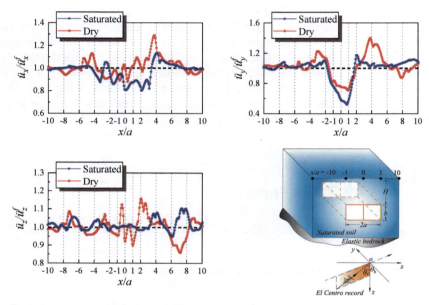

Fig. 2. Ratios of the peak ground accelerations above the rectangular tunnel to those of the free field for the saturated and dry poroelastic half-space cases.

Figure 3 compares the spectral accelerations above the tunnel with those of the free field for the saturated poroelastic half-space case, and in each plot, the grey lines represent the response spectra with damping 5% for the different locations above the tunnel, and the red line represents the response spectra of the free field. It can be seen that the effects of the underground tunnel on the spectral accelerations are limited in the period range of 1 s and are more pronounced in the shorter period range. Compared with the free-field scenario, the modification of the horizontal spectral accelerations by the presence of the tunnel is more significant in the y-direction (parallel to the tunnel longitudinal axis) than in the x-direction (perpendicular to the tunnel longitudinal axis), and the horizontal spectral acceleration Sa-y can be considerably amplified or de-amplified in the period range of 0.02 to 0.6 s. As for the vertical spectral acceleration Sa-z, large amplification can be found in the period range of 0.02 to 0.4 s.

Figure 4 shows the same type of results as Fig. 3, but for the dry poroelastic half-space. Consistent with the saturated poroelastic half-space case, the effects of the tunnel on the spectral accelerations for the dry poroelastic half-space case are also limited in the period range of 1 s. However, the period contents of the ground accelerations are very different for the two cases. It seems that the saturated soil has a tendency to weaken the shorter period contents while enhance the longer period contents. For the saturated poroelastic half-space case, the two horizontal spectral accelerations, Sa-x and Sa-y, both have large peaks around the period 0.5 s, and the largest peak of the vertical spectral acceleration Sa-z appears around the period 0.6 s, while for the dry poroelastic half-space case, the spectral accelerations in the three directions tend to be larger around the period 0.25 s.

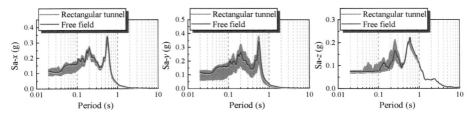

Fig. 3. Comparison of the spectral accelerations above the rectangular tunnel with those of the free field for the saturated poroelastic half-space case.

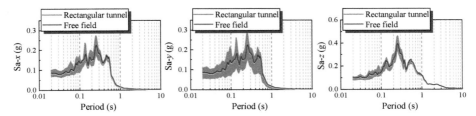

Fig. 4. Comparison of the spectral accelerations above the rectangular tunnel with those of the free field for the dry poroelastic half-space case.

3.2 Effects of Partial Saturation

The soil may not always be fully saturated due to the varying groundwater table, and in particular, partial saturation is common for offshore soils. The partial saturation effect on the ground motion amplification is investigated by a modified two-phase model, considering the air-water mixture as homogeneous pore fluid. The soil parameters of the partial saturation scenario are the same as those of the full saturation scenario in Sect. 3.1, while the bulk modulus of the pore water is modified according to the saturation degree of the soil to account for the air phase [11].

Figure 5 shows the ratios of the peak ground accelerations for different saturation degree values, and the high saturated degree case of S_r = 99.9%, 99%, 98% and 95% is considered such that the modified two-phase model can be justified [9, 11]. It can be seen that even a slight reduction in the saturation degree can lead to significant changes in the amplification ratios, both in terms of the magnitude and distribution. The partial saturation scenario can result in considerably larger amplification ratios than the full saturation scenario, which may be attributed to the increasing compressibility of the air-water mixture with the reduction of the saturation degree.

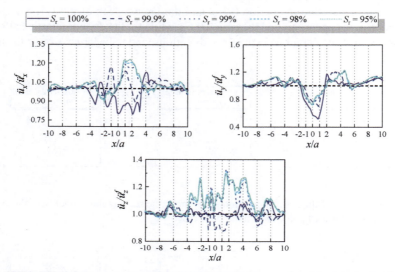

Fig. 5. Ratios of the peak ground accelerations above the rectangular tunnel to those of the free field for different saturation degree values.

Figure 6 shows the ratios of the spectral accelerations above the tunnel to those of the free field for different saturation degree values, and the spectral ratios are for the locations of the maximum ratios of the peak ground accelerations. It can be seen that the saturation degree has a big influence on the period contents of the ground surface acceleration. The partial saturation scenario generally leads to larger spectral ratios than the full saturation scenario when the period is beyond 0.2 s. Moreover, the partial saturation scenario may show an opposite tendency of the change in the spectral ratios compared to the full saturation scenario. For instance, for the case of $S_r = 100\%$, the spectral ratio of $\ddot{u}_x$ the has a general tendency to decrease with the period, and on the other hand, for the cases of $S_r = 99\%$, 98%, and 95%, the spectral ratio of $\ddot{u}_x$ tends to first decrease with the period and then increases, reaching its peak value after the period of 0.2 s.

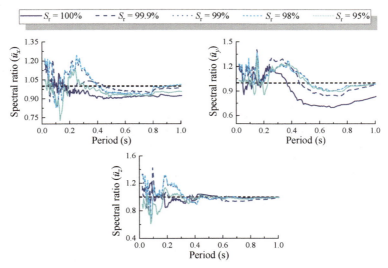

Fig. 6. Ratios of the spectral accelerations above the rectangular tunnel to those of the free field for different saturation degree values.

4 Conclusions

This study presents findings from a numerical investigation on the effects of a rectangular tunnel on the ground motion of a saturated poroelastic half-space for obliquely incident seismic waves. The results show that the rectangular tunnel can significantly modify the ground surface accelerations compared with the free-field site response, and this effect should be taken into account for the seismic design of the aboveground structures around it. For this example, the rectangular tunnel tends to considerably de-amplify the two horizontal ground accelerations right above it, while slight amplification of the two horizontal ground accelerations can be found at a distance from the tunnel. As for the vertical ground acceleration, the amplification effect of the tunnel seems to be more pronounced than the de-amplification effect. Moreover, the influence of the rectangular tunnel on the spectral accelerations is limited in the period range of 1 s and is more prominent in the shorter period range.

The large differences in the ground motion amplification between the saturate and dry poroelastic half-space cases, both in terms of the magnitude and period contents of the ground surface accelerations, indicate the important role of the pore water and highlight the significance of employing a rigorous two-phase model for simulating the seismic response of a saturated poroelastic half-space. In addition, the saturation degree is a critical factor that can significantly influence the ground motion amplification, and the partial saturation scenario may result in larger amplification of the peak ground accelerations than the full saturation scenario.

Acknowledgments. This study was supported by the National Natural Science Foundation of China under Grants 52008013, U1839202, 51978462, and 51878625, which are gratefully acknowledged.

References

1. Gizzi, F.T., Masini, N.: Historical damage pattern and differential seismic effects in a town with ground cavities: a case study from Southern Italy. Eng Geol **88**, 41–58 (2006). https://doi.org/10.1016/j.enggeo.2006.08.001
2. Sgarlato, G., Lombardo, G., Rigano, R.: Evaluation of seismic site response nearby underground cavities using earthquake and ambient noise recordings: a case study in Catania area, Italy. Eng Geol **122**, 281–291 (2011). https://doi.org/10.1016/j.enggeo.2011.06.002
3. Abuhajar, O., El Naggar, H., Newson, T.: Experimental and numerical investigations of the effect of buried box culverts on earthquake excitation. Soil Dyn. Earthq. Eng. **79**, 130–148 (2015). https://doi.org/10.1016/j.soildyn.2015.07.015
4. Baziar, M.H., Moghadam, M.R., Kim, D., Choo, Y.W.: Effect of underground tunnel on the ground surface acceleration. Tunn. Undergr. Sp. Tech. **44**, 10–22 (2014). https://doi.org/10.1016/j.tust.2014.07.004
5. Rabeti Moghadam, M., Baziar, M.H.: Seismic ground motion amplification pattern induced by a subway tunnel: shaking table testing and numerical simulation. Soil Dyn. Earthq. Eng. **83**, 81–97 (2016). https://doi.org/10.1016/j.soildyn.2016.01.002
6. Alielahi, H., Kamalian, M., Adampira, M.: Seismic ground amplification by unlined tunnels subjected to vertically propagating SV and P waves using BEM. Soil Dyn. Earthq. Eng. **71**, 63–79 (2015). https://doi.org/10.1016/j.soildyn.2015.01.007
7. Liang, J., Zhang, J., Ba, Z.: Amplification of in-plane seismic ground motion by group cavities in layered half-space (II): with saturated poroelastic soil layers. Earthq. Sci. **25**(4), 287–298 (2012). https://doi.org/10.1007/s11589-012-0854-2
8. Liang, J., Han, B., Ba, Z.: 3D Diffraction of obliquely incident SH waves by twin infinitely long cylindrical cavities in layered poroelastic half-space. Earthq. Sci. **26**(6), 395–406 (2013). https://doi.org/10.1007/s11589-013-0046-8
9. Degrande, G., De Roeck, G., Van den Broeck, P., Smeulders, D.: Wave propagation in layered dry, saturated and unsaturated poroelastic media. Int. J. Solids Struct. **35**, 4753–4778 (1998). https://doi.org/10.1016/S0020-7683(98)00093-6
10. Han, B., Zdravković, L., Kontoe, S.: Analytical and numerical investigation of site response due to vertical ground motion. Géotechnique **68**, 467–480 (2018). https://doi.org/10.1680/jgeot.15.P.191
11. Yang, J., Sato, T.: Analytical study of saturation effects on seismic vertical amplification of a soil layer. Geotechnique **51**, 161–165 (2001). https://doi.org/10.1680/geot.2001.51.2.161
12. Biot, M.A.: Theory of propagation of elastic waves in a fluid-saturated porous solid. I Low-frequency range. J. Acoust. Soc. Am. **28**, 168–178 (1956). https://doi.org/10.1121/1.1908239
13. Biot, M.A.: Mechanics of deformation and acoustic propagation in porous media. J. Appl. Phys. **18**, 1482–1498 (1962). https://doi.org/10.1063/1.1728759
14. Zhu, J., Li, X., Liang, J.: 3D seismic responses of a long lined tunnel in layered poro-viscoelastic half-space by a hybrid FE-BE method. Eng. Anal. Bound Elem. **114**, 94–113 (2020). https://doi.org/10.1016/j.enganabound.2020.02.007
15. Zhu, J., Liang, J.: Soil pressure and pore pressure for seismic design of tunnels revisited: considering water-saturated, poroelastic half-space. Earthq. Eng. Eng. Vib. **19**(1), 17–36 (2020). https://doi.org/10.1007/s11803-020-0545-2

Foundations and Soil-Structure Interaction

The Effect of Soil Damping on the Soil-Pile-Structure Interaction Analyses in Liquefiable and Non-liquefiable Soils

Ozan Alver(✉) and E. Ece Eseller-Bayat

Civil Engineering Department, İstanbul Technical University, İstanbul, Turkey
alver16@itu.edu.tr

Abstract. Preliminary design of piles is performed by considering the static loads, but the final design must include the dynamic loads, especially in earthquake-prone regions. The soil nonlinearity under seismic loading is evaluated using the modulus degradation curves in the total-stress approach. In this study, two different centrifuge tests were simulated in FLAC^{3D}. The nonlinear elastic method (hyperbolic model) and the elastoplastic Mohr-Coulomb (MC) model were employed in the study. The soil-single pile-structure systems were analyzed under the specific earthquake events, and the soil-pile-structure response was compared. The analyses with the low-intensity input motions show that the superstructure accelerations and the bending moments in the single pile are estimated with reasonable accuracy. However, the superstructure accelerations might be underestimated, especially in the MC model, compared to the centrifuge test results due to the increase in the amplitude of the input motion. The low accelerations can be attributed to the high damping ratios in the perfectly plastic constitutive model. Although the nonlinear elastic model is less complex, closer results might be obtained since the more realistic damping ratios are implemented. The results show that even the less elaborate models, such as the hyperbolic model with an indefinite failure criterion, might give reasonably accurate results in the total-stress approach, thanks to the limited damping ratios. As a result, the responses of the superstructure and the pile in soil-pile-structure interaction problems are highly dependent on the soil damping; in turn, the due account must be given to the selection of the constitutive model.

Keywords: Soil-pile-structure interaction · Damping · FLAC^{3D} · Modulus · Degradation

1 Introduction

Dynamic soil-pile-structure interaction analyses are carried out for piles constructed in a seismically active region. Two common methods are available for the analyses: Three-dimensional finite element (or finite difference) or beam on nonlinear Winkler foundation (BNWF) approaches. In the mentioned methods, the pile and superstructure behavior is assumed to remain in the linear stage. The main uncertainty in the analyses stems from the modeling of soil behavior (Boulanger et al. 1999; Finn and Fujita 2002; Allotey and

El Naggar 2008; Rahmani 2018). This paper aims to show the effect of the hysteretic damping of the soil on the pile-soil-structure system response. For this purpose, two different centrifuge test setups were simulated by three-dimensional models in FLAC^{3D} (Itasca Consulting Group 2009). Gohl's (1991) and Wilson's (1998) studies are used for non-liquefiable and liquefiable sands.

Figure 1 shows Gohl's (1991) and Wilson's (1998) containers in prototype units, which are called Case 1 and Case 2 in this paper. In Case 1, the pile having a diameter of 0.57 m was placed inside a dry Nevada sand with a relative density of $D_R = 40\%$. The single mass was placed on top of the pile extending to 2.0 m from the ground surface.

The diameter and the length of the highly instrumented single pile in Case 2 were 0.67 m and 16.7 m, respectively. A 49 Mg mass placed on the pile created a single degree of freedom system. The free height of the single pile was 3.8 m. The soil in which the piles were embedded was the saturated Nevada sand placed at two different relative densities. The thicknesses of these layers were 9.4 m and 11.3 m, and the relative densities were 55% and 80%, respectively.

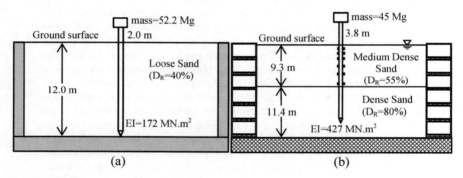

Fig. 1. a) Rigid soil container of Gohl (1991), and b) Laminar soil container of Wilson (1998)

2 Numerical Modeling

The numerical model was created in FLAC3D in this study, as shown in Fig. 2. The model dimensions in x-y-z directions are 16 m × 10 m × 12 m and 20 m × 51 m × 20 m, in Gohl (1991) and Wilson (1998), respectively. The zone sizes must be fine enough to transmit the waves from the base through the model accurately. Finer zones are used in the vicinity of the pile in the lateral direction, and the zone sizes gradually increase using an aspect ratio of 1.1. In the vertical direction, the dimension of each solid zone is constant, but the height of each element must be lower than the limit, which (Δl) is one-tenth of the wavelength (λ) of the input wave motion ($\Delta l < \lambda/10$). The minimum zone size in the vertical direction is selected ($\Delta l < \lambda/10 = 1.0$ m). The pile and the superstructure were modeled with the beam elements in the software. As the nonlinear elastic (hyperbolic) soil model has no ultimate strength, the beam for the pile model has rigidly connected to the soil domain. Therefore, the same approach was followed in the MC model to be consistent.

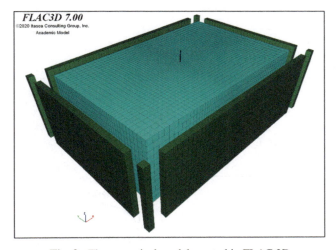

Fig. 2. The numerical model created in FLAC 3D

2.1 Soil Model

Two different soil constitutive models available in FLAC3D have been used to investigate the effect of soil damping ratio on the pile and structure response in soil-pile-structure interaction analyses. The first approach is the hysteretic damping model with the hyperbolic model (Model-1). The soil's nonlinear behavior can be considered using the small-strain shear modulus (G_{max}) and modulus degradation curves. In the second approach, the Mohr-Coulomb model is used with the modulus degradation curves (Model-2). Both approaches adopt Masing's rule for unloading/reloading behavior. However, the hysteretic soil damping differs since the hyperbolic model has no definite ultimate shear strength, while the MC model has constant yield stress.

In this study, the soil's small strain shear moduli (G_{max}) were determined by Seed and Idriss's equation (1970), and the bulk modulus was calculated by elastic theory using the Poisson's ratio. The nonlinear stress-strain behavior of soils can be considered by normalized shear modulus reduction (G/G_{max}) curves in FLAC3D. Several researchers have studied the stress-strain behavior of sands, and reduction curves have been suggested so far (Seed and Idriss 1970; Darendeli 2001). Boulanger et al. (1999), Thavaraj et al. (2010) and Kwon and Yoo (2020) implement the curves of Seed and Idriss (1970). In this study, effective stress-dependent curves of Darendeli (2001) were used in the analyses to better capture the variation with depth (Fig. 3a). The shear modulus reduction and the reference strain equations for cohesionless soils are given in Eqs. (1) and (2), respectively, according to Darendeli (2001). The hysteretic damping command in FLAC^{3D} was utilized for all zones in the model. The reference strain values were assigned to each zone by taking the initial effective stress into account. The stress-strain curves of hyperbolic and Mohr-Coulomb models are shown in Fig. 3b. The hyperbolic model allows larger pressures to occur, especially at the lower confining stresses, and less damping ratios could be provided.

$$\frac{G}{G_{max}} = \frac{1}{1 + \left(\frac{\gamma}{\gamma_r}\right)^{0.919}}$$ (1)

$$\gamma_r = 0.0352 * \left[\frac{\sigma'_m}{p_a}\right]^{0.3483}$$ (2)

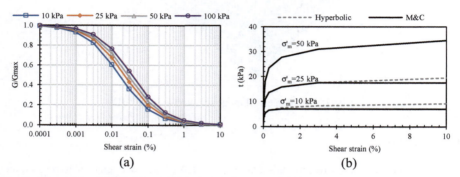

Fig. 3. The soil models (a) Modulus degradation curves of Darendeli (2001) and (b) Stress-strain curves

Figure 3b shows the monotonic stress-strain curves under dynamic loading. However, the unloading/reloading behavior must be defined for the fully nonlinear analyses in the time domain. Even if the FLAC^{3D} implements Masing's rule for the models employed in this study, the resulting damping ratios are different due to the backbone curves. A perfectly plastic MC model causes the damping to be higher than the hyperbolic model since the area of the loop under the cyclic loading is larger.

2.2 Input Motions

The numerical models for Case 1 and Case 2 were created in the FLAC^{3D} and analyzed under the same input motions as in the centrifuge tests. In Case 1, a random input motion having 0.15g acceleration amplitude was applied to the rigid container base by Gohl (1991). The acceleration-time history of the motion is given in Fig. 3a (provided by Dr. Amin Rahmani and Dr. Mahdi Taibeat). In Case 2, two real earthquake records (Loma Prieta, 1989-Santa Cruz station and Kobe, 1995-Port Island station) were used in the centrifuge tests (Wilson 1998). The centrifuge laboratory of UC Davis provides the acceleration time history of the input motion. Although the soil in Case 2 was saturated sand, the Santa Cruz motion with amax = 0.05 g (Event K) did not cause significant excess pore water pressure. That Event was selected in this study for the non-liquefiable case, in which the soil was dry sand. Event I and Event J are the motions of Santa Cruz and Kobe with amax = 0.49 g and amax = 0.22 g, respectively, in which the medium dense sand completely liquefied under these events in Case 2.

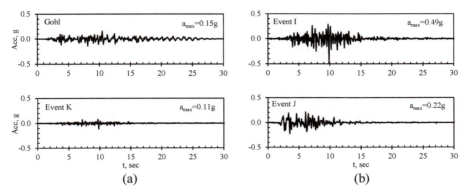

Fig. 4. Earthquake records used in the numerical analysis a) for nonliquefiable cases, b) for liquefiable cases

3 Results

The soil-pile-structure system was subjected to base excitation with the earthquake input motions given in Fig. 4. The structure responses in Case 1 (Gohl and Event K) were compared with the centrifuge test results by the acceleration response spectra, as shown in Fig. 5. MC model results of Gohl are very close to the centrifuge test for both the peak and the spectral accelerations, whereas the hyperbolic model slightly overestimates the accelerations (Fig. 5a). Figure 5b shows the results for Event K. Although the peak accelerations are close, the hyperbolic model better captures the spectral accelerations than the MC, which underestimates the response.

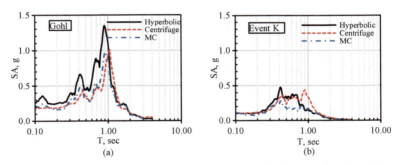

Fig. 5. The acceleration response spectrum of structure accelerations in a) Gohl and (b) Wilson (Event K)

Model responses were compared through the maximum bending moments along with the pile. Figure 6a shows the results of Gohl, where the bending moments were underpredicted and overpredicted, by MC and hyperbolic models, respectively. Although the models underestimate the pile response in Event K, the hyperbolic model is closer to the centrifuge test. The results showed that the hyperbolic model could predict the structure and the pile response with reasonable accuracy. MC model slightly underestimates the structure response in Event K and results in the lower bending moments. The simulation of Gohl with the MC model gives better results in the structure response. However, the pile bending moments are again underpredicted.

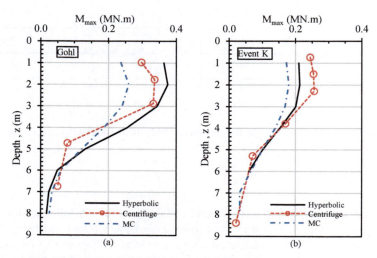

Fig. 6. Bending moments of (a) Gohl and (b) Wilson (Event K)

The soil-pile-structure system response in the liquefiable soils under Event I and Event J (Case 2) were investigated by comparing the analyses with the centrifuge test results. Figure 7a shows the acceleration response spectra for Event I, where the hyperbolic and the MC models capture the general trend. However, the MC model significantly underestimates the accelerations in Event J (Fig. 7b), in which the hyperbolic model shows closer results. The high superstructure response in the centrifuge test under Event J was reported as the reduction in the pore pressures, which results in the high effective stresses (Boulanger, 1999). Since the pore water pressures are not considered in the models employed in this study, the system behavior is mainly affected by the soil

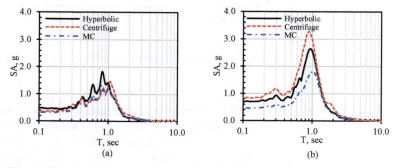

Fig. 7. The acceleration response spectrum of the superstructure acceleration under a) Event I and (b) Event J

damping. MC model causes the damping to be higher than the hyperbolic model due to the perfectly plastic behavior.

Model responses in the liquefiable case were compared through the maximum bending moments along with the pile. Figure 8a shows the results of Event I, where the bending moments were very close in the MC model, whereas the hyperbolic models slightly overpredict the bending moments. Figure 8b shows the pile response under Event J. Both models underpredict the bending moments due to the low superstructure accelerations. However, the hyperbolic model leads to a closer pile response in Event J than the MC model.

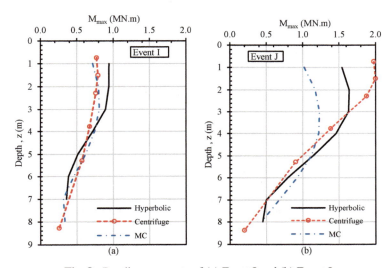

Fig. 8. Bending moments of (a) Event I and (b) Event J

4 Conclusions

The effect of the damping ratio on the soil-pile-structure system response was investigated in this study. Two centrifuge test setups available in the literature were modeled in FLAC^{3D}, and dynamic analyses were performed using the hyperbolic and the Mohr-Coulomb models. Acceleration response spectra for the superstructure and bending moments along the piles were compared with the test results for non-liquefiable and liquefiable cases. The MC model generally underestimates the structure and pile responses, whereas the hyperbolic model with indefinite ultimate stress yields more reasonable and consistent results due to the limited damping ratios. Overall system behavior could be well predicted provided that the soil damping is represented realistically.

Acknowledgment. This research was funded by TUBITAK (The Scientific and Technological Research Council of Turkey) under the Grant No. 119M624 and project title "Development of Lateral Load Resistance-Deflection Curves for Piles in Sands under Earthquake Excitation".

We thank Dr. Amin Rahmani and Dr. Mahdi Taiebat for providing the centrifuge test data given in Gohl (1991).

References

Allotey, N., Naggar, M.H.N.H.E.: A numerical study into lateral cyclic nonlinear soil–pile response. Can. Geotech. J. **45**(9), 1268–1281 (2008). https://doi.org/10.1139/t08-050

Boulanger, R.W., Curras, C.J., Kutter, B.L., Wilson, D.W., Abghari, A.: Seismic soil–pile–structure interaction experiments and analyses. J. Geotechn. Geoenviron. Eng. **125**(9), 750–759 (1999)

Darendeli, M.B.: Development of a new family of normalized modulus reduction and material damping curves. Ph.D. Dissertation, University of Texas at Austin (2001)

Finn, W.D.L., Fujita, N.: Piles in liquefiable soils: seismic analysis and design issues. Soil Dyn. Earthq. Eng. **22**, 731–742 (2002)

Gohl, W.L.: Response of pile foundations to simulated earthquake loading: experimental and analytical results. Ph.D. Dissertation, University of British Columbia (1991)

Itasca Consulting Group: Fast Lagrangian Analysis of Continua in 3-Dimensions, Version 7. Minneapolis, USA (2009)

Kwon, S.Y., Yoo, M.: Study on the dynamic soil-pile-structure interactive behavior in liquefiable sand by 3D numerical simulation. Apl. Sci. **10**, 2723 (2020)

Rahmani, A., Taiebat, M., Finn, W.L., Ventura, C.E.: Evaluation of p-y springs for nonlinear static and seismic soil-pile interaction analysis under lateral loading. Soil Dyn. Earthq. Eng. **115**, 438–447 (2018)

Seed, H.B., Idriss, I.M.: Soil moduli and damping factor for dynamic analyses. Report No. EERC 70-10, Earthquake Engineering Research Center. University of California, Berkeley, California, 40 p. (1970)

Thavaraj, T., Finn, W.D., Wu, G.: Seismic response analysis of pile foundations. Geotech. Geol. Eng. **28**(3), 275–286 (2010)

Wilson, D.W.: Soil pile superstructure interaction in liquefying sand and soft clay. Ph.D. Dissertation, University of California at Davis (1998)

Bearing Capacity of Shallow Strip Foundations Adjacent to Slopes

Orazio Casablanca, Giovanni Biondi, Giuseppe Di Filippo, and Ernesto Cascone(✉)

Department of Engineering, University of Messina, Contrada Di Dio, 98166 Messina, Italy
ecascone@unime.it

Abstract. In this paper the static and seismic bearing capacity factors for shallow strip foundations adjacent to slopes have been evaluated using the method of characteristics, extended to the seismic case by means of the pseudo-static approach. Bearing capacity factors have been evaluated for different values of the slope angle and, under seismic conditions, accounting for the effect of inertia forces arising in the foundation soil and transmitted by the superstructure. These effects are dealt with independently and it was also demonstrated that they can be superimposed without significant error.

The proposed solutions, obtained assuming Hill's and Prandtl's failure mechanisms, have been checked against those obtained through finite element analyses and compared with results already available in the literature.

Closed-form solutions and an empirical formula have been provided to evaluate corrective coefficients accounting for the effect of the sloping ground on bearing capacity that, under static conditions, allow a straightforward application of the bearing capacity trinomial formula.

Keywords: Foundations · Slope stability · Bearing capacity · Method of characteristics

1 Introduction

Under static condition and with reference to the case of horizontal ground surface, the bearing capacity of shallow strip footings subject to vertical and centred loads is generally evaluated using the well-known trinomial formula proposed by Terzaghi:

$$q_{ult} = c' \cdot N_c + q \cdot N_q + \frac{1}{2} \cdot \gamma \cdot B \cdot N_\gamma \tag{1}$$

In Eq. (1) q_{ult} represents the limit load that the soil-foundation system can sustain under the assumption of rigid perfectly plastic behaviour, γ and c' are the unit volume weight and the cohesion of the soil, B is the foundation width, q is the vertical surcharge acting acting on the founding plane aside the foundation and, finally, N_c, N_q e N_γ are the bearing capacity factors, depending on the angle of soil shear strength φ'.

For foundations near or adjacent to a slope, Eq. (1) can be still used introducing suitable bearing capacity factors or, alternatively, corrective coefficients taking into account the slope effect on the bearing capacity.

Many solutions are available in the literature that allow evaluating bearing capacity factors or corrective coefficients considering both static and seismic conditions; these solutions have been obtained through the limit equilibrium method (e.g. [6, 11, 13, 16]), the upper bound theorem of limit analysis (e.g. [1, 17]), the method of characteristics (e.g. [2–5, 8, 12]) and the finite element limit analysis (e.g. [7, 10, 15]).

This paper presents the values of the static and seismic bearing capacity factors for shallow strip footings adjacent to a slope obtained using the method of characteristics (MC), extended to the seismic case through the pseudo-static approach.

Differently from most of the available solutions, the soil inertia effect, due to the seismic wave propagation, and the superstructure inertia effect, due to the structural response, are dealt with independently; however it was demonstrated that these effects can be overlapped without significant error [3].

Finally, in order to validate the results obtained through MC, numerical finite element (FE) analyses have been carried out using the code Plaxis [14].

2 Method of Analysis

MC assumes that the soil behaves as a rigid perfectly plastic medium according to Mohr-Coulomb criterion and that in the soil mass at incipient failure both equilibrium equations and plastic condition must be satisfied.

The resultant equations can be solved to obtain the limit load of the soil-foundation system without arbitrary assumptions about the plastic mechanism.

For plane strain conditions, the equilibrium equations and the plastic condition are:

$$\frac{\partial \sigma'_x}{\partial x} + \frac{\partial \tau_{xy}}{\partial y} = \gamma \cdot k_h$$
$$\frac{\partial \tau_{xy}}{\partial x} + \frac{\partial \sigma'_y}{\partial y} = \gamma \cdot (1 - k_v)$$
(2)

$$\sqrt{\left(\frac{\sigma'_x - \sigma'_y}{2}\right)^2 + \tau_{xy}^2} = \frac{\sigma'_x + \sigma'_y}{2} \sin \varphi' + c' \cos \varphi'$$
(3)

where σ'_x, σ'_y and τ_{xy} are the normal effective stresses and the shear stresses acting on the horizontal and vertical planes and k_h and k_v are the horizontal and vertical seismic acceleration coefficients taking into account for the inertia forces acting in the foundation soil. Substituting the plastic condition (3) into the equilibrium Eqs. (2), a system of two hyperbolic partial differential equations is obtained. Using MC, it is possible to transform the system of partial differential equations in a system of ordinary differential equations in the variables s' and ω, which represent, respectively, the mean stress $s' = (s'_1 + s'_3)/2$ and the angle formed between the maximum principle stress direction σ_1 and the horizontal direction x. The evaluation of bearing capacity through MC is carried out solving three boundary-value problems. Figure 1 shows the Cauchy (C), Riemann (R) and Goursat (G) domains relative to the Hill (Fig. 1a) and Prandtl (Fig. 1b) mechanism, associated to these three problems.

In both cases, the upper boundary of the Cauchy domain is inclined by an angle β in order to simulate the presence of a slope adjacent to the foundation. The evaluation of N_c and N_q does not imply particular numerical difficulties. Conversely, to solve the N_γ problem it was necessary to introduce a small value of the lateral surcharge q equal to $\gamma B \cdot 10^{-9}$, which contribution on the actual value of bearing capacity factor is negligible.

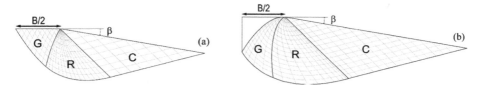

Fig. 1. Goursat, Riemann e Cauchy domains: a) Hill mechanism; b) Prandtl mechanism.

3 Static Bearing Capacity

For a shallow foundation adjacent to a slope, in the case of weightless soil ($\gamma = 0$), MC allows evaluating closed form expressions for bearing capacity factors N_{cg} and N_{qg} where the subscript "g" has been introduced to specify that the ground surface adjacent to the foundation is inclined. If the self-weight of the soil is considered ($\gamma \neq 0$) the values of the bearing capacity factor $N_{\gamma g}$ can be obtained through the MC only by numerical integration of the plastic equilibrium equations. Bearing capacity factors N_{cg}, N_{qg} and $N_{\gamma g}$ (the last obtained assuming both Hill and Prandtl mechanism) are given in Fig. 2, for values of the angle of shear strength φ' in the range $15° \div 45°$ and of the slope angle β up to $45°$. For each curve in Fig. 2b–d, it is possible to evaluate a limit value β_{lim} of the slope angle for which the bearing capacity factor drops to zero. For $N_{\gamma g}$ relative to both smooth and rough foundations and N_{qg}, this condition is verified for $\beta_{lim} = \varphi'$ because the bearing capacity problem degenerates into an infinite slope stability problem and the numerical solution cannot be achieved. As far as N_{cg} is concerned ($c' - \varphi'$ soil with $\gamma = 0$), a larger value of β_{lim} can be found. For example, for $\beta = 90°$ the solving equations degenerate into those related to the stability of a vertical cut with a uniform load applied at the crest. In this case, since the soil is weightless, in order to generate a slip surface, the load applied onto the soil surface needs to be always greater than zero, independently of φ'.

Bearing capacity factors N_{cg}, N_{qg} and $N_{\gamma g}$ obtained for the case of foundations adjacent to a slope can be normalized with respect to their homologous factors (N_c, N_q, N_γ) relative to the case of horizontal ground surface. Thus, the following closed form corrective coefficients can be introduced for the N_c and N_q problems:

$$g_c = \frac{N_{cg}}{N_c} = \frac{(1 + \sin \varphi') \exp[(\pi - 2\beta) \tan \varphi'] - (1 - \sin \varphi')}{(1 + \sin \varphi') \exp[\pi \tan \varphi'] - (1 - \sin \varphi')} \quad (4)$$

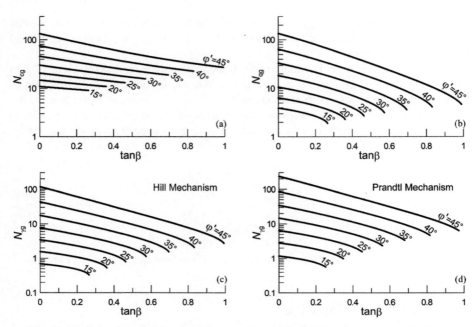

Fig. 2. Static bearing capacity factors N_{cg}, N_{qg} and $N_{\gamma g}$ (Hill and Prandtl mechanism).

$$g_q = \frac{N_{qg}}{N_q} = \frac{(1 - \sin \varphi') \cos^2 \beta}{1 - \sin \varphi' \cos(\Delta_\beta - \beta)} \exp[-(\beta + \Delta_\beta) \tan \varphi'] \quad (5)$$

while for the N_γ problem, a best fit of the numerical values of g_γ is given by the following equation, to be used along with the coefficients listed in Table 1:

$$g_\gamma = \frac{N_{\gamma g}}{N_\gamma} = [1 - (l_1 + l_2 \cot \varphi') \tan \beta]^{m_1/[m_2 + (\tan \varphi')^{m_3}]} \quad (6)$$

Table 1. Coefficients of Eq. (6)

Mechanism	l_1	l_2	m_1	m_2	m_3
Hill	−1.091	1.225	4.735	−0.812	−2.143
Prandtl	−1.047	1.195	4.037	−0.822	−1.852

4 Seismic Bearing Capacity

Bearing capacity factors for shallow strip foundations adjacent to slopes were evaluated also with reference to seismic conditions accounting only for earthquake-induced inertia forces arising in the foundation soil (N^s_{qgE} and $N^s_{\gamma gE}$) or considering only the inertial forces transmitted onto the foundation soil by the superstructure (N^{ss}_{cgE}, N^{ss}_{qgE} and $N^{ss}_{\gamma gE}$). Figure 3 shows the variation of the factors N^s_{qgE} and $N^s_{\gamma gE}$ with k_h, computed via MC for values of the ratio k_v/k_h equal to 0 and to ± 0.5, for $\varphi' = 30°$ and β equal to 10° and 20°; for comparison, the values obtained for horizontal ground surface (β = 0) are also shown. As k_h increases, the seismic bearing capacity factor decreases and drops to zero when the limit value $k_{h,\lim} = \tan(\varphi' - \beta)$ is attained, meaning that the soil shear strength is fully mobilized to resist the shear stresses due to the slope inclination and the inertia forces arising in the soil. It can be also observed that a positive value of k_v (vertical acceleration directed upward) affects significantly the bearing capacity factors because it reduces the vertical component of the surcharge q and the resultant of the body forces. Conversely, negative values of k_v (vertical acceleration directed downward) have a beneficial effect on bearing capacity in comparison to the case $k_v = 0$.

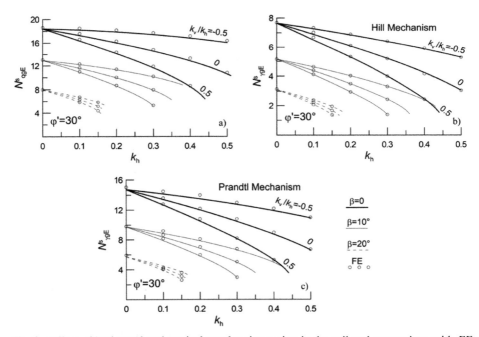

Fig. 3. Effect of horizontal and vertical acceleration acting in the soil and comparison with *FE* results for $\varphi' = 30°$ and β = 0–20°: a) N^s_{qgE}; b) $N^s_{\gamma gE}$ for the Hill mechanism; c) $N^s_{\gamma gE}$ for the Prandtl mechanism.

Figure 3 also depicts the results obtained through *FE* analyses. It can be observed that there exists a good agreement between the results obtained using the two methods of analysis. The comparison is also shown in terms size of the plastic mechanism (Fig. 4) where the characteristic lines network is overlapped to the map of the incremental displacements derived by FE analysis for the N^s_{qgE} (Fig. 4a) and the $N^s_{\gamma gE}$ problems (Fig. 4b and 4c for the Hill and the Prandtl mechanism, respectively).

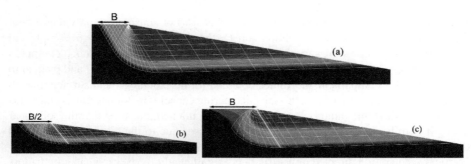

Fig. 4. Comparison between the characteristic lines network and the total incremental displacement contours obtained via *FE* analyses for $\varphi' = 30°$, $k_h = 0.3$, $k_v = 0$ and $\beta = 10°$: a) N^s_{qgE} problem; b) $N^s_{\gamma gE}$ problem for the Hill mechanism; c) $N^s_{\gamma gE}$ problem for the Prandtl mechanism.

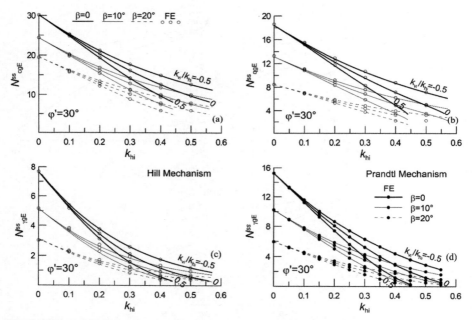

Fig. 5. Effect of horizontal and vertical acceleration transmitted by the superstructure and comparison with *FE* results for $\varphi' = 30°$ and $\beta = 0\text{--}20°$: a) N^{ss}_{cgE} factor; b) N^{ss}_{qgE} factor; c) $N^{ss}_{\gamma gE}$ factor for the Hill mechanism; d) $N^{ss}_{\gamma gE}$ factor for the Prandtl mechanism.

The inertial effect of the superstructure was taken into account in the analyses applying an inclined load onto the foundation with angle $\theta_i = \tan^{-1}[(k_{hi}/(1-k_{vi})]$, where k_{hi} and k_{vi} are respectively the seismic acceleration coefficients involved in the evaluation of the horizontal and vertical inertia forces transmitted by the superstructure.

For the Prandtl mechanism it was not possible to establish the correct boundary condition to be imposed on the Goursat domain; therefore, this case was analyzed only through the FE approach. Figure 5 shows the results obtained for N^{ss}_{cgE}, N^{ss}_{qgE} and

$N^{ss}_{\gamma gE}$ for $\varphi' = 30°$, $\beta = 0\text{–}20°$ and for different values of the ratio k_{vi}/k_{hi}; for comparison, the bearing capacity factors for horizontal ground surface ($\beta = 0$) are also shown in the figure. By increasing k_{hi}, the bearing capacity factors decrease until they reach a limit value for $\theta_{i,\text{lim}}$ (namely the limit inclination of the applied load), which implies a sliding failure mechanism of the foundation. This limit value, independent of β, was derived in closed form by Kezdi [9] for the bearing capacity factor N^{ss}_{cgE} (Fig. 3a) while for N^{ss}_{qgE} and $N^{ss}_{\gamma gE}$ the angle $\theta_{i,\text{lim}}$ is equal to φ'.

From the comparison with the results obtained by FE, in terms of the values of the bearing capacity factors (Fig. 5) and size of the plastic mechanisms (Fig. 6), an excellent agreement can be observed between the results obtained using the two methods of analyses.

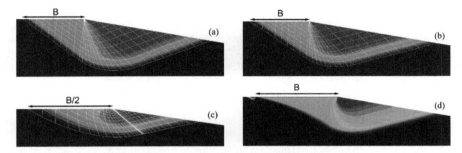

Fig. 6. Comparison between the characteristic lines network and the total incremental displacement contours obtained through *FE* analyses for $\varphi' = 30°$, $k_h = 0.3$, $k_v = 0$ and $\beta = 10°$: a) N^{ss}_{cgE}; b) N^{ss}_{qgE}; c) $N^{ss}_{\gamma gE}$ for the Hill mechanism; d) $N^{ss}_{\gamma gE}$ for the Prandtl mechanism.

5 Conclusions

Static and seismic bearing capacity factors for shallow strip foundations adjacent to slopes were evaluated using the method of characteristics. Results of numerical analyses carried out through the finite element method confirm the correctness of the proposed solutions. Unlike most of the available studies, the effect of inertia forces involved in the plastic mechanism and the effect of inertia forces transmitted by the superstructure were dealt with independently; as shown in other studies ([3]), it is still possible to overlap these effects without producing significant errors. This approach allows to determine the bearing capacity of shallow foundations even in those cases where the acceleration acting on the superstructure significantly differs from that acting in the foundation soil.

Acknowledgements. This work is part of the research activities carried out by the University of Messina Research Unit in the framework of the work package WP16 - Contributi normativi - Geotecnica (Task 16·3: Interazione terreno-fondazionestruttura, Sub-task 16·3·2: Capacità dissipativa dell'interfaccia terreno-fondazione) of a Research Project funded by the ReLuis (Network of Seismic Engineering University Laboratories) consortium (Accordo Quadro DPC/ReLUIS 2019–2021).

References

1. Askari, F., Farzaneh, O.: Upper bound solution for seismic bearing capacity of shallow foundations near slopes. Géotechnique **53**(8), 697–702 (2003)
2. Casablanca, O., Cascone E., Biondi G.: The static and seismic bearing capacity factor Nγ for footings adjacent to slopes. In: Proc. VI Italian Conf. of Researchers in Geotechnical Engineering, Procedia Engineering, vol. 158, pp. 410–415 (2016)
3. Casablanca, O., Cascone, E., Biondi, G., Di Filippo, G.: Static and seismic bearing capacity of shallow strip foundations on slopes. Géotechnique. (2021). https://doi.org/10.1680/jgeot.20.P.044
4. Cascone, E., Biondi, G., Casablanca, O.: Influence of earthquake-induced excess pore water pressures on seismic bearing capacity of shallow foundations. In: Earthquake Geotechnical Engineering for Protection and Development of Environment and Constructions-Proceedings of the 7th International Conference on Earthquake Geotechnical Engineering, pp. 566–581 (2019)
5. Cascone, E., Biondi, G., Casablanca, O.: Groundwater effect on bearing capacity of shallow strip footings. Comput. Geotech. **139**, 104417 (2021)
6. Castelli, F., Motta, E.: Bearing capacity of strip footings near slopes. Geotech. Geol. Eng. **28**(2), 187–198 (2010)
7. Chakraborty, D., Kumar, J.: Bearing capacity of foundations on slopes. Geomech. Geoeng. **8**(4), 274–285 (2013)
8. Graham, J., Andrews, M., Shields, D.H.: Stress characteristics for shallow footings in cohesionless slopes. Can. Geotech. J. **25**(2), 238–249 (1988)
9. Kezdi, A.: The effect of inclined loads on the stability of a foundation. In: Proc. 5th ICSMFE, vol. 1, pp. 699–703. Paris (1961)
10. Kumar, J., Chakraborty, D.: Seismic bearing capacity of foundations on cohesionless slopes. J. Geotech. Geoenviron. Eng. **139**(11), 1986–1993 (2013)
11. Kumar, J., Kumar, N.: Seismic bearing capacity of rough footings on slopes using limit equilibrium. Géotechnique **53**(3), 363–369 (2003)
12. Kumar, J., Mohan Rao, V.B.K.: Seismic bearing capacity of foundations on slopes. Géotechnique **53**(3), 347–361 (2003)
13. Narita, K., Yamaguchi, H.: Bearing capacity analysis of foundations on slopes by use of logspiral sliding surfaces. Soils Found. **30**(3), 144–152 (1990)
14. Plaxis V8: Finite Element Code for Soil and Rock Analyses, Plaxis b.v., The Netherlands (2002)
15. Raj, D., Singh, Y., Shukla, S.K.: Seismic bearing capacity of strip foundation embedded in c − φ soil slope. Int. J. Geomech. **18**(7) (2018)
16. Sarma, S.K., Chen, Y.C.: Seismic bearing capacity of shallow strip footings near sloping ground In: 5th SECED Conf., October 1995, Chester, UK, AA Balkema, Rotterdam, pp. 505–512 (1995)
17. Sawada, T., Nomachi, S.G., Chen, W.F.: Seismic bearing capacity of a mounded foundation near a down-hill slope by pseudo-static analysis. Soils Found. **34**(1), 11–17 (1994)

Seismic Response of Anchored Steel Sheet Pile Walls in Dry and Saturated Sand

Alessandro Fusco[1(✉)], Giulia M. B. Viggiani[1], Gopal S. P. Madabhushi[1], Riccardo Conti[2], and Cécile Prüm[3]

[1] University of Cambridge, Cambridge, UK
af649@cam.ac.uk
[2] Università Niccolò Cusano, Rome, Italy
[3] ArcelorMittal, Global Research and Development, Esch-sur-Alzette, Luxembourg

Abstract. Anchored Steel Sheet Pile (ASSP) walls are complex systems whose behaviour during an earthquake depends on the interaction between the soil, the anchor and the wall. The design of ASSP retaining walls is frequently carried out using pseudo-static methods, where the earthquake-induced inertial forces are represented using an equivalent pseudo-static coefficient that is constant in time and space, usually estimated neglecting any effects of soil-structure interaction, whereas their permanent displacements are often estimated using pseudo-dynamic approaches based on Newmark's sliding block method. However, ASSP walls are complex systems and understanding their seismic response is necessary to determine the extent to which the design assumptions of pseudo-static methods are acceptable and the extension of Newmark's method to the prediction of their seismic-induced permanent displacements is sound. This paper presents the results of two dynamic centrifuge tests on reduced scale models of ASSP walls in dry and saturated uniform medium dense sand performed on the Turner Beam Centrifuge at the Schofield Centre, University of Cambridge. Both trains of sinusoidal waves and realistic earthquake motions were applied to the base of the models. Digital image correlation was used to measure the displacements of a cross-section of the models during each applied earthquake. In dry sand, the largest amplification of the horizontal accelerations was observed near the retaining wall, and the main wall accumulated significant outward rotation with limited horizontal displacement of the toe. In saturated sand, the generation of excess pore pressure led to significant de-amplifications and phase lag in the accelerations, and caused larger permanent displacements of the structure.

Keywords: Retaining structures · Physical modelling · Liquefaction · Soil-structure interaction

1 Introduction

Anchored Steel Sheet Pile (ASSP) walls are frequently adopted as retaining structures in ports and waterfront facilities, as they can provide a more efficient and economical solution in comparison to other types of concrete retaining structures. The dynamic response of ASSP walls is, however, a rather complex Soil-Structure Interaction (SSI) problem that is often addressed, in design practice, by relying on a number of rather simplifying assumptions. For instance, according to Eurocode 8 [1], the seismic design of ASSP walls can be carried out using pseudo-static methods, where earthquake-induced inertial forces are represented using an equivalent pseudo-static seismic coefficient that is constant with time and in space. Although building codes normally provide simplified procedures to estimate the seismic coefficient based on the soil characteristics and the height of the retaining wall, the effects of SSI on the seismic actions are usually ignored.

On the other hand, in recent years, performance based design methods have received increasing attention as it was recognised that, rather than catastrophic failures, many quay walls suffered from excessive displacements and deformations that compromised the serviceability of port facilities [2]. The permanent displacements accumulated by gravity retaining walls under seismic actions can be estimated using pseudo-dynamic methods based on Newmark's sliding block analysis [3]. The application of similar solutions to the prediction of earthquake-induced permanent displacements of ASSP walls is still an open issue, because of the multiple factors affecting their seismic capacity, including the multiple failure mechanisms that these can experience [4]. A better understanding of the seismic response of ASSP walls is therefore crucial to determine the extent to which the assumptions at the base of pseudo-static methods are reasonable, and to extend the application of pseudo dynamic approaches based on Newmark's method to the prediction of their permanent displacements.

The complexity of the soil-structure interaction problem under examination, strongly dependent on the mechanical behaviour of the soil in cyclic and saturated conditions, makes dynamic centrifuge modelling an ideal research tool, because of its ability to capture true soil behaviour, and hence the correct failure mechanisms, and any potential liquefaction, without introducing constitutive assumptions.

This paper presents and discusses the results of two dynamic centrifuge tests performed on small scale models of ASSP walls in dry and saturated uniform medium dense sand. Digital image correlation performed during the centrifuge tests provided novel insight in the displacement field of ASSP walls under seismic actions.

2 Experimental Layout

The centrifuge tests were carried out on the Turner beam centrifuge at the Schofield Centre, University of Cambridge. One centrifuge test was performed on small scale models of ASSP walls in dry sand and one in saturated sand. The centrifuge tests were performed at an enhanced gravity of 60 g. Figure 1 shows a typical layout of an ASSP wall and Table 1 summarizes the dimensions at prototype scale of the two layouts investigated.

2.1 Centrifuge Tests Preparation

Preparation of the Models and Input Motion. The anchor wall and the main wall were modelled using 4.7 mm thick aluminum plates, with a bending stiffness at prototype scale, $EI = 123.8$ MNm^2/m, or the same as the bending stiffness of an AZ28 steel profile. The tiebacks were modelled using steel cables, and hinged connections were created between the walls and the tiebacks. Four tiebacks were used in the dry centrifuge test and two in the saturated test. Hostun sand ($G_S = 2.65$, $e_{max} = 1.011$, $e_{min} = 0.555$, $\varphi'_{cs} = 33°$) was used to prepare the models. This was air-pluviated inside the containers at a relative density of approximately 50% using an automatic sand pourer. For the saturated tests, the sand was saturated using a high-viscosity aqueous solution of hydroxypropyl methylcellulose (60 cSt) to overcome the inconsistency between the dynamic and seepage time scales. Soil accelerations and pore pressures were measured using piezoelectric accelerometers and Pore Pressure Transducers (PPTs). The layouts of the instruments for the tests are shown in Fig. 2.

A servo-hydraulic actuator was used to apply input acceleration time histories at the base of the container [5]. The majority of the seismic inputs consisted of trains of nearly sinusoidal waves with increasing peak acceleration and main frequency of 1 Hz. The acceleration time history recorded during the Imperial Valley earthquake, scaled to a peak acceleration of 0.18 g, was applied in Test AF05. Figure 3 shows the acceleration time traces applied in the two tests. Further details about instrumentation, input motions, and model preparation are given by [6].

The Containers and Displacements Measurements. The dry and saturated centrifuge tests were performed in rigid containers with internal dimensions W × H × D of respectively 500 × 360 × 250 mm^3 and 730 × 398 × 250 mm^3 and a Perspex transparent side that allowed taking digital images of a cross section of the model using a high-speed camera during the earthquakes, at a frame rate of 975 Hz. A layer of Duxseal at both ends of the containers reduced energy reflection at the rigid boundaries [7]. The displacements of the soil and of the structural elements were obtained by Particle Image Velocimetry (PIV), conducted using GeoPIV-RG [8].

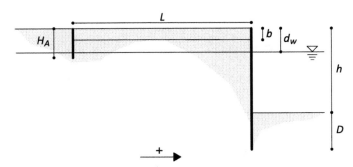

Fig. 1. General layout of an ASSP wall in saturated sand, and sign convention on the accelerations and displacements, positive if towards the excavation.

Table 1. Dimensions of the centrifuge tests at prototype scale [m].

Test	h	D	b	H_A	L	d_w
AF05	8	2	0.5	2	12	–
AF06	8	2	0.5	2	12	1.5

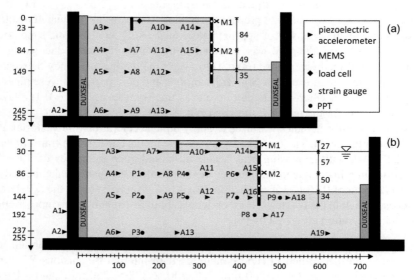

Fig. 2. Layout of the instruments at model scale for (a) Test AF05 and (b) Test AF06.

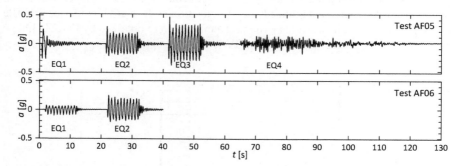

Fig. 3. Input acceleration time histories applied in Tests AF05 and AF06, at porotype scale.

3 Amplification Phenomena of Acceleration Signals

In the following, all dimensions are at prototype scale unless noted otherwise.

Figure 4 presents the horizontal acceleration time histories recorded along an array of piezoelectric accelerometers in the backfill (A9, A10, A11) and at base of the model

(A12) during EQ2 of Test AF05 and during EQ1 and EQ2 of Test AF06, together with the excess pore pressure, ue, measured by transducers P1, P2, P3, in Test AF06. The figure also gives an indication of the estimated vertical effective stress acting at the depth of the pore pressure transducers at the beginning of EQ1. During EQ2, in Test AF05, the amplitude of the seismic waves increased as they travelled towards the soil surface. The same occurred during EQ1 in Test AF06, since the generation of excess pore pressures was limited. On the contrary, during the stronger EQ2 of Test AF06, in which significant excess pore pressures developed, significant de-amplification of accelerations occurred towards the soil surface.

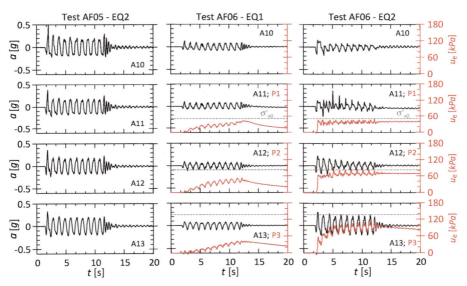

Fig. 4. Time histories of horizontal acceleration and excess pore pressure recorded along an array of instruments in the backfill during: EQ2, in Test AF05; EQ1 and EQ2 in Test AF06.

This is a typical liquefaction-induced phenomenon, caused by the softening of the sand resulting from significant generation of excess pore pressures and consequent reduction of the effective confining stress. During this earthquake, the acceleration time histories recorded by accelerometer A10 showed spikes, which can be attributed to the dilation occurring in the sand.

Figures 5(a) to (d) show the Arias Intensity of the horizontal accelerations recorded during all earthquakes applied in Test AF05 respectively in the far-field (A3, A4, A5), the backfill (A10, A11, A12), the soil immediately behind the wall (A14, A15, A16), and on the main wall (M1, M2). The data are normalized by the Arias Intensity of the base signal (A6). The results for the same instruments alignments in Test AF06 are shown respectively in Figs. 5(e) to (h). Figures 5(i) to (k) show the excess pore pressure ratio, $r_u = u_e/\sigma'_{v0}$, estimated at approximately half the duration of each earthquake applied in Test AF06, without taking into account the cyclic oscillations in the excess pore pressures, respectively in the far-field (P1, P2), in the backfill (P5), and in the soil immediately behind the main wall (P6, P7). Transducer P4 malfunctioned during the test. The ratio r_u at the base of the model, approximately at a depth of 14.2 m, was obtained from PPT P3. The Arias Intensity was preferred to the peak acceleration to discuss amplification phenomena, since the latter was likely to be misleading when high frequency acceleration spikes occurred in liquefied soil.

In dry conditions, in the majority of the cases the acceleration signals amplified as the seismic waves travelled towards the ground surface. In Test AF05, the Arias intensity of the near-surface acceleration signals in the far field was on average 48% larger than at the base. The amplification increased steadily towards the main wall, where the maximum amplification between the base and the near-surface signals was observed, on average 80% in terms of Arias Intensity.

During the relatively weak EQ1 of Test AF06, in saturated conditions but with limited generation of excess pore pressures, similarly to what observed for AF05 in dry conditions, the acceleration signals amplified towards the ground surface. In this case, however, the amplifications were in general smaller than in Test AF05, since the Arias Intensity of the near-surface acceleration increased to a maximum of approximately 1.5 times that of the base signal. During the stronger earthquake EQ2 of Test AF06, a different trend was observed. The large excess pore pressures developed during seismic shaking were responsible for a significant degree of liquefaction in the soil deposit, which caused the soil to become partially isolated from the base motion. Consequently, both in the far-field (Fig. 5(e)) and in the backfill (Fig. 5(f)), the accelerations de-amplified

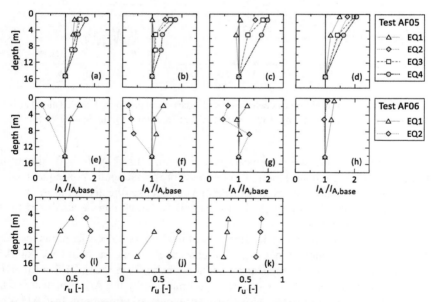

Fig. 5. Arias Intensity of the horizontal accelerations recorded in the far-field, backfill, soil behind the wall, and on the wall, normalized by the Arias Intensity of the base signal, in Test AF05 (Fig. (a) to (d), respectively), and Test AF06 (Fig. (e) to (h), respectively). Ratio r_u calculated in the far-field, the backfill, and the soil behind the wall in Test AF06 (Fig. (i) to (k) respectively).

drastically towards the ground surface. In general, the largest de-amplifications occurred close to the shallower layers of soil, consistently with the tendency of the soil to liquefy first at shallower depths.

In the soil immediately behind the main wall (Fig. 5(g)), despite the large excess pore pressures measured, the de-amplification of the seismic signals was not as severe as further away from the excavation, and, in some cases, the accelerations were amplified. This is likely to be due to the fact that, in the soil close to the main wall, the accelerations were transmitted by the movement of the wall itself. Similarly, the accelerations on the main wall always amplified with respect to the base signals, since they travelled through the retaining structure.

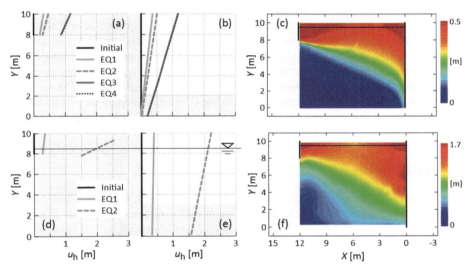

Fig. 6. Position of the (a) anchor wall and the (b) main wall after each earthquake applied in Test AF05, and (c) contours of incremental horizontal displacement during EQ3 in Test AF05. The same is shown for Test AF06 in (d), (e), and (f). The displacement contours in (f) refer to EQ2.

4 Seismic Induced Displacements of ASSP Walls

Figure 6(a) and (b) show respectively the position of the anchor and the main wall at the end of every earthquake applied in Test AF05. For the same test, Fig. 6(c) shows the contours of the incremental horizontal displacement of the soil on a cross section of the model during EQ3. Figure 6(d), (e), (f) report the same information for Test AF05.

At the end of Test AF05, the top of the wall had displaced horizontally by about 1.18 m, whereas the final displacement of the toe was limited, approximately 0.2 m. This corresponded to an outward rotation of the main wall of 9.8%. The final rotation of the anchor wall was approximately 1.5 times the rotation of the main wall, about 15%. Very limited displacements occurred during EQ4, which was applied after several stronger earthquakes. During Test AF05, the horizontal displacements of the soil in the backfill included between the anchor wall and the main wall, and above a surface extending from the toe of the anchor wall to dredge level, on the main wall, was approximately uniform. These characteristics are typical of a global failure [4].

Despite the fewer and weaker earthquakes applied, the displacements observed in Test AF06 were significantly larger than in Test AF05. The final displacements of the top and the toe of the main wall in Test AF06 were 2.2 m and 1.5 m respectively, which corresponds to an outward rotation of the main wall of 7%, slightly lower than that observed in Test AF05. However, in Test AF06, the rotation of the anchor wall was larger than in the other test, this being about 50%, or approximately 7 times the rotation of the main wall. The presence of porous fluid in the soil not only led to larger displacements, but also affected the failure mode of the main wall, causing its toe to experience large horizontal displacements. This was a consequence of soil softening resulting from the generation of excess pore pressures. Similar to what observed in Test AF05, also during EQ2 in Test AF06 the horizontal displacements of the soil between the two walls, and above a surface extending from the toe of the anchor to dredge level, was nearly uniform, again showing features of a global failure mechanism [4]. In saturated conditions, however, the horizontal displacements interested deeper strata of soil in the backfill, as large displacements occurred also on the retained side of the wall, below dredge level.

5 Conclusions

Two dynamic centrifuge tests were conducted on small scale models of ASSP walls in dry and saturated sand. PIV allowed to track the displacements of soil and structure on a cross section of the models during the applied earthquakes. The experimental results showed that, in dry conditions, the seismic waves in general amplified towards the ground surface and the largest amplification of the accelerations occurred near the main wall. Moreover, the main wall experienced significant outward rotation with limited displacement of the toe. In saturated conditions, large development of excess pore pressures caused a de-amplification of the horizontal accelerations in the soil deposit, and the wall suffered both an outward translation and a rotation. The contours of horizontal displacements calculated from PIV revealed that a global failure occurred in both tests, as the displacement of the upper part of the backfill was nearly uniform.

References

1. Eurocode 8, EN1998-5: Design of Structures for Earthquake Resistance-Part 5: Foundations, Retaining Structures and Geotechnical Aspects. European Committee for Standardization, Brussels (2004)
2. PIANC: Seismic design guidelines for port structures. In: Working Group No. 34 of the Maritime Navigation Commission, International Navigation Association (2001)
3. Conti, R., Viggiani, G.M.B., Cavallo, S.: A two-rigid block model for sliding gravity retaining walls. Soil Dyn. Earthq. Eng. **55**, 33–43 (2013)
4. Caputo, V.G., Conti, R., Viggiani, G.M.B., Prüm, C.: Improved method for the seismic design of anchored steel sheet pile walls. J. Geotech. Geoenv. Eng. **147**(2), 04020154 (2021)

5. Madabhushi, G.S.P., Haigh, S.K., Houghton, N.E., Gould, E.: Development of a servo-hydraulic earthquake actuator for the Cambridge Turner beam centrifuge. Int. J. Phys. Model. Geotech. **12**(2), 77–88 (2012)
6. Fusco, A., Viggiani, G.M.B., Madabhushi, G.S.P., Caputo, G., Conti, R., Prüm, C.: Physical modelling of anchored steel sheet pile walls under seismic actions. In: Proceedings of the 7th International Conference on Earthquake Geotechnical Engineering (2019)
7. Steedman, R.S., Madabhushi, G.S.P.: Wave transmission at a multi-media interface. In: Proceedings of the 5th International Conference on Soil Dynamics and Earthquake Engineering, Elsevier (1991)
8. Stanier, S.A., Blaber, J., Take, W.A., White, D.J.: Improved image-based deformation measurement for geotechnical applications. Can. Geotech. J. **53**(5), 727–739 (2015)

Response of Suction Bucket Foundation Subjected to Wind and Seismic Loading During Soil Liquefaction

Bin Gao[1,2], Guanlin Ye[1,2(✉)], Qi Zhang[1,2], and Wenxuan Zhu[1,2]

[1] State Key Laboratory of Ocean Engineering, Shanghai Jiao Tong University, Shanghai 200240, China
`{gaobinsd,ygl}@sjtu.edu.cn`
[2] Shanghai Key Laboratory for Digital Maintenance of Buildings and Infrastructure, Shanghai Jiao Tong University, Shanghai 200240, China

Abstract. Offshore wind turbines (OWTs) are now being constructed in seismic areas as part of a plan to achieve net-zero carbon dioxide emissions. Liquefaction caused by earthquakes poses a serious threat to the operation of OWTs. The dynamic response of the OWT on the suction bucket foundation in liquefiable sand under earthquakes combined wind loads was investigated in the present study. The FE-FD software named DBLEAVES-X with the Cyclic Mobility constitutive model was used in the research. The nonlinear dynamic analyses were carried out using a three-dimensional model of the OWT. The effects of earthquake, the wind load, sand density, and aspect ratio of the suction bucket on the behavior of the OWT were studied. The findings show that OWTs will be permanently tilted and displaced caused by wind, earthquake, and liquefaction. In seismic areas, this numerical model could be used in a comparative sense for the design of the OWT.

Keywords: Liquefaction · Earthquake · Suction bucket foundation

1 Introduction

As many countries begin to implement net-zero CO_2 emission plans in earnest, offshore wind power is showing impressive growth among the world's renewable energy sources. A suction bucket foundation is an effective OWT's foundation design solution with fast construction, low cost, and reusability. The suction bucket foundation resembles an inverted circular bucket, as shown in Fig. 1. The performance of OWTs under long-term wind loads and possible earthquakes is a major consideration in design.

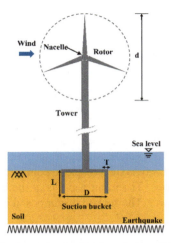

Fig. 1. A suction bucket foundation of OWT.

The performance of the suction bucket foundation under horizontal loads was investigated by model tests, centrifugal tests, and numerical simulations. Small-scale models were conducted out by Byrne and Houlsby [1] to investigate the behavior of suction bucket foundations under cyclic horizontal loads. Wang et al. [2] conducted centrifugal tests to reproduce the performance of suction bucket foundations under horizontal loads. Many studies have established numerical models of suction bucket foundations and investigated the bearing capacity of suction bucket foundations under horizontal loads [3, 4]. Many studies have focused on the performance of suction buckets under static loads, and less on the combined effects of seismic and static loads.

Over the last decades, the great majority of the OWTs have been installed in low seismicity areas. Thus, the earthquake and liquefaction have not been the main concerns in the design of OWTs. New wind farms have to be built in seismically active areas (Fujian and Guangdong, China) due to site constraints and the distribution of wind resources. The offshore seabed consists mainly of silty sands, which is easily liquefied. Earthquake-induced liquefaction assessment is necessary for the construction of OWTs in seismic areas. The seismic performance of piles, mudmat foundations, and caissons were studied by Kaynia [5]. During the design phase, the effects of liquefaction on soil and foundation performance shall be evaluated [6], and procedures for evaluating suction bucket foundations have not been established.

To accurately describe the dynamic response of soil and structure under liquefaction, it is necessary to establish an advanced constitutive model. The poroelastic seabed was used in many previous researches [7, 8] and the plastic deformation of the soil is not well represented. The elastic-plastic model, Cyclic Mobility (CM) constitutive model, is advantageous in simulating sand liquefaction [9, 10]. The CM model is embedded in the FE-FD code to investigate the behavior of OWTs in the liquefiable sand under wind and earthquakes. The effects of earthquake, the wind load, sand density, and aspect ratio of the suction bucket on the behavior of the OWT were studied.

2 Numerical Model of Suction Bucket

2.1 Geometry and Modeling

Based on a 3.5 MW OWT, a 3D OWT on suction bucket foundation was modeled, as shown in Fig. 2. Table 1 shows the essential features of the 3.5 MW OWT. The size of the seabed was 160 m long, 60 m wide, and 70 m high. The diameter of the suction bucket is 21 m, and the length of the skirt is 20 m. This tower stands at a height of 80 m. In the numerical simulation, the model dimensions (6 D and 3.5 D) are sufficient to eliminate boundary effects. All boundaries except for the top are undrained boundaries. The bottom boundary is fixed. Equivalent displacement or periodic boundaries are used in the left and right boundaries. The input wave is at the bottom boundary. The wind load acts on the top of the tower. The seabed is composed of Toyoura sand.

Table 1. The basic properties of 3.5 MW offshore wind turbine.

Type	Hub height (m)	Tower Young's modulus (Pa)	Density of tower (kg/m^3)	Rotor diameter (m)	Nacelle + rotor mass (t_n)
3.5 MW	80	210×10^9	8500	80	220

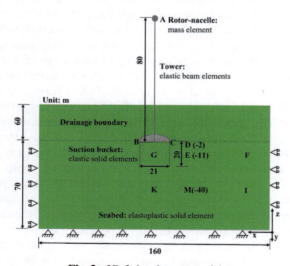

Fig. 2. 3D finite element model.

2.2 Load Conditions

Three working conditions of the OWTs were carried out, including operational and seismic conditions (wind loads and earthquakes), parked and seismic conditions (only earthquakes), operational conditions (only wind loads). Load cases are shown in Table 2. The frequency of input waves is 4 Hz. The sine waves $a = k \cdot \sin(2\pi ft)$ ($k = 0.1$ g, 0.2 g, and 0.4 g indicate seismic intensity VII, VIII, and IX, respectively) were used in the numerical calculation. 0.95 (aspect ratio, L = 20 m and D = 21 m), VIII (seismic intensity), medium sand, and 1600 kN (wind load) were the primary parameters for a typical situation. The time of the earthquake is 20 s.

Table 2. Load cases under different design conditions.

Design conditions	Wind loads (kN)	Earthquake (g)	Soil	Aspect ratio (L/D)
Operational	1600	NA	Medium sand	0.95
Parked and seismic	NA	0.2	Medium sand	0.95
Operational and seismic	1600	0.1	Loose sand	0.54 0.65
		0.2	Medium sand	0.76 0.95
		0.4	Dense sand	1.14 1.33
				1.52 1.71

3 Results

3.1 Effect of Working Condition

The time histories of accelerations at monitoring points A, G, and K under the various OWT conditions are shown in Fig. 3. They are the same in the two conditions. Wind loads play a minor role in the development of accelerations. Before 5.6 s, there are acceleration amplifications at the top of the nacelle (point A), and then the accelerations rapidly decline to near zero after 8.1 s. Similar occurrences can also be found in the foundation (point G). Under the foundation, however, the accelerations do not decay with time (point K). The EPWPR development of monitoring points (G, K, E, and M) in the seabed is compared in Fig. 4 under various OWT conditions. The rise of EPWPRs and the emergence of liquefaction in the earthquake can explain these events. Liquefaction begins in the shallow seabed, then progresses downward and weakens over time. Due to liquefaction, the sandy seabed softens and loses its ability to transfer shear stresses. The acceleration received by the top turbines is reduced as a result of liquefaction in the seabed. The order of EPWPRs is E > G > M > K. At the same depth, the EPWPRs inside and below the suction bucket are lower than those outside the suction bucket.

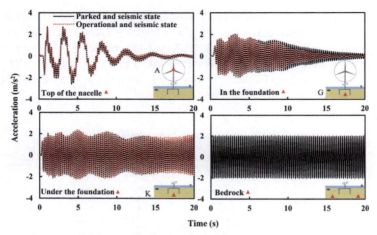

Fig. 3. Time histories of accelerations at monitoring point A, G, and K during different OWT conditions.

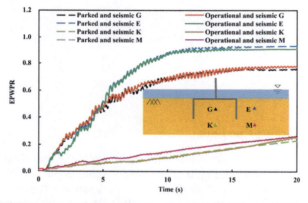

Fig. 4. EPWPRs of monitoring points in seabed during different OWT conditions.

While changes in acceleration and EPWPRs are difficult to detect under various OWT conditions, the variances in horizontal displacement, settlements, and rotation are obvious. The OWT's horizontal displacements, settlements, and rotations are compared in Fig. 5 under three OWT conditions. The only negligible distortion appears in the horizontal displacements, settlements, and rotations under operational conditions. Due to the symmetry of the sine wave, the tilt occurs in the former cycle and recovers in the latter cycle, producing a small residual tilt under earthquake conditions. As can be seen in Fig. 5, the rotation accumulates and exceeds the maximum allowable rotation (0.25°) under operational and seismic conditions.

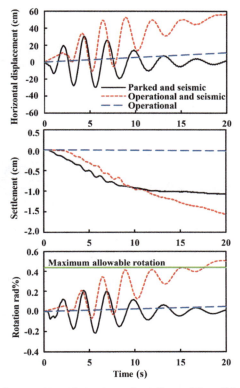

Fig. 5. Horizontal displacements, settlements, and rotations of the offshore wind turbine during different OWT conditions.

3.2 Effect of Sand Density

The time histories of accelerations at monitoring points A, G, and K under varied sand density conditions are compared in Fig. 6. Among the three sands, the acceleration at points A and G in the loose sand decays the earliest and fastest. The acceleration at points A and G in dense sand decreases the latest and slowest among the three sands. Loose sands are the easiest to liquefy, while dense sands are difficult to liquefy, as shown in Fig. 7. The shear waves cannot be transmitted through liquefied soil.

Figure 8 shows the influence of the sand density on the horizontal displacements, settlements, and rotations of the OWT. With decreasing sand density, OWT's horizontal displacement, settlement, and rotation increase. The rotation of the OWT in dense sand is reduced by 87.2% compared with the loose sand. It can be observed that the rotation of the OWT in loose and dense sands exceeds 0.25°.

1090 B. Gao et al.

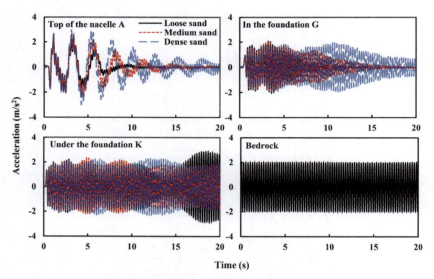

Fig. 6. Accelerations of monitoring points in loose, medium, and dense sand.

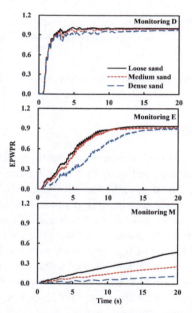

Fig. 7. EPWPRs of monitoring points in loose, medium, and dense sand.

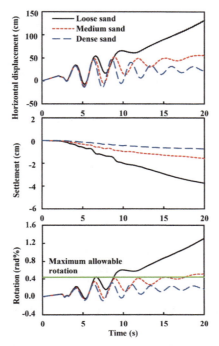

Fig. 8. Horizontal displacements, settlements, and rotations of OWT in loose, medium, and dense sand.

3.3 Effect of Seismic Intensity

Figure 9 represents the effect of seismic intensity on OWT's acceleration. It is self-evident that as the input acceleration rises, so does the OWT's reaction acceleration. This assumption holds true for the first 0–3.4 s of the earthquake. After 8.1 s, however, the acceleration is the polar opposite of the assumption. The variation of OWT acceleration is determined by the liquefaction process in a sandy seabed. The rate of seabed liquefaction increases with increasing input acceleration, as shown in Fig. 10. Differences in the onset of seabed liquefaction result in phase differences in acceleration. As the seismic intensity rises, the OWT's horizontal displacements, settlements, and rotations grow, as shown in Fig. 11. In particular, the remaining rotation exceeds the allowed rotation at 0.2 g and 0.4 g. OWTs in high-intensity areas should be rigorously evaluated for seismic hazards.

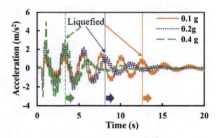

Fig. 9. Influence of seismic intensity on the acceleration of OWT.

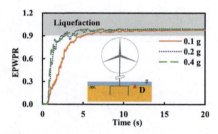

Fig. 10. Influence of seismic intensity on EPWPR.

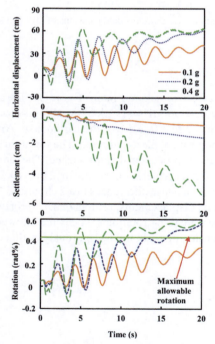

Fig. 11. Influence of seismic intensity on horizontal displacement, settlement, and rotation of OWT.

3.4 Effect of Aspect Ratio

The effect of varying geometric dimensions on horizontal displacements, settlements, and rotations of OWTs are evaluated in Fig. 12. The final rotations and horizontal displacements can be successfully decreased by raising the aspect ratio of the suction bucket. The long skirt increases the area of contact with the unliquefied soil, and the unliquefied soil can provide more resistance. After L/D > 1.3, the final displacements and rotations are stable. These indicate that increasing the skirt length can no longer reduce the rotations and displacements.

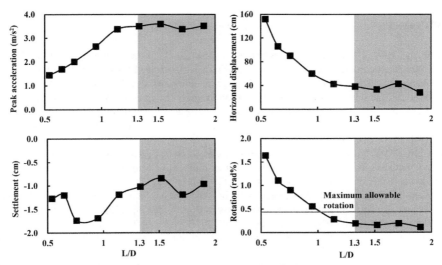

Fig. 12. Influence of aspect ratios on peak acceleration, horizontal displacement, settlement, and rotation of OWT.

4 Conclusions

Under earthquakes combined wind loads, the dynamic response of the OWT on the suction bucket foundation in liquefiable sand was investigated by the DBLEAVES-X. The effects of earthquake, the wind load, sand density, and aspect ratio of the suction bucket on the behavior of the OWT were studied. The main findings are as follows.

- In different conditions, no significant differences were observed for accelerations and EPWPRs, while significant gaps were observed for horizontal displacements, settlements, and rotations of the OWT. The mechanism of the rotation is explained by wind anisotropy and the degradation of soil strength due to liquefaction. OWT's tilt exceeds the limit tilt under winds and earthquakes.
- With decreasing sand density, OWT's horizontal displacement, settlement, and rotation increase. The rotation of the OWT in dense sand is reduced by 87.2% compared

with the loose sand. The rate of seabed liquefaction increases with increasing input acceleration. Differences in the onset of seabed liquefaction result in phase differences in acceleration. As the seismic intensity rises, the OWT's horizontal displacements, settlements, and rotations grow. OWTs in high-intensity areas should be rigorously evaluated for seismic hazards.
- The final rotations and horizontal displacements can be successfully decreased by raising the aspect ratio of the suction bucket. After L/D > 1.3, increasing the skirt length can no longer reduce the rotations and displacements.

References

1. Byrne, B.W., Houlsby, G.T.: Experimental investigations of the response of suction caissons to transient combined loading. J. Geotech. Geoenviron. Eng. **130**(3), 240–253 (2004)
2. Wang, X.F., Yang, X., Zeng, X.W.: Centrifuge modeling of lateral bearing behavior of offshore wind turbine with suction bucket foundation in sand. Ocean Eng. **139**, 140–151 (2017)
3. Ahlinhan, M.F., Houehanou, E.C., Koube, B.M., Sungura, N.: 3D Finite element analyses of suction caisson foundations for offshore wind turbines in drained sand. Int. J. Geotech. Eng. **14**(1), 110–127 (2020)
4. Choo, Y.W., Kim, D.J., Youn, J.U., Hossain, M.S., Seo, J., Kim, J.H.: Behavior of a monopod bucket foundation subjected to combined moment and horizontal loads in silty sand. J. Geotech. Geoenviron. Eng. **147**(5), 04021025 (2021)
5. Kaynia, A.M.: Earthquake geotechnics in offshore engineering. In: Pitilakis, K. (ed.) Recent Advances in Earthquake Engineering in Europe, pp. 263–288. Springer International Publishing, Cham (2018). https://doi.org/10.1007/978-3-319-75741-4_11
6. Det Norske Veritas: Design of Offshore Wind Turbine Structures, Offshore Standard DNV-OS-J101 (2014). [Edition]
7. Ye, J.: Seismic response of poro-elastic seabed and composite breakwater under strong earthquake loading. Bull Earthquake Eng. **10**(5), 1609–1633 (2012)
8. Ye, J.H., Jeng, D.S.: Three-dimensional dynamic transient response of a poro-elastic unsaturated seabed and a rubble mound breakwater due to seismic loading. Soil Dyn. Earthquake Eng. **44**, 14–26 (2013)
9. Bao, X.H., Xia, Z.F., Ye, G.L., Fu, Y.B., Su, D.: Numerical analysis on the seismic behavior of a large metro subway tunnel in liquefiable ground. Tunn. Undergr. Space Technol. **66**, 91–106 (2017)
10. Gao, B., Ye, G.L., Zhang, Q., Xie, Y., Yan, B.: Numerical simulation of suction bucket foundation response located in liquefiable sand under earthquakes. Ocean Eng. **235**, 109394 (2021)

A Class of Thermodynamic Inertial Macroelements for Soil-Structure Interaction

Davide Noè Gorini(✉) and Luigi Callisto

Sapienza University of Rome, Rome, Italy
davideno.gorini@uniroma1.it

Abstract. The seismic performance of structures can be significantly influenced by the interaction with the foundation soils, with effects that depend on the frequency content and the amplitudes of the ground motion. A computationally efficient method to include these effects in the structural analysis is represented by the macroelement approach, in which a geotechnical system is modelled with a single macroelement that describes the generalized force-displacement relationship of the system. While this method has been mainly developed for shallow foundations, the present study proposes a class of macroelements representing the macroscopic response of different foundation types, including abutments, piled and caisson foundations. The generalized force-displacement relationships for these models are elastic-plastic and are derived using a rigorous thermodynamic approach. The plastic responses of the macroelements are bounded by the ultimate capacities of the geotechnical systems, while the inertial effects associated with the soil mass involved in the dynamic response of the structure are simulated by introducing appropriate participating masses. The macroelements are implemented in OpenSees; in this paper they are applied to assess the seismic performance of a tall viaduct showing highly nonlinear features.

Keywords: Dynamic soil-structure interaction · Nonlinear behaviour · Inertial effects · Thermodynamic framework · OpenSees

1 The Response of Geotechnical Systems in Structural Analysis

During the last years the need to include the effects of soil-structure interaction in the assessment of the seismic risk of structures has gained increased interest worldwide. These effects are particularly evident for bridges because of the recurring presence of abutments and deep foundations, that imply a large participation of soil mass to the dynamic response of the structure [1–6].

It is well known that the dynamic behaviour of geotechnical systems is markedly nonlinear starting from small strain levels, showing amplification at the resonance frequencies of the soil-foundation system. This combined nonlinear and frequency-dependent response may increase the ductility demand for the structural members [5], and this effect cannot be evaluated through the usual substructure approach, which is based on linear behaviour. The strength and stiffness properties at the macroscopic scale, that refers to the entire soil-foundation system, are moreover affected by the loading direction. This effect reflects also on the mass participation of the soil, as demonstrated for foundations [7, 8] and bridge abutments [9, 10].

In this paper, a class of macroelements for geotechnical systems is proposed as constitutive relationships able to reproduce the soil-structure interaction phenomena discussed above with a minimal computational effort. The main ingredients of the adopted thermodynamic formulation are described below, and an illustrative example on the use of the macroelements in the structural analysis is finally shown.

2 A Potential-Based Formulation

The macroelements (MEs) developed in this work are aimed at simulating the response of geotechnical systems, such as foundations and bridge abutments, in the numerical analysis of structures. Each ME relates the generalised forces Q_i (three forces and three moments), exchanged between the geotechnical system and the superstructure, to the corresponding displacements and rotations q_j through a second-order tangent stiffness matrix H_{ij}, such that $Q_i = H_{ij} \times q_j$.

The MEs are multi-surface plasticity models with kinematic hardening derived with a consistent thermodynamic approach, using hyperplasticity [11]. The constitutive response requires the definition of two potentials, namely the energy and dissipation functions, formulated to provide a multi-axial, frequency-dependent and hardening response. The dissipative response is based on some primary assumptions, that are the validity of Ziegler's principle [12], the additive decomposition of the elastic and plastic components of deformations and the associativity of the plastic flows, so that the dissipation function can be obtained by the yield surfaces.

The constitutive ingredients needed to compute the potentials are 1) the identification of the mass and stiffness tensors at small displacements (elastic response) and 2) the evaluation of the ultimate limit state surface bounding the plastic domain.

3 Energy Function

The proposed MEs represent a multi-axial generalisation of the model developed in [5] for bridge abutments, shown in Fig. 1. The energy function for all the MEs, expressed by Gibbs free energy g, therefore reads:

$$g\left(Q_i^{(1)}, q_i^{(n)}, m_i^{(n)}\right) = -\frac{1}{2} \cdot C_{ij}^{(0)} \cdot Q_j^{(0)} \cdot Q_i^{(0)} - \sum_{n=1}^{N} Q_i^{(n)} \cdot q_i^{(n)} + \frac{1}{2} \cdot \sum_{n=1}^{N} H_{ij}^{(n)} \cdot q_j^{(n)} \cdot q_i^{(n)}$$

$$-Q_i^{(R)} \cdot q_i^{(R)} - \sum_{n=1}^{N} m_{ij}^{(n)} \cdot \sum_{h=n}^{N} \ddot{q}_j^{(h)} \cdot \sum_{k=n}^{N} q_i^{(k)} + \frac{1}{2} \cdot \sum_{n=1}^{N} m_{ij}^{(n)} \cdot \sum_{h=n}^{N} \dot{q}_j^{(h)} \cdot \sum_{k=n}^{N} \dot{q}_i^{(k)} \quad (1)$$

According to Eq. 1, for sufficiently small displacements q_j the response is purely elastic and is controlled by the elastic free energy $0.5 \cdot C_{ij}^{(0)} \cdot Q_j^{(0)} \cdot Q_i^{(0)}$, that is proportional to the square of the elastic force vector $Q_j^{(0)}$ through the second-order compliance matrix $C_{ij}^{(0)}$. When the total force vector Q_i reaches the innermost yield surface the response becomes elastic-plastic. In this case plastic displacements occur with amplitude and direction that depend on the number of plastic flows activated. During plastic loading, the force $Q_i^{(n)}$ developing in the nth plastic flow spends work in the respective plastic displacement $q_i^{(n)}$ (term $Q_i^{(n)} \cdot q_i^{(n)}$ in Eq. 1); moreover, by virtue of the kinematic hardening the macroelement stores energy through the second order kinematic tensors $H_{ij}^{(n)}$ (term $H_{ij}^{(n)} \cdot q_j^{(n)} \cdot q_i^{(n)} > 0$). The term $Q_i^{(R)} \cdot q_i^{(R)}$ is associated with an additional irreversible displacement vector $q_i^{(R)}$ aimed at reproducing the ratcheting phenomenon, as already proposed in [13] for piles. The energetic contributions described so far define completely the nonlinear behaviour of the ME, while the frequency-dependent features are enclosed in the last two terms of Eq. 1, that are proportional to the second order mass tensors $m_{ij}^{(n)}$ associated with the plastic flows.

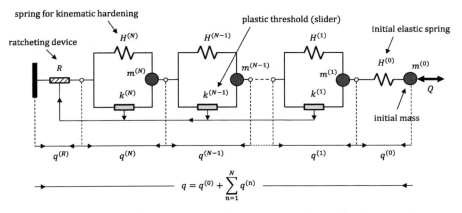

Fig. 1. One-dimensional layout of the inertial, multi-surface hyper-plastic macroelement.

4 Dissipative Response

Energy dissipation occurs when the MEs exhibit a plastic response, that is when the force state is within the plastic domain. In the context of multi-surface plasticity, the plastic response is obtained through a series of yield surfaces $y^{(n)}$ ($n = 1, \ldots, N$) defined in the force space, that provide a multi-linear response up to the attainment of the ultimate conditions of the system, described by the ultimate limit state surface. Since the internal yield surfaces are homothetic to the ultimate locus, the definition of the plastic domain requires only the definition of the ultimate capacity of the geotechnical system under multi-axial loading. The following models of ultimate surface are here taken into consideration, as schematically shown in Fig. 2:

- shallow foundations (Fig. 2a): translated hyper-ellipsoid in the space of the forces, Q_1-Q_2-Q_3, and moments, Q_{R1}-Q_{R2}-Q_{R3}, exchanged between the foundation and the superstructure, as proposed by Martin et al. [14];

- deep foundations (Fig. 2b): hyper-egg with super-elliptical generatrices describing the combinations of forces, Q_1-Q_2-Q_3, and moments, Q_{R1}-Q_{R2}, producing failure of a pile group [15];
- caisson foundations (Fig. 2c): roto-translated hyper-ellipsoid in the generalised force space, Q_i ($i = 1, 2, 3, R1, R2, R3$), relative to the force transfer between foundation and superstructure in presence of sloping ground (developed in this work);
- integral bridge abutments (Fig. 2d): roto-translated hyper-ellipsoid in the generalized force space, Q_i, representing the force exchange at the deck-abutment contact, as an extension of the model by Gorini et al. [16] for semi-integral abutments.

The reader can refer to the papers above for the description of the calibration procedures and to [3, 5] for the derivation of the incremental elastic-plastic response.

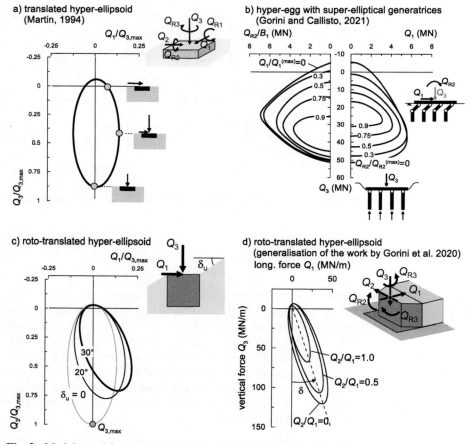

Fig. 2. Models used for the ultimate limit state surface of the macroelements, for the case of a) shallow foundations, b) deep foundations, c) caisson foundations, d) integral bridge abutments.

5 Implementation and Analysis Method

The MEs are implemented in the analysis framework OpenSees [17] as multi-axial materials. Each material can be assigned to a novel zero-length finite element, named *ZeroLength6D*, simulating the full translational-rotational coupling of the response between the two overlapped nodes. In the current version in OpenSees, the response of the new materials does not include inertial effects, hence only the effect of the first mass tensor $m_{ij}^{(0)}$ can be reproduced by assigning the corresponding masses to the soil-structure contact node in the global structural model. As a result, the MEs can reproduce with good accuracy the frequency-dependent response of the geotechnical system from small to medium strain levels, while they may somewhat underestimate the period lengthening towards failure [3, 5].

The analysis procedure is the one proposed in [3, 5], based on which the subsoil is considered as composed of two regions: the near field, intended as the soil zone interacting with the structure whose response is simulated by means of the ME, and the far field not influenced by the presence of the structure. In this view, the propagation of the seismic waves from the bedrock up to the lower boundary of the near field, called effective depth z_{eff}, can be studied separately through a free field site response analysis. The motion computed at z_{eff} represents the seismic input for the MEs in the global structural model to carry out dynamic analyses in the time domain. The effective depth can be taken as $z_{\text{eff}} = 10 \times D$ for deep foundations (D is the pile diameter) and $z_{\text{eff}} = L_{\text{f}}$ or $\max\{L_{\text{f}}, 10 \times D\}$ for bridge abutments with shallow and deep foundations, respectively (L_{f} is the width of the abutment foundation).

6 Illustrative Example and Discussion

The proposed MEs are now employed to analyse the seismic performance of the idealised case study shown in Fig. 3a. The bridge rests on five supports and crosses a V-shaped valley whose subsoil reflects typical layering and mechanical features of the Apennine area of central Italy. The subsoil is composed of four layers, S1 to S4 in the figure, with shear wave velocity, V_{s}, increasing with depth. At the location of abutment A1, V_{s} ranges from 200 m/s at the foundation level to 1000 m/s at the bedrock, the latter encountered at a depth of 95 m. The strength of the soil layers is described by a cohesion of 10 kPa and angle of shearing resistance in the range 24°–26°.

The structural members were designed referring to Italian technical provisions (Italian Building Code, 2018), considering the seismic demand for the site at hand. According to the static scheme of the bridge, in the longitudinal direction the deck is connected only to piers P2 and P3. The abutment A2 is connected longitudinally to the deck using viscous dampers, while relative deck-abutment deformations at A1 are free. In the transverse direction, the deck is connected to all piers and abutments. The foundation piles have diameters ranging between 1.0–1.2 m.

The seismic performance of the bridge was investigated using a numerical representation of the bridge model with the MEs implemented in OpenSees. The results of this analysis are compared with the ones of a dynamic analysis of the structural system in which soil-structure interaction is neglected. As per the calibration of the MEs, here omitted for brevity, one can refer to [5, 10] for the evaluation of the modal characteristics of the soil-abutment system, and for instance to the commercial software DYNA [18] for the identification of the dynamic behaviour of piled foundations; the definition of the plastic domain follows the rationale exposed in Sect. 4. The resulting dynamic responses at small displacements are characterised by fundamental vibration periods of 0.15 s to 0.25 s for the abutments and of 0.05 s to 0.11 s for the pier foundations. The piers were modelled as displacement-based beam elements with hollow sections reproducing the properties of the effective reinforced concrete cross section. The Kent-Scott-Park model

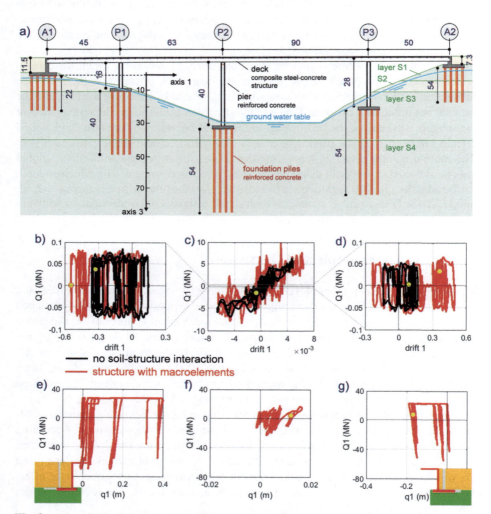

Fig. 3. a) longitudinal section of the reference soil-bridge system (lengths in meters), and representation of the seismic performance in terms of: b,d) shear force-drift responses of the bearing devices on A1 and A2, respectively, c) shear force-drift response at the base of P2, e,f,g) longitudinal, force-displacement responses of the MEs for A1, P2 and A2.

[19] was assigned to the concrete fibers and an elastic-plastic material with kinematic hardening to the steel fibers. The P-δ transformation was used to account for geometric nonlinearity effects. The stiffness and strength of the bearing devices were simulated through the combination of nonlinear rheological elements. The deck was reproduced through equivalent elastic force-based beam elements.

In the numerical model with MEs, a staged analysis procedure was adopted, composed of a first gravitational stage and the subsequent application of a three-component seismic motion following the procedure in Sect. 5. The results shown in the following refer to a no-collapse earthquake scenario. The bridge performance is concisely quantified in terms of the shear force-drift responses of the bearing devices on the abutments A1 and A2, in Figs. 3b and d, the shear force-drift response at the base of pier P2, in Fig. 3c, the longitudinal force-displacement responses of the MEs of abutment A1, pier P2 and abutment A2, in Figs. 3e and f. It is evident that the nonlinear response of the geotechnical systems magnifies the displacements of the superstructure compared to the fixed-base model, causing significant permanent effects that increase the displacement and ductility demands for the bearing devices and the piers members, respectively. The abutments show a more pronounced nonlinear response than the piers foundations because of the higher participation of the soil mass involved in the embankment. The abutments attain the active resistance (force cut-off in Figs. 3e,g) and accumulate irreversible displacements towards the centre of the bridge. However, this permanent effect is less evident for the strong abutment, A2, due to the smaller height and to the interaction with the superstructure.

The above results point to the need to account for a combined frequency-dependent and nonlinear response of the geotechnical systems in the assessment of the structural performance. The proposed macroelements may be employed in large parametric studies and for a direct verification of the preliminary design assumptions, as they require limited computational resources.

Acknowledgements. Part of this research was sponsored by the Italian Department of Civil Protection (Project RINTC 2019/21). The Authors acknowledge the contribution of Prof. P. Franchin and Dr. F. Noto for the development of the structural model of Fig. 3.

References

1. Elgamal, A., Yan, L., Yang, Z., Conte, J.P.: Three-dimensional seismic response of humboldt bay bridge-foundation-ground system. J. Struct. Eng. **134**(7), 1165–1176 (2008). https://doi.org/10.1061/(ASCE)0733-9445(2008)134:7(1165)
2. Stefanidou, S.P., Sextos, A.G., Kotsoglou, A.N., Lesgidis, N., Kappos, A.J.: Soil-structure interaction effects in analysis of seismic fragility of bridges using an intensity-based ground motion selection procedure. Eng. Struct. **151**, 366–380 (2017)
3. Gorini, D.N.: Soil–structure interaction for bridge abutments: two complementary macroelements. PhD thesis. Sapienza University of Rome, Rome, Italy (2019). https://iris.uniroma1.it/handle/11573/1260972
4. Gorini, D.N., Callisto, L.: A coupled study of soil-abutment-superstructure interaction. In: Calvetti, F., Cotecchia, F., Galli, A., Jommi, C. (eds.) Geotechnical research for land protection and development - proceedings of CNRIG 2019, pp. 565–574. Springer, Cham, Switzerland (2020). https://doi.org/10.1007/978-3-030-21359-6_60

5. Gorini, D.N., Callisto, L., Whittle, A.J.: An inertial macroelement for bridge abutments. Géotechnique **72**(3), 247–259 (2020). https://doi.org/10.1680/jgeot.19.P.397
6. Gallese, D., Gorini, D.N., Callisto, L.: On a novel seismic design approach for integral abutment bridges based on nonlinear static analysis. In: 4th International Conference on Performance-based Design in Earthquake Geotechnical Engineering. China (2022)
7. Paolucci, R., Pecker, A.: Seismic bearing capacity of shallow strip foundation on the dry soils. Soils Found. **37**(3), 95–105 (1997)
8. Pane, V., Vecchietti, A., Cecconi, M.: A numerical study on the seismic bearing capacity of shallow foundations. Bull. Earthq. Eng. **14**(11), 2931–2958 (2016). https://doi.org/10.1007/s10518-016-9937-0
9. Kotsoglou, A.N., Pantazopoulou, S.J.: Bridge-embankment interaction under transverse ground excitation. Earthquake Eng. Struct. Dynam. **36**(12), 1719–1740 (2007). https://doi.org/10.1002/eqe.715
10. Gorini, D.N., Callisto, L., Whittle, A.J.: Dominant responses of bridge abutments. Soil Dyn. Earthq. Eng. **148**, 106723 (2021). https://doi.org/10.1016/j.soildyn.2021.106723
11. Collins, I.F., Houlsby, G.T.: Application of thermomechanical principles to the modeling of geotechnical materials. Proc. R. Soc. Lond. Ser. A **453**(1964), 1975–2001 (1997)
12. Ziegler, H.: An introduction to thermomechanics. Amsterdam, the Netherlands: North Holland (1977)
13. Houlsby, G.T., Abadie, C., Beuckelaers, W., Byrne, B.: A model for nonlinear hysteretic and ratcheting behaviour. Int. J. Solids Struct. **120**, 67–80 (2017)
14. Martin, C.M.: Physical and numerical modeling of offshore foundations under combined loads. PhD thesis. University of Oxford (1994)
15. Gorini, D.N., Callisto, L.: Generalised ultimate loads for pile groups. Accepted for publication in Acta Geotechnica **17**, 2495–2516 (2022). https://doi.org/10.1007/s11440-021-01386-4
16. Gorini, D.N., Whittle, A.J., Callisto, L.: Ultimate limit states of bridge abutments. J Geotech Geoenviron Eng **146**(7), (2020). https://doi.org/10.1061/(ASCE)GT.1943-5606.0002283
17. McKenna, F., Scott, M.H., Fenves, G.L.: OpenSees. nonlinear finite-element analysis software architecture using object composition. J. Compu. Civil Eng. **24**, 95–107 (2010). https://doi.org/10.1061/(ASCE)CP.1943-5487.0000002
18. El Naggar, M.H., Novak, M., Sheta, M., El-Hifnawy, L., El-Marsafawi, H.: DYNA6.1 - a computer program for calculation of foundation response to dynamic loads. Geotechnical Research Centre, University of Western Ontario, London, Ontario (2011)
19. Scott, B.D., Park, R., Priestley, M.J.N.: Stress-strain behavior of concrete confined by overlapping hoops at low and high strain rates. ACI J **79**, 13–27 (1982)

Simulation Analyses of Centrifuge Model Tests on Piled Raft Foundation with Deep Mixing Walls

Junji Hamada[1]([✉]), Takehiro Okumura[1], Yoshimasa Shigeno[1], and Yoshihiro Fujiwara[2]

[1] Takenaka Research and Development Institute, Chiba, Japan
hamada.junji@takenaka.co.jp
[2] Mind Inc., Chiba, Japan

Abstract. Three-dimensional finite element analyses were conducted to simulate the dynamic centrifuge model tests on piled raft foundation with deep mixing walls (DMWs). The DMWs is one of the most efficient liquefaction countermeasure methods. After the shaking model tests, the DMWs were observed that several cracks were locally induced in the wall. However the DMWs were quite effective at reducing the sectional force of the piles to an acceptable level under the large earthquake load. The stresses in the DMWs, which was difficult to measure in the tests, were investigated by the numerical analyses. The shear stresses on the DMWs in the analyses were almost consistent with the inertia force of the structure similar to the experimental results. In addition, tensile stresses of about 1.5 MPa were generated, which were more than 0.2 times higher than an unconfined compressive strength of 6.33 MPa, confirming the experimental results of cracking in the DMWs. The model test results were reproduced by the numerical analyses considering the separation from the ground around the pile and at the base of the foundation and the sliding.

Keywords: Piled raft foundation · Ground improvement · Finite element method · Non-linear analysis · Centrifuge model test

1 Introduction

Recently piled raft foundations combined with grid-form cement deep mixing walls (DMWs), which is one of the most efficient liquefaction countermeasure methods, was developed and applied to actual buildings on soft clayey ground or liquifiable ground [1, 2]. The piled raft foundation with DMWs showed a good performance during the 2011 off the Pacific coast of Tohoku Earthquake. And dynamic centrifuge model tests on the piled rafts with DMWs were conducted to investigate the seismic behavior during a large earthquake [3]. The model tests have used real soil-cement models with an unconfined compressive strength of about 6330 kPa in order to investigate the sectional forces on piles during large earthquakes by which the DMWs can be partially failed. After the shaking tests, the DMWs were observed that several cracks were locally induced in the wall. However the DMWs were quite effective at reducing the sectional force of the piles to an acceptable level under the large earthquake load. In general, the local failures of DMWs are not acceptable in the existing method of allowable stress design. If a local failure can be appropriately taken into account for the foundation design, the DMWs and piles enclosed by it can be designed more rationally, using a performance-based design method.

This paper offers three-dimensional finite element analyses that simulate the centrifuge model tests considering the separation from the ground around the pile and at the base of the foundation and the sliding. The stresses in the DMWs, which was difficult to measure in the tests, were investigated and the test results were supported.

2 Centrifuge Model Tests

2.1 Setting Up the Model

Dynamic centrifuge model tests on a piled raft foundation with soil-cement column walls (DMWs) were performed at an acceleration of 50 g [3]. Figure 1 shows a schematic view of the test model. The scaling ratio is 1/50. The inner dimensions of the laminar shear box are 1000 mm long, 300 mm wide and 550 mm high. A 2 mm-thick membrane was attached inside the box to make it watertight. The superstructure was modeled after a 13 story residential building of 40 m high. Model piles reproduced prototype precast concrete piles with a diameter of 0.8 m. Four piles made of aluminum pipes with a diameter of 16.0 mm and a thickness of 1.0 mm were connected to a slab plate through pile caps. The building model was made of steel parts to reproduce axial force on a column (9330 kN), and the natural period of the superstructure was 0.412 s at the 1st mode in a prototype scale.

Cement-treated silica sand with an unconfined strength of about 1400 kPa was used as the bearing stratum under the pile tips. The model ground was made of Kanto loam soil as to be soft cohesive ground, with a shear wave velocity of about 100 m/s. Table 1 shows the physical and mechanical properties of the loam for the centrifuge models [4].

Models of DMWs were made from Iide silica sand, kaolin clay and blast-furnace slag cement type B, with a W/C of 60%. After setting a clayey ground model, continuous holes with a diameter of 18 mm (0.9 m in prototype scale) are created in the model ground by using a drill machine and poured soil-cement into the holes and tamping it. The model longitudinal wall was set at the center of the laminar shear box in order to avoid an effect of the shear box's boundary. The unconfined compressive strength of the soil-cement is 6330 kPa on the test day. Sandpaper was attached beneath the steel structure model (slab plate) to increase the friction between the structure model and the DMWs.

2.2 Earthquake Waves and Test Cases

Shaking table tests were carried out, increasing the acceleration level of the input motion step by step. A series of the shaking table tests were conducted after a small sine sweep motion. Though peak acceleration levels and corresponding waveforms of the waves used for the tests were selected so that the levels might be gradually increased (from the lowest to the highest) step by step.

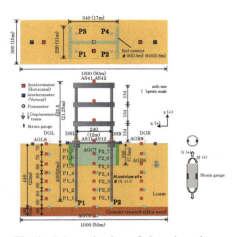

Table 1. Physical and mechanical properties of loam used in centrifuge model.

Wet density, ρ_t (g/cm^3)	1.538
Void ratio, e	1.756
Water content, w (%)	63.1
Degree of saturation, Sr (%)	93.4
Cohesion, c' (kN/m^2)	23.5
Friction angle, ϕ' (deg.)	35.7

Fig. 1. Schematic view of plate shaped building model with DMWs and soft clayey ground.

3 Simulation Analyses of Model Tests

The ground, the piles and the grid-form ground improvement were modelled in as much detail as possible using a three-dimensional finite element method (FEM) and conducted simulation analyses for the centrifuge model tests. The model mesh is divided finely around the piles as shown in Fig. 2. The lateral boundaries for normal direction are periodic boundaries because of using laminar shear box, while the bottom boundary is a fixed boundary. The numerical analysis code is the in-house program called MuDIAN.

3.1 Three-Dimensional Finite Element Model

Modeling of Structure

In order to match the actual layout of the test model, the columns were modelled using beam elements and the floors were modelled using plate elements and the foundation slab was modelled as solid elements. These materials were assumed as linear. The damping ratio, h of the building is a stiffness proportional damping, which is 2% at the first order natural frequency, 1.4 Hz of the coupled system of building and ground.

Modeling of Piles and Grid-Form Deep Mixing Walls (DMWs)

The piles are modeled as the beam elements considering their volume, as shown in Fig. 2(b). The beam nodes and the ground nodes on the circumference of the piles are joined by rigid bars. The joint elements are set the interface nodes controlling the separation between pile and ground. The characteristics of the joint elements are shown in Fig. 3. In the normal direction, only compressive load is transmitted and its stiffness K_n is set to 10^7 kN/m^2/m. On the other hand, in the tangential (in-plane) direction, the stiffness K_s was set to be sufficiently large (10^7 kN/m^2/m) and $\tau_{max} = \sigma_n \tan\phi$ (friction angle, $\phi = 10°$) to represent the slip between the pile and the ground as a bilinear spring.

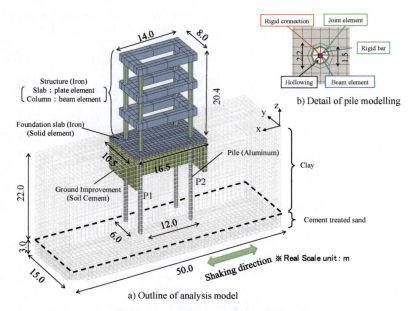

Fig. 2. Schematic view of three-dimentional finite element model.

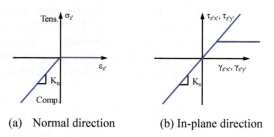

Fig. 3. Characteristics of joint elements employed in simulation analysis.

The grid-form ground improvements (DMWs) are modeled with solid elements as shown in Fig. 2. The Young's modulus of the ground improvements is set at 2530.8 MPa that is assumed at 400 times of the uniaxial compressive strength of 6.33 MPa according to the 28-day strength of the samples. The value corresponds to the shear modulus of 1004 MPa (density, $\rho = 2.1$ t/m^3, shear wave velocity, Vs = 692 m/s, Poisson's ratio of 0.26). The damping ratio was set to 3.0%. Since sandpaper is applied beneath the foundation slab (raft), the stiffness K_s was set to slide beneath it at $\tau_{max} = \sigma_n \tan\phi$ (ϕ = 30°).

Modeling of Ground

Ground material properties and initial shear modulus, G_0 of clay layer are considered to depend on confining pressure, which is confirmed by laboratory test results [4] as shown in Fig. 4. The stiffness and damping of the soils are calculated considering the nonlinear characteristics of the soils, which were obtained from cyclic triaxial tests of the samples. Figure 5 shows the strain dependency characteristics of the soil, (G/G_0-γ and h-γ) used in the three-dimensional analyses.

The G/G_0-γ curve are extrapolated so that the shear stress becomes the yield stress τ_f obtained in Eq. (1) when the shear strain reaches 100%. The confining pressure, σ_m is calculated by the self-weight analysis as the coefficient of earth pressure at rest, $K_0 = 1.0$. The friction angle, $\phi = 35.7°$ and cohesion, $c = 23.5$ kPa are employed from the results of triaxial CD test.

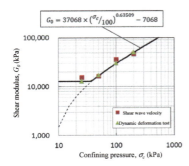
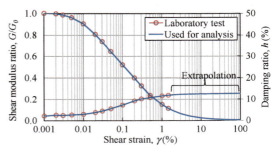

Fig. 4. Initial shear modulus depending on confining pressure

Fig. 5. Variation of shear modulus and damping ratio by shear strain

The employed constitutive laws of the ground are based on the Yoshida model which is expanded of the Ishihara, Yoshida and Tsujino model [5] for the three dimensions. The groundwater table set to be GL–7.5 m. The damping ratio, h of the ground was set to 3% at the first order natural frequency of 1.4 Hz obtained by eigenvalue analysis.

$$\tau_f = \sigma_m \sin\phi + c \cos\phi \qquad (1)$$

4 Results of Simulation Analyses

Figure 6 shows the time histories of the simulated structure and ground accelerations, pile bending moments at pile head and intermediate depth, pile axial forces at pile head and near the pile tip, and lateral load acting on the pile head in case of peak acceleration of 600 Gal at base input (2E) in comparison with the experimental values.

Although the analytical value of the acceleration at the top of the building is a little smaller than the experimental result, the accelerations at the building foundation (1F) and ground surface (GL-0 m) are well reproduced (Fig. 6(a), (b), (c)). The analytical bending moments at piles are different for the left and right piles (P1 and P2), whereas the experimental values are quite similar, indicating the residual bending moments are

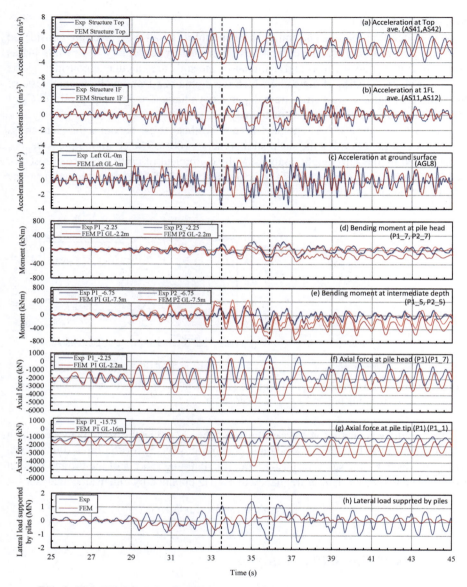

Fig. 6. Time histories of numerical simulations comparing to experimental results

caused by the overestimated residual settlement of the building (Fig. 6(d), (e)). The analytical bending moments at intermediate depth (near the lower end of the DMWs) showed larger values than those at the pile heads, similar to the experimental results (Fig. 6(d), (e)). The axial forces at the pile heads corresponded roughly to the share (about 20%, −2000 kN) from the dead load analysis before the experiment started, and the axial force at the pile tip also expressed well the degree of reduction due to friction

(Fig. 6(f), (g)). However, the fluctuating axial force tended to be larger in the compressive (negative) side at the analysis than that at experiment (Fig. 6(f), (g)). The analysis shows that the DMWs cannot bear the variable axial force very well. The analytical lateral load acting on the piles (sum of the pile head shear forces of the four piles) was smaller than the experimental value (Fig. 6(h)). The reason is thought to be that the DMWs is modeled without consideration of the softening and plasticization.

The dashed lines shown in the figures are the times of peak building inertia force (33.5 and 35.9 s), and the shear force distribution τ_{xz} and principal stress distribution σ_1 (tension: positive) of the longitudinal wall at these times are shown in Fig. 7. The shear stress τ_{xz} in the DMWs was larger in the middle of the back side than in the rocking compression side (front side) of the building.

Assuming that 90% of the building inertia force acts on the longitudinal wall (1 m wide and 17 m long in the analysis), when the lateral load acts on the DMWs with 15 MN × 0.9 = 13.5 MN, the shear stress in the DMWs becomes 794 kPa (13.5 MN/17m²) due to the building inertia force. The stresses in the analysis are generally consistent with the inertia forces, although stresses due to ground deformation are also generated. In

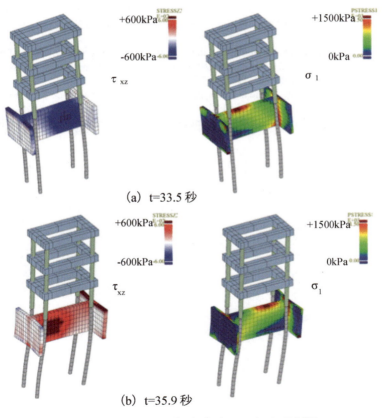

Fig. 7. Shear stress and principal stress in the DMWs

addition, tensile stress of about 1.5 MPa was generated, which is more than 0.2 times higher than the uniaxial compressive strength of 6.33 MPa, confirming the experimental results of cracks in the DMWs.

5 Conclusions

The numerical simulation analyses were conducted to investigate the stresses in the DMWs which was difficult to measure in the experiments. The shear stresses on the DMWs in the analyses were almost consistent with the inertia force of the structure similar to the experimental results. In addition, tensile stresses of about 1.5 MPa were generated, which were more than 0.2 times higher than an unconfined compressive strength of 6.33 MPa, confirming the experimental results of cracking in the DMWs.

References

1. Yamashita, K., Hamada, J., Onimaru, S., Higashino, M.: Seismic behavior of piled raft with ground improvement supporting a base-isolated building on soft ground in Tokyo, Special issue on Geotechnical Aspects of the 2011 off the Pacific coast of Tohoku Earthquake. Soils Found. **52**(5), 1000–1015 (2012)
2. Yamashita, K., Shigeno, Y., Hamada, J., Chang, D.W.: Seismic response analysis of piled raft with grid-form deep mixing walls under strong earthquakes with performance-based design concerns. Soils Found. **58**(1), 65–84 (2018)
3. Hamada, J., Okumura, T., Honda, T.: Centrifuge model tests on seismic behavior of piled raft foundation with soil-cement wall in soft clayey ground. In: Earthquake Geotechnical Engineering for Protection and Development of Environment and Constructions, (7ICEGE), pp. 2787–2794 (2019)
4. Okumura, T., Honda, T., Hamada, J.: Dynamic centrifuge model tests on plate-shape building's pile foundation in clayey ground. In: 16th Asian Regional Conference on Soil Mechanical and Geotechnical Engineering (2019)
5. Tsujino, S., et al.: A simplified practical stress-strain model in multi-dimensional analysis. In: Proc., International Symposium on Pre-Failure Deformation. Characteristics of geomaterials, pp. 463-468 (1994)

Seismic Interactions Among Multiple Structures on Liquefiable Soils Improved with Ground Densification

Yu-Wei Hwang[1(✉)], Shideh Dashti[2], and Juan Carlos Tiznado[3]

[1] National Yang Ming Chiao Tung University, Hsinchu City, Taiwan
yuwei.hwang@nycu.edu.tw
[2] The University of Colorado Boulder, Boulder, CO, USA
[3] Pontificia Universidad Católica de Chile, Santiago, Chile

Abstract. Current guidelines for evaluating the performance of ground densification as a liquefaction countermeasure near buildings are based on free-field conditions. However, particularly in urban areas, where structures are constructed in the vicinity of each other, structure-soil-structure interaction in liquefiable deposits near two adjacent buildings ($SSSI_2$) and multiple buildings (≥ 3) in a building cluster ($SSSI_{3+}$) have been shown to be consequential on key building engineering demand parameters (EDPs). Yet, their potential tradeoffs associated with ground densification are currently not well understood. In this paper, three-dimensional, fully-coupled, nonlinear, dynamic finite element analyses, validated with centrifuge models of $SSSI_2$, were used to explore the influence of ground densification and the building spacing to width ratio (S/W) on key EDPs of buildings experiencing $SSSI_2$ or $SSSI_{3+}$ (particularly for four structures in a square-shape configuration). For the conditions considered, ground densification reduced the permanent settlement of an isolated structure by up to 58% compared to a similar unmitigated structure. Both $SSSI_2$ and $SSSI_{3+}$ had a minor impact on the mitigated structures' average settlement compared to SSI. In contrast, $SSSI_2$ and $SSSI_{3+}$ strongly amplified the permanent tilt of the mitigated structure by up to 6.5 times compared to SSI at S/W < 0.5, due to the enhanced degree of asymmetry in the properties of and demand on the underlying soil. Increasing spacing reduced the permanent tilt of structures under $SSSI_2$ or $SSSI_{3+}$. Overall, the results suggest that $SSSI_2$ and $SSSI_{3+}$ can adversely affect the foundation performance and hence, damage potential of the mitigated soil-foundation-structure system compared to SSI, particularly at S/W < 0.5. Such complexities need to be considered in mitigation planning and design of urban structures.

Keywords: Soil-structure interaction · Structure-soil-structure interaction · Liquefaction · Ground densification · Finite element modeling · Centrifuge modeling

1 Introduction

Previous experimental studies have shown that ground densification can significantly affect soil-structure interaction (SSI) and building performance in potentially liquefiable soils (Housler 2002; Olarte et al. 2018). However, structures are constructed in the vicinity of each other in urban areas. Kirkwood and Dashti (2018, 2019) performed a series of centrifuge tests to explore the influence of ground improvement on adjacent structures, accounting for structure-soil-structure interaction near two buildings ($SSSI_2$). The experiments showed that mitigation around one structure could adversely impact the seismic demand within a neighboring unmitigated structure and its overall performance. In a city, however, multiple buildings (≥ 3) are often constructed in close proximity to each other in a cluster ($SSSI_{3+}$). The mechanics of $SSSI_{3+}$ in relation to SSI and $SSSI_2$ and its influence on mitigated soils and foundation-structure systems are currently not sufficiently understood, neither experimentally nor numerically.

In this paper, 3D, fully-coupled, nonlinear, dynamic finite element analyses, validated with centrifuge models of $SSSI_2$, are used to explore the influence of ground densification and building spacing on key engineering demand parameters (EDPs) of structures experiencing $SSSI_2$ and $SSSI_{3+}$.

2 Centrifuge Testing Program

A series of centrifuge experiments was conducted under a centrifugal acceleration of 70 g at the University of Colorado Boulder's 5.5 m-radius, 400 g-ton centrifuge facility (Kirkwood and Dashti 2018; Hwang et al. 2022), to explore the seismic performance of adjacent shallow-founded, potentially inelastic structures on liquefiable soils with and without ground densification. Figures 1a-b show the centrifuge test layout of $[BA]_{AL-DS}$ with a building spacing (S) of 3 m, as a representative example. The name associated with the centrifuge test was based on the type of structures present in the model container (e.g., A representing one isolated structure like A; BA identifying two adjacent dissimilar structures, B and A). In the subscript of model names, we first identify the building arrangement (e.g., AL referring to two structures in alignment under $SSSI_2$). The second subscript specifies whether the liquefiable layer below a given structure or adjacent structures was densified (DS) or was not mitigated (NM).

Two potentially inelastic, multi-degree-of-freedom (MDOF), shallow-founded, moment-resisting structures were designed based on modern US seismic design codes for high seismic locations in California. Structure A represented a shorter, lighter, 3-story structure (simplified as a 3-DOF model) on a 1 m-thick mat foundation, while Structure B represented a taller and heavier, 9-story structure (simplified as a 2-DOF model) on a 3 m-deep, single-story basement foundation. The foundation footprint size was 9.56 m $\times$ 9.56 m in prototype scale for both structures, but the net bearing pressure of Structure A (i.e., 67 kPa) was notably less than that of B (i.e., 166 kPa). In all experiments, a 10 m-thick (in prototype scale), dense layer of Ottawa sand F65 ($D_{50} = 0.15$ mm, $C_u = 1.56$, $e_{min} = 0.53$, $e_{max} = 0.81$) was prepared by dry pluviation to attain a relative density (D_r) of 90%. Above this layer, a 6 m-thick, loose layer of Ottawa sand was dry pluviated with a $D_r \approx 40\%$. Subsequently, a 2 m-thick layer of Monterey 0/30 sand ($D_{50} = 0.40$ mm, $C_u = 1.3$, $e_{min} = 0.54$, $e_{max} = 0.84$) at a $D_r \approx 90\%$ was dry pluviated as a draining crust. The densification zone was dry pluviated at $D_r \approx 90\%$ below and around the foundation. The densification depth covered the entire thickness of the loose Ottawa sand layer (i.e., 6 m), while the densification width beyond the foundation edge was half of the densification depth (i.e., 3 m), following recommendations from JGS (1998).

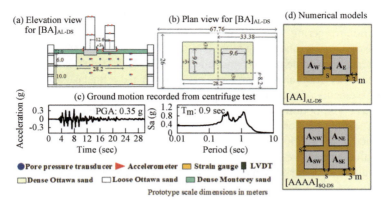

Fig. 1. (a)–(b) Configuration and instrumentation layout of the centrifuge experiments (all units are in prototype scale meters); (c) the acceleration time history and response spectrum (5%-damped) of the base motion (Note: PGA is the peak ground acceleration; T_m is the mean period.); and (d) configuration of the numerical models for numerical sensitivity study.

A methylcellulose solution was prepared to attain a viscosity 70 times that of water for specimen saturation under the centrifuge scaling laws. Subsequently, after the centrifuge model was spun to a nominal centrifugal acceleration of 70 g at the center of the loose Ottawa sand layer, five one-dimensional (1D) horizontal earthquake motions were applied in sequence. Figure 1c shows the mean acceleration time history and response spectrum (5%-damped) of the first major motion (identified as Kobe-L) recorded at the base of the container, which was later used as input to the numerical simulations.

3 Numerical Simulations

A series of three-dimensional (3D) numerical simulations was conducted in a parallel version of the object-oriented, open-source, finite element, OpenSEES platform (OpenSEES MP version 3.0.0 rev 6692; Mazzoni et al. 2006) on the Summit supercomputer at University of Colorado Boulder.

The structural components (e.g., beams and columns) were modelled as the elastic beam elements with the linear elastic material. The reduced section of beam and column fuses was modelled with nonlinear, displacement-based, beam elements with fiber sections (with the uniaxial steel material), to account for nonlinear structural behavior. The superstructure's dynamic properties were calibrated against the fixed-base modal response obtained from hammer impact tests. The key structural properties were summarized by Hwang et al. (2021).

The nonlinear response of saturated, granular soils was numerically simulated using the 20-node brick elements with the u-p formulation, and the pressure-dependent, multi-yield surface, version 2 (PDMY02) soil constitutive model (Elgamal et al. 2002). The fluid bulk modulus was 2×10^6 kPa at atmospheric pressure. The soil model parameters were systematically calibrated and detailed by Hwang et al. (2021) using: (1) a series of drained and undrained, monotonic and cyclic triaxial tests; (2) a free-field seismic site response in the centrifuge with the same soil deposits; and (3) field observations in terms of liquefaction triggering. The soil properties within the improved area were altered to those of dense Ottawa sand with D_r of 90%.

The foundation elements were modelled with 20-node, brick elements with the u-p formulation (its fluid mass density was set as zero) and a linear-elastic material. The foundation elements were connected to the surrounding soil elements through a partially-tied interface. The bottom foundation elements were fully tied (in the x-, y-, and z-directions) to the soil elements. The nodes at the foundation's lateral perimeter were only attached to the soil in the two horizontal directions (both x and y), to allow for its relative settlement with respect to the surrounding soil.

Only half of the centrifuge models across the container width (in the perpendicular direction to shaking) was numerically modelled to reduce computational cost. Hence, out-of-plane displacements (perpendicular to shaking) were restricted at the symmetry face. However, when modelling $SSSI_{3+}$, the soil-foundation-structure systems were simulated in 3D to allow for asymmetric deformations in the direction perpendicular to shaking. The periodic boundary condition was applied to soil domain's lateral boundary nodes located at the same elevation. Additionally, the bottom boundary nodes were fully fixed to represent a nearly rigid container base. The soil domain length in parallel to shaking was 7 W (where W is the foundation width), while the domain width in perpendicular to shaking was 3.5 W. The element size was calculated at each depth based on soil's empirical small-strain shear wave velocity profile and the minimum wavelength of interest. The acceleration time history recorded at the base of the container (see Fig. 1c) in the centrifuge during the Kobe-L motion was applied to the base nodes after the model reached initial static equilibrium.

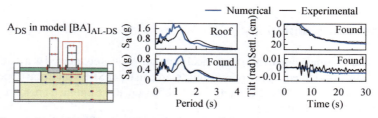

Fig. 2. Comparison of experimental and numerical results in terms of 5%-damped acceleration response spectra on structures and foundations as well as settlement and tilt time histories of foundations of A_{DS} in model $[BA]_{AL-DS}$ with 3 m spacing.

Figure 2 shows the numerical simulation of mitigated Structure A_{DS} in model $[BA]_{AL-DS}$ with 3 m spacing, as a representative example. The numerical model generally captured the accelerations as well as foundation's settlement and tilt response observed experimentally. The mitigated Structure A_{DS} in model $[BA]_{AL-DS}$ tended to settle less compared to its unmitigated isolated counterparts [i.e., 48.5 cm for isolated A_{NM}, according to Hwang et al. (2021)]. This was partly due to the stress increment from the neighbor increased soil's initial shear stiffness and resistance to liquefaction below the two structures and partly due to less contractive tendencies in a densified soil that reduced the magnitude of excess pore pressure generation. Contrary to the trends for settlement, the large bias (due to $SSSI_2$) in terms of initial stiffness, stress fields, and flow paths below the foundations encouraged a greater accumulation of tilt for both adjacent structures. However, ground densification reduced soil deformation bias below the mitigated foundations (due to less soil softening), decreasing their tilt in model $[BA]_{AL-DS}$. For the cases considered and the one ground motion investigated in this paper, the net outcome of these competing mechanisms led to an increase in the magnitude of permanent rotation of A_{DS} in model $[BA]_{AL-DS}$ ($\approx$0.006 rad) with 3 m spacing compared to that of isolated and unmitigated A_{NM} ($\approx$0.002 rad). 3D, fully-coupled, effective-stress, finite element models with a well-calibrated constitutive model and numerical setup (e.g., with use of higher-order elements and sufficient boundary size) could effectively capture these trends observed experimentally.

4 Impact of Ground Densification on $SSSI_2$ and $SSSI_{3+}$

To evaluate the combined influence of ground densification with $SSSI_2$ or $SSSI_{3+}$ on building performance under 1D shaking, we numerically analyzed two configurations: (i) two adjacent structures in alignment parallel to the shaking direction (e.g., $[AA]_{AL}$) representing $SSSI_2$; and (ii) four structures in a square configuration (e.g., $[AAAA]_{SQ}$) representing $SSSI_{3+}$, with spacings of 1, 3, 6, and 9 m (S/W $\approx$ 1/9, 1/3, 2/3, 1). For all models considered, the soil below and around the building cluster was mitigated, as shown in Fig. 1(d). The densification depth covered the entire thickness of the liquefiable layer (i.e., 6 m) below the building cluster, while the lateral extent of densification beyond the edge of the building cluster was kept as half the densification depth. The results from these simulations are compared with the isolated structure when unmitigated (A_{NM}) and mitigated (A_{DS}), which only experienced SSI alone.

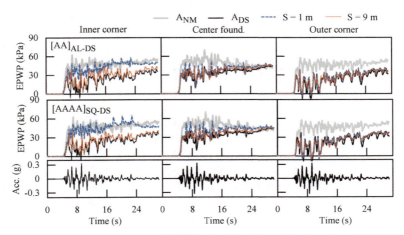

Fig. 3. The excess pore water pressure (EPWP) time histories under A_{DS} in models $[AA]_{AL-DS}$ and $[AAAA]_{SQ-DS}$ with spacings 1 and 9 m in the middle of the critical layer ($z = 5$ m) compared with the isolated unmitigated and mitigated counterparts (A_{NM} and A_{DS}).

Figure 3 shows the numerically computed excess pore water pressure (EPWP) time histories in the middle of the critical layer (z = 5 m) beneath mitigated A_{DS} in models [AA]$_{AL-DS}$ and [AAAA]$_{SQ-DS}$ with spacings of 1 and 9 m. The EPWP responses of unmitigated and mitigated isolated Structure A (i.e., A_{NM} and A_{DS}) are also presented for comparison. Note that the inner corner represents the location near the center of the building cluster. Soil densification limited the generation of EPWP and degree of softening during strong shaking under the center and two corners of mitigated A_{DS} under SSI, SSSI$_2$, and SSSI$_{3+}$ compared to an unmitigated isolated A_{NM}. However, EPWPs below mitigated structures continued to rise after strong shaking due to an inward and upward pore water migration, particularly from the neighboring unmitigated soils. At shorter spacings (e.g., 1 m), mitigated A_{DS} under SSSI$_2$ typically experienced greater EPWPs below its inner corner compared to the outer corner and compared to that of isolated A_{DS} (SSI). This is attributed to greater confining stresses in soil below the foundations' inner corner (overlapping normal stress bulbs), increasing soil's contractive tendencies and capacity for generating larger excess pore pressures. This effect was more pronounced for structures that experienced SSSI$_{3+}$ compared to SSSI$_2$. However, increasing spacing made the response below the foundation's two edges more symmetric under both SSSI$_2$ and SSSI$_{3+}$ (similar to the response below isolated A_{DS}).

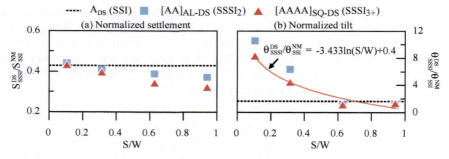

Fig. 4. Numerically computed normalized permanent foundation settlement and tilt of mitigated Structure A_{DS} in models experiencing SSI, SSSI$_2$, and SSSI$_{3+}$ as a function of normalized building spacings (S/W).

Figure 4 shows the settlement (S^{DS}_{SSSI}) and tilt (θ^{DS}_{SSSI}) of mitigated structures experiencing SSSI (with two or more buildings) in models [AA]$_{AL-DS}$ and [AAAA]$_{SQ-DS}$ normalized by those of the unmitigated isolated A_{NM} (i.e., S^{NM}_{SSI} and θ^{NM}_{SSI}). Building spacing (S) is also normalized by the corresponding foundation width (W). The normalized settlement and tilt of isolated A_{DS} (SSI) are also provided for comparison.

For the conditions considered, densification reduced the permanent foundation settlement of the isolated A_{DS} by up to 58% compared to a similar isolated but unmitigated Structure A_{NM}. This is because soil densification reduces the degree of soil softening and the associated shear and volumetric strains below the foundation. For S/W exceeding 0.5, a slight reduction on average settlement of A_{DS} was observed under SSSI$_2$ and SSSI$_{3+}$ compared to isolated A_{DS} (SSI) by up to 5% and 8%, respectively. This was partly due to a larger volume of densified soil below foundations that were more widely spaced and partly due to the cancelation of shear stresses below the foundations experiencing $SSSI_2$ and $SSSI_{3+}$ that reduced the contribution of shear deformations associated with bearing capacity failure.

Contrary to the patterns observed for average settlement, densification slightly amplified the permanent rotation of isolated A_{DS} by about 1.5 times that of isolated A_{NM}. This is because soil densification increased the seismic demand on the foundation and superstructure, amplifying the foundation's overturning moment (due to P-Δ effect) and its cumulative rotation, for the conditions considered in this study. In this case, $SSSI_2$ and $SSSI_{3+}$ increased the bias below the foundations at shorter spacings (in terms of soil's initial confining stress, initial shear stiffness, and degree of softening that affects shear stiffness and damping at larger strains). This bias amplified the accumulation of biased shear and volumetric strains and hence tilt of A_{DS} under $SSSI_2$ and $SSSI_{3+}$ by up to 8 and 6 times that under SSI, respectively, for S/W < 0.5. This effect became negligible for S/W > 0.5 (due to a more symmetric stress and strain state). In general, densification reduced the soil deformations below the outer corners of mitigated structures under either $SSSI_2$ or $SSSI_{3+}$, while the cancellation of shear stresses was more pronounced below the inner corners of structures that experienced $SSSI_{3+}$ in a square cluster compared to two adjacent buildings ($SSSI_2$). A more consistent reduction of strains under the inner and outer corners of A_{DS} in $[AAAA]_{SQ-DS}$ slightly reduced its bias and tilt compared to A_{DS} in $[AA]_{AL-DS}$ for S/W < 0.5.

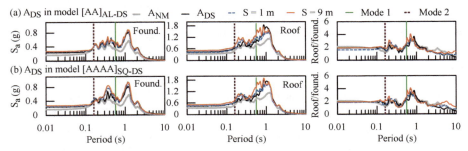

Fig. 5. The numerically computed 5%-damped foundation and roof response spectra as well as roof/foundation spectra ratios of mitigated Structure A_{DS} in models $[AA]_{AL-DS}$ and $[AAAA]_{SQ-DS}$ (Note: Modes 1 through 2 are the fixed-base, natural periods of A obtained numerically).

As shown in Fig. 5, densification amplified the foundation and roof transverse accelerations of A_{DS} experiencing SSI, $SSSI_2$, and $SSSI_{3+}$ for all spacings considered, compared to that of isolated, unmitigated A_{NM}. This was expected due to less soil softening (in an average manner) in a larger volume of soil below the mitigated building cluster. In general, all structures showed amplification of transverse accelerations from their foundation to roof at a period slightly greater than its own first or second mode due to period-lengthening. These results also indicate that $SSSI_{3+}$ slightly increased the seismic demand transferred to the foundation compared to more simplistic cases of SSI and $SSSI_2$, potentially amplifying overall damage within the superstructure.

5 Concluding Remarks and Engineering Implications

In this paper, 3D, fully-coupled, effective-stress, finite element simulations with a carefully calibrated, nonlinear soil constitutive model, validated with centrifuge experiments of SSI and $SSSI_2$, are used to evaluate the seismic performance of multiple structures on liquefiable soils improved with ground densification.

The numerical simulations indicate that building spacing and seismic coupling under $SSSI_2$ and $SSSI_{3+}$ strongly influence the effectiveness of ground densification in terms of foundation settlement, tilt, and accelerations within the superstructure. Ground densification, in general, reduced the foundation average settlement for all spacings and building configurations considered. The presence of double or multiple neighboring structures slightly reduced the average settlement of a mitigated structure compared to an isolated counterpart when $S/W > 0.5$ (particularly under $SSSI_{3+}$). However, for the conditions considered, both $SSSI_2$ and $SSSI_{3+}$ notably amplified the permanent tilt of A_{DS} (by up to a factor of 6.5) compared to SSI at $S/W < 0.5$ (especially $SSSI_2$).

Further research is needed to comprehensively evaluate the impact of seismic coupling in a city block on the performance of mitigated soil-foundation-structure systems with more variations in the soil profile, building properties and arrangement, and ground motion characteristics (including multi-dimensional shaking). However, the preliminary results presented in this paper indicate that the beneficial effects of $SSSI_2$ and $SSSI_{3+}$ on average settlement need to be considered together with their strong adverse impact on foundation's tilt and transverse accelerations transferred to the mitigated superstructure compared to the simpler case of isolated SSI. These complexities can influence the planning and design of ground improvement in urban settings.

References

Elgamal, A., Yang, Z., Parra, E.: Computational modeling of cyclic mobility and post-liquefaction site response. Soil Dyn. Earthquake Eng. **22**(4), 259–271 (2002). https://doi.org/10.1016/S0267-7261(02)00022-2

Hausler, E.A.: Influence of ground improvement on settlement and liquefaction: a study based on field case history evidence and dynamic geotechnical centrifuge tests. Ph.D. Dissertation. University of California, Berkeley (2002)

Hwang, Y.W., et al.: Seismic interaction of adjacent structures on liquefiable soils: insight from centrifuge and numerical modeling. J. Geotech. Geoenviron. Eng. **147**(8), 04021063 (2021)

Hwang, Y.W., Dashti, S., Kirkwood, P.: Impact of ground densification on the response of urban liquefiable sites and structures. J. Geotech. Geoenviron. Eng. **148**(1), 04021175 (2022)

Kirkwood, P., Dashti, S.: Considerations for the mitigation of earthquake-induced soil liquefaction in urban environments. J. Geotech. GeoEnviron. Eng. **144**(10), p. 04018069 (2018)

Kirkwood, P., Dashti, S.: Influence of prefabricated vertical drains on the seismic performance of similar neighbouring structures founded on liquefiable deposits. Geotechnique **69**, 971–985 (2019). https://doi.org/10.1680/jgeot.17.P.077

Mazzoni, S., McKenna, F., Scott, M., Fenves, G.: Open System for Earthquake Engineering Simulation User Command-Language. Network for Earthquake Engineering Simulations, Berkeley, CA (2006)

Olarte, J., Dashti, S., Liel, A.: Can ground densification improve seismic performance of the soil-foundation-structure system on liquefiable soils. Earth. Eng. Struct. Dyn. **47**, 1193–1211 (2018). https://doi.org/10.1002/eqe.3012

Numerical Simulation of Real-Scale Vibration Experiments of a Steel Frame Structure on a Shallow Foundation

Marios Koronides[1](✉) ⓘ, Stavroula Kontoe[1] ⓘ, Lidija Zdravković[1] ⓘ, Athanasios Vratsikidis[2] ⓘ, Dimitris Pitilakis[2] ⓘ, Anastasios Anastasiadis[2] ⓘ, and David M. Potts[1] ⓘ

[1] Imperial College London, London SW7 2AZ, UK
marios.koronides17@imperial.ac.uk
[2] Aristotle University of Thessaloniki, 54124 Thessaloniki, Greece

Abstract. Dynamic Soil-Structure-Interaction (DSSI) phenomena can considerably affect the structural response under dynamic loading. Time domain finite element analysis allows to study these phenomena in depth, but several computational challenges need to be addressed first to achieve rigorous modelling of both the structure and soil domain. Within this context, this study presents three-dimensional FE analyses of real-scale forced vibration tests of a steel frame structure founded on a shallow foundation. The field tests were carried out with the real-scale prototype structure of EUROPROTEAS at the Euroseistest experimental facility located in the Mygdonian Valley in Northern Greece. The structure has outer dimensions of 3 × 3 × 5 m and it is placed in an area whose geotechnical properties have been well documented by previous studies. The study focuses on the modelling of the soil-foundation interface, which is one of the major challenges of such complex SSI simulations. A novel approach to model potential contact imperfections at the interface is proposed, showing very good agreement with the field data and hence improving the reliability of numerical predictions. The analyses show that contact imperfections affect considerably the predicted motion in both the structure and the soil.

Keywords: Dynamic-soil-structure interaction · Finite element analysis · Soil-foundation interface modelling · Real-scale forced vibration tests · Shallow foundation

1 Introduction

Dynamic Soil-Structure Interaction (DSSI) is a complex phenomenon with important implications for the design of structures. The foundation of a vibrating structure transmits waves into the soil beneath, engaging it in the system response, while concurrently the soil response affects the structural response. Quantification of this interaction can be challenging, especially when foundation rocking and sliding take place.

A popular approach to consider these phenomena is to represent the soil domain with springs and dashpots at the soil-foundation interface [1, 2], without explicitly modelling it. Some studies (e.g. [3]) offer modifications to consider soil-nonlinearity, radiation damping and foundation detachment. Macro-element models are very efficient but have limitations in the simulation of wave propagation in layered deposits and in capturing the soil constitutive behaviour. Time domain Finite Element analysis of DSSI phenomena poses several challenges, but offers a rigorous approach with explicit modelling of the structure and the soil domain (e.g.[4, 5]).

This study presents three-dimensional (3D) FE analyses of real-scale forced vibration tests of a steel frame structure founded on a shallow foundation, placing particular emphasis on modelling the soil-foundation interface. Foundations are not always in full adhesion with the soil beneath, forming interface imperfections, a condition that affects the response of the soil-structure system significantly. A novel approach to model potential contact imperfections at the interface is proposed, showing very good agreement with the field data.

2 Experimental Campaign

2.1 Structure and Instrument Set-Up

The field experiment discussed herein is part of a wider testing programme of forced and free vibrations on the prototype structure EUROPROTEAS [6] in Northern Greece. Figure 1a depicts the structural configuration examined in this paper. EUROPROTEAS is a 5 m tall symmetric frame structure with steel columns (SHS 150 × 150 × 10 mm) and cross-bracing system (L-shape 100 × 100 × 10 mm), founded on a shallow foundation. The foundation and superstructural mass consist of one (9Mgr) and two (18Mgr) reinforced concrete slabs respectively, with dimensions 3 × 3 × 0.4 m each. A vibrator was mounted on the top slab and applied unidirectional sinusoidal loads of various magnitudes and frequencies; each one was held for a certain time period until the system reached steady state. The SSI system was well instrumented (Fig. 2), allowing the calibration of the numerical model.

2.2 Soil Conditions

EUROPROTEAS is located at the centre of the well-documented Euroseistest valley, for which extensive geotechnical and geophysical surveys are available (e.g. [6, 7]). The soil stratigraphy and shear wave velocity profiles resulted from past surveys are shown in Fig. 1b and c. The shallow soil layers consist mostly of silty sand and low plasticity clay of very low stiffness, a condition that accentuates SSI effects.

3 Numerical Model

3.1 Problem Geometry and Boundary Conditions

Numerical analyses were undertaken with the Imperial College Finite Element Program (ICFEP) [8]. Figure 3 illustrates the 3D FE model, including dimensions and the static boundary conditions. Exploiting the symmetry of the structure, only half of the problem is modelled. The soil domain dimensions are 15 × 7.5 × 6 m which is divided into two layers whose elastic properties are presented in Table 1. The soil domain and slabs are modelled with 20-noded brick elements, while 3-noded beam elements were used for the steel columns and braces. The beam elements of columns are extended into the brick elements of slabs to achieve a moment connection. The elastic properties of all structural elements, which are selected according to the Eurocodes [9, 10], are shown in Table 2, including Young's modulus (E), Poisson's ratio (ν), cross sectional area (A), moment of inertia (I), torsional constant (J) and target Rayleigh damping (ξ).

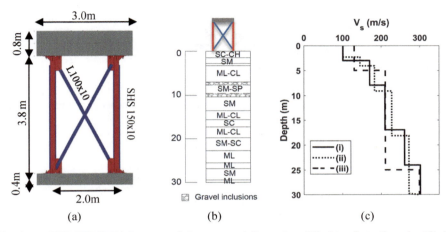

Fig. 1. (a) EUROPROTEAS structural elements and dimensions [6], (b) soil stratigraphy [6], (c) shear wave velocity profiles as interpreted from (i) [7], (ii) [11] and (iii) [12]

The bottom boundary edges of the soil domain are fully fixed, while the nodes at the plane of symmetry are fixed in the out-of plane direction. Dashpots and springs [13] are applied in all three directions (two tangential and one normal) on the remaining boundary nodes of the soil domain, with the exception of the ground surface.

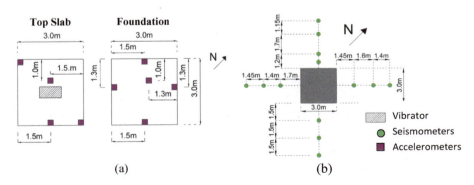

Fig. 2. Plan views of the instrument configurations at the (a) structure and (b) soil

All analyses presented in this paper are linear elastic. Although not shown herein, analyses with non-linear soil were conducted and suggested that soil non-linearity was not important for the presented experiment.

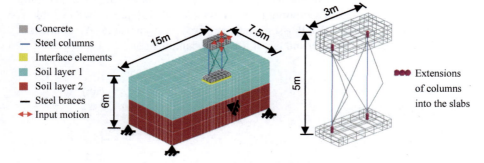

Fig. 3. Numerical model dimensions and some of the boundary conditions

Table 1. Elastic soil properties

	Vs (m/s)	v	ξ (%)
Layer 1	31	0.2	5
Layer 2	200	0.25	5

Table 2. Elastic properties of the structural elements

	E (GPa)	v	A (cm^2)	I (cm^4)	J (cm^4)	ξ (%)
Slabs	31	0.2	–	–	–	1
Braces	200	0.25	19.2	177	6.97	2
Columns	200	0.25	54.9	1770	2830	2

3.2 Input Motion

The input motion is applied at a node on the top slab of the structure as prescribed horizontal and vertical displacement time histories. These were interpreted from accelerations recorded during the experiments by the instrument closest to the geometrical centre of the top slab. The experiment discussed herein has an input frequency of 5 Hz and amplitude of 11 kN. As Fig. 4a shows, during this experiment, the structure oscillated along the loading direction only (in-plane), while the out-of plane motion was very weak, allowing the use of a half geometry model. The initial part of the employed in-plane steady state motion was tapered, see Fig. 4b.

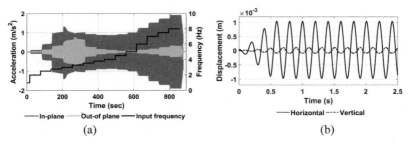

Fig. 4. (a) Recorded in and out of plane top slab acceleration time histories; (b) input motion

3.3 Interface Modelling

One of the most challenging aspects of SSI numerical models is the simulation of the interface between foundations and soil, to allow sliding and rocking of shallow foundations. In this study, zero thickness interface elements [14] are employed at the soil-foundation interface. The relative displacement between the soil surface and the foundation base is controlled by the shear (Ks) and normal (Kn) elastic stiffness of the interface elements as well as by their plastic behaviour. All the analyses of this study adopt high stiffness of interface elements, Ks = Kn = 1e8 kN/m^3, coupled with an elastoplastic Mohr-Coulomb failure surface which adopts cohesion $c' = 0$ and angle of shearing resistance $\phi' = 30°$ (EPint model). These interface elements are capable of opening when a mobilized tensile normal stress exceeds the tensile strength set as $c'/\tan \phi'$, which in the current model equals zero. While the interface element remains open it has no stiffness and hence no normal or shear stresses are transferred from the foundation to the soil, while any strains occurring are accumulating as normal and shear plastic strains, respectively. The former strains result from differential normal displacements and the latter from differential tangential displacements between the corresponding nodes on the two sides of an interface element, hence having the unit of length.

The EPint model is applicable when the foundation is in full contact with the soil prior to the application of dynamic loading. However, during the experiments, it was observed that, under static loads, the foundation, at its edges, was not in full adhesion ouwith the soil beneath. A novel method to simulate these imperfections is proposed.

The suggested interface model initiates the detachment (gap) of the shallow foundation from the soil by prescribing a non-zero plastic normal strain ($\varepsilon_{n,pl}$), which is the hardening parameter of the Mohr-Coulomb interface model, to the selected interface elements. These elements behave as "open", simulating a gap and lack of contact, while those with zero $\varepsilon_{n,pl} = 0$ are "closed", simulating full contact conditions. After the application of the self-weight of the structure and during the vibration of the structure, some initially "opened" elements at the interface may "close" regaining their stiffness and "re-open" at a subsequent increment, which is enabled by a complex contact-tracing algorithm in ICFEP [14]. If the initially "opened" interface elements never close during the excitation, they retain zero stiffness. With respect to the latter, it is attempted here to model the gaps that never close by adopting a relatively low stiffness (e.g. Ks = Kn = 1e2 kN/m^3) of elastic interface elements (LSELint).

Although not shown herein for brevity, different interface gap configurations have been investigated. This investigation showed that the impact of the gaps is minimal when a gap is prescribed at an inner part of the foundation surrounded by contact areas. On the contrary, gaps have a significant effect when they are prescribed at the foundation edges. Figure 5 presents the interface contact configuration which was found to be applicable for the modelled experiments and used in the current numerical model. Numerical analyses presented employ the following interface modeling: (a) EPint without gaps, (b) EPint with initial gaps of 2 mm (EPint_2 mm) or (c) gaps of 0.6 mm (EPint_0.6 mm), and (d) LSELint.

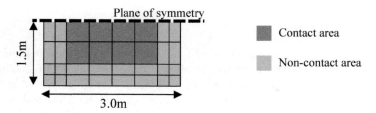

Fig. 5. Initial contact conditions between the foundation and soil in plan view

4 Results and Discussion

The experimental and numerical steady state motions are compared at the following locations: (a) top slab edge, (b) foundation edge, (c) soil surface node located 1.20 m far from the foundation edge in the loading direction.

Figure 6 compares top slab experimental and numerical acceleration time histories and Fourier spectra, from three analyses with different boundary conditions. The first analysis simulates the structure with a fully fixed foundation (fixed-base), representing the case where SSI effects are neglected. The second and third analyses model the soil domain, according to Fig. 3, with or without elastoplastic interface elements (EPint) at the soil-foundation interface. The fixed-base approximation underestimates the vertical motion component (better shown from Fourier spectra), indicating the need of explicit soil modelling. The remaining two analyses predict similar top slab accelerations, very close to the experimental ones. They are also identical to each other, implying that the dynamic load is not large enough to force interface elements to "open". This highlights the need to model imperfections at the interface which can trigger foundation rocking, as this was visually observed during the experiments.

Figure 7 compares numerical results of analyses with different interface modelling at various parts of the SSI system (top slab, foundation, soil). The horizontal structural motion is well-captured by all analyses. The existence of gaps (EPint_2 mm, EPint_0.6 mm) increases the foundation vertical motion (rocking) and reduces the soil motion considerably (better shown from Fourier Spectra), in comparison with the analyses without gaps (EPint). The former analyses approach much better the field-data providing further evidence of the existence of interface imperfections. It is also observed that the EPint_2 mm analysis gives identical results to the LSEint analysis, implying that the EPint_2 mm initially "opened" interface elements do not close at any point during the motion. When the initial gaps are smaller though (EPint_0.6 mm), the elements "close" and "re-open" at different increments during the analysis, producing the high frequency content observed in the vertical foundation motion.

For the specific experiment simulated here, it is demonstrated that the gaps at the interface affect the foundation rocking and soil motion considerably. Other experiments that were simulated, but not presented herein for brevity, have shown that gaps can affect the SSI response more pronouncedly, especially at a frequency close to the resonant one which is also affected (e.g. free vibration motion). Following this conclusion, potential gaps below prefabricated shallow foundations should be carefully considered in design.

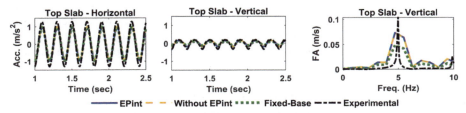

Fig. 6. Top slab acceleration time histories and Fourier Spectra for different interface boundary conditions

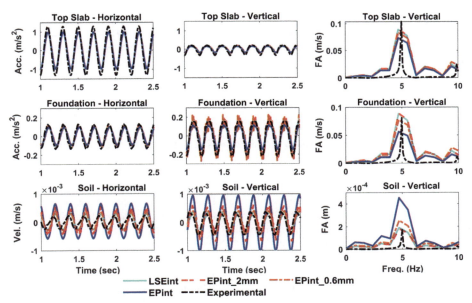

Fig. 7. Numerical results of analyses with different interface imperfections at different parts of the SSI system (top slab, foundation, soil)

5 Conclusions

This paper presented linear elastic 3D finite element simulations of real-scale forced vibration tests where SSI phenomena were prominent, showing very good agreement with the field data. Emphasis was placed on the novel interface modelling between the shallow foundation and soil, aiming to capture imperfections in the adhesion between the foundation and the soil which were observed in the field.

The proposed numerical model uses zero-thickness elastoplastic interface elements at the soil-foundation interface and initiates contact imperfections (gaps) through the formulation of the interface constitutive model, by prescribing initial plastic normal strains to the interface elements of the non-contact areas. It was found that the value of the initial strain (gap) affects the contact area during the excitation and therefore, the numerical results. For the studied case, when the gaps were 0.6mm, the elements "open" and "close" during the excitation, while they remain always "opened" when the gaps are 2mm. The latter case could also be simulated with linear elastic interface elements of much reduced stiffness.

It was shown that interface imperfections can increase foundation rocking and decreases soil motion significantly. A parallel work which has been conducted, but not presented herein for brevity, has shown that the effect can be more pronounced at resonance. Hence, gaps which are common below prefabricated shallow foundations should be carefully considered in design.

Acknowledgements. The experimental campaign was funded by SERA (Seismology and Earthquake Engineering Research Infrastructure Alliance for Europe). The first author wishes to thank Skempton Scholarship, A.G. Leventis Foundation and Cyprus State Scholarship Foundation for their support.

References

1. Gazetas, G.: Formulas and charts for impedances of surface and embedded foundations. J. Geotech. Eng. **117**(9), 1363–1381 (1991)
2. Gazetas, G.: Analysis of machine foundation vibrations: State of the art. Int. J. Soil Dyn. Earthq. Eng. **2**(1), 2–42 (1983). https://doi.org/10.1016/0261-7277(83)90025-6
3. Pelekis, I., McKenna, F., Madabhushi, G.S.P., DeJong, M.J.: Finite element modeling of buildings with structural and foundation rocking on dry sand. Earthq. Eng. Struct. Dyn. **50**(12), 3093–3115 (2021). https://doi.org/10.1002/eqe.3501
4. Amorosi, A., Boldini, D., di Lernia, A.: Dynamic soil-structure interaction: a three-dimensional numerical approach and its application to the Lotung case study. Comput. Geotech. **90**, 34–54 (2017). https://doi.org/10.1016/j.compgeo.2017.05.016
5. Gazetas, G.: 4th Ishihara lecture: Soil-foundation-structure systems beyond conventional seismic failure thresholds. Soil Dyn. Earthq. Eng. **68**, 23–39 (2015). https://doi.org/10.1016/j.soildyn.2014.09.012
6. Pitilakis, D., Rovithis, E., Anastasiadis, A., Vratsikidis, A., Manakou, M.: Field evidence of SSI from full-scale structure testing. Soil Dyn. Earthq. Eng. **112**, 89–106 (2018). https://doi.org/10.1016/j.soildyn.2018.04.024

7. Pitilakis, K., Raptakis, D., Lontzetidis, K., Tika-Vassilikou, T., Jongmans, D.: Geotechnical and geophysical description of euro-seistest, using field, laboratory tests and moderate strong motion recordings. J. Earthq. Eng. **3**(3), 381–409 (1999). https://doi.org/10.1080/13632469909350352
8. Potts, D.M., Zdravković, L.: Finite Element Analysis in Geotechnical Engineering: Theory. Thomas Telford, London (1999)
9. British Standards Institution: BS EN 1992-1-1:2004. Eurocode 2. Design of concrete structures. General rules and rules for buildings (2004)
10. British Standards Institution: BS EN 1993-1-1:2005, Eurocode 3. Design of steel structures. General rules and rules for buildings (2005)
11. Raptakis, D., Chávez-García, F.J., Makra, K., Pitilakis, K.: Site effects at Euroseistest-I. Determination of the valley structure and confrontation of observations with 1D analysis. Soil Dyn. Earthq. Eng. **19**(1), 1–22 (2000). https://doi.org/10.1016/S0267-7261(99)00025-1
12. Raptakis, D., Makra, K.: Multiple estimates of soil structure at a vertical strong motion array: understanding uncertainties from different shear wave velocity profiles. Eng. Geol. **192**, 1–18 (2015). https://doi.org/10.1016/j.enggeo.2015.03.016
13. Kontoe, S.: Developement of time integration schemes and advanced boundary conditions for dynamic geotechnical analysis (2006)
14. Day, R.A., Potts, D.M.: Zero thickness interface elements—numerical stability and application. Int. J. Numer. Anal. Methods Geomech. **18**(10), 689–708 (1994). https://doi.org/10.1002/nag.1610181003

Experimental Behavior of Single Pile with Large Mass in Dry Medium Sand Under Centrifuge Shaking Table Test

Longyu Lu[1], Chunhui Liu[1(✉)], Mengzhi Zhang[1], and Tiqiang Wang[2]

[1] Yantai University, Yantai, Shandong, China
lch_hit@163.com
[2] Institute of Engineering Mechanics, China Earthquake Administration, Yantai, Heilongjiang, China

Abstract. In this study, a single pile centrifuge shaking table model tests was conducted. The foundation soil was dry medium sand with 50% relative density which was made of pouring method. In prototype, the pile diameter was 1.0 m, and a mass of 312.5 ton was fixed on pile head which was 5 m above the ground surface. The strain gauges were uniformly instrumented along the pile to measure the bending moment response of the pile, the accelerometers were instrumented to obtain the acceleration response of soil and pile. The input seismic excitation was the sinusoidal wave of 0.3 g. The results show that: the obvious acceleration amplification effect was observed in the soil above 15 m; the maximum bending moment of pile occurred in 1.65 m depth, and the residual bending moment in the lower part of the pile was observed.

Keywords: Centrifuge test · Soil-pile interaction · Acceleration amplification effect

1 Introduction

In the decades, the clean power has become an essential trend to protect our earth village, and wind energy was one of the best ideal clean powers. The large diameter single pile was used to be the foundation of wind turbines to sustain kinds of lateral loads, such as wind, waves and earthquakes [1]. However, most study focused on the pile behavior under long-term load caused by wind or wave. Limited study was conducted on the larger diameter pile under earthquake motion, especially in the dry sand ground [2–10]. In addition, the response of single pile has a basic function for the development of pile foundation engineering.

To investigate the dynamic response of single pile in the dry medium sand (D_{50}=0.52 mm), a single degree-of-freedom (SDOF) centrifuge shaking table test was carried out. The mass of 312.5 ton fixed on the pile head in prototype, which was enough to produce huge inertial action in the pile, the acceleration response of soil and bending moment response of pile were analyzed.

2 Model Tests

The centrifuge shaking table test was performed in the Institute of Engineering Mechanics, China Earthquake Administration. The centrifuge shaking table system is electro-hydraulic driven, as shown in Figure 1. This system is able to reach a maximum centrifugal acceleration of 100 g, and a maximum vibration acceleration of 30g. A laminar shear container, as show in Figure 1, was used during the shaking to absorb possible reflecting waves, which can better mimic the boundary conditions in the field. The laminar container has the dimension of 0.6 m in height, 0.5 m in width, and 1.2 m in length.

Fig. 1. (a) Centrifuge shaking table, (b) The laminar shear container

2.1 Model-Prototype Relations

According to the centrifuge scaling laws [6], the scale factors N used in the present study were shown in the Table 1, where N is the scaling ratio of the model to the prototype. In this test, the centrifuge acceleration of 50 g was adopted.

Table 1. Scale factor

Items	Physical laws	Scale factor
Geometrical dimension, L	C_l	$1/N$
acceleration, a	C_a	N
Mass density, ρ	C_ρ	1
Young's modulus, E	C_E	1

(*continued*)

Table 1. (*continued*)

Items	Physical laws	Scale factor
Mass, m	$C_m = C_\rho C_l^3$	$1/N^3$
Internal friction angle, φ	C_φ	1
Bending moment, M	$C_M = C_a C_\rho C_l^4$	$1/N^3$
Stress, ε	$C_\varepsilon = C_\rho C_l C_a / C_E$	1
Strain, σ	$C_\sigma = C_\varepsilon C_E$	1
Dynamic time, t_l	$C_{tl} = (C_a / C_l)^{1/2}$	$1/N$

2.2 Experimental Setup

2.2.1 Soil

The soil used in the centrifuge test is Fujian sand. The sand properties were presented in Table 2. The particle size distribution curve was shown in the Figure 2. The soil model was prepared using the conventional sand pouring method. The relationship between relative density and falling distance was shown in Figure 3. According to Figure 3, the dry sand model with 50% of relative density was constructed using pouring method. Dry medium sand was air poured and the accelerometers in the soil were placed at the specific depths during model construction for centrifuge test, and except for the bottom accelerometer, the vertical interval of other accelerometers was 2.8 m, the accelerometers were represented as a1 to a10 in the Figure 4. Besides, in the Figure 4, the settlement of soil (LT1) was recorded with LVDT sensor, the LW1 to LW3 were applied to measure the lateral displacement of the soil in different depth, AW1 was applied to measure the acceleration of table-board, AW2 was used to measure the acceleration corresponding to its position of LW2.

Table 2. The properties of dry sand

Parameter	Units	Medium sand
Specific gravity	–	2.630
Mean grain size, D_{50}	(mm)	0.52
Effective grain size, D_{10}	(mm)	0.29
Coefficient of uniformity, C_u	–	0.84
Maximum dry unit weight, γ_{max}	(kN/m^3)	16.87
Minimum dry unit weight, γ_{min}	(kN/m^3)	14.00

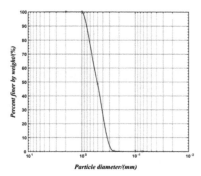

Fig. 2. Grain size distribution curve

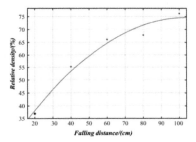

Fig. 3. The relationship between relative density and falling distance

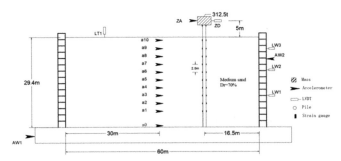

Fig. 4. The layout of sensors

2.2.2 Single Pile and Superstructure

An aluminum tubular pipe was used to represent the pile in the centrifuge model, the outer diameter, thickness and length of the aluminum tubular pipe are 20 mm, 2 mm and 650 mm. The strain gages were pasted along the pile to record the bending moment response of pile, as shown in Figure 4. A solid steel fixed in the head was used to mimic the superstore in model test, the size of the mass is 85 mm × 85 mm × 45 mm. The properties of pile and superstructure in prototype scale and model scale were listed in the Table 3. In the superstructure, the displacement and acceleration were recorded with LVDT and MEMS sensor, which show as ZD and ZA in Figure 4.

Table 3. Properties of pile and superstructure

Physical parameters	Model scale	Prototype scale
Pile		
Material	Aluminum	Concrete
Outer Diameter, D (mm)	20	1000
Thickness, t (mm)	2	---
Pile length, L (mm)	680	34×10^3
Young's Modulus, E (kPa)	70×10^6	27×10^6
Superstructure		
Mass (kg)	2.5	3.125×10^6

The input waves which applied in the bottom of model were sinusoidal wave, with peak amplitude of 0.3 g. The acceleration-time curves and its Fourier spectrum were shown in the Figure 5.

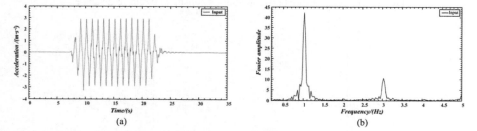

Fig. 5. (a) The input sinusoidal wave of 0.3g, (b) the fourier spectrum of input wave

3 Experimental Results and Discussion

3.1 Acceleration of Soil

The acceleration versus time curve is shown in Fig. 6. It can be observed from the figure that the waveform is very regular, basically maintaining the shape of the input wave, however, the peak acceleration of soil varies with the depth of the soil layer, i.e., as the increase of depth, the peak acceleration decreases.

In order to clearly observe the relationship between peak acceleration and depth, the acceleration profile was plotted in Fig. 7, it can be obviously obtained that the acceleration peak gradually increases and becomes linear as the decrease of depth. The peak acceleration amplification factor of the soil surface was about 1.63. However, the peak acceleration varies around 3 m/s² when the depth is more than 15 m, no obvious acceleration amplification effect was observed for the soil below 15 m.

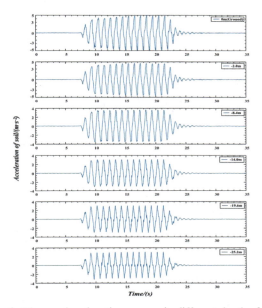

Fig. 6. The acceleration-time curves in different depth of soil

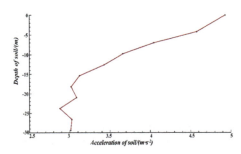

Fig. 7. The distribution of the peak acceleration with the soil depth

3.2 Bending Moment of Pile

The bending moment time history is shown in Fig. 8. In order to facilitate the analysis of the length of the pile with the depth of the soil, the zero point on the pile position is the surface of the soil layer. Therefore, the part of the pile body inserted below the ground gradually decreases with the buried depth of the soil, and the part above the ground gradually increases, for example, -1 m represents the position where the pile is in the soil depth of 1 m. The bending moment of the pile reaches the maximum value at 1.65 m below the soil surface, and the bending moment gradually decreases and stabilizes with the increasing depth. In addition, the residual bending moment appeared in the lower part of the pile, which may be caused by collapse of sand around the pile foundation.

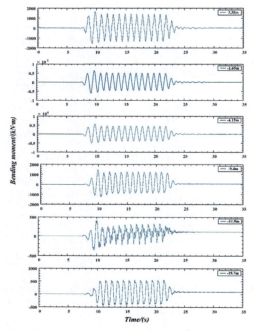

Fig. 8. The bending moment-time curves in the different depth of pile

4 Conclusion

According to the centrifuge shaking table test, the seismic behavior of large mass SDOF pile-structure was analyzed in this study. The conclusions are listed as follow;

(1) The distribution of the peak-acceleration amplification factor had the different behavior with the changing of soil depth. The obvious acceleration amplification effect was observed in the soil above 15 m.
(2) The maximum bending moment of pile occurred in 1.65 m depth, and the residual bending moment in the lower part of the pile was observed.

References

1. Liu, J., Wan, Z., Dai, X., Jeng, D., Zhao, Y.: Experimental study on whole wind power structure with innovative open-ended pile foundation under long-term horizontal loading. Sensors (Basel), **20**(18) (2020)
2. Goit, C.S., Saitoh, M., Igarashi, T., Sasaki, S.: Inclined single piles under vertical loadings in cohesionless soil. Acta Geotech. **16**(4), 1231–1245 (2020). https://doi.org/10.1007/s11440-020-01074-9
3. Chatterjee, K., Choudhury, D., Rao, V.D., Poulos, H.G.: Seismic response of single piles in liquefiable soil considering P-delta effect. Bull. Earthq. Eng. **17**(6), 2935–2961 (2019). https://doi.org/10.1007/s10518-019-00588-2
4. Durante, M.G., Di Sarno, L., Mylonakis, G., Taylor, C.A., Simonelli, A.L.: Soil-pile-structure interaction: experimental outcomes from shaking table tests. Earthq. Eng. Struct. D **45**(7), 1041–1061 (2016)
5. Saeedi, M., Dehestani, M., Shooshpasha, I., Ghasemi, G., Saeedi, B.: Numerical analysis of pile-soil system under seismic liquefaction. Eng. Fail Anal. **94**, 96–108 (2018)
6. Ubilla, J., Abdoun, T., Dobry, R.: Centrifuge scaling laws of pile response to lateral spreading. Int. J. Phys. Modell. Geotech. **11**(1), 2–22 (2011)
7. Dobry, R., Abdoun, T., O'Rourke, T.D., Goh, S.H.: Single piles in lateral spreads: field bending moment evaluation. J. Geotech. Geoenviron. Eng. **129**(10), 879–889 (2003)
8. Lim, H., Jeong, S.: Simplified p-y curves under dynamic loading in dry sand. Soil Dyn. Earthq. Eng. **113**, 101–111 (2018)
9. Tombari, A., El Naggar, M.H., Dezi, F.: Impact of ground motion duration and soil non-linearity on the seismic performance of single piles. Soil Dyn. Earthq. Eng. **100**, 72–87 (2017)
10. Brandenberg, S.J.: Behavior of pile foundations in liquefied and laterally spreading ground, p. 340. University of California, Davis (2005)

Numerical Investigation on Dynamic Response of Liquefiable Soils Around Permeable Pile Under Seismic Loading

Chi Ma[1,2], Guo-Xiong Mei[2], and Jian-Gu Qian[1(✉)]

[1] Department of Geotechnical Engineering, Tongji University, Shanghai 200092, China
qianjiangu@tongji.edu.cn
[2] College of Civil Engineering and Architecture, Guangxi University, Nanning 530004, China

Abstract. Permeable piles have been widely adopted as a new anti-liquefaction treatment measure in the liquefiable site. The dynamic numerical analysis to study the anti-liquefaction effects of liquefiable soils around the permeable pile under the seismic wave was carried out to estimate the excess pore water pressure ratio (EPWPR), pile settlement, and the soil settlement. Numerical results show that the EPWPR has exceeded the critical level of liquefaction at 2 m far from the pile at the top surface of the liquefiable soil layer. The drainage effect of the soil around the pile weakens as the depth of the soil layer increases, that is, the shear stress and strain are being strong rapidly. Additionally, the effect of pile permeability on dynamic responses of liquefiable soil indicates that the increase of pile permeability will considerably weaken dynamic responses. Furthermore, the failure of liquefaction never takes place in the site if the permeability ratio of the pile to that of soil exceeds the critical value of 10.

Keywords: Seismic response · Liquefiable soil · Permeable pile

1 Introduction

Liquefaction is one of the significant and complex issues in geotechnical engineering and is also the main cause of damage and failure in piles under earthquakes [1]. The liquefaction takes place due to the accumulation of pore pressure in loose sand deposits under seismic loading, which in turn produce an important reduction in the strength of the underlying soils. Therefore, the reliable and economical anti-earthquake design of piles requires that the pile can not only resist the seismic loading but also promote soil drainage.

In general, desaturation is a method to mitigate liquefaction of sand soil by generating gas or air in the pores of fully saturated sands to prevent liquefaction. However, this method also influences the superstructure deformation [2]. Stone columns composite foundation in liquefiable soil is helpful to reduce the surface peak acceleration and beneficial to improve the compactness of sand soil. Stone columns have a restraint effect on the deformation of the site and bear more distributed stress of the soil around the piles when an earthquake occurs. However, the strength of this pile is not enough to resist high seismic loading [3]. The strength composite pile is a kind of prefabricated concrete pipe pile inserted into the cement-soil pile, but this method cannot solve the liquefiable of the soil under seismic loading [3]. Permeable piles which can effectively drain water in time and improve the bearing capacity of piles have been adopted as a new anti-liquefaction treatment measure in liquefiable sites by Mei [4]. The effect of drainage hole blockage on drainage under seismic loading is studied.

2 Numerical Model

2.1 Soil Mesh

In this study, the soil volume dimensions considered are 40 × 20 × 10 m. The upper and bottom layers of the model are dense sand (Dr = 80%) and thick loose sand (Dr = 40%) respectively (Fig. 1) [6]. The properties of the sand are presented in Table 1.

Table 1. Sand properties

Properties	Unit	Dr = 40%	Dr = 80%
Poisson's ratio		0.45	0.45
Shear modulus	kPa	2.7×10^4	4.47×10^4
Bulk modulus	kPa	2.16×10^5	4.32×10^5
$(N_1)60$		7.2	30
Permeability	cm/s	6.6×10^{-2}	3.2×10^{-5}
Friction angle	°	33	39.5
Soil density	kg/m³	2962	2047

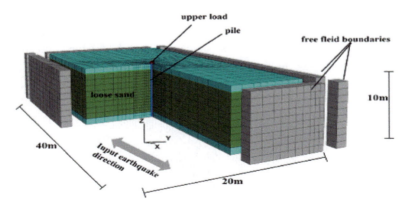

Fig. 1. Geometry of the numerical model

2.2 Pile Mesh

The permeable pile model is established by referring to the previous study results. The self-adaptive decompression permeable pile (Fig. 2) is composed of a pipe pile body, at least one row of plug holes which are arranged on the pipe pile body and water-soluble plugs which are arranged in the plug holes. The cross-section of the permeable pile is circular with a pile diameter of 0.6 m(D) and a pile length of 10 m. The material parameters of the pile are shown in Table 2.

Table 2. Parameters of pile

Properties	Bulk modulus	Shear modulus	Diameter	Density
Pile	1.67 GPa	0.76 GPa	0.6 m	2400 kg/m^3

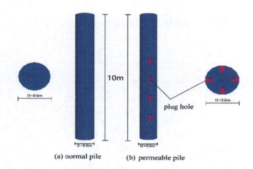

Fig. 2. Sketch of permeable pile

2.3 Input Motions

The seismic wave input duration is 20 s. The peak acceleration increases to 0.2g in 2 s and decreases to 0 at 17 s (Fig. 3). This seismic wave is applied in the x-direction at the bottom of the model (Fig. 1).

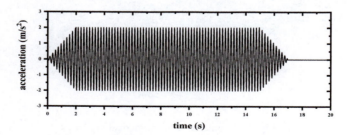

Fig. 3. Input acceleration history of calculation

3 Dynamic Response of Liquefiable Soils Around Permeable Pile

The EPWPR is defined as the ratio of the difference of pore pressure (u) at a certain moment and initial pore pressure (u_0) over the initial effective stress (σ_0). If the total stress is kept constant during seismic loading, the true liquefaction will occur when EPWPR is equal to 1 theoretically. In fact, the soil will undergo large deformation when the EPWPR exceeds 0.8 and the soil has already lost its bearing capacity [7, 8]. Therefore, it is considered that the liquefaction is triggered when the EPWPR exceeds 0.8 in this study.

As shown in Fig. 4, the horizontal distance is analyzed in units of one pile diameter (D). For example, the coordinate of point A is (1.5, 0, 4), which is 1.5 m from the pile core and 1.2 m from the pile toe.

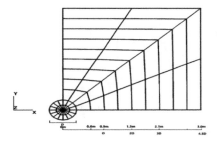

Fig. 4. Schematic diagram of analysis distance

To analyze the effective area of the permeable pipe pile, Fig. 5 to 12 plot the time history of EPWPR of the soil at four locations around the normal pile and permeable pile respectively. The trend of EPWPR with time for the normal pile at the depth of 4 m during the seismic loading is shown in Fig. 5(a). The seismic loading reaches the peak acceleration at 2 s. The EPWPR rises rapidly during this period because the pore water is too difficult to drain. With the excitation of seismic loading, the pore water pressure gradually and slowly increases until it stabilizes after liquefaction. Additionally, it is also found that the farther away from the pile, the faster the pore water pressure increases and the earlier it enters the liquefaction state. Compared with the normal pile, the EPWPR decreases after the use of permeable piles in Fig. 5(b). Therefore, the soil still can resist the load when subjected to the seismic loading. When the length is 2.0D, the EPWPR is all less than 0.8. When the length is 2.5D, a small part of the EPWPR exceeds 0.8. However, the soil liquefaction will occur when the length exceeds 2.5D and the EPWPR exceeds 0.8 for a long time. The influence range of the pile is 2.5D at a depth of 4m, and the drainage effect is the best when the pile length is 2.0D.

The development law of the time history of EPWPR at depths of 5 m and 6 m is similar to that at a depth of 4 m, as shown in Fig. 6 and Fig. 7 respectively. The influence range of the pile is 2.0D at a depth of 5 m, and the drainage effect is the best when the pile length is 1.5D. Similarly, the influence range of the pile is 2.0D at a depth of 6 m, and the drainage effect is the best when the pile length is 1.5D.

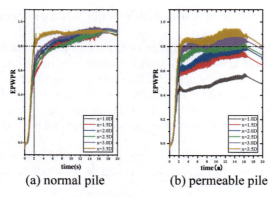

(a) normal pile (b) permeable pile

Fig. 5. The EPWPR at a depth of 4 m

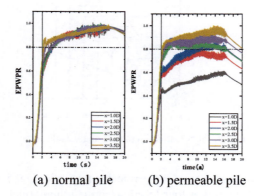

(a) normal pile (b) permeable pile

Fig. 6. The EPWPR at a depth of 5 m

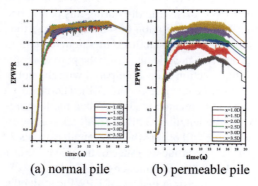

(a) normal pile (b) permeable pile

Fig. 7. The EPWPR at a depth of 6 m

The development law of the time history of EPWPR at depths of 7 m is shown in Fig. 8. The influence range of the pile is 1.5D at a depth of 7 m, and the drainage effect is the best when the pile length is 1.0D. At the top surface of the liquefiable soil layer, the EPWPR has exceeded the critical level of liquefaction at 2.5D (about 2 m) far from the pile. Additionally, the drainage effect of the soil around the pile weakens as the depth of the soil layer increases.

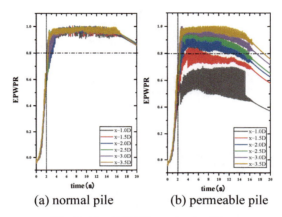

(a) normal pile (b) permeable pile

Fig. 8. The EPWPR at a depth of 7 m

4 Influence of Clogging of Drainage Holes on Drainage

In actual engineering, the drainage through the pile will drive the movement of soil particles. The drainage holes will be blocked owing to the long-term drainage effect. The influence of the drainage hole blockage on the drainage of the permeable pile is studied in this section. Based on the analysis results in Sect. 4.1, the EPWPR of points A (1.5, 0, 4), B (1.2, 0, 5), C (1.2, 0, 6), and D (0.9, 0, 7) are selected for analysis respectively. The position of each point is shown in Fig. 9.

Figure 10 shows the EPWPR of the four points with different permeability coefficients of the drainage holes. The permeability of 1.0×10^{-1} cm/s is consistent with the effect of complete drainage in the permeable area. Furthermore, the maximum pore water pressure is concentrated around 0.8 when the permeability is 5.0×10^{-2} cm/s, while exceeding 0.8 at 8 s when the permeability is 1.0×10^{-2} cm/s. There are similar trends at points B and D. Therefore, it is concluded that it belongs to a completely drained state when the permeability is 1.0×10^{-1} cm/s, and when the anti-liquefaction effect of the permeable pile disappears when the permeability is 1.0×10^{-2} cm/s. The failure of liquefaction never takes place in the ground if the pile-soil penetration ratio exceeds the critical value of 10.

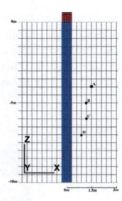

Fig. 9. Measuring point location

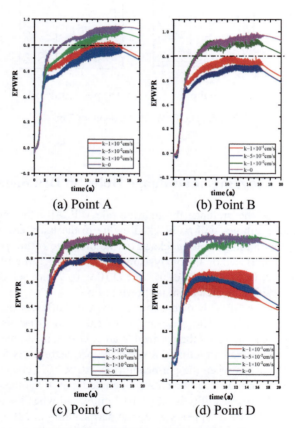

(a) Point A (b) Point B (c) Point C (d) Point D

Fig. 10. The EPWPR caused by fouling effect

5 Conclusion

Seismic analysis of permeable pile improved liquefiable site was conducted in this study. The calculation results show that the permeable pipe installation can improve the anti-liquefaction effect of the foundation. Additionally, the permeable pile can drain the pore water in time under the action of seismic loading and reduce EPWPR within a certain range. Horizontal reinforcement area is about 2–3 times its diameter.

The study on the clogging effect of the permeable pile hole found that the drainage effect is the best when the permeability is 1.0×10^{-1} cm/s. and The anti-liquefaction effect of the permeable pile disappears when the permeability is reduced to 1.0×10^{-2} cm/s. The failure of liquefaction never takes place in the site if the permeability ratio of the pile to that of soil exceeds the critical value of 10.

References

1. Bhattacharya, S., Ma, D.B.S.: A critical review of methods for pile design in seismically liquefiable soils **16**(3), 407–446 (2008)
2. Darby, K.M., Hernandez, G.L., Dejong, J.T., Boulanger, R.W., Wilson, D.W.: Centrifuge model testing of liquefaction mitigation via microbially induced calcite precipitation. J. Geotech. Geoenviron. Eng. **145**(10) (2019)
3. Zou, Y., Wang, R., Zhang, J.: Analysis on the seismic response of stone columns composite foundation in liquefiable soils. J. Rock Soil Mech. **40**(6), 2443–2455 (2019)
4. Liu, S.H., Wang, A.H., Zhang, D.W., Lin, W.L.: Seismic response of piles in cement-improved soil in a liquefiable soil. J. Civil Environ. Eng. **2**(5), 99–110 (2021)
5. Zhou, X.P., Mei, G.X.: Finite element simulation of permeable pipe pile driving considering consolidation process. Rock Soil Mech. **35**(2), 676–682 (2014)
6. López Jiménez, G.A., Dias, D., Jenck, O.: Effect of the soil–pile–structure interaction in seismic analysis: case of liquefiable soils. Acta Geotech. **14**(5), 1509–1525 (2018). https://doi.org/10.1007/s11440-018-0746-2
7. Koutsourelakis, S., Jean, H.P., Deodatis, G.: Risk assessment of an interacting structure–soil system due to liquefaction. Earthquake Eng. Struct. Dynam. **31**(4), 851–879 (2010)
8. Liu, Z.Y., Qian, J.G., Mohammadjavad, Y., Xue, J.F.: The effects of initial static deviatoric stress on liquefaction and pre-failure deformation characteristics of saturated sand under cyclic loading. Soil Dyn. Earthquake Eng. **149** (2021)

Rotation of a Cantilevered Sheet-Pile Wall with Different Embedment Ratios and Retaining a Liquefiable Backfill of Various Relative Densities

Satish Manandhar, Seung-Rae Lee, and Gye-Chun Cho(✉)

Korea Advanced Institute of Science and Technology (KAIST), Daejeon, Republic of Korea
gyechun@kaist.edu

Abstract. Damage to retaining structures at the waterfront during major earthquakes can be attributed to the development of positive excess pore water pressure in the retained or foundation soils, leading to liquefaction. As part of Liquefaction Experiments and Analysis Project (LEAP), dynamic centrifuge tests of a cantilevered sheet-pile wall model floating in dense sand and retaining a liquefiable backfill were carried out. The embedment ratio (embedment depth / excavation height) and the relative density (D_r) of the backfill were varied to observe the rotation of the sheet-pile during a tapered 1 Hz sinusoidal input base motion. The D_r of the backfill affected the wall rotation during gravity loading and the initial static shear stress in the backfill; smaller D_r backfill led to a larger wall rotation and initial static shear. During seismic loading, the model with a higher initial static shear had a dilative pore pressure response which increased the wall stability. On the other hand, the model with a smaller initial static shear showed a contractive response leading to initial liquefaction in the active failure wedge, which resulted in a large increase of wall rotation. Moreover, the model with a higher embedment ratio had a smaller wall rotation during initial loading cycles, but the wall rotation increased rapidly after the soil in front of wall softened due to positive excess pore water pressure buildup. These results showed the importance of considering the initial static shear stress prior to earthquake loading and the vulnerability of wall stability due to excess pore pressure increase in front of the wall.

Keywords: LEAP · Liquefaction · Sheet-pile wall · Wall rotation · Centrifuge Test

1 Introduction

Extensive damage to retaining structures near the waterfront with saturated backfill has been observed in past earthquakes ([1, 2]). The failure of retaining structures (sheet-pile wall, quay wall etc.) could be structural failure due to high stresses and/or loss of serviceability due to large deformations. For the performance-based design, the prediction of the deformation of these structures during design earthquake motion is critical.

Limit-equilibrium analysis based on Mononobe-Okabe method can predict wall stability under seismic loads but it cannot predict the magnitude of wall deformations. Also, numerical simulation using effective stress analysis method have a varying degree of success in predicting wall displacements depending on the formulation and calibration of soil constitutive models, especially in the presence of excess pore water pressure buildup and liquefaction in the backfill ([3, 4]). In this context, centrifuge model tests have proven to be useful in understanding the deformation mechanism of retaining structures with liquefiable backfills and in validating numerical simulation procedures ([5, 6]).

This paper presents the results of a series of dynamic centrifuge tests performed on a cantilevered sheet-pile wall model retaining a liquefiable backfill sand. These tests were part of an international collaborative project on soil liquefaction and its consequences, called the Liquefaction Experiments and Analysis Project (LEAP). The main objectives of the work were to understand the mechanism of wall rotation and to study the effect of backfill relative density (D_r) and embedment ratio of the wall on the seismic performance of the wall.

2 Dynamic Centrifuge Tests

Three dynamic centrifuge tests were performed at a centrifugal acceleration of 40-g using a 5 m radius beam-type centrifuge at KAIST and KAIST Analysis Center for Research Advancement (KARA). The test model consisted of a submerged cantilevered sheet-pile wall model floating in middle of dense sand ($D_r = 90\%$) and retaining a liquefiable backfill. Two tests (LEAP2020_T2 and LEAP2020_T3) had an embedment ratio of 0.5 and target backfill D_r of 65% and 55%. One test (LEAP2021_T1) has an embedment ratio of 1.0 and a target backfill D_r of 65%. Embedment ratio is the ratio of wall embedment depth to the excavation height (Fig. 1). The model wall was made of aluminum and had a thickness of 0.112 m in the prototype scale. For this study, the backfill represented the sand layer (in front and back of the wall) above the dense sand.

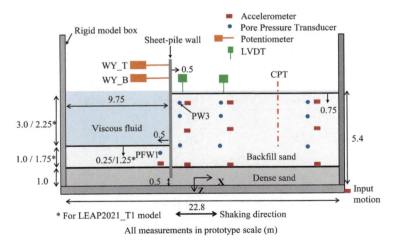

Fig. 1. Model schematic and instrumentation for LEAP2020_T2 and LEAP2020_T3 models. Changes are indicated for the LEAP2021_T1 model.

The model geometry along with the instrumentation is shown in Fig. 1. All measurements are reported in the prototype scale [7] unless stated otherwise.

The models were prepared following the LEAP-RPI-2020 specifications. The sand used for the test was Ottawa F-65 sand, which has been used extensively in previous LEAP exercises [8]. The sand was dry-pluviated into a rigid model container from a calibrated drop height to achieve a target D_r. Various instrumentation such as accelerometers, pore pressure transducers, LVDTs, and potentiometers were installed during model construction to measure the dynamic response of the model. The models were fully saturated with viscous fluid to satisfy the centrifuge scaling law.

Before the seismic event, CPT was performed in the backfill to obtain the cone tip resistance (q_c) profile. The measured q_c value at a depth of 2 m was used to obtain the D_r of the backfill sand based on a correlation by [9]. A tapered 1 Hz sinusoidal motion with five consecutive cycles at peak ground acceleration (PGA) of different amplitudes was used as an input base excitation (Fig. 2). The details of the three tests and the input base motions are presented in Table 1.

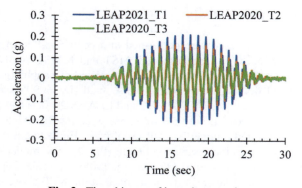

Fig. 2. Time-history of input base motions

Table 1. Details of the test models and input base motions

Model	q_c at 2 m (MPa)	Backfill D_r (%)	PGA_{eff} (g)	FS_{liq}
LEAP2021_T1	3.06	67.9	0.183	0.84
LEAP2020_T2	2.72	63.7	0.144	0.91
LEAP2020_T3	2.15	56.4	0.12	0.87

The PGA_{eff} is the effective PGA of the input base motion defined based on [10]. Also, factor of safety against liquefaction triggering (FS_{liq}) for the backfill soil is presented in Table 1. The cyclic stress ratio (CSR) is calculated based on the simplified procedure [11] and the cyclic resistance ratio (CRR) is based on the q_c-based liquefaction triggering curve from [12].

Although detailed dynamic response of the model was obtained from various sensors, the measured wall rotation during static (centrifuge spin-up) and seismic loadings will be presented in this paper. The wall rotation was calculated as the arctangent of the ratio of the difference in horizontal displacements recorded by two potentiometers at top of the wall to their vertical separation (Fig. 1). Furthermore, the excess pore pressure response in the active failure wedge (PW3) and in front of the wall (PFW1) will also be discussed.

3 Results and Discussion

The wall rotation during the spinning of the centrifuge from 1-g to the target centrifuge g-level of 40-g is shown in Fig. 3. The final wall rotation was largest in the LEAP2020_T3 (0.88°) model while it was smallest in the LEAP2021_T1 model (0.19°). For the cantilevered sheet-pile, the wall rotation depends on the embedment ratio and the friction angle (relative density) of the backfill soil [13]. The rotational stability of the wall is provided by the passive resistance from the soil in front of the wall. So, the wall with a larger embedment ratio had a higher rotational stability. Meanwhile, for the same embedment ratio, the model with a relatively looser backfill (LEAP2020_T3) had a greater wall rotation than the model with a denser backfill (LEAP2020_T2). This is because the looser backfill exerts a comparatively higher driving force (active pressure) and a smaller stabilizing force (passive pressure) on the wall due to a smaller friction angle.

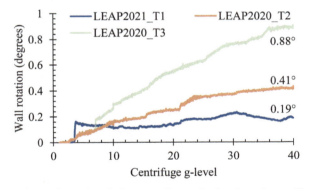

Fig. 3. Wall rotation during centrifuge spinning from 1-g to 40-g

The wall rotation during dynamic loading for the three models is shown in Fig. 4. The final wall rotation of the LEAP2020_T3 model was the smallest (1.94°), while the final wall rotation of LEAP2020_T2 and LEAP2021_T1 models were 3.35° and 3.55°, respectively. The mechanism of wall rotation and the effects of the backfill D_r and the embedment ratio on the wall rotation during dynamic loading are discussed below.

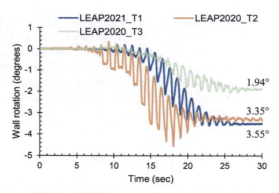

Fig. 4. Wall rotation during dynamic loading

3.1 Mechanism of Wall Rotation

At the beginning of the input motion (t < 15 s), the stress waves propagated through the backfill and caused an increase in horizontal stresses on the wall due to the inertia force from the active failure wedge. As a result, the wall rotated towards the excavation. At this stage, the excess pore pressure ratio (r_u), which is defined as the excess pore water pressure divided by the vertical effective stress, alternated between positive and negative values ($r_u = \pm 0.5$) but did not develop a net increase in excess pore pressure (Fig. 5). This is because the backfill soil near the wall was under the influence of the initial static shear stress, which prevented stress reversals during loading and unloading cycles of the base motion.

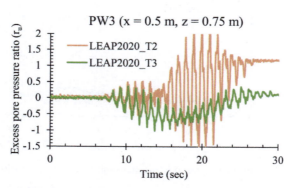

Fig. 5. Excess pore pressure ratio (r_u) in the active failure wedge (PW3) in the LEAP2020_T2 and LEAP2020_T3 models.

Once the backfill soil experienced the peak stress cycles (t = 15 to 20 s), the rotation of the wall increased at a higher rate towards the excavation due to a large inertia force from the active failure wedge. The excess pore water pressure also increased rapidly and approached initial liquefaction ($r_u = 1$). At this moment, the soil underwent cyclic mobility phenomena in which alternating cycles of positive (contractive) and negative (dilative) pore pressures were observed and with a net increase in excess pore pressure. This caused the backfill soil to gradually soften and lose its ability to transmit stress waves. As the stress waves could not propagate through the soil, the inertia force from the active failure wedge and the rate of accumulation of wall rotation reduced. Hence, towards the end of the motion (t > 20 s), the wall rotated at a much slower rate than at the beginning of the motion (t < 15 s) for similar input shaking intensity.

3.2 Effect of Backfill Relative Density

The effect of backfill D_r can be analyzed by observing the seismic response of LEAP2020_T2 and LEAP2020_T3 models. The initial wall rotation was twice larger in the LEAP2020_T3 model than in the LEAP2020_T2 model (Fig. 3). Hence, the soil in the active failure wedge was under a higher initial static shear stress in the LEAP2020_T3 model than in the LEAP2020_T2 model.

The cyclic resistance ratio of medium-dense and dense sands increases under the influence of initial static shear [14]. This can be observed from the excess pore pressure response in the active failure wedge behind the wall (Fig. 5). At the beginning of the input motion (t < 15 s), the excess pore pressure response alternated between positive and negative values with a net zero change in excess pore pressure in the LEAP2020_T2 model, while a net dilative response was observed in the LEAP2020_T3 model. During the peak cycles (t = 15–20 s), the excess pore pressure increased rapidly and reached initial liquefaction with cyclic mobility in LEAP2020_T2 model, while the soil response was mostly dilative in the LEAP2020_T3 model. As a result, the wall rotation increased rapidly under the action of a larger inertia force from the liquefied active failure wedge in the LEAP2020_T2 model. On the other hand, wall rotation, which was driven by inertia force from the non-liquefied active failure wedge, increased at a slower rate in the LEAP2020_T3 model.

Although the shaking intensity was smaller in the LEAP2020_T3 model, liquefaction did trigger in the backfill (Table 1). So, the observed pore pressure response in the active failure wedge is due to the effect of initial static shear and not due to the smaller level of shaking intensity.

3.3 Effect of Embedment Ratio

The effect of embedment ratio can be studied by comparing the seismic responses of the two models: LEAP2020_T2 and LEAP2021_T1. At the beginning of the cyclic loading, the sheet-pile in the LEAP2020_T2 model rotated more than in the LEAP2021_T1 model. This is due to a higher passive resistance in the front of the wall for the model with a higher embedment ratio. During the peak stress cycles (t = 15–20 s), the wall rotation increased rapidly in both the LEAP2020_T2 and LEAP2021_T1 models. However, the rate of increase in wall rotation was higher in the LEAP2021_T1 model. This is because once positive excess pore pressure developed in the passive failure wedge (Fig. 6), the greater passive resistance in the front of the wall due to a higher embedment ratio diminished and the wall rotated under the action of the inertia force from the active failure wedge. Since, the shaking intensity was larger in the LEAP2021_T1 model, the inertia force was also greater and caused the observed higher rate of wall rotation.

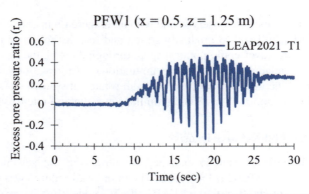

Fig. 6. Excess pore pressure ratio (r_u) in front of the wall (PFW1) in LEAP2021_T1 model.

4 Conclusions

This paper presented the results of three dynamic centrifuge tests on the behavior of a cantilevered sheet-pile wall embedded in a submerged backfill sand. The embedment ratio and the relative density of the backfill were varied to observe the wall rotation and the excess pore pressure response near the wall.

The rotation of the cantilevered sheet-pile wall model was driven by the inertia force from the active failure wedge. The excess pore pressure response in the active failure wedge depended on the initial static shear stress. The model which had a higher initial static shear (smaller D_r and higher initial wall rotation) experienced a dilative pore pressure response in the active failure wedge. This increased the wall stability and resulted in a smaller wall rotation. On the other hand, the model with a higher D_r (smaller initial wall rotation) showed contractive response leading to initial liquefaction during peak stress cycles. As a result, wall rotation increased rapidly. Moreover, greater passive resistance in the model with a higher embedment ratio was lost after the soil softened due to excess pore pressure buildup in front of the wall.

Acknowledgements. This research was supported by the KAIST Analysis Center for Research Advancement (KARA).

References

1. Bureau of Ports and Harbors: Earthquake resistant design for quay walls and pier in Japan. Ministry of Transport, Tokyo, Japan (1989)
2. Kamon, M., et al.: Geotechnical disasters on the waterfront. Soils Found. **36**, 137–147 (1996)
3. Iai, S., Kameoka, T.: Finite element analysis of earthquake induced damage to anchored sheet pile quay walls. Soils Found. **33**(1), 71–91 (1993)
4. Madabhushi, S.P.G., Zeng, X.: Simulating seismic response of cantilever retaining walls. J. Geotech. Geoenviron. Eng. **133**(5), 539–549 (2007)
5. Dewoolkar, M.M., Ko, H.Y., Pak, R.Y.S.: Seismic behavior of cantilever retaining walls with liquefiable backfills'. J. Geotech. Geoenviron. Eng. **127**(5), 424–435 (2001)

6. Madabhushi, S.P.G., Zeng, X.: Seismic response of gravity quay walls. II: numerical modeling. J. Geotech. Geoenviron. Eng. **124**(5), 418–427 (1998)
7. Garnier, J., et al.: Catalogue of scaling laws and similitude questions in geotechnical centrifuge modelling. Int. J. Phys. Modell. Geotech. **7**(3), 1–23 (2017)
8. El Ghoraiby, M., Park, H., Manzari, M.T.: Physical and mechanical properties of Ottawa F65 sand. In: Kutter, B.L., Manzari, M.T., Zeghal, M. (eds.) Model Tests and Numerical Simulations of Liquefaction and Lateral Spreading, pp. 45–67. Springer International Publishing, Cham (2020)
9. Carey, T.J., Gavras, A., Kutter, B.L.: Comparison of LEAP-UCD-2017 CPT results. In: Kutter, B.L., Manzari, M.T., Zeghal, M. (eds.) Model Tests and Numerical Simulations of Liquefaction and Lateral Spreading, pp. 117–129. Springer International Publishing, Cham (2020)
10. Kutter, B.L., et al.: LEAP-GWU-2015 experiment specifications, results, and comparisons. Soil Dyn. Earthq. Eng. **113**, 616–628 (2018)
11. Seed, H.D., Idriss, I.M.: Simplified procedure for evaluating soil liquefaction potential. J. Soil Mech. Found. Div. **97**(9), 1249–1273 (1971)
12. Idriss, I.M., Boulanger, R.W.: Soil liquefaction during earthquakes. Earthquake Engineering Research Institute MNO-12 (2008)
13. Madabhushi, S.P.G., Chandrasekaran, V.S.: Rotation of cantilever sheet pile walls. J. Geotech. Geoenviron. Eng. **131**(2), 202–212 (2005)
14. Youd, T.L., et al.: Liquefaction resistance of soils: summary report from the 1996 NCEER and 1998 NCEER/NSF workshops on evaluation of liquefaction resistance of soils. J. Geotech. Geoenviron. Eng. **127**(10), 817–833 (2001)

Foundation Alternatives for Bridges in Liquefiable Soils

Juan Manuel Mayoral[✉], Daniel De La Rosa, Mauricio Alcaraz, Nohemi Olivera, and Mauricio Anaya

Institute of Engineering at National Autonomous University of Mexico, Mexico City, Mexico
JMayoralV@iingen.unam.mx

Abstract. Ground failure during major seismic events associated with soil liquefaction can lead to major structural damage in both the columns and bridge upper deck, due to large seismic-induced displacements in the support foundation. Liquefaction driven ground motion incoherence during the dynamic event, and permanent soil deformations are key variables in the observed damage. This paper summarizes a numerical study of an alternative bridge foundation design proposed to reduce support displacements during and after an earthquake, as well as relative settlement associated with partial loss of bearing capacity when the bridge column is founded on a potential liquefiable layer. Three dimensional numerical models were developed with the program FLAC^{3D}. The seismic environment was characterized by a uniform hazard spectrum, UHS, developed for a nearby rock outcrop, considering a return period of 1000 years. Initially, a one-dimensional analysis was performed with the software SHAKE, to evaluate the liquefaction susceptibility. Later, to consider changes in both topography and ground subsurface layering of the site, a two-dimensional model was developed with the program QUAD4M. Based on the results gathered in here, it was concluded that structured cell foundation is a sound alternative to improve the seismic performance of bridges located in liquefiable soils and allows reducing detrimental effects associated with liquefaction-induced ground deformations.

Keywords: Liquefaction · Bridges · Seismic soil-structure interaction · Lateral spreading · Foundations

1 Introduction

Soil liquefaction is defined as the loss of strength associated with the reduction of effective stress that occurs due to excess pore pressure generation during cyclic loading of uncompacted non plastic fine-grained soil, such as silty sands and sandy silts. Although, it occurs mostly in loose, saturated sands, it has also been observed in gravel and non-plastic silts. This gradual loss of strength leads to bearing capacity reduction, which, in turn, may preclude poor structural performance related to permanent ground movements, and lateral spreading. Some examples of poor performance of bridge foundations under seismic loading were observed during 1964 Niigata, Anchorage and Alaska earthquakes.

Similarly, during the 1989 Loma Prieta, 1994 Northridge and 1995 Kobe earthquakes. In Mexico, during the 2010 Mexicali earthquake, a section of the San Filipito bridge collapsed, due to the lateral displacement of the soil foundation due to ground liquefaction. In bridge structures, piles have been used extensively in both liquefiable and non-liquefiable soils. In liquefiable soils, the progressive build-up of pore water pressure can result in a loss of strength and stiffness, resulting in large bending moments and shear forces in the pile [1]. The behavior mechanism of piles in liquefiable soils has been studied by various researchers in recent decades [2]. In this work, a structured cell foundation alternative is proposed for bridge supports founded in potentially liquefiable soil deposits and its behavior is evaluated using three-dimensional finite-difference models with FLAC3D software. To represent the behavior of the liquefiable stratum, a constitutive model of the generation of excess pore pressures is used. Finally, a comparison is made based on lateral displacements using footing, piles, and structured cell foundations.

2 Case Study

2.1 Geotechnical Conditions

To characterize the geotechnical subsoil conditions where supports A-1 to A-6 are placed, 21 standard penetration tests with selective undisturbed sample recovery were conducted. Based on this field investigation, it was established that the soil profile deposit is mainly comprised by sands, some clays and andesitic rock. Figure 1 presents the geotechnical soil profile obtained based on the aforementioned exploration along the bridge. The control points indicated in the profile (i.e. A-1, A-2, A-3, A-4, A-5, A-6) correspond to the bridge support locations. The geotechnical properties obtained are presented in Table 1. As can be seen, the geotechnical unit No. 16 corresponds to a loose sand, just below the support A-4.

2.2 Dynamic Soil Properties

Due to the lack of experimental information regarding the soil dynamic properties of the materials found at the site, the shear wave velocity distribution was obtained from several correlations with the Standard Penetration Test [3–5]. The normalized modulus degradation and damping curves were estimated based on the proposed by Vucetic & Dobry [6] for plastic fine materials, as a function of plasticity index, PI considering the information gathered from index properties. Regarding the sand and silt layers, the curves proposed by Seed & Idriss [7] were deemed appropriate. For gravels and rock, the curves proposed by Seed et al. [8] and Schnabel [9] were used. Figure 2 present the shear wave velocity profile obtained for the analysis of support A-4.

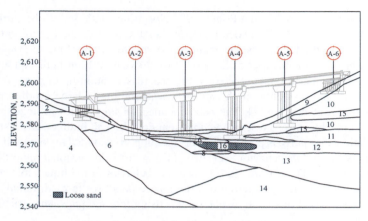

Fig. 1. Geotechnical profile and bridge.

Table 1. Soil properties used in the analysis.

Geotechnical unit	γ (kN/m³)	c (kPa)	ϕ (°)	Description
UG-1	17.0	30	27	Low plasticity clay
UG-2	18.0	30	30	Low and high plasticity clays of hard consistency
UG-3	18.0	30	30	Low and high plasticity clays
UG-4	20.0	112	24	Agglomerate, RQD (0–100%)
UG-5	17.0	24	25	Low and high plasticity clays of soft to medium consistency
UG-6	18.8	38	36	Sequence of clayed sands with gravels
UG-7	20.0	355	31	Basalt, RQD (0–46%)
UG-8	23.0	172	36	Rhyolite, RQD (24%)
UG-9	16.7	116	34	Silty and clayed sand of medium to very dense compactness
UG-10	19.5	685	32	Silty sand with gravel and clay
UG-11	19.6	152	33	Rhyolite, RQD (0%)
UG-12	20.9	73.6	36	Agglomerate, RQD (0–46%)
UG-13	19.0	355	31	Gravel
UG-14	21.5	72.8	17	Agglomerate, RQD (25–57%)
UG-15	21.0	72.1	17	Agglomerate, RQD (20–71%)
UG-16	18.0	9.3	25	Loose sand

2.3 Seismic Environment

The seismic environment was characterized by a uniform hazard spectrum, UHS, developed for a return period of 1000 years (Fig. 3b), considering the seismogenic sources associated to subduction, normal (intermediate depth), and local (shallow) events. To develop an acceleration time history which response spectrum reasonably matches the design response spectrum, the selected time history, usually called seed ground motion, was modified using the method proposed by Lilhanand & Tseng [10] as modified by Abrahamson [11]. The seed ground motion was selected from those recorded at SLPA seismological station, which is located in firm soil, approximately 13.5 km from the bridge. Equivalent linear properties were deemed appropriated for this sensitivity study.

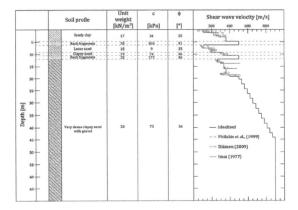

Fig. 2. Shear wave velocity profile corresponding to support A-4.

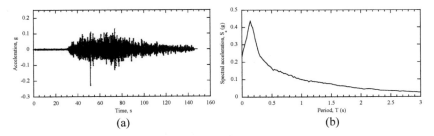

Fig. 3. (a) Synthetic time history and (b) uniform hazard response spectra.

3 Assessment of Liquefaction Potential

To determine the liquefaction potential, the empirical method presented by Youd et al., [12], was used. This method is evaluated using a safety factor against liquefaction. The cyclic stress ratio to cause liquefaction was evaluated using the equation proposed by

[13]. In order to compute the value of the magnitude scaling factor the equation proposed by [14] was used. Based on the geotechnical characterization shown in Fig. 2, models were made for one-dimensional site response analysis using the computer code SHAKE [15] which performs an equivalent linear analysis in the frequency domain. Figure 4 shows the results of Standard Penetration Test, SPT, N_{60}, CSR and safety factor obtained for the soil profile of the support A-4. As can be seen, the localized granular soil stratum at 6.30–8.10 m, is potentially liquefiable; that is, the cyclic stress ratio induced by the earthquake exceeds the cyclic resistance ratio.

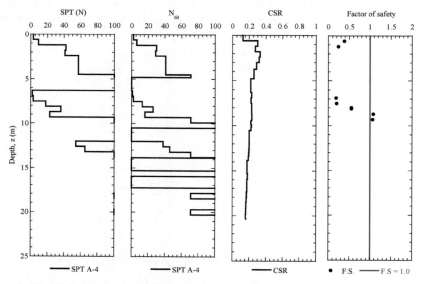

Fig. 4. Safety factor obtained from one-dimensional site response analysis.

3.1 Two-Dimensional Site Response Analysis

To account for potential topographic effects, a model was developed using QUAD4M [16], which is a finite element program that allows two-dimensional site response analysis in the time domain, using the equivalent linear method. To perform the analysis in QUAD4M, a model with 548 quadrilateral zones, 290 m wide and variable height from 107 m to 137.4 m was developed. Absorbent boundaries were used on the sides to minimize the reflection of the seismic waves, while at the base of the model to represent the half-space, viscous boundaries developed by [17] were used. Figure 5a shows the acceleration time history and Fig. 5b shows response spectrum of the bridge support A-4, calculated with SHAKE and the two-dimensional model made with QUAD4M. Similarity is observed between the results. The peak ground acceleration at support A-4 was equal to 0.28 g and maximum spectral acceleration was 0.87 g for a period of 0.22 s. Table 2 shows the results of one-dimensional and two-dimensional site response analysis.

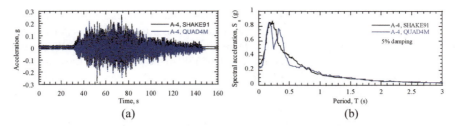

Fig. 5. (a) Calculated time history and (b) response spectra.

Table 2. Results of one-dimensional and two-dimensional site response analysis.

	PGA (g)	Sa_{max} (g)	Ts_{max}
SHAKE	0.28	0.87	0.22
QUAD4M	0.23	0.84	0.16

4 Three-Dimensional Numerical Model

To analyze the behavior of each alternative foundation, three-dimensional finite-difference models were developed with the FLAC3D software [18], (Fig. 6). To represent the elastoplastic behavior of the soil, the Mohr-Coulomb failure criterion was used, while for the liquefiable soil, the Finn [19] and Byrne [20] model was implemented. The foundations considered in the analysis are shown in Fig. 6b to 6d. The concrete strength at 28 days, f'c, was 25 MPa.

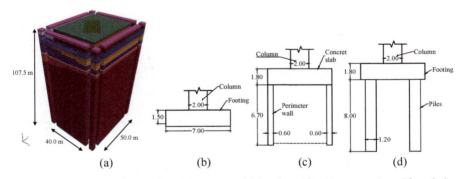

Fig. 6. (a) Three-dimensional finite difference model developed for the case study and foundations used in the analysis: (b) Footing, (c) structured cell and (d) piles.

Figure 7a presents the excess pore water pressure ratio obtained at different depths of the soil profile, r_u, where $r_u = \Delta u/\sigma'$. Δu, is the excess pore pressure, and σ', is initial effective confining pressure. When $r_u = 0.9$ or greater, significant liquefaction effects are expected to occur. It is observed that for a depth of 5.90 m the excess pore pressure ratio is as high as 0.9, after 60 s. Also, Fig. 7b presents the total and effective stresses, and the pore pressure histories during the earthquake obtained at 5.90 m. Clearly, there is a decrease in effective stresses at this depth. Figure 8 presents the history of lateral displacements obtained in the center of each foundation, to quantify the contribution to the reduction of lateral displacements. As can be seen, with the structured cell, the displacements during, and at the end of the earthquake, are substantially less than those obtained for the footing, and the piles. Table 3 presents a summary of lateral displacements obtained at the end of the earthquake for each foundation.

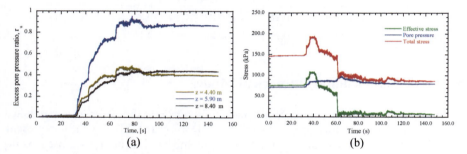

Fig. 7. (a) Excess pore water pressure ratio obtained at different depths of the soil profile and (b) stresses histories obtained at 5.9 m depth

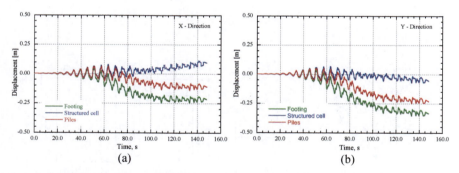

Fig. 8. (a) Displacement in x direction and (b) y direction at the center of each foundation.

Table 3. Summary of lateral displacements for each foundation.

Foundation	Lateral displacement (m)	
	X direction	Y direction
Footing	0.23	0.34
Structured cell	0.09	0.06
Piles	0.11	0.23

5 Conclusions

A numerical study of the response of structured cells as a feasible alternative for improving the seismic performance of the foundation of a 230 m long-bridge, located in potentially liquefiable soil was conducted. Sets of three-dimensional finite difference models were developed. Free filed response was calibrated against two and one dimensional analyses conducted with the programs QUAD4M and SHAKE respectively. Three foundation alternatives were analyzed, footing, piles, and structured cell foundations. As can be seen, the structural cell is capable of reducing the lateral displacements up to 40% with respect to the conventional foundations. Thus, it was concluded that structured cell foundation is a sound alternative to improve the seismic performance of bridges located in liquefiable soils and allows reducing detrimental effects associated with liquefaction-induced ground deformations.

References

1. Kramer, S.L., Arduino, P., Shin, H.: Using OpenSees for performance-based evaluation of bridges on liquefiable soils. Pacific Earthquake Engineering Research Center (2008)
2. Elgamal, A., Yan, L., Yang, Z., Conte, J.P.: Three-dimensional seismic response of Humboldt Bay bridge-foundation-ground system. J. Struct. Eng. **134**(7), 1165–1176 (2008)
3. Pitilakis, K., et al.: Geotechnical and geophysical description of Euro-Seistests, using field and laboratory test, and moderate strong ground motions. J. Earth Eng. **3**, 381–409 (1999)
4. Dickmen, U.: Statistical correlations of shear wave velocity and penetration resistant for soils. J. Geophys. Eng. **6**, 4167 (2009)
5. Imai, T.: P and S wave velocities of the ground in Japan. In: Proceedings of IX International Conference on Soil Mechanics and Foundation Engineering, vol. 2, pp. 127–132 (1977)
6. Vucetic, M.Y., Dobry, R.: Effect of soil plasticity on cyclic response. J. Geotech. Eng. ASCE, **114**(1), 89–107 (1991)
7. Seed, H.B., Idriss, I.M.: Soil Moduli and Damping Factors for Dynamic Response Analysis. (UBC/EERC-70/10). University of California, Berkeley (1970)
8. Seed, H.B., Wong, R.T., Idriss, M., Tokimatsu, K.: Moduli and damping factors for dynamic analyses of cohesionless soils. Geot. Eng. Div. **112**(11), 1016–1032 (1986)
9. Schnabel, P.B.: Effects of Local Geology and Distance from Source on Earthquake Ground Motions. University of California, Berkeley, USA (1973)
10. Lilhanand, K., Tseng, W.: Development and application of realistic earthquake time histories compatible with multiple damping response spectra. In: 9th WCEE Vol II, pp. 819–824 (1988)

11. Abrahamson, N.: State of the practice of seismic hazard evaluation. In: Proceeedings of GeoEng 2000, vol. 1, pp. 659–685 (2000)
12. Youd, T.L., Idriss, I.M.: Liquefaction resistance of soils. J. Geotech. Geoenviron. Eng. **127**(4) (2001)
13. Idriss, I.M., Boulanger, R.W.: Semi-empirical procedures for evaluating liquefaction potential during earthquakes. In: Proceeding of 11th International Conference on Soil Dynamic and Earthquake (2004)
14. Andrus, R., Stokoe, K.: Liquefaction resistance based on shear wave velocity. Nat. Ctr Earthq. Eng. Res. 89–128 (1997)
15. Schnabel, P.B., Lysmer, J., Seed, H.B.: A computer program for earthquake response analysis of horizontally layered sites. Rep. No. EERC 72–12, UC Berkeley (1972)
16. Hudson, M., Idriss, I.M., Beikae, M.: QUAD4 M-A. In: Center for Geotechnical Modeling, Dep of Civil and Env Eng, University of California, Davis, CA., Davis, CA (1994)
17. Lysmer, J., Kuhlemeyer, R.L.: Finite dynamic model for infinite media. J. Eng. Mech. **95**(EM4), 859–877 (1969)
18. Itasca Consulting Group, FLAC3D Fast Lagrangian Analysis of Continua, Ver. 5.0 User's Manual, Minneapolis, USA (2008)
19. Finn, W., Martin, G., Lee, K.: An effective stress model for liquefaction. J. Geotech. Eng. Div. **103**(6), 517–533 (1975)
20. Byrne, P.M.: A cyclic shear-volume coupling and pore-pressure model for sand. In: Proceedings of the 2nd International Conference on Recent Advances in Geotechnical Earthquake Engineering and Soil Dynamics, Issue 1, pp. 47–55 (1991)

A Case Study of Seismic Design of Pile Foundation Subject to Liquefaction, Cyclic Softening, and Lateral Spreading

Yasin Mirjafari[✉] and Malcolm Stapleton

Babbage Consultants Limited, Auckland, New Zealand
yasin.mirjafari@babbage.co.nz

Abstract. Pile foundations are frequently subjected to the lateral load imposed by soil movement during and after an earthquake event. This lateral load transfer from the soil to the pile is a complex soil-structure interaction problem that involves an iterative process and requires effective communication between structural and geotechnical engineers. Several methodologies for the design of laterally loaded pile foundations have been proposed by researchers. However, limited practical research papers are available which explain the design procedure of piles subject to lateral ground movement due to liquefaction and cyclic softening. This paper presents the details of the methodology that was implemented for a structure located on an area subject to the risk of kinematic ground movement due to cyclic softening and liquefaction during an earthquake event and lateral spreading after the earthquake shaking. The foundation comprises 82 piles in very soft cohesive soil. Separate design scenarios representing the ground and structure behavior at different stages of earthquake shaking have been considered. Structural software ETABS is used to check the buckling load capacity of the pile-supported laterally by soil springs.

Keywords: Soil-structure interaction · Pile · Liquefaction · Cyclic softening

1 Introduction

A piled structure located on an area with high seismicity prone to soil liquefaction or cyclic softening can be subject to inertia loads (base shear), cyclic ground displacement, and lateral displacement during or after an earthquake event. These actions separately or in combination may influence the performance of the pile foundation. To design the foundation, effective collaboration between geotechnical and structural engineers is required to first estimate the magnitude of the ground movement and then assess the impact of the movement on the behavior of the structure during and after an earthquake event. This paper discusses the evaluation of different ground movement scenarios and their impact on a piled structure during and after an earthquake.

2 Ground Condition

2.1 Soil Profile

A building with heavy equipment was proposed to be constructed on a site with challenging soft soil conditions that precluded more conventional foundation solutions. The platform for the structure is underlain by hardfill to a depth of approximately 0.9 m to 1.5 m overlying peat/highly organic silt to a depth of 3.5 to 5 m. The overall depth below the ground level to the underlying weak sedimentary rock is approximately 25 m but varies from west to east beneath the platform. The materials above that are predominantly low strength and compressible. Alluvial sand is however present which is reported to have included gravels at the western part of the building footprint and north of the site. The site is affected by a high groundwater table all year round with groundwater close to the ground surface in winter.

The general soil profile of the site is summarized in Table 1. The undrained shear strength and friction angle were estimated from CPT results undertaken within the footprint of the proposed structure.

Table 1. Stratigraphic profile

Soil thickness (m)	Description	Su peak (kPa)	Friction angle
1.2	Hardfill	–	40
2.8	Peat	15	–
2	Very soft clay	10	–
6	Soft clay	18	–
2.5	Firm clay	30	–
2	Sand/Gravel	–	32
4.5	Stiff silt	70	–
3.5	Very stiff silt	150	–
–	Weak mudstone	600	–

2.2 Liquefaction and Cyclic Softening

Cyclic softening and liquefaction potential of the soil were evaluated using CPT data. A total number of six CPT were undertaken at the subject site. The site is expected to have a very low to negligible susceptibility to liquefaction in response to Serviceability Limit State (SLS) seismic activity. Under Ultimate Limit State (ULS), seismic loading, liquefaction is however predicted to occur at depth in the central part of the structure footprint. Free field liquefaction-related settlement is predicted to be less than 100 mm. The soft alluvial deposits including the organic silts in the upper 11 m of the soil profile are expected to experience cyclic softening in response to a ULS event.

2.3 Lateral Spreading

The maximum general slope across the site and surrounding area was estimated to be not more than 3% down towards a drainage channel that marks the eastern and southern sides of the proposed development area and property boundaries. The order of unrestrained lateral displacement following an Ultimate Limit State design event is estimated to be less than 150 mm.

2.4 Kinematic Soil Displacement

The cyclic ground displacement has been assessed using the method developed by Tokimatsu & Asaka (1998) [5] and Tsai et al. (2014) [8]. Based on a pile founding level of −25 m bgl, a cyclic ground displacement of 200 mm has been estimated in the ULS earthquake event at the ground level. The estimated cyclic ground displacement has been incorporated in the pile analysis.

2.5 Negative Skin Friction

The presence of liquefied soils reduces the support around and apparent adhesion available to support piles carrying vertical and lateral loads. Therefore, any skin friction and associated bearing capacity derived from the soils predicted to liquefy has been ignored. The overlying non-liquefied soils above the pile founding level and the soils inferred to be susceptible to seismically induced settlement have the potential to settle and induce down-drag. This effect has been considered in the pile bearing capacity analysis.

2.6 Site Period

For the purpose of pile design, the site period of the proposed development has been estimated using the CPT data. By assuming an average shear wave velocity of 120 m/s, the site period of 0.7 s has been calculated.

3 Soil-Structure Interaction

A pseudo-static approach based on soil-structure interaction was used in the design process of the pile foundation.

The interaction between pile and soil is influenced by the relative stiffness of the two components. A flexible pile can relatively bend with the soil movement; on the other hand, soil can displace past the piles if the piles are relatedly stiff. In other words, the interaction between a pile and surrounding soil depends on the relative stiffness and relative movement of the two components.

The other important faction which can influence the interaction between soil and pile is the direction or phase of the soil and pile movement relative to each other. In the seismic design of a foundation, shaking of the supported building (base shear loading) can be in-phase or out-of-phase with the shaking of the ground (kinematic loading due to cyclic displacement); or somewhere between these two extremes. The natural period of the structure and ground are the important factors dictating the phase of the relative movement of the soil and structure. Further information on this aspect is provided by (Tokimatsu et al. 2005) and (Tamura and Tokimatsu 2005) [6, 4].

The described factors are considered in a number of different design scenarios according to the different load demands applied to the structure. The load demand and different scenarios are described in the following sections.

3.1 Loading Demand

The considered load demands for the proposed pile foundation are as follow:

1. Gravity load from the structure, equipment, and foundation weight (Static Design)
2. Seismic load from the structure (Seismic design).
3. Soil-structural interaction load due to soil lateral spreading (Seismic design).
4. Soil structure interaction load due to kinematic ground movement (Seismic design).

Three soil-structural action load cases for the seismic design of the foundation are considered. These cases are described as follow:

Seismic Case 1- Kinematic load – No ground displacement:
In Case 1, the building movement is assessed by modelling the full structural load under ULS seismic conditions without applying any soil load on the piles arising from soil displacement relative to the piles. The soil resistance against the piles was modelled as non-linear springs. (Cubrinovski et al. 2006) [1] provides some guidance on the selection of these parameters. No liquefaction and cyclic softening were allowed in this scenario, therefore lateral resistance of the piles was modeled using the nonlinear spring stiffness derived based on full strength material. The soil resistance against or lateral support for the piles was modelled as horizontal springs. In this scenario, computer modeling using Etabs software estimated the maximum pile head movement to be less than 80 mm. No piles failed for the considered load combinations.

Seismic Case 2- Kinematic load and kinematic ground displacement:
To incorporate the effects of kinematic ground movement, the second case was analysed using a spring value generated based on reduced strength of the soil due to cyclic softening in the upper 11 m. The reduced soil strength of the materials was estimated based on the methodology developed by Idriss and Boulanger 2007 and 2008 [1, 2]; hence, the spring stiffness values were calculated based on reduced soil strength parameters and a conservative assumption of 300 mm of total building and soil movement. For this case, a distributed load was applied to the upper 11 m of the piles to generate the 300 mm expected building movement using the ETABS model.

An important aspect of scenario 2 is to identify whether the shaking of the building (base shear loading) be in-phase or out-of-phase with the shaking of the ground (kinematic loading due to cyclic displacement); or somewhere between these two extremes. Based on the estimated site period of 0.7 s, and the natural period of the structure, the movement of the structure is considered to be in-phase with the direction of the ground movement.

Seismic Case 3- No kinematic load- lateral spreading:
In case 3, conditions representing a lateral spreading scenario after the earthquake shaking was modelled. Lateral spreading was assumed to occur after shaking. In this case, only vertical gravity loads from the building and floor live load was applied. In this case, similar to case 2, soil spring stiffness was estimated based on reduced strength of soil due to cyclic softening and liquefaction. All piles are subjected to a UDL load from 0 m to 5 m below ground based on the assessed depth of soil subject to soil lateral spreading. This UDL pile load combination was scaled to force the foundation to reach a lateral movement of 150 mm. This movement stimulates the post-earthquake soil movement, i.e., without earthquake load. This load condition is not considered as a governing load case in the pile design.

The structural software ETABS was used to accurately simulate the building loads and piles foundation under different load combinations for the described loading cases.

3.2 Horizontal Spring Stiffness

Confirming the appropriate lateral stiffness/spring value is an iterative process. The steps described below were followed to achieve convergence of the analyses.

1- An initial stiffness value was selected based on a predicted deflection.
2- The Etabs model was run and the deflection calculated in that model was determined in various locations across the building footprint.
3- The spring value was re-evaluated based on the maximum deflection assessed in step 2.
4- Steps 2 and 3 were repeated until the lateral displacement predicted in the ETABS model was consistent with the stiffness/soil spring value used in the model.

The structural and geotechnical modelling predicts the maximum pile deflection will occur at the pile head. As noted above, modelling indicated piles below 11 m are affected by very small lateral deflection. For small movement, stiffer springs consistent with the specified construction methodology have been used in the model.

At the depth from 11 m to 16 m below the pile head, the spring stiffness we have used has been based on pile deflection of 10 mm. No deflection is predicted to occur below 16 m. Hence below this depth, the tangent stiffness has been taken directly from the linear elastic part of the p-y curve for materials below 16 m.

4 Structural Analysis

Steel Circular Hollow Section (CHS) piles are considered to have the most efficient cross-section for addressing the risk of pile buckling issue due to the soil lateral spreading, in addition to vertical loading gravity + seismic). The layout of piles and ground beams is shown in Fig. 1.

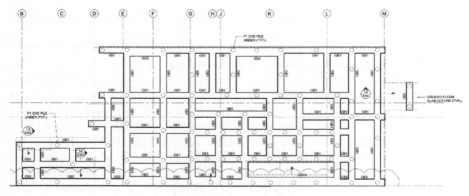

Fig. 1. Ground beam and pile layout

The pile moment demand is calculated as the combination of the gravity/live load plus seismic load (100% in one direction, +30% in other direction). For pile buckling capacity check, seismic loads with gravity load was used to check the p-delta effect on the pile. Pile buckling analysis was carried out to calculate pile axial capacity with the lateral soil spring support. Then, the axial capacity was used in the design check of the pile axial load and moment combination demand in accordance with NZS 3404 [7].

For structural design verification, NZS 3404 was used to check the combined action of axial load plus flexural moments (including P-Delta effects) on the CHS piles. ETABS was used to determine the maximum load applied to the pile-supported laterally by soil springs that were then used to assess the occurrence of buckling.

To validate the foundation design, an equivalent pseudo-static analysis in the geotechnical computer model LPile V2019 was undertaken. In this method similar load scenarios were adopted; the building load was applied as an equivalent point load at the top of a pile with free head connection as of the ETABS model. Similar to the ETABS analysis, all the loads were within the capacity of the piles.

5 Conclusions

The structure was proposed to be constructed on a site with challenging soft soil conditions that precluded more conventional foundation solutions. Steel CHS piles were considered to have the most efficient cross-section for addressing the potential for pile buckling as a result of the lateral soil pressure, seismic loading, and vertical dead and live loads. For site conditions such as these, detailed soil-structure interaction analyses are required which necessitate effective communication between geotechnical and structural engineers. Careful consideration of all possible loading conditions and an iterative design process are required to address the challenges posed by complex design projects such as this one.

Acknowledgments. The authors would like to acknowledge Dr Victor Lam and Cong Liu for their collaboration and involvement in the structural design of the foundation in this project.

References

1. Boulanger, R.W., Idris, I.M.: Evaluation of cyclic softening in silts and clays. J. Geotech. Geoenviron. Eng. ASCE **133**(6), 641–652 (2007)
2. Boulanger, R.W., Idriss, I.M.: Soil liquefaction during earthquakes, Oakland, Calif. Earthquake Engineering Research Institute (2008)
3. Cubrinovski, M., Kokusho, T., Ishihara, K.: Interpretation from large-scale shake table tests on piles undergoing lateral spreading in liquefied soils. Soil Dyn. Earthq. Eng. **26**(2–4), 275–286 (2006)
4. Tamura, S., Tokimatsu, K.: Seismic earth pressure acting on embedded footing based on large-scale shaking table tests. ASCE Geotech. Special Pub. **145**, 83–96 (2005)
5. Tokimatsu, K., Asaka, Y.: Effects of liquefaction-induced ground displacements on pile performance in the 1995 Hyogoken-Nambu earthquake. Soils Found. **138**(Special), 163–177 (1998)
6. Tokimatsu, K., Suzuki, H., Sato, M.: Effects of inertial and kinematic interaction on seismic behaviour of pile with embedded foundation. Soil Dyn. Earthq. Eng. **25**, 753–762 (2005)
7. Standards Association of New Zealand. Steel Structures Standard. NZS 3404 Parts 1 and 2 (1997)
8. Tsai, C., Mejia, L., Meymand, P.: A strain-based procedure to estimate strength softening in saturated clays during earthquakes. Soil Dyn. Earthq. Eng. **66**(2014), 191–198 (2014)

Dynamic Centrifuge Model Tests on Plate-Shaped Building Supported by Pile Foundation on Thin Load-Bearing Stratum Overlying Soft Clay Layer

Takehiro Okumura(✉) and Junji Hamada

Takenaka R&D Institute, Ohtsuka 1-5-1, Inzai-shi, Chiba, Japan
okumura.takehiro@takenaka.co.jp

Abstract. In Japan, many plate-shaped buildings are built as residences in coastal areas. As a result, some of them are sustained by the pile foundation embedded on the thin load-bearing stratum overlying the soft clay layer because a continuous load-bearing stratum is too deep to construct economically. During a large earthquake, plate-shaped buildings cause large lateral force and varying axial force in their pile foundations and make their seismic behavior more complicated because they each have a high aspect ratio and single-span composed of two columns and several beams. In addition, the thin load-bearing stratum overlying soft clay layer makes the seismic behavior of a plate-shaped building more complex because of the less stiffness of soft clay. In this study, two dynamic centrifuge model tests in a 50 g field were carried out to compare the seismic behavior of the plate-shaped building and pile foundation on the thin load-bearing stratum overlying soft clay layer with that on the continuous load-bearing stratum. From the several shaking tests, it was observed that the large instantaneous acceleration of the superstructure occurred when the gap between the pile tip and the load-bearing stratum was closed in the model on the continuous load-bearing stratum. On the other hand, the bending moment at the pile head in the thin load-bearing stratum was almost equal even though the inertial force differed. However, the pile foundation on the thin load-bearing stratum did not settle critically, although the peak axial force was almost double the preload.

Keywords: Thin load-bearing stratum · Pile foundation · Plate-shaped building · Centrifuge motel test · Seismic behavior

1 Introduction

The seismic behavior of pile foundations in the soft ground is complicated since the soft ground is intensively nonlinear in mechanical properties and gives large relative displacements to the pile foundations during an earthquake. Plate-shaped buildings cause large lateral force and varying axial force in their pile foundations and make their seismic behavior more complicated because they each have a high aspect ratio and a single span composed of two columns and several beams.

The dynamic centrifuge model test is an effective experimental method to investigate the seismic behavior of the building-ground system during an earthquake. Many experimental studies have been carried out on the soft ground, which has a longer natural period than the superstructures. However, most of them were focused on the liquefaction phenomenon in sandy soil, and there were few experimental cases on the building and foundation structures in the soft clayey ground [1–3]. Besides, those experiments focused on the building loads as horizontal force and did not include the overturning moment, and the effect of the overturning moment on the seismic behavior of superstructures and the pile stress was not known.

The recent studies, in which the soft clayey ground came to be easily produced in a short time by using Kanto loam as the ground material in a centrifugal model test, confirmed that the clayey ground has a low range of shear wave velocity from 100 to 200 m/s in a 50 g centrifugal field [4]. In addition, the shaking table tests were carried out using the model simulating a plate-shaped building supported on a pile foundation. The results of these tests confirmed that when the axial force of the piles reaches zero (0) in a large earthquake, the pile tip ground and the piles are separated, and the rotation angle of the building rapidly increases.

In this study, two centrifuge model tests have been carried out to study the effect of the thin load-bearing stratum on the seismic behavior of plate-shaped buildings supported by pile foundations in the soft clayey ground during large earthquakes. In the centrifuge tests, the different load-bearing stratum conditions were tested.

2 Centrifugal Experiment

2.1 Structure Model

Figure 1 shows a superstructure's prototype with a pile foundation and the schematic diagrams of the centrifuge models in a 50 g field. As shown in Fig 1 a), a prototype of the building model was assumed to be a plate-shaped building (13 stories, 40.0 m height, with the first natural period of 0.412 s). The dimensions of the building in the long and short directions are 12.0 m and 6.0 m, respectively. In the superstructure modeling, the weight, the first natural period, and the moment of inertia were combined with the prototype as a single-layer truss structure, according to the similarity law in a 50 g field [5]. The piles were modeled after a cast-in-place concrete belled pile (with the shaft diameter of φ 1.5 m and the bell diameter of φ 2.2 m) using aluminum pipes (30 mm in diameter and 1.5 mm in thickness) aluminum alloy was used for their tips.

2.2 Ground Model

As shown in Fig. 1b) and c), models A and B had two and three layers made in a laminar shear box, respectively. Kaolin clay was used as the bottom clay layer in Model B, and a vacuum mixing apparatus stirred kaolin clay powder and water. The agitated material was charged into a soil tank, and consolidation was carried out at 500 kPa for 14 days to withstand an assumed loading pressure of 439 kPa from a pile in a 50 g field. The physical test results of the bottom clay layer and the triaxial compression test (CUb) results are shown in Table 1. This table also shows the results of the elastic wave velocity tests on the specimens, which were consolidated under the confining pressure of 500 kPa and then returned to the confining pressure of 150 kPa, simulating the conditions in the centrifugal field. Iide silica sand No. 4 (relative density (Dr) = 90%) with 100 kg/m^3 of cement (W/C = 100%) was used as the load-bearing stratum which was strong (unconfined compression strength (q_u) = 1.4 N/mm^2) and stiff (shear wave velocity (Vs) = approximately 550 m/s). The clayey layer was a reconstructed sample of Kanto loam collected in Tokyo and prepared by wet compaction at an optimum water content (w = 63.1%) to achieve the maximum dry density (ρ_d = 0.943 g/cm^3) to facilitate consolidation in the centrifuge. Water was used for the pore fluid because the permeability of the Kanto loam is very low at 9.0×10^{-5} cm/s. Thus, we consider that the effect of drainage during shaking would be small enough. The physical properties of this clayey ground are shown in Table 2 and described in detail about the dynamic deformation characteristics in the literature [4].

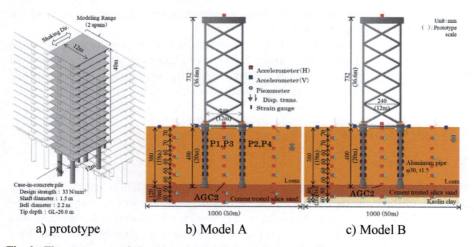

Fig. 1. The prototype of the plate-shaped building supported by a pile foundation and the schematic diagram of the centrifugal models

Table 1. Physical and mechanical properties of kaolin clay as the bottom clay layer

Soil Particle density (g/cm^3)	2.693	Cohesion (kN/m^2)	c	25.9
Gravel fraction (%)	0.0		c'	17.4
Sand fraction (%)	0.0	Internal friction angle (°)	ϕ	11.5
Silt fraction (%)	39.8		ϕ'	23.4
Clay fraction (%)	60.2	S-wave velocity (m/s)		184
Liquid limit (%)	53.0	P-wave velocity (m/s)		1,616
Plastic limit (%)	29.1			
Plasticity index	23.9			

Table 2. Physical and mechanical properties of Kanto loam as the upper clayey layer

Soil Particle density (g/cm³)	2.599	Wet density (g/cm³)		1.538
Gravel fraction (%)	0.0	Void ratio		1.756
Sand fraction (%)	31.0	Water content (%)		63.1
Silt fraction (%)	43.7	Degree of saturation (%)		93.4
Clay fraction (%)	25.3	Cohesion (kN/m²)	c'	23.5
		Internal friction angle (°)	φ'	35.7

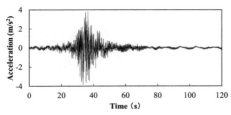

Fig. 2. The input motion for the shaking table test for the largest input case

Table 3. Peak acceleration used in shaking table tests and case indexes

Case Number	Peak acceleration (m/s²)		Case index	
	Outcrop (2E)	Within (E+F)	Model A	Model B
0	(Sweep wave)		A-0	A-0
1	0.50	0.35	A-1	B-1
2	2.00	1.23	A-2	B-2
3	3.50	2.14	A-3	B-3
4	4.50	2.96	A-4	B-4
5	6.00	3.85	A-5	B-5
6	0.50	0.35	A-6	B-6

2.3 Input Motion

The shaking table tests were conducted after confirming that the clayey ground surface settlement and pore water pressure were stabilized. However, an input motion composed of the upward and downward waves (E+F) was necessary to control the shaking table. Therefore, seismic response analyses were executed using the one-dimensional equivalent linear analysis program. The parameters of the ground analytical model (the shear stiffness considered confining pressure and the dynamic deformation characteristics) were determined from the laboratory tests [4]. For the outcrop motion (2E) used in these analyses as an input motion, the acceleration response spectrum characteristic defined by the notification [6] was used. Also, its phase characteristic was observed in Kobe Marine Observatory (NS component) during the Hyogo-ken Nanbu Earthquake in 1995. We used the waveforms (E+F) obtained from these analyses using outcrop motions with changing the peak acceleration of 0.5, 2.0, 3.5, 4.5, and 6.0 m/s² as the input motion of the shaking table tests. Figure 2 shows the example of the input motion for the shaking table test (E+F) calculated by 1-D analysis with the input motion (2E) as the peak acceleration of 6.0 m/s². Table 3 shows the relationship between the peak acceleration of 2E waves and E+F waves. The case indexes are also shown in Table 3. The shaking table test results are shown in the prototype scale calculated by the similarity law [5].

3 Test Results

3.1 Superstructure's Response

To compare the inertial force generated in the superstructure, Fig. 3 shows the relationship between the peak inertial force and overturning moment and the peak acceleration observed at AGC2, which position is indicated in Fig. 1. Figure 3 shows that the peak inertial force and overturning moment increased with the input intensity. However, the peak inertial force and overturning moment of Case B-5 were almost equal to those of B-4 because the bottom clay layer yielded, and the seismic wave was not transmitted from the shaking table to the load-bearing stratum. In addition, the relationship between the peak inertial force was linear under the peak acceleration was 2.0 m/s^2. However, in the larger input cases in which the peak acceleration was over 2.0 m/s^2, the peak inertial force in Model A was larger than Model B. The relationship between the peak overturning moment and the peak acceleration was similar to the inertial force's.

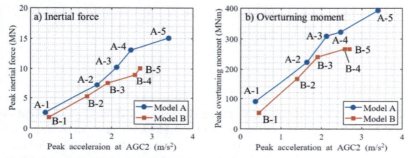

Fig. 3. The relationship between the peak values of inertial force and overturning moment of the superstructure and the peak acceleration observed at AGC2

To consider why the inertial force and the overturning moment in Model A were larger than Model B in the larger input cases, Fig. 4 shows the time histories of acceleration observed in Case A-5 and Case B-5. As shown in Fig. 4 a3) – a5) and b3) – b5), the ground response's difference was little. On the other hand, the superstructure's responses differed between Model A and Model B, as shown in Fig. 4 a1) – a2) and b1) – b2). Especially, the footing's acceleration observed in Case A-5 had a large instantaneous amplitude than Case B-5. Also, the superstructure top's acceleration observed in Case A-5 was instantaneous larger than Case B-5. From the above results, we considered that the large instantaneous acceleration in Case A-5 made the peak inertial force and overturning moment of the superstructure was larger than Case B-5.

Figure 5 shows the relationship between the residual settlement amount and the residual inclination angle for each test. Figure 5 a) and b) show that the superstructure's settlement at the time of centrifugal loading (swing-up) was small in Model A and large in Model B. However, in both models, the amount of ground settlement was larger than the settlement of the superstructure. The ground settlement increased with each vibration, while superstructure settlement hardly increased. From Fig. 5 c) and d), the residual inclination angle was very small and similar regardless of the load-bearing stratum.

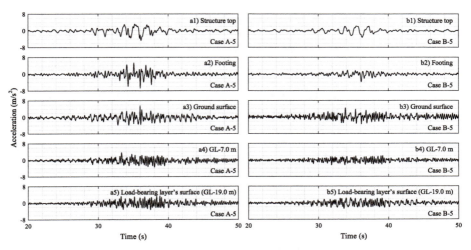

Fig. 4. The time histories of acceleration observed in Case A-5 and Case B-5

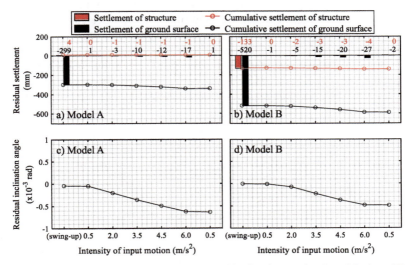

Fig. 5. The history of residual settlement and inclination angle of models A and B

3.2 Pile Stress

Axial force and bending moment of the pile were calculated from the strain gauge inside the pile shaft. Figure 6 shows the relationship between the axial force of the pile at GL-6.75 m depth considering the effect of bending deformation and the vertical displacement of the pile head obtained by the displacement transducers. From Fig. 6 a), in Case A-5, the large vertical displacement of the pile head was generated upward within the range where the pulling force was generated in a pile and that the critical pulling resistance of the pile exists in 0–2 MN. In addition, the vertical rigidity of the pile was smaller in

Case B-5, as indicated by the black dashed line in Fig. 6. It is considered because the bottom clay layer was less rigid than the load-bearing stratum. From the above results, the piles in Case A-5 were pulled out, and it is considered that the gap between the pile tip and load-bearing stratum was generated. We considered that the large instantaneous acceleration was generated when the gap between the pile tip and the load-bearing stratum was closed. Also, the gap did not occur in Case B-5 because the settlement rigidity in the model on the thin load-bearing stratum was small, and the superstructure settled rather than floated. By the way, the preload pressure on the bottom clay stratum was designed to withstand the statical axial force. However, the peak compression axial force in Case B-5 was double the statical axial force. In other words, the superstructure over the thin load-bearing stratum did not settle critically (shown in Fig. 5 b)) even though the bottom clay layer suffered the compression double the preload pressure.

Figure 7 shows the relationship between the inertia force of the superstructure and the bending moment at the pile heads (GL-1.25 m). Figure 6 shows a good correlation between the pile head bending moment and the inertial force. The peak of bending moment generated at the pile head was about -12 MNm in Case A-5 and about 12 MNm in Case B-5. In Case B-5, the inertia force and bending moment were linear. On the other hand, in Case A-5, the inertial force decreased a little more than the proportional relationship at a large inertia force. We confirmed that the decrease in the bending moment was caused by the overturning moment being added to the pile head because of the axial force becoming constant as the pile was pulled out [7]. From the above results, it is considered that the condition that the piles were pulled out or not pulled out influenced the fact that the peak axial force and the peak bending moment were almost equal to each model even though the inertial force and the overturning moment were different.

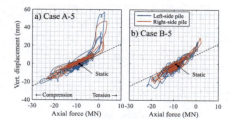

Fig. 6. The relationship of pile axial force and verti-cal displacement of pile head

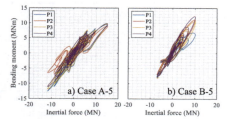

Fig. 7. The relationship of the inertial force of super-structure and bending moment of pile heads

4 Conclusion

We carried out a series of centrifuge experiments to investigate the seismic behavior of a plate-shaped building supported by a pile foundation in the soft clayey ground during a large earthquake. Two ground models with different conditions of the load-bearing stratum were subjected to a shaking table test with large input motions in a 50 g centrifuge field. As a result, the following conclusions were obtained:

1. The inertial force and overturning moment generated in the superstructure on the continuous load-bearing stratum were larger because the large instantaneous acceleration occurred when the gap between the pile tip and the load-bearing stratum was closed.
2. The superstructure over the thin load-bearing stratum did not settle critically, although the peak axial force was almost double the preload pressure. Furthermore, the residual inclination angle of the superstructure was very small, although the input acceleration was about 6.0 m/s^2.
3. The peak axial force and bending moment generated in the piles were almost equal to each model because the condition in which the piles were pulled out or not pulled out influenced the stress even though the inertial force and overturning moment of the superstructure were different.

Lastly, we point out that the input motions used in this research were only in one waveform. We would study the performance-based design from finite element analyses using different input waveforms.

References

1. Kohama, E., Sugano, T.: Centrifugal model tests on the dynamic behavior of pile foundation structures in cohesive soil during earthquakes. In: 39th Japan National Conference on Geotechnical Engineering, pp. 1791–1792 (2004). (in Japanese)
2. Zhang, L., Goh, S.H., Liu, H.: Seismic response of pile-raft-clay system submitted to a long-duration earthquake: centrifuge test and final element analysis. Soil Dyn. Earthq. Eng. **92**, 488–502 (2017)
3. Garala, T.K., Madabhushi, G.S.P.: Seismic behavior of soft clay and its influence on the response of friction pile foundations. Bull. Earthq. Eng. **17**, 1919–1939 (2019)
4. Okumura, T., Honda, T., Hamada, J.: Dynamic centrifuge model tests on plate-shape building's pile foundation in clayey ground. In: Proceedings of 16th Asian Regional Conference on Soil Mechanics and Geotechnical Engineering (2019)
5. Kazama, M., Inatomi, T.: Application of centrifuge model testing to dynamic problems. J. Jpn. Soc. Civ. Eng. **477**(I-25), 83–92 (1993). (in Japanese)
6. Building Standard Law of Japan: Notification No. 1461. The Japanese Ministry of Construction (2000). (in Japanese)
7. Okumura, T., Hamada, J.: Dynamic centrifuge model tests on plate-shaped building's pile foundation in the soft clayey ground; effect of building's moment of inertia. In: Proceedings of 20th International Conference on Soil Mechanics and Geotechnical Engineering (2022, in press)

The Behaviour of Low Confinement Spun Pile to Pile Cap Connection

Mulia Orientilize[1](✉) , Widjojo Adi Prakoso[1] , Yuskar Lase[1], and Carolina Kalmei Nando[2]

[1] Universitas Indonesia, Kampus UI Depok, Depok City, Jawa Barat 16424, Indonesia
`mulia@eng.ui.ac.id`
[2] Universitas Indonesia, Depok City, Indonesia

Abstract. An experimental and numerical study of the low confinement spun pile to pile cap connection typically used in Indonesia was performed. A cyclic loading test on two spun pile connections was performed. One was an empty spun pile (SPPC01) and another was filled with reinforced concrete (SPPC02). No shear failure was observed during the test. Filling the reinforced concrete into the spun pile increase the capacity by 54% and also improve the ductility from 4.26 to 4.70. A numerical study was performed to simulate the effect of the soil. A similar spun pile with a fixed connection to the pile cap was embedded in clay. Soil affects the strength and ductility of the connection, but it is not significant. Stiffer soil results in higher bending capacity. The capacity could be approximated based on the P-M interaction. To conclude, the connection of the spun pile to pile cap perform well though the spun pile has insufficient confinement.

Keywords: Low confinement spun pile · Bending capacity · Ductility

1 Introduction

A spun pile is a circular hollow precast prestress pile which is commonly used for bridges and ports since it has several advantages such as convenience in construction, lower cost, higher bearing capacity, and more reliable quality. The connection between pile to pile cap is a critical part since the change of area, stress, and stiffness occurs in this region suddenly [1]. The design of the seismic resistance of the foundation in Indonesia is guided by article 7.13.6 of SNI 1726-2019. For fixed connection, prestressed bars from the pile are embedded into a pile cap with a specified length. To accommodate higher curvature on the connection region, concrete and mild rebar could be added.

The design code of the bottom structure in Indonesia, SNI 8640, is still based on the elastic concept. The code does not allow the foundation to suffer damage. The lateral displacement is restricted to 25 mm for a severe earthquake. In contrast, several detailings of the foundation should meet ACI 318-19 which are intended for upper structural components which applies the performance based design. One of the requirements is the number of transverse reinforcements to ensure that the concrete has sufficient confinement during the inelastic stage. The spun pile in Indonesia is mostly produced with low confinement. It is known that inadequate confinement results in low ductility. However, it should not be a problem if the structure is designed to behave elastically.

A post-severe-earthquake observation reported by Kaneko, et al. [2] found a several damages occur on the pile to pile cap connections which indicate that damage is unavoidable. Several research has been conducted to study the inelastic behaviour of the spun pile to pile cap connections under seismic load [3–6]. More insight about the performance based design of the bottom structure has been gained. Even though the research are still carried out as indicated by international journal articles until 2020, the design of bridges and wharves have been implemented in the code [7].

Research of piles with low confinement was reported by Park and Falconer [8, 9]. The amount of confinement was 30–60% than NZ-code which were 1% out of 1.2%. The study found that the connection still performed ductile. A study of low confinement of the spun pile in Indonesia has been conducted by Irawan [10]. The amount of transverse reinforcement was 0.24% which is about 21% of the minimum requirement. The experimental study found that the confinement is insufficient to resist the explosion of the pile's concrete at the ultimate state due to compression. Another study found that ductility of limited confinement of two different spun piles was 2.50 and 4.50 [11].

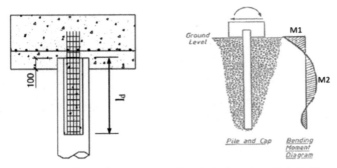

Fig. 1. The typical spun pile – pile cap connection in Indonesia which is illustrated as fixed connection and its associated bending moment diagram

Indonesia should move to the performance based design of bottom structures since the seismic demand based on the latest seismic risk maps produced in 2017 tend to increase. More research of spun pile to pile cap connections based on common practice in Indonesia are required. The past studies focused on the spun pile and none of them have been studied the behaviour of the spun pile connection.

A typical detailing of the spun pile to pile cap connection in Indonesia is shown in Fig. 1. An experimental study of low confinement of spun pile connection based on this detail has been performed. Soil effects are neglected during the test. Further numerical investigation was conducted to see the behaviour of the pile embedded in the soil. OpenSees software was employed to simulate a single spun pile surrounded by one layer of clay. Push over analysis was performed to investigate the inelastic behaviour and the results were presented as strength and ductility.

2 Research Methodology

2.1 The Experimental Study

The experimental study of low confinement spun pile connected to pile cap was conducted on two full scales specimens which were tested until failure. The pile was 450 mm in diameter with a wall thickness of 80 mm. It was made of 52 Mpa of concrete, 10@7.1 mm PC wires and it was confined by a 40 mm spiral with a pitch of 120 mm. The volumetric ratio of the spiral was 0,113% which was about 10% of the minimum requirement set by ACI 318-19. SPPC01 was the specimens where the spun pile was empty whereas SPPC02 was filled by 6D19 reinforced concrete.

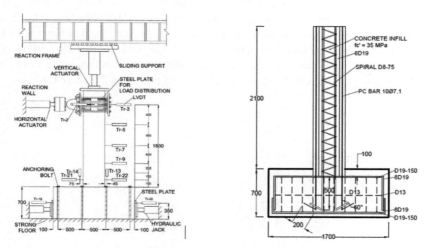

Fig. 2. The experimental set-up and the specimens

Figure 2 shows the setup and the specimen's detail. The spun pile was connected to the pile cap through 10 PC wires. To prevent the slip failure, end of the wires was bent about 30° and hence the total embedment length was 700 mm. Mild rebar, 6D19, was also similarly embedded in the pile cap fr SPPC02. A reverse cyclic lateral load according to ACI 374.2r-13 was applied with a constant vertical force as $0.1fc'Ag$. Eight horizontal and two vertical LVDTs were employed to monitor the displacement.

2.2 The Numerical Study

Further numerical investigation was conducted to see the contribution of soil to pile connection. The study was conducted by OpenSees software. A similar pile used in the experiment was modelled as Wrinkler beam which is supported by spring. The single pile with 7.5 m of length was attached to one layer of clay with Su was varied from 20 to 100 kPa. Push over analysis was carried out to investigate the inelastic stage.

To ensure that the FE model could represent the actual condition, results from the experimental study was used as a reference to validate the model. Figure 3 shows the geometric and the fibre section used in OpenSees. Concrete07 and Steel01 were chosen to present the properties of concrete and steel. The prestressed was modelled through the initial stress as 1080 Mpa which was 80% of the initial prestressed. Fully bonded was assumed between the spun pile and the concrete infill. The pile was assumed as a cantilever structure connected to a pile cap with fixed support with some degree of rotation due to strain penetration. The rotation depends on bond-slip between prestressed wires or mild rebars and the concrete of the pile cap. The slip was calculated based on the equations proposed by Zhao and Sritharan [12]. The zero-length element of OpenSees was used to simulate the inelastic angle in the connecting region.

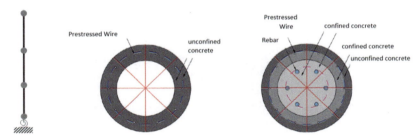

Fig. 3. Fiber element

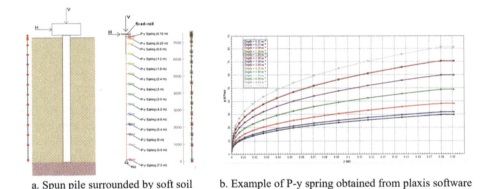

a. Spun pile surrounded by soft soil   b. Example of P-y spring obtained from plaxis software

Fig. 4. The numerical model of the spun pile embedded in soil

Figure 4 shows the numerical model and the p-y spring of the soil. The spring stiffness was determined from Plaxis software. The pile was divided into 12 elements and the spring was attached to each element. The model was named as SPPCAnt which represents the spun pile connection that is commonly used in Indonesia as shown in Fig. 1. Concrete and rebars are filled to the spun pile with an approximate depth of one meter to obtain a fixed restrain to the pile cap. Hence, SPPCAnt consisted SPPC01 except the connecion region which was SPPC02. The 6D19 rebars were anchored to the pile cap together with the 10@7.1 PC wires. A monotonic push over analysis on the horizontal direction was performed where $0.1fc'Ag$ dead load was constantly applied in the vertical direction.

3 Results and Discussion

3.1 The Experimental Results

The two specimens failed due to flexural which was revealed from the crack pattern as shown in Fig. 5. There was no shear failure detected. The crack of SPPC01 spread on the connection region to 450 mm of height measured from pile cap surface and 650 mm of height for SPPC02. Approximately, the crack which is represented as plastic hinges has a length of 1D to 1.5D. All wires of SPPC01 were fractured meanwhile SPPC02 still had one unfractured wire. As can be seen in Fig. 5, the fracture occur at a depth of $+10$ mm to -30 mm from the pile cap surface. It indicated that there was no slip between the wire and the pile cap's concrete since it could develop the ultimate strength.

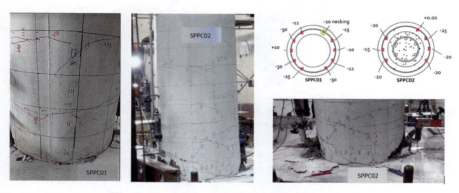

Fig. 5. The failure mode of SPPC01 and SPPC02

The hysteretic curves are shown in Fig. 6. The maximum lateral capacity of SPPC01 and SPPC02 are 95.47 kN and 146.59 kN, respectively. The reinforced concrete in SPPC02 improve the strength by 54% compared to SPPC01. Referring to hysteretic curve and the spread of crack, SPPC02 has better energy absorption. The bending capacity of SPPC01 is 179 kNm. The value is slightly higher than that determined from the P-M interaction which is 171.04 kNm. Meanwhile, SPPC02 has a bending capacity of 284 kNm which is 10% higher than P-M interaction at 258.7 kNm.

Ductility is defined as an ability of a structure to undergo large amplitude cyclic deformation in the inelastic range without a substantial reduction in strength. Displacement ductility is determined as a ratio of the ultimate displacement (du) over the yield displacement (dy). Yield displacement (dy) is defined based on an equivalent elasto-plastic system at reduced stiffness as 75% of the ultimate lateral load. Meanwhile, the ultimate displacement (du) is the displacement when the load-carrying capacity decrease approximately 15%. Based on these definitions, the ductility of SPPC01 and SPPC02 are 4.26 and 4.7, respectively.

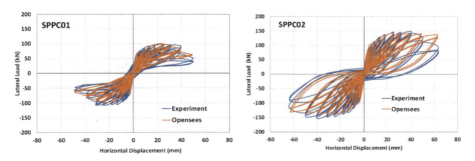

Fig. 6. The hysteretic curves from the experiment and Opensees's results

3.2 The Numerical Results

Prior to the numerical study, the validation was conducted on the FE model based on the experimental results and the comparisons are shown in Fig. 6 Strain penetration effect was considered in the FE model. Since the spun pile was anchored to the pile cap through PC Wire and 6D19, the slip was affected by bar diameter (db), its yield stress (Fy) and the concrete strength (fc') [12]. Three FE models were studied, one was without slip, and the other two models were slipped with different anchored bars, based on PC wires and 6D19. The study found that all models have similar results. Hence, to simplify the analysis, in the numerical study it is assumed as fixed restraint with no slip.

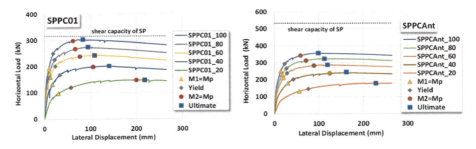

Fig. 7. The load – displacement curves of SPPC01 and SPPCAnt

Figure 7 presents the results of the pushover analysis. Two different spun piles were compared, SPPC01 and SPPCAnt. Similar to the experiments, both connections fail due to flexural. The graph shows the stage when the value of M_1 and M_2 start to exceed the plastic moment (Mp), the yield and the ultimate state. M_1 and M_2 are the maximum bendings which are located on the connection and in the ground as shown in Fig. 1. The plastic moment of SPPC01 and SPPC02 are 166.52 kNm and 236.33 kNm, respectively. The study found that the stiffer the soil, the higher the capacity of the connection. Filling the reinforced concrete into the spun pile on the connection region (SPPCAnt) effectively improve the strength of the connection by 18–20%. Meanwhile, the use of SPPC02 increase the strength by 25% to 30%.

The maximum forces of all connections and the ductility are shown in Fig. 8. As predicted, SPPC02 has the highest strength. However, the strength of SPPCAnt which represent the real condition is not much different from SPPC02. The ductility is not affected much by the soil, except SPPC02. Meanwhile, for SPPCAnt, the ductility increases faintly when surrounded soil is stronger. The correlation between soil strength and ductility is found scattered for SPPC01.

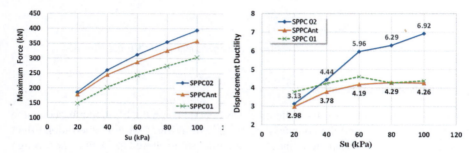

Fig. 8. The maximum force and ductility with the change of soil shear strength Su

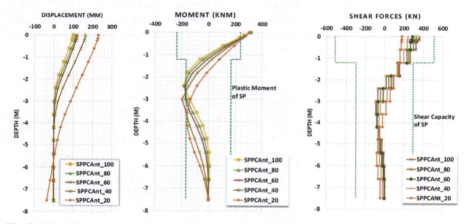

Fig. 9. Distribution of displacement, bending moment and shear force along the pile depth of SPPCAnt at the ultimate states

The distribution of displacement, shear force and bending moment along the pile at the ultimate state is shown in Fig. 9. The transition from positive to negative bending moment occurs at depth of -1.00 to -1.40 measured from the connection meanwhile for pile in soft clay (Su 20kPa), it locates on -1.75 m. The diagram shows that the maximum moment M_1 and M_2 exceed the plastic moment (Mp) of the spun pile, except those in Su 20 kPa. The maximum M_2 is noticed at 0.24–0.32 L for stiff clay, 0.32–0.4 L for medium clay and 0.8 L for soft clay. The shear diagram further confirms that there is no shear failure occur. Similarly, there is no soil failure observed since the soil stress represented by springs forces are still within soil capacity.

Table 1 presents the maximum value of M1 at different soil strengths at the ultimate state. To get more insight about soil contribution to the connection, the value is normalized by the moment obtained from the test, which is 179 kNm for SPPC01 and 284 kNm for SPPC02. Based on the ratio it can be concluded that soil contribution to the strength of the connection is varied linearly with the soil strength. SPPCAnt has a ratio within the range of 1.06 to 1.12. It demonstrates that the bending capacity of the spun pile embedded in clay is about 10% higher than spun pile without surrounding soil. When M1 is normalized by moment obtained from P-M interaction, the ratio for SPPCAnt is from 1.17 (Su = 20 kPa) to 1.23 (Su = 100 kPa). It seems that the bending capacity could be approximately determined based on the P-M interaction.

Table 1. Maximum M_1 and its ratio to Mu obtained from the test

Su (kPa)	Axial load (P)	Max M_1 – kNm (Max M_1/Mu test)		
		SPPC01	SPPC02	SPPCAnt
20	0.1fc'Ag	196.9 (1.10)	305.5 (1.08)	301.8 (1.06)
40	0.1fc'Ag	199.3 (1.11)	319.1 (1.12)	310.7 (1.09)
60	0.1fc'Ag	201.7 (1.13)	326.4 (1.15)	311.3 (1.10)
80	0.1fc'Ag	203.3 (1.14)	334.2 (1.18)	316.2 (1.11)
100	0.1fc'Ag	204.9 (1.14)	340.0 (1.20)	317.3 (1.12)

4 Conclusions

The experimental study of low confinement spun pile connection found that there is no shear failure detected during the test and the connections failed due to flexure. Filling the spun pile with reinforced concrete improves the strength by 54% and increase the ductility from 4.27 to 4.70. Better energy absorption was observed on SPPC02. The connection did not experience a slip and the prestressed suffered fracture. The bending capacity is 5 to 10% higher than the capacity determined by the P-M interaction.

Based on the numerical study, SPPCAnt, a typical spun pile connection in Indonesia, which was embedded in clay demonstrated good performance. A small amount of reinforced concrete filled into the spun pile on the connection region effectively improve the connection behaviour. The shear failure did not detect and its bending capacity was slightly higher than the test. Compared to the test result, the ratio was in the range of 1.06 to 1.12 and 1.17 to 1.23 when it is compared to P-M interaction. Stiffer soil results in higher connection capacity. The bending moment obtained from the P-M interaction could be used to approximate the bending capacity of the spun pile connection. The ductility of SPPCAnt is in the range of 2.98 to 4.26 which is slightly affected by the soil strength. The amount of confinement on the spun pile should be studied.

Acknowledgement. The research is supported by Indonesia Ministry of Research and Universitas Indonesia through Applied Research, contract number NKB-254/UN2.RST/HKP.05.00/2021. The authors would like to acknowledge: PT Wijaya Karya Beton, Tbk for the research collaboration.

References

1. Bang, J.-W., Oh, S.-J., Lee, S.-S., Kim, Y.-Y.: Pile-cap connection behavior dependent on the connecting method between PHC pile and footing. J. Korea Inst. Struct. Maint. Insp. **20**(3), 25–32 (2016). https://doi.org/10.11112/jksmi.2016.20.3.025
2. Kaneko, O., Kawamata, S., Nakai, S., Sekiguchi, T., Mukai, T.: Analytical study of the main causes of damage to pile foundations during the 2011 off the Pacific coast of Tohoku earthquake. Jpn. Archit. Rev. **1**(2), 235–244 (2018). https://doi.org/10.1002/2475-8876.12033
3. Wang, T., Yang, Z., Zhao, H., Wang, W.: Seismic performance of prestressed high strength concrete pile to pile cap connections. Adv. Struct. Eng. **17**(9), 1329–1342 (2014). https://doi.org/10.1260/1369-4332.17.9.1329
4. Yang, Z., Wang, W.: Experimental and numerical investigation on the behaviour of prestressed high strength concrete pile-to-pile cap connections. KSCE J. Civ. Eng. **20**(5), 1903–1912 (2016). https://doi.org/10.1007/s12205-015-0658-8
5. Yang, Z., Li, G., Nan, B.: Study on seismic performance of improved high-strength concrete pipe-pile cap connection. Adv. Mater. Sci. Eng. **2020** (2020). https://doi.org/10.1155/2020/4326208
6. Guo, Z., He, W., Bai, X., Chen, Y.F.: Seismic performance of pile-cap connections of prestressed high-strength concrete pile with different details. Struct. Eng. Int. **27**(4), 546–557 (2017). https://doi.org/10.2749/222137917X14881937845963
7. ASCE 61-14: Seismic design of piers and wharves (2014). https://doi.org/10.1061/9780784413487
8. Joen, P.H., Park, R.: Flexural strength and ductility analysis of spirally reinforced prestressed concrete piles. PCI J. **35**(4), 64–83 (1990)
9. Joen, P.H.: Seismic performance of prestresses concrete piles and pile - pile cap connections, p. 319 (1987)
10. Irawan, C., Djamaluddin, R., Raka, I.G.P., Faimun, Suprobo, P., Gambiro: Confinement behavior of spun pile using low amount of spiral reinforcement - an experimental study. Int. J. Adv. Sci. Eng. Inf. Technol. **8**(2), 501–507 (2018). https://doi.org/10.18517/ijaseit.8.2.4343
11. Irawan, G.C., Raka, I.G.P., Djamaluddin, R., Suprobo, P.: Ductility and SEismic performance od spun pile under constant axial and reverse flexural loading. In: International Symposium on Concrete Technology (ISCT 2017), pp. 35–44 (2017)
12. Zhao, J., Sritharan, S.: Modeling of strain penetration effects in fiber-based analysis of reinforced concrete structures. ACI Struct. J. **104**(2), 133–141 (2007)

Rocking Pilegroups Under Seismic Loading: Exploring a Simplified Method

Antonia Psychari[✉], Saskia Hausherr, and Ioannis Anastasopoulos

Institute for Geotechnical Engineering, ETH Zurich, Zürich, Switzerland
antonia.psychari@igt.baug.ethz.ch

Abstract. Recent research on the rocking response of bridges founded on pilegroups has demonstrated the potential benefits of allowing full mobilization of pile bearing capacity during seismic loading. Especially for the retrofit of existing bridges, allowing strongly nonlinear foundation response may allow avoiding foundation retrofit, which can be a challenging and costly operation. However, this calls for 3D numerical analysis of the bridge–foundation system with adequately sophisticated constitutive models, which can be time-consuming. To promote practical application of such performance-based design philosophy, this study explores the efficiency of a simplified analysis method, where the pilegroup is replaced by an assembly of nonlinear (rocking) and linear (horizontal and vertical) springs and dashpots. Such a simplified approach has been shown to offer reasonable predictions for rocking shallow foundations, and it is therefore of interest to explore its applicability to pilegroups. A bridge pier founded on a 2 × 1 pilegroup in homogeneous clay is chosen as an illustrative example. The bridge-foundation system is initially modelled with 3D finite element (FE) analysis, accounting for soil and structural nonlinearity of the reinforced concrete (RC) piles. The system is subjected to monotonic and cyclic pushover loading, based on which the nonlinear rotational spring and the corresponding dashpot are calibrated. The simplified model is subjected to dynamic time history analyses and compared to the full 3D model. The comparison yields promising preliminary results, but also offers insights on the reasons of the observed discrepancies and the unavoidable limitations of such simplified technique.

Keywords: Pilegroup · Bridge · Finite element modelling · Simplified method

1 Introduction

The rapid growth of urban areas has led to the need for retrofit and expansion of existing infrastructure. Especially in the fields of transport infrastructure, widening of existing road and rail bridges in bottleneck areas becomes unavoidable. Commonly founded on pilegroups, such bridges do not satisfy current seismic design standards. While strengthening of the superstructure is relatively straightforward, strengthening of the foundation system can be challenging, requiring non-trivial excavation and addition of piles, leading to increased costs and network interruption.

An alternative seismic design approach is explored. Following the concept of "rocking isolation" [1–3], full mobilization of foundation bearing capacity is promoted, aiming to exploit strongly nonlinear soil-foundation response in defense of the superstructure. A "do nothing" approach may be feasible for the foundation in some cases, taking advantage of the inherent pile design conservatism of past decades. By allowing soil yielding and full mobilization of pile bearing capacity, the foundation may act as a "fuse", absorbing seismic energy [1]. Recent studies on various types of foundations [2, 4, 5] have demonstrated the advantages of such a concept, while case studies on piled foundations have shown that foundation strengthening may lead to inferior seismic performance compared to the "do-nothing" approach [5].

However, allowing such strongly nonlinear foundation response is unavoidably related to increased settlement [4]. The application of such performance-based design in practice requires 3D numerical analysis of the bridge–foundation–soil system, which can be challenging for everyday design practice. The scope of this study is to explore the potential of a simplified analysis method, aiming to replace the time consuming and demanding 3D finite element (FE) analyses, facilitating the application of such new concept in practice. Inspired by a simplified method that was developed for rocking foundations by Anastasopoulos & Kontoroupi [6], an assembly of nonlinear (rocking) and linear (horizontal and vertical) springs and dashpots is explored as an alternative to the full 3D FE analysis of the soil–foundation system. A parametric study is conducted to explore the efficiency of the simplified method and address its limitations.

2 Simplified Model: Description

Figure 1 illustrates the full 3D FE model of a 2×1 pilegroup (Fig. 1a) and the corresponding simplified model (Fig. 1b). Considering the transverse direction, a single bridge pier is modelled as a single degree of freedom (SDOF) system. The height H corresponds to the pier height and the cap height. The dead and live loads of the bridge deck are represented by a concentrated mass m at the top of the pier. An idealized 2 x 1 pilegroup with piles of length L and diameter D is used as an illustrative example. Consistent with practice, a typical pile spacing $s/D = 3$ is selected to avoid group effects, while the piled foundation lies in an idealized homogeneous clay layer of undrained shear strength S_u.

In the case of the simplified model, the soil and the piles are replaced by an assembly of springs and dashpots, calibrated to match the nonlinear pilegroup response. Since the response of the system is rocking dominated, the horizontal (K_h and C_h) and vertical (K_v and C_v) springs and dashpots are assumed elastic, and are calibrated according to published solutions. The rocking response of the foundation is represented by a nonlinear rotational spring and a linear rotational dashpot, calibrated on the basis of cyclic displacement-controlled pushover analysis of the full 3D model. The simplified model can be easily implemented in commercial FE programs, or even programed in a Matlab code.

Monotonic and cyclic pushover analyses of the full 3D FE model are conducted for calibration of the simplified model. More specifically: (a) the nonlinear rotational spring K_R is defined by the moment–rotation relation ($M - \vartheta$), calculated through monotonic,

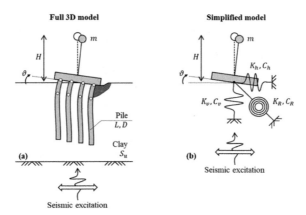

Fig. 1. Schematic illustration of: (a) full 3D FE model; and (b) simplified model.

displacement-controlled pushover analysis of the full 3D FE model; (b) the coefficient C_R of the linear rotational dashpot is calculated through $M - \vartheta$ loops, from cyclic analyses; and (c) the settlement–rotation ($w - \vartheta$) response (not discussed herein) is computed from the cyclic response of the 3D FE model. This paper is a first attempt to explore the feasibility of such a simplified approach, which can facilitate practical application without the need for advanced 3D FE analyses.

3 Numerical Analysis Methodology

The RC piles of the analyzed 2×1 pilegroup have a length $L = 15$ m, diameter $D = 1$ m and spacing $s = 3$ m. C25/30 concrete is assumed, using a typical reinforcement ratio of 1%. The pier height H_{pier} is parametrically varied from 6 m to 20 m, being representative of an overpass or viaduct bridge, respectively. The pilecap is 6 m $\times$ 3 m with a height of 1.5 m. The deck mass is also parametrically varied, so as to achieve safety factor against static vertical loading $FS_v = 3$ to 10. An idealized 21 m thick clay layer of $S_u = 100\,kPa$ is considered, allowing a distance of $6D$ below the pile tip. The lateral boundaries of the model are at a distance of $10D$ from the edge of the piles in each direction to avoid boundary effects.

The soil is modelled with 8-node brick elements, assuming a thoroughly validated kinematic hardening model, with a Von Mises failure criterion and associated flow rule [7]. The model has been validated against centrifuge test results. Its calibration is based on the undrained shear strength S_u, the small-strain shear modulus G_o, and stiffness degradation ($G - \gamma$ and $\xi - \gamma$ curves). The piles are modelled with 3D nonlinear continuum elements, employing the Concrete Damage Plasticity model (CDP) model, which is available in Abaqus [8]. The key parameters of the model are derived for the uniaxial stress-strain response of concrete, while the inelastic strain (under tension or compression) is associated to damage parameters to capture the effect of cracking or crashing. Model parameters are calibrated based on the basis of widely used publications [9–11]. The reinforcement is modelled with shell elements, embedded in the continuum elements that represent the concrete. The detailed modelling of the RC piles enables the definition of separate material laws for the confined core of the cross-section and the unconfined cover concrete, as well as the consideration of the interaction between axial force (N) and bending moment (M) and cumulative damage during consecutive load cycles. Figure 2 offers an overview of the constitutive models used for the soil and the piles. Tensionless interfaces with appropriate friction coefficient μ are introduced between the soil and the foundation (piles and pilecap).

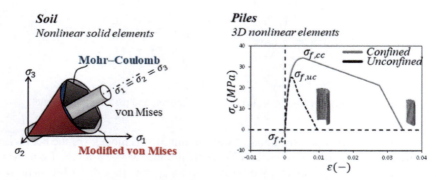

Fig. 2. Soil modelling with kinematic hardening model and RC pile modelling with CDP.

The modelling of the soil–pile system has been verified against analytical solutions for single pile capacity. Figure 3a compares the FE-computed axial resistance of a single pile in tension and compression to analytical solutions, according to the α-method assuming $\alpha = 0.75$ [12]. Similarly, Fig. 3b compares the FE-computed lateral response of a single pile to the analytical solution of Broms [13], assuming a frictionless interface, and the solution of Randolph and Houlsby [14], which accounts for friction.

The full 3D FE model is subjected to monotonic and cyclic pushover loading to derive the moment–rotation ($M - \vartheta$) response. Cyclic pushover analysis is conducted to obtain hysteretic response. The results are used to calibrate the simplified model. Then, both models are subjected to dynamic analysis, using 5 records (Fig. 4).

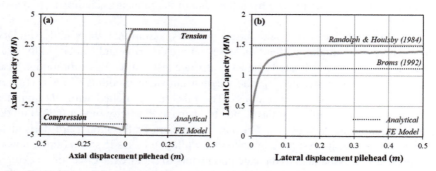

Fig. 3. 3D FE model verification for a single pile under: (a) axial; and (b) lateral loading.

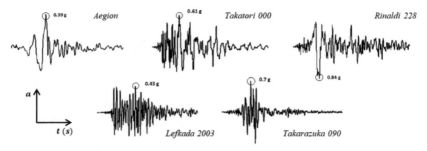

Fig. 4. Seismic records used as seismic excitation for the dynamic time history analyses.

4 Implementation of Simplified Model

In the simplified model, the soil–foundation system is replaced by a nonlinear rotational spring and a linear dashpot, calibrated on the basis of monotonic and cyclic pushover analysis of the 3D FE model, and linear horizontal and vertical springs and dashpots. The vertical and horizontal springs are calibrated using published solutions. The vertical stiffness of a single pile is given by Gazetas [15]:

$$K_{v,s} = \frac{2\pi\, LG}{\ln(4L/D)} \quad (1)$$

where: L is the pile length; D is the pile diameter; and G is the shear modulus of soil.

Accordingly, the horizontal stiffness of a single pile is given by [15]:

$$K_{h,s} \cong E_s D (E_p/E_s)^{0.21} \quad (2)$$

where: E_s and E_p is the elasticity modulus of soil and pile, respectively.

The springs corresponding to the 2 × 1 pilegroup are obtained by multiplying by 2 (group effects are not accounted for, as they are negligible for the specific configuration). The nonlinear rotational spring is obtained directly from the pushover analysis of the 3D FE model. Figure 5 shows an indicative example for $H_{pier} = 6\,\text{m}$ and $FS_v = 3$.

The damping coefficients of the dashpots are calculated in function of the corresponding stiffness K, the damping ratio ξ and a characteristic frequency ω:

$$C \approx \frac{2K\xi}{\omega} \quad (3)$$

However, the fundamental frequency is not constant, but a function of rotation ϑ [16]. As a simplification, the initial frequency of the rocking system is used herein:

$$T_{n,0} = 2\pi \sqrt{\frac{mH^2}{K_{R,0} - mgH}} \quad (4)$$

where $K_{R,0}$ is the initial rocking stiffness ($\vartheta = 0$); m is the mass of the oscillator; and H is total height of the oscillator.

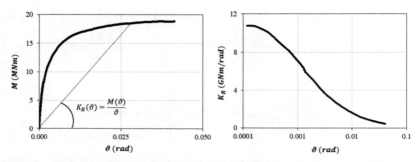

Fig. 5. Nonlinear rotational spring and secant stiffness for $H_{pier} = 6$ m and $FS_v = 3$.

For the vertical and the horizontal dashpots, a damping ratio $\xi = 5\%$ is assumed (gross simplification). For the rotational dashpot, the hysteretic damping ratio is computed through the $M - \vartheta$ loops of the cyclic pushover analysis of the full 3D FE model. As shown in the example calculation of Fig. 6 (for $H_{pier} = 6$ m and $FS_v = 3$), the damping coefficient is a function of rotation ϑ. Therefore, a nonlinear dashpot would be more appropriate. To maintain simplicity a linear dashpot is employed, selecting the maximum value of the bell-type curve.

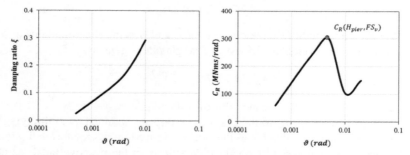

Fig. 6. Cyclic pushover analysis ($H_{pier} = 6$ m, $FS_v = 3$): calculation of rotational dashpot.

5 Simplified vs. Full 3D Model

Indicative comparisons are presented, focusing on deck acceleration and foundation moment-rotation ($M - \vartheta$) response. Figure 7 offers such a comparison for a system with $H_{pier} = 20$ m and: (a) $FS_v = 10$, subjected to the Aegion record; and (b) $FS_v = 3$, subjected to the Rinaldi record. Subjected to the moderate intensity Aegion record, the lightly loaded ($FS_v = 10$) system exhibits a rather elastic response (Fig. 7a). The simplified model predicts reasonably well the response, both in terms of acceleration time history and $M - \vartheta$ loops. Evidently, the comparison is not perfect. The simplified model underestimates the maximum acceleration and, consequently, the maximum overturning moment that is transmitted to the pilegroup. The opposite is observed for the heavily loaded ($FS_v = 3$) system subjected to the Rinaldi excitation (Fig. 7b). As revealed by the $M - \vartheta$ loops, the response is strongly nonlinear. The simplified model is in reasonable agreement with the full 3D model, this time overestimating the response.

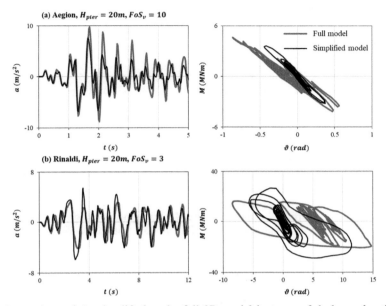

Fig. 7. Comparison of the simplified to the full 3D model in terms of deck acceleration time histories (left) and pilegroup moment–rotation ($M - \vartheta$) response for a system with $H_{pier} = 20$ m and: (a) $FS_v = 10$ subjected to Aegion; and (b) $FS_v = 3$ subjected to Rinaldi record.

Interestingly, the simplified model under-predicts ϑ in both cases. Looking at the second case, it appears that the under-prediction of ϑ is due to over-estimation of damping. As previously discussed (Eq. 3), the calculation of the damping coefficient requires an assumption of period $T_{n,0}$ (Eq. 4), which is calculated on the basis of the initial rocking stiffness $K_{R,0}$ ($\vartheta = 0$). The validity of this assumption is challenged with the increase of ϑ, and this is why the damping is significantly over-estimated in the second case, where the system exhibits strongly nonlinear response (Fig. 7b).

6 Conclusions

The current study presents preliminary results of a simplified methodology to analyse the seismic rocking response of pilegroups, aiming to facilitate implementation in practice of an alternative seismic design method that allows full mobilization of foundation bearing capacity. The simplified model replaces the pile–soil system with an assembly of a nonlinear rotational spring and linear dashpot, accompanied by vertical and horizontal linear springs and dashpots. The results are not perfect but promising. The presented results indicate that the key limitation of the simplified model is the assumption of a linear rotational dashpot.

Acknowledgement. This work was supported by the Swiss National Science Foundation (SNSF) under Grant 200021_188459. The authors gratefully acknowledge this funding. The methods, results, opinions, findings, and conclusions presented in this report are those of the authors and do not necessarily reflect the views of the funding agency.

References

1. Anastasopoulos, I.: Beyond conventional capacity design: towards a new design philosophy. In: Orense, R.P., Chow, N., Pender, M.J. (eds.) Soil–Foundation–Structure Interaction, pp. 213–220. Taylor & Francis Group (2010)
2. Anastasopoulos, I., Gazetas, G., Loli, M., Apostolou, M., Gerolymos, N.: Soil failure can be used for seismic protection of structures. Bull. Earthq. Eng. **8**(2), 309–326 (2010)
3. Mergos, P.E., Kawashima, K.: Rocking isolation of a typical bridge pier on spread foundation. J. Earthq. Eng. **9**(2), 395–414 (2005)
4. Agalianos, A., Psychari, A., Vassiliou, M.F., Stojadinovic, B., Anastasopoulos, I.: Comparative assessment of two rocking isolation techniques for a motorway overpass bridge. Front. Built Environ. **3**(47) (2017). https://doi.org/10.3389/fbuil.2017.00047
5. Sakellariadis, L., Marin, A., Anastasopoulos, I.: Widening of existing motorway bridges: pile group retrofit versus nonlinear pile–soil response. J. Geotech. Geoenviron. Eng. **145**(12), 04019107 (2019)
6. Anastasopoulos, I., Kontoroupi, T.: Simplified approximate method for analysis of rocking systems accounting for soil inelasticity and foundation uplifting. Soil Dyn. Earthq. Eng. **56**, 28–43 (2014)
7. Anastasopoulos, I., Gelagoti, F., Kourkoulis, R., Gazetas, G.: Simplified constitutive model for simulation of cyclic response of shallow foundations: validation against laboratory tests. ASCE J. Geotech. Geoenviron. Eng. **137**(12), 1154–1168 (2011)
8. ABAQUS 2019: Standard user's manual. Dassault Systèmes Simulia Corp., Providence, RI, USA (2019). ETH Zürich
9. Chang, G.A., Mander, J.B.: Seismic energy based fatigue damage analysis of bridge columns: Part 1 – evaluation of seismic capacity, section 3. Technical report NCEER-94-0006 (1994)
10. Mander, J.B., Priestley, M.J.N., Park, R.: Theoretical stress-strain model for confined concrete. J. Struct. Eng. **114**(8), 1804–1826 (1988)
11. Psychari, A., Agalianos, A., Sakellariadis, L., Anastasopoulos, I.: On the effect of RC moment–axial load interaction in the moment capacity of a pile group. In: Proceedings of 2nd International Conference on Natural Hazards & Infrastructure, Chania, Greece (2019)
12. Tomlinson, M.J.: Pile Design and Construction Practice. Viewpoint Publications, London (1977)
13. Broms, B.B.: Lateral resistance of piles in cohesive soils. J. Soil Mech. Found. Div. **90**(2), 27–63 (1964)
14. Randolph, M.F., Houlsby, G.T.: The limiting pressure on circular pile loaded laterally in cohesive soil. Geotechnique **34**(4), 613–623 (1984)
15. Gazetas, G., Anastasopoulos, I., Garini, E., Gerolymos, N.: Interaction Between Soil, Foundation and Structure. Tsotras, Athens (2015)
16. Housner, G.W.: The behavior of the inverted pendulum structures during earthquakes. Bull. Seismol. Soc. Am. **53**(2), 403–417 (1963)

Remediation of Structure-Soil-Structure Interaction on Liquefiable Soil Using Densification

Shengwenjun Qi[1(✉)] and Jonathan Adam Knappett[2]

[1] State Key Laboratory of Coastal and Offshore Engineering, Dalian University of Technology, Dalian 116024, China
sqi@dlut.edu.cn
[2] University of Dundee, Dundee DD1 4HN, UK
j.a.knappett@dundee.ac.uk

Abstract. Earthquake induced soil liquefaction has drawn significant attention in urban areas where buildings are closely spaced due to the extensive damage it has caused on buildings in recent earthquakes. This study investigated the effect of targeted soil densification as a remediation method on one of a pair of adjacent buildings with different storey levels on liquefiable soil using dynamic centrifuge modelling. The effects of remediation on structural and foundation performance of both treated and untreated buildings are presented. It was demonstrated that localised soil densification can be effective and beneficial in treating detrimental foundation behaviours of the treated building on the liquefiable soil and an untreated adjacent building can also be beneficially affected. From a structural perspective, soil densification can reduce the structural demand on the treated building with negligible detrimental effect or even some beneficial effect on the untreated building on the liquefiable soil. In the case studied, localised soil densification when building new structures adjacent to existing ones can be an effective remediation method in improving seismic performance on liquefied soil. This study also suggests that the decision of remediating any building in an urban setting should be made after considering effects of SSSI.

Keywords: Structure-soil-structure interaction · Liquefaction · Ground remediation · Centrifuge modelling

1 Introduction

Earthquake-induced soil liquefaction can cause significant damage to low-rise buildings in urban centres, as seen in the 2010 Maule and 2011 Christchurch Earthquakes, which will become exacerbated due to population growth and urbanisation worldwide.

This study investigates the effectiveness of targeted soil densification as a remediation method on a pair of adjacent buildings in consideration of structure-soil-structure (SSSI) effects on liquefiable soil. A case of an 'existing' two-storey building (B1) with

---

S. Qi—Formerly University of Dundee, UK.

© The Author(s), under exclusive license to Springer Nature Switzerland AG 2022
L. Wang et al. (Eds.): PBD-IV 2022, GGEE 52, pp. 1193–1200, 2022.
https://doi.org/10.1007/978-3-031-11898-2_99

separated strip foundations adjacent to a taller 'new' four-storey building (B2) with similar foundations was taken as the benchmark. Densification was conducted only beneath B2 to a limited depth to (i) reflect this being incorporated into the seismic design of B2; and (ii) due to the fact that the ownerships of adjacent buildings are most likely to be different, such that B2 should mitigate its impacts on B1 where possible and not require remedial works to B1. The comparative foundation performance and structural response of cases with treated and untreated B2 were considered. The liquefiable cases were conducted using dynamic centrifuge modelling.

2 Centrifuge Modelling

Two centrifuge tests (benchmark non-densified and remediated densified case) were conducted using the Actidyn Systèmes C67 3.5 m radius beam centrifuge facility at 1:40 scale tested at 40 g condition at the University of Dundee, UK. All data are shown at prototype scale unless stated otherwise.

2.1 Model Structures

The prototype of the B1 (existing) structure was designed as an idealised single bay, steel moment resisting frame with concrete slab floors standing on separated concrete strip foundations. A taller (new) structure B2 with the same foundation and storey properties but greater number of storeys was placed adjacent to B1 with 1.2 m edge-to-edge spacing to enable SSSI effects. The design process of B1 involved firstly setting the targeted fundamental natural fixed-base period to approach the result of Eq. 1:

$$T_n = 0.1N \qquad (1)$$

where N is the number of storeys, so T_n should approach 0.2 s for B1.

Then, the mass of each concrete floor slab was calculated, so that the targeted equivalent elastic stiffness can be derived by combining Eq. (1) with Eqs. (2)–(4):

$$T_n = 2\pi \sqrt{\frac{M_{eq}}{K_{eq}}} \qquad (2)$$

$$M_{eq} = \sum_{i=1}^{N} M_i \bar{y}_i^2 \qquad (3)$$

$$K_{eq} = \sum_{i=1}^{N} K_i (\bar{y}_i - \bar{y}_{i-1})^2 \qquad (4)$$

For B1 ($N = 2$), the normalised modal coordinates of the fundamental mode were $\bar{y}_1 = 0.45$ and $\bar{y}_2 = 0.89$ determined from an elastic eigenvector analysis. By selecting the closest available steel Universal Column size to provide sufficient bending stiffness EI (UC 203 × 203 × 86) for the four columns in each storey, the fixed-base natural period was $T_{n1} = 0.21$ s. B2 shared the same values of M_i and K_i as B1, so the fix-base natural period of B2 was calculated as $T_{n2} = 0.36$ s based on the normalised modal coordinates

Table 1. Building properties (prototype scale)

Parameter	B1	B2	units
Total height (storey height)	6 (3)	12 (3)	m
Slab floor dimensions	3.6×3.6×0.5		m
Strip foundation dimensions	4.8×1.2×0.5		m
M_1-M_4	16.5×10³		kg
K_1-K_4	37.1×10⁶		N/m
Bearing pressure	50	88	kPa
Static factor of safety	3	1.7	

of $\overline{y_1} = 0.18$, $\overline{y_2} = 0.37$, $\overline{y_3} = 0.55$ and $\overline{y_4} = 0.73$ when $N = 4$. The detailed properties of the two building structures are provide in Table 1.

The prototype design of the building structures was fabricated at 1:40 reduced scale based on centrifuge scaling laws [1]. The model frame and the foundation were fabricated from solid 6082-series square-section aluminium alloy rods, and the slab floors were fabricated with same aluminium plates with additional steel plate bolted on the top to account for the storey mass (shown in Fig. 1).

2.2 Model Preparation and Soil Properties

The soil layouts of the densified (test SQ14) is shown in Fig. 1 where the densified area reached $D_r = 80\%$ (within the practical range of densities achieved in field cases) beneath B2 to a limited depth of 3 m, while the surrounding soil was medium dense ($D_r = 55\%$–60%). The layout of the non- densified (test SQ10) case was similar to the densified (test SQ14), while without the densified zone. The instrumentation of ADXL-78 uniaxial MEMS accelerometers, HM-91 and PDCR-81 pore pressure transducers and linear variable differential transformers (LVDTs) used to record soil and building performance are also shown in Fig. 1. Properties of HST95 silica sand used in the tests are shown in Table 2. The soil was saturated with hydroxyl-propyl methyl-cellulose (HPMC) pore fluid (40 times viscosity of water) to achieve the same permeability as saturated soil in the prototype.

Table 2. Soil properties of HST95 silica sand

Properties	Value	Units
Specific gravity, G_s	2.63	-
D_{10}	0.09	mm
C_u(uniformity) and C_z (curvature)	1.9 and 1.06	-
e_{max} and e_{min}	0.769 and 0.467	-
ϕ_{crit}	32	degrees

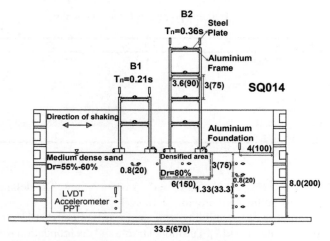

Fig. 1. Centrifuge model configuration: dimensions at prototype scale are shown in m; dimensions at model scale are given in mm in brackets ().

2.3 Dynamic Excitations

The selected input ground motions were recorded from the New Zealand Canterbury Series of 2010–2011, followed by the 2011 Japan Tohoku Earthquake (time histories of each earthquake are shown in Fig. 2). The first and last motions were the 'mainshocks' aiming to generate full liquefaction; the three 'aftershocks' in between aimed to generate relatively smaller amounts of excess pore water pressure rise after the first mainshock. The data for each earthquake were recorded for 4 min at model scale after the start of each earthquake to ensure full dissipation of excess pore water pressure (EPWP) before applying a subsequent earthquake in all tests. Therefore, the dashed lines in Fig. 2 indicate periods where there was no seismic input while EPWP dissipated.

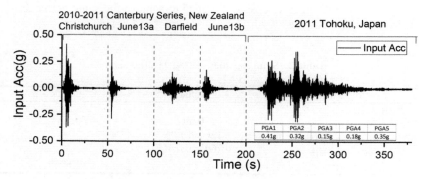

Fig. 2. Input acceleration time history and peak ground acceleration (PGA)

3 Results and Discussion

This section demonstrates the ground response to shaking, and the effect of remediation on building performance in consideration of SSSI.

3.1 Liquefaction Condition

The maximum EPWP ratio r_u in the free-field and beneath the foundations are shown in Fig. 3. During the 'mainshocks' of EQ1 and EQ5, the soil in the free-field reached full liquefaction at all depths; during the aftershocks from EQ2 to EQ4, the less intensive EQ2 was still strong enough to generate full liquefaction at full depth, while EQ3 and EQ4 only induced full liquefaction at very shallow depth (i.e. limited downwards propagation of the liquefaction front). The variability in peak r_u profile between tests SQ10 and SQ14 in EQ3 and 4 were thought be caused by the presence of the densified area beneath structure B2and localised density variation in the free-field during model preparation.

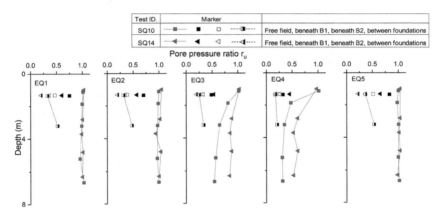

Fig. 3. EPWP ratio r_u in the free field and beneath foundations in the liquefied cases

In Fig. 3: (i) the r_u values beneath the foundations were generally lower than the free-field due to the extra loading from the buildings' foundations; (ii) the r_u values beneath B2 were generally lower than B1 in untreated cases similarly, due to the higher foundation loading caused by the additional storeys of B2; (iii) The r_u values between the foundations were lower than beneath the foundations in the untreated case because of the overlapping stress fields from both buildings; (iv) The r_u values in the densified zone were lower than in the non-treated cases at the same location in the mainshocks.

3.2 Effect of Remediation on Foundation Performance

Foundation accumulative settlement and post-earthquake rotation (tilt) magnitudes were determined to represent the foundation performance of B1 and B2 and are shown in Figs. 4 and 5, respectively. The initial settlements and any initial rotation due to centrifuge spin-up (which were unavoidable) are shown as '40 g' values in Figs. 4 and 5.

For the treated building B2, densification was an effective way of remediation where it was possible to achieve reductions in settlement and rotation by more than or close to 50%. For the untreated adjacent building B1, densification beneath its neighbour had similar beneficial effects on both foundation settlement and rotation. As a result, densification as a remediation method can be generally beneficial both to the building being treated as part of its seismic design, while also not detrimentally (even beneficially) affecting its neighbour's seismic response considering SSSI.

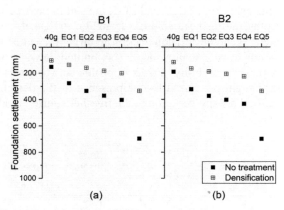

Fig. 4. Foundation settlement of (a) B1 and (b) B2 in the liquefiable soil

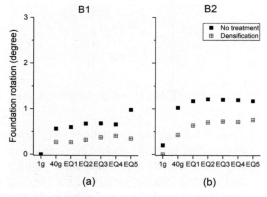

Fig. 5. Magnitude of foundation rotation for (a) B1 and (b) B2 in liquefiable soil

3.3 Effect of Remediation on Structural Performance

The structural performance in terms of the peak storey acceleration and inter-storey drift across the first storey of each building (where the maximum structural deformations occurred) are shown in Figs. 6 and 7. The peak horizontal storey acceleration is a measure of the peak response including both structural and rocking mechanisms, and the inter-storey drift indicates pure structural distortion of the columns. The latter was derived by removing the rotation-induced displacement from total displacement.

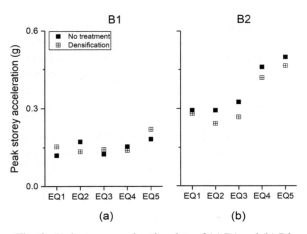

Fig. 6. Peak storey acceleration data of (a) B1 and (b) B2

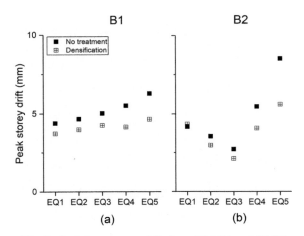

Fig. 7. Peak inter-storey drift data of (a) B1 and (b) B2

For the treated B2, densification beneath can lower structural demand (both storey acceleration and inter-storey drift) on the liquifiable soil with nearly negligible detrimental effect on the untreated B1 in terms of peak storey acceleration and some limited beneficial effects in terms of storey distortion. However, the improvements to structural performance are smaller proportionally than those observed for the foundation response, so densification can be considered mainly a technique for improving foundation response that can be applied with low/positive impact on surrounding buildings.

4 Conclusion

This paper presented the effect of localised densification as a remediation method beneath one of two adjacent buildings on liquefiable soil using centrifuge modelling. The effect of remediation in terms of foundation performance and structural demand on both the treated and untreated adjacent building were demonstrated in consideration of SSSI.

It was shown that localised soil densification beneath a ('new') taller building with longer natural period can benefit this structure in both foundation performance and structural demand, while not detrimentally affecting a neighbouring shorter ('existing') building, even offering it some protection in some of earthquake events simulated. This suggests that soil densification can be chosen as an appropriate method of improving overall building performance in an urban area when soil is liquefiable.

This paper also suggests that the decision of remediating any building in an urban setting should be made after considering effects of SSSI on different soil, i.e. the impacts of such a decision on the seismic performance of existing adjacent structures should be considered.

Reference

1. Wood, D.M.: Geotechnical Modelling. Spon Press, Taylor and Francis, London (2004)

Seismic Responses Analysis on Basements of High-Rise Buildings Considering Dynamic Soil-Structure Interaction

Yan-Jia Qiu[1,2(✉)], Hong-Ru Zhang[2,3], and Zhong-Yang Yu[2]

[1] Changjiang Survey, Planning, Design and Research Co., Ltd., 430010 Wuhan, China
17115316@bjtu.edu.cn
[2] Key Laboratory of Urban Underground Engineering of Ministry of Education, Beijing Jiaotong University, 100044 Beijing, China
[3] School of Civil Engineering, Beijing Jiao Tong University, 100044 Beijing, China

Abstract. Soil-structure interaction is a research hotspot in the geotechnical and civil engineering, especially with the further accelerated development of urban high-rise buildings. Unlike previous studies focusing on the superstructures, this paper presents a numerical study on the seismic response characteristics of underground structures, i.e., the basements of high-rise buildings. To this end, a comprehensive model considering soil-basement-superstructure system was analyzed firstly, and the dynamic response of basement was divided into the kinematic interaction (KI) effect and inertial interaction (II) effect based on a theoretical framework. Then, a typical 25-story symmetric wall-frame structure with a 3-story basement was selected as the case to verify the founds obtained in theoretical analysis. The results demonstrate that: 1) Unlike the single underground structure, the dynamic response of basement would be significant amplified by the inertial interaction effect of superstructure (averagely enlarger by 1.524 times); and 2) There is a larger KI effect on basement whetever the superstructure exists or not. Therefore, it is necessary for seismic design of building basements to comprehensively consider the KI and II effects. In addition, this paper investigates the possibility on acquiring the KI and II effects with simplified methods according to the characteristics of two effects.

Keywords: Dynamic interaction · Seismic response · Underground subway station · Ground adjacent building · Soil-structure interaction

1 Introduction

As a complex interdisciplinary problem, dynamic soil-structure interaction is progressed rapidly and becomes a research hotspot in the geotechnical and civil engineering [1–5], especially with the further accelerated development of urban high-rise buildings. Zhang [6] and Bilotta et al. [7] studied the effects of soil-structure interaction on seismic behaviour of high-rise buildings via numerical method, they concluded that SSI has a remarkable impact on the seismic behaviour of high-rise frame-core tube structures since it can increase the lateral deflections and inter-storey drifts and decrease storey shear forces of structures. Han [8] analyzed the influences of foundation displacements on the performance of high-rise buildings. The results indicated that the conventional design procedure which assumed the role of the soil-structure interaction is beneficial to superstructure cannot fully represent the real behaviour of high rise buildings, and sometime will lead to a dangerous design.

The effects of dynamic soil-structure interaction on seismic behavior of superstructures are significant and have been widely studied, while the seismic performances of underground parts of high-rise buildings received little attentions. Some researchers simplified buildings basements as the foundations of superstructures and their emphasis are still seismic responses of superstructures. The motions of basements were widely studied as the input excitations for analysis system [9]. Lee et al. [10] and Scarfone et al. [11] examined the seismic response of a high-rise building with the effect of the basement. They reminded that the lateral stiffness of high-rise building may be significantly overestimated if the basement is ignored in analytical model (viz., ignoring the KI effect), which would lead to a shorter natural periods of vibration. Borghei et al. [12] and Kim et al. [13] observed that with the effect of kinematic interaction, the high-frequency components of foundation motion are more suppressed than those of the free-field.

Obviously, little attention has been paid to the aseismic performances and dynamic responses of basements with the effects of dynamic soil-structure interaction. As an essential component of underground structure infrastructure, the safety of basements is significant for urban high-rise building system. On the one hand, with further accelerating of urban construction, more and more underground shopping malls and underground garages are constructed under the high-rise buildings. Thus, the basements bear important commercial functions; On the other hand, as the foundations of superstructures, the collapse of basements would lead to a severe damage of upper high-rise buildings.

Therefore, focusing on the seismic responses of basements which are connected with high-rise superstructures and embedded into the soil layers, the research objective of this study is to figure out the mechanisms and the characteristics of dynamic performances of basements. Specifically, the mechanisms of seismic responses of basements are revealed by theoretical methods, and regularities founded in theoretical analysis as well as the characteristics of dynamic performances of basements are investigated in 3D numerical simulation.

2 Dynamic Soil-Basement-Structure Interaction Analysis

2.1 Dynamic Governing Equations

The high-rise building can be divided into over-ground superstructure (S) and underground basement according to ground surface, as shown in Fig. 1.

The dynamic governing equation of soil-basement-superstructure system under an earthquake action in frequency-domain is given by:

$$\left\{ \begin{array}{ccc} [S_{SS}] & [S_{SI}] & [0] \\ [S_{IS}] & [S_{II}] & [S_{IB}] \\ [0] & [S_{BI}] & [S_{BB}] \end{array} \right\} \left\{ \begin{array}{c} \{u_S\} \\ \{u_I\} \\ \{u_B\} \end{array} \right\} = \left\{ \begin{array}{c} \{P_S\} \\ \{P_I\} \\ \{P_B\} \end{array} \right\} \quad (1)$$

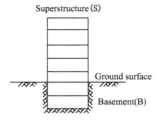

Fig. 1. Dynamic soil-basement-superstructure interaction (SBSI) model.

where subscript I denotes the boundary interface between superstructure and basement; Vector $\{P\}$ is the external load on analysis model; Matrix $[S]$ represents the dynamic stiffness in frequency-domain,

$$[S] = [K] + i\omega[C] - \omega^2[M] \qquad (2)$$

where matrixes $[K]$, $[C]$ and $[M]$ express the mass, damping and stiffness, respectively. The only external load is the dynamic earth pressure on the basement, which is given by:

$$\{P_B\} = -\left[S_{BB}^D\right]([u_B] - \left[u_B^D\right]) \qquad (3)$$

where D represents a cavity field, and matrix $\left[S_{BB}^D\right]$ denotes the ground impedance around the basement [14].

$$\left[S_{BB}^D\right] = \left[K_{BB}^D\right] + i\omega\left[C_{BB}^D\right] \qquad (4)$$

According to Eq. (1) and (3), we obtain,

$$\begin{Bmatrix} [S_{SS}] & [S_{SI}] & [0] \\ [S_{IS}] & [S_{II}] & [S_{IB}] \\ [0] & [S_{BI}] & [S_{BB}] + \left[S_{BB}^D\right] \end{Bmatrix} \begin{Bmatrix} \{u_S\} \\ \{u_I\} \\ \{u_B\} \end{Bmatrix} = \begin{Bmatrix} 0 \\ 0 \\ \left[S_{BB}^D\right]\left[u_B^D\right] \end{Bmatrix} \qquad (5)$$

Then, as shown in Fig. 2, if we replace the basement with a same volume soil, Eq. (3) would be changed to [15]:

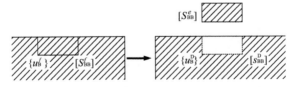

Fig. 2. Replacing the basement with a same volume soil.

$$\{P_B^f\} = \left[S_{BB}^e\right]\{u_B^f\} = -\left[S_{BB}^D\right](\{u_B^f\} - \{u_B^D\}) \qquad (6)$$

$$[S_{BB}^D]\{u_B^D\} = ([S_{BB}^e] + [S_{BB}^D])\{u_B^f\} = [S_{BB}^f]\{u_B^f\} \qquad (7)$$

in which, superscripts e and f denote the original soil with a same volume of basement and free-field, respectively. So, the Eq. (7) can be simplified as

$$\begin{cases} [S_{SS}] & [S_{SI}] & [0] \\ [S_{IS}] & [S_{II}] & [S_{IB}] \\ [0] & [S_{BI}] & [S_{BB}] + [S_{BB}^D] \end{cases} \begin{Bmatrix} \{u_S\} \\ \{u_I\} \\ \{u_B\} \end{Bmatrix} = \begin{Bmatrix} 0 \\ 0 \\ [S_{BB}^f]\{u_B^f\} \end{Bmatrix} \qquad (8)$$

According to the first and third formulas in Eq. (8), the seismic response of basement is given by

$$\{u_B\} = [S_{BB}^f]\{u_B^f\}([S_{BB}] + [S_{BB}^D])^{-1} + [S_{BI}][S_{SI}]^{-1}[S_{SS}]\{u_S\}([S_{BB}] + [S_{BB}^D])^{-1} \qquad (9)$$

Obviously, the seismic response of basement can be divided as two parts,

$$\{u_B\} = \{u_B\}^f + \{u_B\}^S \qquad (10)$$

in which,

$$\begin{cases} \{u_B\}^f = [S_{BB}^f]\{u_B^f\}([S_{BB}] + [S_{BB}^D])^{-1} \\ \{u_B\}^S = [S_{BI}][S_{SI}]^{-1}[S_{SS}]\{u_S\}([S_{BB}] + [S_{BB}^D])^{-1} \end{cases} \qquad (11)$$

Therefore, the seismic responses of basements which are connected with high-rise superstructures and embedded into the soil layer, can be divided into two parts: the kinematic interaction effect $\{u_B\}^f$ induced by the motion of field $\{u_B^f\}$; and the inertial interaction effect $\{u_B\}^S$ existential for the vibration of superstructure $\{u_S\}$, as shown in Fig. 3.

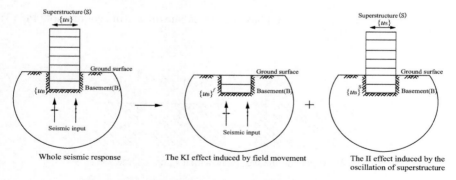

Fig. 3. Decomposition of the dynamic response of basement

2.2 Characteristic Analysis

It should be noted that the dynamic stiffness matrixes $\left[S_{BB}^f\right]$, $[S_{BB}]$, $\left[S_{BB}^D\right]$ would be constant matrixes after the site and the structure are determined. Then, the kinematic interaction (KI) effect of basement is determined by the seismic response of free-field, $\left\{u_B^f\right\}$.

In addition, due to the product of $[S_{SS}]$ and $\{u_S\}$ is the inertial force of superstructure produced during earthquake, the external load on basement induced by the inertial force of superstructure, $\{P\}_B^S$, can be given by,

$$\{P\}_B^S = [S_{BI}][S_{SI}]^{-1}[S_{SS}]\{u_S\} \tag{12}$$

Then, the inertial interaction (II) effect can be rewriten by,

$$\{u_B\}^S = \{P\}_B^S \left([S_{BB}] + \left[S_{BB}^D\right]\right)^{-1} \tag{13}$$

Therefore, inertial interaction (II) effect of building basement is determined by $\{P\}_B^S$, after the site and the structure are determined.

According to the above characteristic analysis, we simplify the theoretical formulas of KI and II effects of basement as follows,

$$\begin{cases} \{u_B\}^f = \alpha\left\{u_B^f\right\} \\ \{u_B\}^S = \beta\{P\}_B^S \end{cases} \tag{14}$$

In which, parameters α and β are the kinematic interaction coefficient and inertia interaction coefficient, which are all depended on the dynamic stiffness of site and basement.

3 Numerical Analysis

3.1 Finite Element Model

A typical high-rise building with basement is selected as case to verify the results acquired in theoretical analysis and illustrate KI and II effects on seismic responses of basements. Three-dimensional schematic diagram of basement as well as ground frame construction plane of superstructure are shown in Figs. 4 and 5, respectively. The mechanical and physical properties of basement and superstructure are presented in Table 1.

The numerical simulation is conducted in commercial software, i.e., ABAQUS. In which, the beams and columns of basement and superstructure are simulated by linear beam elements (b31) and the other components are simulated by reduced integral shell element (s4r). Elastic constitutive model is utilized to simulated the stress-strain relationship of structures. Table 2 presents the physical properties of ground and underground structures.

Table 1. Geometrical properties of basement and superstructure

Structure	Roof slab (m)	Medium slab (m)	Base slab (m)	Side wall (m)	Longitudinal beam (m)	Cross beam (m)	Column (m)
Basement	0.5	0.35	0.7	0.6	0.6 × 1.5	0.5 × 1.2	0.8 × 0.8
Superstructure	0.4	0.3	/	0.4	0.5 × 1	0.4 × 0.8	0.6 × 0.8

Fig. 4. Three-dimensional model of the basement

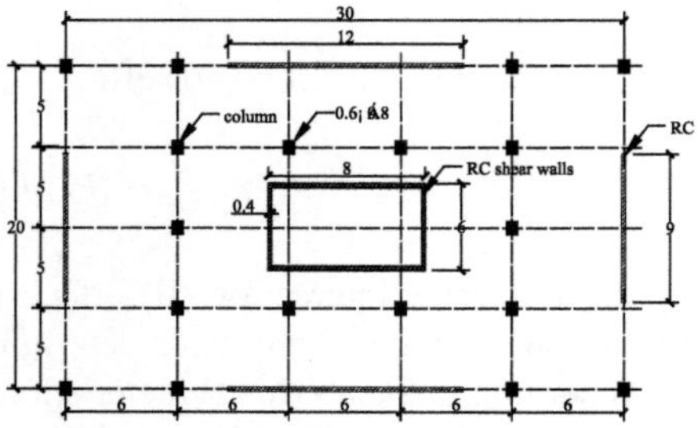

Fig. 5. Plan of the superstructure (unit: m).

Table 2. The physical properties of structures

Components	Concrete number	Density (kg/m^3)	Poisson's ratio	Young's modulus (MPa)
Column	C45	2600	0.2	33500
Others	C35	2600	0.2	31500

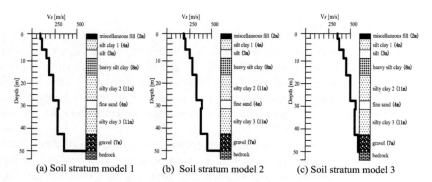

(a) Soil stratum model 1   (b) Soil stratum model 2   (c) Soil stratum model 3

Fig. 6. Shear-wave velocity profile and stratigraphic profile

The shear-wave velocity profiles and stratigraphic profiles of soil stratum models in the case study are presented in Fig. 6. The soil element is simulated by eight nodes linear hexahedron element (c3d8) and it is considered as a continuous, isotropic, and nonlinear medium. The equivalent-linear dynamic constitutive model [16] is adopted in simulation to reflect the nonlinear characteristic of soil. Specifically, the stiffness degradation of soil under a dynamic load is achieved by iterative method. The principle and property of the equivalent-linear dynamic constitutive model are introduced by Qiu et al. [17]. Three typical soils shear modulus degradation and damping curves acquired by Seed and Idriss [18], as shown in Fig. 7, are adopted in this simulation, and Table 3 present the material parameters of each soil layer.

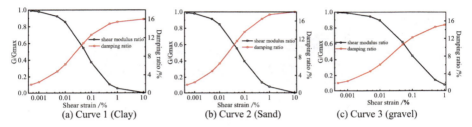

(a) Curve 1 (Clay) (b) Curve 2 (Sand) (c) Curve 3 (gravel)

Fig. 7. Shear modulus and damping ratio versus shear strain curves.

Table 3. Physical properties of the soil layer

Soil type	Density (kg/m^3)	Poisson's ratio	Degradation curves type
Miscellaneous fill	1900	0.44	Curves 2
Silty clay 1	1950	0.4	Curves 1
Silt	1950	0.42	Curves 2
Heavy silt clay	2050	0.38	Curves 1
Silty clay 2	2100	0.4	Curves 1
Fine sand	2150	0.37	Curves 2
Silty clay 3	2150	0.4	Curves 1
Gravel	2150	0.31	Curves 3

3.2 Boundary Conditions, Seismic Motion Inputs and Other Settings

Under the premise of both computational efficiency and accuracy, the tied degrees of freedom (TDOF) boundary is applied in model [19]. A fixed contact condition with no slip between soil and the underground structure of high-rise building is assumed. In addition, coupling contact conditions are assumed between the beams, columns and the walls they connect to. According to the studies of Abate et al. [20], in order to ensure the effectively reflect the propagation of seismic wave, the mesh element of soil should be chosen in order to be 1/6–1/8 of the minimum wavelength in simulation. Therefore, the mesh element of soil layer near the ground are more refined for its smaller shear wave velocity. In addition, a more refined mash is set around the high-rise building in order to obtain a more accurate response of building basement. As for the element of beam and shell, the sizes of them are 0.2 m and 0.2 m × 0.2 m, respectively.

A 3D dynamic finite element model (as shown in Fig. 8 and 9) with 530 m in X direction, 250 m in Z direction and thickness 50 m (Y direction) is established according to aforesaid settings and processes.

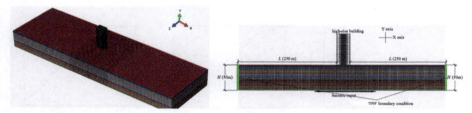

Fig. 8. Mesh of the 3D finite element model.

Fig. 9. Settings in finite element model.

Four natural earthquakes and an artificial wave are selected as the input motions. The original acceleration histories of earthquakes are shown in Fig. 10. In the simulation, the amplitude of each input wave is adjusted to a same value (i.e., 0.1g, g is acceleration of gravity), and the seismic accelerations are applied at the bottom of the model in X direction, as shown in Fig. 9. The elastic response spectras of five earthquake waves (after the adjustment) are presented in Fig. 10 (**f**).

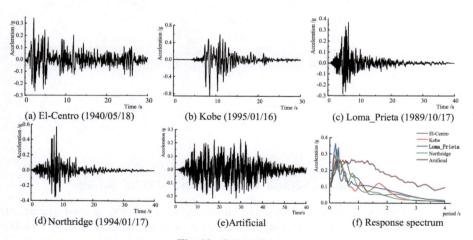

Fig. 10. Seismic inputs

4 Results and Discussion

4.1 Seismic Characteristics of Building Basement

Figure 11 present the displacement difference time-history of the basement with a high-rise superstructure (the seismic input is Artificial wave) and the Fourier spectrum curve calculated by fast Fourier transform (FFT) algorithm. For comparison, Fig. 12 presents the results of single basement, i.e., the basement without superstructure.

It can be seen from the figures that: 1) The presence of superstructure would augment the dynamic response of basement (the maximum displacement differences of basements enlarger by 1.497 times, 1.514 times and 1.562 times for soil stratum models 1, 2 and 3); 2) There are two peaks in spectrum curve of seismic response of the basement with a high-rise superstructure, while that is one peak for the basement without superstructure. The presence of superstructure would significantly augment the seismic response of basement at the first-order natural frequency of superstructure. Obviously, this is the II effect of superstructure on the seismic response of basement; 3) There is a peak in spectrum curve of seismic response of the basement when the transverse axis approaches to the natural frequency of site, whatever the superstructure exists or not. Correspondingly, this is the KI effect of surrounding field on the seismic response of basement. The presence of superstructure has less effect on the value of this peak, i.e., the II effect do not affect the value of KI effect.

Obviously, it is necessary for seismic design of building basements to comprehensively consider the KI and II effects because both of them have great influences on dynamic performances of basements.

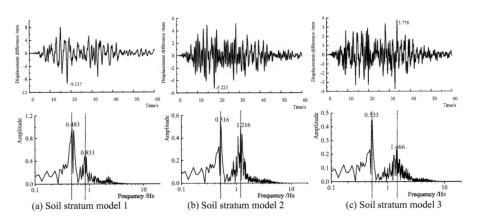

Fig. 11. Seismic response of the basement with a high-rise superstructure.

1210   Y.-J. Qiu et al.

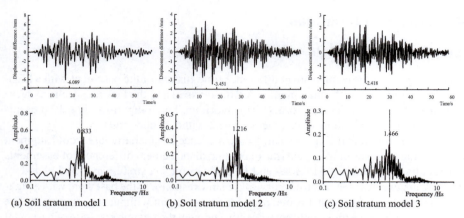

(a) Soil stratum model 1   (b) Soil stratum model 2   (c) Soil stratum model 3

Fig. 12. Seismic response of the single basement (without superstructure).

4.2 Analyses on the KI Effect and Kinematic Interaction Coefficient

Figure 13 compares the maximum displacement differences of basement induced by the KI effect under different calculation conditions. As shown in Figure that the seismic responses have significant difference under various input earthquakes, especially in soil stratum model 1. The results under Artificial wave are always larger than those under other earthquakes, because the acceleration response spectra is bigger than other seismic inputs at most frequencies. In addition, the KI effect would be reduced with the site getting harder (the average KI effects are 5.244, 3.110, and 2.002 mm for soil stratum models 1to 3, respectively).

The KI effect is highly affected by the frequency property of earthquakes. However, regular would be different when we investigate the ratio of KI effect to free-field response. Figure 14 compares the time-histories of the KI effect and the free-field response under Kobe wave. It shows that the trend of the KI effect is basically consistent to the free-field response, and the KI effect can be obtained by the product of the free-field response and a similarity ratio. Certainly, this similarity ratio is the kinematic interaction coefficient. As

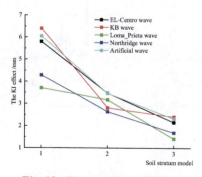

Fig. 13. The maximum KI effect

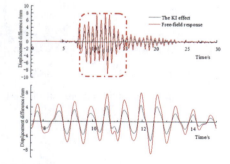

Fig. 14. Time-histories of the KI effect and free-field response

for other earthquakes, Table 4 summaries the maximum displacement difference of the soil within the region of roof and floor of basement in the free-field condition, i.e., $\{u_B^f\}$, and the ratios of the KI effects to free-field responses are presented in Fig. 15. Although the absolute values of the KI effect and free-field responses under various earthquakes are different, the radios of two responses are always constant. Moreover, the kinematic interaction coefficient would be enlarged with the soil stratum becoming harder.

Table 4. The maximum displacement differences of free-field

Seismic input	Soil stratum model		
	Model 1 (mm)	Model 2 (mm)	Model 3(mm)
El-Centro wave	9.7	3.41	1.51
Kobe wave	10.79	2.74	1.71
Loma_Prieta wave	6.22	3.1	1.03
Northridge wave	7.24	2.57	1.23
Artificial wave	10.33	3.37	1.75

4.3 Analyses on the II Effect and the Inertial Interaction Coefficient

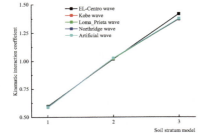

Fig. 15. Kinematic interaction coefficient analysis

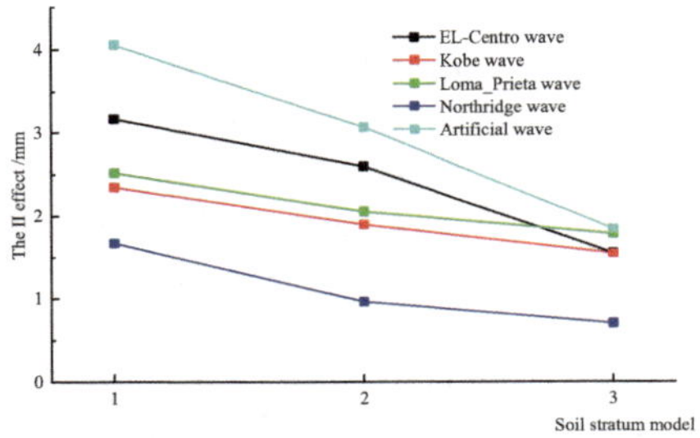

Fig. 16. The maximum II effects

Figure 16 presents the maximum displacement difference of basement produced by the vibration of superstructure (i.e., the II effect). The II effects are significant different under the various seismic signals. In which, the results under artificial wave are always bigger than others, too.

Evidently, the absolute value of the II effect is also affected by the frequency properties of earthquakes. Then, Fig. 17 compares the time histories of II effect and the seismic shear on basement transmitted from superstructure. The trend of the II effect is basically consistent to the seismic force on the basement. Therefore, the II effect can be obtained by the product of the seismic force on the basement transmitted from superstructure and a flexibility coefficient. This flexibility coefficient is the inertial interaction coefficient.

As for other earthquakes, the results are summarized in Fig. 18. Same to the characteristics of the kinematic interaction coefficient, although the absolute values of the II effect under various earthquakes are different, the inertial interaction coefficients are always constants. With the soil stratum becoming harder, the inertial interaction coefficient would be lessened (the inertial interaction coefficients equal to 0.293 mm/MN, 0.149 mm/MN, and 0.098 mm/MN for soil models 1 to 3, respectively).

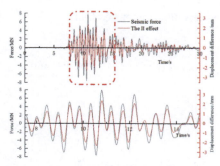

Fig. 17. Time-histories of the II effect and seismic force

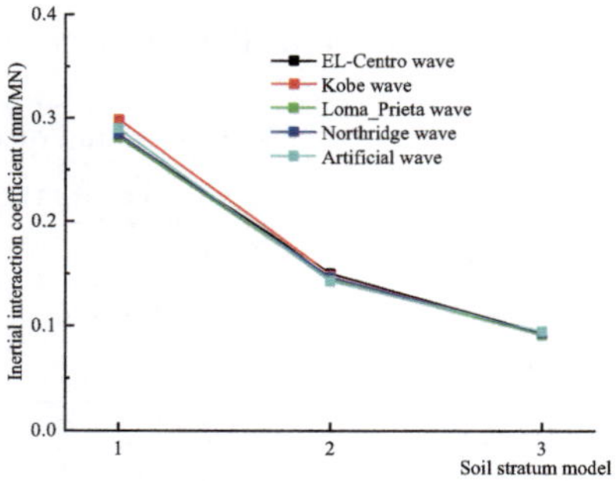

Fig. 18. Inertial interaction coefficient

5   Conclusion

1) The seismic response of the basement with high-rise superstructure can be distinguished as the KI and II effects. In which, the KI effect is induced by the field movement, and the II effect is induced by the vibration of superstructure;
2) Due to the presence of superstructure would significantly augment the seismic response of basement at the first-order natural frequency of superstructure, the dynamic response of the basement with a high-rise superstructure would be amplified comparing to the single basement, e.g., the maximum displacement differences of basement enlarger by 1.497 times, 1.514 times and 1.562 times for soil stratum models 1, 2 and 3, respectively;
3) There is larger KI effect on the seismic response of basement whatever the superstructure exists or not. In addition, the presence of II effect has less effect on the value of the KI effect;
4) The KI effect on the seismic response of basement equal to the product of free-field response and kinematic interaction coefficient, and the II effect equal to the product of the seismic force on the basement transmitted from superstructure and inertial interaction coefficient. With the harder of soil stratum model becoming, the kinematic interaction coefficient is enlarged while the inertia interaction coefficient is lessened.

It is necessary for seismic design of building basements to comprehensively consider the KI and II effects because both of them have great influences on dynamic performances of basements. This paper has investigated the characteristics of KI and II effects, and proved that two effects can be acquired by simplified methods.

Acknowledgment. The research of this paper is supported by National Natural Science Foundation of China (52078033).

References

1. Nasab, M.S.E., Chun, S., Kim, J.: Soil-structure interaction effect on seismic retrofit of a soft first-story structure. Structures **32**, 1553–1564 (2021)
2. Taghizadeh, M., Gholhaki, M., Rezaifar, O.: A study on effect of soil-structure interaction on performance of strong-back structural system subjected to near and far-field earthquakes. Structures **32**, 116–136 (2021)
3. Tabatabaiefar, H.R., Fatahi, B.: Idealisation of soil–structure system to determine inelastic seismic response of mid-rise building frames. Soil Dyn. Earthq. Eng. **66**, 339–351 (2014)
4. Kuzniar, K., Tatara, T.: The ratio of response spectra from seismic-type free-field and building foundation vibrations: the influence of rockburst parameters and simple models of kinematic soil-structure interaction. Bull. Earthq. Eng. **18**(3), 907–924 (2020)
5. Mylonakis, G., Gazetas, G.: Seismic soil-structure interaction: beneficial or detrimental? J. Earthq. Eng. **4**(3), 277–301 (2000)
6. Van Nguyen, D., Kim, D., Nguyen, D.D.: Nonlinear seismic soil-structure interaction analysis of nuclear reactor building considering the effect of earthquake frequency content. Structures **26**, 901–914 (2020)
7. Bilotta, E., De Sanctis, L., Di Laora, R., D'onofrio, A., Silvestri, F.: Importance of seismic site response and soil-structure interaction in dynamic behaviour of a tall building. Géotechnique **65**(5), 391–400 (2015)
8. Han, Y.: Seismic response of tall building considering soil-pile-structure interaction. Earthq. Eng. Eng. Vib. **1**(1), 57–64 (2002)
9. NEHRP Consultants Joint Venture. Soil-structure interaction for building structures NIST report 2012; 12-917-21, Washington, D.C
10. Lee, D.G., Kim, H.S.: Efficient seismic analysis of high-rise buildings considering the basements. In: Proceedings of NZSEE c2001 (2001)
11. Scarfone, R., Morigi, M., Conti, R.: Assessment of dynamic soil-structure interaction effects for tall buildings: a 3D numerical approach. Soil Dyn. Earthq. Eng. **128**, 105864 (2020)
12. Borghei, A., Ghayoomi, M.: The role of kinematic interaction on measured seismic response of soil-foundation-structure systems. Soil Dyn. Earthq. Eng. **125**, 105674 (2019)
13. Kim, S., Stewart, J.P.: Kinematic soil-structure interaction from strong motion recordings. J. Geotech. Geoenviron. Eng. **129**, 323–335 (2003)
14. Akira, T.: A study on seismic analysis methods in the cross section of underground structures using static finite element method. Struct. Eng. Earthq. Eng. **122**(1), 41–54 (2005)
15. Wolf, J.P.: Dynamic Soil-Structure-Interaction. Prentice-Hall, Englewood Cliffs (1985)
16. Kramer, S.L., Paulsen, S.B.: Practical use of geotechnical site response models. In: Proceedings of International Workshop on Uncertainties in Nonlinear Soil Properties and Their Impact on Modeling Dynamic Soil Response, p. 10. Univ. of California, Berkeley (2004)
17. Qiu, Y.J., Zhang, H.R., Yu, Z.Y., et al.: A modified simplified analysis method to evaluate seismic responses of subway stations considering the inertial interaction effect of adjacent buildings. Soil Dyn. Earthq. Eng. **150**, 106896 (2021)

18. Seed, H.B., Idriss, I.M.: Soil moduli and damping factors for dynamic response analyses. Rep. No. EERC-70/10, Earthquake Engineering Research Center, Univ. of California at Berkeley, Berkeley, CA (1970)
19. Li, W., Chen, Q.: Effect of vertical ground motions and overburden depth on the seismic responses of large underground structures. Eng. Struct. **205**, 110073 (2020)
20. Abate, G., Massimino, M.R.: Numerical modelling of the seismic response of a tunnel–soil–aboveground building system in Catania (Italy). Bull. Earthq. Eng. **15**(1), 469–491 (2016). https://doi.org/10.1007/s10518-016-9973-9

Centrifuge and Numerical Simulation of Offshore Wind Turbine Suction Bucket Foundation Seismic Response in Inclined Liquefiable Ground

Xue-Qian Qu[1], Rui Wang[1], Jian-Min Zhang[1], and Ben He[2(✉)]

[1] Tsinghua University, Beijing 100084, China
[2] Power China Huadong Engineering Corporation Limited, Hangzhou 311122, China
he_b2@ecidi.com

Abstract. With the rapid development of offshore wind farms, suction bucket foundation is gaining attention due to installation convenience, cost efficiency, and reusability. Currently, increasing number of offshore wind turbines are built in seismically active areas, the seismic stability of bucket foundation is becoming a major concern, especially in liquefiable soil. By comparing the results of centrifuge shaking table test with solid-fluid coupled dynamic finite difference analysis, the seismic response of suction bucket foundations in liquefiable ground is analyzed. A unified plasticity model for the large post-liquefaction shear deformation of sand is used to provide high fidelity representation of the behavior of saturated sand under seismic loading. The soil-bucket-superstructure dynamic interaction mechanism is investigated, via analysis of the rotation and translation of the bucket foundation and the acceleration and displacement of the superstructure. The results show that excess inclination angle of the bucket foundation and amplified turbine response during the earthquake in the inclined liquefiable foundation should be considered in design.

Keywords: Centrifuge modelling · Offshore wind turbine · Bucket foundation · Earthquake loading · Liquefiable soil · Soil-bucket-superstructure interaction

1 Introduction

Offshore wind turbines have been planned or built in seismic activity zones (Asareh et al. 2016; Smith et al. 2016; Kaynia et al. 2019). The suction bucket foundation, which has the advantages of reusability, large bearing capacity and low cost, have been increasingly used for offshore wind turbines. Liquefaction of seabed soil may occur under earthquakes, which can cause damage to offshore wind turbine suction bucket foundations. Especially in mildly inclined ground, where offshore wind turbines are often founded on (Wen 2016).

Research on the seismic response of offshore wind turbine bucket foundation has been carried out using both centrifuge tests (Bhattacharya et al. 2013; Yu et al.

2014; Wang et al. 2017; Ueda et al. 2020; Qu et al. 2021) and numerical simulation (Asheghabadi et al. 2019; Esfeh et al. 2020). However, study on the seismic response of wind turbine bucket foundations in mildly inclined liquefiable ground is limited.

In this study, the seismic behavior of offshore wind turbine bucket foundations in mildly inclined liquefiable ground are analyzed via centrifuge shaking table test and numerical simulation.

2 Centrifuge Shaking Table Test

A centrifuge test was performed in this study under 40 g centrifugal acceleration. The model layout is plotted in Fig. 1. The wind turbine model consisted of bucket foundation, turbine tower, and a lumped mass used to simulate the nacelle and blades. The wind turbine model is made of stainless steel with Poisson's ratio of 0.3, Young's modulus of 210 GPa and density of 7.9×10^3 kg/m^3. Chinese Fujian standard sand with 44% relative density and 1.61×10^3 kg/m^3 dry density was used. Sand was saturated using 0.35% hydroxyl propyl methyl cellulose (HPMC) solution for consistency in modelling scales for time. The basic properties of the standard sand are listed in Table 1. The inclination angle of the ground soil was 0.85°.

Table 1. Basic properties of the standard sand in test.

Maximum dry density ρ_{dmax} (kg/m^3)	Minimum dry density ρ_{dmin} (kg/m^3)	Specific gravity Gs	Permeability coefficient k (m/s)
1.81×10^3	1.48×10^3	2.65	1.25×10^{-3}

For consistency, all measurements mentioned in this study are in prototype scale. The input Parkfield seismic motion is shown in Fig. 2.

3 Numerical Analysis

Numerical simulation of the centrifuge shaking table test is achieved using the finite difference code FLAC^{3D}. A unified plasticity model for the large post-liquefaction shear deformation of sand (CycLiq) developed by Wang et al. (2014) is used to simulate the ground soil in this study. The constitutive model can simulate the mechanical behavior of sand in different states from pre- to post-liquefaction under cyclic loading, which has been implemented in FLAC^{3D} (Zou et al. 2020). The model has 14 parameters, as listed in Table 2. The parameters used in this study modified within a reasonable range based on relevant research (Wang et al. 2014; Zou et al. 2020). The simulation results of a typical undrained cyclic torsional shear test of the sand using the constitutive model with the parameters are shown in Fig. 3 (a) and (b).

The numerical model is shown in Fig. 4. The boundary nodes at the left and right sides of the same height are tied to simulate free field conditions The pore pressure of the soil surface is set to 0 for simulation free drainage condition. The bucket foundation

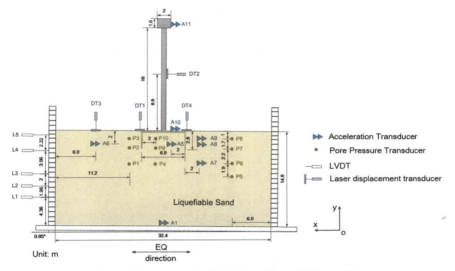

Fig. 1. Model cross-section for the centrifuge shaking table test.

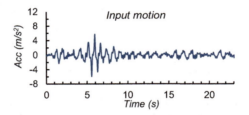

Fig. 2. The acceleration time history of input motion in centrifuge shaking table test.

Table 2. Constitutive model parameters used in the numerical analysis.

G_0	κ	h	$d_{re,1}$	$d_{re,2}$	d_{ir}	α	$\gamma_{d,r}$	n^p	n^d	M	λ_c	e_0	ξ
200	0.008	1.8	0.35	30	0.75	10	0.05	1.1	7.8	1.35	0.019	0.934	0.7

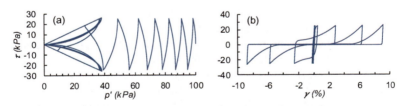

Fig. 3. The simulation results of a typical undrained cyclic torsional shear test of the sand: (a) effective stress path; (b) stress-strain relation.

is simulated by shell elements with Young's modulus of 210 GPa, Poisson's ratio of 0.3. To simulate the interface between the bucket foundation and the soil, interface with the friction angle of 15° is set. Turbine tower and lumped mass are modeled using linear elasticity with bulk modulus of 175 GPa, Shear modulus of 80.8 GPa.

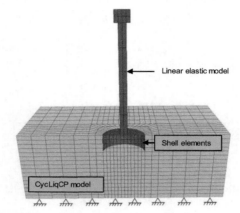

Fig. 4. Numerical dynamic analysis model in FLAC^{3D}.

4 Results and Discussion

Figure 5 presents the acceleration and excess pore pressure time histories at various locations within the model from both the centrifuge test and numerical simulation. Good overall agreement is achieved for both acceleration and excess pore pressure. This shows the effectiveness of using the model and numerical platform in performing such simulation.

Compared with the acceleration of the near field soil and far field soil, the bucket skirt causes reflection seismic wave in the bucket, resulting in the increase of acceleration amplitude of the inside-bucket soil, as shown in Fig. 5 (b). The acceleration amplitude of soil near the soil surface is smaller than the input motion, which is caused by the softening of soil. Due to the influence of the vertical load provided by the turbine weight, the excess pore pressure of the inside-bucket soil is greater than of the near field soil and far field soil at the same depth, visible in Fig. 5 (d)–(f).

Simulations of the soil and bucket displacement also shows good agreement with test results. The horizontal displacement of the soil surface and bucket foundation are similar (Fig. 6 (a) and (c)). The accumulation of horizontal displacement is mostly due to the inclination of the ground. The peak and residual settlement of bucket foundation are greater than that at the soil surface (Fig. 6 (b) and (d)), with the bucket foundation settlement being significantly greater. Rotation of the bucket foundation is also observed in Fig. 6 (e).

The horizontal displacement and inclination of the bucket may lead to the failure of the wind turbine during the earthquake. In Fig. 6 (c), the test result shows that the peak

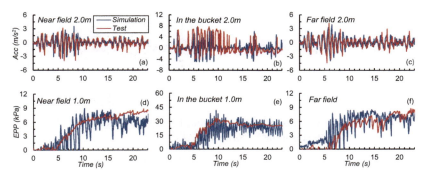

Fig. 5. The test and simulation results of acceleration time histories at: (a) near field 2.0m depth; (b) in the bucket 2.0 m depth; (c) far field 2.0 m depth, and excess pore pressure time histories at: (d) near field 1.0 m depth; (e) in the bucket 1.0 m depth; (f) far field 1.0 m depth.

horizontal displacement of the bucket foundation is smaller than the failure horizontal displacement (Hesar et al. 2003). The numerical and test peak inclination angles of bucket foundation are 0.63° and 0.49°, respectively, which are both greater than the design restrictions in China (FD-003 2007) and in UK (Peire et al. 2009), as shown in Fig. 6 (e). It can be seen that excess foundation inclination caused by the seismic load in liquefiable ground should be considered in design.

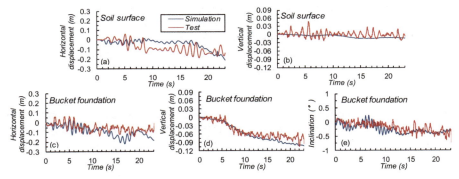

Fig. 6. The test and simulation results of soil surface response time histories: (a) horizontal displacement; (b) vertical displacement, and bucket foundation response time histories of: (c) horizontal displacement; (d) vertical displacement; (e) inclination.

The numerical and test peak accelerations of wind turbine are 2.96 m/s² and 3.71 m/s² respectively, which are slightly amplified than that of the soil surface (2.83 m/s² in simulation and 3.23 m/s² in test) before 15 s, as shown in Fig. 7. It is worth noting that the accelerations of the soil surface exhibit significant spikes after 15 s, especially in centrifuge test. The numerical and test maximum accelerations of wind turbine are 2.66 m/s² and 5.47 m/s² respectively, which are 2.0 and 4.1 times of maximum input motion after 15 s, respectively. This phenomenon is caused by the dilatancy of the soil, as analyzed by Qu et al. (2021).

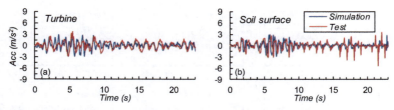

Fig. 7. Numerical and test results of acceleration time histories of the turbine and at soil surface: (a) turbine acceleration; (b) soil surface acceleration.

5 Conclusions

In this study, the seismic response of offshore wind turbine bucket foundation in inclined liquefiable ground is investigated via numerical simulation and centrifuge shaking table test. The results can be summarized as follows:

1. The CycLiq model and solid-fluid coupled dynamic finite difference analysis method can reflect the dynamic response of the ground soil and offshore wind turbine bucket foundation.
2. The excess pore pressure of the inside-bucket soil is greater than of the near field soil and far field soil at the same depth due to the vertical load provided by the turbine weight. The bucket skirt causes reflection of the seismic wave in the bucket, resulting in the increase of acceleration amplitude of the inside-bucket soil.
3. Excess inclination of the bucket foundation due to earthquake caused by the inclined liquefiable ground should be considered in design.
4. The acceleration of the wind turbine may be amplified during the earthquake and further analysis should be conducted.

Acknowledgement. The authors gratefully acknowledge the financial support provided by National Natural Science Foundation of China (51909249) and Zhejiang Provincial Natural Science Foundation (LQ19E090001).

References

1. Asareh, M.A., Schonberg, W., Volz, J.: Effects of seismic and aerodynamic load interaction on structural dynamic response of multi-megawatt utility scale horizontal axis wind turbines. Renewable Energy **86**, 49–58 (2016)
2. Asheghabadi, M.S., Jebeli, A.J.: Seismic behavior of suction Caisson foundations. Int. J. Geotech. Geol. Eng. **13**(2), 30–36 (2019)
3. Bhattacharya, S., Cox, J.A., Lombardi, D., Wood, D.M.: Dynamics of offshore wind turbines supported on two foundations. P I Civil Eng-Geote. **166**(2), 159–169 (2013)
4. Esfeh, P.K., Kaynia, A.M.: Earthquake response of monopiles and Caissons for offshore wind turbines founded in liquefiable soil. Soil Dyn. Earthq. Eng. **136**, 106213 (2020)
5. FD-003.: Design regulations on subgrade and foundation for wind turbine generator system. China Water Power Press, Beijing (2007)

6. Hesar, M.: Geotechnical design of the Barracuda and Caratinga suction anchors. In: Offshore Technology Conference (2003)
7. Kaynia, A.M.: Seismic considerations in design of offshore wind turbines. Soil Dyn. Earthq. Eng. **124**, 399–407 (2019)
8. Peire, K., Nonneman, H., Bosschem, E.: Gravity base foundations for the thornton bank offshore wind farm. Terra et Aqua **115**(115), 19–29 (2009)
9. Qu, X.Q., Zhang, Z.T., Hu, J., Wang, R., Zhang, J.M.: Centrifuge shaking table tests on offshore wind turbine bucket foundation in mildly inclined liquefiable seabed. Soil Dyn. Earthq. Eng. **151**, 107012 (2021)
10. Smith, V., Mahmoud, H.: Multihazard assessment of wind turbine towers under simultaneous application of wind, operation, and seismic loads. J. Perform. Constr. Facil. **30**(6), 04016043 (2016)
11. Ueda, K., Uzuoka, R., Iai, S., Okamura, T.: Centrifuge model tests and effective stress analyses of offshore wind turbine systems with a suction bucket foundation subject to seismic load. Soils Found. **60**(6), 1546–1569 (2020)
12. Wang, R., Zhang, J.M., Wang, G.: A unifified plasticity model for large post-liquefaction shear deformation of sand. Comput Geotech. **59**, 54–66 (2014)
13. Wang, X.F., Yang, X., Zeng, X.W.: Seismic centrifuge modelling of suction bucket foundation for offshore wind turbine. Renewable Energy **114**, 1013–1022 (2017)
14. Wen, F.: Developments and characteristics of offshore wind farms in China. Adv. New Renewable Enengy **4**(2), 152–158 (2016)
15. Xia, Z.F., Ye, G.L., Wang, J.H., Ye, B., Zhang, F.: Fully coupled numerical analysis of repeated shake-consolidation process of earth embankment on liquefiable foundation. Soil Dyn. Earthq. Eng. **30**(11), 1309–1318 (2010)
16. Yu, H., Zeng, X.W., Lian, J.J.: Seismic behavior of offshore wind turbine with suction caisson foundation. In: Geo-Congress, pp. 1206–1214 (2014)
17. Zou, Y.X., Zhang, J.M., Wang, R.: Seismic analysis of stone column improved liquefiable ground using a plasticity model for coarse-grained soil. Comput. Geotech. **125**, 103690 (2020)

Unconventional Retrofit Design of Bridge Pile Groups: Benefits and Limitations

L. Sakellariadis[✉], S. Alber, and I. Anastasopoulos

Department of Civil, Environmental and Geomatic Engineering, ETH, Zürich, Switzerland
lampros.sakellariadis@igt.baug.ethz.ch

Abstract. This paper investigates an example problem of bridge retrofit due to widening, focusing on the foundation. Conventional foundation retrofit is compared to an unconventional "do-nothing" approach, where the existing pile group is maintained, allowed to fully develop its moment capacity. The two systems are comparatively assessed employing the finite element (FE) method, accounting for all sources of material and geometric nonlinearities. The soil is modelled with the hypoplastic model for sand, while the Concrete Damaged Plasticity (CDP) model is used for all reinforced concrete (RC) members (piles and bridge pier). The soil-pile interface is modelled with frictional interface elements. Both foundation systems (retrofitted and existing) are initially subjected to pushover loading to comparatively assess their moment capacities. Subsequently, the widened bridge with and without foundation retrofit is subjected to dynamic time history analysis using excitations representative of Swiss seismicity. The performance of the two approaches is comparatively assessed, considering structural and geotechnical performance criteria. The paper provides useful insights regarding the benefits and limitations of unconventional retrofit design of piled foundations.

Keywords: Seismic design · Soil-structure interaction · Pile groups · Nonlinear analysis

1 Introduction

The study is motivated by the on-going upgrade of existing motorway infrastructure in Switzerland (and worldwide). One such example is the construction of additional traffic lanes in critical areas, aiming to enhance capacity of the roadway network and eliminate bottlenecks. Such process often involves widening of bridges, often founded on piles. Both the structure and the foundation need to be evaluated and retrofitted (if necessary), in order to accommodate the increased loads. Whereas pier retrofit is relatively straightforward, pile group strengthening can be a challenging, costly, and time-consuming operation, calling for optimized solutions. An unconventional "do-nothing" approach is explored, where the existing pile group foundation is maintained without any retrofit. In contrast to conventional design, under seismic loading the non-retrofitted pile group is allowed to sustain strongly nonlinear response, fully mobilizing its moment capacity. The energy dissipated through soil yielding that can potentially improve the seismic performance of the soil–foundation–structure system. Such a concept, termed "rocking isolation", was outlined by [2] and explored primarily for shallow footings, showing superior seismic performance.

The feasibility of applying such an approach to bridge pile groups is explored herein for dense sand implementing advanced numerical methods, which reproduce the non-linear pressure-dependent behavior of sand and the axial load-dependency of the reinforced concrete (RC) piles. A simplified yet representative prototype problem is outlined and used to comparatively assess the performance of an idealized widened bridge with two foundation alternatives: (a) retrofitted according to current seismic codes; and (b) un-retrofitted allowing nonlinear response. The benefits and limitations of the two approaches are critically discussed.

2 Problem Definition

The definition of the prototype problem follows an inverse procedure, briefly described. Based on the developed database of existing bridges in Switzerland, a fundamental yet representative pile group typology is selected, consisting of a 2 × 2 group of end-bearing bored piles of diameter $D = 1$ m, length $L = 15$ m, and spacing $s = 3D = 3$ m. Perth sand of relative density $D_r = 80\%$ is assumed, to allow model calibration and validation against soil element and centrifuge model tests conducted at ETH Zurich (ETHZ). A frictional soil-pile interface is assumed, calibrated after interface tests. Installation effects are not accounted for assuming the piles wished-in-place.

In most European countries, the majority of existing bridges were built before the 90's on the basis of global safety factors, with no (or transitional) seismic design, while their current upgrade requires seismic assessment according to current codes. To that end, based on the selected foundation, the dead load of the structural system prior to widening is selected to correspond to a representative static factor of safety, $FS_v = 3$. This requires calculation of the pile group bearing capacity, which is conducted using traditional bearing capacity theory:

$$V_{tip} = \pi D^2 N_q \sigma'_{v,tip} = 10.5 \, MN \tag{1}$$

$$V_{shaft} = \pi D K \tan\delta \int_0^{tip} \sigma'_v = 3.2 \, MN \tag{2}$$

where: N_q is the factor of [4]. The calculation uses the critical state friction angle $\varphi_{cv} = 29.6°$, derived after drained triaxial tests. The calculation of shaft resistance uses the earth pressure coefficient, $K = 0.7$, after [7] for bored piles, while the interface friction angle, $\delta = 37°$,, is based on the peak value derived from interface shear tests. The total dead load, W_{tot}, of the foundation and the structural system (simplified as SDOF), W_{tot} is selected so that:

$$FS_v = \frac{n(Q_{tip} + Q_{shaft})}{W_{tot}} = 3 \tag{3}$$

where: n is the number of piles. This calculation neglects group effects and the contribution of the pile cap (typical design assumption). The selected initial system (prior to widening is depicted in Fig. 1.

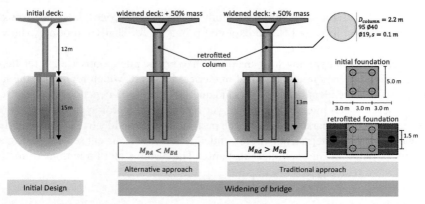

Fig. 1. Problem definition and overview of the studied problem; structural and foundation details.

2.1 Deck Widening Scenario

In order to consider deck widening, the concentrated mass representing the bridge deck is increased by 50%. The evaluation of structural and foundation elements and their potential retrofit design is performed following the current Swiss code, adopting partial safety factors. The critical design action is the overturning moment, M_{Ed}. The structural system satisfying the design requirements is presented in Fig. 1. With respect to the foundation, the moment capacity of the initial foundation ($M_{Rd,i}$) is calculated based on the procedures described by [7], treating the applied moment as eccentric vertical load undertaken by the piles via axial loading. The acting moment is larger than the capacity of the initial foundation ($M_{Rd,i} < M_{Ed}$), which is retrofitted accordingly. An overview of this evaluation is given in Table 1.

Table 1. Initial pile group and retrofit design for the seismic loading scenario

M_{Ed} (MNm)	$M_{Rd,i}$ (MNm)	$M_{Rd,r}$ (MNm)
35.1	26.3	40.1

3 Analysis Methodology

The two foundations are analysed employing the FE method. The key attributes of the FE model are shown in Fig. 2a. The RC pier and piles are modelled using the CDP model of [1]. Separate material laws are implemented for the confined core and the unconfined cover, describing the uniaxial stress-strain response under tension and compression after [8] and [4]. The longitudinal reinforcement is modelled using elastic-perfectly plastic surface elements, embedded in the concrete. The performance of the CDP model is assessed against the results of RC section analysis, confirming its capability to reproduce the axial force–bending moment interaction of RC (Fig. 2b).

The soil is modelled using the hypoplastic model, which is suitable to reproduce the nonlinear, pressure- and density-dependent behaviour of sand. Our study uses the version of [11], enhanced to incorporate the intergranular strain concept of [9]. The calibration of the 8 parameters of the basic model requires oedometer and triaxial compression tests, while the 5 extra parameters of the enhanced model used the results of cyclic simple shear tests (Table 2). The performance of the model are shown in Fig. 2c. The interfaces between the foundation and the surrounding soil are frictional - tensionless. The friction coefficient, $\mu = 0.75$, is selected on the basis of interface shear tests (adopting rough conditions). Finally the pile cap is modeled elastic, assuming the un-cracked modulus of RC. The numerical analysis methodology has been validated against monotonic and cyclic pushover tests performed at the ETHZ drum centrifuge (Fig. 3).

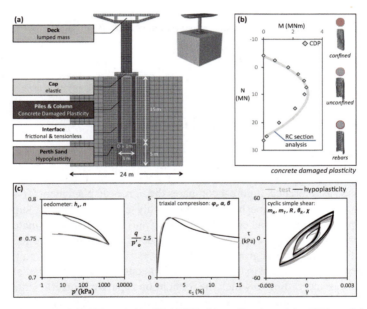

Fig. 2. (a) Overview of the FE model in ABAQUS; (b) verification of the CDP model against RC section analysis; and (c) calibration of the hypoplastic model against soil element tests

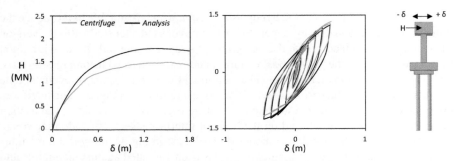

Fig. 3. Validation of numerical analysis against monotonic and cyclic pushover tests performed at the ETHZ drum centrifuge.

Table 2. Hypoplastic model parameters for Perth sand

φ_c (°)	h_s (MPa)	n	e_{d0}	e_{c0}	e_{i0}	a	β	m_R	m_T	R	β_R	χ
29.6	3644	0.44	0.47	0.82	0.98	0.14	0.25	5	3.5	10^{-4}	0.9	1

4 Retrofitted vs. Un-retrofitted Foundation

4.1 Static Loading

The two alternatives are initially assessed under static pushover loading. The fixed-base pier is also used as reference. Figure 4 compares the Force–displacement ($H - u$) results. Furthermore, the deformed mesh of the three analyses is shown, with superimposed contours of plastic points at the concrete members and the soil. The ultimate resistance is similar for all three cases, controlled by the structural capacity of the RC pier. This is expected for the retrofitted group, which complies with capacity design (foundation over-designed to guide failure to the structure), but contradicts to the calculations of the un-retrofitted case according to which the foundation capacity is insufficient, calling for retrofit. This result indicates the conservatism of current design practice.

Based on these results, retrofit is not justified since the same resistance is mobilized for both alternatives, however, considering the entire $H - u$ response, the differences become non-negligible. The retrofitted system is almost equally stiff to that the fixed-base pier. On the contrary, the un-retrofitted foundation is more flexible, experiencing more than double displacement at the ultimate state. The FE deformed plots of Fig. 4 show extensive strains on the soil, indicating highly nonlinear response.

The structural performance of the piles is also different. In the un-retrofitted case, the piles yield, tending to form plastic hinges at a certain depth. However, the failure of the pier prevents full formation of plastic hinges at the piles, or mobilisation of the pull-out resistance of the trailing pile. Thus, the moment resistance of the pile group is not fully reached. In the retrofitted case, the piles remain elastic, with the maximum bending moments developing at their connection to the cap. This is attributed to the increased rotational restraint due to the added pile row. Overall, under static loading, the "do-nothing" approach is proven feasible, provided that the deformations under the design seismic actions, remain within tolerable serviceability limits and the piles maintain their structural integrity. Plastic hinges are allowed by some modern seismic codes, but their acceptance is a matter of design decision.

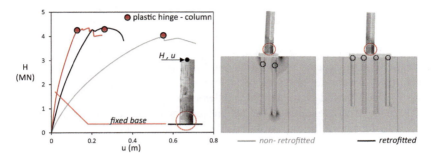

Fig. 4. Pushover analysis: H – u curves, deformed mesh with contours of plastic points.

4.2 Seismic Loading

The retrofitted and the un-retrofitted pile group are further subjected to nonlinear time history analyses. The analysis uses input excitations, compatible with the assumed Swiss seismicity. Figure 5 shows the target design used in this study. The selection of real seismic records is performed using Seismoselect [10], based on the NGA West 2 Ground motion Database (PEER). Due to the computational cost of 3D analyses, three different ground motions were selected. These are scaled on peak ground acceleration (PGA) automatically by Seismoselect, matching on average the target spectrum. Since the target spectrum corresponds to rock outcrop, the selected records were scaled by 50% and used as input applied at the base of the model.

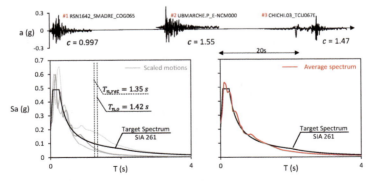

Fig. 5. Scaled time histories of input motions used for the analyses, along with elastic acceleration response spectra matching on average the target design spectrum.

For the dynamic analysis, the fixed boundaries used for the static pushover, are replaced by periodic boundaries. These are materialized by imposing kinematic constraints between the edge nodes of the model at each level. These boundaries allow the model to mimic free-field soil response. The results are compared indicatively for the scaled Chi-chi record in Fig. 6. The performance is first assessed in terms of total deck drift (δ_{total}) time histories (Fig. 6a). The two systems experience similar δ_{total}, which can be partially explained on the basis of the corresponding spectral demand. As shown in Fig. 5, with pier nonlinearity governing the response of both systems, their predominant period (after eigenfrequency analysis) is quite similar.

To further explain the observed behaviour, δ_{total} time history is decomposed to its rocking (δ_r, due to foundation rotation) and flexural (δ_f, due to deflection of the pier) drift (Figs. 6b, c). In all cases examined, the swaying component is limited, and is therefore not presented. In the case of the un-retrofitted pile group, the rocking component is dominant (about 75% of the total drift). On the contrary, in the retrofitted foundation the two components are of similar magnitude. Thanks to the increased rocking response of the un-retrofitted pile group, the flexural drift of the pier is reduced. The latter is directly related to structural damage, implying superior performance of the under-designed foundation. Finally, for the example studied herein, the residual rotation (θ_{res}) and settlement (w_{res}) of the two groups is similar (Fig. 6d) and limited.

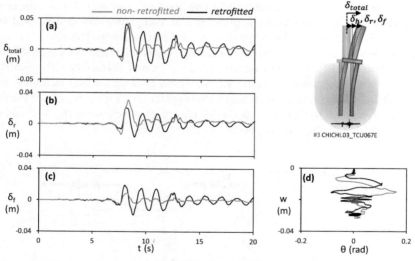

Fig. 6. Comparison for for Chi-chi record. Time histories of: (a) total drift (δ_{total}); (b) rocking drift (δ_r); (c) flexural drift (δ_f); and (d) settlement-rotation ($w - \theta$) response.

The performance of the two design alternatives is summarized in Table 3, in terms of maximum δ_{total}, δ_r, δ_f, maximum bending moment at the base of the pier M_e, and residual settlement w_{res} and rotation θ_{res}. The un-retrofitted pile group experiences larger rocking drift, leading to a reduction of pier deflection, limiting structural damage. This is also depicted on the significant reduction of the bending moments acting at the base of the pier.

The improved structural performance comes at the cost of slightly increased w_{res} for the un-retrofitted case, which however remains within tolerable limits. In all cases examined, for both foundation alternatives, the pier does not reach its ultimate bending moment capacity, sustaining repairable damage, remaining within fully operational performance limits [7].

Considering that the motions correspond to design levels, the reduced levels of structural damage are attributed partially to the reduction of inertial loads due to the kinematic interaction of the pile group, and partially to the safety factors implemented in the design. For the cases examined, foundation retrofit cannot be justified. However, this conclusion cannot be fully generalized, due to the limited number of excitations used. More analyses are essential, allowing for a wider variety of seismic motions.

Table 3. Summary of seismic performance of the two foundation alternatives.

Record	δ_t (cm)	δ_r (cm)	δ_f (cm)	M_e (MNm)	w_{res} (cm)	θ_{res} (°)	
SMadre	0.9	0.6	0.3	8.5	3.1	0.006	
	1.3	0.7	0.6	17.0	2.5	0.003	
Marche	1.2	0.7	0.4	3.7	2.5	0.017	*un-retrofitted*
	1.4	0.4	2.0	7.0	2.1	0.009	*retrofitted*
Chi-chi	4.2	3.1	0.9	8.5	3.1	0.016	
	4.1	2.1	1.9	16.9	3.0	0.02	

5 Conclusions

Two retrofit approaches of existing pile groups were compared: conventional retrofit; and a "do nothing" approach (the existing foundation is maintained and allowed to fully mobilize its moment capacity). Under static pushover loading, the un-retrofitted group is equally efficient in terms of ultimate resistance of the bridge, being however more flexible. The ultimate state was controlled by the structural capacity of the pier.

Under seismic loading, using motions compatible with the design spectrum, the performance of the two approaches was similar. The nonlinear response of the un-retrofitted pile group increases settlements (still tolerable) and rocking drift. Associated with increased energy dissipation, the latter leads to a pronounced decrease of flexural drift of the pier (less structural damage). Both systems perform equally well regarding structural (plastic hinging of the RC members) as well as geotechnical performance criteria (residual settlement, rotation).

Summarizing, by means of a validated numerical analysis methodology, the pros and cons of the two design alternatives were revealed. Considering the substantial effort required to upgrade an existing pile group, it is concluded that under static and seismic loading, within the design limits, such challenging operation is not justified.

Acknowledgements. The authors would like to acknowledge the contribution of Manuela Fehr in the evaluation of the results of the FE analyses presented in this study.

References

1. ABAQUS: Standard user's manual. Dassault Systemes Simulia Corp., Prov., RI, USA (2019)
2. Anastasopoulos, I., Gazetas, G., Loli, M., Apostolou, M., Gerolymos, N.: Soil failure can be used for seismic protection of structures. Bul. of Earthq. Eng. **8**(2), 309–326 (2010)
3. Berezantzev, V.G.: Load bearing capacity and deformation of piled foundations. In: Proceedings of the 5th International Conference, ISSMFE, Parisn, 1961, vol. 2, pp. 11–12 (1961)
4. Chang, G., Mander, J.B.: Seismic energy based fatigue damage analysis of bridge columns: Part I Evaluation of seismic capacity. NCEER Tech. Rep. No. 94-0006, Buffalo (1994)
5. FEMA, P.: 750 (2009) NEHRP recommended seismic provisions for new buildings and other structures, Washington, DC, USA (2009)
6. Fleming, K., Weltman, A., Randolph, M., Elson, K.: Piling engineering. CRC press (2008)
7. Ghobarah, A.: Performance-based design in earthquake engineering: state of development. Eng. Struct. **23**(8), 878–884 (2001)
8. Mander, J.B., Priestley, M.J.N., Park, R.: Theoretical stress-strain model for confined concrete. J. Struct. Eng. **114**(8), 1804–1826 (1988)
9. Niemunis, A., Herle, I.: Hypoplastic model for cohesionless soils with elastic strain range. Mech. Cohesive-Frict. Mater. Int. J. Exp. Model. Comput. Mater. Struct. **2**(4), 279–299 (1997)
10. Seismosoft: SeismoSelect 2021 – A computer program for the selection and scaling of ground motion records (2021). https://seismosoft.com/
11. Von Wolffersdorff, P.A.: A hypoplastic relation for granular materials with a predefined limit state surface. Mech. Cohesive Frict. Mater. Int. J. Exp. Model. Comput. Mater. Struct. **1**(3), 251–271 (1996)

Shaking Table Tests on Level Ground Model Simulating Construction of Sand Compaction Piles

Hiroshi Yabe[1(✉)], Junichi Koseki[2], Kenji Harada[1], and Keiichi Tanaka[1]

[1] Fodo Tetra Corporation, Tokyo, Japan
hiroshi.yabe@fudotetra.co.jp
[2] University of Tokyo, Tokyo, Japan

Abstract. The sand compaction pile (SCP) method is a ground improvement method that increases the density of the ground by enlarging the diameter of the sand pile through penetration and repeated withdrawal/re-driving of the casing pipe. To confirm the improvement effect of the SCP method by model tests, it is desirable to reproduce the actual construction as accurately as possible, taking into account the effect of stress history during construction. The authors compacted horizontal model ground using a newly-developed sand pile driving apparatus that simulates the static SCP construction, and conducted shaking table tests on the compacted ground. In these tests, the effect of casing rotation for the ground behavior during sand pile driving was investigated, and the behavior of the ground models was measured by the changes in the driving force, the earth and water pressure gauges installed into the ground models when performing ground improvement with the SCP driving system. The relationship between excess pore water pressure and acceleration during shaking was investigated after improvement. From these measurements, it was confirmed that the driving force and earth pressure were smaller and the ground was denser in the model compacted with rotation of casing than in the model without rotation. The behavior of the improved model ground, which was made to simulate the real construction process, was studied based on the measured records of acceleration, pore water pressure, and settlement. The results showed that the pore water pressure and the amount of settlement were larger in the order of the unimproved ground, the improved ground without rotation of casing, and with rotation. In addition, a simple dynamic cone penetration test was conducted as a sounding test before and after shaking table tests, and from the results, the ground conditions before and after shaking and the improvement effect were confirmed.

Keywords: Sand compaction piles · Liquefaction · Shaking table test · Sounding tests

1 Introduction

The sand compaction pile (SCP) method is a ground improvement method used to densify sandy soil deposits for liquefaction during earthquakes. SCP method has the repeated process of inserting a sand-filled casing, withdrawing the casing, and re-inserting the casing newly filled with sand to enlarge the diameter of the sand pile (column) thus formed into the ground [1]. There are two types of SCPs: vibratory and non-vibratory, but recently the non-vibratory type is more commonly used because it has less noise and vibration. If the effectiveness of the Non-vibratory SCP method is to be verified by model testing, the actual ground improvement work should preferably be realistically simulated by taking into consideration factors such as the shear history effect. The authors modified an existing sand piling apparatus designed to simulate the Non-vibratory SCP construction, placed sand piles in the saturated level ground models in a gravity field, and evaluated the ground-improving effect through shaking-table testing and sounding. During the sand piling process, the authors also investigated the influence of casing rotation on the behavior of the improved ground models. This paper reports on those tests.

2 Sand Piling Apparatus and Experiment Method

Photo. 1 shows an overall view of the sand piling apparatus utilized for the testing. The apparatus was developed by modifying a previous apparatus developed by Kusakabe et al. [2]. And adding a rotation mechanism mimicking the mechanism utilized in the static SCP method. The casing pipe shown in Photo. 2 consists of an outer pipe 40 mm in diameter and an inner pipe used to remove the sand from the casing and enlarge the sand pile diameter. The diameter of the outer pipe was a 1/10th-scale model of

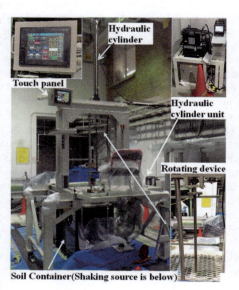

Photo. 1. Sand piling apparatus

Photo. 2. Casing pipe

a typical casing pipe (400 mm diameter) used in the SCP method. The sand piling apparatus was designed to be capable of being moved around over the soil container and placing a sand compaction pile at any location along the plane surface. An operator's console mounted on top of the apparatus allows controlling conditions such as the penetration depth and pile construction steps from a touch panel controller. Sand piling was carried out in the same manner as in the tests conducted by Kusakabe et al. [2], by following the procedure shown in Fig. 1. Casing rotation was simulated at 20 rpm (vertical penetration rate: approximately 10 mm/s) at Steps 1, 4, and 5 in which the outer pipe moves vertically.

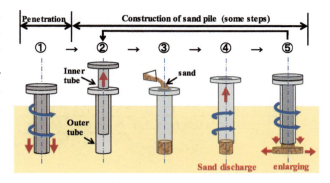

Fig. 1. Sand pile construction process in experiment

The level ground model was made in a rigid soil container, which consists of the zones compacted without/with rotation of casing and the un-improved zone, as shown in Fig. 2. It was prepared by the method of air pluviation in which Gifu silica sand No. 7 (ρ_s = 2.646 g/cm³, ρ_{dmax} = 1.565 g/cm³, ρ_{dmin} = 1.235 g/cm³) was discharged vertically in the air from a perforated small container into the soil container. In all cases, the lowest 100-mm-thick layer was prepared as a non-liquefied layer with a relative density Dr of 80%, and the overlying 440-mm-thick layer as a liquefiable layer with a relative density Dr of about 60% (designed to simulate the pre-improvement condition). After saturation by water flow under the soil container, a total of eight sand piles were placed at depths of GL-40 to GL-440 mm at 200 mm spacing (improvement ratio a_s = 9.6%). Two types of improved ground models were prepared with a rotating or non-rotating casing and step by step shaking from 85 to 515 gal was carried out with 20 cycles of sinusoidal waves at 5 Hz. Table 1 shows the measured input acceleration at each shaking step.

Table 1. Input acceleration at each shaking step

STEP	Acceleration (gal)
1	85
2	120
3	166
4	195
5	235
6	255
7	305
8	319
9	342
10	410
11	445
12	515

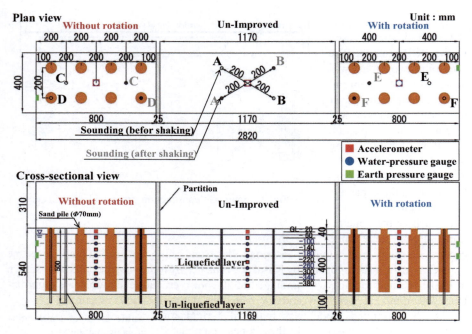

Fig. 2. Outline drawing of ground models in the soil container

3 Ground Model Behavior During Sand Piling

3.1 Measurements During Sand Piling

Figure 3 gives examples of ground models behavior measurements during sand piling that show the time histories of the measured values of hydraulic cylinder pressure (driving force), increases in Earth pressure, and the excess pore water pressure ratio during the 8th (last) sand pile placement, along with the rotation/non-rotation data. Comparison of without-rotation (red-line) and with-rotation (blue-line) hydraulic pressure measurement data at the top of Fig. 3 reveals that the with-rotation values tend to be smaller than without-rotation values although the measuring range (up to 2.5 MPa) of the pressure

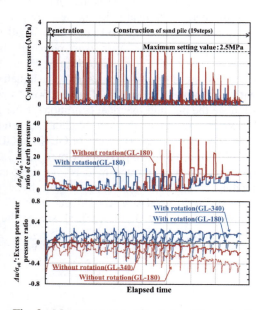

Fig. 3. Measurement results during 8th sand pile construction

gauge was exceeded in some cases. The diagram at the middle of Fig. 3 shows the results of measurement of changes in earth pressure (lateral pressure) (= $\Delta\sigma_h/\sigma_{v0}'$) under initial effective overburden pressure. This diagram also shows that with the elapse of time, constant earth pressure kept acting continuously in the with-rotation case, while in the without-rotation case the amount of increase in earth pressure increased during both penetration and the latter half of the pile construction process. The diagram at the bottom, which shows the changes in water pressure under initial effective overburden pressure, $\Delta u/\sigma_{v0}'$, shows that the values in the without-rotation case tended to decrease over time.

This type of water pressure change resembles the type of water pressure change that can be observed during the expansion of sandy soil in undrained shear conditions. Noda et al. [3] analyzed the behavior of sandy soil elements during non-vibratory SCP construction and found similar expansive behavior to this experiment, but the analysis was done without rotation, so it is inferred that large penetration loads and expansive behavior occurred. Therefore, these behaviors indicate that the casing rotation reduces pressure during penetration and prevents an excessive increase in earth pressure and, from the fact that negative excess pore water pressure, which indicates expansive undrained shear behavior, decreased, it can be inferred that the force exerted by rotation greatly contributes to penetration.

3.2 Measurements at Each Sand Pile Driving Step

Figure 4, based on the values obtained from all sand piles before and after placement and the maximum and minimum values extracted during sand piling, shows the changes over time in the earth pressure increase ratio and the excess pore water pressure ratio during the sand piling process. As the sand piling progressed, the piling location approached the earth pressure gauge. Consequently, the amount of earth pressure increase gradually increased, and the influence of rotation (with or without) was significantly great only during the placement of the eighth sand pile, which was the pile closest to the earth pressure gauge. The diagram at the bottom shows that excess pore water pressure fluctuates significantly and that the measured values

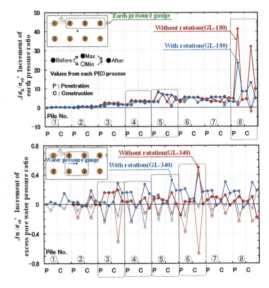

Fig. 4. Change of earth and water pressure at each sand pile driving step

for water pressure change considerably during sand piling in the vicinity of the water pressure gauge, indicating that positive excess pore water pressure tends to occur more in the with-rotation case than in the without-rotation case.

Similar earth pressure and water pressure measurements were conducted at other sites such as reported in Harada et al. [4]. Although changes under the influence of the number of sand piles cannot be compared because of differences from the reported conditions (piling locations and measurement points), similar rises and falls in excess pore water pressure and earth pressure were reported. Although the stresses investigated in the laboratory tests reported in Harada et al. [5] were based solely on forces in the longitudinal profile (depth) direction in the without-rotation case, it can be inferred that in actual situations, three-dimensional stress resultants including the force due to rotation in-plane surface occur and contribute to the compaction mechanism.

3.3 Geometry of Sand Piles and the Amount of Pile Heave

Average diameter of sand piles: 71mm (Without rotation), 73mm (With rotation)

Heave of ground after sand pile driving: 13mm (Without rotation), 5mm (With rotation)

Photo. 3. Formed diameter of sand pile and heave of ground

After the sand piling, the surface of the ground models into the container was measured, and the finished shape sizes of the sand piles were observed at GL-100 mm through GL-300 mm when the apparatus and materials used in the shaking test were removed. Photo. 3 shows how the diameter of the sand piles was measured as well as the pile heave after the sand piling, and Fig. 5 shows the changes in relative density during sand piling. The measured diameters of the sand piles showed little difference between the with-rotation case and the without-rotation case, but the amount of pile heave tended to be greater in the without-rotation case. Comparison of the amount of increase in relative density (ΔDr) resulting from sand piling reveals that the amount of change in relative density from the initial ground condition tended to be greater and the amount of pile heave due to sand piling was smaller in the with-rotation case than in the without-rotation case.

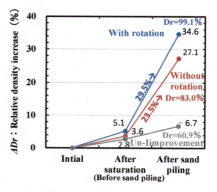

Fig. 5. Changes in relative density during sand piling

From this change, it can be inferred that the force exerted by the rotation of the casing helped reduce the pile heave in the surrounding ground surface to achieve a higher soil density into the ground model.

4 Ground Model Behavior During Shaking

Step-by-step shaking of the ground model shown in Fig. 2 was carried out at the accelerations shown in Table 1. Figure 6 shows changes in the maximum excess pore water pressure ratio ($\Delta u/\sigma_{v0}'$) at different shaking steps. Figure 7 shows changes in the amount of ground surface settlement. The water pressure ratio used here is the average of the values for each layer (L-2 to L-5) shown in the cross section in Fig. 2.

Both the excess pore water pressure ratio and the amount of subsidence began to change after the input acceleration exceeded 300 gal and changed considerably at the shaking step of 410 gal. Comparison under different conditions reveals that the amounts of change in the pre-improvement, without-rotation improvement, and with-rotation improvement cases were greater in that order. In the with-rotation case in which the highest density was observed, the maximum excess pore water pressure ratio approached closer to the values under the other conditions at the last step (515 gal).

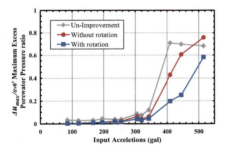

Fig. 6. Relationship between input acceleration and maximum excess pore water pressure ratio

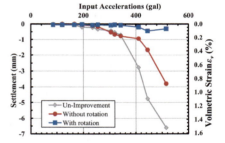

Fig. 7. Relationship between input acceleration and ground surface settlement

At the preceding step (445 gal), however, the rate of increase in water pressure was kept lower than in the other cases, and the amount of subsidence also remained lower (within 0.5 mm) than in the other cases until the end of the test. This behavior is thought to be due mainly to the dependence on the density differences resulting from the compaction shown in Fig. 5 (ground improvement effect). As reported in Tsukamoto et al. [6], however, two-way loading including loading in the direction of rotation tended to cause greater negative dilatancy and a higher degree of compaction than one-way loading in the vertical direction (i.e. without casing rotation). Loading in other directions may also possibly contribute to the stabilization of the soil particle structure.

5   Sounding Before and After Shaking

This section describes the results of sounding obtained by simple dynamic cone testing conducted before and after shaking. Figure 8 [7] shows the simple lightweight dynamic cone penetrometer (product name: PANDA; hereinafter referred to as "dynamic cone tester") used for this study. In a penetration test conducted by use of the dynamic cone penetrometer, the anvil is hammered consecutively, and the dynamic penetration resistance q_d-values received by the anvil is recorded each time by the data logger. Penetration is performed at approximate intervals of 10 mm, and the penetration depth is recorded, by use of a dedicated wire, together with the q_d-values. This sounding method was developed originally in France and has some advantages including that it can be conducted by a single person in a short time and a minimum of space, test results can be checked immediately, and arbitrary hammer impacts can be used for evaluation.

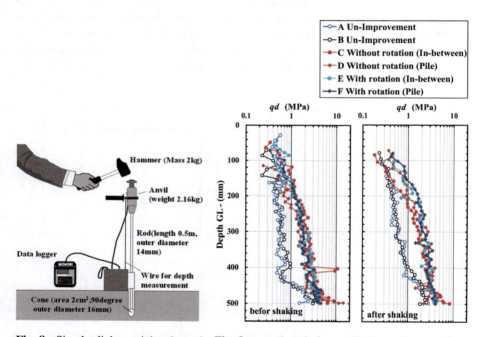

Fig. 8. Simple light weight dynamic cone penetration test (Reproduction from [6])

Fig. 9. q_d-values before and after shaking (Refer to Fig. 2 for location of soundings before and after shaking)

Figure 9 shows the dynamic cone test results. Hammer impact energy was adjusted so that each hammer blow would cause penetration of approximately 10 mm, and hammering was continued until a depth of 500 mm, which is deeper than the improved model layer, was reached. The uppermost part of the figure has been adjusted to eliminate the influence of natural subsidence. As shown, both pre-shaking and post-shaking results show higher q_d-values at improved ground-locations than at unimproved ground- locations, and the differences are discernible at depths between GL-200 mm and GL-440 mm. Comparison between pre-shaking and post-shaking results reveals that q_d-values at unimproved locations tend to be low at depths shallower than GL-200 mm, that a similar tendency can be seen at locations between piles in the without-rotation case, and that q_d-values at locations between piles and at the centers of piles did not decrease even after shaking in the with-rotation case. These results indicate a sufficient ground-improving effect.

6 Conclusion

In the present study, level ground model behavior during sand piling and the effect of casing rotation were investigated by using a sand piling apparatus designed to simulate the static SCP construction. The measurement results confirm that casing rotation reduces the pressing compressive load during sand piling, prevents an excessive increase in earth pressure, and increases the soil density. The ground model behavior observed during shaking and the sounding test results from before-and-after shaking also indicate that the sand compaction pile method is sufficiently effective for preventing liquefaction and that casing rotation enhances the effectiveness of the SCP method.

References

1. Kitazume, M.: The Sand Compaction Pile Method, A.A.BALKEMA PUBLISHERS ISBN: 0415372127 (2005)
2. Kusakabe, S., Koseki, J., Harada, K.: Shaking table test on horizontal ground model simulating sand compaction piling. In: 55th Annual Meeting of the Japan National Conference on Geotechnical Engineering, 22–11–1–05 (2020) (in Japanese)
3. Noda, T, Yamada, E., Yamada, S., Asaoka, A.: A Soil-water Coupled Analysis on Compaction of Sandy Ground with Static Cavity Expansion, Geotechnical Special Publication No.143 (Eds: J. A. Yamamuro and J. Koseki), pp. 269–285 (2005)
4. Harada, K., Yabe, H., Hashimoto, N., Ito, T.: Measurement of horizontal stress during placement of multiple sand compaction piles. In: Japan Association for Earthquake Engineering Annual Meeting 2020, B-3-5 (2020) (in Japanese)
5. Harada, K., Obayashi, J., Yabe, H.: Study on densification mechanism of a compaction method involving penetration and pile construction and its ground improvement effect. In: Proceedings of the 12th National Symposium on Ground Improvement, pp. 235–242 (2016) (in Japanese)
6. Tsukamoto, Y., Ishihara, K., Yamamoto, M., Harada, K., Yabe, H.: Soil densification due to statics and pile installation for liquefaction remediation. Soils Found. **40**(2), 9–20 (2000)
7. Langton, D.D.: The Panda lightweight penetrometer for soil investigation and monitoring material compaction, Ground Engineering September, 33–34 (1999)

3D Numerical Lateral Pushover Analysis of Multiple Pile Group Systems

Amelia Yuwono(✉), Widjojo A. Prakoso, and Yuskar Lase

Universitas Indonesia, Depok 16424, Indonesia
amelia.yuwono@ui.ac.id

Abstract. In the development of a performance-based design approach for pile groups, it is important to identify aspects that may have significant effects. To achieve that objective, a series of 3D numerical lateral pushover analyses of pile groups is performed, and the results are subsequently evaluated and synthesized. The effects of pile structural capacity, soil conditions, and foundations systems (single pile groups and multiple pile group systems) are examined explicitly. The highlighted observations include: 1) the structural performance criteria of piles as structural elements alone may not be directly applicable for pile group systems, 2) the performance criteria for the pile group systems would need to consider explicitly the soil conditions, 3) the behavior of pile groups in softer soils appears to be displacement-controlled, while that in stiffer soils appears to be force-controlled, and 4) the progression of plastic hinge development would not most likely lead to a sudden decrease in the pile group system stiffness.

Keywords: Pile groups · Clays · Pushover analysis · Performance based design

1 Introduction

The performance of pile groups and pile group systems would be critical to ensure that structures are to perform within the acceptance criteria. However, the performance criteria for pile groups and pile group systems are not yet well established [e.g., 1], and major research programs are still active worldwide in this area [e.g., 2–4]. In Indonesia, similar research programs are also active. At the element level, some studies could be cited [e.g., 5, 6], while at the system level, the publications include [7].

The primary objective of this paper is to identify important aspects to be considered in the development of a performance-based design approach for pile groups. A series of 3D numerical lateral pushover analyses of pile groups is to be performed, including the effects of pile structural capacity (elastoplastic bending moment models), soil conditions (shear strength and modulus), and foundations systems (single pile groups (1PC9) and multiple pile group systems (9PC9)). It is noted that the latter has not been widely explored in the literature, and is the main focus of this paper. The results are then to be synthesized and used to develop some notes on the identified important aspects. The additional objective includes the evaluation of a conventional analytical model to estimate pile lateral resistance.

2 Numerical Models

The geotechnical conditions considered consisted of two soil layers. The thickness of the upper soil layer was 12 m, while that of the lower soil layer was 8 m thick. To evaluate the soil strength effect on the behavior of laterally loaded pile groups, the upper soil layer undrained shear strength S_u was varied; the undrained soil modulus E_u was taken as $150 \times S_u$. The S_u values used were 20 kPa, 60 kPa, and 200 kPa, representing soft clays, medium clays, and very stiff clays, respectively. The S_u and E_u of the lower soil layer were 200 kPa and 50 MPa, respectively. The undrained friction angle of both soil layers was zero, and the groundwater was not modeled explicitly. The soil constitutive model used was the Mohr-Coulomb model.

The piles were 500 mm square piles with an effective length of 12 m, and the pile tip was allowed to rotate freely on top of the lower soil layer. The pile modulus was 30 GPa, and the pile ultimate shear capacity was 580 kN. The pile ultimate bending moment capacity was 400 kN-m at zero axial load, but the pile was observed to experience a plastic hinge at about 300 kN-m.

The pile groups were 3-by-3 pile groups (1PC9) with piles spaced 1.5 m center-to-center. The pile cap was 4.5 m × 4.5 m, and the concrete pile cap stiffness was determined based on a thickness of 1.0 m. Multiple pile group systems (9PC9) consisting of 8 m spaced, nine 1PC9 pile groups were also examined. There were 0.25 m × 0.75 m concrete tie-beams connecting the pile caps, and there were 0.25 m thick concrete floor slabs connecting the pile caps and tie-beams. The pile caps, tie-beams, and slabs were not in contact with the upper layer soil. The basic models are shown as Fig. 1.

The analyses were performed using Plaxis 3D [8], and the total stress analysis option was used. The soil elements were modeled using 10-node tetrahedral elements. The embedded piles were modeled using 3-node beam elements and used the elastoplastic constitutive model. The tie-beam elements were modeled using 3-node beam elements, while the pile cap and floor slab elements were modeled using 6-node thin shell elements. The connection between the soil and a pile was modeled using interface elements with predetermined maximum capacities, facilitated by additional nodes in the soil elements. The connection between a pile and the pile cap was a rigid connection. All the elements were automatically generated by Plaxis 3D; the number of nodes was between 127,043 and 169,061 nodes, while the number of soil elements was between 85,600 and 103,795 elements.

The cyclic lateral displacement was applied at the pile cap level at 3 mm interval to 240 mm, and the resulting pushover forces at pile heads were recorded. In this paper, only the backbone curves are examined.

3 Results and Discussion

3.1 No Soil Conditions

The pile groups 1PC9 and 9PC9 were analyzed without the upper soil layer. The objective of these analyses was to observe how the two pile groups would response structurally to the pushover lateral displacement. Figure 2 shows the load-lateral displacement curves for both pile groups. The behavior is essentially an elastic-perfectly plastic behavior, following the constitutive model used for the piles.

The responses appeared to be very similar, with maximum average pushover force per pile of about 31.4 kN. This maximum force was due to the pile ultimate bending moment capacity, while this force was well below the pile shear capacity. The lateral displacement to reach this maximum force for both pile groups was about the same. The coefficient of variation (= standard deviation / mean, COV) of the average force was about 10%; however, it was also observed that the COV increased with increasing lateral displacement.

The dashed line is approximately the limit of the pile groups having no plastic hinges. The difference in displacement between this limit and the plastic conditions appears to be not too significant. The lateral displacement limit is about 0.1 m.

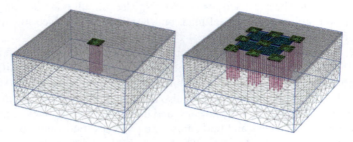

Fig. 1. Basic models for 1PC9 and 9PC9.

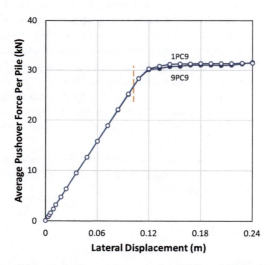

Fig. 2. Load-lateral displacement curves for 1PC9 and 9PC without soils.

3.2 General Behavior

The results were validated by comparing them to the results of test results reported in the literature [e.g., 9]. The lateral pile force distribution among the leading piles, trailing piles and middle piles in 1PC9 was found within the reported values. As this study considered a wide range of soil strengths, the actual force distribution varied depending on the soil strength considered. This is extensively reported elsewhere [10].

The load-lateral displacement curves for both 1PC9 and 9PC9 having different soil parameters are shown as Figs. 3 and 4, respectively. The left-hand side figures show the overall curves, while the right-hand side figures are the details of the initial part of the curves. It is noted that the y-axis shows the average pushover force per pile. For both pile groups, as the soil conditions become stiffer, the initial stiffness of curves and the maximum force increase.

By comparing the left-hand side figures of Figs. 3 and 4 with Fig. 2, the significant effect of soil mass on the overall load-lateral displacement of pile groups as expected could be observed clearly. It is highlighted that no elastic-perfectly plastic behavior was present, even though plastic hinges have developed in the pile groups. This indicates the complex interactions between the soil mass and the structural pile groups. Furthermore, this also may indicate that the structural performance criteria of piles as structural elements alone may not be directly applicable for pile group systems, as these criteria were developed from pure structural standpoints.

The dashed lines on the right-hand side figures are approximately the limits of the pile groups having no plastic hinges. For both pile groups, a pile group in softer soils would have a limit with a lower average force and a higher lateral displacement. In addition, the limits for 1PC9 appear to be higher than those for 9PC9. The limits for both pile groups are much lower than the limit for pile groups without soils (Fig. 2), indicating the limits based on structural capacities may not again be directly applicable for pile groups.

The load distribution among the pile caps in 9PC9 is also examined. The average pushover force per pile for each pile group is shown in Fig. 5 for different soil parameters. Also shown is the corresponding average pushover force per pile for 1PC9. The distribution is quite wide, but the distribution appears to become more uniform for stiffer soils. Nevertheless, the distribution of the pile pushover forces of 1PC9 is much narrower than that of 9PC9, as shown in Fig. 6 ($S_u = 200$ kPa, lateral displacement $= 0.024$ m). All these indicate that the performance criteria for the pile group system would need to consider explicitly the soil conditions.

3.3 Pile Structural Degradation

The detailed progression of structural degradation of pile groups is shown in Fig. 7. The notation ① indicates the last lateral displacement at which no initial pile plastic hinges have developed, while the notation ② indicates the last lateral displacement at which no piles have reached their ultimate bending moment capacity. The notation ③ indicates the last lateral displacement at which no more than 5% of piles have resisted pushover force greater than 580 kN. It is highlighted that the lateral displacement ② is less than the lateral displacement ③ for all cases, indicating the pile bending moment capacity would be the predominant controlling pile group structural aspect.

The progression from the lateral displacement ① to the lateral displacement ② represents the progression of plastic hinge development. It can be observed that there is no sudden decrease in the pile group system stiffness in this progression; clearer curves are also shown on right hand side of Figs. 3 and 4. The effect of this progression would be rather gentle, albeit the elastic-perfectly plastic behavior observed in Fig. 2.

The average pushover force is compared with the pile ultimate shear capacity in Fig. 8. It is noted that the pile model used does not include the shear capacity model. It can be seen that the average pushover force of the pile groups in soft clays was well below the shear capacity, while that of the pile groups in medium clays was about the same as the capacity. The force of the pile groups in stiff clays was well above the shear capacity. All these suggest that the behavior for softer soils appears to be displacement-controlled, while that for stiffer soils appears to be force-controlled. The actual boundary for these two behaviors needs to be examined further.

It is of interest to compare the lateral displacement ③ in Fig. 7 and the pile ultimate shear capacity in Fig. 8. When the average pushover force is used, it appears that the use of 75% of the pile ultimate shear capacity would match better the systemic initiation of pile ultimate shear capacity. This is particularly true for the variability of pile pushover forces as shown in Figs. 5 and 6.

The load-lateral displacement curves are also used to examine the analytical pile ultimate lateral resistance model proposed by Broms [e.g., 11]. The "long" condition was assumed in the analysis. The comparison is shown in Fig. 8. For both 1PC9 and 9PC9, the analytical model appears to provide rather conservative estimates; the model appears to be more conservative for stiffer soils.

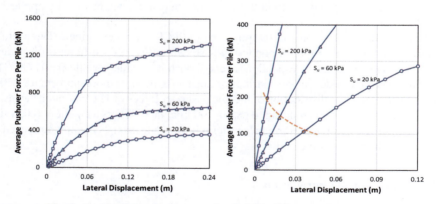

Fig. 3. Load-displacement curves for 1PC9 in different soil conditions.

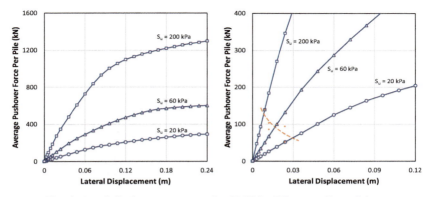

Fig. 4. Load-displacement curves for 9PC9 in different soil conditions.

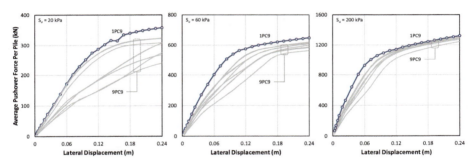

Fig. 5. Pile cap average pushover force distribution for 9PC9 in different soil conditions.

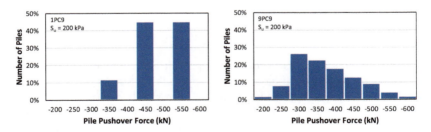

Fig. 6. Distribution of pile pushover forces.

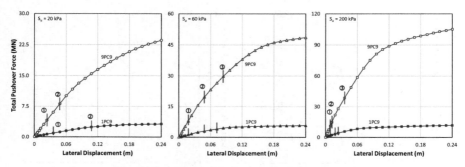

Fig. 7. Progression of pile structural degradation.

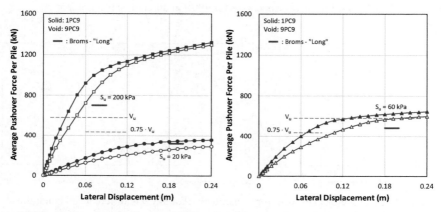

Fig. 8. Evaluation of pile ultimate shear capacity and pile ultimate lateral resistance.

4 Conclusions

The objective of this paper is to identify important aspects to be considered in the development of a performance-based design approach for single pile groups and pile group systems. It is noted that the latter has not been widely explored in the literature, and was highlighted in this paper. The results of a series of numerical lateral pushover analyses of pile groups using Plaxis 3D were evaluated and synthesized. The effects of pile structural capacity (elastoplastic bending moment models), soil conditions, and foundations systems (single pile groups (1PC9) and multiple pile groups (9PC9)) were examined explicitly. The highlighted observations include the following:

- The structural performance criteria of piles as structural elements alone may not be directly applicable for pile group systems.
- The performance criteria for the pile group systems would need to consider explicitly the soil conditions.
- The behavior of pile groups in softer soils appears to be displacement-controlled, while that in stiffer soils appears to be force-controlled.

- The pile bending moment capacity would be the predominant controlling pile group structural aspect. The progression of plastic hinge development would not most likely lead to a sudden decrease in the pile group system stiffness.
- The use of 75% of the pile ultimate shear capacity is suggested when the evaluation is based on the average lateral force.
- The analytical pile ultimate lateral resistance model proposed by Broms appears to provide rather conservative estimates.

Overall, the behavior of single pile groups (1PC9) was found to be different from that of multiple pile group systems (9PC9), and this observation deserves further examination.

Acknowledgment. The work presented was partially funded by the Post Graduate Research Grant Universitas Indonesia 2020 (Contract No: NKB-659/UN2.RST/HKP.05.00/2020).

References

1. American Society of Civil Engineers. Seismic evaluation and retrofit of existing buildings, Standard ASCE/SEI 41–17, ASCE (2017)
2. Abu-Farsakh, M., Souri, A., Voyiadjis, G., Rosti, F.: Comparison of static lateral behavior of three pile group configurations using three-dimensional finite element modeling. Can. Geotech. J. **55**, 107–118 (2018)
3. Liu, T., Wang, X., Ye, A.: Roles of pile-group and cap-rotation effects on seismic failure mechanisms of partially-embedded bridge foundations: quasi-static tests. Soil Dyn. Earthq. Eng. **132**, 106074 (2020)
4. Chiou, J.-S., You, J.-Q.: Theoretical solutions of laterally loaded fixed-head piles in elastoplastic soil considering pile-head flexural yielding. Can. Geotech. J. **57**, 650–660 (2020)
5. Irawan, C., Djamaluddin, R., Raka, I.G.P., Faimun, Suprobo, P., Gambiro: Confinement behavior of spun pile using low amount of spiral reinforcement - An experimental study. Int. J. Adv. Sci. Eng. Inf. Technol. **8**(2), 501–507 (2018)
6. Orientilize, M., Prakoso, W.A., Swastinitya, A.P.: The different effect of cyclic loading protocol on spun pile performance. In: IOP Conference Series: Earth and Environmental Science, vol. 622. no.1 (2021)
7. Yuwono, A., Prakoso, W.A., Lase, Y.: Preliminary 3D numerical pushover analysis of laterally loaded pile groups. In: IOP Conference Series: Materials Science and Engineering, vol. 930 no. 1 (2020)
8. Brinkgreve, R.B.J., Zampich, L.M., Ragi Manoj, N.: Plaxis Connect Edition 2019, Delft (2019)
9. Ilyas, T., Leung, C.F., Chow, Y.K., Budi, S.S.: Centrifuge model study of laterally loaded pile groups in clay. J. Geotech. Geoenviron. Eng. **130**(3), 274–283 (2004)
10. Yuwono, A., Prakoso, W.A., Lase, Y.: Force Distribution in Pile Group Systems Subjected to Lateral Pushover (2022, submitted)
11. Poulos, H.G., Davis, E.H.: Pile Foundation Analysis and Design. Wiley, Hoboken (1980)

Responses of Adjacent Building Pile to Foundation Pit Dewatering

Chao-Feng Zeng[✉], Hai-Yu Sun, Hong-Bo Chen, Xiu-Li Xue, Yun-Si Liu, and Wei-Wei Song

Hunan Provincial Key Laboratory of Geotechnical Engineering for Stability Control and Health Monitoring, School of Civil Engineering, Hunan University of Science and Technology, Xiangtan 411201, Hunan, China
cfzeng@hnust.edu.cn

Abstract. In water-rich area, foundation pit pumping has adverse effects on adjacent buildings. In this study, a series of numerical simulations are carried out, based on a practical dewatering test in an excavation for metro station in Tianjin, to investigate the characteristics of near building pile deformations induced by dewatering. The effect of different pumping depth (H_d) on the pile axial force and lateral friction are revealed. The results indicate that the axial force of the pile foundation experiences a substantial increment; accordingly, the pile shaft resistance becomes negative in the same depth range, and the neutral point appears at the position of 1/2 times the pile length from the pile crown. The research results can provide guidance to the safety assessment of the existing building pile during pumping of adjacent foundation pit.

Keywords: Dewatering · Pile foundation · Deformation · Negative friction

1 Introduction

Due to the limited ground space and the large demand for underground space, the foundation pit engineering is developing towards a larger and deeper trend. In a crowded urban environment, the surrounding area of a foundation pit is often adjacent to existing buildings, roads, and subway tunnels. Studies have shown that the construction of the foundation pit will cause significant deformation of the adjacent environment [1–6], which means that more refined design and construction should be conducted.

At present, domestic related research mainly concentrates on the impact of foundation pit construction on the deformations of the soil and adjacent structure. Zhang et al. [7] deduced a theoretical formula to compute the surrounding soil settlement caused by excavation. Zeng et al. [8] used ABAQUS to establish a three-dimensional fluid-solid coupled finite element model to investigate the wall deflection induced by pre-excavation dewatering under different construction conditions. Additionally, Deng et al. [9] simulated the whole process of excavation construction, and the induced deformations of the adjacent subway station were analyzed by the measured and computed results. Zheng et al. [10] analyzed the development of horizontal deformation of a metro station during the construction process of an adjacent foundation pit, and evaluated the effectiveness of several control measures in limiting the deformation through numerical simulations.

However, only a few scholars have investigated the characteristics of adjacent building pile deformations induced by pumping. For example, Zhang [11] established numerical simulation models, calculated and analyzed the influence of the dewatering and excavation of foundation pits on adjacent pile groups; then, this numerical model was employed to analyze the influence of the ground reinforcement in the passive zone on the adjacent pile groups. Liu [12] presented a procedure to estimate the drawdown, the soil consolidation settlement, the negative friction on the pile shaft and the additional pile settlement, induced by dewatering. However, what the above researchers have done is regarding to investigate the response of piles in a single aquifer system, and did not analyze the behaviour of pile foundation during foundation pit dewatering in a multi-layered aquifer-aquitard system (MAAS). In this MAAS, the water level changes in different aquifers caused by the pumping are not synchronized, and thus, the adjacent building pile deformations would also be different. Therefore, this paper carries out a series of three-dimensional numerical analyses to investigate the responses of adjacent building pile foundation during foundation pit dewatering in a MAAS.

2 Numerical Simulations

2.1 Project Background

Figure 1 presents a plan view of a foundation pit showing the layout of pumping wells and selected monitoring points (only those in the north side were shown) for a metro station in Tianjin, China. The length of the pit was approximately 154 m and the width of that was approximately 40 m. The final excavation depth was 16.9–19.3 m. Diaphragm walls, with a thickness of 0.8 m and depths of 31.5–33.5 m, did not totally cut off the aquifers in the site (refer to Fig. 2).

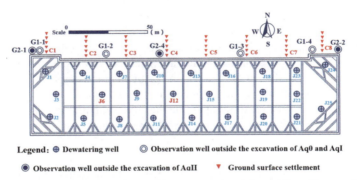

Fig. 1. Plan view of an excavation and instrumentation (adapted from Zeng et al. [13])

The soil layer distribution in this project area is the typical soil layer in Tianjin, and the strata distribution and main soil mechanical parameters are shown in Fig. 2. In the upper 50 m below the ground level (BGL), silty clays, silts and silty sands were the main soil types. There were five aquifers, labeled as Aq0, AqI, AqII, AqIII and AqIV, respectively. 25 pumping wells were arranged inside the excavation and 7 observation wells were installed on the north side of the excavation (in order to monitor the change of water level outside the pit).

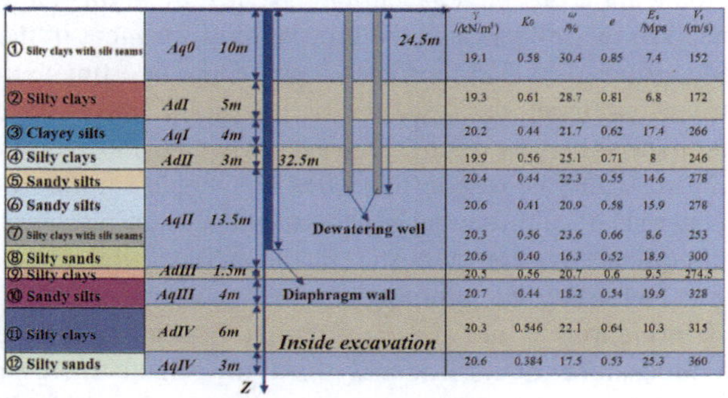

Layer	Aquifer	Thickness		γ /(kN/m³)	K_0	ω /%	e	E_s /Mpa	V_s /(m/s)
① Silty clays with silt seams	Aq0	10m	24.5m	19.1	0.58	30.4	0.85	7.4	152
② Silty clays	AdI	5m		19.3	0.61	28.7	0.81	6.8	172
③ Clayey silts	AqI	4m		20.2	0.44	21.7	0.62	17.4	266
④ Silty clays	AdII	3m	32.5m	19.9	0.56	25.1	0.71	8	246
⑤ Sandy silts				20.4	0.44	22.3	0.55	14.6	278
⑥ Sandy silts	AqII	13.5m	Dewatering well	20.6	0.41	20.9	0.58	15.9	278
⑦ Silty clays with silt seams				20.3	0.56	23.6	0.66	8.6	253
⑧ Silty sands				20.6	0.40	16.3	0.52	18.9	300
⑨ Silty clays	AdIII	1.5m		20.5	0.56	20.7	0.6	9.5	274.5
⑩ Sandy silts	AqIII	4m	Diaphragm wall	20.7	0.44	18.2	0.54	19.9	328
⑪ Silty clays	AdIV	6m	Inside excavation	20.3	0.546	22.1	0.64	10.3	315
⑫ Silty sands	AqIV	3m		20.6	0.384	17.5	0.53	25.3	360

Note γ= unit weight; K_0= the coefficient of soil pressure at-rest; ω=moisture content; e= initial void ratio; E_s=compression modulus; V_s=shear wave velocity

Fig. 2. Strata distribution and the main soil mechanical parameters

After the installation of the diaphragm wall and dewatering wells, 21 days pre-excavation pumping tests (inside the excavation) were carried out to test the pumping capacity of dewatering well and evaluate the impact of pumping on the surrounding environment. In this study, only a typical pumping test of 3.2 days (i.e., 77 h) was shown. In order to reduce the disturbance of the pumping to the enclosure structure, the first level of strut was constructed in advance. Water level declines in different aquifers outside the excavation and surrounding ground surface settlements were monitored during the test. The water level inside the foundation pit decreased by about 15 m, observed by J6, J12; at the same time, the water level of aquifers at different depths outside the pit also decreases, causing apparent soil consolidation settlement (see Sect. 2.3 for details). This shows that the surrounding environment deformation caused by pumping cannot be ignored. In the following sections, a numerical model is developed based on this project, and the measured groundwater drawdown and settlement are employed to validate the model. On this basis, a building pile foundation is added outside the model foundation pit to investigate the influence of pumping on the adjacent building pile foundation.

2.2 Model Building

In this section, two types of models are created. One is that used to simulate the pumping test in Sect. 2.1. The other is that used to simulate the building pile response during pumping. The modeling process of the two types of models is basically the same; and thus, the later one is introduced in detail in this section.

During engineering design and construction, attention is often paid to the maximum deformation of excavation and the surrounding environment. Therefore, the model in this section only simulates the middle section of the long pit in Fig. 1, and the mesh of the finite element model is shown in Fig. 3. The dimensions of the diaphragm wall and dewatering well in the model are basically the same as those in Sect. 2.1. The adjacent building pile foundation consists of a cushion cap with a height of 10 m and 36 piles with a length of 31 m and a diameter of 0.85 m. The distance between piles is 2.2 m. The distance between the foundation pit and the pile foundation is labelled as D. Models under different D are established, and this paper takes the case of $D = 40$ m as an example to investigate the response of adjacent building pile during foundation pit dewatering.

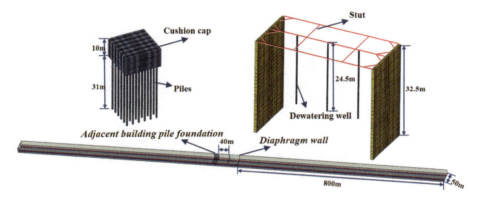

Fig. 3. Mesh of the finite element model

The simulation of the soil is based on the engineering site conditions in Sect. 2.1. The entire soil thickness is set 50 m and the model is divided into 12 soil layers. According to the experience of previous scholars [5, 14], the Mohr-Coulomb model is used to simulate the deformation behavior of the soil during the pumping, and all models consider the fluid-solid coupling; the C3D8P solid element is used to simulate the soil. The soil parameters calculated by the model are shown in Table 1. The enclosure structure is simulated by C3D8I solid element, the pumping well is simulated by S4 shell element, the struts are simulated by B31 linear beam element, and the Young's moduli of the three structure is set 210 MPa, 30 MPa, and 30 MPa, respectively. The coefficient of friction between the soil and structures is set 0.3.

In order to avoid the influence of soil settlement beyond the model boundary during the dewatering, this paper sets the soil boundary at a distance of 800 m from the foundation pit (this is beyond the influence radius of pumping). The four lateral boundaries of the model are all set with horizontal displacement constraint to the direction perpendicular to the corresponding surface; the bottom of the model is set with both horizontal and vertical displacement constraints. Assuming that the initial water level of the model is on the model surface, and set a constant head with fixing head also at model surface on the two outer boundaries with distance of 800 m from the foundation pit. The contact surface between soil and dewatering wells was also set a constant head to simulate pumping process. The range of constant head surface are adjusted to achieve pumping simulation with different depths.

Table 1. Soil parameters used in the model

Hydrogeology	Soil classification	Buried depth/m	K_v (m/d)	K_H (m/d)	E (Mpa)	c' (kPa)	φ'
Aq0	Silty clays with silt seams	0~10	0.03	0.003	43.5	17	25
AdI	Silty clays	10~15	0.025	0.001	56.3	18	23
AqI	Clayey silts	15~19	0.2	0.1	137.6	10	34
AdII	Silty clays	19~22	0.006	0.001	118.6	19	26
AqII	Sandy silts	22~24.5	2.5	0.5	151.8	8	34
	Sandy silts	24.5~29.5	1	0.2	153.3	8	36
	Silty clays with silt seams	29.5~32.5	1	0.16	128	17	26
	Silty sands	32.5~35.5	3	0.6	178.5	7	37
AdIII	Silty clays	35.5~37	0.02	0.004	152.1	19	26
AqIII	Sandy silts	37~41	3	0.9	214.5	10	34
AdIV	Silty clays	41~47	0.005	0.001	198.4	18	27
AqIV	Silty sands	47~50	3.5	1.5	257	7	38

Note: c'= effective cohesion; φ'= effective friction angle; K_H= horizontal hydraulic conductivity; K_V= vertical hydraulic conductivity; E= elastic modulus

2.3 Model Verification

Figure 4 is the comparison curve between the simulated value and the measured value. From Fig. 4(a), it can be found that with the development of dewatering, the measured and computed values of the drawdown increase gradually, and the drawdown inside the pit is about 15 m, which causes the groundwater drawdowns in different aquifers outside the pit (the groundwater drawdowns of the AqI and the AqII are about 3 m and 7 m, respectively), which shows that the calculated results are in good agreement with the measured results. Figure 4(b) is the ground surface settlement during dewatering, and it can be seen that the simulated and measured values of the soil settlement increase with time, and the maximum error is about 30%, which indicates the model calculation can basically reflect the ground response during dewatering.

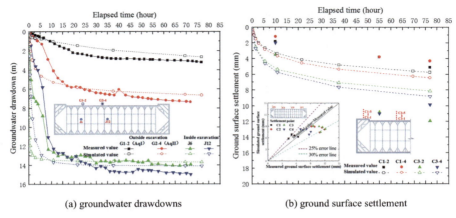

Fig. 4. Computed and measured groundwater drawdown and ground surface settlement (adapted from Zeng et al. [13])

3 Responses of Pie Foundation to Dewatering

3.1 Pile Deformation

Taking the central pile as an example, Fig. 5 shows the relationship between groundwater drawdown at the bottom of the pile (S) and the pile top settlement (δ_V) under different pumping depth (H_d). It can be seen that δ_V increases linearly with S regardless of dewatered depth, which indicates a good correlation between adjacent building pile deformations and the change of groundwater level.

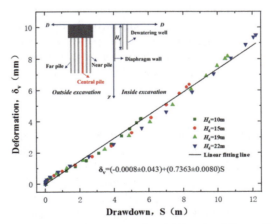

Fig. 5. Relationship between S and δV under different Hd

Figure 6 shows the time-history curves of S and δ_V under different pumping depth (H_d). It can be seen that S and δ_V grow simultaneously during pumping, and the growth was faster in the early stage, while both tend to be stable after 200 h. Additionally, S and δ_V increase as H_d increases, and the stabilization times of S and δ_V do not have great difference under the different H_d.

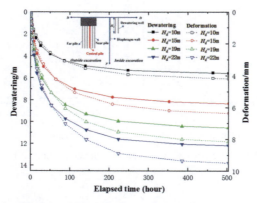

Fig. 6. The time-history curves of S and δ_V under different H_d

3.2 Pile Shaft Resistance

Taking the central pile as an example, Fig. 7 demonstrates the distribution of pile shaft resistance along the depth. The following laws can be obtained:

Negative friction induced by pit dewatering occurs within a certain buried depth (i.e., 1/2 times the pile length), which is due to the fact that the settlement of the shallow soil (outside the pit) induced by dewatering is larger than the pile settlement. In addition, the pumping depth H_d has a great impact on the distribution of pile shaft resistance. Specifically, with the increase of H_d in the pit, the soil outside the pit will have greater settlement, and accordingly, the differential settlement between soil and pile also increases; therefore, the pile shaft resistance tends to increase.

3.3 Pile Axial Force

The curves of pile axial force distribution along the depth are shown in Fig. 8. It can be seen that the pile axial force tends to increase first and then decrease, and the peak value appears at the position of approximately 1/2 times the pile length. This result corresponds to its frictional resistance as exhibited in Fig. 7. In addition, the pile axial force increases with the increase of pumping depth under different pumping conditions.

Obviously, as to the building design, the engineers rarely consider the possible impact of the future construction of adjacent projects, such as the above-mentioned lateral negative friction induced by dewatering in a nearby foundation pit. Therefore, in the structural design of a building resting on pile foundation, a certain safety stock should be considered to prevent additional stresses and deformations caused by future construction disturbances of neighboring projects.

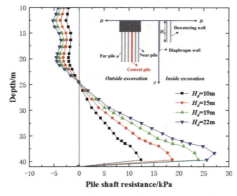

Fig. 7. Curves of pile shaft resistance along the depth

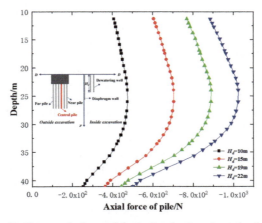

Fig. 8. Curves of pile axial force distribution along the depth

4 Conclusions

In order to analyze the effect of foundation pit pumping on the response of existing buildings pile foundations, a series of three-dimensional numerical models are established to investigate the axial force and deformation of adjacent building pile under different pumping depth (H_d). Based on this research, the following conclusions can be drawn:

1. During pumping, δ_V increases linearly with S, and stabilized after 200 h. S and δ_V increase as H_d increases, and the stabilization times of S and δ_V are generally consistent under different H_d.
2. In terms of pile shaft resistance, the settlement of the shallow soil around the pile is greater than the pile settlement during pumping, which causes the pile to develop negative friction; the neutral point is at the position of about 1/2 times the pile length. With the increase of H_d in the pit, the soil outside the pit will have greater settlement,

and the differential settlement between soil and pile also increases; therefore, the pile shaft resistance tends to increase.
3. In terms of pile axial force, when the negative friction appears, the pile axial force tends to increase and then decrease along the depth, and the peak value appears at the neutral point; under different H_d, both pile axial force and pile shaft resistance increase with H_d increases.

This study could provide guidance for building design, i.e., the design should be forward-looking and consider the secondary effects induced by the future construction of adjacent buildings to prevent the additional stresses and deformation.

References

1. Ong, D.E.L., Leung, C.F., Chow, Y.K.: Behavior of pile groups subject to excavation-induced soil movement in very soft clay. J. Geotech. Geoenviron. Eng. **135**(10), 1462–1474 (2009)
2. Liao, S.-M., Wei, S.-F., Tan, Y., et al.: Field performance of large-scale deep excavations in Suzhou. Chin. J. Geotech. Eng. **37**(03), 458–469 (2015)
3. Zheng, G., Zhu, H.-H., Liu, X.-R., et al.: Control of safety of deep excavations and underground engineering and its impact on surrounding environment. China. Civil Eng. J. **49**(06), 1–24 (2016)
4. Zeng, C.-F., Zheng, G., Xue, X.-L., et al.: Combined recharge: a method to prevent ground settlement induced by redevelopment of recharge wells. J. Hydrol. **568**, 1–11 (2019)
5. Zeng, C.-F., Zheng, G., Xue, X.-L.: Responses of deep soil layers to combined recharge in a leaky aquifer. Eng. Geol. **260**, 105263 (2019)
6. Zeng, C.-F., Zheng, G., Zhou, X.-F., et al.: Behaviours of wall and soil during pre-excavation dewatering under different foundation pit widths. Comput. Geotech. **115**, 103169 (2019)
7. Zhang, S.-M., Jing, F.-W., Huang, Y.-S., et al.: The influence of foundation pit excavation and dewatering to ground surface settlement. J. Civil Environ. Eng. **38**(05), 43–49 (2016)
8. Zeng, C.F., Xue, X.L., Zheng, G., et al.: Responses of retaining wall and surrounding ground to pre-excavation dewatering in an alternated multi-aquifer-aquitard system. J. Hydrol. **559**, 609–626 (2018)
9. Deng, X., Zheng, H., Song, Z.-H., et al.: Deformation analysis of deep excavation adjacent to the new subway station. Chin. J. Undergr. Space Eng. **14**(S1), 270–277 (2018)
10. Zheng, G., Pan, J., Cheng, X.-S., et al.: Passive control and active grouting control of horizontal deformation of tunnels induced neighboring excavation. Chin. J. Geotech. Eng. **41**(07), 1181–1190 (2019)
11. Zhang, R.-J., Zheng, J.-J., Ding, L.-Y., et al.: Influence of dewatering and excavation of foundation pits on adjacent pile groups and its control measures. J. Huazhong Univ. Sci. Technol. (Nat. Sci. Ed.) **39**(07), 113–117 (2011)
12. Liu, Y., Xiang, B., Fu, M.: Influence of dewatering in deep excavation on adjacent pile considering water insulation effect of retaining structures. Geotech. Geol. Eng. **37**(6), 5123–5130 (2019)
13. Zeng, C.-F., Wang, S., Xue, X.-L., et al.: Evolution of deep ground settlement subject to groundwater drawdown during dewatering in a multi-layered aquifer-aquitard system: insights from numerical modelling. J. Hydrol. **603**, 127078 (2021)
14. Zeng, C.-F., Xue, X.-L., Zheng, G., et al.: Responses of retaining wall and surrounding ground to pre-excavation dewatering in an alternated multi-aquifer-aquitard system. J. Hydrol. **559**, 609–626 (2018)

Dynamic Interaction Between Adjacent Shallow Footings in Homogeneous or Layered Soils

Enza Zeolla[1](✉) 📵, Filomena de Silva[2] 📵, and Stefania Sica[1] 📵

[1] Università del Sannio, Benevento, Italy
enza.zeolla@unisannio.it
[2] Università Federico II, Napoli, Italy

Abstract. Seismic design of buildings is generally carried out without accounting for the interaction between adjacent structures through the underlying soil. The increasingly growth of urbanization in modern metropolitan areas or the urgent need for seismic requalification of historical centres with very closely-spaced structures require the Structure-Soil-Structure Interaction problem to be properly investigated and quantified. The paper tries to shed lights on cross interaction phenomena arising between two identical rigid shallow foundations excited by harmonic loads. Through a 3D continuum approach, solved by the finite difference method, the stiffness matrix of a footing in presence of a neighbouring one was obtained. Different values of foundation-foundation spacing and subsoil type (halfspace or layered soil deposit) were considered.

Keywords: Structure-Soil-Structure Interaction · Impedance functions · Footing group

1 Introduction

In most engineering design cases, proximity among buildings is hardly ever considered even though it is well known that a sort of cross interaction between closely spaced structures may exist under both static and dynamic loading conditions. In the dynamic field, any vibrating foundation spreads a wave field that could be seen as a disturbance affecting the adjacent foundation and vice versa. In his comprehensive state-of-art on analysis of machine foundation vibrations, Gazetas [1] cites the early pioneering works on dynamic cross interaction among adjacent foundations, namely Wartburton et al. 1971 [2], Chang-Liang (1974) [3] and Roesset & Gonzales (1977) [4]. Later on, Qian and Beskos (1995) [6] criticized the assertion of the ATC-3 regulations (1984) [5] *"that neglecting coupling effects between footings could lead to conservative results"* by showing that this is not always true for certain bands of frequency.

In this work, the influence of a neighboring footing on the dynamic response of a similar footing (the master one) is evaluated in computing its frequency-dependent matrix of impedance functions. As well-known, foundation impedances $\overline{K}_{ij}$ are the sum of a real part representing the dynamic stiffness and an imaginary part accounting for damping [1]:

$$\overline{K}_{ij} = k_{ij}(a_0)K_{ij} + i\omega(a_0)C_{ij} \qquad (1)$$

where:

- the subscripts i, j indicate that $\overline{K}_{ij}$ links the component i of the vector of the loads transmitted by the foundation into the soil to the component j of the displacement vector;
- the low-frequency stiffness, K_{ij}, and the dashpot coefficient, C_{ij}, depend on the soil shear modulus, G, and Poisson's ratio, v, as well as on a characteristic dimension of the foundation;
- the dynamic coefficients $k_{ij}(a_0)$ and $c_{ij}(a_0)$, depend on the vibration frequency, ω, the characteristic dimension of the foundation, (B), and the soil shear wave velocity, V_S, through the dimensionless frequency, $a_0 = \omega B/V_S$.

In the following, the impedance components of a rigid rectangular footing affected by the presence of a nearby identical foundation will be compared to the corresponding terms of the same footing when considered alone.

2 Performed Numerical Analyses

The numerical study was performed considering the typical subsoil configurations of homogeneous halfspace and layer over halfspace. Both the single and the double foundations were assumed to be rectangular with a base, 2B, equal to 2 m and a length, 2 L, of 10 m. Figure 1 shows the 3D reference schemes of soil and closely-spaced foundations whereas Table 1 lists all the analysed cases. In detail, the spacing, S, between the two foundations (Fig. 1) was varied between 0.5 m and 4 m while the thickness H of the upper layer over halfspace was varied between 3 m and 10 m. The adopted soil unit weight, γ, bulk modulus, K, and shear modulus, G, are reported in Table 2.

Table 1. Analyzed reference schemes

Cases	Subsoil configuration	S/B	H/B
1	Homogeneous halfspace	0.5, 1, 2, 4	∞
2	Stratum over halfspace	1, 2, 4	3
3			5
4			10

The reference schemes shown in Fig. 1 were modelled through the FLAC3D software [7]. The mesh size, Δ, was calibrated according to the rule $\Delta < V_S/8f_{max}$ proposed by Kuhlemeyer and Lysmer (1973) [7], to allow seismic wave frequencies up to $f_{max} = 25$ Hz to be reliably propagated through the soil. The infinite extension of the soil in depth and along the lateral sides was simulated by means of dashpots, providing viscous normal and shear stresses proportional to the volume and shear wave velocities of the connected mesh element. A linear visco-elastic isotropic constitutive law was assigned to all soil elements. Viscous damping was introduced through the Rayleigh formulation and set to a very low value (1%) so that the overall damping encompassed in the numerical simulations may be assumed to be completely generated by wave scattering from the oscillating foundation, i.e., radiation damping.

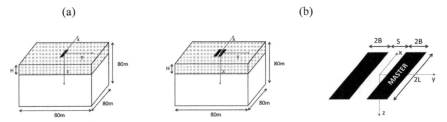

Fig. 1. Reference schemes for single footing (a) and twin footings (b) on stratum over halfspace.

Table 2. Soil parameters adopted in the numerical study

	γ [kN/m^3]	K [MPa]	G [MPa]	V_S [m/s]
Halfspace	19	12.2	5.6	54
Layer	19	38.6	19	100

Since the foundation has been assumed rigid, a uniform displacement field was simulated for all nodes placed along the foundation footprint. To this aim, a harmonic velocity along x-, y- or z-direction was applied to all grid points of the contact area in order to calculate the corresponding translational impedances. The input frequency $f = \omega/2\pi$ was alternatively set equal to 1, 3, 5, 7, 9, 11 and 13 Hz to cover a likely frequency band in the engineering vibration field. In the simulations of the twin foundations, both of them were excited in the same way. During each computation, the displacement in the centroid of the master foundation (Fig. 1b) as well as the stress state at the foundation-soil interface were recorded. The contact stresses were later properly summed to compute the corresponding forces and calculate the real and the imaginary parts of the soil-foundation impedance. Finally, imposing a constant velocity at the nodes of the foundation, also the low-frequency values K_{ij} and C_{ij} in Eq. (1) were obtained. Details on the overall developed procedure and its validation against analytical solutions have been reported in Zeolla et al. (2020) [9].

3 Results

3.1 Low Frequency Stiffness

Table 3 and 4 respectively report the low-frequency stiffness computed for the rectangular footing on homogeneous halfspace and on stratum over halfspace for different values of the foundation-foundation spacing, S, and layer thickness H. Both S and H have been divided by the half-width B of the foundation. By analyzing the numerical results in Table 3 and Table 4, it emerges that, when the nearby foundation is added, all stiffness components of the master foundation are lower than the values corresponding to the same footing isolated. This effect diminishes as the spacing between the two foundations increases. As expected, both single and twin footing systems are stiffer when settled on a stratum over halfspace.

Table 3. Low-frequency translation stiffness components of the master foundation on halfspace.

		Single	S/B = 0.5	S/B = 1	S/B = 2	S/B = 4
H/B = ∞	K_{zz} [MN/m]	96.40	64.34	68.69	73.42	79.15
	K_{yy} [MN/m]	64.66	44.96	45.31	47.67	49.36
	K_{xx} [MN/m]	60.73	42.89	44.35	47.49	49.46

Table 4. Low-frequency translation stiffness of a rigid strip on stratum over halfspace.

		Single	S/B = 1	S/B = 2	S/B = 4
H/B = 3	K_{zz} [MN/m]	134.35	115.26	121.77	125.80
	K_{yy} [MN/m]	75.17	58.42	61.55	64.30
	K_{xx} [MN/m]	68.73	56.84	60.84	63.18
H/B = 5	K_{zz} [MN/m]	118.24	99.31	106.59	112.62
	K_{yy} [MN/m]	69.46	52.98	55.77	58.06
	K_{xx} [MN/m]	63.75	51.31	55.03	57.61
H/B = 10	K_{zz} [MN/m]	103.25	82.45	88.82	95.10
	K_{yy} [MN/m]	64.63	48.18	50.65	52.52
	K_{xx} [MN/m]	59.90	46.93	50.23	52.44

The numerical results corroborate what expected from the simple theory of elasticity and Boussinesq solutions in the static filed. The static stiffness of the master footing in a group is always smaller than that of the same foundation considered isolated ($K_{double} < K_{single}$). In addition, for translational modes the static stiffness of the master footing in a group decreases with decreasing the ratio S/B due to cross interaction effects. Basing on the above numerical outcomes, an interaction coefficient may be defined to quantify the group effect on the stiffness of the master foundation:

$$\alpha_{ij} = \left(K_{ij\ single} - K_{ij\ group}\right)/K_{ij\ single} \qquad (2)$$

where $K_{ij\ single}$ and $K_{ij\ group}$ are the stiffness components of the master foundation in case it is alone or in group.

Figure 2 shows the variation with the S/B ratio of the interaction coefficient associated to the translation along x, y, and z. As expected, the interaction reduces as the spacing ratio S/B increases. The α-values are lower for the stratum-over-halfspace subsoil and converge to the halfspace solution as the ratio H/B increases.

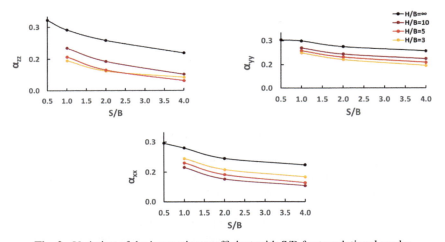

Fig. 2. Variation of the interaction coefficient with S/B for translational modes

3.2 Dynamic Coefficients of a Single Foundation on a Stratum over Halfspace

Figure 3 shows the dynamic stiffness coefficients of the translation impedances along the vertical axis (z) and along the horizontal axis (y) which is parallel to the footing width (Fig. 2b). As reported in previous literature studies [1], the variation of the dynamic stiffness coefficients with frequency is almost smooth when the footing is placed on the halfspace. Conversely, if the same footing is placed on a stratum the dynamic coefficients k exhibits many peaks and valleys. In the latter case, the waves generated by the oscillating foundation are reflected back by the interface between the stratum and the underlying stiffer halfspace. The observed fluctuation of k descends from the interference between the reflected waves and those propagated from the foundation. As a matter of fact, the first valleys of k_{yy} are centered around $a_0 = 0.50, 0.28$ and 0.15 for H/B = 3, 5 and 10, respectively. Being $V_S = 54$ m/s and B = 1 m, such a_{0s} correspond to a frequency equal to 4.3 Hz, 2.4 Hz and 1.3 Hz, which are the first shear frequency of each stratum-over-halfspace configuration as computed through the Strata software [10]. Multiple wave reflections generate P, S and Rayleigh waves making the consecutive valleys to occur around frequencies close but different from the consecutive natural frequencies of the subsoil. The shape of the resonant valleys of k_{zz} and k_{yy} becomes less pronounced with increasing the ratio H/B so that for H/B = 10 the stiffness coefficients $k(a_0)$ approach the curves corresponding to the halfspace solution.

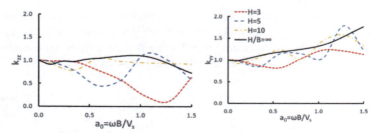

Fig. 3. Stiffness dynamic coefficients for swaying modes (along z and y) of the master foundation (single) on halfspace or stratum over halfspace.

3.3 Group Effect on the Dynamic Stiffness Coefficients

Figure 4 compares the stiffness dynamic coefficients for swaying modes (along z and y) of the master foundation when it is in-group with a nearby identical footing. Again, the two subsoil configurations of halfspace and of a stratum over halfspace have been considered.

As the halfspace solution concerns, the impedances of the master foundation in group (continuous lines) are close to those of the single footing (continuous black line) for low values of the dimensionless frequency a_0. Significant differences may be observed at higher frequencies on the k_{zz} coefficient that shows significant undulations. The dimensionless frequency at which the presence of an additional footing influences the results decreases with increasing S/B. The interference occurs when the wave originated from the clone foundation forces the master one to move out of phase with respect to its original motion. Such effect depends on the forcing frequency and on the S/B ratio. Actually, longer wavelengths, associated to lower frequencies, make very close foundations (e.g., S/B = 0.25) to move together with minor differences with respect to the single footing response.

As already observed in Sect. 3.2 for the single strip, the stratum-over-halfspace case makes the stiffness coefficients (dotted lines) to fluctuate more and more over the analysed range of frequencies due to multiple interference between the wave field generated by the master foundation and that reflected at the interface between the two soil layers. The twin footing system exacerbates such response, especially for the dynamic coefficients k_{yy} that regulates the swaying mode along the alignment of the two footings.

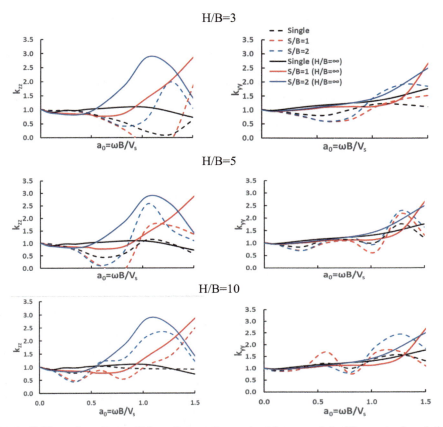

Fig. 4. Stiffness dynamic coefficients for swaying modes (along z and y) of the master foundation (in group) on halfspace or stratum over halfspace

4 Conclusions

The dynamic cross interaction between two closely-spaced shallow foundations placed on a halfspace or on a stratum over halfspace was numerically investigated through a 3D continuum approach solved numerically by the finite difference technique.

For the double footing system, the cross interaction produces a reduction of the low-frequency stiffness for all degrees of freedom of the footing. The interaction effect is higher at smaller distances between the two nearby foundations. Interaction coefficients were introduced to quantify the change in low-frequency stiffness of the master footing in dependence of footing-footing spacing and layer thickness in case of a stratified soil deposit.

The dynamic stiffness coefficients of the master foundation in-group were found almost coincident with those of the single foundation in the low frequency range but this statement is no longer true at higher frequencies of oscillation.

The case of one or more footings on a stratum over halfspace proved to be more challenging to deal with as a wavier trend of the stiffness dynamic coefficients with frequency was found. Peaks and valleys correspond to frequencies almost close to the natural ones of the stratum.

Cross interaction among neighboring foundations can produce a beneficial or detrimental effect depending on oscillation frequency, natural frequencies of the soil deposit and distance between the foundations of the group. For some cases, neglecting these effects could lead to a non-conservative design. The problem is surely worth of deeper insight to propose guidelines for designing structures in densely populated areas.

Acknowledgements. The work has been developed in the framework of the 2019–2021 ReLUIS-DPC research programme funded by the Italian Civil Protection Department, as a contribution to the geotechnical Work Package 'Soil-Foundation-Structure Interaction' (Task 16.3).

References

1. Gazetas, G.: Analysis of machine foundation vibrations: state of the art. Int. J. Soil Dyn. Earthq. Eng. **2**, 2–42 (1983)
2. Warburton, G.B., Richardson, J.D., Webster, J.J.: Forced vibrations of two masses on an elastic half space. ASME J. Appl. Mech. March **38**(1), 148–156 (1971)
3. Chang-Liang, V.: Dynamic response of structure in layered soils. Ph.D. thesis, MIT (1974)
4. Roesset, J.M., Gonzalez, J.J.: Dynamic interaction between adjacent structures. Dyrt Meth. Soil Rock Mech. **1**, 127 (1977)
5. ATC-3: Applied Technology Council, Tentative Provisions for the Development of Seismic Regulations for Buildings, National Science Foundation and National Bureau of Standards, Washington, DC, 1st edn. (1978), 2nd edn. (1984)
6. Qian, J., Beskos, D.E.: Dynamic interaction between 3-D rigid surface foundations and comparison with the ATC-3 provisions. Earthq. Eng. Struct. Dyn. **24**, 419–437 (1995)
7. Itasca Consulting Group, Inc.: FLAC3D User Manual: Version 5.0. USA (2004)
8. Kuhlmeier, R.L., Lysmer, J.: Finite element method accuracy for wave propagation problems. J. Soil Dyn. Div. **99**, 421–427 (1973)
9. Zeolla, E., De Silva, F., Sica, S.: Dynamic cross-interaction between two closely-spaced shallow foundations. In: Compdyn 2021, 8th ECCOMAS Thematic Conference on Computational Methods in Structural Dynamics and Earthquake Engineering, Athens (2021)
10. Kottke, A., Rathje, E.: Technical Manual for Strata (2021)

Effects of Nonliquefiable Crust on the Seismic Behavior of Pile Foundations in Liquefiable Soils

Gang Zheng[1,2,3], Wenbin Zhang[1,2], and Haizuo Zhou[1,2,3](✉)

[1] School of Civil Engineering, Tianjin University, Tianjin 300072, China
hzzhou@tju.edu.cn
[2] Key Laboratory of Coast Civil Structure Safety, Tianjin University, Ministry of Education, Tianjin 300072, China
[3] State Key Laboratory of Hydraulic Engineering Simulation and Safety, Tianjin University, Tianjin 300072, China

Abstract. Pile foundations located in liquefiable soil deposits are vulnerable to liquefaction-induced damage during earthquakes. The constraint conditions of both the pile head and pile tip are vital for the dynamic response of a pile. In this paper, a validated three-dimensional finite element model, where a PressureDependMultiYield constitute model based on multisurface plasticity theory is utilized for soil, is established to study the seismic behavior of a single pile foundation in liquefiable ground with variable nonliquefiable crust thicknesses. The effects of the properties of the nonliquefiable crust and pile head fixity on the dynamic behavior are investigated. Analysis shows that the existence of a nonliquefiable crust significantly affects the behavior of end-bearing piles and floating piles. The effect of a pile head constraint varies with the properties of the nonliquefiable crust.

Keywords: Pile foundation · Liquefaction · Nonliquefiable crust · Pile head constraint

1 Introduction

Pile foundations located in saturated sandy soils are vulnerable to failure during earthquakes [1]. Recent case histories have shown that seismic-induced liquefaction has caused significant damage to pile-supported structures [2–4]. Due to the loss of soil strength and stiffness, the performance of piles in liquefiable soil is more complex than that of piles in nonliquefiable soil. Many researchers have used various methods, such as centrifuge shaking table tests [5–9], dynamic Winkler models [10–12], and three-dimensional finite element models [13–18], to study the seismic behavior of pile foundations located in liquefiable soils. It can be clearly observed that a pile head rotation constraint and overlying nonliquefiable crust significantly affect the dynamic behavior of the soil-pile system. For example, Finn [1] concluded that the existence of a stiff crust would dramatically increase the moment and deflection demands on a pile. Wang et al.

[13] reported that a pile head rotation constraint changes the kinematic and inertial interactions of the piles. A nonliquefiable crust weakly constrains the rotation of a pile head. Rahmani et al. [14] observed that the fixity of a pile head more strongly influences piles embedded in dry soils than that of piles in liquefiable soils. Dash et al. [19] summarized the mechanism of pile failure in liquefiable soils. End-bearing piles and floating piles tend to suffer distinct damage due to the constraint of pile tips. Although researchers have investigated the behavior of single piles in liquefiable soils, limited studies have been conducted to quantitatively compare the effect of nonliquefiable crust and pile head fixity conditions on both end-bearing and floating piles in liquefiable soils.

This paper aims to enhance the understanding of nonliquefiable crust and pile head fixity conditions on the seismic behavior of piles in liquefiable soils. The Open System for Earthquake Engineering Simulation (OpenSees) platform [20] is adopted, and the established model is validated through a well-established centrifuge test. Then, parametric studies are performed to investigate the effect of crust properties and pile head fixity conditions. Finally, a comparison of seismic behavior between end-bearing piles and floating piles is discussed.

2 Numerical Model and Validation

2.1 Numerical Model

All models were calculated using OpenSeesPL software [21]. Figure 1 shows the finite element mesh of the centrifuge model. Due to symmetry, only half of the model is taken perpendicular to the vibration direction for calculation. Saturated sands are simulated using 8-node brickUP elements, and each node includes three translational degrees of freedom and one pore pressure degree of freedom. The PressureDependMultiYield material [22] is used to simulate the material properties of saturated sand. The pile foundation is modeled with elastic materials, and the flexural stiffness EI is 429,000 kNm^2. The ElasticBeamColumn unit is used to simulate the pile foundation. The pile nodes contain three translational degrees of freedom and three rotational degrees of freedom. A rigid beam is set between the pile and cap to connect the soil and the pile. As introduced by some researchers, this is considered to be an effective method for simulating the pile volume and pile-soil connection [16, 23].

In the first step, the weight of the soil is applied in the model to simulate the initial ground stress. The three translational degrees of freedom of the bottom node are all fixed. The nodes on the side are fixed in the normal direction. The pore pressure on the water table surface of the model is set to 0 to simulate drainage conditions. The second step is to install the pile foundation. In this step, the soil occupying the pile volume is removed and replaced with piles and rigid connections, followed by monotonic loading from the self-weight of the structure. The third step is ground motion excitation. The three translational degrees of freedom of the nodes on the boundary perpendicular to the excitation direction are bound to simulate the shearing effect in an earthquake. The base acceleration recorded in the centrifuge experiment is applied to the finite element model by the uniformExcitation command in OpenSees.

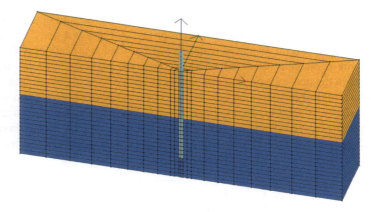

Fig. 1. Model configuration

2.2 Centrifuge Test and Validation

The established numerical model was validated through a dynamic centrifuge experiment (CSP-2F) carried out by Wilson et al. [5]. Figure 2 shows the layout of the centrifuge experiment. The centrifuge model is 51.6 m in length, 20.6 m in width, and 20.5 m in height. The size is consistent with the centrifuge prototype. The main parameters of the material are shown in Table 1. The soil layer is double-layer Nevada sand, including an upper loose sand layer and a bottom dense sand layer. The structure uses steel pipe pile models with diameters and wall thicknesses of 670 mm and 19 mm, respectively. The pile length is 20.6 m, of which 3.8 m is above the soil surface, and the pile top has a

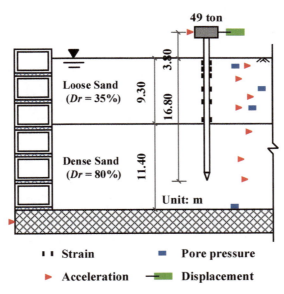

Fig. 2. Centrifuge test configuration [5]

concentrated mass of 49 tons. Kobe waves are input from the bottom of the model, and the ground motion used in this study is shown in Fig. 3. Figures 4 and 5 compare the soil response and structural response, respectively. For the loose sand layer (Fig. 4(d)), the excess pore pressure reaches the effective confining pressure after an initial rise, while the soil in the dense layer does not liquefy (Fig. 4(c)). As shown in these figures, the calculated soil response (i.e., acceleration and excess pore pressure) and structural response (i.e., i.e., superstructure acceleration and pile bending moment) show good consistency with the centrifuge results, indicating that the model established in this paper can capture the seismic response of pile foundations in liquefiable soils.

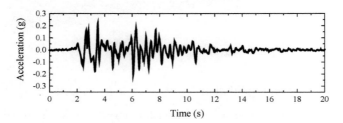

Fig. 3. Input motion

Table 1. Material properties [21]

Parameter	Loose sand	Mid-dense sand	Dense sand
Mass density (t/m^3)	1.7	1.9	2.1
Reference shear modulus (kPa)	55000	75000	130000
Reference bulk modulus (kPa)	150000	200000	390000
Friction angle	29	33	40
Reference confining pressure (kPa)	80	80	80
Peak shear strain	0.1	0.1	0.1
Phase transformation angle (degree)	29	27	27
Contraction parameter	0.21	0.07	0.03
Dilation parameter 1	0	0.4	0.8
Dilation parameter 2	0	2	5
Liquefaction parameter 1	10	10	0
Liquefaction parameter 2	0.02	0.01	0
Liquefaction parameter 3	1	1	0
Permeability coefficient	6.6e-5	6.6e-5	3.7e-5

Effects of Nonliquefiable Crust on the Seismic Behavior of Pile Foundations 1269

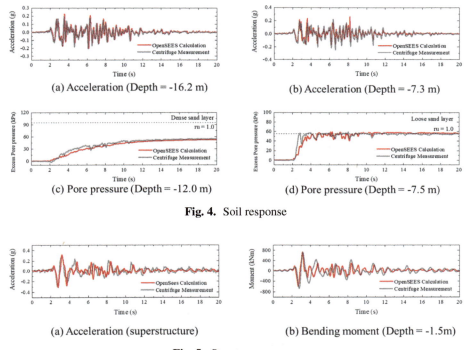

Fig. 4. Soil response

Fig. 5. Structure response

3 Results

3.1 Configuration

To study the dynamic response of a single pile in liquefiable soils, two models were established, including an end-bearing pile model and a floating pile model, as shown in Fig. 6. The effect of the pile head fixity, thickness (T), and relative density (Dr) of the nonliquefiable crust on the dynamic response of the pile foundation are studied. The nonliquefiable crust layer in this paper is represent by dry sand. The research is based on the validated numerical model as described in Sect. 2. The Kobe motion in Fig. 2 is adopted.

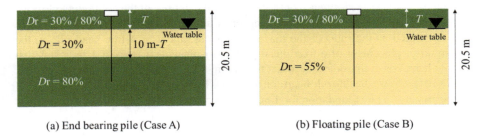

Fig. 6. Studied cases

3.2 Lateral Displacement

Figures 7 (a) and 7 (b) show the maximum lateral displacements of the end-bearing piles and floating piles with/without overlying nonliquefiable crust, respectively. For end-bearing piles, under the fixed head condition, the overlying nonliquefiable layer with a relative density of 30% increases the maximum lateral displacement of the pile, and the dense nonliquefiable layer reduces the lateral displacement. For the free head condition, the overlying nonliquefiable crust is beneficial to reduce the lateral displacement, and the displacement decreases as the density of the overlying nonliquefiable crust increases.

For floating piles, the nonliquefiable crust significantly reduces the lateral displacement of the pile head, but it increases the lateral displacement of the pile body. In addition, the lateral displacement is not sensitive to the density of the nonliquefiable crust.

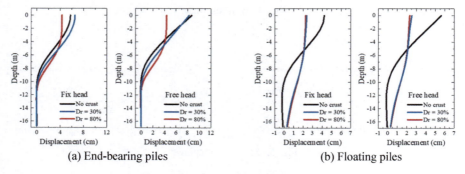

(a) End-bearing piles   (b) Floating piles

Fig. 7. Maximum lateral displacement

Figure 8 shows the influence of the thickness and density of nonliquefiable crust on the maximum lateral displacement of end-bearing piles. The lateral displacement mainly occurs in the liquefiable soil layer. With the increase in the thickness of the nonliquefiable crust, the difference in the maximum lateral displacement between the fixed head cases and the free head cases decreases, indicating that the nonliquefiable crust acts as a constraint on pile rotation. The constraint effect increases as the density of the nonliquefiable layer increases. Figure 9 shows the effect of the nonliquefiable crust thickness and density on the maximum lateral displacement of floating piles. Unlike end-bearing piles, the bottom of the pile is inclined to a certain extent due to the lack of constraint from a dense bottom layer.

Figure 10 shows the influence of the thickness and density of the nonliquefiable crust on the displacement ratio for end-bearing piles and floating piles. The displacement ratio is defined as the ratio of the maximum displacement of a free head pile to the maximum displacement of a fixed head pile. A displacement ratio of 1 indicates that the pile head constraint does not affect the lateral displacement of the pile foundation. Figure 10 (a) shows that the pile head constraint has limited effect on the maximum lateral displacement of a pile foundation for the cases of a crust relative density of 80% and crust thickness greater than 3 m. When the density of the overlying nonliquefiable layer is 30% (Fig. 10 (b)), as the crust thickness increases, the displacement ratio first decreases and then increases. The lateral displacement of the pile foundation can be divided into two components, one in the liquefiable layer and the other in the overlying nonliquefiable crust. When the thickness of the nonliquefiable layer is within 3 m for the floating piles, the displacement of the pile foundation in the liquefiable layer plays a leading role in the maximum lateral displacement of the pile foundation. Increases of the thickness of the nonliquefiable crust reduce the lateral displacement in the liquefiable layer. For the floating pile cases in which the thickness of the loose nonliquefiable crust is greater than 3 m, the lateral displacement of the pile in the overlying nonliquefiable crust dominates the lateral movement of the pile. A pile head constraint affects the lateral displacement more significantly with the increasing thickness of the loose nonliquefiable crust. For end-bearing piles, the valley value of crust thickness is 5 m. The difference between end-bearing piles and floating piles may be because the displacement distribution of end-bearing piles is more concentrated.

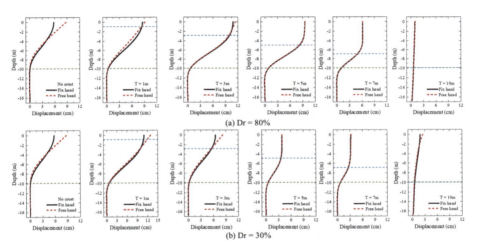

Fig. 8. Maximum lateral displacement for end-bearing piles with varied crust thicknesses

3.3 Bending Moment Envelope

Figures 11 (a) and 11 (b) show the maximum bending moment envelope diagrams of end-bearing piles and floating piles with/without crust. For end-bearing piles, the nonliquefiable crust under a fixed head condition reduces the bending moment of the pile top, especially when the density of the overlying nonliquefiable crust is 80%. However, the existence of a nonliquefiable layer increases the bending moment of the pile body. As the density of the overlying nonliquefiable layer increases, the bending moment of the pile body increases. For floating piles, nonliquefiable crusts of different densities under fixed head conditions significantly reduce the pile head bending moment and pile body bending moment. For a free pile head, an overlying nonliquefiable crust reduces the maximum bending moment of the pile body.

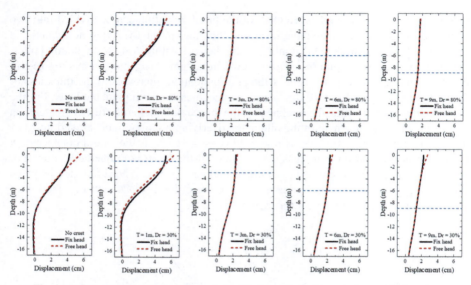

Fig. 9. Maximum lateral displacement for floating piles with varied crust thicknesses

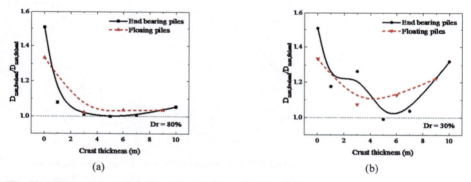

Fig. 10. Effect of crust thickness on the maximum lateral displacement ratio between a fixed head and free head pile with (a) crust relative density = 80% and (b) crust relative density = 30%

Figure 12 displays the maximum bending moment envelope of the end-bearing piles. As the thickness of the nonliquefiable layer increases, the bending moment at the pile head decreases. A maximum bending moment is found at the interface between the liquefiable layer and the nonliquefiable layer, especially for the case with a crust relative density of 80%. When the overlying soil layers are all nonliquefiable (T = 10 m), the bending moment of a pile foundation is significantly reduced, indicating that the kinematic effect of a liquefiable layer on the pile significantly increases the risk of pile damage.

Effects of Nonliquefiable Crust on the Seismic Behavior of Pile Foundations 1273

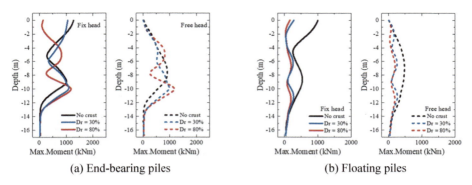

(a) End-bearing piles   (b) Floating piles

Fig. 11. Maximum bending moment envelops

Figure 13 displays the maximum bending moment envelope of the floating piles. As the nonliquefiable crust thickness increases, the bending moment at the pile head and the pile body decreases. When the thickness of the nonliquefiable crust is greater than 3 m, the bending moment at the pile foundation is less significant, indicating that the failure of floating piles may be more likely to be caused by other factors, such as excessive settlement [19].

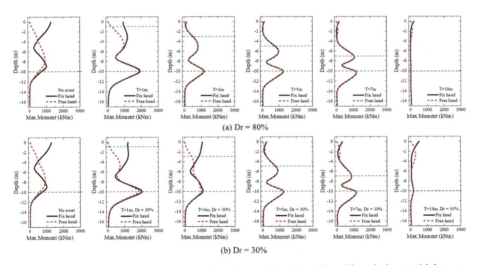

Fig. 12. Maximum bending moment envelopes for end-bearing piles with varied crust thicknesses

Figure 14(a) shows the effect of the nonliquefiable crust thickness on the maximum bending moment of a pile head. For end-bearing and floating piles, the presence of a nonliquefiable crust has a constraining effect on the pile head. As the thickness of the nonliquefiable layer increases, the bending moment of the pile head decreases. Furthermore, for end-bearing piles and a 30% relative density of nonliquefiable crust, the pile head bending moment is unchanged after the crust thickness exceeds 5 m. For other

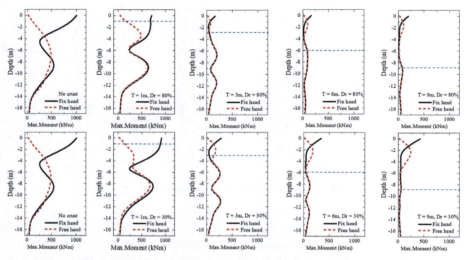

Fig. 13. Maximum bending moment envelopes for floating piles with varied crust

cases, the influence of crust thickness becomes limited when the nonliquefiable crust thickness exceeds 3 m.

The maximum bending moments of a pile body with varied crust thicknesses are plotted in Fig. 14(b). For floating piles, the lateral displacement of the piles decreases with increasing crust thickness, thus decreasing the bending moment in the pile body. For end-bearing piles, the bending moment in the pile body is larger than that in the pile head due to the constraint from the bottom. The bending moment reaches a peak with a crust thickness of 1 m. When the crust thickness rises from 3 m to 7 m, the lateral displacement of the pile drops by approximately 35%. However, the bending moment remains constant. This phenomenon may be attributed to the fact that a thicker crust tends to increase the kinematic force of the soil.

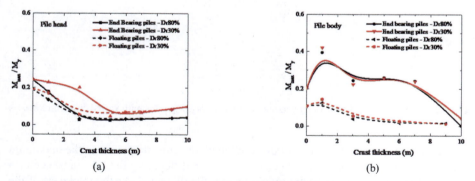

Fig. 14. Effect of crust thickness on the maximum bending moment of the (a) pile head and (b) pile body

4 Conclusion

The main conclusions are as follows:

(1) For end-bearing piles, a nonliquefiable soil layer may reduce both the maximum lateral displacement and the bending moment. A proper crust thickness and density are beneficial for reducing the lateral displacement and bending moment of piles. The presence of a liquefiable soil layer significantly increases the risk of pile damage.
(2) For floating piles, a nonliquefiable crust reduces the bending moment of the pile foundation. Compared with end-bearing piles, the bending moment of a floating pile is smaller due to the lack of a bottom constraint, especially for the cases considered in this work in which the crust thickness is greater than 3 m. The failure of floating piles in a liquefiable soil layer may be more prone to be caused by other factors, such as excessive settlement.
(3) When a thin and loose nonliquefiable crust overlies liquefiable soil, the constrained conditions of a pile head significantly affect the lateral displacement and bending moment of the pile. For the cases in which the crust thickness exceeds 3 m, the pile head fixity conditions have almost no effect on the response of the pile.

Acknowledgments. This research was funded by the National Natural Science Foundation of China (Nos. 52078335, 41630641, and 52078337) and the project of the Natural Science Foundation of Tianjin, China (No. 20JCQNJC01080).

References

1. Finn, W.D.L.: 1st Ishihara lecture: an overview of the behavior of pile foundations in liquefiable and non-liquefiable soils during earthquake excitation. Soil Dyn. Earthq. Eng. **68**, 69–77 (2015)
2. Dash, S.R., Govindaraju, L., Bhattacharya, S.: A case study of damages of the Kandla port and customs office tower supported on a mat–pile foundation in liquefied soils under the 2001 Bhuj earthquake. Soil Dyn. Earthq. Eng. **29**(2), 333–346 (2009)
3. Tokimatsu, K., et al.: Building damage associated with geotechnical problems in the 2011 Tohoku Pacific earthquake. Soils Found. **52**(5), 956–974 (2012)
4. Fujii, S., et al.: Investigation and analysis of a pile foundation damaged by liquefaction during the 1995 hyogoken-nambu earthquake. Soils Found. **38**, 179–192 (1998)
5. Wilson, D.W.: Soil-pile-superstructure interaction in liquefying sand and soft clay. In: Department of Civil & Environmental Engineering. University Of California at Davis (1998)
6. Bhattacharya, S., Madabhushi, S.P.G., Bolton, M.D.: An alternative mechanism of pile failure in liquefiable deposits during earthquakes. Geotechnique **54**(3), 203–213 (2004)
7. Dobry, R., et al.: Single piles in lateral spreads: field bending moment evaluation. J. Geotech. Geoenviron. Eng. **129**(10), 879–889 (2003)
8. Boulanger, R.W., et al.: Seismic soil-pile-structure interaction experiments and analyses. J. Geotech. Geoenviron. Eng. **125**(9), 750–759 (1999)
9. Liu, X., Wang, R., Zhang, J.M.: Centrifuge shaking table tests on 4x4 pile groups in liquefiable ground. Acta Geotech. **13**(6), 1405–1418 (2018)

10. Brandenberg, S.J., et al.: p-y plasticity model for nonlinear dynamic analysis of piles in liquefiable soil. J. Geotech. Geoenviron. Eng.. **139**(8), 1262–1274 (2013)
11. Wang, X.W., et al.: Efficient finite-element model for seismic response estimation of piles and soils in liquefied and laterally spreading ground considering shear localization. Int. J. Geomech. **17**(6), 06016039 (2017)
12. Su, L., et al.: Pile response to liquefaction-induced lateral spreading: a shake-table investigation. Soil Dyn. Earthq. Eng. **82**, 196–204 (2016)
13. Wang, R., Fu, P., Zhang, J.-M.: Finite element model for piles in liquefiable ground. Comput. Geotech. **72**, 1–14 (2016)
14. Rahmani, A., Pak, A.: Dynamic behavior of pile foundations under cyclic loading in liquefiable soils. Comput. Geotech. **40**, 114–126 (2012)
15. Haldar, S., Babu, G.L.S.: Failure mechanisms of pile foundations in liquefiable soil: parametric study. Int. J. Geomech. **10**(2), 74–84 (2010)
16. Zhang, X.Y., et al.: Using peak ground velocity to characterize the response of soil-pile system in liquefying ground. Eng. Geol. **240**, 62–73 (2018)
17. Wang, R., Liu, X., Zhang, J.-M.: Numerical analysis of the seismic inertial and kinematic effects on pile bending moment in liquefiable soils. Acta Geotech. **12**(4), 773–791 (2016). https://doi.org/10.1007/s11440-016-0487-z
18. Wang, R., Zhang, J.M., Wang, G.: A unified plasticity model for large post-liquefaction shear deformation of sand. Comput. Geotech. **59**, 54–66 (2014)
19. Dash, S.R., Bhattacharya, S., Blakeborough, A.: Bending–buckling interaction as a failure mechanism of piles in liquefiable soils. Soil Dyn. Earthq. Eng. **30**(1–2), 32–39 (2010)
20. Mazzoni, S., McKenna, F., Scott, M.H., Fenves, G.L.: OpenSees command language manual. Pac. Earthq. Eng. Res. (PEER) Cent. **264**, 137–158 (2006)
21. Lu, J., Elgamal, A., Yang, Z.: OpenSeesPL: 3D lateral pile-ground interaction, user manual, beta 1.0 (2011)
22. Yang, Z.: Numerical modeling of earthquake site response including dilation and liquefaction. In: Department of Civil Engineering and Engineering Mechanics. Columbia University, New York (2000)
23. Rajeswari, J.S., Sarkar, R.: A three-dimensional investigation on performance of batter pile groups in laterally spreading ground. Soil Dyn. Earthq. Eng. **141**, 106508 (2021)

Numerical Implementation of Ground Behaviors Beneath Super-Tall Building Foundations During Construction

Youhao Zhou(✉) and Takatoshi Kiriyama

Shimizu Institute of Technology, 3-4-17 Ecchujima, Koto-ku, Tokyo, Japan
`y.zhou@shimz.co.jp`

Abstract. Recently, in Japan, the number of super-tall buildings has been increasing, and weight per area for some has exceeded 1000 kN/m^2. The behavior of soil in Japan under such high pressure has not been fully studied, and it is not clear whether such behavior can be simulated using conventional soil constitutive models. As a first step, a series of K0 triaxial tests using Toyoura sand was conducted to study the behavior of the soil beneath the super-tall building foundation during excavation and construction. In this study, a new constitutive model for geomaterials is proposed, where both the confining pressure dependency and strain dependency are considered. Finally, a series of numerical simulations of the CD triaxial test were carried out using the proposed model, along with conventional models for comparison. The proposed model best describes the stress–strain relation obtained from the test because both the confining pressure dependency and strain dependency are properly modelled.

Keywords: Super-tall building · Ground settlement · Triaxial test · Numerical modelling · Numerical simulation

1 Introduction

In recent years, the number of super-tall buildings has increased in Japan. The average weight per area of the building exceeds 1000 kN/m^2 in some cases, and the maximum local contact pressure can be as large as 1500 kN/m^2. Under such high pressure, both total and differential ground settlement during construction can be so large that they greatly impact the structural design of not only the foundation but also the entire building. Therefore, for such super-tall buildings, it is crucial to accurately predict ground settlement with the help of numerical analysis.

Around the world, settlement analysis for super-tall buildings of this level of self-weight, such as Burj Khalifa in Dubai [1], Shanghai Tower in China [2], and Lotte World Tower in Korea [3], have been performed using conventional soil constitutive models. However, owing to the difference in geotechnical conditions, it is unclear whether the same methods can be used for super-tall buildings in Japan. Tamaoki [4] proposed a method for predicting soil stiffness based on case studies of buildings with weight per area up to 500 kN/m² in Japan. Because the weight of upcoming new super-tall buildings is significantly larger, the applicability of this method is questionable.

In this study, as a first step, a series of K0 triaxial tests that simulate the stress path of the soil under a super-tall building foundation during excavation and construction was conducted using dense Toyoura sand. Consolidated drained (CD) triaxial tests were also performed to obtain the strength parameters of the sand. Next, a new constitutive model for soil is proposed, in which both the confining pressure dependency and strain dependency are considered. Finally, a numerical simulation of the CD triaxial test was performed using the proposed model along with conventional models for comparison.

2 Triaxial Compression Tests of Toyoura Sand

2.1 K0 Triaxial Test Using Toyoura Sand

A series of K0 triaxial tests were performed to simulate the stress path of the soil under a super-tall building foundation during excavation and construction. Figure 1 shows schematics of the stress path and timeline of the K0 triaxial test.

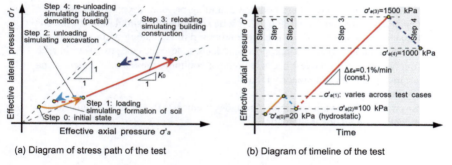

Fig. 1. Schematics stress path and timeline of K0 triaxial test

First, the specimen was subjected to hydrostatic pressure at step 0. Then, the specimen was subjected to four sequential loading and unloading steps. Step 1 simulates the formation of the soil layer by loading; step 2 simulates the excavation of soil above the foundation by unloading; step 3 simulates the construction of the super-tall building by loading; and step 4 simulates building demolition by unloading. Throughout the test, the lateral strain ε_r is constrained to zero ($\varepsilon_r \equiv 0$) by controlling the effective lateral pressure σ'_r in order to maintain the specimen in the K0 condition. Therefore, the test was named 'K0 triaxial test'. The test was performed under a constant axial strain rate of $\Delta \varepsilon_a = 0.1\%/\text{min}$ by controlling the applied axial load.

Table 1 lists the test cases and loading sequences. The effective axial pressure at step 1 $\sigma'_{a(1)}$ is varied across the cases to simulate the effective overburden pressure of soil at different depths. Test case IDs correspond to the value of $\sigma'_{a(1)}$. The axial displacement, axial load, lateral displacement, lateral pressure, and pore water pressure were monitored and recorded. The test results were processed according to JGS 0525-2020 [5]. Toyoura sand was used in the test. The cylindrical specimen was 5 cm in diameter and 10 cm in height. It is prepared by air pluviation to achieve a high relative density of $D_r = 80\%$.

Table 1. List of cases and loading sequence for K0 triaxial test

Case ID	Effective axial pressure σ'_a [kPa]				
	Step 0	Step 1	Step 2	Step 3	Step 4
250	20 (hydrostatic)	250	100	1500	1000
500		500			
1000		1000			
1500		1500			

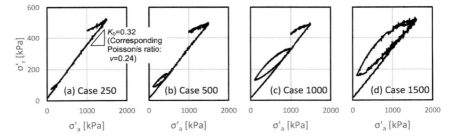

Fig. 2. Relation between effective axial pressure σ'_a and effective lateral pressure σ'_r

Figure 2 shows the relation between effective axial pressure σ'_a and effective lateral pressure σ'_r in all test cases. Regardless of the test case, $\sigma'_a \sim \sigma'_r$ relation appears as a straight skeleton curve during the initial loading (step 1). It then appears as an upward-convex curve during unloading (step 2) and as a downward-convex curve during reloading (step 3), forming a spindle-shaped hysteresis loop. However, after surpassing the unloading point during reloading, it appears as a straight line again, continuing the skeleton curve seen during the initial loading (step 1). This implies the lateral pressure coefficient K_0 remains a constant on the skeleton curve.

Figure 3 shows the relation between deviatoric stress s ($=[\sigma'_a - \sigma'_r]/2$) and deviatoric strain ε_{dev} ($=\varepsilon_a - \varepsilon_r$) for all cases. Note that because $\varepsilon_r \equiv 0$ throughout the K0 triaxial test, $\varepsilon_{dev} = \varepsilon_a$. In the figure, each step of the test is presented in different colors, whereas step 3 is further distinguished before and after surpassing the unloading point (see Fig. 1(b)). The $s \sim \varepsilon_{dev}$ relations are notably in downward-convex curves throughout the test.

To further examine this behavior, the reference stress σ_{ref} ($=\sigma - \sigma_{rev}$) and reference strain ε_{ref} ($=\varepsilon - \varepsilon_{rev}$) were calculated. Here, σ and ε are the stress tensor and strain tensor, respectively. The subscript rev indicates the last reverse point, which is the initial state, unloading point, or reloading point for the corresponding loading step. The subscript ref indicates the stress or strain relative to the last reverse point. Note that after surpassing the unloading point during the reloading step, the reverse point refers to the initial state instead of the reloading point. This is because the stress–strain relation typically returns to the skeleton curve from the hysteresis curve.

Figure 4 shows $s_{ref} \sim \varepsilon_{ref,dev}$ relation for all cases. Remarkably, the stress–strain relations appear in downward-convex curves during loading (steps 1 and 3) despite the strain dependency, indicating that the stiffness increases along with the effective confining pressure. In contrast, the stress–strain relations appear in upward-convex curves during unloading (steps 2 and 4), indicating that the stiffness decreases along with the effective confining pressure. The initial slope in step 2 becomes steeper in cases where $\sigma'_{a(1)}$ (i.e., stress of the unloading point) is higher, indicating a higher unloading stiffness at higher confining pressures. Although omitted due to space limitation, the same trend can be observed in the relation between effective mean stress p' $(=[\sigma'_a + 2\sigma'_r]/3)$ and volumetric strain ε_v $(=\varepsilon_a + 2\varepsilon_r)$.

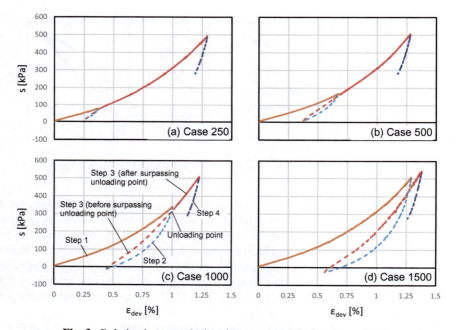

Fig. 3. Relation between deviatoric stress s and deviatoric strain ε_{dev}

The results imply that the confining pressure dependency of Toyoura sand largely affects its behavior during the K0 triaxial test, notably more so than the strain dependency. Therefore, it is important to consider the confining pressure dependency of the soil when predicting the vertical ground displacement during excavation and building construction, in addition to the strain dependency.

2.2 Consolidated Drained (CD) Triaxial Test of Toyoura Sand

A series of CD triaxial tests were conducted to obtain the strength parameters of Toyoura sand. Four test cases are planned with cell pressures of $\sigma = 250, 500, 1000,$ and 1500 kPa, which correspond to each case of the K0 triaxial tests. The tests were performed, measured, and processed according to JGS 0524-2020 [5]. This paper presents the results from cases with $\sigma = 250$ and 1500 kPa.

Figure 5 shows the relationship between the differential stress σ_D ($=\sigma_a - \sigma_r$) and axial strain ε_a. Figure 6 shows the stress path on the deviatoric stress and mean effective stress plane ($s \sim p'$). The dotted line in Fig. 6 represents the failure envelope:

$$q_f = c_s + H \cdot p' \quad (1)$$

where q_f is the failure stress, and c_s and H are the intercept and slope of the failure envelope, respectively.

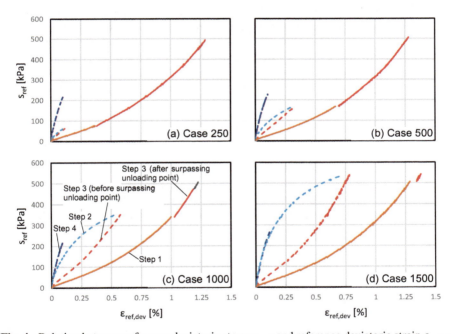

Fig. 4. Relation between reference deviatoric stress s_{ref} and reference deviatoric strain $\varepsilon_{ref,dev}$

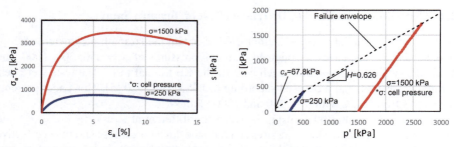

Fig. 5. Relation between differential stress σ_D and axial strain ε_a

Fig. 6. Stress path on s-p' plane

3 Constitutive Model of Soil with Implementation of Both Confining Pressure Dependency and Strain Dependency

A new constitutive model for three-dimensional numerical analysis, in which both the confining pressure dependency and strain dependency are considered, is proposed. As a first step, this paper only shows the modelling of the skeleton curve. A hyperbolic model is chosen to model the strain dependency in this study for simplicity, but any other functions can be applied to the model in a similar manner. To accommodate the loading in an arbitrary direction (i.e., regardless of element coordination), the stress–strain relation is defined in the deviatoric stress–strain ($s \sim \varepsilon_{dev}$) plane:

$$s = f\left(p', \varepsilon_{dev}\right) \tag{2}$$

$$ds = \partial f / \partial p' \cdot dp' + \partial f / \partial \varepsilon_{dev} \cdot d\varepsilon_{dev} \tag{3}$$

where s is the deviatoric stress ($s = [\sigma'_1 - \sigma'_3]/2$), p' is the effective mean stress ($p' = [\sigma'_1 + \sigma'_2 + \sigma'_3]/3$), and ε_{dev} is the deviatoric strain ($\varepsilon_{dev} = \varepsilon_1 - \varepsilon_3$). The strain dependency of the skeleton curve is modelled as follows:

$$s = \left(q_f \cdot G_0 \cdot \varepsilon_{dev}\right) / \left(q_f + G_0 \cdot \varepsilon_{dev}\right) \tag{4}$$

where q_f is the failure stress, as shown in Eq. 1, and G_0 is the initial shear modulus. The confining pressure dependency is modelled as follows:

$$G_0 = G_{0i} * p'^n \tag{5}$$

where G_{0i} is the normalized shear modulus and n is a parameter for confining pressure dependency. Consequently, the incremental form of the skeleton curve is obtained as

$$ds = \frac{(p' \cdot H \cdot G_0 \cdot \varepsilon_{dev} + n \cdot qf^2) \cdot G_0 \cdot \varepsilon_{dev}}{p' \cdot (qf + G_0 \cdot \varepsilon_{dev})^2} \cdot dp' + \frac{qf^2 \cdot G_0}{(qf + G_0 \cdot \varepsilon_{dev})^2} \cdot d\varepsilon_{dev} \tag{6}$$

The first term of Eq. 6 describes the confining pressure dependency, and the second term describes the strain dependency. Note that the incremental effective mean stress dp', which is a part of the incremental stress, is required to obtain the incremental deviatoric stress ds. This means that dp' effectively appears on both sides of Eq. 6. Therefore, iterations are required to obtain the convergent stress at each increment.

4 Numerical Simulation of CD Triaxial Test

To validate the proposed model, a numerical simulation of the CD triaxial test described in Sect. 2.2 is carried out as a first step. Numerical simulations of the K0 triaxial tests will be conducted in the future. The Mohr–Coulomb model (where only the strength depends on the confining pressure and the stress–strain relation is bi-linear) and a standard hyperbolic model (independent of confining pressure and stress–strain relation is in a smooth curve) are chosen for comparison.

Table 2. Parameters for proposed constitutive model

Parameter	Value
Normalized initial shear modulus G_{0i} [kN/m^2]	5300
Poisson's ratio ν	0.24
Parameter for confining pressure dependency n	0.44
Intercept of failure envelope c_s [kPa]	67.8
Slope of failure envelope H	0.626

Table 3. Parameters for Mohr–Coulomb and standard hyperbolic models

Parameter	Value	
	$\sigma = 250$ kPa	$\sigma = 1500$ kPa
Initial shear modulus G_0 [kN/m^2]	62 100	131 400
Poisson's ratio ν	0.24	
Cohesion c^* [kPa]	65.6	
Angle of internal friction φ^* [deg]	31.2	
Reference strain $\gamma_{0.5}^{**}$	3.62×10^{-3}	7.67×10^{-3}

*: for Mohr–Coulomb only **: for standard hyperbolic only

Tables 2 and 3 show parameters for each model. For the proposed model, the parameters were back-calculated from both triaxial tests. For the Mohr–Coulomb model, because the confining pressure dependency for the initial shear modulus G_0 is not considered, G_0 is given different values for each case according to the cell pressure. For the standard hyperbolic model, because the confining pressure dependency is not considered at all, both G_0 and $\gamma_{0.5}$ are given different values for each case according to the cell pressure (i.e., both stiffness and strength are determined by the initial state for the standard hyperbolic model).

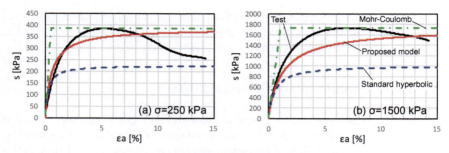

Fig. 7. Numerical simulation of CD triaxial test

Figure 7 shows the relationship between the deviatoric stress s and axial strain ε_a. In the test, the deviatoric stress reaches the failure stress at approximately $\varepsilon_a = 5\%$ and shows the strain softening behavior afterwards. The Mohr–Coulomb model precisely estimates the failure stress because the strength is pressure-dependent. However, owing to its bilinear nature, the Mohr–Coulomb model overestimates the stress leading up to the failure point. The standard hyperbolic model significantly underestimates the stress after the axial strain surpasses 0.5% approximately, because it does not consider the change in either stiffness or strength as the confining pressure increases. The proposed model underestimates the failure stress because it uses a hyperbolic function as its skeleton curve, which approaches asymptotically to but never reaches the failure envelope. Overall, the proposed model gives the closest result to the test among the three, because both the confining pressure dependency and strain dependency are properly modelled.

5 Conclusions

A series of triaxial tests using Toyoura sand were conducted to study the behavior of soil under a super-tall building foundation during excavation and construction, and a new constitutive model that considers both confining pressure dependency and strain dependency is proposed. The conclusions are as follows.

1. The confining pressure dependency of Toyoura sand largely affects its behavior during the K0 triaxial test, which simulates the stress path of soil under the super-tall building foundation during excavation and construction. Therefore, it is important to consider the confining pressure dependency of the soil, in addition to the strain dependency.
2. A new constitutive model using a hyperbolic function as its skeleton curve while also considering the confining-pressure dependency is proposed. The stress–strain relation is defined in the deviatoric stress–strain plane to accommodate the loading in an arbitrary direction (i.e., regardless of element coordination).
3. Compared with the Mohr–Coulomb model and standard hyperbolic model, the proposed model best describes the stress–strain relation obtained from the CD triaxial test, because both the confining pressure dependency and strain dependency are properly modelled.

In the next step, the authors plan to perform laboratory tests using in-situ soil samples to further study the soil behavior, and to add a hysteresis curve to the proposed model in order to accommodate cyclic loading. Numerical simulations of the CD triaxial test are conducted as a preliminary step in this paper, and numerical simulations of the K0 triaxial tests presented in this paper will also be conducted in the future. Furthermore, a variation of the proposed constitutive model based on modified RO model is also planned for additional versatility.

References

1. Poulos, H.G., Bunce, G.: Foundation design for the Burj Dubai–the world's tallest building. In: Proceedings of the 6th International Conference on Case Histories in Geotechnical Engineering, Arlington, VA, Paper, vol. 1 (2008)
2. Xiao, J.H., Chao, S., Zhao, X.H.: Foundation design for the Shanghai Center Tower. In: Advanced Materials Research, vol. 243, pp. 2802–2810. Trans Tech Publications Ltd. (2011)
3. Sze, J., Lam, A.K.: A combo foundation does the job: piled-raft foundation design for a supertall skyscraper. Geo-Strata—Geo Inst. ASCE **23**(4), 42–49 (2019)
4. Tamaoki, K., Katsura, Y., Nishio, S., Kishida, S.: Estimation of Young's Moduli of bearing soil strata. In: Excavation in Urban Areas, KIGForum 1993 (Adachi ed.), pp. 23–33 (1993)
5. The Japanese Geotechnical Society. Japanese Geotechnical Society Standards: Laboratory Testing Standards of Geomaterials (2017)

Slope Stability and Reinforcement

Failure Mechanism Analysis of Loess Slope Under the Coupling Effect of Rainfall and Earthquake Using Shaking Table Test

Jinchang Chen[1], Lanmin Wang[1,2], and Ailan Che[1](✉)

[1] School of Naval Architecture, Ocean and Civil Engineering, Shanghai Jiao Tong University, Shanghai 200040, China
alche@sjtu.edu.cn

[2] Lanzhou Institute of Seismology, China Earthquake Administration, Lanzhou 730000, China

Abstract. Rainfall and earthquake are two main factors that trigger landslides. In practice, landslides would occur more easily under the coupling effect of them and present long sliding distance and large sliding scale characteristics, which brought huge losses and casualties to people. Shaking table test of a model loess slope was performed in the study. The wave of Minxian station in 2013 Minxian-Zhangxian Ms6.6 earthquake with intensities of 50 gal, and 100 to 800 gal (interval 100gal) were exerted on the slope successively. Then, the rainfall was applied after 800 gal by using artificial rainfall device and 1300 gal is applied at last. Based on Hilbert-Huang transform (HHT), we investigated the acceleration response of the slope in time-frequency-energy space. The change characteristics of Hilbert energy and predominate frequency under the single effect of earthquake and the coupling effect of rainfall and earthquake were analyzed based on marginal spectrum. According to the dynamic response results of the slope, failure mechanism and damage evolution law of the landslide under the coupling effect of rainfall and earthquake were clarified. The study shows that the coupling effect greatly decreased the stability of the slope and the final damage presents soil flow. The instantaneous energy and cumulative energy have a great dissipation under the coupling effect, which provide the kinetic energy to the soil flow.

Keywords: Loess slope · Rainfall and earthquake · Shaking table test · HHT · Failure mechanism

1 Introduction

Loess is a fragmented porous multiphase medium. When it is soaked in water or subjected to medium or strong earthquake, its structure will collapse, resulting in the loss of strength completely [1, 2]. So, rainfall and earthquake are two main factors that induced landslides in practice. From a practical point of view, it seems unnecessary to study the landslide induced by the coupling effect of the two factors, because the climatic characteristics of arid and semi-arid areas and the small probability of earthquake events in the loess region. However, the soil or mud flow in Yongguang village induced by the Minxianzhangxian Ms 6.6 earthquake in Gansu Province in 2013 has changed people's inherent understanding. Before the earthquake, there was about a week of rainfall in the local area, and the soil moisture content was very high. After the earthquake, a loess landslide with large volume was induced immediately and 12 people were buried [3].

Studies on the stability of slope under the coupling effect of rainfall and earthquake have been carried out. However, few of them investigate the failure mechanism in energy point. It is concluded that the coupling action will reduce the shear strength of the landslide and cause the instability and collapse of the slope [4]. The stability of hinge block ecological slope under the coupling effect of rainfall and earthquake is studied. It is obtained that the stability of hinge block ecological slope is better than that of soil slope under the coupling effect [5]. The numerical simulation study on the failure of Guantan landslide under the coupling effect is carried out. Result shows that rainfall aggravates the landslide disaster [6]. By means of numerical simulation and laboratory test, the formation mechanism of debris flow under rainfall and earthquake is revealed [7]. Dynamic response characteristics of loess slope under the different rainfall and earthquake conditions were studied based on large-scale shaking table test [8–10]. Above studies focus on the time domain to study the dynamic response and failure mechanism. Hilbert Huang transform (HHT) is a new time-frequency analysis method. It is considered as a major breakthrough in linear and steady-state spectral analysis based on Fourier transform in recent years. It was originally proposed by Huang [11] of NASA in 1998. This method is not limited by the global nature of Fourier analysis. The obtained time-frequency spectrum has high time resolution and frequency resolution. HHT has been applied to seismic mechanics, geophysics and other fields and achieved good results [12–13].

To clarify the dynamic response characteristics and failure mechanism of loess slope under the coupling effect of rainfall and earthquake from the point of time-frequency domain, we carried a large-scale shaking table test of model loess slope. Based on HHT method, we investigated the change rules of Hilbert spectrum and marginal spectrum under different loading sequences. Combined with the final failure phenomenon, we discussed the failure mechanism according to the change characteristics of the instantaneous and cumulative energy.

2 Shaking Table Test

Shaking table test is a scale model test. So, we need to determine the similarity ratio based on similarity theory at first. The geometric similarity is 10 and the geometric similarity ratio, density ratio and acceleration ratio are the fundamental dimensions in the study. The materials of the model loess slope are loess, barite, fly ash, sawdust, and water with ratio 5:2:2.1:0.5:0.4. The main similarity ratio and physical parameters of model slope are showed in Table 1.

Table 1. Similarity ratio and physical parameters

Physical parameters	Density (g/cm³)	Elastic modulus (MPa)	Cohesive strength (kPa)	Internal friction angle (°)	Dry unit weight (kN/m³)	Hydraulic conductivity (cm/s)
Similarity coefficients	1	10	10	1	1	1
Target value	1.28–1.45	5–8	3–6	26–30	11.37–13.52	1.58
Actual value	1.31	6.2	7	25	12.15	1.1

The test was done at the Key Laboratory of Loess Earthquake Engineering, China Earthquake Administration. The model and the rainfall equipment are showed in Fig. 1. The size of model is 2.85 m (length) × 1.4 m (width) × 1.0 m (height). The slope angle is approximately 20°. The model was divided into 10 layers to build. Each layer was compacted uniformly. Rainfall was applied through artificial rainfall equipment. The intensity of the rain was controlled by the valves. 18 acceleration sensors are used. The specific distribution of them is shown in Fig. 2(a). Most of them are distributed at the surface or along elevation direction.

Fig. 1. Test model and artificial rainfall equipment

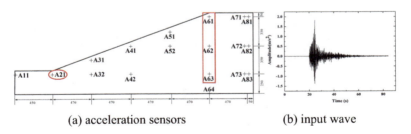

(a) acceleration sensors   (b) input wave

Fig. 2. Distribution of acceleration sensors (units: mm) and the input wave (Minxian-EW)

The input seismic wave form is the recording of Minxian station that triggered by Minxian-Zhangxian Ms 6.6 earthquake, as shown in Fig. 2(b). The wave was scaled and amplified to achieve the loading sequence under different intensities. The specific loading sequence is showed in Fig. 3. First, earthquake with intensity 50 gal and 100–800 gal were applied successively. Then, 20 mm rainfall was exerted before the last seismic loading (1300 gal). This is because that rainfall before the earthquake was around 20 mm at Yongguang landslide. In the test, we used optical measurement technique to measure the displacement of slope surface. However, the rainfall has an impact on the quality of displacement data. So, the rainfall is applied on the slope shoulder of model.

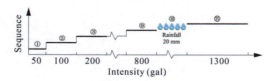

Fig. 3. Loading sequence

3 Method

3.1 Empirical Mode Decomposition (EMD)

The main process of EMD is as follows: 1) According to the upper and lower extreme points of the original signal, the upper and lower envelope lines are drawn respectively; 2) Find the mean value of the upper and lower envelope lines and draw the mean envelope line; 3) The mean envelope of the original signal is subtracted to obtain the intermediate signal; 4) Judge whether the intermediate signal meets the conditions of IMF. If so, the signal is an IMF component. Through EMD, the original signal was divided into a series of intrinsic mode functions (IMFs) and a residue, as shown in Eq. (1).

$$X(t) = \sum_{i=1}^{n} IMF_i(t) + r(t) \quad (1)$$

where X(t) is the original signal, IMF_i are the intrinsic mode functions of original signal, and r(t) is the residue.

The process of EMD has the over decomposition problem. So, there is need to extract the useful components from the IMFs. The basis of extraction is shown in Eq. (2), which is mainly based on the correlation coefficient.

$$\rho = \frac{\max(\mu_i)}{10\max(\mu_i) - 3}, i = 1, 2...n \quad (2)$$

where μ_i is the correlation coefficient and n are the number of IMFs.

3.2 Hilbert Transform

The expression of Hilbert transform is shown in Eq. (3). Hilbert transform is to convolute x (T) with $1/(\pi T)$.

$$y(t) = H[x(t)] = \frac{1}{\pi} P \left(\int_{-\infty}^{\infty} \frac{x(\tau)}{t - \tau} d\tau \right) \quad (3)$$

where P is the Cauchy principal value.

The analytic signal can be expressed as the combination of real part and imaginary part, as shown in Eq. (4). The Hilbert transform is the imaginary part of the analytic signal. According to the Euler formula, the analytic signal can be furtherly expressed by amplitude and phase.

$$z(t) = x(t) + jy(t) = a(t)e^{j\theta(t)} \tag{4}$$

where a(t) is the amplitude of the instantaneous frequency, and θ(t) is the phase function.

The instantaneous frequency is the first derivative of phase to time, as shown in Eq. (5).

$$\omega(t) = \frac{d\theta(t)}{dt} \tag{5}$$

where ω(t) is the instantaneous frequency.

Hilbert spectrum is the combination of amplitude, time and instantaneous frequency of the meaningful IMFs, as shown in Eq. (6).

$$H(t, \omega) = \sum_{1}^{m} a_i(t, \omega_i) \tag{6}$$

where H (t, ω) is the Hilbert spectrum.

Marginal spectrum is the integration of Hilbert spectrum to time, as shown in Eq. (7).

$$h(\omega) = \int_0^T H(t, \omega) dt \tag{7}$$

where h(ω) is the marginal spectrum.

4 Results and Discussion

4.1 Hilbert Spectrum

Hilbert spectrum reflects the distribution characteristics of instantaneous energy (amplitude), frequency and time. The change rules of different sensors under different intensities are investigated in the study. Hilbert spectrum results of A61 under 50 gal, 500 gal, 800 gal and 1300 gal are presented in Fig. 4. Under 50 gal, the instantaneous energy mainly concentrated within 80 Hz. With the increase of intensity, the energy gradually moved toward low frequency. Under 500 gal and 800 gal, the energy is concentrated within 60 Hz and 50 Hz, respectively. After rainfall, the distribution characteristics of the energy have a great change. Under 1300 gal, the energy concentrated within 20 Hz. Before rainfall, the energy increases with the increase of intensity. However, after rainfall, the energy has a great dissipation. When the input seismic intensity increased from 800 gal to 1300 gal, the energy did not increase greatly. The failure phenomenon shows that soil flow occurs under 1300 gal. So, the energy may dissipate by the soil flow.

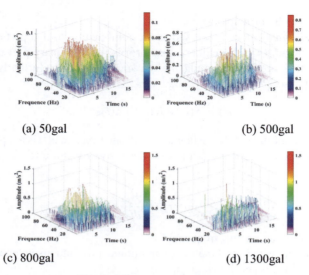

Fig. 4. Hilbert spectrums (A61)

4.2 Marginal Spectrum

Marginal spectrum reflects the distribution characteristics of cumulative energy at different instantaneous frequencies. The shape of marginal spectrum is related to the intensity and elevation. Marginal spectrums of A61, A62, A63 and A21 are shown in Fig. 5. The location of them is shown in Fig. 2. The energy concentrated within 60 Hz. The shape of it presents two peak and with predominate frequency approximately 10 Hz and 20 Hz respectively. The amplitude increases with the increases of intensity. However, the peak energy under 1300 gal is close to it under 800 gal. It indicates that the energy has a dissipation after rainfall. Peak value at same elevation (A63 and A21) is close under different intensities. First predominate frequency is close to the input seismic wave. Second predominate frequency is the natural frequency of slope. With the increases of elevation, the second peak energy also increases. It reflects that elevation has selective amplification effect on high frequency energy.

4.3 Failure Phenomenon

Failure phenomenon of model slope under the effect of rainfall and earthquake is shown in Fig. 6. As it mentioned above, rainfall was applied on the slope shoulder. Under 1300 gal, liquefaction occurs at the slope shoulder and the seismic subsidence is obvious (5–10 cm). The soil flows induced by the liquefication move down from slope shoulder under loading. The failure type is close to the Yongguang county landslides which is induced by the Minxian-zhangxian Ms 6.6 earthquake and rainfall. The dissipated instantaneous energy and cumulative energy mentioned above provided the kinetic energy to the soil flow

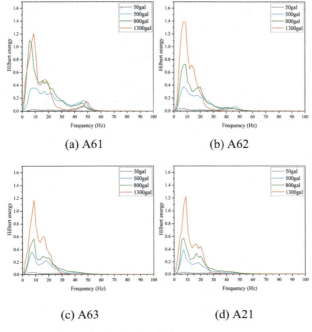

(a) A61   (b) A62

(c) A63   (d) A21

Fig. 5. Marginal spectrums

(a) Before liquefaction   (b) After liquefaction

Fig. 6. Failure phenomenon

5   Conclusions

A large-scale shaking table test of loess slope under the effect of rainfall and earthquake was performed in the study. Based on HHT, we investigated the change characteristics of instantaneous energy and cumulative energy at the monitoring points. Some conclusions are listed as following:

1. The distribution characteristics of instantaneous energy is related to the seismic intensities and rainfall. With the increase of intensity, the energy gradually moved toward low frequency (approximately 20 Hz). The peak energy increases with the increase

of intensity, but the energy has a great dissipation after rainfall. The dissipated energy provides the kinetic energy to the soil flows.
2. The shape of marginal spectrum presents two peak and with predominate frequency approximately 10 Hz and 20 Hz respectively at different monitoring points. The cumulative energy increases with the increases of intensity and the energy has a dissipation after rainfall. The distribution characteristics of the energy are closely related to the elevation. Elevation has selective amplification effect on high frequency energy (approximately 20 Hz).
3. Failure phenomenon of slope under the coupling effect of rainfall and earthquake presents soil flow, which is close to the Yongguang landslides. The instantaneous energy and cumulative energy are decreased at the monitoring points, which provided the kinetic energy to the soil flow.

References

1. Wang, L.: Loess Dynamics, 1st edn. Earthquake Press, Beijing (2003)
2. Sun, J.., Wang, L., Long, P.: Method for evaluation of regional landslide disaster,coupled with earthquake and rainfall. J. Rock Mech. Eng. **30**(4), 752–760 (2011)
3. Xu, S., Wu, Z., Sun, J.: Typical landslide characteristics and induced mechanism of the Ms6.6 earthquake in Zhang County, Minxian County. Earthq. Eng. **35**(3), 471–476 (2013)
4. Hailun, W.: The Stability Study of Landslides Under Coupling Condition of Earthquake and Rainfall: Jiulong Mountain Landslide in Meishan as an Example. Chengdu University of Technology, Chengdu (2015)
5. Yubin, W.: Study on stability of hinge block ecological slope under earthquake-rainfall coupling. Southeast University, Nanjing (2017)
6. tong, X.: Simulation Study on the failure of the Guantan landslide under the action of earthquake and rainfall. Jilin University, Changchun (2016)
7. Wenhui, Z.: Mechanisms of slope instability and the starting of debris-flow under the complex function of rainfall and earthquake. Inner Mongolia University of Technology, Hohhot (2011)
8. Jinchang, C., Lanmin, W., Xiaowu, P., Fuxiu, L.T: Experimental study on the dynamic characteristics of low-angle loess slope under the influence of long- and short-term effects of rainfall before earthquake. Eng. Geol. **273**, 105684 (2020)
9. Pu, X., Wang, L., Wang, P., Chai, S.: Study of shaking table test of seismic subsidence loess landslides induced by the coupling effect of earthquakes and rainfall. Nat. Hazards **103**(1), 923–945 (2020). https://doi.org/10.1007/s11069-020-04019-3
10. Pu, X., Wan, L., Wang, P.: Initiation mechanism of mudflow-like loess landslide induced by the combined effect of earthquakes and rainfall. Nat. Hazards **105**(3), 3079–3097 (2021). https://doi.org/10.1007/s11069-020-04442-6
11. Song, D., Liu, X., Huang, J., Zhang, J.: Energy-based analysis of seismic failure mechanism of a rock slope with discontinuities using Hilbert-Huang transform and marginal spectrum in the time-frequency domain. Landslides **18**, 105–123 (2021)
12. Danqing, S., Zhuo, C., Chao, H., Yutian, K., Wen, N.: Numerical study on seismic response of a rock slope with discontinuities based on the time-frequency joint analysis method. Soil Dyn. Earthq Eng. **133**, 106112 (2020)

Seismic Stability Analysis of Earth Slopes Using Graphical Chart Solution

Hong-zhi Cui[1] and Jian Ji[1,2(✉)]

[1] Geotechnical Research Institute, Hohai University, Nanjing, China
ji0003an@e.ntu.edu.sg
[2] Key Lab of Ministry of Education for Geomechanics and Embankment Engineering, Hohai University, Nanjing, China

Abstract. The stability charts for static and pseudo-static (PS) stability analysis of homogenous slopes have been widely used for preliminary design work. Based on the concept of a graphical stability chart, this study presents a series of improved stability charts solution based on PS analysis of a wide range of homogeneous slope models with inclination angles varying from 10° to 60°, and considering both horizontal and vertical seismic loads. The concept of limit state line (LSL) is presented, and their mathematical expressions are obtained from extensive regression analyses. The LSL-based graphical stability charts' solutions are proposed for PS stability analysis of homogeneous slopes with any given combination of physical parameters. The proposed charts show very good performances for obtaining the PS FoSs accurately and rapidly; the calculation errors are generally within 10%. It is anticipated that the developed charts, which consider the relatively realistic conditions of seismic loads, may serve as a promising tool for practicing engineers to estimate the FoS of homogenous slopes equivalent to PS stability analysis in less time.

Keywords: Slope stability · Graphical stability chart · Limit state line · Seismic coefficient · Limit equilibrium analysis.

1 Introduction

One of the most devastating disasters threatening the safety of human life and properties is the earthquake and its triggered secondary hazards, such as landslides [1]. The seismic instability of slopes located in earthquake-stricken regions that are susceptible to landslides has been an attractive research topic in geotechnical engineering [2, 3]. Transparent evaluation methods are useful tools to geotechnical practitioners for project planning that requires to minimize the losses and to ensure the engineering safety. In this regard, the stability charts are usually applied as a direct approach when a preliminary stability design is involved. Although Taylor [4] firstly proposed the concept of a stability chart which can estimate the factor of safety (FoS), the stability charts suffer the basic limitation of iterative computation of FoS despite that the computation is straightforward by itself. Recently, a new graphical construction method was developed to provide more convenience in obtaining directly the state of slope stability without iterative process, e.g. by presenting the stability charts in terms of limit state lines (LSL) on which the FoS equal to unity [5, 6].

Note that when the seismic effect is to be considered for stability analysis, the pseudo-static (PS) method is usually adopted as the main tool for obtaining FoS under seismic conditions. For example, the chart solutions for slope stability can be easily extended into the PS approach based both on the variational formulation of limit equilibrium and on the strength reduction or the limit analysis [7, 8]. Conventionally, many researchers were devoted to only investigating the horizontal seismic inertia force when discussing the slope stability under earthquake. Some other researches indicated that vertical peak acceleration of the area around the epicenter also plays a significant role in slope stability according to a massive quantity of spectrum records [9–11]. Furthermore, it is an important topic worth discussing that the seismic FoS considering horizontal and vertical earthquake loads can be obtained directly from the stability charts for geotechnical engineers.

In this study, an effort has been made to develop more transparent graphical stability charts for seismic slope stability analysis that is easy to understand and follow by geotechnical practitioners. First, hundreds of homogenous LEM models with the slope angle ranging from 10° to 60° are numerically computed using Morgenstern-Price's method. To characterize the influence of seismic loads, the horizontal and vertical coefficients are employed in the PS analysis. For slopes subject to limit state, i.e., FoS = 1 condition, the numerical cases studies with modified soil properties are implemented thousands of times, which serve as the basis for generating new stability charts under both horizontal and vertical seismic inertia forces. In order for a universal solution, the regression method is also introduced for mathematically expressing these case-specific transparent stability charts. At last, the accuracy and feasibility of the proposed method are demonstrated through two worked examples.

2 Improved Procedure to Estimate FoS for Pseudo-static Stability Analysis

2.1 Graphical Stability Charts Based on Limit State Line

First, the solution of static homogeneous slope stability depends on the slope geometries and soil properties defining the problem, which can be briefly written as:

$$FoS = FoS(c, \phi, \gamma, H, \beta) \tag{1}$$

As a general assumption of soil shear strength for stability analysis, the Mohr-Coulomb failure criterion is adopted as the basis for developing the stability charts. The FoS defining the ratio between shear strength to the mobilized strength required for limit equilibrium is expressed by [5, 12]:

$$FoS = \frac{\tau_f}{\tau_{mob}} = \frac{c + \sigma \tan(\phi)}{c_{mob} + \sigma \tan(\phi_{mob})} = FoS_c = \frac{c}{c_{mob}} = FoS_\phi = \frac{\tan(\phi)}{\tan(\phi_{mob})} \tag{2}$$

where **c** is the soil effective cohesion, φ is the soil effective angle of internal friction, c_{mob} and ϕ_{mob} denote the mobilized strength parameters, respectively. The stability chart is commonly used for preliminary design work due to its simplicity in concept. For stability analysis of homogenous soil slopes, only some basic information including the slope height (H), slope angle (β), the shear strength parameters (c, ϕ), and the unit weight (γ) are required. Combining the parameters in dimensionless form can more effectively reduce the number of design variables and make the stability chart more universal.

In Taylor's chart solution, a stability number N required for the iterative computation of FoS is expressed as follows [4, 5]:

$$N = \frac{c_{mob}}{\gamma H} \tag{3}$$

On the other hand, a new graphical stability chart solution is proposed by Klar et al. [6], and the key feature of the chart is to construct the LSL, namely, the curve reflecting the combination between two dimensionless parameters $c/\gamma H$ and $\tan(\phi)$ at limit state (FoS = 1), such that:

$$FoS = 1 = \frac{\tau'_f}{\tau_{mob}} = FoS_c = \frac{c'}{c_{mob}} = FoS_\phi = \frac{\tan(\phi')}{\tan(\phi_{mob})} \tag{4}$$

$$\tau'_f = \tau_{mob} \tag{5}$$

$$\begin{cases} c' = c_{mob} = \dfrac{c}{FoS} \\ \tan(\phi') = \tan(\phi_{mob}) = \dfrac{\tan(\phi)}{FoS} \end{cases} \tag{6}$$

where $\{\tau'_f, c', \phi'\}$ represent soil parameters at limit state under static condition.

Previous studies have also shown that for homogenous soil slope, the critical failure surface is only related to $c/\gamma H$ and $\tan(\phi)$. At this time, another crucial dimensionless parameter, i.e., $\lambda_{c\phi}$ is defined as follows [13]:

$$\lambda_{c\phi} = \frac{\gamma H \tan(\phi)}{c} \tag{7}$$

Theoretically, the position of each critical slip circle is related only to $\lambda_{c\varphi}$, and a shallower failure mode will be induced with the increase in $\lambda_{c\varphi}$ [14]. By the concept of LSL, it can be directly seen that when a slope has a parameter combination ($c/\gamma H$, $\tan(\phi)$) located in the region above LSL, the corresponding FoS will be greater than unity, that is, the slope remains stable, and vice versa. Further, the graphical computation of FoS can best be illustrated in Fig. 1, where for example, Point A (x_1, y_1) denotes a dimensionless parameter combination of a given slope. Then it is straightforward to show that FoS is simply the ratio between any two companion lengths on the chart, such that:

$$FoS = \frac{l_1}{l_c} = \frac{y_1}{y_c} = \frac{x_1}{x_c} \tag{8}$$

where l_1 is the distance from the origin to the coordinate A (x_1, y_1); l_c is the distance from the origin to its intersection point B (x_c, y_c) on the limit state line. The FoS obtained by Eq. (8) is mathematically identical to that given by Eq. (2).

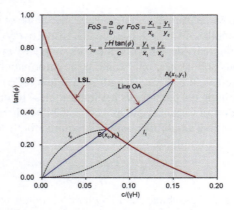

Fig. 1. FoS calculation using the dimensionless chart (adopted from ref. [6])

2.2 Improved Stability Chart Method Under Static Condition

In the framework of graphical stability chart, Tsai et al. [15, 16] adapted the Morgenstern-Price method to carry out a large number of static stability analyses. They summarized the data pairs of $c/\gamma H$ and $\tan(\phi)$ for homogenous slope models with FoS close to unity and plotted the LSL with respect to slope angle (β) ranging from 15° to 60° with 5° intervals. Then, all the LSL's were collectively expressed in a mathematical form as follows [15, 17]:

$$[\tan(\phi)]' = -0.40 \times [\tan(\beta)] \times \ln\left[5.11 \times \left(\frac{c}{\gamma H}\right)' + 0.101\right] \quad (9)$$

where $\tan \phi'$ and $\left(\frac{c}{\gamma H}\right)'$ are two dimensionless parameters of the LSL which are defined as follows:

$$\begin{cases} [\tan(\phi)]' = \dfrac{\tan(\phi)}{FoS} \\ \left(\dfrac{c}{\gamma H}\right)' = \dfrac{\frac{c}{\gamma H}}{FoS} \end{cases} \quad (10)$$

Combining Eq. (9) and (10), FoS for a given homogeneous slope can be directly calculated. In this study, our attempt is made to develop more transparent stability chart solutions for seismic stability analysis as will be elaborated in the next section.

To determine the updated data pairs $(c/\gamma H, \tan(\phi))$ on LSL, 2471 stability evaluations on homogenous slope models as shown in Fig. 2 are performed using the Morgenstern-Price method subjected to the numerical package *Rocscience Slide*. Figure 3 shows the LSL's data pairs recalculated with slope angles (β) ranging from 10° to 60° with a gradient of 5°. The improved expression of the collective data pairs with multiparameter regression analysis is displayed as follows:

$$[\tan(\phi)]' = -0.43061 \times [\tan(\beta)] \times \ln\left[5.2765 \times \left(\frac{c}{\gamma H}\right)' + 0.11248\right] \quad (11)$$

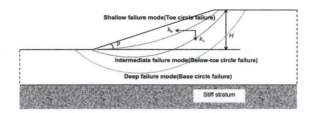

Fig. 2. Geometrically unified simple slope model used in slide

For comparison, the LSL's drawn by the improved regression formula (solid line) have a high degree of consistency with the source data pairs as shown in Fig. 3 (a). The predicted results by the proposed method are mostly within ± 5% difference from those by the *Slide's* slope-stability analysis, as indicated in Fig. 3 (b). On this basis, we further explore the possibility of developing graphical stability chart solution with the consideration of seismic coefficients.

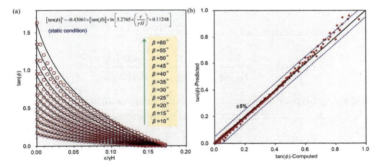

Fig. 3. Stability chart with limit state lines for varying slope angles (a) limit state line in the space of dimensionless variables; (b) accuracy plot

2.3 Selection of Pseudo-static Coefficients

Next, the graphical stability chart solution accounting for both horizontal and vertical seismic loads will be presented. Numerous researchers mainly focused on considering the effect of horizontal earthquake loading [8]. According to Hynes-Griffith and Franklin [18], the seismic coefficient, k, is used to characteristic the seismic loading under the framework of PS analysis. It should be taken as one-half the peak bedrock acceleration when FoS is required to be unity and the actual range of seismic coefficients will be between 0 and 0.375 under the premise. In this work, k_h (horizontal) scenarios are respectively taken as 0.05, 0.10, 0.15, 0.20, 0.25, 0.30, and 0.35.

Note that most codes only consider the influence of horizontal component of seismic load on slope stability, i.e., the k_h. However, to further study the influence of vertical earthquake component, the corresponding relationship of earthquake coefficients can be determined according to relevant design manuals or standards when considering both the horizontal and vertical seismic loads. The U.S. Army Corps of Engineers (USACE,1989) suggested $k_v = (1/2)k_h$ [19, 20]. In this work, the $k_v = 0.5k_h$ is used to investigate the slope stability under seismic loads.

2.4 Improved Regression Formula of Limit State Line Considering Seismic Loads

To construct the LSL considering both horizontal and vertical seismic loads, a series of transformations between soil parameters were derived as follows.

First, the general expression of the seismic factor of safety can be written as follows:

$$FoS_d = FoS_d(c, \phi, \gamma, H, \beta, k_h, k_v) \tag{12}$$

Due to the assumption of constant ratio between k_v and k_h, Eq. (12) can be simplified as:

$$FoS_d = FoS_d(c, \phi, \gamma, H, \beta, k_h) \tag{13}$$

similarly, Eq. (2) can be written as,

$$FoS_d = \frac{\tau_{fd}}{\tau_{mobd}} = \frac{c}{c_{mobd}} = \frac{\tan(\phi)}{\tan(\phi_{mobd})} \tag{14}$$

where τ_{fd} and τ_{mobd} respectively denote the available shear strength and mobilized shear strength and c_{mobd} and ϕ_{mobd} are nominal shear parameters considering the act of seismic load (i.e. k_h).

At the limit state of seismic load, Eq. (14) can be written as

$$1 = \frac{\tau'_{fd}}{\tau_{mobd}} = FoS_{cd} = \frac{c'_d}{c_{mobd}} = FoS_{\phi d} = \frac{\tan(\phi'_d)}{\tan(\phi_{mobd})} \tag{15}$$

where τ'_{fd} denotes the shear strength under seismic load, and $\{c'_d, \phi'_d\}$ represent the shear strength data pair on seismic LSL.

$$\begin{cases} c_{mobd} = c'_d \\ \tan(\phi_{mobd}) = \tan(\phi'_d) \end{cases} \tag{16}$$

On the other hand, consider a homogeneous slope is at limit state with shear strength data pair $\left(c', \tan(\phi')\right)$ as defined previously. When the limit state slope is further exerted with seismic loads, the pseudo-static factor of safety can be expressed as follows:

$$FoS_d = \frac{\tau'_f}{\tau_{mobd}} = FoS_{cd} = \frac{c'}{c_{mobd}} = FoS_{\phi d} = \frac{\tan(\phi')}{\tan(\phi_{mobd})} \tag{17}$$

where FoS_d represents the pseudo-static factor of safety corresponding to the seismically mobilized shear strength τ_{mobd} with nominal shear parameters $\{c_{mobd}, \phi_{mobd}\}$.

Substituting Eq. (16) into Eq. (17), it is straightforward to obtain the following relationships:

$$\begin{cases} c'_d = \dfrac{c'}{FoS_d} \\ \tan(\phi'_d) = \dfrac{\tan(\phi')}{FoS_d} \end{cases} \quad (18)$$

Therefore, the shear strength parameters for a specific slope at the limit state under seismic loads can be obtained by transforming according to Eq. (18) under the static condition. Then, the LSL-based stability chart accounting for the earthquake actions in the framework of the PS method can be expressed by the following nonlinear relationship between $c/\gamma H$, $\tan(\phi)$:

$$[\tan(\phi'_d)] = P_1 \times [\tan(\beta)] \times \ln\left[P_2 \times \left(\dfrac{c'_d}{\gamma H}\right) + P_3\right] \quad (19)$$

where $\{\tan(\phi'_d), (\dfrac{c'_d}{\gamma H})\}$ are now the dimensionless seismic LSL data pairs of shear strength, and they are defined the same way as in Eq. (10) where the FoS is now from PS stability analysis; P_1, P_2, and P_3 are three dimensionless parameters reflecting the contribution of seismic load in terms of the ratio $\tan(\beta)/k_h$, and they are given by Eq. (20) as follows:

$$\begin{cases} P_1 = C_{P_1,1} \cdot \ln\left[C_{P_1,2} \cdot \dfrac{\tan(\beta)}{k_h} + C_{P_1,3}\right] \\ \ln(P_2) = C_{P_2,1} \cdot \mathrm{Exp}\left[C_{P_2,2} \cdot \dfrac{\tan(\beta)}{k_h}\right] + C_{P_2,3} \\ P_3 = C_{P_3,1} \cdot \ln\left[C_{P_3,2} \cdot \dfrac{\tan(\beta)}{k_h} + C_{P_3,3}\right] \end{cases} \quad (20)$$

where $C_{P_i,j}$ represents the jth coefficient of the non-linear regression equation subjected to the ith parameter P_i, and describes the nonlinear relationships between $\tan(\beta)/k_h$ and P_i.

Clearly, Eq. (19) and (20) cover the basic features of the slope under earthquake action, including the slope angle and height, the soil shear strength and unit weight, as well as the seismic coefficient. It is worth pointing out that the coefficients $C_{P_i,j}$ are potentially controlled by horizontal seismic coefficient k_h. To further investigate the relationship between $C_{P_i,j}$ and k_h, computational effort was made to continue the nonlinear regression analysis for an universal mathematical expression. Our extensive study revealed that the optimal curves for different cases could be further expressed by the uniform logarithmic function as follows:

$$C_{P_i,j} = \alpha_{C_{P_i,j},1} \times \ln\left[\alpha_{C_{P_i,j},2} \times (k_h) + \alpha_{C_{P_i,j},3}\right] \quad (21)$$

Note that $\alpha_{C_{P_i,j},k}$ denotes the kth ($k = 1, 2, 3$) coefficient belonging to $C_{P_i,j}$. For each combination, the predicted accuracy is almost close to 1.0, as listed in Table 1.

The predicted values by the proposed Equations are mostly within 10% difference [21–23] with actual values according to statically results. Till now, the Eqs. (19), (20) and (21) work together to provide improved solution for the computation of PS FoS for homogeneous slopes with basic parameters being given.

Table 1. List of values for coefficient C_1, C_2, and C_3

Parameters	Coefficients	$\alpha_{C_{P_i,j},1}$	$\alpha_{C_{P_i,j},2}$	$\alpha_{C_{P_i,j},3}$	R^2
P_1	$C_{P_1,1}$	0.5826	0.7984	1.0911	1.00
	$C_{P_1,2}$	−0.0197	−2.7108	1.1868	0.99
	$C_{P_1,3}$	0.1066	−0.4032	1.0171	1.00
P_2	$\ln(-C_{P_2,1})$	0.3861	3.2876	−0.1074	1.00
	$\ln(-C_{P_2,2})$	0.3107	0.8324	−0.0299	0.98
	$C_{P_2,3}$	−0.4169	0.1224	0.0187	0.99
P_3	$C_{P_3,1}$	−0.0248	4.2706	2.4409	0.97
	$C_{P_3,2}$	−0.1490	−2.5155	1.1648	0.99
	$C_{P_3,3}$	−1.5161	0.1987	0.9936	0.99

2.5 Procedure of Estimating FoS Under Seismic Loads

Based on the improved regression formula of LSL considering seismic loads for seismic slope stability analysis, the PS FoS can be immediately calculated through the detailed procedure below:

Step 1: Determine the coordinate of point A (x_1, y_1). Specifically speaking, the combination of two dimensionless parameters is obtained according to the existing slope geometry and shear strength parameters.

Step 2: Connect the origin and point A will intersect with LSL with point B (x_c, y_c). The formula of LSL can be established by two steps as well when it is not given in the stability chart. First, the coefficients $C_{P_i,j}$ (j = 1, 2, 3) are determined according to Eq. (20) and (21) on the premise known of the design seismic coefficients. Then, the parameters P_i (i.e. P_1, P_2, P_3) can be estimated by using Eq for a specific slope. Hence, the mathematical expression of LSL can be formed based on Eq. (19).

Step 3: Calculate FoS through Eq. (4) as displayed in Fig. 1.

Therefore, the PS FoS under seismic loads can be immediately estimated by numerically performing Eqs. (19), (20) and (21).

3 Application of Proposed Method to Homogenous Earth Slope Involving Seismic Loads and Discussion

3.1 Application

One case which can be solved mathematically is presented to illustrate how to conveniently estimate FoS using the proposed approach. This case is a 4.6-m-tall slope with slope angle $\beta = 33°$ composed of soft clayey soil with an undrained strength of $\phi = 10°$, $c = 23.8$ kPa, and weight $\gamma = 19.2$ kN/m^3. Besides, $k_h = 0.3$, $k_v = 0.15$ to obtain the FoS under both horizontal and vertical seismic loadings.

Step 1: Determining the coordinate of Point A (x_1, y_1)

$$\frac{c}{\gamma H} = \frac{23.8}{19.2 \times 4.6}, \tan(\phi) = \tan(10°) = 0.1763$$

Therefore, the coordinate of A is (0.2695, 0.1763).

Step 2: Producing the LSL

The k_h is used to determine the $C_{P_{i,j}}$ (j = 1, 2, 3) which denotes the coefficient of the non-linear regression equation involving parameters P_i (i.e. P_1, P_2, P_3). Hence, the $C_{P_{i,j}}$ can be solved mathematically according to Eq. (21) and Table 1. The corresponding parameter P_i is estimated based on Eq. (21) when introducing a given slope angle β. Then, the mathematical formula of the LSL is established by using Eq. (19). The final expression can be presented as follows:

$$[\tan(\phi'_d)] = -0.5821 \times [\tan(\beta)] \times \ln\left[2.6115 \times \left(\frac{c'_d}{\gamma H}\right) + 0.051\right] \quad (22)$$

Step 3: Calculating the $\lambda_{c\phi}$
According to Eq. (3),

$$\lambda_{c\phi} = \frac{\gamma H \tan(\phi)}{c} = \frac{19.2 \times 4.6 \times \tan(10°)}{23.8} = 0.65434 \quad (23)$$

To obtain the intersection point B, a straight line passing through the origin with a slope of $\lambda_{c\varphi}$ is constructed, the independent variable is $\frac{c'_d}{\gamma H}$, and the corresponding dependent variable of $[\tan(\varphi'_d)]$, as displayed follows:

$$[\tan(\phi)] = \lambda_{c\phi} \times \left(\frac{c}{\gamma H}\right) \quad (24)$$

Then, the $\frac{c'_d}{\gamma H}$ and $\tan(\phi'_d)$ can be obtained by combining the Eq. (22) and Eq. (24). Therefore, the coordinate of point B is (0.2353, 0.1540).

Step 4: Estimating the FoS

FoS can be calculated according to Eq. (8) and is 1.1453. Furthermore, the FoS is 1.201 by using the *SLOPE/W* which is almost close to the computed result based on the proposed approach in this work.

3.2 Discussion

It is worth pointing out that the dimensionless $c/\gamma H$, instead of a particular combination of parameters, was employed to contain more situations. The chart solutions based on the LSL method were observed to match the iterative process and the kinematic solutions from the graphical approach. Therefore, the applicability of the LSL-based graphical method in slope stability prediction is substantiated. Compared with the iterative process or the chart-based kinematic method, the LSL-based chart solution procedure is more 'user-friendly' for practitioners who are unfamiliar with slope stability analysis. Nevertheless, the main limitation is that chart method is only applicable for homogeneous soil slopes at present[5, 12, 14, 24, 25]. In the future, relevant research should further implement the heterogeneous slope when adopting the chart method.

4 Conclusion

This paper presented an improved graphical stability chart solution for estimating the pseudo-static FoS of homogenous soil slopes with both horizontal and vertical seismic loads. By the expression of two dimensionless parameters ($c/\gamma H$, $\tan(\phi)$), the limit state lines (LSL's) were obtained by analyzing a large volume of pseudo-static stability modelling results. For any given case of the slope angle and horizontal seismic coefficient, the PS FoS can be straightforwardly calculated from LSL-based graphical solution method, without additional iterative calculations. In addition, it has high accuracy within 10% allowable error range. At last, one example was used to demonstrate the engineering applications.

References

1. Ji, J., Wang, C.-W., Gao, Y., Zhang, L.: Probabilistic investigation of the seismic displacement of earth slopes under stochastic ground motion: a rotational sliding block analysis. Can. Geotech. J. **58**, 952–968 (2021)
2. Ji, J., Gao, Y., Lü, Q., Wu, Z., Zhang, W., Zhang, C.: China's early warning system progress. Science **365**, 332 (2019)
3. Jibson, R.W.: Regression models for estimating coseismic landslide displacement. Eng. Geol. **91**, 209–218 (2007)
4. Taylor, D.W.: Stability of earth slopes. J. Boston Soc. Civ. Eng. **24**, 197–247 (1937)
5. Sun, J., Zhao, Z.: Stability charts for homogenous soil slopes. J. Geotech. Geoenviron. Eng. **139**, 2212–2218 (2013)
6. Klar, A., Aharonov, E., Kalderon-Asael, B., Katz, O.: Analytical and observational relations between landslide volume and surface area. J. Geophys. Res. Earth Surf. **116**, 1–10 (2011)
7. Li, A.J., Lyamin, A.V., Merifield, R.S.: Seismic rock slope stability charts based on limit analysis methods. Comput. Geotech. **36**, 135–148 (2009)
8. Baker, R., Shukha, R., Operstein, V., Frydman, S.: Stability charts for pseudo-static slope stability analysis. Soil Dyn. Earthq. Eng. **26**, 813–823 (2006)
9. Du, W.Q.: Effects of directionality and vertical component of ground motions on seismic slope displacements in newmark sliding-block analysis. Eng. Geol. **239**, 13–21 (2018)

10. Nouri, H., Fakher, A., Jones, C.J.F.P.: Evaluating the effects of the magnitude and amplification of pseudo-static acceleration on reinforced soil slopes and walls using the limit equilibrium horizontal slices method. Geotext. Geomembr. **26**, 263–278 (2008)
11. Ling, H.I., Leshchinsky, D., Mohri, Y.: Soil slopes under combined horizontal and vertical seismic accelerations. Earthquake Eng. Struct. Dynam. **26**, 1231–1241 (1997)
12. Steward, T., Sivakugan, N., Shukla, S.K., Das, B.M.: Taylor's slope stability charts revisited. Int. J. Geomech. **11**, 348–352 (2011)
13. Duncan, J., Wright, S.: The accuracy of equilibrium methods of slope stability analysis. Eng. Geol. **16**, 5–17 (1980)
14. Jiang, J.-C., Yamagami, T.: Charts for estimating strength parameters from slips in homogeneous slopes. Comput. Geotech. **33**, 294–304 (2006)
15. Chien, Y.-C., Tsai, C.-C.: Immediate estimation of yield acceleration for shallow and deep failures in slope-stability analyses. Int. J. Geomech. **17**, 1–9 (2017)
16. Tsai, C.C., Chien, Y.C.: A simple procedure to directly estimate yield acceleration for seismic slope stability assessment, Japanese geotechnical society special. Publication **2**, 915–919 (2016)
17. Tsai, H.Y., Tsai, C.C., Chang, W.C.: Slope unit-based approach for assessing regional seismic landslide displacement for deep and shallow failure. Eng. Geol. **248**, 124–139 (2019)
18. M.E. Hynes-Griffin, A.G. Franklin, Rationalizing the seismic coefficient method, in, Army Engineer Waterways Experiment Station Vicksburg Ms Geotechnical Lab, 1984
19. Sahoo, P.P., Shukla, S.K.: Taylor's slope stability chart for combined effects of horizontal and vertical seismic coefficients. Géotechnique **69**, 344–354 (2019)
20. USACE, EM 1110–2–2502: Engineering and design of retaining and flood walls, in, USA: USACE, Washington, DC (1989)
21. Grilli, S.T., Watts, P.: Modeling of waves generated by a moving submerged body. Appl. Underwater Landslides Eng. Anal. Boundary Elem. **23**, 645–656 (1999)
22. Korup, O., Clague, J.J., Hermanns, R.L., Hewitt, K., Strom, A.L., Weidinger, J.T.: Giant landslides, topography, and erosion. Earth Planet. Sci. Lett. **261**, 578–589 (2007)
23. Cui, S.-H., Pei, X.-J., Huang, R.-Q.: Rolling motion behavior of rockfall on gentle slope: an experimental approach. J. Mt. Sci. **14**(8), 1550–1562 (2017). https://doi.org/10.1007/s11629-016-4144-7
24. Bell, J.M.: Dimensionless parameters for homogeneous earth. J. Soil Mech. Found. Div. **92**, 51–65 (1966)
25. Gao, Y., Zhang, F., Lei, G., Li, D., Wu, Y., Zhang, N.: Stability charts for 3D failures of homogeneous slopes. J. Geotech. Geoenviron. Eng. **139**, 1528–1538 (2013)

Influence of Cyclic Undrained Shear Strength Degradation on the Seismic Performance of Natural Slopes

Giuseppe Di Filippo, Orazio Casablanca, Giovanni Biondi, and Ernesto Cascone(✉)

Department of Engineering, University of Messina, Contrada di Dio, 98166 Messina, Italy
gdifilippo@unime.it

Abstract. The consequences of several large earthquakes occurred worldwide, reveal that the seismic stability conditions and the post-seismic serviceability of natural slopes are largely affected by the cyclic behaviour of soils. In fact, failures of natural slopes and large permanent displacements may be triggered by the weakening effects due to cyclic shear strength degradation, which may arise simultaneously with inertial effects. In this paper, effect of the cyclic behaviour of soils on the slope seismic response is accounted for in a displacement-based analysis using a simplified degradation model for the soil undrained shear strength. To assess the seismic performance of the slope, a modified Newmark-type analysis is proposed in which a time-dependent value of the slope yield acceleration coefficient $k_{h,c}$ is assumed. $k_{h,c}$ is expressed as a function of the number of equivalent loading cycles of the input motion, a degradation parameter T and a degradation ratio R, the latter quantifying the reduction of $k_{h,c}$ with respect to its initial value. The results of a parametric analysis were used to derive empirical predictive models for a safe estimate of the influence of cyclic undrained shear strength degradation on the seismic-induced permanent displacements.

Keywords: Slope stability · Undrained shear strength degradation · Permanent displacements

1 Introduction

The damages induced by earthquake-triggered landslides frequently exceed those directly related to the ground shaking and were often ascribed to the effects of the cyclic behaviour of soils. In fact, during strong earthquakes soils develop significant deformations that may affect the stability conditions of natural slopes possibly causing failure and damages to environment, structures and lifelines (e.g. [1, 2]).

A number of studies were developed to account for the effect of soil strength degradation in the evaluation of the slope response (e.g. [3–5]). The proposed approaches were developed using effective stress or total stress analyses. In the first case the available soils shear strength was estimated as a function of the excess pore pressure induced by the cyclic loading; in the other case the cyclic reduction of undrained shear strength is accounted in the analysis.

The present paper deals with a procedure to assess the co-seismic performance of slopes accounting for the possible occurring cyclic degradation of shear strength and for the corresponding reduction of the slope yield acceleration coefficient $k_{h,c}$. In the procedure the cumulative number of equivalent loading cycles N_{eq} was adopted as a robust indicator of the destructive capacity of the ground motion. A set of seismic records were selected and a parametric analysis was carried out using different values of the parameters describing the degradation path of $k_{h,c}$, providing several empirical predictive models for the displacements evaluation.

2 Reference Scheme and Yield Acceleration Coefficient

To analyze the effects of the possible reduction in soil shear strength, the infinite slope scheme, providing the most conservative value of the slope yield acceleration, was considered. This scheme is suitable for slopes characterized by translational failure mechanism, with shallow and wide sliding surface. The stability conditions are evaluated through the limit equilibrium total stress analysis, using a pseudo-static approach and assuming a rigid-plastic behavior of the un-stable soil with a uniform reduction of the soil shear strength along the sliding surface. The slope stability conditions are described by the horizontal component of the yield seismic coefficient $k_{h,c}$:

$$k_{h,c} = \frac{\delta_{Cu} C_{u,0} / \gamma D - \sin \beta}{(\cos \beta \mp \Omega \sin \beta)} \tag{1}$$

In Eq. (1), γ is the soil unit weight and $C_{u,0}$ is the available undrained shear strength along the sliding surface; β and D are the slope angle and the depth of the failure surface, respectively; $\delta_{Cu} = C_{u,c}/C_{u,0}$ is the current value of a cyclic degradation coefficient; $\Omega = k_v/k_h$ is the ratio between the vertical (k_v) and the horizontal (k_h) component of the seismic acceleration coefficient ($\Omega > 0$ for k_v directed upwards).

Cyclic shear strength reduction makes the yield acceleration coefficient k_c decay with time. Thus, the term δ_{Cu} in Eq. (1) implies a reduction of the yield acceleration coefficient $k_{h,c}$, starting from its initial value $k_{h,co}$ (Eq. 2) down to a minimum $k_{h,cmin}$ (Eq. 3), attained when the degradation coefficient δ_{Cu} attains its minimum $\delta_{Cu,min}$.

$$k_{h,co} = \frac{C_{u,0}/\gamma D - \sin \beta}{(\cos \beta \mp \Omega \sin \beta)} \quad k_{h,cmin} = \frac{\delta_{Cu,min} C_{u,0}/\gamma D - \sin \beta}{(\cos \beta \mp \Omega \sin \beta)} \tag{2}$$

3 Proposed Modified Newmark-Type Analysis

The numerical analyses presented in the paper refer to a rigid block sliding on a horizontal plane. Thus, no effect of the soil compliance is accounted for in the analyses which focuses only on the effect of the cyclic strength degradation. It is worth noting that, through suitable shape factors [6], depending on the shape of the actual failure mechanism and on the geometrical and mechanical properties of the slope, the results presented herein can be assumed representative also for other several slope scheme (e.g. [5]) and other geotechnical systems such as sliding gravity (e.g. [7]) and earth-reinforced walls (e.g. [8]) and rock slopes (e.g. [9]).

In the analysis presented in the paper only the horizontal component of the ground motion is considered for simplicity and the initial value $k_{h,co}$ of the slope yield acceleration coefficient was assumed equal to a fraction of the peak horizontal acceleration coefficient $k_{h,max}$ of the ground motion selected for the displacement analysis. Specifically, values of the acceleration ratio $k_{h,co}/k_{h,max}$ in the range $0.1 \div 0.8$ were considered and, for each of the selected seismic records.

In the displacement analyses, the reduction $\Delta k_{h,c} = k_{h,co} - k_{h,cmin}$ of the yield acceleration coefficient was assumed as a datum and a parametric analysis was carried out using different values of a degradation ratio $R = \Delta k_{h,c}/k_{h,co}$. Obviously, a proper evaluation of the expected reduction of $k_{h,c}$ (i.e. proper values of R) is required to apply the proposed procedure and the empirical predictive models presented in next section.

It is known that the number of equivalent loading cycles N_{eq} suitably describes the energy and frequency content and the significant duration of a ground motion. Also, N_{eq} represents a robust indicator of the destructive capacity of a ground motion and allows describing the more relevant effects of the cyclic soil behaviour such as cyclic excess pore pressures and degradation of the undrained shear strength. The conversion procedures proposed in the literature to estimate N_{eq} starting from an acceleration record, aim to detect the loading cycles which mostly affect the cyclic strength degradation and, through a proper weighting procedure, allow establishing the contribute of each cycle to the overall strength degradation. The corresponding degradation path is defined by the time-history of the cumulated weighted cycles $n_{eq}(t)$ culminating (as a summation) in the total number of cycles N_{eq}.

For each of the acceleration time-history selected for the displacement analysis carried out in this paper, $n_{eq}(t)$ and N_{eq} were evaluated using the conversion procedure proposed by [10]. According to these Authors, starting from an acceleration time-history, assumed proportional to the time-history of the earthquake-induced shear stress, $n_{eq}(t)$ and N_{eq} can be computed through a conversion procedure in which the cycles that significantly affect the soil shear strength reduction are detected, using a suitable cycle-counting method, and weighted through appropriate weighting factors derived from a large set of cyclic laboratory test results.

Starting from these statements, the time-history of the cumulated weighted cycles $n_{eq}(t)$ was adopted herein to describe the degradation path of the slope yield acceleration coefficient.

According to [11], using a degradation parameter T, the degradation coefficient δ_{Cu} was related to the number of loading cycles whose time-history was assumed coincident with that of the cumulated weighted cycles $n_{eq}(t)$. Thus, $\delta_{Cu} = (n_{eq})^{-t}$ and $\delta_{Cu,min} = (N_{eq})^{-t}$ were assumed in the analysis and the following current (time-dependent) value of the slope yield acceleration coefficient was considered in the displacements evaluation:

$$k_{h,c}(t) = k_{h,co} \left\{ 1 - \frac{\left[1 - n_{eq}^{-T}\right]}{\left[1 - N_{eq}^{-T}\right]} R \right\} \qquad (3)$$

As a result, the reduction of the yield acceleration coefficient is assumed to occur only in the time interval relevant for the conversion procedure and, thus, it depends on the cumulation of the number of those cycles whose amplitude is effective in producing the reduction of the soil shear strength.

4 Results and Proposed Predictive Models

The displacement analyses were carried out using the horizontal acceleration time-histories selected by [12]. The corresponding time histories of the slope yield acceleration coefficient were computed using different values of the degradation ratio R, varying in the range $0.1 \div 0.8$) and different values of the acceleration ratio $k_{h,co}/k_{h,max}$, varying in the range $0.1 \div 0.8$.

For each of the selected record, both positive and negative signs were considered and the larger of the two computed permanent displacements, hereafter denoted as d, was assumed as representative of the slope seismic performance.

The condition $R = 0$ (i.e. $\Delta k_{h,c} = 0$), corresponding to the conventional Newmark-type analysis, with yield acceleration constant over the time ($k_{h,cmin} = k_{h,co}$), was assumed as a reference to estimate the increment in permanent displacements due to the reduction of $k_{h,c}$ associated to the soil shear strength degradation. To assess the influence of the cyclic shear strength degradation on the slope permanent displacement, the final displacement d_o computed through the conventional Newmark-type analysis ($R = 0$) was assumed as a reference for each of considered value of the acceleration ratio $k_{h,co}/k_{h,max}$. Accordingly, the displacement ratio d/d_o can be regarded as a corrective coefficient of the earthquake-induced permanent displacements that accounts for the reduction of the yield acceleration coefficient due to the occurrence of soil shear strength degradation during the ground motion.

The results of the displacement analyses which neglect (d_o) or account for (d) the degradation of $k_{h,c}$ are represented in Figs. 1 and 2, respectively, for the cases $T = 0.15$, $R = 0.2, 0.4, 0.6$ and 0.8. The figures show also the average values of the dataset of results together with the 84th and the 90th percentile of the data distribution.

The above-described procedure, aimed to estimate the corrective coefficient d/d_o, was proposed to be used in conjunction with the displacement predictive models recently proposed by [12] to assess d_o (starting from a conventional Newmark-type analysis). However, due to its generality, the values of the corrective coefficient d/d_o computed herein can be applied also the cumulated permanent displacement d_o computed using other predictive models available in the literature.

Furthermore, as previously described, the use of proper shape factors allows applying the proposed corrective coefficient d/d_o also to predict the actual permanent displacement d of several slope schemes and other sliding geotechnical systems.

The values of the corrective coefficient d/d_o computed herein are plotted in Fig. 3 for each value of the degradation ratio R considered in the analyses; the average trends are also potted in the figure together with the 84th and the 90th percentile of the dataset distribution.

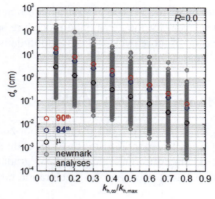

Fig. 1. Values of the displacement d_0 compute for $R = 0$ (conventional Newmark-type analysis).

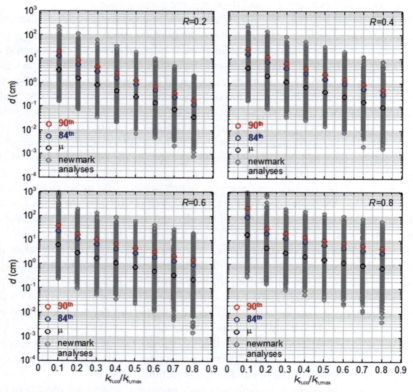

Fig. 2. Permanent displacements d computed for different values of the degradation (R) and the acceleration ($k_{h,co}/k_{h,max}$) ratios and $T = 0.15$.

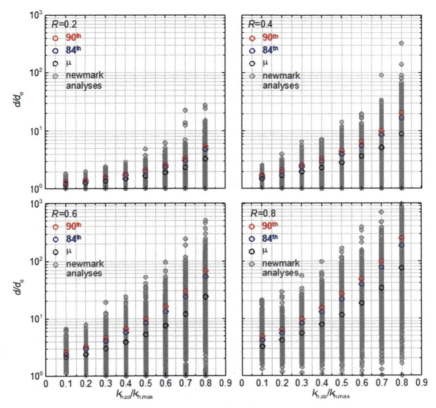

Fig. 3. Displacement ratio d/d_o computed for different values of the degradation (R) and the acceleration ($k_{h,co}/k_{h,max}$) ratios and $T = 0.15$.

In order to provide a simple and practical tool for the displacement prediction, capable to account for the effects of the shear strength degradation, regression analyses of the computed corrective coefficients d/d_o were carried out using proper functional forms to explain the distribution of the data shown in Fig. 3. Specifically, a polynomial and a S-shaped functional form have been selected to fit the average values of the computed corrective coefficients. Only the S-shaped model are presented herein. Figure 4 shows the results of the regression analyses carried out for each of the selected values of the degradation (R) and acceleration ($k_{h,co}/k_{h,max}$) ratios. The proposed S-shaped model is described by the following equation:

$$\log\left(\frac{d}{d_0}\right) = A \log\left(1 - \frac{k_{h,co}}{k_{h,max}}\right) + B \log\left(\frac{k_{h,co}}{k_{h,max}}\right) + C \qquad (4)$$

where the regression coefficients A, B and C depend on the values of the degradation ratio R according to the expressions given in Fig. 4.

$$A = 2.31R^2 - 4.25R, \quad B = 1.20R^2 - 0.79R, \quad C = 1.52R^2 - 0.58R \qquad (5)$$

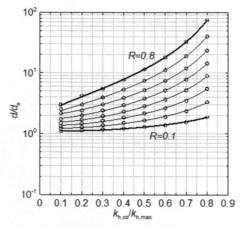

Fig. 4. Empirical predictive models proposed for $T = 0.15$ to assess the displacement corrective coefficient d/d_o as a function of the degradation (R) and acceleration ($k_{h,co}/k_{h,max}$) ratios.

5 Conclusions

The paper describes a simplified procedure to perform a modified Newmark-type analysis to account for the reduction of the slope yield acceleration coefficient $k_{h,c}$ due to the possibly occurring reduction of the soil shear strength. The reduction $\Delta k_{h,c}$ of $k_{h,c}$ (starting from its initial value $k_{h,co}$) is innovatively expressed as a function of the number of equivalent loading cycles N_{eq}, computed using an effective conversion procedure, and depends on a degradation ratio $R = \Delta k_{h,c}/k_{h,co}$ whose proper evaluation is preliminary required. The results of a large set of displacement analyses show that the reduction of $k_{h,c}$ remarkably affects the computed permanent displacements d which are generally larger than those (d_o) evaluated neglecting such effect (conventional Newmark-type analysis). The ratio d/d_o, representing a displacement corrective coefficient, quantifies the effect of the reduction of $k_{h,c}$ on the increase in permanent displacements and was computed for a large set of acceleration records and several empirical predictive models was derived, giving the displacement corrective coefficient d/d_o as a function of the ratios R and $k_{h,max}/k_{h,co}$, $k_{h,max}$ being the peak horizontal acceleration of the input ground motion. Due to its generality, using appropriate shape factors, the proposed procedure and solutions can actually be applied to any geotechnical system that under seismic conditions behaves as a sliding block.

Acknowledgements. This work is part of the research activities carried out by the Research Group of the University of Messina in the framework of the work package WP2-Slope Stability of a Research Project funded by the ReLuis (Network of Seismic Engineering University Laboratories) consortium (DPC/ReLUIS 2019-2021).

References

1. Harp, E.L., Jibson, R.W.: Landslides triggered by the 1994 Northridge, California, earthquake. Bull. Seismol. Soc. Am. **86**(1B), S319–S332 (1996)
2. Biondi, G., Maugeri, M.: Seismic response analysis of Monte Pohill (Catania). Adv. Earthq. Eng. **14**, 177–195 (2005)
3. Ingegneri, S., Biondi, G., Cascone, E, Di Filippo, G.: Influence of cyclic strength degradation on a Newmark-type analysis. In: Earthquake Geotechnical Engineering for Protection and Development of Environment and Constructions, Rome, 17–20 June 2019, pp. 2996–3004. CRC Press/Balkema, London (2019). ISBN: 978-0-367-14328-2
4. Di Filippo, G., Biondi, G., Cascone, E.: Influence of earthquake-induced pore-water pressure on the seismic stability of cohesive slopes. In: Earthquake Geotechnical Engineering for Protection and Development of Environment and Constructions, Rome, 17–20 June 2019, pp. 2136–2144. CRC Press/Balkema, London (2019). ISBN: 978-0-367-14328-2
5. Bandini, V., Biondi, G., Cascone, E., Rampello, S.: A GLE-based model for seismic displacement analysis of slopes including strength degradation and geometry rearrangement. Soil Dyn. Earthq. Eng. **71**, 128–142 (2015)
6. Madiai, C., Vannucchi, G.: Potenziale sismico distruttivo e stabilità dei pendii: abachi e tabelle per la stima degli spostamenti. Riv. Ital. Geotecn. 3–4 (1997)
7. Biondi, G., Cascone, E., Maugeri, M.: Displacement versus pseudo-static evaluation of the seismic performance of sliding retaining walls. Bull. Earthq. Eng. **12**(3), 1239–1267 (2013)
8. Gaudio, D., Masini, L., Rampello, S.: A performance-based approach to design reinforced-earth retaining walls. Geotext. Geomembr. **46**(4), 470–485 (2018)
9. Di Filippo, G., Bandini, V., Biondi, G., Cascone, E.: A two-wedge approach for the evaluation of seismic-induced rock-slides. In: Earthquake Geotechnical Engineering for Protection and Development of Environment and Constructions, Rome, 17–20 June 2019, pp. 2128–2135 (2019). CRC Press/Balkema, London (2019). ISBN: 978-0-367-14328-2
10. Biondi, G., Cascone, E., Di Filippo, G.: Affidabilità di alcune correlazioni empiriche per la stima del numero di cicli di carico equivalente. Riv. Ital. Geotecn. (2012)
11. Idriss, I.M., Dobry, R., Singh, R.D.: Nonlinear behaviour of soft clays during cyclic loading. ASCE **104**(12), 1427–1447 (1978)
12. Gaudio, D., Rauseo, R., Masini, L., Rampello, S.: Semi-empirical relationships to assess the seismic performance of slopes from an updated version of the Italian seismic database. Bull. Earthq. Eng. **18**(14), 6245–6281 (2020)

Dynamic Analysis of Geosynthetic-Reinforced Soil (GRS) Slope Under Bidirectional Earthquake Loading

Cheng Fan, Kui Cai, and Huabei Liu[✉]

School of Civil and Hydraulic Engineering, Huazhong University of Science and Technology, 1037 Luoyu Road, Wuhan 430074, Hubei, China
hbliu@hust.edu.cn

Abstract. For the past few decades, geosynthetic-reinforced soil slopes (GRS slopes) have been increasingly used in geotechnical, hydraulic and geoenvironmental engineering applications, due to their great earthquake resistance. Near-field strong ground motion usually involves a vertical component, which is very large in some cases. However, existing design guidelines do not provide a clear approach of earthquake resistant design for GRS slopes subjected to combined horizontal and vertical accelerations. In this study, a nonlinear Finite Element procedure was further validated by a centrifuge shaking table test, and then employed to investigate the seismic responses of a GRS slope model considering a large range of bidirectional earthquake loadings, based on a highway project located in Xinjiang. The results showed that the vertical acceleration had a great effect on permanent displacement, if the corresponding horizontal acceleration was large, and it also played an important role on the stiffness and natural resonant frequency of the soil due to soil compaction. There existed good correlations between the earthquake intensity parameter a_{rs} at the center of gravity of active wedge and seismic responses of the GRS slope after horizontal earthquake loading.

Keywords: GRS slope · Bidirectional ground motion · Finite element analysis · Ground motion intensity parameter

1 Introduction

China is a country with frequent earthquakes, and geosynthetic-reinforced soil (GRS) slopes have a great potential of extensive application in areas of high seismicity. Near-field ground motions containing vertical components are of interest in the fields of earthquake engineering. However, vertical acceleration is rarely taken into account in the existing design of reinforced soil structures.

There exist limited studies aiming to analyze the seismic responses of GRS structures under bidirectional earthquake loadings. It was reported that several GRS retaining walls failed during near-field earthquakes, in which the ratio of vertical to horizontal acceleration was very large [1, 2]. A series of large-scale shaking table tests of GRS retaining walls were carried out by Ling et al. [3], and actual records of bidirectional ground motion from the 1995 Kobe earthquake was considered in one of these tests. It was found that the vertical ground motion had little influence on the facing displacement, but led to higher reinforcement loads. Using the Finite Element procedure validated against the aforementioned test [3], Fan et al. [4] analyzed the influence of bidirectional earthquake loadings on the seismic response of GRS retaining walls, and the interplay between the wall responses with the Arias intensity of the horizontal seismic motion was also shown, a transfer function defining the reinforcement loads based on the Arias intensity was proposed. Ling et al. [5] analyzed the residual displacement of a GRS slope subjected to combined horizontal and vertical earthquake loadings with a log-spiral failure mechanism, the importance of vertical excitation to the sliding displacement analysis was discussed.

The objective of this study is to investigate the seismic performance of GRS slopes subjected to combined horizontal and vertical accelerations. With a nonlinear Finite Element procedure validated by a centrifuge shaking table test, a series of numerical models of a GRS slope considering a large range of bidirectional earthquake loadings were established, based on a highway project located in Xinjiang. The influence of vertical acceleration on the seismic response of GRS slopes was analyzed. A good correlations between the earthquake intensity parameter a_{rs} at the center of gravity of active wedge and seismic responses of the GRS slope was proposed, and the intensity parameter a_{rs} can be employed in a pseudo-static analytical method to determine the maximum reinforcement load.

2 Finite Element Procedure and Its Validation

PLAXIS [6] (Finite Element method) is widely used in geotechnical engineering applications including GRS structure systems. In this study, The backfill soil was modeled by Hardening Soil model with small-strain stiffness(HSS model) [6]. A parameter of 0.67 was considered to describe the reduced shear strength and stiffness of the reinforcement-soil interface. The geosynthetic reinforcements were modeled by one-dimension linear elastic elements. The acceleration time-history of the ground motion was input at the base of the model. The viscous damping was set as 5% in the dynamic analysis. Further details can be found in Cai [7].

A centrifuge shaking table test at a centrifugal acceleration of 20 g considering bidirectional seismic excitation was employed to validate the Finite Element procedure for the dynamic analysis. The wrap-faced GRS retaining wall model configurations and the test results are shown in scale-down units. Figure 1 shows the Finite Element mesh. The model height was 200 mm, the reinforcement length was 210 mm, and the vertical spacing of the reinforcement layers was 50 mm. The model parameters of the backfill soils are presented in Table 1, which were calibrated from cyclic triaxial compression tests on Fujian standard sand (relative density Dr = 70%) as shown in Fig. 2. The 100 mm thick foundation used the same sand but at a Dr of 90%. The reinforcement layers were one type of PP geogrid with a stiffness of 35 kN/m, which represented the secant stiffness at a strain of 2%. The model wall was subjected to several consecutive bidirectional sinusoidal shakings. In the first shaking, the peak horizontal acceleration was 1 g, while the peak vertical acceleration was 0.67 g(the vertical to horizontal (V/H) ratio = 0.67), and there was no phase difference. The input accelerations in the later shaking were approximately 2, 4, 8, 10 times of those in the first one with the same V/H ratio. The frequency of input motions was 66.7 Hz. Further details of the input motions can be found in Cai [7].

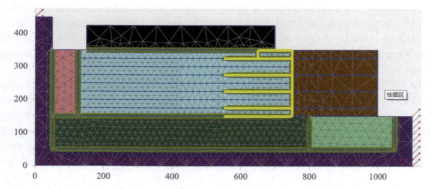

Fig. 1. Finite Element mesh of the centrifuge shaking table test (unit: mm)

Table 1. Backfill parameters of the GRS retaining wall model

Parameter	Unit	Value
Unit weight, γ	kN/m^3	15.1
Cohesion, c	kPa	2
Friction angle, φ	°	36.6
Dilatancy angle, ψ	°	6.6
Secant stiffness in standard drained triaxial test, E_{50}^{ref}	kPa	2.45E4
Tangent stiffness for primary oedometer loading, E_{oed}^{ref}	kPa	2.5E4
Unloading–reloading stiffness, E_{ur}^{ref}	kPa	7.35E4
Power for stress-level dependency of stiffness, m	–	0.5
Shear strain, $\gamma_{0.7}$	–	0.0004
Reference shear modulus at very small strains (e < 10E6), G_0^{ref}	kPa	9E4
Strength reduction factor in the interface, R_{inter}	–	0.67

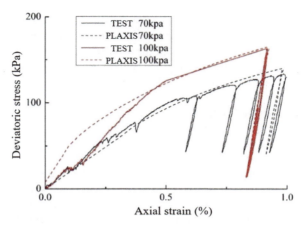

Fig. 2. Cyclic triaxial compression tests on Fujian standard sand (Dr = 70%)

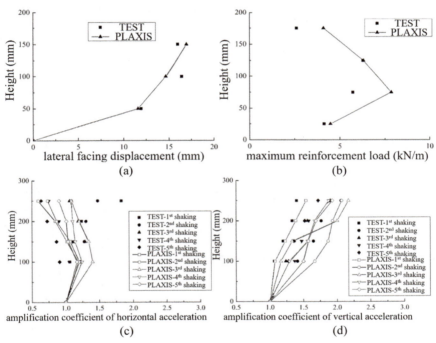

Fig. 3. Comparisons between the results from the test and the Finite Element method: (a) lateral facing displacement (b) maximum reinforcement load of each layer (c) amplification coefficient of horizontal acceleration of the reinforced soil (d) amplification coefficient of vertical acceleration of the reinforced soil

Figure 3 shows the comparisons between the results of the centrifuge shaking table test and the Finite Element method. It shows that the Finite Element procedure satisfactorily reproduced the seismic responses of the wrap-faced GRS retaining wall model. The validated procedures can then be used for analyzing the seismic responses of a GRS slope model considering bidirectional earthquake loadings.

3 Finite Element Models

The seismic responses of a gabion-GRS slope model with a height of 10.1 m were investigated, based on a highway project located in Xinjiang, as shown in Fig. 4. The gradient of the slope facing and its back were 1:0.75 and 1:1.5, respectively. The vertical spacing of the reinforcement layers was 0.6 m. The foundation soil was assumed to be stiff rock, hence fixed boundaries were used on the base, and roller boundaries were used on the sides. The unit weight of gabion facings was 24 kN/m^3, which was the same as Xu [8]. Parameters of the backfill soil are shown in Table 2, and further details can be found in Xiong [9]. The reinforcements were modeled by one-dimension linear elastic elements with a stiffness of 4000 kN/m. The other parameters of the Finite Element model can be found in Cai [7].

Table 2. Backfill parameters of the Gabion-GRS slope model

Parameter	Unit	Value
Unit weight, γ	kN/m^3	20
Cohesion, c	kPa	12
Friction angle, φ	°	41.7
Dilatancy angle, ψ	°	0
Secant stiffness in standard drained triaxial test, E_{50}^{ref}	kPa	5.421E4
Tangent stiffness for primary oedometer loading, E_{oed}^{ref}	kPa	5.421E4
Unloading–reloading stiffness, E_{ur}^{ref}	kPa	1.63E5
Power for stress-level dependency of stiffness, m	–	0.5
Shear strain, $\gamma_{0.7}$	–	4.5E−4
Reference shear modulus at very small strains (e < 10E6), G_0^{ref}	kPa	1.05E5
Strength reduction factor in the interface, R_{inter}	–	0.67

After construction simulation, the base excitations were input from bottom of the model. Altogether 30 strong bidirectional ground excitations taken from the records of 8 actual earthquake were employed in the dynamic analyses. The peak acceleration of all the input horizontal excitations were scaled to be 0.4 g, and the corresponding vertical excitations were scaled proportionally according to the actual vertical to horizontal (V/H) ratio.

Further details can be found in Cai [7].

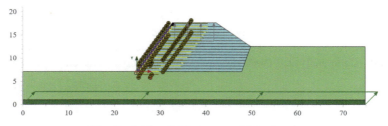

Fig. 4. Gabion-GRS slope model (unit: m)

4 Results and Discussions

Figure 5(a) shows the comparisons between the maximum residual lateral facing displacement δ_{max} after bidirectional earthquake loading and that after horizontal loading only. It can be seen that the vertical acceleration had a great effect on permanent displacement, if the corresponding horizontal acceleration was large.

The comparisons between the sum of maximum reinforcement loads $\sum T_{max}$ after bidirectional earthquake loading and that after horizontal loading only are summarized in Fig. 5(b). It shows that the vertical acceleration led to smaller reinforcement loads with larger corresponding horizontal acceleration, which mean it played an important role on the stiffness and natural resonant frequency of the soil due to soil compaction.

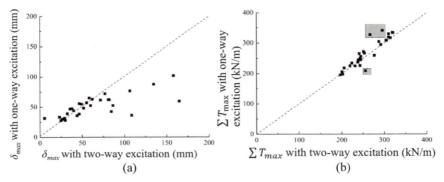

Fig. 5. Influence of vertical acceleration on (a) the maximum residual lateral facing displacement δ_{max} (b) the sum of maximum reinforcement loads $\sum T_{max}$

The earthquake intensity parameter a_{rs} is defined as:

$$a_{rs} = \left(\frac{1}{t_0}\int_0^{T_d} a(t)^2 dt\right)^{0.5} \quad (1)$$

Here t is time, $a(t)$ is the time history of the horizontal excitation, T_d is the duration of the horizontal earthquake motion, and t_0 is a unit time.

Figure 6 shows that there existed good correlations between the earthquake intensity parameter a_{rs} at the center of gravity of active wedge and reinforcement loads of the GRS slope after horizontal earthquake loading. The intensity parameter a_{rs} can be employed in a pseudo-static analytical method to determine the maximum reinforcement load.

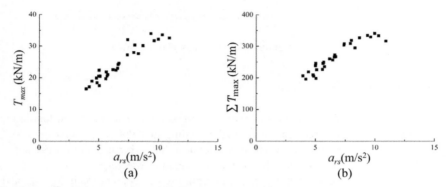

Fig. 6. Relationship between a_{rs} and (a) maximum reinforcement load T_{max} (b) the sum of maximum reinforcement loads $\sum T_{max}$ after horizontal earthquake loading

5 Conclusion

In this study, a nonlinear Finite Element procedure was further validated by a centrifuge shaking table test, and then employed to investigate the seismic responses of a GRS slope model considering a large range of bidirectional earthquake loadings, based on a highway project located in Xinjiang. The results showed that:

(1) The vertical acceleration had a great effect on permanent displacement if the corresponding horizontal acceleration is large, and it also played an important role on the stiffness and natural resonant frequency of the soil due to soil compaction;
(2) There existed good correlations between the earthquake intensity parameter a_{rs} at the center of gravity of active wedge and seismic responses of the GRS slope after horizontal earthquake loading; The intensity parameter a_{rs} can be employed in a pseudo-static analytical method to determine the maximum reinforcement load.

References

1. Sandri, D.: Retaining walls stand up to the Northridge earthquake. In: Geotechnical Fabrics Report. IFAI, St. Paul, vol. 12, no. 4, pp. 30–31 (1994)
2. Ling, H.I., Leshchinsky, D.: Failure analysis of modular-block reinforced-soil walls during earthquakes. J. Perform. Constr. Fhorizontal and verticaacil. **19**(2), 117–123 (2005)
3. Ling, H.I., Mohri, Y., Leshchinsky, D., et al.: Large-scale shaking table tests on modular-block reinforced soil retaining walls. J. Geotech. Geoenviron. Eng. **131**(4), 465–476 (2005)

4. Fan, C., Liu, H.B., Cao, J.Z., et al.: Responses of reinforced soil retaining walls subjected to horizontal and vertical seismic loadings. Soil Dyn. Earthq. Eng. **129**, 105969 (2020). https://doi.org/10.1016/j.soildyn.2019.105969
5. Ling, H.I., Leshchinsky, D., Mohri, Y.: Soil slopes under combined horizontal and vertical seismic accelerations. Earthq. Eng. Struct. Dyn. **26**(12), 1231–1241 (1997)
6. PLAXIS. Reference Manual, 2D – Version 9.0, PLAXIS. Delft University of Technology, Delft (2008)
7. Cai. Dynamic Analysis of Reinforced Soil Slope Under Bidirectional Earthquake Loading. Huazhong University of Science and Technology, Wuhan (2020)
8. Xu, G.Y.: Experimental Study on Uniaxial Compressive Peak Strength of Gabion and Influence Factors. Southwest Jiaotong University, Chengdu (2019)
9. Xiong, K.J.: Study on Seismic Performance of Back-to-Back Geosynthetic-Reinforced Soil (GRS) Road Embankments with Different Cross-sections. Huazhong University of Science and Technology, Wuhan (2019)

Seismic Performance of Slopes at Territorial Scale: The Case of Ischia Island

Francesco Gargiulo[✉], Giovanni Forte, Anna d'Onofrio, Antonio Santo, and Francesco Silvestri

University of Naples Federico II, Napoli, Italy
francesco.gargiulo6@unina.it

Abstract. Ischia is an active volcanic island in the gulf of Naples (Italy), historically hit by several earthquakes which caused extensive structural damage as well as landslides, especially localized in the north-western area of the island. The aim of this study is to assess the seismic performance of slopes at territorial scale in the three municipalities which were mostly damaged by a recent earthquake occurred in 2017. The main results are presented through different maps individuating the areas potentially unstable in terms of earthquake-induced slope displacements. Validation of the results was carried out comparing such areas with those historically affected by earthquake-induced landslides, as reported by the chronicles and the literature. The comparison confirmed the effectiveness of the proposed approach, while the resulting maps may provide a significant planning tool for managing the emergency after a strong-motion event and defining a priority scale of interventions to mitigate the instability risk. Finally, they permitted to identify the most critical areas where further site-specific investigations as well as simplified to advanced numerical analyses will be carried out.

Keywords: Seismic slope stability · Geographic information system · Displacement-based methods

1 Introduction

On August 21st 2017 the volcanic island of Ischia in the Gulf of Naples (Southern Italy) was struck by a M_w 4.0 seismic event, which caused more damage than would be expected from an earthquake of this magnitude. The subsequent Seismic Microzonation studies for the municipalities of Casamicciola Terme, Lacco Ameno and Forio [1] have been aimed at defining homogeneous zones in terms of amplification, but left out about 48% of the urban areas classified as unstable on the basis of official national and regional landslides inventories. As a matter of fact, the area affected by the 2017 earthquake is characterized by several landslide phenomena triggered by historical earthquakes [2], as well as by high susceptibility to weather-induced instability phenomena, e.g. flow slides and debris flows in pyroclastic soils occurred in 2006 (Vezzi Mt.) and 2009 (Casamicciola) respectively [3, 4].

Several past studies aimed at the risk assessment of Ischia territory in terms of earthquake-induced landslides were mainly limited to seismic hazard and susceptibility evaluation by empirical-qualitative approaches, not adequately supported by mechanical characterization of the pyroclastic soils [5]. This paper is part of a wide research project aimed at zoning the whole territory affected by the 2017 earthquake in terms of slope instability induced by seismic motion, taking into account the complex hazard characterizing the island due to the combination of volcanic, seismic and hydrogeological factors, by adopting an up-to-date multi-level approach.

2 Outline of the Methodology

Several authors in literature (e.g. [6–9]) have adopted territorial scale approaches to assess earthquake-induced displacements. The latter can be implemented in a Geographic Information System (GIS) by combining hydrogeological, geotechnical, topographic, and seismological data as shown in the flowchart of Fig. 1.

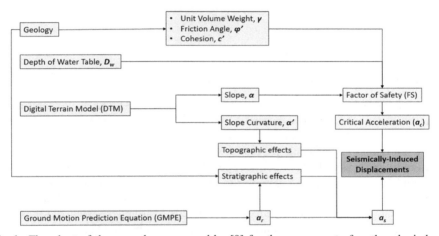

Fig. 1. Flowchart of the procedure proposed by [9] for the assessment of earthquake-induced displacements

The mapping scale usually varies between 1:25.000 and 1:5.000, depending on the reference geological survey, with a cell grid size determined by the resolution of the Digital Terrain Model. At these territorial scales, the seismic performance of slopes can be evaluated using the displacement-based approach. Following the well-known method by Newmark (1965) [10], the potential landslide is modelled as a rigid friction block sliding on an inclined plane. The triggering occurs when the acceleration overpasses the 'critical acceleration' threshold, a_c, which can be calculated as:

$$a_c = (FS - 1)g \sin \alpha \quad (1)$$

where g is the gravity acceleration, α is the slope angle and FS is the factor of safety. The latter for an infinite slope can be calculated as:

$$FS = \frac{c'}{\gamma D \sin \alpha} + \frac{\tan \varphi'}{\tan \alpha} - \frac{m \gamma_w \tan \varphi'}{\gamma \tan \alpha} \quad (2)$$

where:

- γ is the unit volume weight of the soil;
- γ_w is the water unit weight;
- c' and φ' are the effective cohesion and friction angle;
- D is the depth of the sliding surface;
- m is the part of the slope above shear surface which is saturated.

Next, the combination of seismological data and critical acceleration a_c leads to the evaluation of the permanent slope displacements. The latter, in a dynamic simplified analysis, can be obtained by double integrating the difference between an accelerogram and a_c until the relative velocity becomes zero. This approach is not straightforward for application at the territorial scale, thus it is more usual to adopt predictive relationships expressing the slope displacements as a function of the critical acceleration and synthetic ground motion parameters, e.g. the peak surface acceleration, a_s.

3 Input Data

3.1 Geological and Geomorphological Setting

The island of Ischia represents the top of an active volcano constituted by volcanic rocks, by landslide deposits and, subordinately, by sedimentary rocks, which derive from the accumulation and the cementation of fragments of pre-existing rocks, dismembered by erosive processes. The volcanic rocks present on the island come from lava flows and domes and explosive products, which have generated tuff formations and extensive tephra of ash and lapilli. The extreme complexity of the formations was already highlighted during the studies of Seismic Microzonation of the municipalities of Casamicciola, Lacco Ameno and Forio, mostly affected by the 2017 earthquake. Hence, it has been appropriately simplified in the geolithological map shown in Fig. 2 [1].

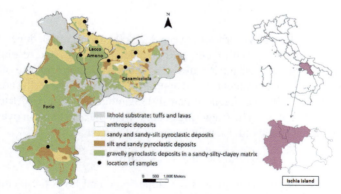

Fig. 2. Geolithological map of the N-W sector of Ischia island in the gulf of Naples (Italy)

The patterns of groundwater flow are particularly complex, due to the volcano-tectonic events that affected the island. Figure 3 shows a map of the seasonal mean groundwater table depth from ground level obtained from [11].

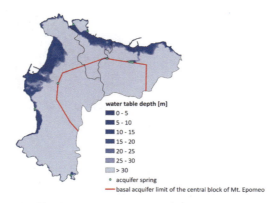

Fig. 3. Groundwater table of the study area

Table 1 summarizes the statistical values of unit weight and strength parameters assigned to each of the units identified in Fig. 2, as derived from the collection of 24 laboratory tests (20 direct shear and 4 triaxial tests) available from 12 previous borehole investigations collected in the area.

Table 1. Unit weight and strength parameters for the main formations identified in Fig. 2

X	$\bar{X}$	sx	CoV (%)	p^{16}	p^{50}	p^{84}
γ	15.7	2.3	0.1	13.1	16.0	17.6
c'	13.2	16.0	1.2	0.0	8.4	19.9
φ'	32.3	5.3	0.2	27.7	31.3	35.6
γ	14.9	7.7	0.5	14.4	15.1	15.3
c'	10.3	12.5	1.2	0.0	7.2	19.3
φ'	32.4	4.4	0.1	29.1	31.6	37.5
γ	15.0					
c'	10.0	colspan: no data available				
φ'	30.0					
γ	14.0					
c'	0.0	colspan: no data available				
φ'	25.0					

X=sample
$\bar{X}$=sample mean
sx=standard deviation
CoV=coefficient of variation
p^{16}=16th percentile
p^{50}=50th percentile
p^{84}=84th percentile

Starting from a Digital Terrain Model (DTM) with 5m resolution (Fig. 4a), a slope angle map has been obtained (Fig. 4b).

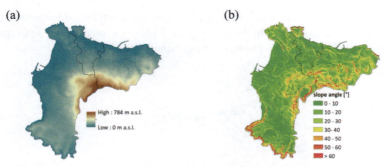

Fig. 4. Digital Terrain Model (a) and slope map (b) of the study area

3.2 Seismological Data

The seismic action has been evaluated considering a Deterministic Seismic Hazard Analysis (DSHA), i.e. simulating the Maximum Historical Earthquake (MHE) occurred at Casamicciola in 1883. To this end, an *ad hoc* Ground Motion Prediction Equation (GMPE) for volcanic areas [12] has been used to simulate the distribution of acceleration at bedrock (Fig. 5a), by considering a point-source and isotropic attenuation model starting from the position of the epicenter and the magnitude of the event reported by the CPT15 catalogue [13]. The reference acceleration, a_r, has been modified to obtain the acceleration at surface, a_s (Fig. 5b), as follows:

$$a_s = S_S \cdot S_T \cdot a_r \tag{3}$$

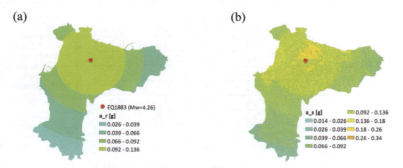

Fig. 5. Maps of acceleration at bedrock (a) and surface (b)

In Eq. 3, S_S is the nonlinear stratigraphic amplification factor, evaluated as variable between 1.3 and 1.9 with the relationships proposed by [14] (Fig. 6b). It was assigned on the basis of the map in Fig. 6a, obtained by integrating the Grade I Seismic Microzonation map of the area with the soil type classification map proposed by [15] for the whole Italian territory. S_T is the topographic factor, evaluated as variable between 0.5 and 1.4 from the slope curvature (Fig. 6c) to take into account the amplification or attenuation of the seismic motion in case of convex or concave geometry (e.g. [9, 16]).

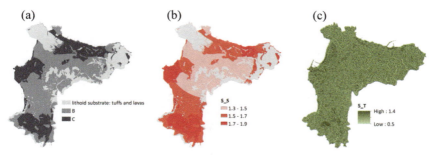

Fig. 6. Soil type classification (a), stratigraphic (b) and topographic (c) amplification factors maps

4 Results

The first step in evaluating seismically-induced displacements is the definition of the critical acceleration beyond which the slope will slide, evaluated through Eq. 1, by assuming an infinite slope model. Three different maps of critical acceleration have been obtained considering the statistical variability of the physical and mechanical properties synthesised in Table 1 (Fig. 7).

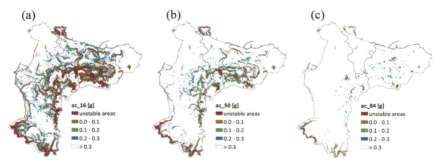

Fig. 7. Maps of critical acceleration for 16th (a), 50th (b) and 84th (c) percentile

Finally, by combining the critical acceleration maps with the ground motion distribution (Fig. 5b), the permanent displacements could be calculated with the semi-empirical relationship proposed by [17] (see Fig. 8):

$$d = B \cdot e^{-A \frac{a_c}{a_s}} \qquad (4)$$

where A and B are coefficients depending on the soil classes (Fig. 6a) and on the range of values of the acceleration expected at surface (Fig. 5b).

The main results of the proposed methodology are presented in the maps of Fig. 8 in terms of spatially distributed displacements. Such displacements are classified according to 5 different classes, from very low (white zones) to unstable areas (red zones) corresponding to as many susceptibility classes. When considering the outcomes associated to the 16th percentile input parameters (Fig. 8a), about 10% of the total area is

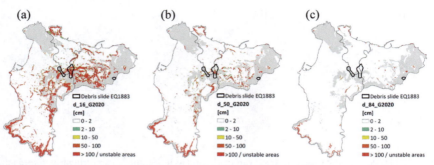

Fig. 8. Maps of earthquake-induced displacements for 16th (a), 50th (b) and 84th (c) percentile

characterized by induced displacements greater than 100 cm. Less conservative results are obtained considering the 50th or 84th percentile (Fig. 8b–c), for which the unstable areas drop to the less than 4% and 1%, respectively. Therefore, the amount of territory potentially unstable in seismic conditions appears less extended than that mapped in the official landslide inventories.

These maps can be useful to manage the emergency after a strong-motion event and planning a priority scale of interventions to mitigate the instability risk individuating the potentially unstable areas. Leaving aside the unstable areas located south of the municipality of Forio consisting of coastal cliffs overlooking the sea, the areas where attention needs to be focused are located between the municipalities of Casamicciola and Lacco Ameno, in the same zones where earthquake-induced landslides during the 1883 event occurred [2], as shown by the black contours drawn on the maps in Fig. 8 in the area of 'Fango', reported by the chronicles [18]. Moreover, the results addressed an ongoing program of site-specific investigations which will support more advanced dynamic numerical analyses, in order to assess and refine the predictions.

References

1. Alleanza, G.A., et al.: Valutazione della risposta sismica locale di alcune aree dell'Isola di Ischia per la pianificazione di interventi di mitigazione del Rischio Sismico. In: XXVII Convegno Nazionale di Geotecnica "La geotecnica per lo sviluppo sostenibile del territorio e per la tutela dell'ambiente" (in Italian), Reggio Calabria, 13–15 July 2022 (2021)
2. Del Prete, S., Mele, R.: Il contributo delle informazioni storiche per la valutazione della propensione al dissesto nell'Isola di Ischia (Campania). Rend. Della Soc. Geol. Ital. **2**, 29–47 (2006)
3. Nocentini, M., Tofani, V., Gigli, G., Fidolini, F., Casagli, N.: Modeling debris flows in volcanic terrains for hazard mapping: the case study of Ischia Island (Italy). Landslides **12**(5), 831–846 (2014). https://doi.org/10.1007/s10346-014-0524-7
4. Santo, A., Di Crescenzo, G., Del Prete, S., Di Iorio, L.: The Ischia island flash flood of November 2009 (Italy): phenomenon analysis and flood hazard. J. Phys. Chem. Earth (2012). https://doi.org/10.1016/j.pce.2011.12.004
5. Caccavale, M., Matano, F., Sacchi, M.: An integrated approach to earthquake-induced landslide hazard zoning based on probabilistic seismic scenario for Phlegrean Islands (Ischia, Procida and Vivara), Italy. Geomorphology **295**, 235–259 (2017). https://doi.org/10.1016/j.geomorph.2017.07.010

6. Jibson, R.W., Harp, E.L., Michael, J.A.: A method for producing digital probabilistic seismic landslide hazard maps. Eng. Geol. **58**(3–4), 271–289 (2000). https://doi.org/10.1016/S0013-7952(00)00039-9
7. Kaynia, A., Skurtveit, E., Saygili, G.: Real-time mapping of earthquake-induced landslides. Bull. Earthq. Eng. **9**, 955–973 (2011). https://doi.org/10.1007/s10518-010-9234-2
8. Anna Nowicki, M., Wald, D.J., Hamburger, M.W., Hearne, M., Thompson, E.M.: Development of a globally applicable model for near real-time prediction of seismically induced landslides. Eng. Geol. **173**, 54–65 (2014). https://doi.org/10.1016/j.enggeo.2014.02.002
9. Silvestri, F., Forte, G., Calvello, M.: Multi-level approach for zonation of seismic slope stability: experiences and perspectives in Italy. Panel report. In: Picarelli, L., Scavia, C., Aversa, S., Cascini, L. (eds.) 12th International Symposium on Landslides, Napoli, Italy, 12–19 June 2016 'Landslides and Engineered Slopes. Experience, Theory and Practice', vol. 1, pp. 101–118 (2016)
10. Newmark, N.M.: Effects of earthquakes on dams and embankments. The V Rankine Lecture of the British Geotechnical Society. Géotechnique **15**(2), 139–160 (1965)
11. Piscopo, V., Lotti, V., Formica, F., Lana, F., Pianese, L.: Groundwater flow in the Ischia volcanic island (Italy) and its implications for thermal water abstraction. Hydrogeol. J. **28**, 1–23 (2019)
12. Lanzano, G., Luzi, L.: A ground motion model for volcanic areas in Italy. Bull. Earthq. Eng. **18**(1), 57–76 (2019). https://doi.org/10.1007/s10518-019-00735-9
13. Rovida, A., Locati, M., Camassi, R., Lolli, B., Gasperini, P., Antonucci, A.: Catalogo Parametrico dei Terremoti Italiani (CPTI15), Versione 3.0. Istituto Nazionale di Geofisica e Vulcanologia (2021)
14. Tropeano, G., Soccodato, F., Silvestri, F.: Re-evaluation of code-specified stratigraphic amplification factors based on Italian experimental records and numerical seismic response analyses. Soil Dyn. Earthq. Eng. (2018). https://doi.org/10.1016/j.soildyn.2017.12.030
15. Forte, G., Chioccarelli, E., Falco, M., Cito, P., Santo, A., Iervolino, I.: Seismic soil classification of Italy based on surface geology and shear-wave velocity measurements. Soil Dyn. Earthq. Eng. **122**, 79–93 (2019). https://doi.org/10.1016/j.soildyn.2019.04.002
16. Torgoev, A., Havenith, H.B., Lamair, L.: Improvement of seismic landslide susceptibility assessment through consideration of geological and topographic amplification factors. In: JAG 2013, 17–18/09 Grenoble, France (2013)
17. Gaudio, D., Rauseo, R., Masini, L., Rampello, S.: Semi-empirical relationships to assess the seismic performance of slopes from an updated version of the Italian seismic database. Bull. Earthq. Eng. **18**(14), 6245–6281 (2020). https://doi.org/10.1007/s10518-020-00937-6
18. Johnston-Lavis, H.J.: Monograph of the earthquakes of Ischia (1885)

Numerical Simulation of Seismic Performance of Road Embankment Improved with Hybrid Type Steel Pile Reinforcement

Chengjiong Qin[1(✉)], Hemanta Hazarika[1], Divyesh Rohit[1], Nanase Ogawa[2], Yoshifumi Kochi[3], and Guojun Liu[4]

[1] Kyushu University, Fukuoka 819-0395, Japan
qinchengjiong@yahoo.co.jp
[2] Giken Ltd., Tokyo 135-0063, Japan
[3] K's Lab Inc., Yamaguchi 753-0212, Japan
[4] Changshu Institute of Technology, Suzhou 215500, China

Abstract. Highway embankments are part of strategic infrastructures that play a critical role in connecting and transporting critical rescue components and disaster relief materials during natural disasters such as earthquakes, tsunamis, typhoons, and rainstorms. The failure of such structures during disasters can exacerbate the extent of damage to human lives due to delays in transporting rescue services. The 2017 NEXCO report points to a recent example of embankment damage in the town of Mashiki in the Kumamoto region as a result of the 2016 Kumamoto earthquake. Therefore, it is important to develop a sustainable and economical geotechnology applicable to existing and new highway embankments. This research proposes a new seismic mitigation technique that uses hybrid reinforcement to protect highway embankments and mitigate the associated ground subsidence. The technique aims to reduce lateral and vertical deformation of the embankment and the buildup of excess pore water pressure by reinforcing the underlying liquefiable soil with two types of steel piles deployed around and below the embankment foundation. The performance of the proposed technique is evaluated through a dynamic effective stress analysis using the LIQCA FEM program.

Keywords: Dynamic effective stress analysis · Ground subsidence · Highway embankments · Hybrid pile retrofit technique

1 Introduction and Background of Study

The 2016 Kumamoto earthquake with a magnitude of Mw 7.0 led to considerable loss of life and structural damage in the Kumamoto Prefecture, Japan. It was observed that multiple sections of the Kyushu Expressway could not support the rescue activities in Kumamoto Prefecture due to damage of embankments during the earthquake [1]. A typical illustration of highway embankment failure in the Mashiki town during the 2016 Kumamoto earthquake is shown in Fig. 1 [2]. It can be seen that the underlying loose sand layer under the embankment liquefied, causing excessive settlement and the collapse of the embankment making it unusable. In order to prevent such failures during future disasters, the authors of this research have developed a novel low-cost hybrid type steel pile reinforcement technique using vertical and slanted piles which can be placed around and under the existing embankments and new embankments, thereby mitigating the liquefaction induced excessive deformations in the highway embankments. The effectiveness of the hybrid type reinforcement technique was evaluated in Qin et al. [3, 4] through 1g shaking table tests. It is observed that the proposed technique is effective in suppressing excess pore water pressure developed during dynamic shaking leading to minimizing the excessive settlements in the embankment. In the current research, the emphasis is placed on evaluating the above-mentioned technique through 2-dimensional dynamic FEM analysis using LIQCA coding program. Here, the technique's effectiveness is evaluated through dynamic effective stress analysis by comparing the unreinforced model, conventional reinforcement (only vertical piles), and new hybrid type reinforcement (vertical piles and slanted piles).

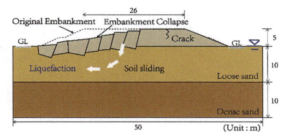

Fig. 1. Schematic diagram of road embankment failure during the 2016 Kumamoto earthquake

2 Numerical Approach

In this research, LIQCA liquefaction analysis code is used, in which a soil-water coupled problem is formulated based on a $u - p$ formulation for a porous media [5–7]. In this formulation, the finite element method (FEM) is utilized for spatial discretization of the equilibrium equation, while the finite volume method is applied for spatial discretization of the pore water pressure in the continuity equation. The numerical approach has been previously validated by Oka et al. [5] by comparing numerical results and analytical solutions for transient response of saturated porous solids. The constitutive equation used for sand elements is a cyclic elasto-plastic model. Several studies can be referred to know the details of the formulation and constitutive model [5–7]. In this research, a series of numerical simulations were performed to access the effectiveness of the hybrid type reinforcement in settlement reduction and liquefaction mitigation in highway embankment. Through the numerical results obtained, we hereby discuss the accuracy of the prediction with liquefaction analyses and observed pile behavior in a prototype scale, in particular the lateral deformation of embankment and foundation caused by the increase of excess pore water pressure in the liquefied ground.

2.1 Numerical Model and Boundary Conditions

The soil elements in the numerical models are assigned Toyoura sand properties. The ground surface below the embankment is modelled in two separate layers. The bottom dense layer has relative density (Dr) of 90%, while the upper loose layer has relative density (Dr) of 60%. The overlying embankment is modelled with a relative density (Dr) of 60%, and is modelled as the dry element. In addition, the underlying soil layers are saturated. Table 1 shows the properties of foundation and embankment materials. The slanted pile is modeled as the linear elastic beam element. Figure 2 shows the hybrid beam element [8] for the vertical pile, consisting of the conventional beam and the solid elastic elements, to represent the pile volume. The joint elements are added along with all interfaces between the soil and structures to share identical displacements in the horizontal direction. The bottom of the model is set to have a rigid boundary condition. When the soil layer composition at the lateral boundary is different, the free boundary should be used. The lateral and bottom boundaries are assumed to be impermeable, whereas the surface of the loose sand layer is a permeable boundary. Dynamic analysis is performed for three different configurations of highway embankment and foundation as shown in Fig. 3: (Case 1) unreinforced foundation of the embankment, (Case 2) only two rows of piles to support embankment, and (Case 3) hybrid type reinforcement.

Table 1. Material parameters of the constitutive model

Material parameters	Unit	Loose layer Dr = 60%	Dense layer Dr = 90%	Embankment
Density	ρ	1.88	2.00	1.88
Coefficient of permeability	k	0.01	0.01	0.01
Initial void ratio	e_0	0.738	0.659	0.738
Compression index	λ	0.02	0.0004	0.02
Swelling index	κ	0.0005	0.00008	0.0005
Failure stress ratio	M_f^*	1.300	1.466	1.300
Phase transformation stress ratio	M_m^*	0.980	0.765	0.980
Initial shear modulus Ratio	G_0/σ_m^*	2343	1133	2343
Dilatancy parameter	D_0^*	0.5	0.12	0.5
Hardening parameter	n	5.0	4.0	5.0
	B_0^*	6550	54000	6550
	B_1^*	65.5	5400	65.5
Reference strain parameter	γ_r^{P*}	0.002	0.03	0.002
	γ_r^{E*}	0.008	0.36	0.008

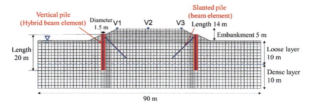

Fig. 2. Finite element model of the entire model for Case 3

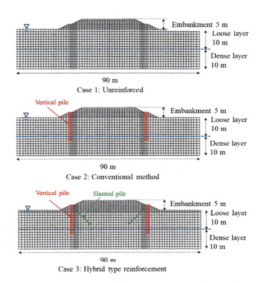

Fig. 3. Layouts of models: (a) Case 1 - unreinforced foundation of the embankment, (b) Case 2 - only two rows of steel piles reinforcement (c) Case 3 - hybrid type reinforcement

2.2 Initial Stress Conditions and Input Motions

For the initial stress analysis, Mohr-Coulomb model was applied to all soil layers of the FEM model. Gravity force is applied to the entire model under the condition that stress analysis is in 100 divisions. The initial stress analysis expresses the stress state of the ground and embankment, and the deformation here is reset to zero. The foreshock and mainshock of the 2016 Kumamoto earthquake in the East-West direction were recorded in the ground surface of KMMH16 station (Mashiki town). The time histories of earthquake are depicted in Fig. 4. A mainshock that followed a foreshock is imparted to the model. Consolidation analysis of 6 h is considered between foreshock and mainshock to ensure maximum dissipation of developed excess pore water pressure during the earthquake. The final stress state after foreshock and consolidation is used as the initial stress state for dynamic analysis for mainshock, whereas the deformed coordinates after foreshock are used for the mainshock.

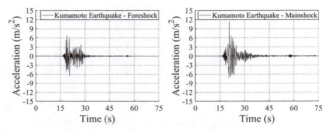

Fig. 4. Seismic input motion applied in the analysis (Strong-motion Seismograph Networks)

3 Results and Discussion

The distribution of excess pore water pressure ratio at the 20th second, 40th second, 60th second, and 75th second of foreshock and mainshock are depicted in Figs. 5, 6, 7, and 8, respectively. As a result, the excess pore water pressure ($r_u = \Delta u/\sigma_v'$) around the foot of the embankment in Case 1, Case 2, and Case 3 reaches 1.0 in the 40th second of the foreshock. The increase of r_u in a partial area underneath the embankment is observed in Fig. 7a and Fig. 8a, whereas there is no significant r_u underneath the embankment in hybrid type reinforcement at the end of foreshock (Fig. 7e). The hybrid type reinforcement limited excess pore water pressure during foreshock by increasing the shear strength of the entire embankment. Since the stress states after foreshock and consolidation are used for analysis in the mainshock simulation. The r_u of the entire free field in each case reaches 1.0 in Case 1 and Case 2 at the 20th second of the mainshock. The excess pore water pressure (Δu) increases on the foundation among the vertical pile foundation, which is illustrated in Fig. 5d, due to the vertical pile elements being set to the undrained condition in the 2D simulation. However, the central area on the foundation, which has a minor increment of excess pore water pressure, can support the embankment used by hybrid type reinforcement during the mainshock.

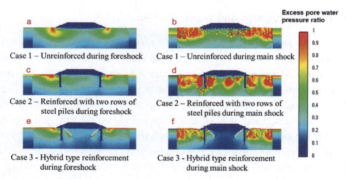

Fig. 5. Computed distribution of excess pore water pressure ratio at the 20th second

Numerical Simulation of Seismic Performance of Road Embankment Improved 1337

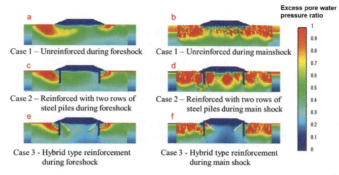

Fig. 6. Computed distribution of excess pore water pressure ratio at the 40th second

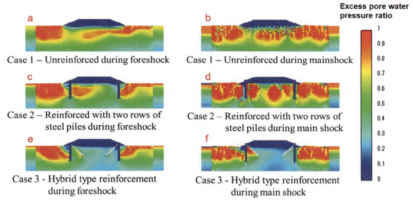

Fig. 7. Computed distribution of excess pore water pressure ratio at the 60th second

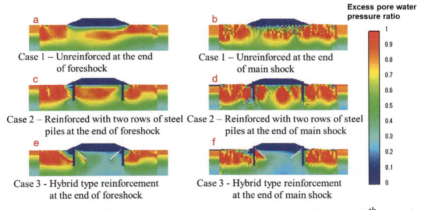

Fig. 8. Computed distribution of excess pore water pressure ratio at the 75th second

The time histories of settlement on the embankment crest and center of the embankment are illustrated in Fig. 9. In the unreinforced case, the considerable embankment failure of the embankment occurs due to the two earthquakes. The loose layers significant deformation is observed in both vertical and lateral directions underneath the embankment. However, the only small settlement is noted in the embankment improved by the hybrid type reinforcement after earthquake loadings. The time histories of settlement, which are recorded from V2, show the stability of the entire embankment, which proves the effectiveness of the hybrid type reinforcement by increasing the shear strength of the whole embankment. Furthermore, the pile foundations inserted in the middle of the embankment slope cannot restrain the movement of the embankment crest during strong earthquakes. The insignificant settlement in the central of the embankment can provide the function of embankment after earthquakes.

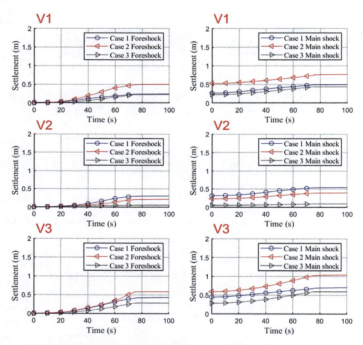

Fig. 9. The time histories of settlement during the earthquakes

4 Conclusions

This study has evaluated the performance of the highway embankment on a liquefiable foundation layer subjected to earthquake loadings. The authors have developed a new reinforcing technique for highway embankment to mitigate the damage induced by the earthquake. The following conclusions can be derived based on the numerical results:

1) In the case of unreinforced embankment, large settlement and horizontal deformation occur even in the foreshock.
2) The loose layer's relatively small buildup of excess pore water pressure improved by hybrid type reinforcement shows that the new deformation countermeasure is effective in the liquefaction mitigation. Therefore, the seismic performance of the proposed method hybrid type countermeasure is superior to that of the existing method using only vertical steel piles.
3) Ground settlement is one of the pivotal issues which occur during earthquakes. The new hybrid type reinforcement is significantly effective in reducing settlement. The numerical results show that slanted piles can reduce the lateral deformation of vertical pile and help in maintaining structural integrity against liquefaction.

In summary, the embankment improvement using the hybrid type reinforcement helps enhanceing the highway embankment's stability during earthquakes, mitigating the seismic effects on the soils, and reducing the loose soil's liquefaction potential. Nevertheless, to achieve a more practical and efficient design for the hybrid type method, it is required to perform further detailed studies such as the mechanism of the hybrid type reinforcement.

Acknowledgment. Financial aids for this study provided by NEXCO (Nippon Expressway Company Limited) and affiliated organizations, Japan and Kyushu Regional Management Service Association, Fukuoka, Japan are gratefully acknowledged.

References

1. Hazarika, H., et al.: Geotechnical damage due to the 2016 Kumamoto earthquake and future challenges. Lowland Technol. Int. **19**(3), 189–204 (2017)
2. West Nippon Expressway Company Limited: Kumamoto Earthquake Response Committee WG2 Report. About earthquake resistance performance of highway embankment (2017)
3. Qin, C.J., et al.: Performance of hybrid pile-nailing supported highway embankments on liquefiable foundation subjected to dynamic loadings. In: Proceedings of the Technical Forum on Mitigation of Geo-disasters in Asia, Kumamoto, pp. 136–141 (2019)
4. Qin, C., et al.: Evaluation of hybrid pile supported system for protecting road embankment under seismic loading. In: Hazarika, H., Madabhushi, G.S.P., Yasuhara, K., Bergado, D.T. (eds.) Advances in Sustainable Construction and Resource Management. LNCE, vol. 144, pp. 733–744. Springer, Singapore (2021). https://doi.org/10.1007/978-981-16-0077-7_62
5. Oka, F., Yashima, A., Shibata, T., Kato, M., Uzuoka, R.: FEM-FDM coupled liquefaction analysis of a porous soil using an elasto-plastic model. Appl. Sci. Res. **52**(3), 209–245 (1994)
6. Oka, F., Yashima, A., Tateishi, A., Taguchi, Y., Yamashita, S.: A cyclic elasto-plastic constitutive model for sand considering a plastic-strain dependence of the shear modulus. Geotechnique **49**(5), 661–680 (1999)
7. Oka, F., Uzuoka, R., Tateishi, A., Yashima, A.: A cyclic elasto-plastic model for sand and its application to liquefaction analysis. In: Constitutive Modelings of Geo-Materials Selected Contributions from the Frank L. DiMaggio Symposium, pp. 75–99 (2002)
8. Zhang, F., Kimura, M., Nakai, T., Hoshikawa, T.: Mechanical behavior of pile foundations subjected to cyclic lateral loading up to the ultimate state. Soils Found. **40**(5), 1–17 (2000)

Distribution of Deformations and Strains Within a Slope Supported on a Liquefiable Stratum

Zhijian Qiu[1(✉)] and Ahmed Elgamal[2]

[1] Xiamen University, Xiamen 361005, Fujian, China
ZhijianQiu@Xmu.edu.cn
[2] University of California San Diego, La Jolla, CA 92093-0085, USA

Abstract. Ground slopes with an underlying liquefiable layer are particularly prone to failure due to seismic excitation. For an embedded foundation, the extent of ground deformation at its specific location within the slope, will dictate the level of detrimental consequences. As such, knowledge about the spatial configuration of expected ground deformation along the slope's length and height will provide insights towards assessment and mitigation of the consequences. For that purpose, calibrated Finite element (FE) simulations of ground slopes are conducted, and derived insights from the seismic response analyses are gleaned, where properties of upper crust layer and thickness of the liquefiable layer are varied. Generally, lower levels of deformation are to be expected with distance away from the crest and toe of the sloping zone, and the study aims to quantify this effect. In addition, it is shown that properties of the upper crust may have a significant influence on the pattern and level of accumulated downslope permanent deformation.

Keywords: Seismic · Liquefaction · Lateral spreading · Finite element · Slope

1 Introduction

Liquefied soil layers may result in considerable accumulated permanent downslope deformations during earthquakes, causing severe damage to overlying structures and deep foundations. Recent earthquakes continue to demonstrate the detrimental effects of liquefaction on bridge abutments, supporting columns, and pile foundations. Key related references are provided in [1].

To gain more insights towards assessment and mitigation of the detrimental consequences, knowledge about the spatial configuration of ground deformation in the slope vicinity plays an important role. In general, lower levels of deformation are to be expected with distance away from the crest and toe of the sloping zone. As such, this study attempts to quantify such distribution of ground deformations within a slope supported on a liquefiable stratum. A total of 9 Finite Element (FE) slope simulations are conducted and the results are systematically discussed, where soil properties of an upper clay layer and thickness of the liquefiable layer are varied.

As such, the following sections of this paper present the: 1) computational framework, 2) specifics and model properties of the ground slope, and 3) computed response and insights derived from the study. Finally, a number of conclusions are presented and discussed.

2 Computational Framework

The Open System for Earthquake Engineering Simulation (OpenSees) framework [2] was employed to conduct the nonlinear slope analyses subjected to seismic excitation. OpenSees is developed by the Pacific Earthquake Engineering Research (PEER) Center and is widely used for simulation of geotechnical systems and soil-structure interaction applications [3, 4]. The OpenSees elements and materials used in this FE model are briefly described below. Further details about the computational framework are discussed in reference [1].

1) Two-dimensional (2D) four-node quadrilateral elements (OpenSees quadUP) following the u-p formulation [5] were employed for simulating saturated soil response, where u is displacement of the soil skeleton and p is pore-water pressure.
2) The employed soil constitutive model was developed based on the multi-surface plasticity theory [6, 7]. In this model, the shear stress-strain backbone curve is represented by the hyperbolic relationship. As such, soil is simulated by the implemented OpenSees materials PDMY02 [6, 7] and PIMY [7, 8].

3 Finite Element Model

For illustration, a 2D Finite Element (FE) mesh (Fig. 1) is created to represent the ground slope, comprising 5,334 nodes and 5,156 quadrilateral elements. This slope is approximately 300 m long and the inclination angle is about 27° (2 Horizontal: 1 Vertical). The subsurface conditions consist of clay (PIMY in OpenSees; [7, 8]), loose sand, and dense sand (PDMY02 in OpenSees; [6, 7]). The water table is prescribed at an elevation of 33 m, at the top of the loose sand layer (Fig. 1). Along the left- and right-side mesh boundaries (Fig. 1), 2D plane strain soil columns of large size (not shown) are included [1, 9]. These soil columns (away from the slope to minimize boundary effects), efficiently reproduce the desired shear beam free-field response at these locations (Fig. 1).

For simplicity, nodes were fixed along the FE mesh base against longitudinal and vertical translation. The seismic motion (Fig. 1) was taken as that of the 1940 Imperial Valley earthquake ground surface El Centro Station record (Component 180). Finally, dynamic analysis was conducted by applying this acceleration time history to the FE model base (elevation = 0.0 m). In this representation, free-field motion along both the left- and right-side side mesh boundaries is seamlessly generated by the included boundary soil columns mentioned above.

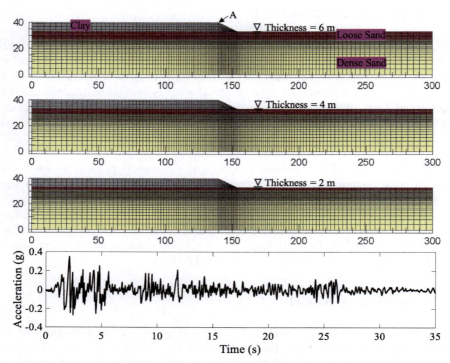

Fig. 1. Two-dimensional FE model with different thickness of loose sand layer (i.e., 6 m, 4 m, 2 m) subjected to a base input motion (1940 El Centro, Component 180).

4 Computed Response

To study the spatial configuration of ground deformation along the slope's length and height, a total of 9 simulations were performed, where properties of upper clay layer and thickness of the liquefiable loose sand layer are varied. On this basis, the computed results are systematically presented and discussed below.

4.1 Longitudinal Deformation at End of Shaking

Figure 2 depicts the longitudinal relative displacement contours at end of shaking. It can be seen that the maximum ground displacement occurs near the crest and toe of the sloping zone, reaching as much as 2.2 m, 1.7 m and 0.7 m, respectively. With distance away from the slope, lower levels of lateral deformation are clearly observed (Fig. 3). In the vicinity of slope, the width of deformed zone and the amount of local downslope ground displacement are significantly affected by thickness of the liquefiable loose sand layer. With the increase of liquefiable layer thickness, permanent ground deformation and the deformed zone are larger. At both side boundaries, (X = 0 m and X = 300 m in Fig. 3), the displacements were minimal, when compared to the large downslope deformation within the slope.

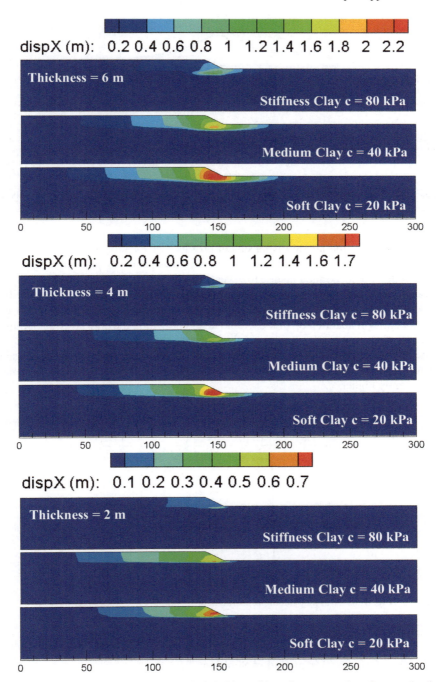

Fig. 2. Lateral ground deformation at end of shaking with various properties of upper clay layer and thickness of the liquefiable loose sand layer.

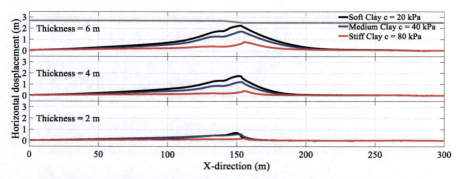

Fig. 3. Lateral displacement along ground surface at end of shaking (grey line represents the outline of ground surface).

For illustration, the lateral displacement profile near location A (Fig. 1) at end of shaking is shown in Fig. 4. It can be seen clearly that lateral displacement increased sharply within the loose sand layer. However, relative deformation within the overlying clay varies along depth (i.e., not behaving as a sliding rigid block as modeled by simplified methods [10]). As seen in Fig. 3, the displacement starts decreasing in elevation (upward) and is not uniform near the top of loose sand layer and within the clay layer. In addition, properties of the upper clay layer and thickness of the liquefiable sand layer play a significant role in dictating the overall lateral ground deformation. With the decrease in cohesion (shear strength) and increase in loose sand layer thickness, lateral displacement is consistently higher.

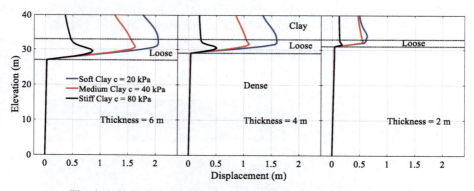

Fig. 4. Lateral displacement profile near location A at end of shaking.

4.2 Vertical Deformation at End of Shaking

In addition to lateral deformation, vertical displacement (Fig. 5) is an important component, providing more insights towards assessment and mitigation of the detrimental consequences. As seen in Fig. 5, peak settlement occurred near the slope crest and showed slumping of as much as 0.55 m, 0.29 m, 0.11 m for soft, medium, and stiff clay, respectively. Similar to the lateral counterpart, peak value of vertical displacement is greater, with decrease in the upper clay curst cohesion. In addition, the vertical displacement is affected by thickness of the liquefiable loose sand layer.

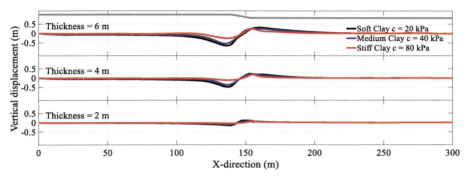

Fig. 5. Vertical displacement along ground surface at end of shaking (grey line represents the outline of ground surface).

4.3 Liquefaction and Accumulated Shear Strains

Figure 6 shows time histories of effective confinement p' divided by the initial value p'_0 (before shaking) at loose sand layer base near location A (Fig. 1). The ratio p'/p'_0 reaching 0.0 indicates loss of effective confinement due to liquefaction. As seen in Fig. 6, the ratio p'/p'_0 gradually decreases towards 0.0 for all scenarios and this loose stratum attains its low specified residual shear strength of 2 kPa. In accordance with the above-mentioned observed lateral deformations, peak shear strain within the loose sand layer becomes larger with decrease of the overlying clay crust cohesion (Fig. 7).

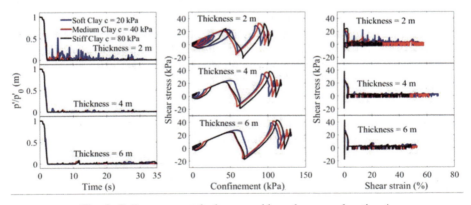

Fig. 6. Soil response at the loose sand layer base near location A.

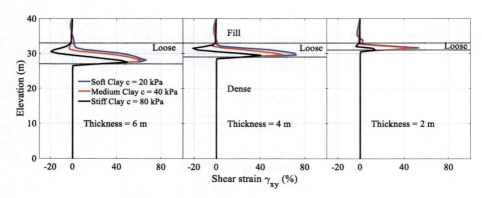

Fig. 7. Shear strain profile near location A at end of shaking.

5 Conclusions

This study aims to quantify the distribution of ground deformations within a slope supported on a liquefiable stratum. As such, a total of 9 Finite Element (FE) slope simulations are performed and the computed results are systematically presented and discussed, where soil properties of the upper clay layer and thickness of the liquefiable layer are varied. Specific observations and conclusions include:

1) With distance away from the crest and toe of the sloping zone, lower levels of resulting ground deformation are achieved as expected.
2) Knowledge about the spatial configuration of ground deformation within the slope provides more insight towards assessment and mitigation of the detrimental consequences due to liquefaction.
3) Peak strength of overlaying clay layer above the liquefied loose sand stratum plays an important role in overall deformation of the sloping ground. With the increase of upper clay layer cohesion, permanent ground displacement was lower. Such a deformation mechanism differs from the simplified method of [10], in which the sliding mass is considered to be rigid.
4) Spatial distribution of deformations around the slope appears to be highly dependent on thickness of the liquefiable layer. It is shown that both the amount of permanent deformation and extent of the deformed zone are larger, with the increase of liquefiable layer thickness.

Acknowledgements. This research was supported by the California Department of Transportation (Caltrans) under Contract No. 65A0548 with Dr. Charles Sikorsky as the project manager, and Fundamental Research Funds for the Central Universities of China (22070103963, 20720220070).

References

1. Qiu, Z., et al.: Aspects of bridge-ground seismic response and liquefaction-induced deformations. Earthquake Eng. Struct. Dynam. **49**(4), 375–393 (2020)

2. McKenna, F., Scott, M.H., Fenves, G.L.: Nonlinear finite-element analysis software architecture using object composition. J. Comput. Civ. Eng. **24**(1), 95–107 (2010)
3. Yang, Z., Elgamal, A.: Influence of permeability on liquefaction-induced shear deformation. J. Eng. Mech. **128**(7), 720–729 (2002)
4. Lu, J., Elgamal, A., Yan, L., Law, K.H., Conte, J.P.: Large-scale numerical modeling in geotechnical earthquake engineering. Int. J. Geomech. **11**(6), 490–503 (2011)
5. Chan, A.H.C.: A unified finite element solution to static and dynamic problems in geomechanics. PhD Thesis, University College of Swansea (1988)
6. Elgamal, A., Yang, Z., Parra, E.: Modeling of cyclic mobility in saturated cohesionless soils. Int. J. Plast **19**(6), 883–905 (2003)
7. Yang, Z., Lu, J., Elgamal, A.: OpenSees soil models and solid-fluid fully coupled elements. In: User's Manual: Version 1. University of California, San Diego, La Jolla (2008)
8. Elgamal, A., Yan, L., Yang, Z.: Three-dimensional seismic response of Humboldt Bay bridge-foundation-ground system. J. Struct. Eng. **134**(7), 1165–1176 (2008)
9. Qiu, Z.: Computational modeling of ground-bridge seismic response and liquefaction scenarios. PhD Thesis, UC San Diego (2020)
10. Newmark, N.M.: Effects of earthquakes on dams and embankments. Geotechnique **15**(2), 139–160 (1965)

Probabilistic Seismic Hazard Curves and Maps for Italian Slopes

Fabio Rollo(✉) and Sebastiano Rampello

Dipartimento di Ingegneria Strutturale e Geotecnica, Università di Roma La Sapienza,
via Eudossiana 18, 00184 Roma, Italy
fabio.rollo@uniroma1.it

Abstract. The seismic performance of an earth slope is commonly evaluated through the permanent displacements developed at the end of an earthquake. In this paper a probabilistic approach is adopted to assess the displacement of the slope for a given hazard level using an updated database of ground motions recorded during the earthquakes occurred in Italy. The results are presented in terms of hazard curves, showing the annual rate of exceedance of permanent slope displacement evaluated using ground motion data provided by a standard probabilistic hazard analysis and a series of semi-empirical relationships linking the permanent displacements of slopes to one or more ground motion parameters. The probabilistic approach permits to take into account synthetically the characteristics of the slope through the yield seismic coefficient, the aleatory variability of the ground motions and the different subsoil classes of the recording stations. Finally, the procedure has been extended on a regional scale to produce seismic landslide hazard maps for Irpinia, one of the most seismically active regions in Italy. Seismic landslide hazard maps are very attractive for practitioners and government agencies for a screening level analysis to identify, monitor and minimise damages in zones that are potentially susceptible to earthquake-induced slope instability.

Keywords: Slopes · Earthquake-induced displacements · Probabilistic analysis · Displacement hazard curves · Hazard maps

1 Introduction

Earthquakes often produce instability in natural slopes, causing severe human and economic losses. Therefore, several efforts have been devoted in the past decades by the geotechnical community to study the seismic response of slopes aimed at mitigating and preventing such catastrophic events.

The seismic performance of a slope can be synthetically assessed by the evaluation the permanent displacements developed at the end of the seismic event, which can be quantified through several methods. In this context, the well-known displacement-based Newmark's method [1] represents a good compromise between simplicity and accuracy of the results, in which the slope is assimilated to a rigid block sliding on a horizontal plane that undergoes permanent displacements only when the acceleration of the input motion is greater than a critical value, the latter depending on slope resistance. The integration of a set of acceleration time histories permits to link the permanent displacements to the yield seismic coefficient, representing the slope seismic resistance, and to one or more ground motion parameters, leading to a series of semi-empirical relationships [2–4]. These relationships are useful for a preliminary estimate of the expected seismic-induced displacement of slopes but they cannot account for the aleatory variability of earthquake ground motion and displacement prediction. Therefore, many researchers developed fully probabilistic-based approaches stemming from the above semi-empirical relationships to introduce the concept of hazard associated to the calculated displacement [5–7]. The results of the probabilistic approach are typically illustrated in terms of displacement hazard curves, which provide the mean annual rate of exceedance λ_d (or the return period $T_r = 1/\lambda_d$) for different levels of permanent displacements.

The present study aims to contribute towards the evaluation of earthquake-induced landslides hazard for Italy, which is characterised by a high level of seismicity. The displacement hazard curves are developed for the Italian territory stemming from the seismic databased recently updated by Gaudio et al. (2020) [8]. Both scalar and vector probabilistic approaches are employed: in the first case the peak ground acceleration (*PGA*) is used as ground motion parameter, while the combination of *PGA* with the peak ground velocity (*PGV*) is chosen for the vector approach. Moreover, the probabilistic approach has been implemented on a regional scale using the ground motion hazard information to evaluate landslides hazard maps, that allow to identify the zones that are more susceptible to earthquake-induced slope instability and the probability of occurrence of a displacement level in a specific time interval. These results provide a useful tool for practitioners and government agencies for a preliminary evaluation of the seismic performance of slopes, stemming from the updated seismic database and the results of the semi-empirical relationships adopted in the study.

2 Probabilistic Approach

The displacement models predict the natural log of permanent horizontal displacement d as a function of the natural log of one or more ground motion parameters (*GM*) or earthquake magnitude.

For the single ground motion parameter model, the expression proposed by Ambraseys & Menu (1988) [9] is adopted:

$$\ln(d) = a_0 + a_1 \ln\left(1 - \frac{k_y}{PGA}\right) + a_2 \ln\left(\frac{k_y}{PGA}\right) \tag{1}$$

where a_0, a_1 and a_2 are regression coefficients and d is expressed in cm. The efficiency of the semi-empirical relationships is described by the standard deviation of the natural log of displacement $\sigma_{\ln}$. Equation (1) complies with the conditions $d \to \infty$ for $k_y/PGA = 0$ and $d = 0$ for $k_y/PGA = 1$ expected for the assumption of a rigid sliding block.

The results of a probabilistic analysis are synthesised in terms of seismic displacement hazard curves, providing the mean annual rate of exceedance λ_d for different levels of displacements given a yield seismic coefficient and a specific site. According to [7], for the single ground motion *PGA* displacement model the annual rate of exceedance λ_d can be computed as:

$$\lambda_d(x) = \sum_i P[d > x | PGA_i] \times P[PGA_i] \qquad (2)$$

where $P[d > x | PGA_i]$ is the probability of the displacement exceeding a specific value x, given a peak ground acceleration PGA_i, and $P[PGA_i]$ is the annual probability of occurrence of the ground motion level PGA_i obtained through a probabilistic seismic hazard analysis (PSHA).

Ambraseys and Menu (1988) [9] suggest to modify Eq. (1) to account for the effects of other ground motion parameters. Therefore, along this track, the following semi-empirical relationship is proposed for the case of two ground motion parameters:

$$\ln(d) = a_0 + a_1 \ln\left(1 - \frac{k_y}{PGA}\right) + a_2 \ln\left(\frac{k_y}{PGA}\right) + a_3 \ln(PGV) \qquad (3)$$

Equation (3) allows to develop a vector probabilistic approach in terms of the ground motion parameters *PGA*, *PGV*, according to Rathje & Saygili (2008) [7]:

$$\lambda_d(x) = \sum_i \sum_j P[d > x | PGA_i, PGV_j] \times P[PGA_i, PGV_j] \qquad (4)$$

Details concerning the calculation of Eq. (4) can be found in [10].

3 Results of the Proposed Approach

The results of the probabilistic approach are synthesised in terms of displacement hazard curves, plotting the mean annual rate of exceedance λ_d for different levels of permanent displacements and in terms of hazard maps, showing the distribution of the return period $T_r = 1/\lambda_d$ for a given permanent displacement and a given yield seismic coefficient. Hazard curves and maps are developed for different zones of the Italian territory employing both the scalar and the vector approaches.

The permanent displacements are computed with the rigid sliding-block model for the simple scheme of an infinite slope assuming a constant shear strength, that is a constant k_y, during earthquake loading. Computation have been carried out for different values of the yield seismic coefficient, using an updated version of the Italian strong motion database. In comparison to the work by Rollo & Rampello (2021) [10], an extended range of the yield seismic coefficient is considered here ($k_y = 0.04$, 0.06, 0.08, 0.1, 0.12, 0.15) to include a higher number of slopes scenarios.

Table 1 reports the regression coefficients of the semi-empirical relationships evaluated for computed displacements greater than 0.0001 cm. Despite very small values of displacements are meaningless from an engineering perspective, they have been considered to develop semi-empirical relationships capable to fit satisfactory the data as the *PGA* approaches k_y. The coefficients of the semi-empirical relationships are further specialised to account for the influence of the subsoil class on the probabilistic approach.

Table 1. Coefficients of the semi-empirical relationships

Subsoil class	GM parameter	a_0	a_1	a_2	a_3	$\sigma_{\ln}$
All	*PGA* (g)	−1.667	2.017	−2.127	–	1.103
	PGA (g), *PGV* (cm/s)	−2.959	2.178	−0.809	1.322	0.579
A	*PGA* (g)	−2.550	1.799	−2.709	–	1.042
	PGA (g), *PGV* (cm/s)	−3.147	2.142	−1.016	1.267	0.592
B	*PGA* (g)	−1.778	1.975	−2.060	–	1.052
	PGA (g), *PGV* (cm/s)	−2.945	2.184	−0.745	1.315	0.574
C	*PGA* (g)	−1.252	2.124	−2.003	–	1.134
	PGA (g), *PGV* (cm/s)	−2.885	2.169	−0.852	1.308	0.567

As reported in Table 1, the use of the two parameters semi-empirical relationship reduces significantly the standard deviation associated to the Newmark's displacements as compared to the scalar approach: an expected result as the couple of ground motion parameters *PGA*, *PGV* are more representative of the strong motion database than the *PGA* only.

The computed permanent displacements and the semi-empirical relationships are plotted in Fig. 1 against the ratio k_y/PGA for the whole database and the three subsoil classes (A, B and C). The vector predictive equation is plotted for $PGV = 8$ cm/s, representing the average value associated to the recorded ground motions. It is seen that the adopted models not only predict correctly the conditions at the extrema but also nicely reproduce the non-linear variation of the permanent displacements with k_y/PGA. Moreover, Rollo & Rampello (2021) [10] show that the vector approach predicts about halved values of the standard deviations with respect to the scalar one, with a trend characterised by negligible biases for increasing values of k_y/PGA and no relevant variation with k_y/PGA, hence demonstrating that the regressions adopted in this study are appropriate for the Italian ground motion database.

The displacement hazard curves shown in this paper refer to the Italian site of Amatrice, afflicted by the 2016 Central Italy seismic sequence. Information pertaining to the seismic hazard of the specific site have been extracted from the INGV interactive seismic hazard maps (http://esse1.mi.ingv.it/d2.html) that permits to query probabilistic seismic hazard maps of the Italian territory on a regular grid spaced by 0.05°.

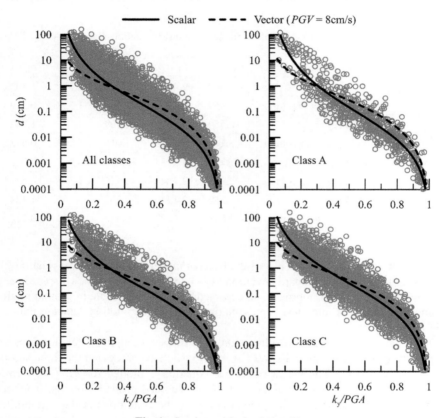

Fig. 1. Semi-empirical relationships

Figure 2 shows the displacement hazard curves obtained through the scalar and vector approaches for different values of the yield seismic coefficient. As expected, the computed displacement hazard curves show a decreasing annual rate of exceedance with increasing permanent displacements and prove that the more stable is the slope (i.e. greater is k_y), the lower is the annual rate of exceedance associated to a given displacement. Furthermore, it is worth noting that the vector approach always predicts smaller values of λ_d in the whole range of permanent displacements considered in the analysis, thus reducing the hazard estimate. This result is consistent with the outcomes of [6, 7, 10] and with the smaller standard deviation and median displacements computed through the vector approach as compared with the scalar one.

Figure 3 illustrates the displacement hazard curves computed with the scalar and vector approaches for the Amatrice site, distinguishing the three subsoil classes and using $k_y = 0.08$.

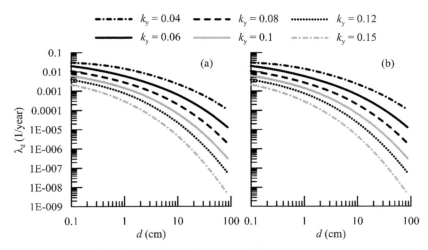

Fig. 2. Effect of k_y for (a) scalar and (b) vector approaches

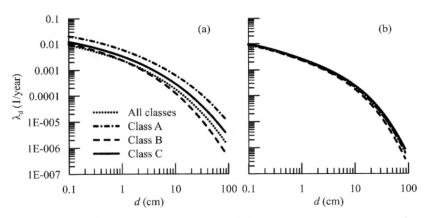

Fig. 3. Effect of subsoil classes for $k_y = 0.08$ for (a) scalar and (b) vector approaches

It is seen that higher annual rate of exceedance λ_d is computed for soft (class C) subsoil as compared to that of stiff (class B) subsoil or, in other terms, for class C sites the model predicts the largest displacements at fixed rate of exceedance. This is consistent with the fact that the ground motions recorded on soft soils are richer in low frequencies than those recorded on stiff soils, hence leading to larger displacements. It is less clear the trend observed for the case of rock-like (class A) subsoil, for which the smallest displacements are expected, while the obtained hazard curve lies above those computed for classes B and C. As discussed by Rollo & Rampello (2021) [10], this could be attributed to the fact that for rock-like subsoils about 50% of the records with $PGA > 0.2\ g$ are characterised by high values of the mean period $T_m > 0.4$ s, that could explain the unexpectedly high displacements computed for subsoil class A, leading the hazard displacement curve to the right of that obtained for the other subsoil classes. An analogous trend is observed for the vector approach, despite the effect of the subsoil classes is less pronounced than that observed for the scalar approach.

Finally, the probabilistic approach is used in the following to develop the hazard maps for the district of Irpinia, in the South Italy, about 50 km East of Naples and characterised by a severe seismic hazard. The hazard maps depict the contours of the return periods T_r associated to different levels of seismic-induced displacement and yield seismic coefficient. Figure 4 shows the contour maps obtained from the vector approach for a threshold displacement $d_y = 2$ cm and different values of the yield seismic coefficient.

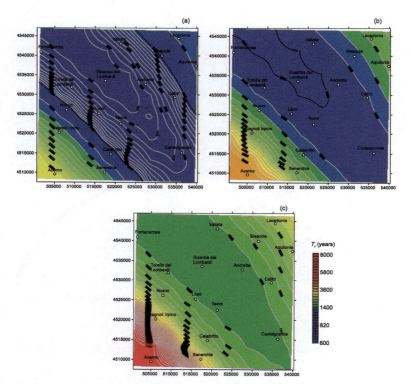

Fig. 4. Hazard maps for the Irpinia district for $d_y = 2$ cm: (a) $k_y = 0.1$, (b) $k_y = 0.12$, (c) $k_y = 0.15$

As expected, the return periods associated to a given value of d_y increase with k_y. The variability of T_r is clearly associated to the PGA hazard curves and disaggregation information of the region, that is more severe for the zone of the Apennines extending from North-Western to South-Eastern corners of the map. Therefore, the contour lines are nearly parallel to the NW-SE diagonal of the map, with larger return periods in the South-Western area, where hills take place of the mountains.

4 Conclusions

The results presented in this paper show that the probabilistic approach represents a useful tool for a preliminary evaluation of the seismic performance of slopes. The displacement hazard curves provide site-specific information, whereas the hazard maps are suitable for the analysis and the evaluation of the seismic risk of natural slopes at a regional scale. The probabilistic nature of these maps also enables them to be combined with other hazards for a more complete hazard analysis, useful for disaster management to minimise damages caused by earthquake-induced landslides.

Acknowledgements. The research work presented in this paper was partly funded by the Italian Department of Civil Protection under the ReLUIS research project – Working Package 16: *Geotechnical Engineering* – Task Group 2: *Slope stability*.

References

1. Newmark, N.M.: Effects of earthquakes on dams and embankments. Geotechnique **15**(2), 139–160 (1965)
2. Saygili, G., Rathje, E.M.: Empirical predictive models for earthquake-induced sliding displacements of slopes. J. Geotech. Geoenviron. Eng. **134**(6), 790–803 (2008)
3. Song, J., Gao, Y., Rodriguez-Marek, A., Feng, T.: Empirical predictive relationships for rigid sliding displacement based on directionally-dependent ground motion parameters. Eng. Geol. **222**, 124–139 (2017)
4. Cho, Y., Rathje, E.M.: Displacement hazard curves derived from slope-specific predictive models of earthquake-induced displacement. Soil Dyn. Earthq. Eng. **138**, 106367 (2020)
5. Rathje, E.M., Saygili, G.: Probabilistic seismic hazard analysis for the sliding displacement of slopes: scalar and vector approaches. J. Geotech. Geoenviron. Eng. **134**(6), 804–814 (2008)
6. Du, W., Wang, G.: A one-step Newmark displacement model for probabilistic seismic slope displacement hazard analysis. Eng. Geol. **205**, 12–23 (2016)
7. Macedo, J., Bray, J., Abrahamson, N., Travasarou, T.: Performance-based probabilistic seismic slope displacement procedure. Earthq. Spectra **34**(2), 673–695 (2018)
8. Gaudio, D., Rauseo, R., Masini, L., Rampello, S.: Semi-empirical relationships to assess the seismic performance of slopes from an updated version of the Italian seismic database. Bull. Earthq. Eng. **18**(14), 6245–6281 (2020). https://doi.org/10.1007/s10518-020-00937-6
9. Ambraseys, N.N., Menu, J.M.: Earthquake-induced ground displacements. Earthq. Eng. Struct. Dynam. **16**(7), 985–1006 (1988)
10. Rollo, F., Rampello, S.: Probabilistic assessment of seismic-induced slope displacements: an application in Italy. Bull. Earthq. Eng. **19**(11), 4261–4288 (2021). https://doi.org/10.1007/s10518-021-01138-5

New Soil-Pile Spring Accounting for a Tree-Root System in the Evaluation of Seismic Slope Stability

Yoshikazu Tanaka[1](✉), Kyohei Ueda[2], and Ryosuke Uzuoka[2]

[1] Graduate School of Bioresources, Mie University, 1577 Kurimamachiya-cho, Tsu, Mie 514-8507, Japan
ytanaka@bio.mie-u.ac.jp

[2] Disaster Prevention Research Institute, Kyoto University, Gokasho Uji Kyoto 611-0011, Japan
{ueda.kyohei.2v,uzuoka.ryosuke.6z}@kyoto-u.ac.jp

Abstract. Tree-root system plays an important role in the stabilization of soil slopes. Many failures of soil slopes covered with trees were observed during the Hokkaido Eastern Iburi earthquake in 2018. Tree vibration may have induced the slope failure. The response of trees on the slope is not well understood under seismic conditions. The influence of a tree-root system on the slope failures may not be ignored under the seismic event. Therefore, it is important to evaluate the seismic behavior of a tree-root system including surrounding soil. The 2D FEM analysis for the slope stability was performed by adding the macroscopic properties such as resistance strength (i.e., cohesion) and stiffness directly to the soil, which simplifies the modeling. The cohesion predicting is influenced by the definition of the rooted zone since the shape of the root differs depending on the tree species. This study aims to investigate the soil-pile spring accounting for tree-root systems using the 2D FEM analysis. The tree-root system is simply modeled as a connection spring between the soil and the pile. The root-soil spring is validated with the tree pull-down tests. The simulation revealed that the root-soil spring differs depending on the tree species. The soil-pile spring could be applied for evaluating the tree-root system. The concept of the soil-pile spring could be applied for evaluating the tree-root system including the effect of the horizontal root growth.

Keywords: Soil-pile spring · Tree-root system · FEM analysis · Earthquake

1 Introduction

Tree-root system plays an important role in the stabilization of soil slopes [1–3]. On the other hand, many failures of soil slopes covered with trees were observed during the Hokkaido Eastern Iburi earthquake in 2018. When the tree vegetation was compared between collapsed and non-collapsed slopes in some small watersheds, the tree vegetation of the collapsed slope was found to be different from the non-collapsed slope [4]. The tree vibration may have induced the slope failure. The response of trees on the slope is not well understood under seismic conditions. The influence of a tree-root system on the slope failures may not be ignored under the seismic event. Therefore, it is important to evaluate the seismic behavior of a tree-root system, including surrounding soil.

It is difficult to consider the interaction of the tree root system with the surrounding soil layers, with the properties of both soil and a tree root remaining highly uncertain. Generally, two different ways have been used in numerical analysis to incorporate the root system effect. One is to use an accurate 3D FEM model to replicate the ideal tree-root system [5, 6], which is computationally costly but precise. Another way is to add the macroscopic properties such as resistance strength (i.e., cohesion) and stiffness directly to the soil, which simplifies the modeling but is relatively less accurate. The 2D FEM analysis was carried out to evaluate the slope stability with cohesion predicting model based on the centrifuge modeling experiments using a 3D root model [7]. The cohesion predicting is influenced by the definition of the rooted zone since the shape of the root differs depending on the tree species. This study aims to investigate that the tree-root system is modeled using the concept of the soil-pile spring accounting for tree-root systems based on the field test. The tree pull-down tests are summarized from the previous field testing results. A horizontal loading analysis is carried out in order to obtain the parameters for the root-soil spring element based on the results of the tree pull-down test. The root-soil springs are adjusted to represent the ideal root conditions.

2 Numerical Simulation Methods

Figure 1 shows the layout of the finite element mesh of the tree root and soil. The root depth and the soil strength was not investigated in the most previous pull-down tests. According to the investigation of the tree vegetated near the coast by Kawasaki [8], the root system grew to around 2 m depth, and the soil strength was SPT N-value from 1 to 4. The model consisted of four layers with SPT N-value decreasing towards the ground surface based on the investigation near the coast. The layout of the finite element mesh around the root is shown in Fig. 2. The connection between a root-soil system is idealized for two conditions: a rigid tied is considered when the relationship between root and soil is strong, while another way is to define a new root-soil spring. The analysis was performed under the linear conditions, with both soil and tree models defined considering linearity since trees may not fall down during an earthquake. The tree model represented a trunk and a root (1 m wide) modeled using the tree properties. The root-soil spring was calculated using the following equations based on the Recommendations for Design of Building Foundations (2014) [9]. In this study, the coefficient a considering the tree root system was added to the original equation as shown in Eq. (1).

$$k_\mathrm{h} = a 80 E_\mathrm{S} \overline{B}^{\left(-\frac{3}{4}\right)} \tag{1}$$

$$E_\mathrm{S} = 700N \tag{2}$$

$$K_\mathrm{H} = k_\mathrm{h} D \Delta L \tag{3}$$

where k_h indicates the coefficient of subgrade reaction (kN/m³), a indicates adjustment factor to consider the influence of the tree root system. E_S means deformation coefficient of the ground (kN/m²), $\overline{B}$ means dimensionless pile diameter, N means SPT-N value. K_H means horizontal soil spring (kN/m). D means tree diameter (m). ΔL represents the length shared by elements (m). The analysis was performed using a soil spring which considers the influence of the root system as a root-soil spring.

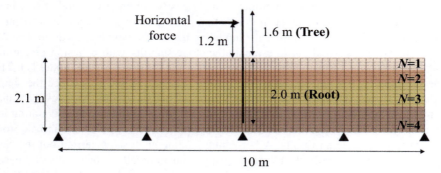

Fig. 1. The layout of the finite element mesh of the tree root and soil on the horizontal loading analysis.

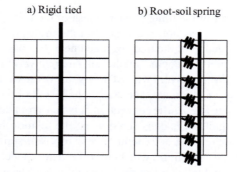

Fig. 2. The layout of the finite element mesh around root: a) Rigid tied, b) Root-soil spring

3 Numerical Simulation Based on the Field Test

3.1 Testing Data Based on the Literature

Tree pull-down test have been conducted to investigate the tree loading resistance for tsunami and strong wind [8, 10–13]. These previous studies observed the rotation angle and the horizontal force of a tree before it falls down. In this study, the horizontal displacement of trees during an earthquake is not large enough to fall down, so that the numerical analysis was performed targeting the linear region of the relationship between the rotating angle and the horizontal force in the literature. The pull-down test cases in the previous studies are shown in Table 1. The literature summarizes 15 cases involving different tree types having different diameters.

Table 1. The pull-down test cases in the previous studies.

Tree species	Author	Diameter
		m
1_Black pine	Kawasaki (2017)	0.16
2_Black pine		0.18
3_Black pine		0.16
4_Black pine		0.15
5_Black pine	Harada et al. (2018)	0.14
6_Black pine		0.15
7_Black pine		0.22
8_Larch	Torita (2009)	0.16
9_Larch		0.14
10_Larch		0.13
11_Camphor	Tachibana et al. (2018)	0.27
12_Wild cherry blossoms		0.19
13_Quercus myrsinifolia		0.10
14_Robinia pseudoacacia	Koizumi et al. (2007)	0.32
15_Robinia pseudoacacia		0.29

3.2 Relationship Between Rotating Angle and Horizontal Force by the Horizontal Tree Loading Analysis

Horizontal loading simulation was carried out for two cases: one in which the connection between the root and the soil was defined as a rigid tied, and the other in which the root-soil spring was used. The relationships between tree diameter and M/θ are shown in Fig. 3. The data from 15 samples is shown in Table 1. The numerical simulation was carried out for 11 cases (Diameter: 0.10 m, 0.13 m, 0.14 m, 0.15 m, 0.16 m, 0.18 m, 0.19 m, 0.22 m, 0.27 m, 0.29 m, 0.32 m). The tree shear modulus was set to 3.65×10^6 kPa, and the root depth was set to 2 m. Since the simulation condition was defined as linear, the M/θ is obtained by dividing the rotating moment by the rotating angle.

The M/θ of the rigid tied case increased as the tree diameter increased. Comparing the testing results with the rigid tied, the results for the 0.1 m diameter were almost identical, but for the 0.3 m diameter, the difference between the rigid tied case and the testing was large. The rigid tied represented an unrealistic condition. In the case of a root-soil spring, the relationship between diameter and M/θ was almost identical with the testing results. The results of adjustment factor are shown in Table 2. Black pine and larch are classified as conifers, and others are classified as hardwoods. The root shape of conifers tends to predominate in the vertical direction, and the root shape of hardwoods tends to predominate in the horizontal direction. The adjustment factor of conifers ranged from 0.01 to 0.03, and that of hardwoods ranged from 0.01 to 1.0. The case of hardwood became larger than that of conifers. The adjustment factor seems to be affected by the horizontal roots growth.

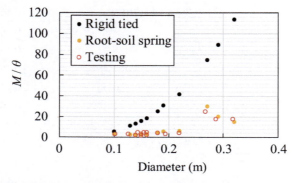

Fig. 3. The relationship between tree diameter and rotating moment M/rotating angle θ in the results of the rigid tied and the root-soil spring, the testing.

Table 2. The results of adjustment factor a

Tree species	Author	Diameter (m)	Adjustment factor a
1_Black pine	Kawasaki (2017)	0.16	0.01
2_Black pine		0.18	0.01
3_Black pine		0.16	0.01
4_Black pine		0.15	0.01
5_Black pine	Harada et al. (2018)	0.14	0.01
6_Black pine		0.15	0.01
7_Black pine		0.22	0.01
8_Larch	Torita (2009)	0.16	0.03
9_Larch		0.14	0.1
10_Larch		0.13	0.03
11_Camphor	Tachibana et al. (2018)	0.27	0.1
12_Wild cherry blossoms		0.19	0.01
13_Quercus myrsinifolia		0.10	1
14_ Robinia pseudoacacia	Koizumi et al. (2007)	0.32	0.01
15_ Robinia pseudoacacia		0.29	0.02

4 Conclusion

This study aims to investigate the soil-pile spring accounting for tree-root systems using the 2D FEM analysis. The tree-root system is simply modeled as a connection spring between the soil and the pile. The root-soil spring is validated with the tree pull-down tests. The rigid tied between root and soil was overestimated compared to the horizontal loading test results. The results of adjustment factor were classified by the tree species such as the conifer and hardwoods. Then, the adjustment factor was affected by the horizontal root growth. The concept of the soil-pile spring could be applied for evaluating the tree-root system including the effect of the horizontal root growth.

Acknowledgements. This paper was supported by Association for Disaster Prevention research grant support.

References

1. Abe, K.: In-situ direct shear tests for study on tree root function of preventing landslides. J. Jpn. Soc. Reveget. Technol. **22**(2), 95–108 (1996) [in Japanese]
2. Nilaweera, N.S., Nutalaya, P.: Role of tree roots in slope stabilization. Bull. Eng. Geol. Env. **57**, 337–342 (1999)
3. Loades, K.W., Bengough, A.G., Bransby, M.F., Hallett, P.D.: Planting density influence on fibrous root reinforcement of soils. Ecol. Eng. **36**, 276–284 (2010)
4. Tanaka, Y., Ueda, K., Uzuoka, R.:The Characteristics of the Vegetation Distribution Related to the Slope Failure Caused by the Earthquake, Understanding and Reducing Landslide Disaster Risk, pp. 479–483 (2021)
5. Dupuy, L.X., Fourcaud, T., Lac, P., Stokes, A.: A genetic 3D finite element model of tree anchorage integrating soil mechnics and real root system architecture. Am. J. Bot. **94**(9), 1506–1514 (2007)
6. Kim, Y., Rahardjo, H., Tsen-Tieng, D.L.: Stability analysis of laterally loaded trees based on tree-root-soil interaction. Urban Forest. Urban Green. **49**, 126639 (2020)
7. Liang, T., Knappett, J.A., Leung, A., Carnaghan, A., Bengough, A.G., Zhao, R.: A critical evaluation of predictive models for rooted soil strength with application to predicting the seismic deformation of rooted slopes. Landslides **17**(1), 93–109 (2019). https://doi.org/10.1007/s10346-019-01259-8
8. Kawasaki, N.: Geotechnical evaluation of resistance against overturning for Pinus thunbergii, Master thesis of Tokushima University, pp. 1–56 (2017) [in Japanese]
9. Architectural Institute of Japan. Recommendations for Design of Building Foundations, Architectural Institute of Japan, Japan (2014) [in Japanese]
10. Harada, K., Mineta, J., Seo, N., Kinpara, T.: Field experiments on coastal tree damages against tsunamis-case study at Yaizu-Tajiri coast-. J. Jpn. Soc. Civil Eng. **74**(4), I_897–I_905 (2018) [in Japanese]
11. Torita, H.: Mechanical evaluation of the Japanese larch for wind damage. J. Jpn. Forest Soc. **91**, 120–124 (2009). [in Japanese]
12. Tachibana, D., Naoki, S., Imai, K.: Evaluation on the wind resistance of transplanted tall trees. AIJ J. Technol. Design **24**(58), 1325–1330 (2018) [in Japanese]
13. Koizumi, A., Hirai, T., Ryu, K., Nakahara, R., Araya, K., Shimizu, H.: Wind damage resistance of Robinia pseudoacacia planted on the roadsides. Res. Bull. Hokkaido Univ. Forests, **64**(2), 105–112 (2007) [in Japanese]

Numerical Study on Delayed Failure of Gentle Sloping Ground

Tetsuo Tobita[1](✉), Hitomi Onishi[2], Susumu Iai[3], and Masyhur Irsyam[4]

[1] Kansai University, Osaka 564-8680, Japan
tobita@kansai-u.ac.jp
[2] Kiso-Jiban Consultants, Co, Osaka, Japan
[3] FLIP Consortium, Osaka, Japan
[4] Bandung Institute of Technology, Bandung, Indonesia

Abstract. The 2018 Sulawesi, Indonesia, earthquake (M_w7.5) triggered massive flow slide on very gentle slopes of 1% to 5%. Thousands of casualties and missing persons were reported due to such unprecedented disaster. Although detailed mechanism of the flow has yet been speculative, liquefaction is identified as a possible suspect. In the present study, cause of such a flow slide is numerically investigated using a 2D finite element method by simulating the dynamic behavior of the 30 m thick and 200 m long gentle sloping ground (slope angle 2°) of alternating silt and sand layers with the recorded acceleration as an input motion. Flow slide was simulated by varying the permeability of silt (5 m thick) located on top of the saturated loose sand (5 m thick). When the permeability of silt (k_{silt} = 1 × 10^{-6} m/s) is set lower than that of the saturated loose sand (k_{sand} = 1 × 10^{-4} m/s), the surface layer above the boundary between the silt and sand starts to flow long after the ground shaking ended. In total 1,000 s of simulation, the ground displacement of about 40 m was obtained. Close observation at the boundary elements revealed that the excess porewater pressure is gradually increasing in the silt layer while that in the liquefied layer dissipates. Mechanism of this type of failure has been studied as the void redistribution mechanism during liquefaction or the formation of water film, which may cause delayed failure of the gentle sloping ground. Although this study successfully simulates the delayed failure in 2D model, there is room for further research into seeking causes of the flow slides occurred in specific areas in Palu, Sulawesi, Indonesia.

Keywords: The 2018 Sulawesi Indonesia · Flow slide · Liquefaction · Delayed failure · Numerical analysis

1 Introduction

On September 28, 2018, at 18:02 (local time), a large earthquake (M_w7.5, D = 10 km) struck Palu City, Sulawesi Island, Indonesia (Fig. 1). The seismic motion was triggered by the Palu-Koro fault, a left strike-slip over 150 km, which runs north-south in the center of Sulawesi Island [1]. In Palu, which is located about 80 km south of the epicenter, large-scale lateral ground flow causing enormous damage occurred in several places due to the earthquake motion. The lateral flow occurred in gentle slope areas, and 2,081 people were killed due to a wide range of ground flow over several hundred meters to several kilometers. There were 4,438 injured and 1,309 missing [2].

Palu city area is located on a basin sandwiched between mountains which run from north to south (Fig. 2). From these mountains alluvial fans of various sizes are formed. All of the damaged sites are located on old alluvial fans, and the surface layer is covered with deposits from the Palu river, forming alternating layers of fine sand and silt.

As shown in Fig. 2, large-scale lateral flow occurred in Palu city at four locations in the eastern part of the Palu river, which flows through the center of the plain, and one location in the western part, for a total of five locations [6–9]. All of them are on gentle slopes with an inclination angle of about 0.6° to 3°, and from the testimony of the residents and the videos taken, the ground flow started a few minutes after the seismic motion had ended with the muddy ground containing a large amount of water which swallowed villages and paddy fields while waving up and down.

In addition, of the five sites where large-scale ground flow occurred, four sites in the eastern part of the Palu river were all damaged with the irrigation canal at the top. Trace of the canal is shown with a red curved line in Fig. 2. It has been pointed out that the inflow of irrigation water into the ground caused damage because the irrigation canal itself collapsed due to the flow, in addition to the possibility that constant water leakage from the irrigation canal changed the environment of the surrounding ground. However, since the same large-scale ground flow is occurring in the Balaroa area on the west side where there is no irrigation canal (Fig. 2), it has not been concluded that it is a direct cause of the massive flow generation.

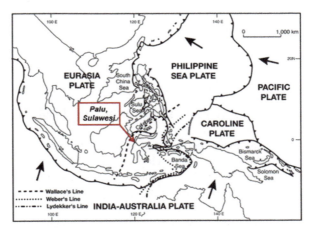

Fig. 1. Location of Palu, Sulawesi, Indonesia and oceanic plates in the area (after [3]).

It is thought that this phenomenon was caused by liquefaction, because the testimony of the residents that the groundwater level was originally shallow and the traces of sand boils were confirmed after the disaster. However, it is extremely rare that such a vast area flows due to liquefaction, and the mechanism of its occurrence has not yet been elucidated. Moreover, this phenomenon is different from the ground deformation that occurs during or immediately after the seismic motion, and the ground rupture is "delayed" due to the change in the stress state over time. Hence, it is thought that the flow is caused by the delayed failure mechanism [5].

In this study, a gentle sloping ground model that imitates the local ground environment is numerically modelled, and elucidate the mechanism of large-scale lateral flow and delayed failure mechanism by a two-dimensional effective stress analysis.

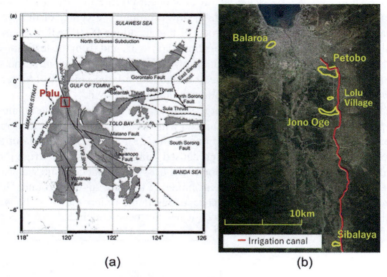

Fig. 2. Location of Palu city in Sulawesi Island and location of flow slides and irrigation canal (after [4]).

2 Numerical Analysis of Flow Failure on Gentle Slopes

Two-dimensional finite element effective stress analysis program FLIP-ROSE (Ver. 7.4) [10, 11] is used as the analysis method. It adopts u-p formulation [12] as the governing equation, and uses the multiple simple shear mechanism for the stress-strain relationship of granular soil, and the cocktail glass model [11] for dilative behavior of soil. The formulation is based on the infinitesimal strain theory.

In this study, a model ground is produced based on the borehole BH-P5A in Petobo [13] (Fig. 3). The borehole site was not affected but close to the flow slide area. The damaged area has a total length of about 2 km and an area of 1.4 km^2, and the residential area from the middle to the lower end of the flow area was filled with thick sediments due to mud flow. As in other areas, the inclination angle of the ground surface is as small as about 2°. In the area where the damage occurred, multiple boring surveys have been conducted after the disaster.

Fig. 3. Location of the borehole, BH-P5A, in Petobo area [13].

From Fig. 4, it can be seen that the soil layers at BH-P5A consists of an alternating layers of silty sand and sandy silt. Table 1 shows the simplified layering system assumed in the present study. With the information, a simplified finite element mesh shown in Fig. 5 is generated. As shown in Table 1, a layer from G.L. (ground level) −5 to −10 m with a SPT - N value of 4 and fines content of 38% is assumed to be the liquefiable layer (layer 2). Because deformation is expected to be larger from the ground surface to a depth of G.L. −10 m, the mesh is divided into 1 m intervals only in the vertical direction (Fig. 5(b)). The mesh size of the ground deeper than G.L. −15 m is set to be 5 m × 5 m. Although the actual flow range in the Petobo area was 2 km in the east-west direction, for the sake of simplicity, the analysis will be performed in the range of 200 m. The water table was set to G.L. −2 m as a trial. The inclination angle of the ground is set to 2°.

The model parameters of each layer were set as shown in Table 2 by referring to the borehole data shown in Fig. 4 and Table 1. To transfer in-situ parameter values into model parameter values, empirical relationship proposed by Mikami et al. [14], which is commonly used in the simulation code employed in this study. Here, since in-situ permeability of the silty sand and the unit weight of soil of each layer are unavailable, they were determined with reference to the values of Ishihara [15] and Japan Road Assoc. [16], respectively. As for the cohesion of sandy silt and silty sand, the cohesion of all layers was set to 0.0 kPa because the test results have not been obtained at this time. Tanaka et al. [17] obtained a liquefaction strength curve by conducting undrained cyclic triaxial tests on samples collected from the liquefaction layer of Sibalaya (Fig. 6) located about 23 km south of Petobo. Thus it is likely to have soil characteristics differ from that of Petobo. Here, due to the lack of experimental data in Petobo, the dilatancy parameters were determined by targeting the experimental results (Table 3). The shear strength q_{us} of the liquefiable layer was determined with reference to the undrained strength from previous studies [8, 18].

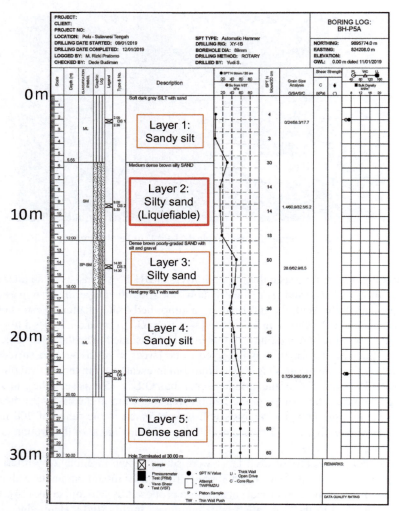

Fig. 4. Borehole profile and assigned layers for the numerical analysis in the present study based on the borehole log of NH-P5A [13].

In the analysis, the self-weight analysis is first performed to obtain the initial stress field in the inclined ground. The gravity and input acceleration are vector-decomposed in the directions parallel and perpendicular to the slope to express the inclination of the ground. Then, seismic response analysis is performed. The duration time for seismic response analysis is 1,000 s. Both self-weight and seismic response analyses are performed under the drainage condition that allow the inflow and outflow of pore water. In the self-weight analysis, the degree of freedom of displacement at the bottom boundary is fixed in both horizontal and vertical directions. At the lateral boundary, it is fixed in the slope direction (deformation in the sloping direction is restricted), while it is kept

Table 1. Simplified layering model for numerical analysis obtained from the borehole BH-P5A [13].

Depth G.L. (m)	Soil type	SPT-N	Fines content, F_c (%)
0 to -5	Sandy Silt	4	76
-5 to -10	Sillty Sand	4	38
-10 to -15	Silty Sand	50	71
-15 to -25	Sandy Silt	45	70
-25 to -30	Sand w/ Gravel	60	-

free in the direction perpendicular to the sloping direction. On the other hand, in the seismic response analysis, the degree of freedom at the bottom boundary is fixed, and the degree of freedom in the vertical direction at both lateral boundaries is set with a multi-point constraint (MPC), and the direction perpendicular to the sloping ground is set free. Here, in general, to simulate an infinite slope deformation, degree of freedom of nodes at both lateral boundary should be set with the MPC. Because this study is preliminary, the boundary condition is set as described above. Although not presented in this paper, a separate study conducted with the MPC shows almost identical results in the central array of the elements. Therefore, in what follows, results in the central array will be presented.

The input seismic motion is an acceleration record obtained at the Indonesian Meteorological Agency located near the Balaroa where a large flow failure occurred in the western side of Palu [13]. In this analysis, the EW component of the recorded motion for 40 s (Fig. 7) is input according to the flow direction in the Petobo area.

To verify whether the alternating layer structure of gently sloping sandy soil and silty soil can be a factor in the generation of delayed flow failure, the permeability of upper silty layer is varied (Table 4). In this analysis, the permeability of the liquefied layer is kept constant as $k_{sand} = 1 \times 10^{-4}$ (m/s), and that of the surface silt layer (G.L. 0 to -5 m) is varied either $k_{silt} = 1 \times 10^{-6}$ (m/s) in Case 1 or in Case 2, $k_{silt} = 1 \times 10^{-4}$ (m/s).

3 Results of the Numerical Analysis

3.1 Ground Displacements

First, looking at the displacement time history in the slope direction (Fig. 8(a)) for Case 1 in which the surface silt layer is set as relatively low permeable layer. The ground displacement at the end of the shaking indicated by "EQ end" in the figure is about 1 m, while the ground shallower than 5 m moved about 4 m in 250 s. After that, large displacement in the vicinity of the ground surface shallower than the silt layer of G.L. $-$ 4 m is observed. The maximum ground surface displacement at the end of the analysis is about 37 m in 1,000 s. From this observation, it can be said that the delayed failure phenomenon is reproduced because a large displacement gradually occurs long after

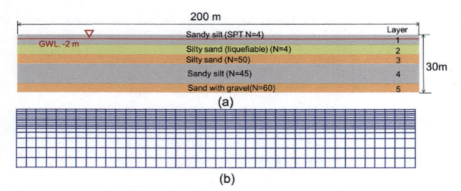

Fig. 5. Finite element mesh: (a) given soil types and (b) analyzed mesh.

Table 2. Model parameters

Layer (Soil type)	1 (Sandy silt)	2 (Silty sand) (Liquefiable)	3 (Sility sand)	4 (Sandy silt)	5 (Sand w/ Gravel)
Depth of boundary	G.L. 0 to -5 m	-5 to -10 m	-10 to -15 m	-15 to -25 m	-25 to 30 m
Wet density, ρ (t/m^3)	1.792	1.930	1.792	1.792	1.792
Reference initial shear modulus, G_{ma} (kPa)	44,701	87,215	240,764	224,433	271,881
Reference volumetric modulus, K_{ma} (kPa)	116,573	227,443	627,874	585,285	709,023
Reference confining stress, σ_{ma}' (kPa)	16.32	51.6	82.32	87.91	92.98
Power index for elastic moduli, m_G & m_K	0.5	0.5	0.5	0.5	0.5
Porosity, n	0.621	0.467	0.562	0.562	0.562
Internal friction angle, ϕ_f (°)	37.20	41.84	39.92	39.92	39.92
Cohesion, c (kN/m^2)	0.0	0.0	0.0	0.0	0.0
Max. damping coef., h_{max}	0.24	0.24	0.24	0.24	0.24
Shear strength, q_{us} (kPa)	-	2.992	-	-	-
Permeability, k (m/sec)	10^{-6}	10^{-4}	10^{-4}	10^{-6}	10^{-4}

Table 3. Dilatancy parameters for layer 2

ϕ_p	ε_d^{cm}	$r_{\varepsilon\,d}^c$	r_{ed}	r_k	l_k
28.0	0.2	5.0	1.0	0.5	2.0

q_1	q_2	q_4	S_1	c_1	q_{us}
10.0	2.0	0.5	0.005	1.0	2.99

the earthquake motion is ended. However, the measured velocity of the ground flow is reported 15 to 20 km/h [7], while the computed one is merely 0.14 km/h. This fact may suggest requirements of other conditions which is not considered in the present analysis, such as water supply from confined aquifer, to produce such a large scale ground flow with high speed.

Table 4. Permeability assigned for each test case.

Case	1	2
k_{silt} (m/s)	1×10^{-6}	1×10^{-4}
k_{sand} (m/s)	1×10^{-4}	1×10^{-4}

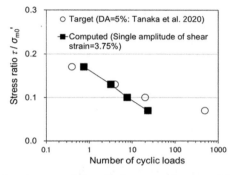

Fig. 6. Measured and computed liquefaction strength curve of the soil (after [17]).

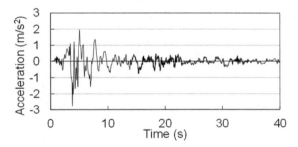

Fig. 7. Recorded earthquake motion (EW motion) [13].

Next, looking at the analysis results of Case 2 in which the permeability of the surface silt layer is set as the same as that of the sand layer, as shown in Fig. 8(b), the displacement other than the ground surface becomes constant at the end of vibration. The displacement of the ground surface at the end of the analysis is about 6 m, because the pore water is discharged to the ground surface as the excess pore water pressure dissipates, so the low effective stress near the ground surface may cause large deformation in the element of surface layer.

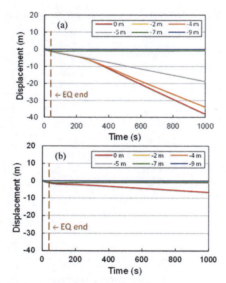

Fig. 8. Time histories of displacement in the sloping direction of each layer: (a) Case 1 and (b) Case 2.

3.2 Effective Stress and Excess Pore Water Pressure

Looking at the time history of excess pore water pressure for Case 1 (Fig. 9(a)), in the liquefied layers G.L. −5 m to G.L. −9 m, the increased excess pore water pressure converges to a constant value over time. On the other hand, in the surface silt layer (G.L. 0 m to −4 m), the excess pore water pressure, which was originally low, increased as the pore water pressure in the lower liquefied layer being decreased, and became constant in about 300 s. From this, it can be inferred that the dissipation of excess pore water pressure generated in the sand layer is blocked by the silt layer. Note that at G.L. − 4 m (Layer 1) in which dilatancy parameters are not assigned, negative pressure (min. −15 kPa) appears at 4 s during shaking. Cause of this can be some numerical errors and is under investigation.

Time histories of excess pore water pressure for Case 2 (Fig. 9(b)) show that those of all layers dissipated in 400 s. The residual excess pore water pressure at the end of the analysis is about 10 kPa. From a separate analysis, it is confirmed that the excess pore water pressure becomes 0 kPa when the computational time step is increased up to 25,000 s.

In order to confirm the transition of excess pore water pressure in depth, the depth profile of excess pore water pressure up to 300 s after the start of analysis is plotted in Fig. 10. The dashed line is the initial effective confining stress. First, regarding Case 1, from Fig. 10(a), the excess pore water pressure dissipates over time in the sand layer deeper than G.L. −6 m, and at 300 s, it shows a constant value of about 65 kPa regardless of the depth. On the other hand, the excess pore water pressure gradually increases in the silt layer shallower than G.L. −5 m. This indicates that the excess pore water pressure generated in the sand layer flows into the silt layer over time. In G.L. −5 m, the excess pore water pressure is almost equal to the initial effective confining stress after about 250 s, and the ground loses its bearing capacity. Hence, the boundary between the sand and the silt layer of G.L. −5 m may become a slip surface.

Next, in the water pressure distribution of Case 2 shown in Fig. 10(b), the excess pore water pressure disappears over time. In this case, the pore water due to the extinction of excess pore water pressure did not stay at the layer boundary and quickly escaped to the upper layer, so the effective stress did not decrease and no large displacement occurred. However, the excess pore water pressure near the ground surface at 300 s was about 10 kPa due to continuous inflow of pore water.

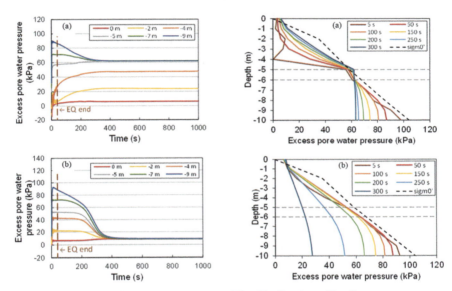

Fig. 9. Time histories of excess pore water pressure at each layer: (a) Case 1 and (b) Case 2.

Fig. 10. Depth profile of excess pore water pressure dissipation: (a) Case 1 and (b) Case 2.

3.3 Effective Stress Path

The effective stress paths of the two elements that sandwich the boundary between the surface silt layer and the sand layer are shown in Fig. 11 (silt layer) and Fig. 12 (sand layer). First, looking at the silt layer, in Case 1 (Fig. 11(a)), the stress path does not reach the failure line at the end of excitation. However, as time passes, the effective stress decreases and the failure line is reached due to the inflow of pore water from the lower layer. In this situation, in the time transition of the average effective stress contour shown in Fig. 13(a), the range where the average effective stress is small (blue) moves from the sand layer (layer 2) to the silt layer (layer 1) with time. In Case 2 shown in Fig. 11(b), the effective stress path touches the failure line during excitation. Then recovers to more than the initial value with the passage of time.

Next, looking at the sand layer shown in Fig. 12(a), in Case 1, the effective stress path reaches the failure line during excitation and then approaches the origin along the failure line. While in Case 2 shown in Fig. 12(b), it was liquefied by shaking and the effective stress path reached the failure line. However, as in Fig. 12(b), the effective stress recovers with time. From the transition of the average effective stress contour shown in Fig. 13(b), the effective stress once decreased by liquefaction gradually recovers from the lower sand layer to the upper silt layer.

It is confirmed that the excess pore water pressure in the silt layer is dissipated and the effective stress is restored when the duration of the analysis is continued more than 1,000 s.

From the above, in Case 2 where the permeability of the surface layer is the same as that of the liquefaction layer, the pore water is drained to the ground surface as the excess pore water pressure disappears. Therefore, the delayed failure due to the decrease in shear strength does not occur. This suggests that the existence of a layer with a small hydraulic conductivity on top of the liquefaction layer, such as Case 1, is an important condition for the delayed failure mechanism.

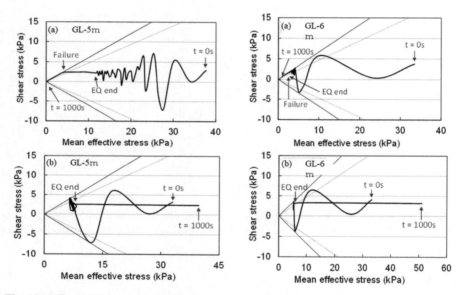

Fig. 11. Effective stress path at G.L. −5 m in silt layer (layer 1): (a) Case 1 and (b) Case 2.

Fig. 12. Effective stress path at G.L. −6 m in sand layer (layer 2): (a) Case 1 and (b) Case 2.

4 Delayed Failure Mechanism of Gentle Slopes

From the analysis results described in Sect. 3, the mechanism of occurrence of the delayed failure of a gently sloping ground is explained as follows. The excess pore water pressure generated in the sand layer due to the earthquake motion creates a hydraulic gradient in the soil and moves the pore water to the ground surface. When a layer with a small hydraulic conductivity such as a silt layer underlain by a liquefied sand layer, the pore

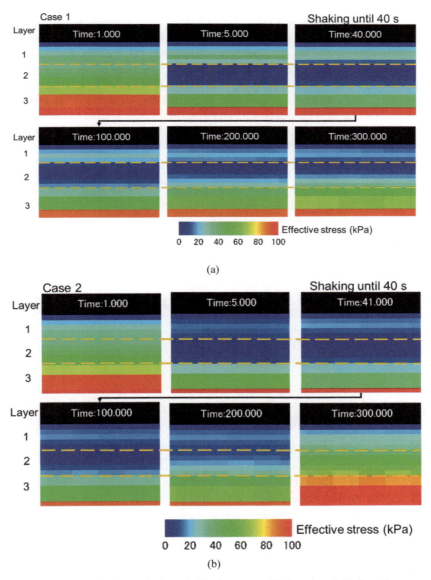

Fig. 13. Transient variation of effective stress: (a) Case 1 and (b) Case 2.

water gradually stays near the layer boundary because the path to the ground surface is obstructed (formation of water film) [19]. For this reason, the top of a sand layer expands (void re-distribution) [20] and the effective stress gradually decreases, which may lead degradation of the shear strength of the sand layer.

This phenomenon can be explained using the compression curve shown in Fig. 14. That is, considering a soil element located at the top of the liquefied layer, the initial effective confining stress p_{us0}' of the ground in the initial state (A and A' in Fig. 14) decreases due to the inflow of pore water into this element. The undrained shear strength at the initial state is q_{us0}. If pore water flows in further, volume expansion occurs and the effective stress path asymptotically approaches to the steady state line (B and B'), and the average effective confining stress p' decreases further. Along with this, the shear strength of the soil element q_{us} decreases, and if it falls below the initial static shear stress, τ_{st}, $q_{us} < \tau_{st}$, then flow occurs (C and C'). This is presumed to be the mechanism of the delayed failure in which the flow occurs after the earthquake motion converges, unlike the flow that occurs immediately due to the earthquake motion.

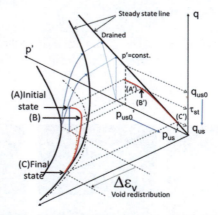

Fig. 14. Schematic view of the stress path of the element in which shear strength decreases with the increase of volumetric strain due to inflow of pore water.

5 Conclusions

Conclusions on the mechanism and causes of large-scale lateral flow on gentle slopes obtained by this study are summarized below.

1) Two-dimensional effective stress analysis was able to reproduce the delayed failure phenomenon of sloping ground due to liquefaction. Excess pore water pressure due to liquefaction creates a hydraulic gradient in the soil. Normally, the pore water is drained to the surface by this hydraulic gradient. However, when a layer with a small hydraulic conductivity such as silt exists in the upper part, the pore water gradually begins to stay near the layer boundary. Along with this, the ground near the layer boundary gradually expands, and the effective stress and shear strength decrease. At some point, the shear stress in the sloping direction due to gravity exceeds the shear strength and then lateral flow occurs. This is called the delayed failure.
2) It was shown that a large-scale flow may occur due to the alternating layer structure of gently sloping sand and silt without considering the effects of water leakage from the irrigation canal and the effects of confined aquifer. However, the measured velocity

of the ground flow is about 100 times faster than computed one in the present study. This fact may suggest requirements of other conditions which is not considered in the present analysis, such as water supply from confined aquifer, to produce such a large scale ground flow with high speed.
3) In Palu, which is surrounded by high-altitude mountain ranges from north to south and formed by alluvial fans of various sizes, alternating layers of sand and silt are widely distributed. It is no wonder that similar flows occur in areas where the water table was high, but there are some locations where large-scale flows did not occur. The reason for this is unknown and should be studied further in the future.

The flow slides with such an enormous size occurred in Palu might be a rare case. However, locations having similar geological subsurface conditions may exist all over the world.

Acknowledgements. This research was supported by Grants-in-Aid for Scientific Research B (JP18H01523: Principal Investigator Yusuke Ono) and (JP20H02244: Principal Investigator Hazarika Hemanta).

References

1. European Commission JRC. Mw 7.5 Earthquake in Indonesia, 28 September 2018 (2018). https://www.gdacs.org/Public/download.aspx?type=DC&id=75. Accessed 2021
2. AHA Centre. M7.4 Earthquake & Tsunami, Indonesia, Situation update No. 13 (2018). https://reliefweb.int/sites/reliefweb.int/files/resources/AHA-Situation_Update-no13-Sulawesi-EQ.pdf. Accessed 23 2021
3. Cipta, A., Robiana, R., Griffin, J.D., Horspool, N., Hidayati, S., Cummins, P.R.: A probabilistic seismic hazard assessment for Sulawesi, Indonesia. Geolog. Soc. Lond. Spec. Publ. **441**(1), 133–152 (2017)
4. Metcalfe, I.: Late Palaeozoic and Mesozoic tectonic and palaeogeographical evolution of SE Asia. Geol. Soc. Lond. Spec. Publ. **315**, 7–23 (2009)
5. Iai, S., Tobita, T.: Nonlinear dynamic analyses for evaluating seismic ground deformation—delayed flow failure. In: Proceedings of the 20th International Conference on Soil Mechanics and Geotechnical Engineering, Sydney 2021 (2021) (Submitted)
6. Mason, H.B., et al.: The 28 September 2018 M7.5 Palu-Donggala, Indonesia Earthquake (version 1.0; 3 April 2019). GEER report (2019)
7. Kiyota, T., Furuichi, H., Hidayat, R.F., Tada, N., Nawir, H.: Overview of long-distance flow-slide caused by the 2018 Sulawesi earthquake, Indonesia. Soils Found. **60**(3), 722–735 (2020)
8. Okamura, M., Ono, K., Arsyad, A.: Large-scale flow-slide in Sibalaya caused by the 2018 Sulawesi earthquake. Soils Found. **60**(4), 1050–1063 (2020)
9. Hazarika, H., et al.: Large scale flow-slide at Jono-Oge due to the 2018 Sulawesi Earthquake, Indonesia. Soils Found. **61**(1), 239–255 (2021)
10. Iai, S., Matsunaga, Y., Kameoka, T.: Strain space plasticity model for cyclic mobility. Soils Found. **32**(2), 1–15 (1992)
11. Iai, S., Tobita, T., Ozutsumi, O., Ueda, K.: Dilatancy of granular materials in a strain space multiple mechanism model. Int. J. Numer. Anal. Meth. Geomech. **35**, 360–392 (2011)
12. Zienkiewics, O.C., Bettess, P.: Soil and other saturated media under transient, dynamic conditions, general formulation and the validity various simplifying assumptions. In: Soil Mechanics – Transient and Cyclic Loads. Wiley (1982)

13. Himawan, A.: Personal communication (2019)
14. Mikami, T., Ozutsumi, O., Nakahara, T., Iai, S., Ichii, K., Kawasaki, T.: Easy method for setting parameters of liquefaction analysis program FLIP. In: Proceedings of the 46th Annual Meeting of the Japan National Conference on Geotechnical Engineering, vol. 813 (2011) (in Japanese)
15. Ishihara, K.: Soil Mechanics, 3rd edn, p. 320. Maruzen (2018)
16. Japan Road Association. Specifications for Highway Bridges, Part I Common (2012)
17. Tanaka, R., Okamura, M., Ono, K.: Cyclic and monotonic strength characteristics of sand at Sibalaya, Sulawesi Indonesia, where massive flowslide occurred. In: The 55th Annual Meeting of the Japan National Conference on Geotechnical Engineering, Japanese Geotechnical Society, 21-11-5-07 (2020) (in Japanese)
18. Ishihara, K., Acacio, A.A., Towhata, I.: Liquefaction-induced ground damage in Dagupan in the July 16, 1990 Luzon Earthquake. Soils Found. **33**(1), 133–154 (1993)
19. Kokusho, T.: Water film in liquefied sand and its effect on lateral spread. J. Geotech. Geoenviron. Eng. ASCE **125**(10), 817–826 (1999)
20. Boulanger, R.W., Truman, S.P.: Void redistribution in sand under post-earthquake loading. Can. Geotech. J. **33**(5), 829–833 (1996)

Seismic Fragility Assessment for Cohesionless Earth Slopes in South Korea

Dung Thi Phuong Tran, Hwanwoo Seo, Youngkyu Cho, and Byungmin Kim(✉)

Ulsan National Institute of Science and Technology, 50 UNIST-gil, Ulju-gun, Ulsan 44919, Republic of Korea
byungmin.kim@unist.ac.kr

Abstract. Seismic performances of earth slopes have been primarily assessed from the predictive models of a horizontal slope displacement as a function of characteristics of slope and ground shaking, based on Newmark or Newmark-type sliding block analyses. A seismic fragility curve of earth slopes has been barely investigated, but it could be useful for resilient analysis (e.g., repair cost, recovery time, and loss of life and property). This study aims to develop the seismic fragility curves of earth slopes in South Korea using finite element (PLAXIS 2D) simulations. We selected a single height of slope (10 m) and three slope angles of 20°, 30°, and 40° defining slope geometries that are representative of slope conditions in South Korea with cohesionless soil types. These conditions resulted in a total of 6 slope models. We considered 280 ground motions calculated by one-dimensional site response analyses for various site conditions. The fragility curves were constructed for three damage states (i.e., minor, moderate, and major) using the computed displacements from the simulations. It turned out that the fragility curves were highly dependent on slope angle and ground motion characteristics.

Keywords: Seismic fragility analyses · Earth slopes · Finite element simulations

1 Introduction

Slopes exist in varied types such as natural slopes, embankments of roadway and railway, dams and levees that are classified as geotechnical structures. Many case histories for severe slope damages by earthquakes have been reported such as a destruction of expressway structure in Japan by the 1995 Kobe earthquake [1] or the 2004 Niigata Chuetsu earthquake [2]. Empirical relationships between the earthquake-induced slope displacement and characteristics of slope and ground shaking have been widely developed to assess the seismic performance of slopes [3–6]. However, the fragility analysis has barely been investigated that expresses the failure probability exceeding a certain damage state of a slope given a level of earthquake intensity in Korea.

This study aims to assess the earthquake-induced slope failure through developing a seismic fragility curve of slope displacement. Toward this end, we characterized the slope conditions and ground motions representative of site classes in South Korea and conducted parametric PLAXIS 2D simulations to compute suites of earthquake-induced slope displacements associated with the ground motions and slope conditions.

The resulting displacements were used for generating the seismic fragility curve of slope displacement for the slopes comprising cohesionless soils across three slope angles.

2 Ground Motion Database

The importance of a stochastic feature in the ground shaking has been shown to significantly impact the responses of the geotechnical structures over a few decades, leading to the use of various ground motion records in the analysis [7]. Table 1 lists the seven pairs of ground motion records (East-West and North-South) for rock outcrop condition to obtain the ground shakings compatible with site conditions in South Korea through one-dimensional site response analyses (DEEPSOIL ver. 7.0 [8]) where the select motions were entered into elastic bedrock. Figure 1 shows the 5% damped acceleration response spectra of the collected motions which are comparable with the Korean design response spectrum for 2400 years return periods. The site response analyses produced surface motions corresponding to site classification system of South Korea (MPSS, 2017) for four site classes (S2, S3, S4, and S5). The shear wave velocities (V_S) larger than 260 m/s and 180 m/s represent shallow (S2) and deep (S4) sites of stiff soils, respectively, whereas the opposite ranges of the V_S values are representative of shallow (S3) and deep (S5) sites of soft soils. The resulting surface ground motions are classified into four types of motion such as M2, M3, M4, and M5 pertaining to each site class S2, S3, S4 and S5, respectively. Those motions were scaled to five levels of peak ground acceleration (PGA) that are 0.1 g, 0.3 g, 0.5 g, 0.7 g, and 0.9 g. The combination of surface motion

Table 1. The collected seven pairs of rock outcrop motions used in this study.

Name	M	Component	PGA (g)	R_{epi} (km)	Mean period (s)
Pohang	5.4	EW	0.24	9.5	0.17
		NS	0.29		0.33
Gyeongju	5.5	EW	0.41	9.1	0.13
		NS	0.34		0.11
Hokkaido	6.7	EW	0.13	34	0.22
		NS	0.12		0.36
N.Palm Springs	6.1	EW	0.13	17.0	0.12
		NS	0.11		0.13
Kozani Greece-01	6.4	EW	0.21	19.5	0.23
		NS	0.14		0.22
Sierra Madre	5.6	EW	0.19	10.4	0.16
		NS	0.28		0.19
L'Aquila Italy	5.6	EW	0.13	15.0	0.27
		NS	0.11		0.26

types and the scaled levels leads to a total of 280 motions (= 14 original motions × four site classes × five PGA levels) for the simulations.

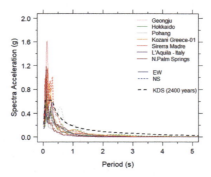

Fig. 1. 5% damped acceleration response spectra of the collected motions and the Korean design response spectrum for 2400 years return periods.

3 Numerical Simulations

A suite of nonlinear finite element analyses (PLAXIS 2D, Bentley [9]) was conducted to produce suites of the permanent earthquake-induced slope displacement for developing the seismic fragility curve of slope displacement. The Hardening Soil model with small-strain stiffness (HSsmall) was used for describing stress-strain responses of soils of the slope models subjected to the earthquake motions. The finite element mesh of the slope models comprises triangular elements with 15 nodes in two-dimensional plane strain condition.

3.1 Earth Slope Modeling

Slope Geometry
A homogeneous geometry was created for the slope model having a height (H-slope) of 10 m and a single slope angle (β) that varies from 20° to 40° shown in Fig. 2(a). Lee [10] investigated the relationship between a slope angle and a slope thickness from a slope surface to firm layer for two sites in South Korea, leading to that the slope thicknesses are generally less than 10 m. Park and Cho [11] performed the seismic slope stability analyses showing that a shallow depth sliding could be regarded as a common failure pattern for potential earthquake events in South Korea. Thus, the selected values of H-slope and β for the geometrical parameters of the slope models are based on the previous studies. Each slope geometry was divided into four sublayers to consider the effect of confining pressure on the V_S and shear strength shown in Fig. 2(b).

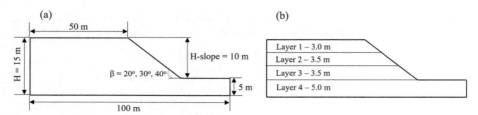

Fig. 2. (a) Slope geometry considered in the finite element analyses, and (b) sublayers of the slope model.

Material Properties

The engineering properties of a slope are generalized in terms of the mechanical behaviors associated with the soil type. This study considered the cohesionless soils (i.e., sand-like behavior, c-ɸ analysis) based on the Mohr-Coulomb failure criterion for the slope simulations. The friction angles ($\phi = 30°$ and $40°$) for the loose and dense sand were determined from the empirical relationship of Terzaghi and Peck [12] between SPT-N value and relative density of sands to consider loose and dense states while keeping the cohesion (c) value set to 10 kPa for both friction angles. The two values of the V_S (=150 m/s and 300 m/s) were associated with the smaller and larger strengths of the loose and dense sands. These values were inferred from the empirical relationship between the V_S and the SPT-N ($V_S = 65.64 \cdot N^{0.407}$ for all soil types) in Korean sites [14].

The four types of motions (M2, M3, M4, and M5) were assigned to the slope models corresponding to the associated V_S. The M2 ($V_S \geq 260$ m/s) and M4 ($V_S \geq 180$ m/s) were applied to the slope models with $V_S = 300$ m/s that are dense sands. The loose sands with $V_S = 150$ m/s were subjected to all the motion types (M2 and M3 for $V_S < 260$ m/s, and M4 and M5 for $V_S < 180$ m/s). Table 2 lists the adopted engineering properties of the slope models and ground motion characteristics of the select earthquake records along with slope geometries.

Boundary Condition

Two types of boundary conditions were used for the finite element analyses. The pinned and roller supports were specified along the bottom and both side edges of the slope model for a gravitational loading analysis with the linear elastic model to provide the initial states of strains and stresses prior to ground shaking. For the transient loading stage, the free-field boundaries were applied to the lateral boundaries of the slope model to prevent the reflected waves at the model edges from going into the slope part. The pinned supports at the model bottom were replaced with the compliant base boundary consisting of the viscous dashpot scheme of Lysmer and Kuhlmeyer [15] to absorb the downward-propagating waves reflected from the top of the slope geometry while allowing for the upward propagation of the incoming waves.

3.2 Model Verification

The finite element analysis was verified through comparing the acceleration response at the free-field location of the slope with that computed from a soil column with the

Table 2. Engineering properties for the slope models used in this study.

H-slope (m)	Slope angle (°)	Soil type	V_S (m/s)	Layer No.	ϕ (°)	Cohesion (kPa)	Applied ground motions
10	20° or 30° or 40°	Cohesionless	150 (Loose)	1	30	10	M2, M3, M4, M5
				2			
				3			
				4			
			300 (Dense)	1	40		M2, M4
				2			
				3			
				4			

same height (15 m of the left side in Fig. 2(a)), sublayers in Fig. 2(b), and engineering properties in DEEPSOIL. The elastic half space condition was chosen in DEEPSOIL that is compatible with the compliant base boundary in PLAXIS 2D The Gyeongju-NS motion (PGA = 0.3 g) was selected for both simulations of PLAXIS 2D and DEEPSOIL. Figure 3 shows the comparisons of the acceleration response in terms of time and frequency domains for both simulations, agreeing well with each other; thus, our PLAXIS 2D simulations can reliably predict the permanent earthquake-induced slope displacements depending on the soil types and ground motion records without being affected by the waves reflected from the lateral boundaries.

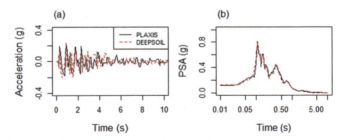

Fig. 3. Comparison of acceleration response between PLAXIS 2D and DEEPSOIL at the top of each simulation: (a) time series and (b) pseudo spectral acceleration (PSA).

4 Fragility Analysis

Using the displacement dataset generated from the previous section, the seismic fragility curves of slope displacement were derived to investigate the effect of slope angles and site class condition on the earthquake-induced slope failure. Building the fragility curves requires to specify a damage state and an associated threshold displacement; thus, we chose three damage states along the thresholds and used PGA for the engineering demanding parameter in the fragility analysis.

4.1 Selection on Damage State and Threshold Level

A damage state of slopes by the earthquakes is represented by a level of slope displacement being exceeded (i.e., threshold). The threshold of the slope displacement against the earthquakes were suggested by California Geological Survey (CGS) [16], which does not distinguish the thresholds for soil slopes from those for rock slopes. These threshold values have been widely applied for various locations in the USA such as Anchorage, Alaska and Mississippi Valley [17, 18], as well as other countries (e.g., South Korea [19], Serbia [20], and Iran [21]). Table 3 lists the damage states by CGS comprising three states (i.e., minor, moderate, and major) associated with three values of the slope displacements (i.e., 5 cm, 15 cm, and 30 cm), respectively.

Table 3. Thresholds for three damage states of the permanent earthquake-induced slope displacement.

Damage state	Minor	Moderate	Major
Threshold for D (cm)	5.0	15.0	30.0

4.2 Seismic Fragility Assessment of Slope Displacement

Figure 4 represents the computed earthquake-induced slope displacements from multiple PLAXIS 2D simulations for the slopes at three slope angles and ground motion conditions corresponding to the site classes. In general, the permanent displacements calculated using the ground motions computed for soft soil sites (S3 and S5) are larger than those for the stiff soil sites (S2 and S4). The displacements for the three slope angles (20°, 30°, and 40°) under the ground motions computed for a S5 site condition with a PGA of 0.5 g ranged from 0.1 to 0.45 m, from 0.11 to 0.55 m, and from 0.12 to 0.69 m, respectively. These minor differences may be attributed to the use of the cohesion = 10 kPa. The permanent displacements were assumed to be lognormally distributed in this study. We calculated the failure probability (P) for given PGA levels using the results of numerical analyses and threshold values based on Eq. 1 [22].

$$P[\text{Demand} > \text{Capacity}|\text{PGA}] = 1 - \Phi\left[\frac{\ln(x_i) - \lambda}{\varsigma}\right] \quad (1)$$

where $\Phi(\cdot)$ is standard normal cumulative distribution function, s_i is a threshold of damage state, $\varsigma^2 = \ln(1 + \delta^2)$ and $\lambda = \ln(\mu) - 0.5\varsigma^2$ in which δ is a ratio of σ to μ, σ/μ (μ and σ are mean standard deviation of permanent displacements corresponding with PGA level).

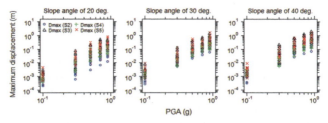

Fig. 4. Permanent earthquake-induced slope displacement for the three slope angles and cohesionless soil.

Figure 5 shows the seismic fragility curve of slope displacement for the four site classes across the three slope angles. The failure probabilities for the M3 and M5 (soft soil motions) are larger than those for the M2 and M4 (stiff soil motions), indicating that the soft slopes are more vulnerable than the stiff one against the ground shakings. This is consistent with the fact that the seismic waves propagating through the soft medium are generally more amplified than the stiff one. Of course, the failure probabilities increase as the slope angle gets larger for all the motion conditions (M2, M3, M4, and M5), which conforms to the fact that the factor of safety for stability analysis of slopes decreases with the slope angle ascending.

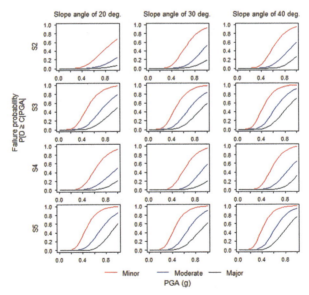

Fig. 5. Fragility curves of three slope angles for cohesionless soil subject to the ground motions computed for the four site classes.

5 Conclusions

The seismic fragility curve of slope displacement in South Korea was investigated. Multiple slope geometries consisting of a single height (10 m) and three slope angles (20°, 30°, and 40°) were developed along with the cohesionless soils (i.e., loose and dense sands) to run the parametric PLAXIS 2D simulations that calculate suites of the permanent slope displacement. The surface motions suited to the Korean ground conditions were characterized by one-dimensional site response analyses with the selected seven pairs of earthquake motion records. The combination of the slope conditions with the developed surface motions produced 1680 permanent slope displacements that were applied to fitting the fragility curves based on the maximum likelihood estimation. The failure probabilities for the slopes subjected to the soft soil motions are less than those for the stiff soil motions, which is attributed to the site amplification effect. Also, the failure probabilities were high for the larger slope angle and the motions of the soft soils with other conditions held same. In this study, we focused on the slope with slope height of 10 m based on other studies for South Korea. We plan to extend the research for more various slopes in the future.

References

1. Kawashima, K.: Seismic performance of RC bridge piers in Japan: an evaluation after the 1995 Hyogo-ken nanbu earthquake. Prog. Struct. Eng. Mater. **2**, 82–91 (2000). https://doi.org/10.1002/(sici)1528-2716(200001/03)2:1%3c82::aid-pse10%3e3.0.co;2-c
2. Honda, R., Aoi, S., Morikawa, N., et al.: Ground motion and rupture process of the 2004 Mid Niigata Prefecture earthquake obtained from strong motion data of K-NET and KiK-net. Earth Planets Space **57**, 527–532 (2005). https://doi.org/10.1186/BF03352587
3. Argyroudis, S., Kaynia, A.M., Pitilakis, K.: Development of fragility functions for geotechnical constructions: application to cantilever retaining walls. Soil Dyn. Earthq. Eng. (2013). https://doi.org/10.1016/j.soildyn.2013.02.014
4. Wu, X.Z.: Development of fragility functions for slope instability analysis. Landslides **12**, 165–175 (2014). https://doi.org/10.1007/s10346-014-0536-3
5. Maruyama, Y., Yamazaki, F., Mizuno, K., et al.: Fragility curves for expressway embankments based on damage datasets after recent earthquakes in Japan. Soil Dyn. Earthq. Eng. **30**, 1158–1167 (2010). https://doi.org/10.1016/j.soildyn.2010.04.024
6. Fotopoulou, S.D., Pitilakis, K.D.: Predictive relationships for seismically induced slope displacements using numerical analysis results. Bull. Earthq. Eng. **13**(11), 3207–3238 (2015). https://doi.org/10.1007/s10518-015-9768-4
7. Bray, J.D., Travasarou, T.: Pseudostatic coefficient for use in simplified seismic slope stability evaluation. J. Geotech. Geoenviron. Eng. **135**, 1336–1340 (2009). https://doi.org/10.1061/(asce)gt.1943-5606.0000012
8. Hashash, Y.M.A.: Version 7.0. 1–170 (2020)
9. Plaxis. Tutorial Manual. Plaxis 2D Connect Ed V20 6–30 (2004)
10. Lee, J., Han, J.-T., Baek, Y., et al.: Development of prediction method considering geometrical amplification characteristics of slope II: construction of landslide hazard map during earthquakes in Seoul. J. Korean Soc. Hazard Mitig. **14**, 85–92 (2014). https://doi.org/10.9798/kosham.2014.14.5.85
11. Park, N.-S., Cho, S.-E.: Development of fragility curves for seismic stability evaluation of cut-slopes. J. Korean Geotech. Soc. **33**, 29–41 (2017)

12. Terzaghi, K., PRB: Soil mechanics in engineering practice (1996)
13. Trask, P.D., Terzaghi, K., Peck, R.B.: Soil mechanics in engineering practice. J. Geol. (1949). https://doi.org/10.1086/625679
14. Sun, C.G., Cho, C.S., Son, M., Shin, J.S.: Correlations between shear wave velocity and in-situ penetration test results for Korean soil deposits. Pure Appl. Geophys. **170**, 271–281 (2013). https://doi.org/10.1007/s00024-012-0516-2
15. Lysmer JKRLI of T, Traffic Engineering U of CBSM, Bituminous Materials Research Laboratory U of CBGEG. Finite dynamic model for infinite media. Dept. of Civil Engineering, Univ. of California, Institute of Transportation and Traffic Engineering, Soil Mechanics Laboratory, Berkeley, Calif (1969)
16. Saygili, G., Rathje, E.M.: Probabilistically based seismic landslide hazard maps: an application in Southern California. Eng. Geol. **109**, 183–194 (2009). https://doi.org/10.1016/j.enggeo.2009.08.004
17. Jibson, R.W., Michael, J.A.: Maps Showing Seismic Landslide Hazards in Anchorage, Alaska. USGS Sci Investig Map 3077 (2009)
18. Jibson, R.W., Keefer, D.K.: Analysis of the seismic origin of landslides: examples from the New Madrid seismic zone. Geol. Soc. Am. Bull. (1993). https://doi.org/10.1130/0016-7606(1993)105%3c0521:AOTSOO%3e2.3.CO;2
19. Kwag, S., Hahm, D.: Development of an earthquake-induced landslide risk assessment approach for nuclear power plants. Nucl. Eng. Technol. (2018). https://doi.org/10.1016/j.net.2018.07.016
20. Garevski, M., Zugic, Z., Sesov, V.: Advanced seismic slope stability analysis. Landslides **10**(6), 729–736 (2012). https://doi.org/10.1007/s10346-012-0360-6
21. Jafarian, Y., Lashgari, A., Haddad, A.: Predictive model and probabilistic assessment of sliding displacement for regional scale seismic landslide hazard estimation in Iran. Bull. Seismol. Soc. Am. (2019). https://doi.org/10.1785/0120190004
22. Chiou, J.S., Chiang, C.H., Yang, H.H., Hsu, S.Y.: Developing fragility curves for a pile-supported wharf. Soil Dyn. Earthq. Eng. **31**, 830–840 (2011). https://doi.org/10.1016/j.soildyn.2011.01.011

Liquefaction-Induced Lateral Displacement Analysis for Sloping Grounds Using Long-Duration Ground Motions

Qiang Wu, Dian-Qing Li, and Wenqi Du[✉]

State Key Laboratory of Water Resources and Hydropower Engineering Science, Institute of Engineering Risk and Disaster Prevention, Wuhan University, Wuhan 430072, Hubei, China
wqdu309@whu.edu.cn

Abstract. Earthquake-induced liquefaction occurring in saturated granular deposits would generally yield permanent deformation of sloping grounds. Many previous studies have highlighted the important role of ground motion duration in triggering liquefaction, yet its effect on liquefaction-induced lateral displacement (LD) of sloping grounds remains ambiguous. In this study, a one-dimensional gently sloping ground model containing a loose-sand liquefiable layer is implemented in OpenSees. A total of 126 pairs of long-duration and spectrally equivalent short-duration ground motions are selected from recently occurred giant earthquakes and the NGA-West2 database, respectively. Numerical analyses are then conducted for the sloping ground using the selected ground motions as input. The results indicate that the long-duration ground motion has a higher potential to cause larger LD than the spectrally equivalent short-duration one. The median of the LDs for the long-duration suite is about 10 times larger than that of the short-duration one, attributing to the much more energy contained for the long-duration ground motions. Moreover, it is indicated that cumulative absolute velocity (CAV) is highly correlated with LD for the long-duration ground motions. Consequently, a simple CAV-based empirical model is proposed to predict LD caused by long-duration ground-motion records.

Keywords: Liquefaction consequence · Long-duration ground motions · Lateral displacement · Sloping ground · CAV

1 Introduction

Liquefaction phenomena usually occurs in saturated granular deposits when subjected to cyclic loading, manifesting as a great reduction of the strength and stiffness of the soil deposits. As a result, the granular soils behave as 'liquid', causing damage to the overlying buildings, earth dams, and other geotechnical structures [1, 2]. Earthquake-induced liquefaction usually causes permanent lateral displacement (LD) of sloping grounds. For example, noticeable lateral spreads were observed during the 1995 Kobe earthquake, which caused many concrete caisson quay walls of the port to move 5 m to the sea side [3].

Some evolutionary intensity measures (IMs), such as Arias intensity (I_a) and cumulative absolute velocity (CAV), are commonly employed to estimate the liquefaction-induced LD [4]. Based on appropriate IMs and site condition parameters (i.e., liquefiable layer thickness and slope inclination), predictive equations for LD have been proposed in literatures [e.g., 5]. Other predictive models of LD were proposed based on field case history data, in which the earthquake magnitude was regarded as a measure related to earthquake duration [6]. However, the influence of duration on liquefaction-induced LD has not been fully investigated.

Recently, long-duration ground motions have attracted increasing attention due to its specific characteristic. According to the incremental dynamic analyses (IDA) for a steel moment frame model, Chandramohan et al. [7] revealed that the estimated median collapse capacity induced by long-duration records is 29% lower than that induced by spectrally equivalent short-duration ones. Fairhurst et al. [8] concluded that the maximum inter-story drifts of reinforced concrete shear wall buildings were sensitive to ground-motion duration. In addition, many other studies also investigated the influence of duration on various structures [9–11]. However, there is currently few study investigating the influence of ground motion duration on liquefaction-induced LD.

This paper thus aims at quantitatively investigating the effect of ground-motion duration on the liquefaction-induced LD of a sloping ground containing liquefiable layers. The remaining part of this paper is structured as follows. First, the selected long-duration and spectrally equivalent short-duration ground-motion suites are presented, which have the similar shape of response spectra at periods in the range of 0 to 6 s. Second, one-dimensional sloping grounds modeled in OpenSees are introduced. Dynamic analyses of each sloping ground subjected to the selected ground motion suites are then conducted; the significance of the effect of duration on post-liquefaction response of sloping grounds is quantitatively analyzed. Finally, a predictive model based on regression analysis is presented and conclusive remarks are provided.

2 Ground Motion Database

During the past two decades, several giant earthquakes, including the 2008 Wenchuan earthquake (moment magnitude Mw of 7.9), 2010 Maule earthquake (Mw 8.8), and 2011 Tohuko earthquake (Mw 9.0), have occurred globally. These giant earthquakes produced a large number of available long-duration ground motions. Taking significant duration D_{s5-75} of 25 s as the threshold [7], the ground motions having D_{s5-75} larger and smaller than 25 s are identified as long- and short-duration ground motions, respectively. Since this study focuses on the post-liquefaction response evaluation, ground motions having PGA smaller than 0.1 g or PGV smaller than 10 cm/s are excluded. As a result, a total of 126 long-duration ground motions are selected from databases such as Kik-Net [12], CESMD [13], NEDC [14], and the NGA-West2 databases [15]. To isolate the effect of amplitude and frequency-content characteristics of ground motions, 126 spectrally equivalent short-duration ground motions, which have comparable spectral accelerations with the long-duration records, are subsequently selected from NGA-West2 database. Specifically, the response spectrum of each long-duration motion is divided into 120 sections, which correspond to spectral accelerations at periods form 0.05 s to 6.0 s in intervals of 0.05 s designated as $L_1, L_2, L_3,\ldots L_{120}$, with mean value of $\bar{L}$. Analogously, the response spectrum of each candidate short-duration ground motion is also divided into 120 sections termed as $\bar{S}_1, S_2, S_3,\ldots S_{120}$, with mean value of S. The scale factor k is expressed as $\bar{L}/\bar{S}$. The sum of squared errors (SSE) of the spectral acceleration at periods of 0 to 6 s between the long-duration and scaled short-duration motions is used to quantify the mismatch of each pair of ground motions, which is expressed as:

$$SSE = \sum_{120}^{i=1} (L_i - kS_i)^2 \tag{1}$$

The one with the smallest SSE is thus selected. The response spectra of these two ground motion suites are schematically showed in Fig. 1.

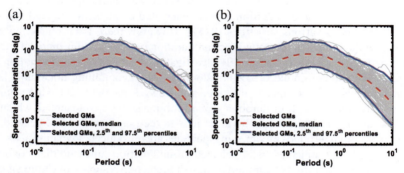

Fig. 1. Response spectra of (a) long-duration ground motions (b) spectrally equivalent short-duration ground motions.

3 Numerical Sloping Ground Model and Dynamic Analysis

3.1 Numerical Sloping Ground Model

A one-dimensional 20-m-high infinite sloping ground containing a loose-sand liquefiable layer was modeled in OpenSees [16]. As schematically shown in Fig. 2, the slope inclination is assigned as 2%. The whole soil deposits consist of a 2-m-thick dry loose-sand layer at surface, a 6-m-thick saturated loose-sand layer in the medium, and a 12-m-thick saturated dense-sand layer at the base. Note that the whole soil deposit is underlain by a bedrock layer. The water level is located at 2 m below the ground surface.

The numerical sloping ground was modeled by SSPbrickUP element, which is a three-dimensional element with 8 nodes. Each node has 4 degrees of freedom (DOFs), namely, two horizontal and one vertical translational DOFs, and one pressure DOF. The former three DOFs were used to monitor the displacements in x, y, and z directions, and the last one was adopted to monitor the pore pressure. The whole soil deposits are composed of 40 cube elements, and the side length of each element is 0.5 m.

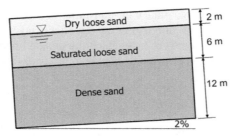

Fig. 2. Schematic representation of the sloping ground.

3.2 Constitutive Model and Boundary Condition

The dynamic behavior of all soil layers in this study is simulated by the pressure-dependent multi-yield-surface constitutive model (abbreviated as the PDMY02 model hereafter) [17]. The PDMY02 model is appropriate to model the dilatancy and cyclic mobility behavior of pressure-sensitive soil materials, e.g., loose and dense sand. The soil property parameters assigned are tabulated in Table 1, which are determined based on recommended values from OpenSees manual.

Periodic boundary was applied to the lateral sides of the soil column, and the horizontal and vertical displacements for nodes having the same elevation were tied together. A Lysmer-Kuhlemeyer [18] dashpot was added at the base of the model to simulate the semi-infinite elastic body, i.e., bedrock. Based on the boundary condition assigned, each ground motion was applied at the bottom of model to conduct dynamic analysis. Rayeigh damping was used to consider the hysteresis behavior of soil with the damping ratio of 0.02.

Table 1. Parameters assigned to each soil layer.

Soils	Density (g/cm^3)	Relative density	Initial void ratio	Shear modulus (kPa)	Bulk modulus (kPa)	Friction angle (°)	Peak shear strain (%)
Loose sand	1.80	40%	0.77	9.0×10^4	2.2×10^5	32	10
Dense sand	2.10	75%	0.55	1.3×10^5	2.6×10^5	36.5	10

3.3 Dynamic Analysis

Gravity analysis was firstly carried out to equilibrate the initial ground stress of the numerical model. The 126 pairs of long-duration and short-duration records were then applied to the model to conduct dynamic analyses. In each dynamic analysis, the time step was set to a small value to guarantee the convergence of the computational process. The excess pore water pressure ratio (r_u) in the liquefiable layer and the LD at ground surface were recorded and monitored as engineering demand parameters (EDPs). The occurrence of liquefaction was identified when the r_u exceeds 0.8.

4 Results

4.1 Dynamic Analysis Results

The ground-motion pair, namely the long-duration record with record sequence number RSN of 4782 in the 2008 Wenchuan earthquake and spectrally equivalent short-duration RSN 1812 in the 1999 Hector Mine earthquake were taken as the demonstrated example herein. As stated above, the short-duration record was scaled to match the response spectrum of the long-duration counterpart. The acceleration-time histories of both long- and short-duration ground motions are shown in Fig. 3a, with the D_{s5-75} ordinates being 43.02 s and 13.84 s, respectively. As shown in Fig. 3b, the response spectra of both records are broadly comparable at periods of 0 to 6 s. The ground motions were then applied to the base of sloping ground model to conduct the dynamic analysis in OpenSees.

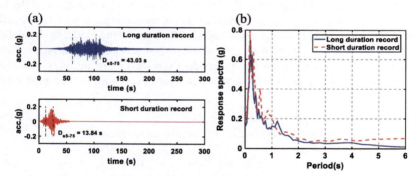

Fig. 3. (a) Acceleration-time histories of the long-duration and spectrally equivalent short-duration ground motions; (b) response spectra of the long-duration and spectrally equivalent short-duration ground motions.

As shown in Fig. 4a, when the sloping ground is subjected to the long-duration record, the time interval of r_u larger than 0.8 is about 150 s, while for spectrally equivalent short-duration one, the $r_u > 0.8$ time interval is only 30 s. This observation indicates that, although liquefaction occurred earlier when subjected to the short-duration record, the long-duration one containing more energy can sustain r_u in high level for much longer time. As a result, the post-liquefaction response is more remarkable when the slope ground is subjected to the long-duration records. Figure 4b demonstrates that the LD at ground surface are 4.67 m and 0.61 m for the long- and spectral equivalent short-duration records, respectively. Because the pore pressure cannot dissipate quickly for the long-duration case, the great reduction of the stiffness and strength of the loose-sand layer inevitably yields the much larger displacement.

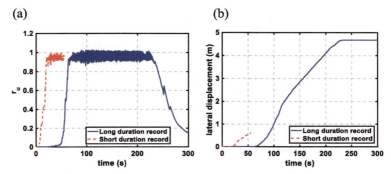

Fig. 4. (a) Excess porewater pressure ratio r_u and (b) Lateral displacement time history curves of long duration and spectrally equivalent short duration ground motions.

Analogous to the abovementioned example, the other 125 pairs of long- and short-duration records were employed to conduct the dynamic analyses. The cumulative probability of the LD ordinates is showed in Fig. 5a. The median LD values induced by the long-duration and short-duration suites are about 3.8 m and 0.3 m, respectively. Thus, it is clear that the long-duration suite yields the LDs about 10 times larger as compared to the short-duration suite.

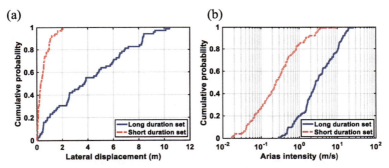

Fig. 5. Plots of the cumulative probabilities of the (a) lateral displacement and (b) Arias intensity for the two ground motion suites considered.

The long-duration ground motions contain much more energy than the short-duration motions, as shown in Fig. 5b. In addition, the long-duration ground motions can trigger liquefaction relative earlies than the short-duration records, so the energy in the post-liquefaction period tends to yield stronger nonlinear ground response. Therefore, for liquefiable sites subject to long-duration ground motion excitations, more attention should be paid to the evaluation of the enlarged liquefaction-induced LDs.

4.2 Regression Analysis Results

Prediction equations similar to the formula proposed by Bardet and Liu [19] were subsequently used to describe the liquefaction-induced LD of long duration set using evolutionary intensity measures, namely I_a and CAV. Both of them are parameters representing the accumulative energy of the ground motions. An efficient IM can reduce the dispersion of the IM-EDP relation, which is important for an appropriate prediction of the dynamic response of an engineering system. As shown in Figs. 6a and b, the coefficient of determination R^2 of CAV and I_a are 0.766 and 0.629, respectively.

Therefore, CAV is a more efficient IM in predicting LD of the long duration suite. The simple CAV-based prediction equation is showed in Eq. 2, with the $\sigma_{log10(LD)}$ of 0.206.

$$\log_{10}(LD + 0.01 \text{ m}) = 2.43 \times \log_{10}(CAV) - 2.78 \tag{2}$$

in which the LD denotes the liquefaction-induced lateral displacement in unit of m, and CAV is the cumulative absolute velocity of the input ground motions in unit of m/s. Note that CAV is the only predictor variable employed in this model. Our future work will study a more complicated formula for predicting LD of the long-duration records by considering the influence of soil and site parameters (e.g., the thickness of liquefiable layer).

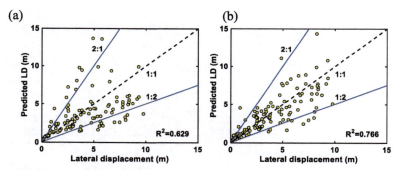

Fig. 6. Comparison between numerical lateral displacement and lateral displacement calculated by regression equation (predicted LD) with (**a**) I_a or (**b**) CAV of long-duration ground motion sets as variable

5 Conclusion

To investigate the influence of the ground motion duration on liquefaction-induced lateral displacement, a one-dimensional sloping ground containing liquefaction layer was built in OpenSees. The slope inclination of the sloping ground was 2%; the relative density and thickness of the liquefiable layer were assigned as 40% and 6 m, respectively. A total of 126 pairs of long-duration and spectrally equivalent short-duration ground motions were selected as input motions for conducting dynamic analyses. Comparative results indicated that the long-duration ground motion suite have a higher potential to yield larger LD than the short-duration suite. Specifically, the median value of LD for the long-duration suite is about 10 times larger than that of the short-duration one. The two evolutionary intensity measures, namely I_a and CAV, were adopted to conduct efficiency evaluation in predicting LDs; it was found that CAV correlated more closely with LD. Finally, a simple CAV-based model is proposed to predict the LDs for the long-duration ground motions.

References

1. Wu, Q., Li, D.-Q., Liu, Y., Du, W.: Seismic performance of earth dams founded on liquefiable soil layer subjected to near-fault pulse-like ground motions. Soil Dyn. Earthq. Eng. **143**, 106623 (2021)
2. Arboleda-Monsalve, L.G., Mercado, J.A., Terzic, V., Mackie, K.R.: Soil–structure interaction effects on seismic performance and earthquake-induced losses in tall buildings. J. Geotech. Geoenviron. Eng. **146**(5), 04020028 (2020)
3. Soga, K.: Soil liquefaction effects observed in the Kobe earthquake of 1995. Proc. Inst. Civil Eng. Geotech. Eng. **131**(1), 34–51 (1998)
4. Araujo, W., Ledezma, C.: Factors that affect liquefaction-induced lateral spreading in large subduction earthquakes. Appl. Sci. **10**(18), 6503 (2020)
5. Greenfield, M., Kramer, S.: The effects of long-duration ground motion on soil liquefaction. In: Eleventh U.S. National Conference on Earthquake Engineering (2018)
6. Bardet, J.-P., Tobita, T., Mace, N., Hu, J.: Regional modeling of liquefaction-induced ground deformation. Earthq. Spectra **18**(1), 19–46 (2002)
7. Chandramohan, R., Baker, J.W., Deierlein, G.G.: Quantifying the influence of ground motion duration on structural collapse capacity using spectrally equivalent records. Earthq. Spectra **32**(2), 927–950 (2016)
8. Fairhurst, M., Bebamzadeh, A., Ventura, C.E.: Effect of ground motion duration on reinforced concrete shear wall buildings. Earthq. Spectra **35**(1), 311–331 (2019)
9. Wang, W., Li, D.-Q., Liu, Y., Du, W.: Influence of ground motion duration on the seismic performance of earth slopes based on numerical analysis. Soil Dyn. Earthq. Eng. **143**, 106595 (2021)
10. Zengin, E., Abrahamson, N.A., Kunnath, S.: Isolating the effect of ground-motion duration on structural damage and collapse of steel frame buildings. Earthq. Spectra **36**(2), 718–740 (2020)
11. Raghunandan, M., Liel, A.B.: Effect of ground motion duration on earthquake-induced structural collapse. Struct. Saf. **41**, 119–133 (2013)
12. Bahrampouri, M., Rodriguez-Marek, A., Shahi, S., Dawood, H.: An updated database for ground motion parameters for KiK-Net records. Earthq. Spectra **37**(1), 505–522 (2020)
13. Haddadi, H., et al.: Report on progress at the center for engineering strong motion data (CESMD). In: 15th World Conference on Earthquake Engineering, Lisbon, Portugal, pp. 19074–19080 (2012)
14. National earthquake data center (NEDC) Homepage. https://data.earthquake.cn/. Accessed 31 Oct 2021
15. Ancheta, T.D., et al.: NGA-West2 database. Earthq. Spectra **30**(3), 989–1005 (2014)
16. Mazzoni, S., McKenna, F., Scott, M.H., Fenves, G.L.: OpenSees Command Language Manual. University of California, Berkely (2006)
17. Yang, Z., Elgamal, A., Parra, E.: Computational model for cyclic mobility and associated shear deformation. J. Geotech. Geoenviron. Eng. **129**(12), 1119–1127 (2003)

18. Lysmer, J., Kuhlemeyer, A.M.: Finite dynamic model for infinite media. J. Eng. Mech. Div. **95**, 859–877 (1969)
19. Bardet, J.P., Liu, F.: Motions of gently sloping ground during earthquakes. J. Geophys. Res. **114**(F2), F02010 (2009)

Performance of Slopes During Earthquake and the Following Rainfall

Jiawei Xu(✉), Kyohei Ueda, and Ryosuke Uzuoka

Disaster Prevention Research Institute, Kyoto University, Gokasho, Uji, Kyoto 611-0011, Japan
{xu.jiawei.4n,ueda.kyohei.2v,uzuoka.ryosuke.6z}@kyoto-u.ac.jp

Abstract. Earthquake is a major cause of landslides in a lot of mountainous areas throughout the world. However, when struck by earthquake, some slopes with certain topographies or geological conditions only experienced a certain loss of stability, resulting in tensile cracks on the surfaces rather than a complete failure. These slopes with shaking-induced cracks become vulnerable to rainfall due to the reduced soil strength and water infiltration. Previous studies also suggested that, rainfall that followed an earthquake could cause a significant failure of the slope with cracks. Therefore, in areas where earthquake and rainfall are common, the prediction and mitigation of slope failure caused by earthquake and rainfall are of great importance, which calls for a thorough investigation into the performance of slopes to earthquake and rainfall. This study aimed to examine the performance of slopes to post-earthquake rainfall through the finite element method. First, the slope deformation caused by earthquake was simulated. Although the generation of cracks were not incorporated, the main features of slope deformation such as the location of slip surface and the deformation scale were reproduced. After that, the effects of permeability change and damage due to shaking were included in the analysis of slope deformation during rainfall. Simulation results showed that, the proposed method performed well to reproduce the slope deformation caused by post-earthquake rainfall.

Keywords: Slopes · Performance · Earthquake · Cracks · Rainfall

1 Introduction

In some areas in the world, slopes with certain topographies or geological conditions have adequate stability under normal static conditions but become unstable when struck by horizontal forces from earthquakes, incurring tension cracks or fissures on their surfaces rather than complete failures, making them vulnerable to the following possible rainfall. Increasing attention is being paid to this phenomenon since it has been reported in the past several earthquakes. For instance, in the 1995 Kobe Earthquake [1], 1999 Chichi Earthquake [2], and 2008 Sichuan Earthquake [3], plenty of slopes did not display failure or slide after completion of earthquakes and only fissures and cracks were found on their surfaces, causing landslides during the next rainy seasons or posing a great risk of slide remobilization or landslides because of possible rainfall later on.

The above findings about the connection between slope failure or landslide with post-earthquake rainfall were mainly supported by satellite or aerial images and the landslide triggering mechanism was still not clarified in detail. It was speculated that post-earthquake rainfall led to the failure of slopes with fissures or cracks, accounting for landslides that entailed casualties and damage to different extents. Given that physical modeling of this kind of slope failure has been conducted [4], this study intends to apply FEM (Finite Element Method) to simulate the response of slopes subjected to post-earthquake rainfall.

2 Constitutive Model

2.1 Stress-Strain Relationship

Slope is usually made of unsaturated soil resting on rock base. Since unsaturated soil is a three-phase material, consisting of solid, water, and air, the constitutive model [5] for describing its hydro-mechanical response is different from that for saturated model. The effective stress for unsaturated soil is

$$\boldsymbol{\sigma}' = \boldsymbol{\sigma} - u_a \boldsymbol{I} + s_w(u_a - u_w)\boldsymbol{I} \tag{1}$$

where $\boldsymbol{\sigma}'$ is the effective stress tensor and $\boldsymbol{\sigma}$ is the total stress tensor; u_a and u_w are the pore air and pore water pressures, with their difference being the matric suction; $\boldsymbol{I}$ is the second-order unit tensor.

The yield surface f is defined by

$$f = \sqrt{3/2}\|\boldsymbol{\eta} - \boldsymbol{a}\| - k = 0 \tag{2}$$

where $\boldsymbol{\eta}$ is the stress ratio between deviator stress $\boldsymbol{s}$ and mean effective stress p'; k is a constant that defines the elastic region. The evolution of the back stress $\boldsymbol{a}$ is described by a non-linear kinematic hardening model

$$\dot{\boldsymbol{a}} = a\left(\frac{2}{3}b\dot{\boldsymbol{e}}^p - \boldsymbol{a}\dot{\varepsilon}_d^p\right) \tag{3}$$

where a and b are material parameters; $\dot{\boldsymbol{e}}^p$ is the plastic deviatoric strain rate tensor; $\dot{\varepsilon}_d^p$ is equal to $\sqrt{2/3}\|\dot{\boldsymbol{e}}^p\|$.

The parameter a evolves as the equivalent strain ε_d^p changes, and their relationship is

$$a = a_0 - \frac{a_0 - a_1}{1 + (a_0 - a_1)\exp(-C_{ref}\varepsilon_d^p)} \tag{4}$$

where a_0, a_1, and C_{ref} are parameters that reproduce the cyclic tests on soil sample through element simulations.

The plastic potential function g has a Cam-clay-type formula written as

$$g = \sqrt{3/2}\|\boldsymbol{\eta} - \boldsymbol{\alpha}\| + M_m \ln(\sigma'/\sigma_0') \tag{5}$$

where M_m is the critical state ratio; σ' is the effective stress and σ'_0 is the value of σ' when $\|\eta - \alpha\| = 0$.

A non-associated flow rule is generalized by

$$\dot{e}^p = 2\lambda_2 \frac{\partial g}{\partial s} \tag{6}$$

$$\dot{\varepsilon}^p_v = (3\lambda_1 + 2\lambda_2) \frac{\partial g}{\partial \sigma'} \tag{7}$$

$$D = \frac{3\lambda_1 + 2\lambda_2}{2\lambda_2} \tag{8}$$

where coefficients a and b are state parameters, and D is the coefficient of dilatancy controlling the ratio of the plastic deviatoric strain increment to the plastic volumetric strain increment.

The evolution rule of dilatancy D [5] is related to suction p^c and satisfies

$$D = \begin{cases} D_1 + (D_0 - D_1) exp\left(-\frac{p^c}{p^c_{\text{ref}}}\right) & p^c > 0 \\ D_0 & p^c \leq 0 \end{cases} \tag{9}$$

where D_0 and D_1 are the initial and minimum dilatancy values; p^c_{ref} is the reference suction that adjusts the dilatancy.

The elastic relationship was described using another two parameters K^e and G^e.

$$K^e = -Kp'; \quad G^e = -Gp' \tag{10}$$

where K^e and G^e are the elastic bulk moduli and K and G are the dimensionless elastic moduli.

2.2 Soil Water Retention

The soil water retention curve describes the relationship between the degree of saturation and suction of unsaturated soil. One widely utilized model is called van Genuchten model [6], with an expression of

$$s_e = [1 + (as)^n]^m \tag{11}$$

where s_s and s_r are the maximum and minimum degrees of saturation; a, n, and m are material parameters, which are equal to 0.42 kPa^{-1}, 2.10, and 0.52 for the unsaturated soil in this study.

The permeability coefficients of unsaturated soil are expressed by

$$k^w_r = k^w_s (s_e)^\xi \left[1 - \left\{1 - (s_e)^{1/m}\right\}^m\right]^2 \tag{12}$$

$$k^a_r = k^a_s (1 - s_e)^\eta \left[1 - (s_e)^{1/m}\right]^{2m} \tag{13}$$

where k_r^w and k_r^a are the relative water and air permeability coefficients and k_s^w and k_s^a are the corresponding saturated values; ξ and η are material parameters, which are equal to 0.5 and 0.33; s_e is the effective degree of saturation, with an expression of

$$s_e = \frac{s_w - s_r}{s_s - s_r} \tag{14}$$

where s_s and s_r are the saturated and residual degrees of saturation, equaling 0.25 and 0.91.

3 Performance of Slope During Earthquake

3.1 Slope Deformation Caused by Earthquake

The numerical model slope is presented in Fig. 1, where the soil is unsaturated, the base is non-deformable, and the void with special water retention features [7] is used for the convenience of smooth water infiltration during rainfall. The finite element model has a total of 503 elements. The horizontal boundary of the model was fixed in terms of both horizontal and vertical displacement; the right-hand side boundary of the model was fixed in terms of vertical displacement. The displacement of all void elements was fixed. The self-weight analysis was conducted before seismic analysis to achieve the initial condition of the simulation. The initial degree of saturation of the soil was 0.35 and void ratio of 0.75. Some basic parameters for soil are given in Table 1.

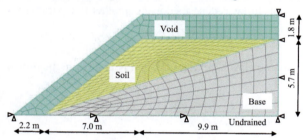

Fig. 1. Finite element slope model

Table 1. Basic parameters of soil for the simulation.

Parameter	Value
Non-dimensional bulk moduli K and G	1500 and 600
Kinematic hardening parameters, a, a_{min}, and a_{max}	1300, 48, and 3000
Kinematic hardening parameters, C_{ref}, p_{ref}^c, and b	7000, 1 kPa, and -1.63
Elastic region, k	0.01
Critical stress ratio, M	1.63
Dilataqncy, D_0 and D_1	0.1 and 0.1

The distribution of effective stress in slope after the self-weight analysis is shown in Fig. 2. The seismic analysis was conducted right after the self-weight analysis.

Fig. 2. Distribution of effective stress in slope after self-weight analysis (unit: kPa)

The input wave to simulate earthquake is given in Fig. 3, which is the same as the one used in centrifuge model test [5].

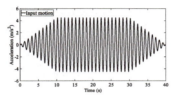

Fig. 3. Input wave simulating earthquake loading

The shaking-induced slope deformation is plotted in Fig. 4 in terms of incremental equivalent and volumetric strains.

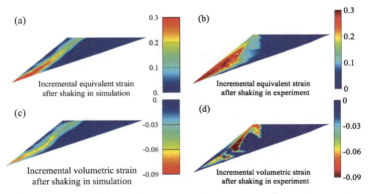

Fig. 4. Incremental equivalent and volumetric strains caused by shaking in simulation and experiment

Despite the inability of the simulation (Fig. 4 a and c) to reproduce the generation of cracks as shown in the crest soil near the slope shoulder in experiment (Fig. 4 b and d), the simulated slope deformation could reproduce the location of the slip surface within the slope caused by the earthquake and the magnitude of deformation was also comparable to the experimental result.

To consider the effect of earthquake on the response of slope to the subsequent rainfall, the influence of deformation should be included, such as the impact of volumetric strain on permeability coefficient reduction of soil in the slip surface and the impact of cracks on the permeability coefficient enhancement of surface soil near the slope shoulder.

3.2 Permeability Change and Damage Due to Earthquake

To include the impact of earthquake on permeability of the soil along the slip surface of the slope, this study adopted the Kozeny–Carman equation to correlate permeability and volumetric strain of the deformed soil caused by seismic loading, and applied the damage concept to consider the strength reduction and the permeability enhancement due to crack due to surface cracks.

The permeability reduction due to volumetric strain is mathematically obtained with reference of the Kozeny–Carman equation below

$$\frac{k'_s}{k_s^w} = \frac{[e_0 + \varepsilon_v(1+e_0)]^3}{e_0^3(1+\varepsilon_v)} \qquad (15)$$

where k'_s is the permeability after deformation; e_0 is the initial void ratio; ε_v is the accumulated volumetric strain after earthquake.

The relationship between permeability of deformed soil and the initial permeability was empirically described using Eq. (15). The distribution of permeability after shaking is then plotted in Fig. 5.

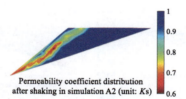

Fig. 5. Permeability distribution in slope due to earthquake-induced volumetric strain

Figure 5 shows the permeability distribution in the slope after earthquake once the Eq. is applied. It is clearly that due to the compression deformation caused by earthquake, the soil along the slip surface exhibits reduction in permeability, with the lowest value 0.76 times the original permeability.

Apart from the permeability reduction effect caused by earthquake, the impact of surface cracks should also be considered. A commonly used method is to adopt the damage concept to indicate the degree of cracks. It is assumed that the permeability enhancement due to cracks will degrade as the depth increases, meaning that the surface soil has the largest increase in permeability and as the depth grows, the enhancement effect gradually reduces. By reference to [8], this study defines the damage m_d as

$$m_d = 1 - \frac{1}{1+e^{-k(h-h_0)}} \qquad (16)$$

where k is parameter controlling the shape of the curve, which is taken as 2; h is the depth of soil with permeability enhancement due to cracks, which is chosen as 1.5 m and h_0 is a constant parameter controlling the location of the turning point of the curve, which is chosen as 0.75 m.

The permeability and friction angle of the surface soil are calculated using the following equations

$$K'_s = \frac{K_s}{(1 - m_d)}; \varphi' = \varphi(1 - m_d) \qquad (17)$$

where K'_s and K_s are the permeability after and before cracks are generated; φ' and φ are friction angle after and before cracks are generated.

The overall distribution of permeability is presented in Fig. 6, where the saturated permeability coefficient of the soil within the depth of 1.5 m has a distribution ranging from 5.5 times the original saturated permeability coefficient to the original permeability according to Eqs. 16 and 17. The soil below the depth of 1.5 m has a distribution of reduced saturated permeability coefficient in the colored figure.

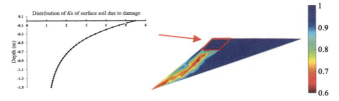

Fig. 6. Permeability distribution in slope due to earthquake

With the updated finite element model with new permeability and strength considering the impacts of volumetric strain and damage, the numerical simulation of the response of the slope subjected to the post-earthquake rainfall was carried out.

4 Performance of Slope During Post-earthquake Rainfall

After earthquake, the rainfall with an intensity of 30 mm/h was turned on and water infiltration started. The flux was given to the top horizontal nodes of the void elements to keep the same infiltration condition as the one in the experiment [4]. As water infiltrated into the soil, seepage was generated and the degree of saturation within the slope began to rise and deformation started to take place. The deformation in terms of incremental equivalent strain caused by post-earthquake rainfall obtained from simulation and experiment are plotted in Fig. 7.

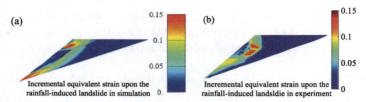

Fig. 7. Deformation caused by post-earthquake rainfall in simulation and experiment (experimental data from [4])

Comparison between the simulated deformation in Fig. 7 a and the experimental deformation in Fig. 7 b shows that, the large deformation near the slope shoulder shaking-induced cracks could be reproduced and the location of slip surface was well simulated. Due to the reduction in soil strength near the shoulder after shaking, rainfall-caused deformation was simulated near the slope shoulder. With the consideration of damage effect around the slope shoulder and permeability change as a result of seismic loading, the slope deformation caused by the following rainfall could be investigated.

5 Conclusions

The performance of slope subjected to earthquake and the following rainfall was investigated using finite element modeling in this study. The slope deformation caused by shaking was reproduced. The shaking-induced crack near the slope crest was not directly simulated but described by adopting the damage concept. The effects of damage and permeability change as a result of seismic loading were considered in the numerical simulation of slope deformation during the subsequent rainfall. Result showed that with the combined effects of damage and permeability change after shaking, the slope deformation caused by rainfall could be largely simulated as the main features such as the deformation magnitude and slip surface location were well reproduced. The approach adopted in this study could be used as a preliminary assessment of slope deformation to the post-earthquake rainfall. Through investigation into other slope responses such as pore water pressure is needed to further evaluate the method.

References

1. Tomita, Y., Sukurai, W., Naka, N.: Study on the extension of collapse caused by rainfall after the earthquake in Rokko Mountain Range. J. Japan Soc. Erosion Control Eng. **48**(6), 15–21 (1996)
2. Lin, C., Liu, S., Lee, S., Liu, C.: Impacts of the Chi-Chi earthquake on subsequent rainfall-induced landslides in central Taiwan. Eng. Geol. **86**(2–3), 87–101 (2006)
3. Tang, C., Zhu, J., Qi, X., Ding, J.: Landslides induced by the Wenchuan earthquake and the subsequent strong rainfall event: a case study in the Beichuan area of China. Eng. Geol. **122**(1–2), 22–33 (2011)
4. Xu, J., Ueda, K., Uzuoka, R.: Evaluation of failure of slopes with shaking-induced cracks in response to rainfall. Landslides **19**(1), 119–136 (2021). https://doi.org/10.1007/s10346-021-01734-1

5. Matsumaru, T., Uzuoka, R.: Three-phase seepage-deformation coupled analysis about unsaturated embankment damaged by earthquake. Int. J. Geomech. **16**(5), C4016006 (2016)
6. van Genuchten, M.T.: A closed-form equation for predicting the hydraulic conductivity of unsaturated soils. Soil Sci. Soc. Am. J. **44**, 892–898 (1980)
7. Mori, T., Uzuoka, R., Chiba, T., Kamiya, K., Kazama, M.: Numerical prediction of seepage and seismic behavior of unsaturated fill slope. Soils Found. **51**(6), 1075–1090 (2011)
8. Gao, Q.F., Zeng, L., Shi, Z.N.: Effects of desiccation cracks and vegetation on the shallow stability of a red clay cut slope under rainfall infiltration. Comput. Geotech. **140**, 104436 (2021)

Seismic Stability Analysis of Anti-dip Bedding Rock Slope Based on Tensile Strength Cut-Off

Qiangshan Yu, Dejian Li, and Yingbin Zhang(✉)

Department of Geotechnical Engineering, School of Civil Engineering,
Southwest Jiaotong University, Chengdu 610031, China
yingbinz719@swjtu.edu.cn

Abstract. For the anti-dip bedding rock slopes (ABRSs), tensile stresses are likely to be generated near the top of the slope, and the results obtained by using the classical Mohr-Coulomb yield criterion are often on the safe side, because the criterion over-estimates the tensile strength of the rock containing joints. In this paper, based on the basic principle of limit analysis upper limit method, a computational failure mode for stability analysis of ABRSs considering tensile strength cut-off is proposed. From the viewpoint of power, the stability coefficient expressions of ABRSs under seismic effects are derived by combining the limit analysis upper limit method and tensile strength cut-off theory. Analysis of the results shows that parameters such as slope angle, horizontal seismic force coefficient, inclination of structural plane and spacing of structural plane have important effects on slope stability coefficient and failure modes.

Keywords: Anti-dip bedding rock slope (ABRS) · Limit analysis · Tensile strength cut-off · Seismic load

1 Introduction

According to the relationship between the inclination of dominant sets of joints and the face of the slope, rock slopes can be divided into two types: bedding rock slopes (BRSs) and anti-dip bedding rock slopes (ABRSs) [1]. Once the ABRS fails, it is usually an instantaneous that breaks through the deep part of the slope [2]. In the past few decades, various methods have been used to study the stability of ABRSs. The limit equilibrium method was first applied to investigate the stability of slopes [3, 4]. Research on ABRSs was not carried out effectively until Goodman [5] divided the failure modes of ABRSs. Most of the current research has investigated the mechanism of ABRSs, which has focused on the effects of gravity, seismic, water and so on [6, 7]. Seismic load is the indispensable factor in ABRSs stability analysis. A pseudo-static approach is commonly used when analyzing the stability of a slope subject to seismic excitation [8]. That is, the problem is simplified by treating the earthquake load as a static load that acts on the centroids of the rock layers. Some scholars [9] studied the dynamic response behavior and mechanisms in the ABRSs using shaking table tests. They found that ABRSs under earthquake load present a failure mode in which tension fracturing occurs at the top. In the terms of the yield of criterion, the yield criterion used in most analyses of slopes is the Mohr-Coulomb criterion. The use of the M-C criterion could lead to an overestimation of the tensile strength of rock slope. There are two main methodologies for considering the impact of tensile strength cut-off. One is to introduce tension cracks [10], and the other is to adjust the mechanism according to the flow rule associated with the strength envelope with tension cut-off [11]. Both these methods are useful for considering the impact of tension cut-off, but the latter is considered in this paper.

In this paper, based on the improved M-C yield criterion considering the tensile strength cut-off and using the work-energy balance equation, a stability coefficient of ABRS is deduced. A parametric study is conducted to investigate the impact of different parameters on the stability of ABRSs. The main contribution of this study is to consider the effects of structural plane parameters and tensile strength truncation on stability of the anti-dip bedding rock slopes.

2 Failure Mode Construction and Stability Calculation

2.1 Failure Mode

Using the limit analysis upper bound method and based on the tensile strength cut-off, the failure model of ABRSs under seismic action is established, as shown in Fig. 1.

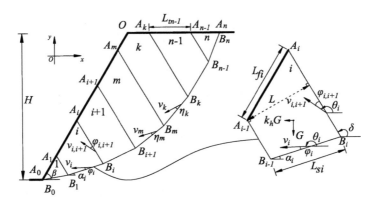

Fig. 1. Failure mode for ABRSs under seismic action

In Fig. 1, the slope height is H, the slope angle is β, the rock weight is γ, and k_h is the horizontal seismic acceleration coefficient. The slope is divided into n slip units, where the number of slip units to the top of the slope O is k. For the ith slip unit, the length of the slip surface at the bottom is $B_{i-1}B_i$, the length of the top is $A_{i-1}A_i$, and the angle between the bottom of the ith slip unit and the horizontal surface is α_i. The length of the structural surface between the ith and the $(i+1)^{th}$ slip units is A_iB_i, and the internal friction angle and cohesion between the structural surfaces are $\varphi_{i,i+1}$, $c_{i,i+1}$. The inclination of the structure plane is δ and the spacing of the structure plane is L.

Calculation of Geometric Size. Assume that the distance of the 1st structural surface closest to the slope angle from the foot of the slope is ζL, where $0 < \zeta < 1$, the distance between the two points where the two adjacent structural surfaces intersect with the slope OA_0 (OB_0) is L_f.

$$L_{f1} = \zeta L / \sin(\delta - \beta) \tag{1}$$

$$L_{fi} = L / \sin(\delta - \beta)(i = 2, ...k - 1), \quad L_{fk} = H / \sin\beta - L_{f1} - (k - 2)L_{f2}, \tag{2}$$

The distance between two points where two adjacent structural surfaces intersect with the slope top surface OA_n (OB_n) is L_t.

$$L_{tk} = (L_{f2} - L_{fk}) \cdot \sin(\delta - \beta) / \sin\delta \tag{3}$$

The distance between two points where two adjacent structural surfaces intersect with the slip surface is L_s.

$$L_{s1} = \zeta L / \sin(\delta - \alpha_i), \quad L_{si} = L / \sin(\delta - \alpha_i)(i = 2, ...n - 1) \tag{4}$$

According to the geometric relationship, the coordinate value of the point B_i (x_{Bi}, y_{Bi}) on the sliding surface can be found as

$$x_{Bi} = x_{B0} + \sum_1^i L_{si} \cos\alpha_i (i = 1, ..., n-1), \quad y_{Bi} = y_{B0} + \sum_1^i L_{si} \sin\alpha_i (i = 1, ..., n-1) \tag{5}$$

$$x_{Bn} = x_{Bn-1} + (-y_{Bn-1})/\tan\alpha_n, \quad y_{Bn} = 0 \tag{6}$$

The coordinates of the point A_i on the slope are

$$x_{Ai} = x_{A0} + \sum_{i=1}^i L_{fi} \cos\beta (i = 1, ..., k-1), \quad y_{Ai} = y_{A0} + \sum_{i=1}^i L_{fi} \sin\beta (i = 1, ..., k-1) \tag{7}$$

The coordinates of the point A_i on the top surface OA_n of the side slope is

$$x_{Ak} = L_{tk}, \quad x_{Ai} = x_{Ak} + \sum_{i=k+1}^i L_{ti} (i = k+1, ...n-1) \tag{8}$$

$$x_{An} = x_{Bn}, \quad y_{Ai} = 0 (i = k, ..., n) \tag{9}$$

where A_0 coincides with B_0 and A_n coincides with B_n.

The length of the bottom surface of the ith block can be expressed as

$$\left|\overrightarrow{B_{i-1}B_i}\right| = \sqrt{(x_{Bi} - x_{Bi-1})^2 + (y_{Bi} - y_{Bi-1})^2} (i = 1, ..., n) \tag{10}$$

Calculation of the Area of the Slide Unit. The area S_i of the ith slip unit can be expressed as

$$S_i = \frac{1}{2}\left(\left|\overrightarrow{A_{i-1}A_i} \times \overrightarrow{A_{i-1}B_{i-1}}\right| + \left|\overrightarrow{B_iA_i} \times \overrightarrow{B_iB_{i-1}}\right|\right)(i = 2, \ldots, n-1) \quad (11)$$

Since the 1st and n^{th} slip units are triangles, their areas can be expressed as

$$S_1 = \frac{1}{2}\left|\overrightarrow{A_0A_1} \times \overrightarrow{A_0B_1}\right|, \quad S_n = \frac{1}{2}\left|\overrightarrow{A_{n-1}A_n} \times \overrightarrow{B_{n-1}A_n}\right| \quad (12)$$

Since the kth block is pentagonal, its area S_k can be expressed as

$$S_k = \frac{1}{2}\left(\left|\overrightarrow{OA_{k-1}} \times \overrightarrow{OA_k}\right| + \left|\overrightarrow{A_{k-1}A_k} \times \overrightarrow{A_{k-1}B_{k-1}}\right| + \left|\overrightarrow{B_{k-1}B_k} \times \overrightarrow{B_kA_k}\right|\right) \quad (13)$$

Calculation of Velocity Field

Calculation of the Shear Area Velocity Field. Considering the coordination of deformation between adjacent slip units, whose velocity polygons are to be vectorially closed, the velocity field is shown in Fig. 2.

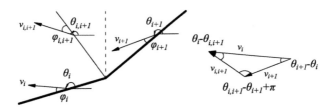

Fig. 2. Schematic diagram of velocity compatibility between slide units

The correlation joint flow law requires that each V_i is at an angle φ_i with the bottom slip surface and requires that the relative velocity $V_{i,i+1}$ of the structural surface of the adjacent slip unit is at an angle $\varphi_{i,i+1}$ with the structural surface. According to Fig. 2, it can be deduced that

$$\frac{V_{i+1}}{V_i} = \frac{\sin(\theta_i - \theta_{i,i+1})}{\sin(\theta_{i+1} - \theta_{i,i+1})}, \quad \frac{V_{i,i+1}}{V_i} = \frac{\sin(\theta_{i+1} - \theta_i)}{\sin(\theta_{i+1} - \theta_{i,i+1})} \quad (14)$$

Further, it can be calculated that

$$V_{i+1} = \prod \frac{\sin(\theta_i - \theta_{i,i+1})}{\sin(\theta_{i+1} - \theta_{i,i+1})} V_1, \quad V_{i,i+1} = \frac{\sin(\theta_{i+1} - \theta_i)}{\sin(\theta_{i+1} - \theta_{i,i+1})} \prod \frac{\sin(\theta_i - \theta_{i,i+1})}{\sin(\theta_{i+1} - \theta_{i,i+1})} V_1 \quad (15)$$

In Eq. (14) and Eq. (15), θ_i, θ_{i+1}, $\theta_{i,i+1}$ are the angles of V_i, V_{i+1}, $V_{i,i+1}$ with the x-axis, respectively, defined as the positive start of the x-axis, with counterclockwise rotation as positive, and with $0 \leq \theta \leq 2\pi$. The specific derivation process is described in [12].

2.2 Stability Calculation

The Principle of Virtual Power. According to the principle of limit analysis upper limit method, when the slope is in the limit state the work power of external force, $W_{external}$, is equal to the internal energy dissipation power, $W_{internal}$.

$$W_{external} = \sum_{i=1}^{m-1} \gamma S_i V_i \sin(\alpha_i - \varphi_i) + \sum_{i=m}^{n} \gamma S_i V_i \sin(\alpha_i - \eta_i)$$
$$+ \sum_{i=1}^{m-1} k_h \gamma S_i V_i \cos(\alpha_i - \varphi_i) + \sum_{i=m}^{n} k_h \gamma S_i V_i \cos(\alpha_i - \eta_i) \quad (16)$$

$$W_{internal} = \sum_{i=1}^{m-1} c_i V_i l_{si} \cos \varphi_i + \sum_{i=m}^{n} c_i v_i l_{si} \left(\cos \varphi \frac{1 - \sin \eta_i}{1 - \sin \varphi} + 2\mu_i \frac{\sin \eta_i - \sin \varphi}{\cos \varphi} \right)$$
$$+ \sum_{i=1}^{n-1} c_i V_{i,i+1} \left| \overrightarrow{A_i B_i} \right| \cos \varphi_{i,i+1} \quad (17)$$

Optimal Calculation of Stability Coefficient. From Eq. (16) and Eq. (17), it is deduced that

$$\frac{\gamma H}{c} = \frac{H \left[\sum_{i=1}^{m-1} V_i l_{si} \cos \varphi_i + \sum_{i=m}^{n} v_i l_{si} \left(\cos \varphi \frac{1-\sin \eta_i}{1-\sin \varphi} + 2\mu_i \frac{\sin \eta_i - \sin \varphi}{\cos \varphi} \right) + \sum_{i=1}^{n-1} V_{i,i+1} \left| \overrightarrow{A_i B_i} \right| \cos \varphi_{i,i+1} \right]}{\sum_{i=1}^{m-1} S_i V_i \sin(\alpha_i - \varphi_i) + \sum_{i=m}^{n} S_i V_i \sin(\alpha_i - \eta_i) + \sum_{i=1}^{m-1} k_h S_i V_i \cos(\alpha_i - \varphi_i) + \sum_{i=m}^{n} k_h S_i V_i \cos(\alpha_i - \eta_i)}$$
$$(18)$$

3 Case Analysis

3.1 Analysis of the Effect of Stability Coefficient

The Effect of k_h and φ on $\gamma H/c$. $H = 50$ m, $\gamma = 25$ kN/m^3, $c = 50$ kN/m^3, $\delta = 120°$, $\eta = 25°$, $\mu = 0.25$, $L = 2$ m. k_h is taken as 0.0, 0.1, 0.3, 0.5, φ is taken as 10°, 15°, 20°, 25°, 30°, $\beta = 45°$–90°, and the influences of k_h and φ on the $\gamma H/c$ is analyzed.

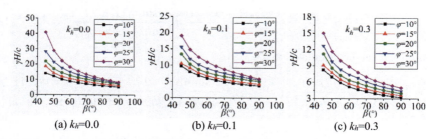

Fig. 3. Variation curves of $\gamma H/c$ with β for different k_h and φ.

In Fig. 3 the $\gamma H/c$ decreases sharply with the increase of β. When the β and the k_h are determined, $\gamma H/c$ increases with the increase of φ. When β and φ are determined, $\gamma H/c$ decreases with the increase of k_h.

The Effect of β and L on $\gamma H/c$. $H = 50$ m, $\gamma = 25$ kN/m^3, $c = 50$ kN/m^3, $\varphi = 20°$, $k_h = 0.2$, $\eta = 25°$, $\mu = 0.5$. The β is taken as 60°, 70°, 80°, 90°, and the L is taken as 2 m, 4 m, 6 m, $\delta = 110°$–160°, respectively, to analyze the influence of β and L on the $\gamma H/c$.

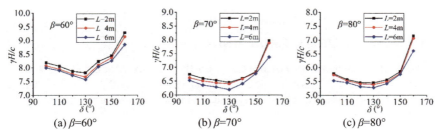

Fig. 4. The variation curves of $\gamma H/c$ with δ for different β and L.

In Fig. 4, the $\gamma H/c$ decreases first and then increases with the δ, and when $\delta = 130°$, the $\gamma H/c$ is the smallest, indicating that 130° is the most unfavorable inclination angle of the structure surface; when the β and the δ are determined the $\gamma H/c$ decreases with the L; when δ and L are determined $\gamma H/c$ decreases with the increase of the β.

The Effect of δ and μ on $\gamma H/c$. $H = 50$ m, $\gamma = 25$ kN/m^3, $c = 50$ kN/m^3, $k_h = 0.2$, $\varphi = 20°$, $\eta = 50°$, $L = 2$ m. The δ is taken as 100°, 120°, 140°, 160°, and the μ is taken as 0.00, 0.25, 0.50, 0.75, 1.00, $\beta = 45°$–90°, to analyze the influence of δ and μ on the $\gamma H/c$.

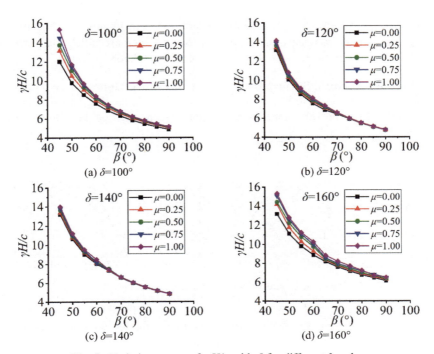

Fig. 5. Variation curves of $\gamma H/c$ with β for different δ and μ

In Fig. 5, the $\gamma H/c$ decreases sharply with the increase of β; when the β and the δ are determined, the $\gamma H/c$ increases with the increase of the μ. When the β gradually increases to 90°, the influence of μ on the $\gamma H/c$ shows a decreasing trend; the δ has an influence on the $\gamma H/c$.

3.2 Analysis of the Effect of Failure Modes

The Effect of Dimensionless Parameter μ on the Failure Modes. $H = 50$ m, $\gamma = 25$ kN/m³, $\beta = 60°$, $\delta = 130°$, $c = 50$ kN/m³, $\varphi = 20°$, $k_h = 0.2$, $\eta = 25°$. The μ is taken as 0.25, 0.50 and 1.00, respectively, and the influence of μ on the failure modes is analyzed.

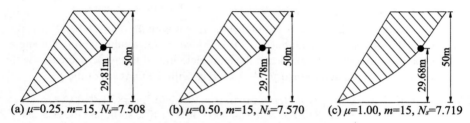

(a) μ=0.25, m=15, N_s=7.508   (b) μ=0.50, m=15, N_s=7.570   (c) μ=1.00, m=15, N_s=7.719

Fig. 6. The effect of dimensionless parameter μ on the failure modes.

In Fig. 6 the effect of μ on the failure modes is not significant, with the increase of μ, the transition point between shear zone and tension zone slowly moves to the bottom of the slope, so with the increase of the μ, the tension zone increases relatively gradually; the $\gamma H/c$ increases gradually with the increase of μ, from 7.508 to 7.719.

The Effect of Structural Surface Inclination δ on Failure Modes. $H = 50$ m, $\gamma = 25$ kN/m³, $\beta = 60°$, $c = 50$ kN/m³, $\varphi = 20°$, $k_h = 0.2$, $\eta = 25°$, $\mu = 0.5$. The δ is taken as 110°, 130° and 150° respectively, and the influence of δ on the failure modes is analyzed.

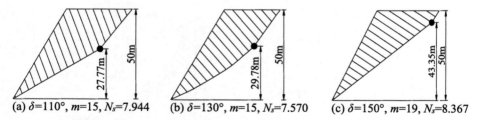

(a) δ=110°, m=15, N_s=7.944   (b) δ=130°, m=15, N_s=7.570   (c) δ=150°, m=19, N_s=8.367

Fig. 7. The effect of structural surface inclination δ on failure modes.

In Fig. 7, the δ has a greater influence on the failure modes, with the increase of δ, the transition point between the shear area and the tension area moves to the top of the slope, so with the increase of the δ, the tension area increases relatively gradually; the $\gamma H/c$ decreases first and then increases with the increase of the δ, and when $\delta = 130°$, the $\gamma H/c$ of the slope is the smallest.

The Effect of Structural Surface Spacing L on Failure Modes. $H = 50$ m, $\gamma = 25$ kN/m^3, $\beta = 60°$, $\delta = 130°$, $c = 50$ kN/m^3, $\varphi = 20°$, $k_h = 0.2$, $\eta = 25°$, $\mu = 0.5$. The L is taken as 2 m, 4 m and 6 m respectively, and the influence of L on the failure modes is analyzed.

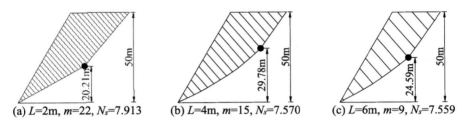

(a) L=2m, m=22, N_s=7.913   (b) L=4m, m=15, N_s=7.570   (c) L=6m, m=9, N_s=7.559

Fig. 8. The effect of structural surface spacing L on failure modes.

In Fig. 8, the L has a greater impact on the failure modes, with the increase of L, the transition point between the shear zone and the tension zone moves up first and then down, so with the increase of the L, the tension zone gradually increases relatively first and then decreases; the $\gamma H/c$ decreases with the increase of L, from 7.913 to 7.559.

4 Conclusion

Based on the limit analysis upper bound method and combined with the tensile strength cut-off theory, the ABRSs are analyzed and the corresponding calculation mode is proposed. Through the imaginary power equation of limit analysis, the formula of stability coefficient of ABRSs is deduced.

The analysis of calculation cases shows that parameters such as slope angle, horizontal seismic force coefficient, tensile strength cut-off coefficient, inclination of structural plane, and spacing of structural plane have effects on slope stability coefficient, and summarizes its change law. In addition, the effects of the above parameters on the slope failure modes are also analyzed by the calculation case.

References

1. Liu, T.T.: Stability analysis of anti-dip bedding rock slopes using a limit equilibrium model combined with bi-directional evolutionary structural optimization (BESO) method. Comput. Geotech. **134**, 104116 (2021)
2. Huang, R.Q.: Formation, distribution and risk control of landslides in China. J. Rock Mech. Geotech. Eng. **2011**(2), 97–116 (2012)

3. Bishop, A.W.: The use of the slip circle in the stability analysis of slopes. Geotechnique **5**(1), 7–17 (1955)
4. Sarma, S.K.: Stability analysis of embankments and slopes. Geotechnique **23**(3), 423–433 (1973)
5. Goodman, R.E.: Toppling of rock slopes. In: Speciality Conference on Rock Engineering for Foundations and Slopes (1976)
6. Amini, M.: Stability analysis of rock slopes against block-flexure toppling failure. Rock Mech. Rock Eng. **45**(4), 519–532 (2012)
7. Che, A.L.: Wave propagations through jointed rock masses and their effects on the stability of slopes. Eng. Geol. **201**, 45–56 (2016)
8. Baker, R.: Stability charts for pseudo-static slope stability analysis. Soil Dyn. Earthq. Eng. **26**(9), 813–823 (2006)
9. Huang, R.Q.: Shaking table test on strong earthquake response of stratified rock slopes. Chin. J. Rock Mech. Eng. **32**(5), 865–875 (2013)
10. Michalowski, R.L.: Stability assessment of slopes with cracks using limit analysis. Can. Geotech. J. **50**(10), 1011–1021 (2013)
11. Utili, S.: Investigation by limit analysis on the stability of slopes with cracks. Geotechnique **63**(2), 140–154 (2013)
12. Tang, G.P.: Inclined slices method for limit analysis of slope stability with nonlinear failure criterion. Rock Soil Mech. **36**(7), 2063–2072 (2015)

Seismic Stability Analysis of High Steep Slopes Considering Spatial Variability of Geo-Materials Based on Pseudo Static Method

Wengang Zhang[1,2,3(✉)], Fansheng Meng[1], Jianxin Li[4], and Changjie He[5]

[1] School of Civil Engineering, Chongqing University, Chongqing 400045, China
zhangwg@cqu.edu.cn
[2] Key Laboratory of MOE of New Technology for Construction of Cities in Mountain Area, Chongqing University, Chongqing 400045, China
[3] National Local Joint Engineering Research Center of Geohazards Prevention in the Reservoir Areas, Chongqing University, Chongqing 400045, China
[4] 3rd Construction Co., Ltd. of China Construction 5th Engineering Bureau, Changsha 410004, China
[5] China Construction Fifth Engineering Division Corp., Ltd., Changsha 410004, China

Abstract. In recent years, with development of underground space and further emphasis on the ecological environment protection, some engineering projects aimed at the ecological restoration of underground abandoned mines have emerged. The Changsha Xiangjiang Happy City project is based on an abandoned mine pit, which was excavated due to quarrying, and the loadings from the super-structure acts directly on the rock mass of the pit wall. Therefore, the stability of the high steep slope of the mine is of great importance for safety of the construction projects. Considering the potential earthquake action and the inherent spatial variability of physical and mechanical parameters of rock and soil, the stability analysis of the high steep slope of the mine under self-weight from the super-structuresis carried out. The pseudo-static method is used to simulate the earthquake action, and the random field method is adopted to characterize the spatial variability of the cohesion c and the internal friction angle φ. The results show that the seismic action significantly affects the stability of the slope, and the failure probability of the slope gradually increases with increase of seismic coefficient. The influence of spatial variability on slope failure probability cannot be ignored, and neglecting spatial variability may lead to overestimation of failure probability.

Keywords: Abandoned mine pit · Slope stability · Spatial variability · Pseudo static method · Random field

1 Introduction

While mining and digging out mineral resources, a large number of mines have been left behind. The high-steep side slope formed by the abandoned mine pit, not only destroys the original ground stress balance and produces stability problems, but also affects the esthetics of the ecological environment. Nowadays, with advancement of urbanization, China's construction land is becoming more and more limited, the development of underground space has been paid more and more attention, and the abandoned industrial and mining areas left over by history have also become the key reconstruction objects. Reasonable planning, ecological restoration and development of abandoned mines have important socio-economic value for commitment to building a resource-saving and environment-friendly society.

Slope stability analysis has always been a key topic in the field of geotechnical engineering [1, 2]. In the process of reconstruction and development of deep waste pit, slope stability will directly affect the safety of construction, so it is of great significance to analyze the stability of mine high and steep slope. Slope stability will be affected by a variety of uncertain factors, including the inherent spatial variability of rock and soil parameters. Spatial variability means that the rock and soil exhibit different properties at different spatial locations due to the different depositional conditions and physical and chemical effects experienced during the historical process of formation. Existing studies have shown that neglecting the influence of the spatial variability of the physical and mechanical parameters of rock and soil will cause the calculation results to deviate from the objective reality [3]. Based on the probability density evolution method for slope dynamic reliability analysis, Huang et al. considered the spatial variability of soil parameters by simulating the relevant non normal random field, and found that the failure probability increased with the increase of correlation coefficient between soil parameters [4].

The spatial variability of rock and soil parameters is one of the internal factors affecting stability, while earthquake is one of the main external factors. The commonly used methods for slope stability analysis under earthquake include quasi-static method, Newmark sliding block method, probability method, etc. As an effective and easy to operate seismic effect analysis method, quasi-static method has been widely used in the engineering field and has been incorporated into the corresponding specifications. Liu et al. analyzed the stability of the Dagangshan dam abutment slope based on the quasi-static method, and pointed out the variation law of the key point displacement of the slope with the height of the slope under geological conditions of different sections and different intensities [5].

In view of this, based on the Changsha Xiangjiang Happy City project, this paper aims at the stability of high and steep slopes in the reconstruction of deep waste mines, and comprehensively considers the spatial variability of the physical and mechanical parameters of the rock and soil and the effects of earthquakes. The pseudo-static method is used to simulate the earthquake action, the random field method is adopted to characterize the spatial variability, and the Morgenstern-Price method is adopted to calculate and analyze the slope stability. Through extensive parameter analysis of the proposed calculation framework, the conclusions obtained can provide certain reference value for the analysis of slope seismic stability of the Changsha Xiangjiang Happy City Project.

2 Numerical Model and Methods

2.1 Numerical Model

The original landform of the Changsha Xiangjiang Happy City project is the alluvial accumulation terraces of the Xiangjiang River. Later, artificial quarrying has formed an approximately elliptical rocky pit with a soil layer covering the top of the pit. According to the geological survey report, select typical sections for numerical modeling. In order to facilitate the calculation and analysis, the slope shape is simplified as assuming that the various layers are approximately horizontally distributed. The established mine slope model is shown in Fig. 1. Considering the heavy load of the upper building structure borne by the pit wall, a concentrated force of 500 kN/m is simulated for each step of the slope. The Mohr-Coulomb elastoplastic constitutive model is adopted to characterize the properties of rock and soil. Through survey reports and specifications, the rock and soil parameters of each layer are determined as shown in Tables 1 and 2, among which the statistical parameters are derived from published literature [6, 7].

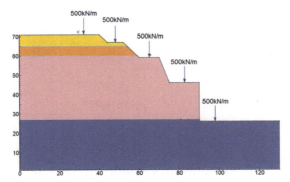

Fig. 1. Slope calculation model.

Table 1. Statistical mechanical parameters of soil layer

Layer	Parameters	Mean value	COV	Distribution	Scale of fluctuation	Correlation coefficient
Miscellaneous fill	Cohesion $c/$ (kPa)	8	0.3	Lognormal	$\delta_h = 20$ m $\delta_v = 2$ m	$\rho_{c,\varphi} = -0.5$
	Friction angle $\varphi/(°)$	10	0.2	Lognormal	$\delta_h = 20$ m $\delta_v = 2$ m	
	Unit weight $\gamma/(\text{kN/m}^3)$	18.8	–	–	–	–

(continued)

Table 1. (*continued*)

Layer	Parameters	Mean value	COV	Distribution	Scale of fluctuation	Correlation coefficient
Silty clay	Cohesion $c/$ (kPa)	50	0.3	Lognormal	$\delta_h = 20$ m $\delta_v = 2$ m	$\rho_{c,\varphi} = -0.5$
	Friction angle $\varphi/(°)$	19	0.2	Lognormal	$\delta_h = 20$ m $\delta_v = 2$ m	
	Unit weight $\gamma/(\text{kN/m}^3)$	19	–	–	–	–

Table 2. Statistical mechanical parameters of rock layer

Layer	Parameters	Mean value	COV	Distribution	Scale of fluctuation	Correlation coefficient
Limestone IV	Cohesion $c/$ (kPa)	450	0.3	Lognormal	$\delta_h = 40$ m $\delta_v = 4$ m	$\rho_{c,\varphi} = -0.5$
	Friction angle $\varphi/(°)$	33	0.2	Lognormal	$\delta_h = 40$ m $\delta_v = 4$ m	
	Unit weight $\gamma/(\text{kN/m}^3)$	24.5	–	–	–	–
Limestone III	Cohesion $c/$ (kPa)	1100	0.3	Lognormal	$\delta_h = 40$ m $\delta_v = 4$ m	$\rho_{c,\varphi} = -0.5$
	Friction angle $\varphi/(°)$	44.5	0.2	Lognormal	$\delta_h = 40$ m $\delta_v = 4$ m	
	Unit weight $\gamma/(\text{kN/m}^3)$	26.5	–	–	–	–

2.2 Stability Analysis Method

Morgenstern-Price Method. The calculation of the safety factor is the focus of slope stability analysis. Morgenstern-Price method is a strict section method, which does not require the shape of the sliding surface, and is more suitable for the sliding analysis of rock slopes. This method satisfies the balance of forces and moments at the same time, so it is generally considered that the calculation result is more accurate. Therefore,

Morgenstern-Price method is selected for slope stability analysis in this paper. Based on SLIDE software and Morgenstern price method, the relationship between vertical force and lateral force is assumed to be $Y = \lambda f(x)X$. Using the balance condition of the bending moment and the horizontal force, the safety factor of the corresponding sliding surface can be obtained by iterative calculation. The expression is as follows [8]:

$$FS = \frac{\sum(c\Delta LR + RN\tan\varphi)}{\sum WL_W - \sum NL_N} \qquad (1)$$

$$FS = \frac{\sum(c\Delta LR\cos\alpha + RN\tan\varphi\cos\alpha)}{\sum N\sin\alpha} \qquad (2)$$

$$N = \frac{W + \lambda f(x)\left(\frac{c\Delta L\cos\alpha}{FS}\right) - \frac{c\Delta L\sin\alpha}{FS}}{\left(\cos\alpha + \frac{\sin\alpha\tan\varphi}{FS}\right) - \lambda f(x)\left(\frac{\cos\alpha\tan\varphi}{FS} - \sin\alpha\right)} \qquad (3)$$

where λ is the variation coefficient of force between soil strips; $f(x)$ is the variation function of force between soil strips; ΔL is the length of the soil strip on the sliding surface; L_W is the length of the force arm from the centroid of the soil strip to the center of the sliding surface; L_N is the distance from the midpoint of the soil strip at the sliding surface to the corresponding normal; α is the included angle between the tangent line of the soil strip and the horizontal plane; R is the length of the moment arm of the circle center; N is the normal force of the sliding surface on the soil strip.

Pseudo-static Method. Since Terzaghi [9] first applied the pseudo-static method to slope seismic stability analysis, the pseudo-static method has been widely adopted and incorporated into the corresponding codes. The pseudo-static method is chosen in this paper because of its simple theory and convenient operation. The principle of pseudo-static method is to simulate the seismic action as a constant acceleration acting in the horizontal and vertical directions, and apply the acceleration to the center of gravity of potentially unstable sliding body. Numerically, the pseudo-static method transforms the seismic action into seismic coefficients in the horizontal and vertical directions. The seismic coefficient represents the magnitude of the earthquake and its value is equal to the ratio of horizontal or vertical acceleration to gravitational acceleration. The horizontal seismic coefficient and seismic force are defined as:

$$k_h = \frac{PGA}{g} \qquad (4)$$

$$F_i = k_h \cdot G_i \qquad (5)$$

where PGA is the ground peak acceleration, and G_i is the gravity of the slider.

Random Field Method. Vanmarcke [10] first proposed the random field method to simulate the spatial variability of rock and soil parameters. Compared with the traditional random variable model, the random field model regards the rock and soil parameters as a random function related to the spatial coordinate position, which has autocorrelation structural characteristics. Therefore, it can more reasonably simulate the spatial distribution of rock and soil parameters. The random field is simulated by two-dimensional Markov theory autocorrelation function, and the expression is as follows:

$$R(\tau_x, \tau_y) = \exp\left\{-\sqrt{\left(\frac{2\tau_x}{\delta_h}\right)^2 + \left(\frac{2\tau_y}{\delta_v}\right)^2}\right\} \tag{6}$$

where $R(\tau_x, \tau_y)$ is the autocorrelation function, τ_x and τ_y are the relative distances in the horizontal and vertical directions of any two points in the spatial position, respectively. δ_h and δ_v are the horizontal and vertical Scale of fluctuation of any two points in the spatial position, respectively.

3 Result Analysis

3.1 Influence of Spatial Variability

The horizontal seismic coefficient is set to 0.2. By comparing the random variable model and the random field model, Fig. 2 shows the influence of spatial variability on the failure probability of slope under earthquake. For the convenience of comparison, make the coefficient of variation of c and φ equal, i.e. $\text{COV}_c = \text{COV}_\varphi$, with a variation range of 0.1–0.5. The other statistical parameters of the random field model are shown in Tables 1 and 2. The results show that whether it is a random variable model or a random field model, with the increase of the COV_c and COV_φ, the failure probability shows an increasing trend. For any coefficient of variation, the failure probability calculated by random field model is smaller than that calculated by random variable model, indicating that spatial variability has a significant impact on the stability of deep waste pit slope. If the spatial variability characteristics of rock and soil are ignored, the failure probability of slope will be overestimated.

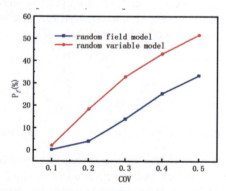

Fig. 2. Influence of spatial variability on failure probability.

Influence of Coefficient of Variation. Figure 3 shows the influence of the variation degree of cohesion c and internal friction angle φ on the slope failure probability, and the variation range of the coefficient of variation of c and φ is 0.1–0.5, respectively. Studies [6, 7] have shown that the degree of variation of φ in rock and soil parameters is generally lower than that of c. Therefore, the coefficient of variation of φ is taken as a fixed value of 0.2 when considering the variability of c, and the coefficient of variation of c is equal to 0.3 when analyzing the variability of φ. The results show that the COV_c and COV_φ have an effect on the failure probability, and the failure probability increases with the increase of the COV_c or COV_φ. However, compared with c, the failure probability is more affected by the coefficient of variation of φ. Therefore, more attention should be paid to the influence of variability in slope stability analysis.

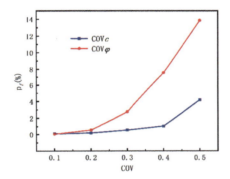

Fig. 3. Influence of COV_c and COV_φ on failure probability.

3.2 Influence of Seismic Effect

By setting the horizontal seismic coefficient k_h to vary from 0.1 to 0.3, the influence of k_h on the stability of spatially variable slope under different earthquake levels is explored. The vertical seismic coefficient k_v is set to half of the k_h, and the influence of k_v on the slope stability is studied by comparing the working conditions of whether the k_v exists. The statistical parameters of the slope rock and soil are shown in Tables 1 and 2, and the obtained results are shown in Fig. 4. The results show that k_h has a significant effect on the slope failure probability, and the slope failure probability increases with the increase of k_h. Make the vertical seismic coefficient is half of the horizontal seismic coefficient, and the failure probability on the contrary decreases after the vertical earthquake is superimposed. This shows that the slope stability is improved when considering k_v, and ignoring the k_v will lead to overestimation the failure probability.

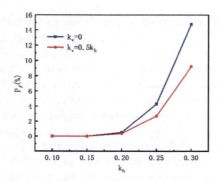

Fig. 4. Influence of seismic on failure probability

4 Summary and Conclusions

Via adopting quasi-static method, characterizing spatial variability by random field method, and calculating slope stability by Morgenstern price method, the effects of spatial variability and the seismic effect on stability of high steep mine slope are analyzed. The conclusions are as follows:

(1) The spatial variability of rock and soil parameters has a significant impact on slope stability, and ignoring it may lead to overestimation of failure probability.
(2) Compared with cohesion, the failure probability is more significantly affected by the coefficient of variation of the internal friction angle. Therefore, special attention should be paid to its accuracy when analyzing slope stability.
(3) The slope failure probability increases with the increase of the horizontal seismic coefficient. The vertical seismic coefficient also has an important influence on the failure probability, and ignoring the vertical seismic coefficient will also lead to an overestimation of the slope failure probability.

References

1. Wang, L.Q., Yin, Y.P., Huang, B.L., et al.: Damage evolution and stability analysis of the Jianchuandong dangerous rock mass in the Three Gorges Reservoir area. Eng. Geol. **265**, 105439 (2020)
2. Zhang, W.G., Meng, F.S., Chen, F.Y., et al.: Effects of spatial variability of weak layer and seismic randomness on rock slope stability and reliability analysis. Soil Dyn. Earthq. Eng. **146**, 106735 (2021)
3. Cho, S.E.: Effects of spatial variability of soil properties on slope stability. Eng. Geol. **92**(3), 97–109 (2007)
4. Huang, Y., Zhao, L., Li, X.R.: Slope-dynamic reliability analysis considering spatial variability of soil parameters. Int. J. Geomech. **20**(6), 04020068 (2020)
5. Liu, J., Li, J.L., Zhang, Y.D., et al.: Stability analysis of Dagangshan dam abutment slope under earthquake based on pseudo-static method. Chin. J. Rock Mech. Eng. **28**(8), 1562–1570 (2009)

6. Lv, Q., Xiao, Z.P., Zheng, J., et al.: Probabilistic assessment of tunnel convergence considering spatial variability in rock mass properties using interpolated autocorrelation and response surface method. Geosci. Front. **9**(6), 1619–1629 (2018)
7. Jiang, S.H., Wei, B.W., Yao, C., et al.: Reliability analysis of soil slopes at low-probability levels considering effect of probability distributions. Chin. J. Geotech. Eng. **38**(6), 1071–1080 (2016)
8. Zhang, L.Z., Qu, J.W., Deng, X.B., et al.: Research on the application about limit equilibrium of Morgenstern-Price method and finite element analysis of ABAQUS method in slope stability analysis. J. Chongqing Univ. Technol. Nat. Sci. **27**(6), 23–27 (2013)
9. Terzaghi, K.: Mechanism of landslides, application of geology to engineering practice (Berkey Volume). In: Geological Society of America, pp. 83–123 (1950)
10. Vanmarcke, E.H.: Probabilistic modeling of soil profiles. J. Geotech. Eng. Div. **103**(11), 1227–2124 (1997)

Liquefaction and Testing

Urban Scale Fragility Assessment of Structures Considering Soil-Structure-Interaction

C. Amendola(✉) and D. Pitilakis

Department of Civil Engineering, Aristotle University of Thessaloniki, Thessaloniki, Greece
{chiaamen,dpitilakis}@civil.auth.gr

Abstract. Fragility curves for structures are typically calculated considering fixed-base structures, i.e., neglecting the interaction between soil, foundation, and structure (SFSI). The state-of-the-art literature proves that considering foundation flexibility, especially for structures resting on soft soil, may lead to different fragility or loss estimates than the fixed-base-on-rock assumption. Including these effects on the city-scale vulnerability analysis is considered a challenging task due to the high exposure concentration and complexity of all the interacting urban systems. For this reason, large-scale analyses are commonly carried out applying existing fragility curves which may have been assessed not correctly accounting for the variation in frequency and amplitude contents imposed by each site's local geotechnical and topographic conditions. To this aim, a new simplified methodology is proposed in this study to perform an urban-scale vulnerability assessment of structures considering the influence of SFSI and local site-effects. The applicability of the proposed approach is based on globally available data regarding the soil parameters, the foundation, and the building taxonomy. The main findings demonstrate that the conventional way of calculating fragility curves may lead to an incorrect evaluation of the seismic risk, especially in soft soil formations.

Keywords: Soil-foundation-structure-interaction · Site-effects · Fragility curves · City-scale risk analysis

1 Introduction

The seismic risk framework requires the definition of fragility curves which define the probability of exceedance of a predefined limit state. The complexity related to the characterization of the soil-foundation system and the common belief in the beneficial effects associated with the interaction between the soil, foundation, and structure led over the years to develop fragility functions considering fixed-base structures. Despite this, a series of effective research attempts [1–7] recognized the modification of the fragility functions of structures founded on soft soil with respect to the typical fixed-base assumption. For a thorough overview of up-to-date literature on fragility curves, including SFSI, we refer the reader to the recent report published by [8]. These studies reveal that the shift of fragility functions or, more in general of the demand with respect to the fixed-base reference case is expected to be significant in deformable soil conditions, leading to either beneficial or unfavorable effects, depending on the dynamic properties of the soil, the foundation, the structure and the characteristics (frequency content, amplitude, significant duration) of the input motion. Even though the results of such studies provided the scientific community with valuable knowledge at site-specific vulnerability assessment, the reliability at a large scale is assessed with certain limitations. Indeed, despite all the previous investigations and efforts, not all the possible SFSI scenarios have been covered so far, making the existing fragility functions accounting for SFSI inadequate for large-scale analyses. Moreover, for the extensive applications, soil conditions may be considered in the seismic hazard assessment adopting either code- or research-based amplification factors or through very advanced models (physical-numerical simulations, microzonation analysis). They are not correctly considered in the fragility curves. Thus, large-scale analyses are commonly carried out to reduce the computation effort applying existing fragility curves [9]. If existing fragility curves are analytically derived, they may have been assessed applying records not accounting for variations in frequency and amplitude contents imposed by each site's local geotechnical and topographic conditions. To this aim, we propose a new methodological framework for developing generalized fragility functions considering SFSI and local site effects applicable to different reinforced concrete and masonry buildings for a great variety of soil-foundation systems encountered in the urban environment. To this aim, the present study intends to start filling the existing gap by developing a systematic methodology for the estimation of fragility curves of different classes of buildings considering SFSI and local site effects, applicable to different reinforced concrete and masonry buildings for a great variety of soil-foundation systems that can be encountered in an urban environment.

2 Proposed Framework for the Development of Fragility Curves Including SFSI and Site Effects

This section aims to propose and quantify an analytical methodology to assess the fragility functions for different building classes founded on shallow or embedded foundations taking into account SFSI and site-effects [10]. Figure 1 summarizes the main steps of the methodological framework. All the analyses are conceived to be implemented in the open-source OpenSees software [11]. To formally consider the aleatoric uncertainties related to the so-called record-to-record variability, a large set of input ground motions recorded on outcropping rock/firm-soil is selected to perform all the dynamic cloud analyses [12]. In order to avoid developing procedures that are excessively structure-site-foundation-dependent, the problem is decoupled through the substructure approach. Accordingly, the modification of the selected records due to the local site-effects is quantified by performing 1D numerical simulations of seismic site response performed on virtual stratigraphic profiles conceived to pertain to different soil types. Following the substructure approach, for structures resting on surface foundation the free-field motion is directly applied as input for the dynamic analyses. While for embedded foundations, the free-field motion is further modified to consider the kinematic interaction before being applied at the base of the superstructure. The latter is modeled following the equivalent single degree of freedom (ESDoF) system approximation [13]. At the same time, the compliance of the foundation subsoil is considered using the Beam-on-Nonlinear-Winkler-Foundation (BNWF) concept [14]. The results of the dynamic analysis are processed to calculate the probability of exceeding for four different limit states (ranging from slight to complete damage state) given the intensity measure, *IM*.

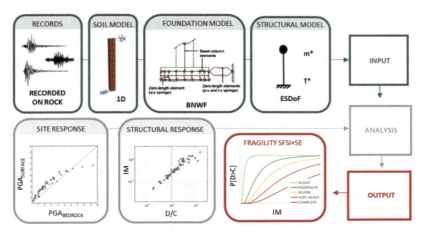

Fig. 1. Flow chart of the proposed methodology for the fragility assessment of structures considering SFSI and site-effects at an urban scale.

2.1 Input

One of the most significant uncertainties in studying the problem of soil-structure-interaction, especially in large-scale applications, is the definition of the main features defining the whole interacting system. With this in mind, the applicability of the proposed approach is based on globally available input data regarding the soil parameters, the foundation, and the building taxonomy, thus making it easily applicable for risk assessment at different cities.

Records. To fully describe structure performance under future earthquakes, the definition of the responses associated with different levels of ground shaking must be assessed. An extensive set of unscaled actual ground motions mostly recorded on stiff rock outcrop is selected to perform all the dynamic cloud analyses. The record selection was carried out following the general recommendation suggested in [12]. Nevertheless, the selected records could not lead the structure to pass the onset of the near-collapse limit state for some structural typologies. In those cases, some records were scaled up to a factor of 2 to avoid the unrealistic and undesired modifications of scaled signals. Different intensity measures are selected in this study for the fragility computation: (i) Peak ground acceleration, *PGA*, (ii) pseudo-spectral acceleration at periods close to the fundamental period of the structure, $Sa(T)$, and (iii) average spectral acceleration, *AvgSa*. The latter is of particular interest in SFSI studies since it compares fragility functions developed for different compliant systems and the reference curves considering the fixed-base assumption. All the chosen intensity measures are computed for the set of input records (i.e., recorded on rock/very stiff soil) and from the free-field motions resulting from the site response analysis, referred to in the following as PGA_R, $Sa(T)_R$, $AvgSa_R$ and PGA_{ff}, $Sa(T)_{ff}$ and $AvgSa_{ff}$ respectively.

Soil. To consider the local site-amplification and define the soil's representative features in the SFSI modeling is necessary to define representative soil profiles. To this aim, seven different representative clayey soil profiles were modeled in the Opensees software. The selected soil profiles were conceived considering different average propagation velocities of the shear waves within the first 30 m of soil depth, $V_{S,30}$ (i.e., ranging from 150 to 450 m/s as reported in Table 1) thus of soil types B, C, and D according to EC8 classification [15]. The adopted shear stiffness profile varies with depth. A simplified distribution is considered in this study to describe the distribution of the soil shear wave velocity with depth as follows:

$$V_s(z) = V_{s,z=30}\left(\left(\frac{V_{s,z=0}}{V_{s,z=30}}\right)^{\frac{1}{a}} + \frac{z}{30}\left(1 - \left(\frac{V_{s,z=0}}{V_{s,z=30}}\right)^{\frac{1}{a}}\right)\right)^a \quad (1)$$

where z stands for the depth measured from the soil surface, while $V_{s,z=0}$ and $V_{S,z=30}$ are the soil shear modulus at the ground surface and a depth of 30 m, respectively. The $V_{S,z=0}$ and $V_{S,z} = 30$ are randomly selected to ensure a $V_{S,30}$ equal to the values reported in Table 1. Some further restrain are adopted for the random selection of the parameters as ensuring values not unrealistically approach zero values near the ground surface and guarantee stiffness parameters vary in the range embedded in the empirical relationships relating the minor strain stiffness, G_0, and the geostatic stress state and history, i.e., according to soil index and state properties [16]. To this aim, the a coefficient was selected equal to 0.25.

Table 1. Main soil parameters selected to characterize the soil profile

Soil parameters							
$V_{s,30}$ (m/s)	150	180	250	300	360	400	450
c_u (kPa)	28	33	58	75	85	110	150
γ (kN/m^3)	15	17	18	19	20	20	21

The numerical model for the site response analysis is performed for a single column of soil deposit 30 m long and constituted of quad elements. The site response analysis is implemented in Opensees using total stress analysis. The nonlinear behavior of the soil has been modeled by assigning a nonlinear material to quad elements "nDMaterial PressureIndependMultiYield" (nonlinear constitutive law based on Von Mises criterion for clays) by adopting the reference parameters suggested in [17].

Foundation. The compliance of the foundation subsoil is considered using the Beam-on-Nonlinear-Winkler-Foundation (BNWF) concept. The advantage of this model is the possibility to directly account for nonlinear soil-foundation behavior, which is expected to occur mainly at higher intensity measures levels. This model accounts for soil-structure interaction phenomena using one-dimensional nonlinear springs distributed along with the soil-foundation interface used in conjunction with gapping and damper elements allowing to consider both geometric and material nonlinearities. To each spring is assigned a backbone curve representing the nonlinear response at the soil-shallow footing interface. The stiffness used to calibrate the springs is selected as a function of a V_S mobilized considering an interaction volume equal to the total foundation length. For the BNWF modeling, to cover different scenarios of foundation systems that can be encountered in an urban environment, the parameters primarily affecting the interaction problem [18], such as the slenderness ratio, H/B (where B is the characteristic foundation half-length), the soil to structure relative ratio, σ, and the structure to soil relative inertia, δ, are parametrically investigated. In this specific application, the parameters are defined based on data gathered after a literature review of previous studies on SFSI available for the case study of Thessaloniki, northern Greece [3, 4, 6].

Structure. All the dynamic analyses are performed following the equivalent single degree of freedom (ESDoF) systems approximation for the superstructure [13]. To reduce the computational effort (as implicitly demanded from seismic fragility assessment of building portfolios), it is convenient to classify buildings by combining a few attributes such as force resisting mechanism, height, and code level. This classification, also known as taxonomy (e.g., GEM taxonomy, [13]), is justified considering that the structures with similar characteristics are more prone to experience similar behavior when subjected to seismic forces. Different "average" structural types (from now on building classes) are selected as representative of the exposure model assigning specified constitutive laws to each building class. Accordingly, each building class is modeled with a single degree of freedom system characterized by a nonlinear hysteretic behavior. The parameters taken to define the specific hysteretic low are defined from the nonlinear backbone curve (capacity curve) available in the literature for different building classes in the GitHub repository (https://github.com/lmartins88/global_fragility_vulnerability, [19]). Even being a simplified approach (e.g. neglecting the contribution of higher modes), the ESDoF approximation has been proved to provide a good estimate of the overall seismic fragility of structures in large-scale analyses [7, 19].

2.2 Results

The proposed methodology results are fragility functions developed for building classes belonging to different SFSI scenarios investigated by changing the dimensionless parameters most influencing the response of structures founded on soft soil profiles. Compares fragility functions developed for a mid-rise regularly infilled structure designed with low-code prescriptions (namely CR-LFINF-DUL-H4 following the GEM taxonomy by changing the H/B ratio, the δ ratio, and the $V_{S,30}$ for all the predefined limit states. All in all (see, for example, Fig. 2), the result of the analyses for the flexible foundations, i.e., considering SFSI and site-effects (dashed lines), produce a shift to the left of the fragility curves compared to the fixed-base case on rock (continuous lines), thus resulting into an increase of the structural fragility. The fragility shift is more pronounced for very soft soil profiles; see Fig. 2 left, developed for the virtual soil profile corresponding to $V_{S,30}$ 180 m/s, compared to Fig. 2 right for $V_{S,30}$ 360 m/s. Such an overall increase is due to the combined effect of site amplification and the higher demand caused my period lengthening. The latter mainly occurs when the period moves to a region of the spectrum with more energy.

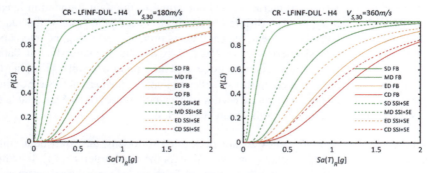

Fig. 2. Comparison between fragility functions in terms of Sa(T)R developed for one reference building class considering the structure fixed at its base (continuous lines) and SFSI and site-effects for one BNWF system characterized by $H/B = 1$, $\delta = 0.05$ and (left) $V_{S,30} = 180$ m/s and (right) $V_{S,30} = 360$ m/s (dashed lines)

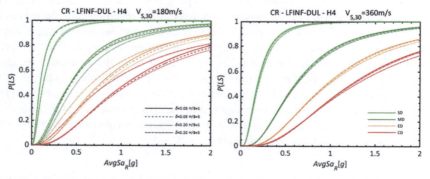

Fig. 3. Fragility functions in terms of $AvgSa_R$ developed for one reference building class considering SFSI and site-effects accounting for different hypotheses on the BNWF system, i.e., by $H/B = 1$ and $\delta = 0.05$ (continuous lines), $H/B = 3$ and $\delta = 0.05$ (dashed lines), $H/B = 1$ and $\delta = 0.2$ (dotted lines) and $H/B = 3$ and $\delta = 0.2$ (dashed-dot lines) for (left) very soft soil profile characterized by $V_{S,30} = 180$ m/s and (right) soft profile characterized by $V_{S,30} = 360$ m/s.

When comparing the fragility functions developed for the selected building class resting on the same soft soil profile but by accounting for different hypotheses on the BNWF systems, i.e., by varying the slenderness and structure-soil relative inertia ratio, it is possible to appreciate the variability associated with SFSI phenomenon in the fragility computation (see Fig. 3). This variability is likely more pronounced for high damage states due to the nonlinear soil-foundation phenomenon occurring for high *IM* values. The seismic fragility curves, including site-effects and SFSI, show up to now, i.e., where the given intensity measure refers to the analysis input records (i.e., as recorded on rock/stiff soil) can be used when the hazard scenario is referring to the underlying bedrock or generally adopted to gain insights into the differences with respect to the standard assessment practice which considers "fixed-base-on-rock" structures and neglects the modification of the input motion due to the deformability of the soil profile (as in the case of Fig. 2), and site amplification effects.

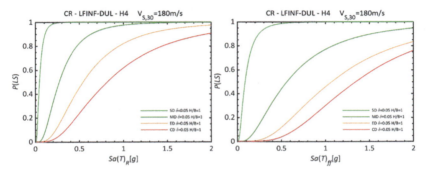

Fig. 4. Fragility functions developed for one reference building class considering SFSI and site-effects for one BNWF system characterized by $V_{S,30} = 180$ m/s, $H/B = 1$, $\delta = 0.1$ in terms of (left) $Sa(T)_R$ and (right) $Sa(T)_{ff}$

On the other hand, the fragility curves as a function of intensity measures defined from the free-field motions (Fig. 4) can also be used in the framework of a risk assessment where the hazard includes site effects adopting either code- or research-based amplification factors or moreover where the hazard scenario comes directly from physics-based numerical simulations [20].

3 Conclusions

A new methodology is proposed to perform an urban-scale vulnerability assessment of structures considering the influence of SFSI and SE. One of the greatest uncertainties in studying soil-structure-interaction is the definition of the main features defining the foundation system. With this in mind, the applicability of the proposed approach is based on globally available data regarding the soil parameters, the foundation, and the building taxonomy, thus making it easily applicable for risk assessment at different cities. This study also provides the first attempt to show how SFSI and site-effects can modify the fragility analysis of structures founded on soft soil. The numerical simulation confirmed that, especially in soft soil formations, the conventional way of calculating fragility curves, i.e., fixed-base structures, may lead to an incorrect evaluation of the seismic risk.

Moreover, the uncertainties associated with the definition of the soil-foundation system can further affect the results, and including all these effects will provide researchers or stakeholders with a correct quantification of the potential fragility or loss estimates, which constitute essential elements in the risk assessment.

References

1. Sáez, E., Lopez-Caballero, F., Modaressi-Farahmand-Razavi, A.: Effect of the inelastic dynamic soil-structure interaction on the seismic vulnerability assessment. Struct. Saf. **33**(1), 51–63 (2011)
2. Rajeev, P., Tesfamariam, S.: Seismic fragilities of non-ductile reinforced concrete frames with consideration of soil structure interaction. Soil Dyn. Earthq. Eng. **40**, 78–86 (2012)
3. Pitilakis, K.D., Karapetrou, S.T., Fotopoulou, S.D.: Consideration of aging and SSI effects on seismic vulnerability assessment of RC buildings. Bull. Earthq. Eng. **12**(4), 1755–1776 (2013). https://doi.org/10.1007/s10518-013-9575-8
4. Karapetrou, S.T., Fotopoulou, S.D., Pitilakis, K.D.: Seismic vulnerability assessment of high-rise non-ductile RC buildings considering soil-structure interaction effects. Soil Dyn. Earthq. Eng. **73**, 42–57 (2015)
5. de Silva, F.: Influence of soil-structure interaction on the site-specific seismic demand to masonry towers. Soil Dyn. Earthq. Eng. **131**, 106023 (2020)
6. Petridis, C., Pitilakis, D.: Fragility curve modifiers for reinforced concrete dual buildings, including nonlinear site effects and soil–structure interaction. Earthquake Spectra **36**(4), 1930–1951 (2020)
7. Cavalieri, F., Correia, A.A., Crowley, H., Pinho, R.: Dynamic soil-structure interaction models for fragility characterisation of buildings with shallow foundations'. Soil Dyn. Earthq. Eng. **132**, 106004 (2020)
8. Amendola, C., Pitilakis, D.: State-of-the-art fragility curves of typical RC and masonry structures considering SFSI and aging effects. Project number 813137 URBASIS-EU New challenges for Urban Engineering Seismology (2021)
9. De Risi, R., Penna, A., Simonelli, A.L.: Seismic risk at urban scale: the role of site response analysis. Soil Dyn. Earthq. Eng. **123**, 320–336 (2019)
10. Amendola, C., Pitilakis, D.: A new methodology for seismic risk assessment of urban areas including soil-structure-interaction and local site effects. Submitted (2022)
11. Mazzoni, S., Mckenna, F., Scott, M.H., Fenves, G.L.: The Open System for Earthquake Engineering Simulation (OpenSEES) User Command-Language Manual (2006)
12. Jalayer, F., Ebrahimian, H., Miano, A., Manfredi, G., Sezen, H.: Analytical fragility assessment using unscaled ground motion records. Earthquake Eng. Struct. Dynam. **46**(15), 2639–2663 (2017)
13. D'ayala, D.F., Vamvatsikos, D., Porter, K.: GEM guidelines for analytical vulnerability assessment of low/mid-rise buildings, vulnerability global component project (2014)
14. NIST. Soil-structure interaction for building structures. Technical report, US Department of Commerce, Washington, DC (2012)
15. CEN Eurocode 8: Design of structures for earthquake resistance. Technical report. European Committee for Standardization (2005)
16. d'Onofrio, A., Silvestri, F.: Influence of micro-structure on small-strain stiffness and damping of fine-grained soil and effects on local site response. In: International Conference on Recent Advances in Geotechnical Earthquake Engineering and Soil Dynamics, vol. 15, pp. 0–6. (2001)

17. Yang, Z., Lu, J., Elgamal, A.: OpenSees soil models and solid-fluid fully coupled elements. User's Manual. Ver. **1**, 27 (2008)
18. Veletsos, A.S., Meek, J.W.: Dynamic behaviour of building-foundation systems'. Earthquake Eng. Struct. Dynam. **3**(2), 121–138 (1974)
19. Martins, L., Silva, V.: Development of a fragility and vulnerability model for global seismic risk analyses. Bull. Earthq. Eng. **19**(15), 6719–6745 (2020). https://doi.org/10.1007/s10518-020-00885-1
20. Paolucci, R., Mazzieri, I., Smerzini, C., Stupazzini, M.: Physics-based earthquake ground shaking scenarios in large urban areas. In: Ansal, A. (ed.) Perspectives on European earthquake engineering and seismology. GGEE, vol. 34, pp. 331–359. Springer, Cham (2014). https://doi.org/10.1007/978-3-319-07118-3_10

Seismic Performance Assessment of Port Reclaimed Land Incorporating Liquefaction and Cyclic Softening

Ioannis Antonopoulos[1(✉)], Alex Park[1], and Grant Maxwell[2]

[1] Level 2, 2 Hazeldean Road, Addington, Christchurch 8024, New Zealand
`ioannis.antonopoulos@stantec.com`
[2] Level 1, 66 Oxford Street, Richmond, Nelson 7020, New Zealand

Abstract. The seismic performance of a reclaimed land part of a New Zealand Port facility was examined through Nonlinear Dynamic Analyses (NDA), under the Operating (OLE) and the Contingency Level (CLE) seismic performance levels described in ASCE/COPRI 61-14 (2014), using advanced tools to investigate the resilience of the asset under the various performance levels and evaluate the seismic hazard, investigate the failure mechanisms leading to ground deformations, and identify key factors contributing to earthquake-induced ground deformations. The geotechnical models were implemented in the finite difference software FLAC and the behaviour of the sand-like and clay-like soils was simulated using the PM4Sand and PM4Silt constitutive models, respectively. Results from the NDA suggest that a compounded effect of both liquefaction of sand-like soils and cyclic softening of clay-like soils may lead to excessive ground deformations that can have detrimental effects on this asset. Liquefaction starts to become a dominant phenomenon under the CLE NDA. Shear failures associated with ratcheting of the strains being developed within the adjacent soils are developed immediately under the revetments. This phenomenon is initiated under the OLE and becomes also dominant under the CLE performance level simulations. Although liquefaction presents a significant challenge, the seismic response of the clay-like soils is indicated to pose a much more significant problem across the site.

Keywords: Port facilities · Liquefaction · Strain-softening · Numerical analyses

1 Introduction

Ports are essential connecting points between countries and are a lifeline for local and global economies. New Zealand, being a new country, started building port facilities as early as the 1860s to assist commerce, transportation of goods, and people.

Their expansion was organic and gradual over the decades of their operation and comprised seawalls of various configurations, reclamation, stone walls, and bunds. All sorts of materials were used for their construction and a main source for the reclamation materials was the dredged marine soils.

The developing understanding of the high seismic risk of the country, not considered at the time of the selection of the port locations, and the need for ever-larger ships to transport an increasingly higher volume of goods poses significant pressure to these old infrastructures.

This requires a rigorous seismic resilience assessment of the old to guide the upgrades required to address today's needs. Such an example is presented in this paper focusing on a particularly interesting area of a port in which the old reclamation and wharves are commonly prone to liquefaction and cyclic softening.

2 Seismic Demands

The area comprises complex geology and a diverse set of geological terranes with mountain areas of uplifted basement rocks (horsts) separated by lower-lying areas in the plains and valleys comprised of younger sedimentary rocks and alluvium (grabens).

The region has several prominent north-easterly trending, mostly active, faults that form the boundaries of these crustal blocks. The regional seismic setting is assessed according to ASCE/COPRI 61-14 (2014) to the below earthquake demands per seismic hazard and performance levels (Table 1).

Table 1. Summary of earthquake demands per ASCE/COPRI 61-14 seismic hazard and performance levels.

Combination	Operating level earthquake (OLE) high (1/72 years)	Contingency level earthquake (CLE) high (1/475 years)
Magnitude, M	6	7.5
Unweighted peak horizontal ground acceleration, α_{max} (g)	0.18	0.40

3 Methodology

3.1 General

The NDA two-dimensional (2D) time history analyses are carried out using the PM4Sand, for the sand-like materials, and PM4Silt, for the clay-like materials, advanced plasticity constitutive models that allow for pore pressure generation with the software package FLAC 8.1 (Itasca Consulting Group, Inc. 2019).

This type of analysis provides a far better insight on the liquefaction and strain softening development and the resulting deformation magnitudes considering the 2D geometry and the soil-foundation-structure-interactions than more simplified pseudo-static analyses.

The NDA were set up, the boundary conditions assigned, and the loading conditions and seismic excitations applied as recommended in the FLAC manual and the work published to date by Boulanger and Ziotopoulou (engr.ucdavis.edu/ dedicated website). The time history analyses were run under total stress conditions. A summary of the modelling procedure is given below:

- Selection of strong-motion records and one dimensional (1D) soil amplification were carried out to select the most demanding strong-motion record for the OLE and CLE events respectively to continue with the 2D NDA.
- Calibration of the PM4Sand and PM4Silt constitutive models for all soil elements to their appropriate stiffness and strength to allow for liquefaction and strain-softening, respectively.
- Hysteretic damping is applied to all other materials.
- Installation of structural elements.
- Application of the selected OLE and CLE records as a time history at the base of the model which now includes the wharf structure, any ground improvement, and any additional surface structures located on the reclamation.
- Monitoring of the accelerations at various depths and ground surface, pore pressure ratio, r_u, displacements, liquefaction and strain softening development, bending moments, shear forces, etc.

3.2 Ground Conditions

Several invasive site investigations (i.e., boreholes and cone penetration tests) were carried out across the site coupled with Multichannel Analysis of Surface Waves (MASW) geophysical surveys. The MASW records were calibrated with the cone penetration tests and soil classification laboratory test results.

The ground materials and stratigraphy were found to be consistent across most of the sites; the local stratigraphy being no different. A typical stratigraphy, identified from the assessment of the available site investigation data and the geophysical investigation report is shown in the following figure (Fig. 1).

- Reclamation fill that comprises mostly coarse-grained – sand-like materials. A denser top layer is found across the site.
- Riprap placed on the revetment sides of the reclamation.
- Marine sediments, mostly coarse-grained – sand-like.
- Marine sediments, mostly fine-grained – clay-like.
- Marine sediments, older – consolidated, mostly fine-grained – clay-like.
- Alluvium sediments, very dense, coarse-grained – sand-like.
- Bedrock that comprises conglomerate and schists.

Fig. 1. Site stratigraphy: ▫Reclamation fill, ▪Rip rap, ▪Marine sediments coarse-grained, ▫Marine sediments fine-grained, ▪Marine sediments intermediate, ▪Alluvium, ▪Bedrock.

The soil materials are also classified in sand-like and clay-like types to assign proper constitutive models for the NDA simulations (Fig. 2).

Fig. 2. ■Sand-like, □Clay-like.

A timber wharf is located on the right-hand side and a building with piled foundations is located on the left-hand side. Both reclamation slopes are armoured with rip rap.

3.3 Material Properties

The geotechnical properties of soil materials encountered are summarised in the below table (Table 2).

Table 2. Summary of geotechnical material properties.

Layer	Dry density, ρ_{dry} (mg/m^3)	Constant volume angle of internal friction, φ'_{cv}	Dilation angle, ψ	Relative density, D_R	Small strain shear modulus, G_{max} (kPa)
Reclamation fill dense	1.42	31	11	0.66	51600
Reclamation fill	1.03	28			40800
Marine sediments coarse-grained	1.34	30	8	0.53	50700
Marine sediments fine-grained	0.96	24			32400
Marine sediments intermediate	1.13	28			59250
Alluvium	1.62	34	10	0.75	80260
Bedrock	1.63	34	10		81400

3.4 Shear Wave Velocity

The typical shear wave velocity profile and the linear amplification function (as both rock outcrop and borehole outputs) of the area are shown in the below figure. The dominant natural frequency is $f_0 \approx 4.4$ Hz, which corresponds to a natural period of $T_0 \approx 0.23$ s.

The time-averaged shear wave velocity over the first 30 m of depth, V_{s30}, from the ground surface was found to range between 236 m/s and 425 m/s. This range of V_{s30} corresponds to a subsoil Class C according to the New Zealand Standard NZS1170.5:2004 (Fig. 3).

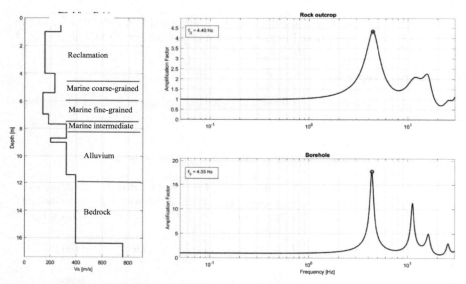

Fig. 3. Typical shear wave velocity profile and amplification function.

4 Selection of Strong Motion Records

Strong motion records from shallow crustal earthquakes were obtained from the PEER NGA database. The search criteria used for both the OLE and the CLE High performance levels are listed in the below table. The selected motions were numerically scaled to the corresponding α_{max} values. The strong motion record for the DE event was selected to be the same as the one for the CLE albeit being numerically scaled higher (Table 3).

From the 1D site response analysis, the most demanding events (in terms of seismic source, the developed shear strains within the soil and the duration of the event) is considered the N. Palm Springs (1986) for the OLE High and Montenegro (1979) for the CLE High performance levels.

Table 3. Strong motion record search criteria.

Performance level	Magnitude	Rupture distance, R_{rup} (km)	Joyner-Boore distance, R_{jb} (km)	V_{s30} (m/s)
OLE high	5.9–6.1	1–10	1–10	236–425
CLE high	7–8			

5 Calibration of Constitutive Models

5.1 General

The NDA are carried out using the plasticity models for sand, PM4Sand, for all sand-like soil layers, and the plasticity model for silt, PM4Silt, for all clay-like (silty/clayey) soil layers. All other materials were simulated using the Mohr-Coulomb constitutive model for their respective shear strength with the below G/G_{max} vs. shear strain backbone curves.

- Bedrock with a weak rock G/G_{max} vs. shear strain backbone curve (EPRI 1993).
- The rip-rap layers with a rockfill G/G_{max} vs. shear strain backbone curve (Gazetas and Dakoulas 1991).

The plasticity constitutive models PM4Sand and PM4Silt were calibrated following the recommendations stipulated in the respective manuals, Boulanger & Ziotopoulou, PM4Sand Version 3.1 (2018) and Boulanger & Ziotopoulou, PM4Silt Version 1 (2018), and the guidelines provided by the authors Boulanger & Ziotopoulou, VGS One-Day Short Course on PM4Sand and Nonlinear Dynamic Modeling of Liquefaction during Earthquakes (2017).

5.2 PM4Sand

The cyclic stress ratio versus the number of equivalent uniform cycles in undrained DSS loading to cause a single amplitude shear strain of 3% with vertical effective consolidation stress of 1 atm are shown in the below figure (Fig. 4).

5.3 PM4Silt

The cyclic stress ratio versus the number of equivalent uniform cycles in undrained DSS loading to cause a single amplitude shear strain of 3% with vertical effective consolidation stress of 1atm are shown in the below figure (Fig. 5).

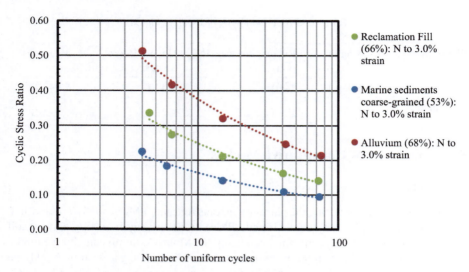

Fig. 4. PM4Sand cyclic stress ratio versus the number of equivalent uniform cycles.

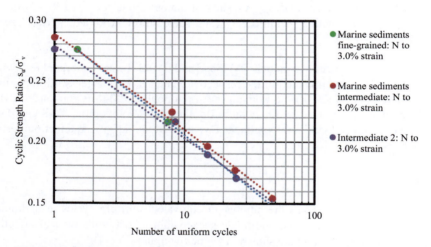

Fig. 5. PM4Silt cyclic stress ratio versus the number of equivalent uniform cycles.

6 NDA Simulations

6.1 OLE High NDA Results

The below figure shows the development of excess pore pressures, shear strains, and total displacements at the end of the OLE High excitation (Fig. 6).

Both revetment slopes tend to move outwards. The lateral extents are approximately 20 m for the left and 50 m for the right. Only the coarse-grained Marine sediments develop the potential for liquefaction.

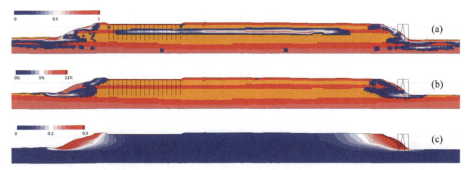

Fig. 6. OLE High dynamic response: (a) excess pore pressure ratio, (b) shear strains, (c) total displacements (in metres).

The piled building (at the left-hand side) is affected by the lateral displacement of the reclamation's revetment slope. The piles supporting the building do not seem to have proper end bearing into the Alluvium or bedrock. They are floating piles and consequently themselves and the building they support tend to move with the ground deformations.

The coarse-grained Marine sediments exhibit the development of excess pore pressures but not of significant shear strains. The below figure shows the stress-strain plot at the middle of that soil layer. Although liquefaction is triggered, the shear strains do not exceed 1%, hence the liquefaction effects are minor. No strain softening is developed within the fine-grained soils (Fig. 7).

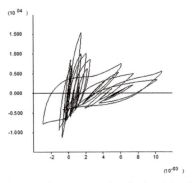

Fig. 7. Strain (horizontal axis) vs. shear stress (vertical axis) of the Marine coarse-grained sediments during the OLE High simulation.

6.2 CLE High NDA Results

The below figure shows the development of excess pore pressures, shear strains, and total displacements at the end of the OLE High excitation (Fig. 8).

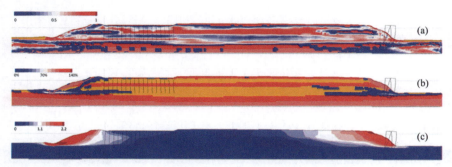

Fig. 8. CLE High dynamic response: (a) excess pore pressure ratio, (b) shear strains, (c) total displacements (in metres).

Both revetment slopes displace by a large magnitude, exceeding 2 m. The lateral extents are approximately 70 m for the left and 110 m for the right. Extensive areas develop liquefaction and strain softening.

It is evident that, as under the OLE High simulation, the piled building is affected by the lateral displacement of the reclamation's revetment slope. However, under the CLE High performance level, the mobilised soil mass extends further than the immediate revetment slope.

Because the piles supporting the building are floating and they show that they do provide some soil pinning effect under a higher seismic shaking level, the piles and the building they support tend to move with the ground deformation. Because of the pinning they provide, they tend to rotate both seawards and landwards resulting in a bending curvature at the foundation level that may affect the structural performance of the building.

It is important to note that a shear surface is being developed at the pile bases under the seaward half of the building's footprint allowing the lateral displacement of that half-block towards the revetment slope.

The high shear stresses associated with the CLE High excitation appear to trigger or exacerbate the nonlinear behaviour of soils (extent of liquefaction and strain softening) and ground deformations. The Marine sediments exhibit the development of strain-softening within their fine-grained (clay-like) phase and liquefaction within their coarse-grained (sand-like) phase. The below figure shows the stress-strain plots at the middle of these soil layers (Fig. 9).

6.3 Horizontal Elastic Spectra

The below figure shows the spectral acceleration vs. period captured at the base of the model (applied excitation) and the middle of the foundation slab of the existing building. The NZS1170.5 Class C spectral plot is also included for comparison. Due to the presence of the two revetment slopes forming the lateral boundaries of the reclamation, the soil mass seems to freely vibrate under the OLE High excitation. This results in the significant amplification of the input motion for periods between 0.5 s and 1.5 s.

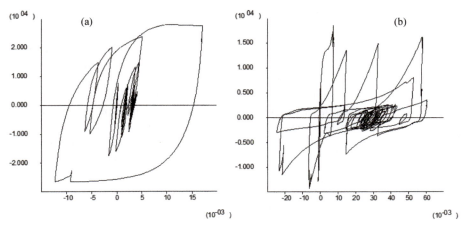

Fig. 9. Strain (horizontal axis) vs. shear stress (vertical axis) of the Marine fine-grained (a) and coarse-grained (b) sediments during the CLE High simulation.

However, when the input seismic energy is much higher, like under the CLE High excitation, the nonlinear behaviour of the soils dominates resulting in large ground deformations. This leads to a significant de-amplification of the input motion for periods between 0.2 s and 1 s (Fig. 10).

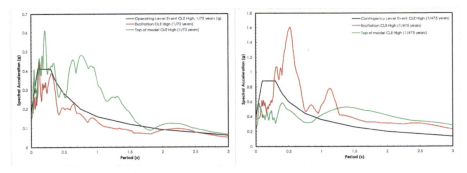

Fig. 10. Spectral acceleration, S_a, vs. period.

7 Discussion

The results of the NDA show that the high shear stresses associated with the CLE excitations appear to trigger or exacerbate the nonlinear behaviour of soils (extent of liquefaction and strain softening) and ground deformations. The below the reclamation marine sediments exhibit the development of strain-softening within their fine-grained (clay-like), phase and liquefaction within their coarse-grained (sand-like) phase.

It was found that because the fine-grained soils maintain a consistent thickness across the examined area, they significantly amplify the ground motions when subjected to the lower energy OLE performance level and develop "free" vibration responses. The main difference between the OLE High and Cle High NDA simulations is that liquefaction requires the build-up and accumulation of seismic cycles under the OLE High excitation, whereas the energy provided by the CLE High excitation is large enough to trigger that phenomenon early on.

Shear failures associated with ratcheting of the strains being developed within the adjacent soils are developed immediately under the revetments. This phenomenon is initiated under an OLE High NDA and becomes dominant under the CLE High performance level simulation. Although liquefaction presents a significant challenge, the seismic response of the clay-like soils is indicated to pose a much more significant problem across the site.

The NDAs showed that even under the OLE High performance level the piled building is affected by the lateral displacement of the reclamation's revetment slope. The piles supporting the building do not seem to have proper end bearing into the Alluvium or bedrock. They are floating piles and consequently themselves and the building they support tend to move with the ground deformations.

Under the CLE High NDA, the deformations increase considerably because the piles supporting the building are floating and they show that they do provide some soil pinning effect under a higher seismic shaking level, themselves, and the building they support tend to move with the ground deformation. Similarly, the wharf is affected by the lateral displacement of the reclamation's revetment slope and is expected to fail under a CLE High performance level.

References

American Society of Civil Engineers, Coasts Oceans, Ports & Rivers Institute: ASCE/COPRI 61-14 ASCE Seismic Design of Piers and Wharves. ASCE (2014)
Antonopoulos, I., McDermott, C.: SH1 realignment – performance-based design of MSEW seawalls. In: 7th International Conference on Earthquake Geotechnical Engineering, pp. 1154–1161. Associazione Geotecnica Italiana, Rome (2019)
Asimaki, D., Shi, J.: SeismoSoil User Manual, v1.3. GeoQuake Research Group, California Institute of Technology (2017)
Boulanger, R.W., Ziotopoulou, K.: VGS One-Day Short Course on PM4Sand and Nonlinear Dynamic Modeling of Liquefaction during Earthquakes. Retrieved from Vancouver Geotechnical Society, 27 October 2017. http://v-g-s.ca/pm4s-short-course
Boulanger, R.W., Ziotopoulou, K.: PM4Sand (Version 3.1). A sand plasticity model for earthquake engineering applications. Davis: Center for Geotechnical Modelling, Department of Civil & Environmental Engineering, College of Engineering, University of California (2018)
Boulanger, R.W., Ziotopoulou, K.: PM4Silt (Version 1). A silt plasticity model for earthquake engineering applications. Davis: Center for Geotechnical Modelling, Department of Civil & Environmental Engineering, College of Engineering, University of California (2018)
EPRI: Guidelines for Determining Design Basis Ground Motions. EPRI, Palo Alto, California (1993)
Gazetas, G., Dakoulas, P.: Aspects of seismic analysis and design of rockfill dams. In: Second International Conference on Recent Advances in Geotechnical Earthquake Engineering and Soil Dynamics, pp. 1851–1888. Missouri University of Science and Technology, St. Louis, Missouri (1991)

Idriss, I.M., Boulanger, R.W.: Soil liquefaction during earthquakes. Earthquake Engineering Research Institute (2008)

Itasca Consulting Group, Inc.: FLAC — Fast Lagrangian Analysis of Continua, Ver. 8.1. Itasca, Minneapolis, USA (2019)

New Zealand Standard. Structural Design Actions NZS 1170, Part 5: Earthquake actions, incorporating amendment No. 1 (2004)

Pretell, R., Ziotopoulou, K., Davis, C.A.: Liquefaction and cyclic softening at Balboa Boulevard during the 1994 Northridge earthquake. J. Geotech. Geoenviron. Eng. **147**(2) (2021)

Shi, J., Asimaki, D.: From stiffness to strength: formulation and validation of a hybrid hyperbolic nonlinear soil model for site-response analyses. Bull. Seismol. Soc. Am. **107**(3), 1336–1355 (2017)

An Attempt to Evaluate In Situ Dynamic Soil Property by Cyclic Loading Pressuremeter Test

Keigo Azuno[1(✉)], Tatsumi Ishii[2], Youngcheul Kwon[3], Akiyoshi Kamura[2], and Motoki Kazama[2]

[1] Chuo Kaihatsu Corporation, Tokyo, Japan
azuno@ckcnet.co.jp
[2] Tohoku University, Sendai, Japan
[3] Tohoku Institute of Technology, Sendai, Japan

Abstract. Laboratory experiments, such as cyclic triaxial tests, are generally conducted to evaluate the dynamic behavior of soil for earthquake-proof design. The disturbance and stress release of soil specimens caused by sampling considerably affect the testing results. To avoid sampling disturbance effect, the authors attempted to measure the dynamic soil properties under cyclic loading by directly implementing an in situ pressuremeter test in the borehole. In the feasibility test on loose sand, it was confirmed that cyclic loading was achieved in situ successfully. Besides, in the tests on medium dense sand, hardening behaviors of surrounding soil were observed.

Keywords: Liquefaction · Pressuremeter · In situ testing · Cyclic loading

1 Introduction

Assessing the dynamic behavior of the ground during earthquake is important in the design of foundation ground against liquefaction. In general, laboratory soil tests are performed to evaluate the dynamic behavior of soil. However, many researchers are confronted with the problem of disturbance during sampling and transportation of sand and gravel. Freeze sampling [1], which can minimize the effects of disturbance, is costly and cannot be applied to all investigations. Even less disturbed samples sustain the adverse effect of stress release. Therefore, even if the number of test results is increased, these results do not reach their true values. As a result, ground improvements are sometimes over-constructed especially in liquefaction countermeasures. Accordingly, the authors believe that the development of a technique for evaluating the in situ dynamic soil property is required.

To avoid the effects of sample disturbance, conducting in situ tests is efficient. The standard penetration test (SPT) has been widely researched for liquefaction assessment. The current liquefaction assessment is based on the relationship between undrained cyclic shear strength and N-value that has been derived by Seed (1979) [2] and Seed et al. (1983) [3]. However, the SPT cannot be applied to gravel because of the overestimation problem caused by gravel strike. The cone penetration test (CPT) has also been widely studied because it is more convenient than the SPT. Seed and Alba (1986) [4] and Shibata and Teparaksa (1988) [5] showed the relationship between undrained cyclic shear strength and cone penetration resistance. The piezocone penetration test (CPTu) can measure pore water pressure in addition to cone penetration. Juang et al. (2008) [6] related the CPTu results to liquefaction history through artificial neural network learning, and Ching and Juang (2011) [7] showed the relationship between the probability of liquefaction occurrence and safety factor. The piezo drive cone test (PDC) [8, 9], which measures the pore water pressure generated by the cone tip in the ground during dynamic penetration, can evaluate the liquefaction safety factor, F_L, and settlement from the measured values. However, CPT, CPTu, and PDC cannot penetrate the hard ground with N-value $\geq$ 20. Although the in situ tests are the most promising method, currently, no investigation method can evaluate the dynamic soil properties, such as stiffness degradations whether liquefaction will occur, regardless of soil type.

This study focuses on the pressuremeter test, which has no problems in testing both hard ground and soft ground. The pressuremeter test [10] is an in situ test method to determine the initial pressure (horizontal stress) and shear modulus of the ground from the relationship between the pressure and displacement of the borehole wall by loading the hole with a probe. Many researchers utilize the pressuremeter test to assess the static properties of the ground. Therefore, we have advanced this technique by modifying the conventional monotonic loading test to the cyclic loading test to enable the pressuremeter test to assess the dynamic soil properties. Thus far, few studies have been conducted on pressuremeter tests with cyclic loading. Reiffsteck et al. (2016) [11] endeavored to implement pressuremeter tests with cyclic loading on sandy to clayey soils at approximately 15 sites. Although Reiffsteck et al. (2016) observed strain increase due to cyclic loading in some sections of the ground where liquefaction is a concern, the study did not reach liquefaction assessment. In addition, Kamura and Kazama (2021) [12] conducted the displacement-controlled cyclic pressuremeter test on loose sand where liquefaction is a concern and obtained the softening behavior and cyclic mobility of soil.

Although a study on cyclic loading pressuremeter tests has been conducted, evidence regarding the dynamic behavior of soil using these tests is insufficient. Accordingly, this paper presents the possibility of assessing in situ dynamic soil property using cyclic pressuremeter tests in relation to the potential for judgment of liquefaction occurrence.

2 Pressuremeter Devices

There are two types of cyclic loading paths for pressuremeter tests: the pressure-controlled and displacement-controlled methods. The former can be widely applied to various soil properties. In contrast, the latter cannot be employed because it may not provide adequate pressure to generate sufficient strain in stiff soil. This section describes the test devices for pressure control.

The schematic of pressuremeter devices for cyclic loading is shown in Fig. 1. This device expands a probe with a membrane by supplying water from a controlled automatic pressurizer. The change in the volume of the probe is measured from the water level change in the standpipe. Other than the automatic pressure device, the other instruments utilized are the same as those in a standard pressuremeter test.

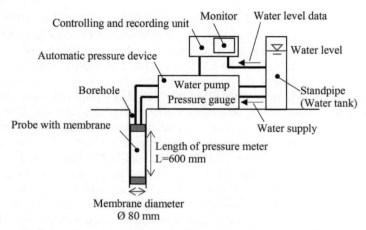

Fig. 1. Schematic of pressuremeter devices for cyclic loading.

3 Feasibility Test in an Outdoor Pit

3.1 Installation and Field Conditions

The authors conducted the feasibility test in a large soil tank made of reinforced concrete. The tank was approximately 2.5 m in height, 9.0 m in width, and 10.0 m in length. It was filled with Ube silica sand No. 6 (relative density: about 60%; permeability coefficient: 7.4×10^{-5} m/s). The probe was installed with pre-burying means into the soil tank. Earth pressure sensors, water pressure sensors, and inclinometers are placed around the probe.

3.2 Result of Cyclic Loading Test Using Pressuremeter

Representative results of the relationship between pressure and radial displacement are shown in Fig. 2, the time history of earth pressure is shown in Fig. 3. The calibrated plot in Fig. 2 indicates the results except for the probe membrane tension. The earth pressure sensor corresponding to the Fig. 3 was located at the center depth as the probe and distance of 30 cm in the horizontal direction. By cyclic loading, the relationship in Fig. 2 exhibits a virtually constant loop shape. As the number of loading increases, the cavity is gradually expanded; the radial displacement is more significant with higher pressure. The earth pressure reacts linearly with the cyclic loading in Fig. 3. The earth pressure only measured less than 1/10 of the loading prove pressure. Accordingly, the authors could confirm that the pressure-controlled cyclic loading had been achieved by using the pressuremeter equipment. In addition, it was found that the spatial extent of influence observed via the cyclic pressuremeter test was very small and limited.

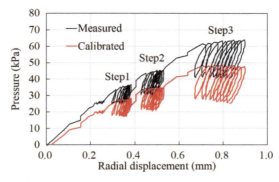

Fig. 2. Relationship between pressure and radial displacement in an outdoor pit.

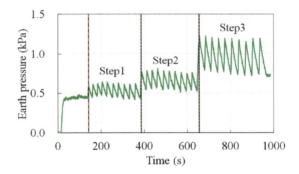

Fig. 3. Time history of earth pressure located at the center depth as the probe.

4 Cyclic Pressuremeter Tests in the Actual Field

4.1 Field Conditions

The authors conducted the cyclic pressuremeter tests to observe the in situ dynamic behavior on the non-liquefiable ground. The authors selected the alluvial lowland site in Sendai, Miyagi, Japan. The site is located on a flooded plain called Sendai Plain, the N-values are relatively large in Fig. 4. The F_C and CSR mean fine grain content and cyclic stress ratio respectively, which were carried out with the specimens sampled from the same borehole.

The cyclic pressuremeter tests were carried out at three depths with different densities and soil types: medium dense sand with $N = 14$, dense sand with $N = 29$, and clay with $N = 3$. These layers have a low potential for liquefaction.

4.2 Testing Procedure

The probe was installed a pre-formed hole means with mud water. The initial probe pressure was corresponding to 1.5 times the effective horizontal stress, which was calculated from the unit weight of soils and the coefficient of earth pressure at rest K_0 ($K_0 = 0.5$), was applied. This was done to prevent the disturbance and fracture of the borehole and to make initial stress condition. The water injection rate to the probe was set to 1.2 L/min with the loading rate of 0.2–0.5 Hz. The number of cyclic loading was 29 cycles, and the pressure amplitude was set to three cases to obtain the shear modulus in the different stress levels. The loading conditions for each depth are shown in Fig. 5.

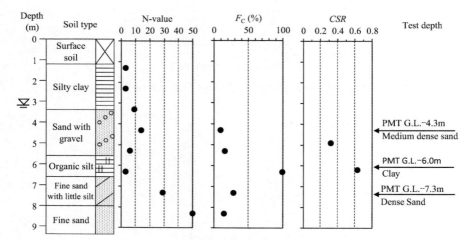

Fig. 4. Field conditions and test depths at the target site in Sendai, Miyagi, Japan.

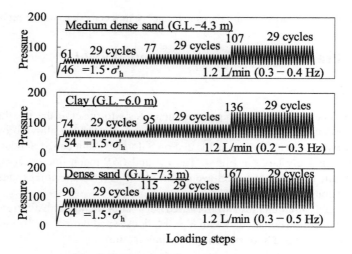

Fig. 5. Loading conditions for each depth.

4.3 Calculation of Shear Strain and Shear Modulus in Assuming an Elastic Material

The shear modulus near the borehole surface is calculated by $G = 1/2 \cdot (\delta p / \delta \varepsilon_c)$, and the radial strain is calculated by $\varepsilon_c = y_c / \rho_0$, based on the thick-walled cylinder theory for elastic materials by Mair et al. (1987) [10]; where G is shear modulus, p is probe pressure, ε_c is radial strain, y_c is radial displacement, ρ_0 is initial radius respectively. In this study, to evaluate the strain level of the obtained shear modulus, the shear strain near the borehole surface is required. Since the circumferential and radial strain are equal and opposite at the borehole wall, probe pressure equals circumferential shear stress [10]. Therefore, the shear strain is calculated from $G = 1/2 \cdot (\delta p / \delta \varepsilon_c)$ and Hooke's law [13] as $\gamma = 2 \cdot \varepsilon_c$, where γ is shear strain.

4.4 Results of Cyclic Loading Test Using Pressuremeter

The relationship between pressure and radial displacement is shown in Fig. 6. The calibrated values mean except for probe membrane tension results. The radial displacement is defined as the relative displacement from the initial step. The set values are the pressure amplitude of the cyclic loading shown in Fig. 5.

The pressure and displacement relationship in Fig. 6 exhibited an almost constant loop shape by cyclic loading in each constant pressure loading step. In the case of clay, as the pressure amplitude in a step increases, the cavity is gradually expanded; the increase of radial displacement is more significant with higher pressure. On the other hand, dense sand showed no significant displacement with increasing pressure amplitude.

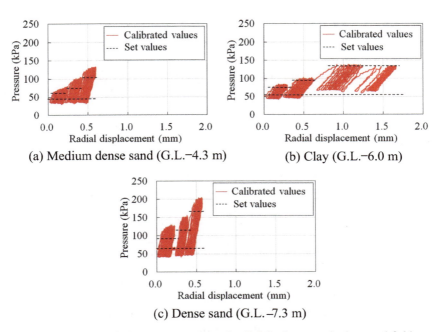

Fig. 6. Relationship between pressure and radial displacement in the actual field.

The relationships between shear modulus and shear strain at certain loading cycles are shown in Fig. 7. The largest shear modulus is obtained on dense sand, followed by medium dense sand, clay, and depends on the N-value and soil type. In addition, as the pressure amplitude increases, the shear strain is expanded in order. The shear strains were obtained in the range of 0.2 to 1.3%, and the shear modulus was obtained at the strain magnitude corresponding to liquefaction occurrence level.

The results of clay showed slight softening behaviors in which the shear modulus decreased with increasing pressure amplitude. On the contrary to this, medium dense sand showed hardening behaviors in which the shear strain and shear modulus increased with increasing pressure amplitude and loading cycles. Furthermore, the shear modulus of dense sand did not change significantly with increasing pressure amplitude. The hardening behaviors indicate that the surrounding soil may have become denser due to cyclic loading under partial drainage condition.

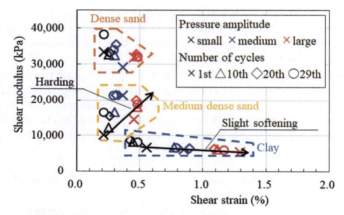

Fig. 7. Relationship between shear modulus and shear strain.

5 Conclusions

In this paper, we reported the results of an attempt to evaluate the in situ dynamic behaviors by cyclic pressuremeter tests. The results and conclusions are as follows.

- The pressure-controlled cyclic loading had been achieved by using the pressuremeter equipment.
- The spatial range of influence observed via the cyclic pressuremeter test was very small and limited, less than 30 cm from the surface of the pressuremeter probe.
- Clay showed slight softening behaviors with increasing pressure amplitude. On the other hand, medium dense sand showed hardening behaviors with increasing pressure amplitude. The latter indicates that the surrounding soil may have become denser due to cyclic loading under partial drainage condition.

In order to increase the applicability, there are the following technical issues:

- Theoretical interpretation of dynamic behavior of ground by cyclic pressuremeter test considering the soils as an elastoplastic material.
- Confirmation of the excess pore water pressure generated by cyclic loading on the liquefied ground.

Acknowledgments. The project presented in this paper was supported by JSPS KAKENHI (Grant No. JP20H02237). The authors express their gratitude for the opportunity to perform field investigations at National Research Institute for Earth Science and Disaster Resilience. The pressuremeter tests were conducted with the support of OYO Geo-monitoring Service Corporation.

References

1. Yoshimi, Y., Hatanaka, M., Oh-oka, H.: Undisturbed sampling of saturated sands by freezing. Soils Found. **18**(3), 59–73 (1978)
2. Seed, H.B.: Soil liquefaction and cyclic mobility evaluation for level ground during earthquakes. ASCE J. Geotech. Eng. Div. **105**(2), 201–255 (1979)
3. Seed, H.B., Idriss, I.M., Arango, I.: Evaluation of liquefaction potential using field performance data. J. Geotech. Eng. **109**(3), 458–482 (1983)
4. Seed, H.B., De Alba, P.: Use of SPT and CPT tests for evaluating the liquefaction resistance of sands. Use In-Situ Tests Geotech. Eng. Geotech. Special Publ. **6**, 281–302 (1986)
5. Shibata, T., Teparaksa, W.: Evaluation of liquefaction potentials of soils using cone penetration tests. Soils Found. **28**(2), 49–60 (1988)
6. Juang, C.H., Chen, C.-H., Mayne, P.W.: CPTu simplified stress-based model for evaluating soil liquefaction potential. Soils Found. **48**(6), 755–770 (2008)
7. Ching, J., Juang, C.H.: Selection among CPTu-based liquefaction models. Procedia Eng. **14**, 2576–2584 (2011)
8. Sawada, S., Tsukamoto, Y., Ishihara, K.: Method of dynamic penetration with pore pressure transducer Part 2 results of chamber test. Japan Natl. Conf. Geotech. Eng. **39**, 1927–1928 (2004)
9. Sawada, S., Yoshizawa, D., Hiruma, N., Sugano, T., Nakazawa, H.: Evaluation of differential settlement following liquefaction using Piezo Drive Cone. In: Proceedings of the 17th International Conference on Soil Mechanics and Geotechnical Engineering: The Academia and Practice of Geotechnical Engineering, vol. 2, pp. 1064–1067 (2009)
10. Mair, R.J., Wood, D.M.: Pressuremeter Testing Methods and Interpretation (1987)
11. Reiffsteck, P., Fanelli, S., Desanneaux, G.: Evolution of deformation parameters during cyclic expansion tests at several experimental test sites. In: Proceedings of the 5th International Conference on Geotechnical and Geophysical Site Characterisation, ISC 20161, pp. 791–796 (2016)
12. Kamura, A., Kazama, M.: Asessment of stiffness degradation of soil by in-situ cyclic loading using pressuremeter. In: 6th International Conference on Geotechnical and Geophysical Site Characterization, ISC2020-153 (2021)
13. Timoshenko, S.P., Goodier, J.N.: Theory of Elasticity (1934)

Effect of Refinements to CPT-Based Liquefaction Triggering Analysis on Liquefaction Severity Indices at the Avondale Playground Site, Christchurch, NZ

John R. Cary[1], Armin W. Stuedlein[1(✉)], Christopher R. McGann[2], Brendon A. Bradley[2], and Brett W. Maurer[3]

[1] Oregon State University, Corvallis, USA
armin.stuedlein@oregonstate.edu
[2] University of Canterbury, Christchurch, New Zealand
[3] University of Washington, Seattle, USA

Abstract. The 2010–11 Canterbury Earthquake Sequence (CES) resulted in significant and widespread liquefaction of the soils underlying Christchurch, New Zealand, offering an opportunity to evaluate factors contributing to seismic ground failure. This study presents the results of a field campaign undertaken to examine the subsurface stratigraphy at a selected liquefaction case history site in Christchurch with cone penetration tests (CPTs) at an unparalleled horizontal resolution. The field data is used to evaluate the role of selected refinements to liquefaction triggering procedures on commonly-used liquefaction severity indices for comparison to observed liquefaction severity. Subsurface cross-sections were constructed at the alluvial Avondale Playground site using 18 closely-spaced CPTs. Refinements to the selected CPT-based liquefaction triggering procedure included the use of a site-specific fines content correlation, corrections for the increased cyclic resistance of partially-saturated soils, and rigorous correction for thin layer and layer transition effects on CPT data. Five liquefaction severity indices (LSIs) were evaluated and compared against observed liquefaction severity. In general, application of the refinements to the liquefaction triggering analyses resulted in improved correspondence between the calculated and observed LSIs by reducing over-prediction of severity when evaluated on a site-wide basis. However, when considering individual soundings, the presence of local and deeper fine-grained lenses appear to explain the lack of observed surficial expression of liquefaction which could not have been anticipated from the magnitude of computed LSIs alone.

Keywords: Liquefaction · Cone penetration tests · Earthquakes · Settlement · Variability

1 Introduction

The Canterbury Earthquake Sequence (CES) consisted of a series of four earthquake events starting with the Darfield earthquake on September 4, 2010 and ending with the

December 23, 2011 event, that occurred along the Pacific-Australian plate boundary on the South Island of New Zealand (Bradley 2014). These events caused an estimated $15B in damage due to liquefaction-induced phenomena in and near Christchurch, NZ (Cubrinovski et al. 2011). Since the CES, numerous geotechnical studies have investigated the widespread liquefaction that contributed to the earthquake-induced damage (e.g., Bastin et al. 2017; Beyzaei et al. 2018), as well as related procedures and models which had produced mixed success capturing the scope and severity of the occurrence and effects of liquefaction (Maurer et al. 2015; van Ballegooy et al. 2015; Markham et al. 2016; McLaughlin 2017; Cubrinovski et al. 2019). These studies illustrated the importance of understanding site-specific responses to seismic ground motions, effects of thin layers to restrict excess pore water redistribution and obfuscate CPT measurements, and the limitations of simplified models to predict liquefaction severity. This paper presents an extensive investigation conducted at the Avondale Playground site following the CES using various liquefaction severity indices (LSIs) and the effects of selected refinements to the liquefaction triggering analyses proposed by Boulanger and Idriss (2015) on the accuracy of liquefaction severity. Liquefaction triggering and its severity are evaluated for the two largest events: the magnitude, M_w, 7.1 September 2010 (Event 1) and M_w 6.2 February 2011 (Event 2) earthquakes using 18 closely-spaced cone penetration tests (CPTs) and supporting laboratory analyses and geophysical measurements to identify opportunities for improvement in the assessment of liquefaction severity.

2 Subsurface Characterization of Avondale Playground Site

2.1 Geological Setting

Christchurch sits upon the boundary between the interacting Pacific and Australian tectonic plates responsible for the emergence of the New Zealand landmass (Begg et al. 2015). The contemporary geologic setting at Christchurch is the product of several long-acting depositional and erosional forces, most notably the cyclically rising and falling sea level and drainage from the Southern Alps. The former was responsible for the deposition of dune sand underlying much of eastern Christchurch to the edge of the Canterbury Plains. Drainage from the topographic relief of the west, currently manifested in the Waimakariri and Avon rivers, avulsed older soil deposits while depositing alluvial sediments (Markham et al. 2016). These meandering and anastomosing rivers deposited, eroded, and redeposited sediments across modern day Christchurch, to raise the land mass above sea level and out into the shallow Pegasus Bay. This resulted in isolating marine and estuarine deposits while forming lagoons and swamps (Begg et al. 2015). The result of these geological processes is a complex system of alluvial silts and sands and isolated peat swamps adjacent to and overlaying dune and estuarine sands in highly variable stratigraphies (Bastin et al. 2017; Beyzaei et al. 2018).

2.2 Subsurface Investigation and Characterization

Assessments of liquefaction triggering and severity require a thorough understanding of the geologic setting, stratigraphic identification, and characterization of the subsurface.

This work leverages significant prior work to characterize the subsurface in Christchurch, including detailed case histories developed for tens of sites in collaboration with members of the University of Canterbury (Cubrinovski et al. 2019), University of Texas, Austin (Cox et al. 2017), and Tonkin & Taylor, select subsurface information of these sites reported by McLaughlin (2017), and information contained within the New Zealand Geotechnical Database (NZGD). Examples of the subsurface information available at the Avondale Playground site included SCPT 57354 and CPT 62768, sonic core borehole BH 57217 with continuous sampling and standard penetration testing (SPT), and direct push crosshole test location DPCH 57062 (Fig. 1). Additional laboratory analyses were conducted on samples retrieved from BH 57217 and sixteen new CPTs were performed along a uniform azimuthal bearing using approximately 1.5 m horizontal spacing to facilitate the construction of a high-resolution subsurface cross-section A-A' along portions of the site that did and did not exhibit manifestation of liquefaction during various earthquakes comprising the CES (Fig. 1). The spacing of the new explorations was selected to balance the need to obtain high-resolution stratigraphy with the prevention of overlapping plastic deformation fields generated during the cavity expansion process associated with cone penetration testing (Salgado and Prezzi 2007).

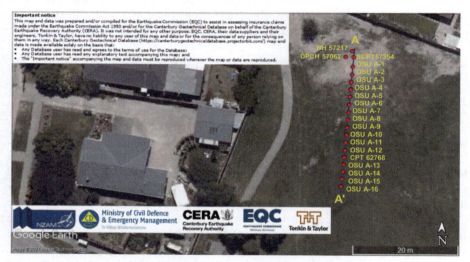

Fig. 1. Liquefaction manifestation at Avondale Playground following the December 2011 event indicating subsurface cross-section A-A' (image from Google Earth via the NZGD).

Figure 2 compares subsurface data collected at and near SCPT 57354, including laboratory indices from samples retrieved from BH 57217 and compression, V_p, and shear wave velocities, V_s, determined from DPCH 57062 (McLaughlin 2017). Figure 2a presents the depth-varying grain sizes D_{10}, D_{30} and D_{60} indicating that the grain sizes generally increase with depth. When considering the soil behavior type index (Robertson 2009), fines content, FC, plasticity index, PI, and corrected cone tip resistance, q_t, in Figs. 2b and 2c, the upper five meters is inferred to consist of highly-layered lenses of very loose to loose silty and clayey sand with FC generally ranging from 20 to 30%, and very soft to medium stiff, low and highly plastic clays. From five meters to the termination of testing at 15 m exists well-graded and silty sands and clean sands interlayered with occasional thin lenses of silt, with relative density, D_r, increasing and FC generally decreasing with depth. Figures 2d and 2e present the variation of cross-hole V_p and V_s to 10 m below ground surface, the former of which varied from 310 to 760 m/s from the ground surface to the groundwater table (GWT), followed by a reduction below the ground GWT to reach ~1,500 m/s at approximately three meters below ground surface and greater for points deeper, indicating fully-saturated conditions below 3 m depth. The corresponding profile of V_s indicates a thin, desiccated crust with $V_s = 145$ m/s, reducing to averages of about 126 and 140 m/s from the depths of 1 to 3 m and 3 to 5 m, respectively. Thereafter, V_s exhibits a stepwise increase to ~170 m/s and increases linearly with depth. The stepwise increase in V_s at 5 m depth correlates to the transition from the loose and soft, highly-interbedded deposits to the relatively consistent sand deposits below.

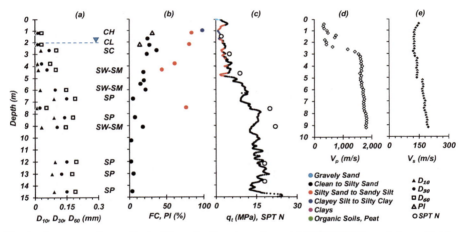

Fig. 2. Summary of geotechnical investigations at and around SCPT 57354: variation of (a) key particle sizes, (b) fines content and plasticity index, (c) cone tip and SPT penetration resistance, (d) compression wave velocity, and (e) shear wave velocity with depth. Note: FC and q_t are color-coded by CPT-based soil behavior type.

2.3 Subsurface Cross-Section A-A'

Figure 3 presents two interpretations of the subsurface stratigraphy along Section A-A' at the Avondale Playground site: one using the field measured and u_2-corrected q_t and sleeve friction, f_s (upper figure), and that using the same data corrected using inverse filtering and thin-layer corrections following the procedure Boulanger and DeJong (2018), described in Sect. 3 below. The standard interpretation of Section A-A' (Fig. 3, upper) indicates the presence of two main stratigraphic layers (Units 1 and 2). The first unit extends to approximately 5 m depth consisting of interbedded, very soft to medium stiff sandy silt and clay interwoven with loose to medium dense silty sand, with typical q_t and I_c of 0.5 to 6 MPa and 1.5 to 2.95, respectively. The second characteristic layer (Unit 2) extends to the termination test depth of approximately 15 m and consists of medium dense to dense clean and silty sand with occasional thin lenses of loose sand and soft to medium stiff silts and clays, with q_t typically ranging from 10 to 20 MPa and the typical I_c of 2.05 or smaller. Within these two characteristic soil layers, two separate and distinct fine-grained lenses may be tentatively identified: the first, noted in CPTs OSU A-7 and A-8 from the depths of ~2 to 4 m, is a very soft to soft, clayey silt to silty clay and clay (A; Fig. 3). The second, present in CPTs OSU A-8 to A-13 and again in OSU A-15 and A-16, at depths of 8 to 10 m, is a very soft to soft, sandy silt to silty clay, and clay, with traces of organic soils (B). These two lenses are expected to exhibit markedly lower hydraulic conductivities than the immediately adjacent soils, with implications for potential migration of excess pore pressures during and following shaking.

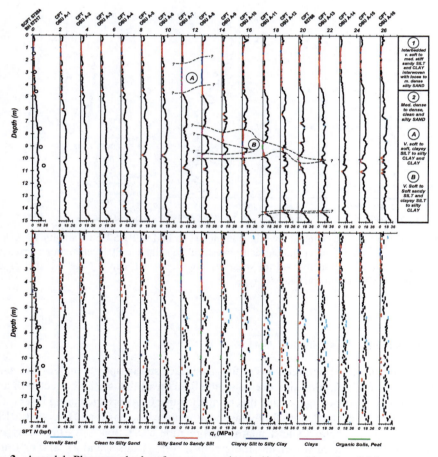

Fig. 3. Avondale Playground subsurface cross section A-A': (upper) based on field measurements of q_t and f_s, and (lower) inverse-filtered and thin-layer corrected q_t, and f_s.

The inverse filtering procedure proposed by Boulanger and DeJong (2018) to account for the presence of thin layers to locally increase or decrease the measured q_t at Section A-A' is shown in the lower portion of Fig. 3. The presence of occasionally contiguous, thin interbeds within the upper 5 m is more clearly observed following inverse filtering, accompanied by a general increase in q_t for the sandier lenses of ~1 to 2 MPa. This correction also facilitates the identification of thin gravelly sand lenses observed at several depths within the range of 7 to 11 m in OSU A-7 through A-16. These thin gravelly sand layers exhibit q_t that are approximately 20% larger, on average, than that measured as a result of inverse-filtering process, with implications for the computation of cyclic resistance using CPT-based liquefaction triggering procedures and corresponding LSIs, described below.

3 Selected Refinements to Liquefaction Triggering Procedures

Following similar, prior studies (e.g., Cox et al. 2017; McLaughlin 2017; Yost et al. 2019), several refinements to CPT-based liquefaction triggering analyses were made to evaluate potential improvements of the accuracy of LSIs relative to the field observations made at the Avondale Playground site. Five individual liquefaction triggering analyses using Boulanger and Idriss (2015) were evaluated for each liquefaction severity index (LSI), including that computed: (1) without site-specific adjustments (the base case), (2) using a site-specific, CPT-based *FC* correlation (only), (3) in consideration of partial saturation below the GWT (only), (4) considering thin-layer effects using the rigorous inverse-filtering procedure of Boulanger and DeJong (2018) described below (only), and (5) a combination of all three refinements.

The first refinement consisted of implementation of a site-specific CPT-based FC-I_c correlation using the functional form proposed by Boulanger and Idriss (2015), in which the *FC* obtained from grain size analyses (from samples retrieved from BH 57217) were matched to the corresponding geometric mean of I_c (from SCPT 57354) over the depth of sample recovery. The geometric mean of I_c was fit to *FC* by fitting:

$$I_c = \frac{FC + 137}{80} - C_{FC} \tag{1}$$

to the *FC* using ordinary-least-squares (OLS) regression, where C_{FC} = the site-specific fitting parameter (set to zero for general application in the absence of sufficient site-specific data; Boulanger and Idriss 2015). Figure 4 presents a comparison between the geometric mean of I_c, measured *FC*, and the resulting site-specific correlation with C_{FC} = 0.179 for the Avondale Playground site. The site-specific correlation represents an improvement to the general CPT-based FC-I_c correlation and exhibits good accuracy on average, with a mean bias (defined as the ratio of calculated and correlated I_c) of 1.02, but exhibits significant variability in accuracy, with a coefficient of variation (COV) in bias of 19%, and a mean absolute error of 18% in *FC*. Although not shown here for brevity, Eq. (1) tended to capture *FC* fairly well for depths ranging from the ground surface to 9 m, and poorly in the clean sands below, indicating silty sand rather than clean sand, with implications for clean sand corrections to cyclic resistance.

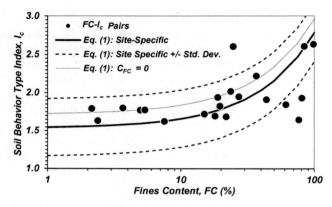

Fig. 4. Comparison of laboratory- and CPT-derived FC using correlation to I_c (Eq. 1).

The second refinement consisted of inverse-filtering CPT measurements to correct for thin-layer effects. This is done following the procedure developed by Boulanger and DeJong (2018); space limitations prevent a presentation of this correction. The third single refinement considered is used to account for the effects of partial saturation of soil below the GWT. This is done by increasing the cyclic resistance ratio, CRR (Hossain et al. 2013):

$$CRR_{corrected} = CRR * K_S \qquad (2)$$

for those depths below the GWT for which $V_p < 1{,}400$ m/s using a partial saturation adjustment factor, K_S. Refer to Hossain et al. (2013) for details. Due to the availability of a single crosshole test, the V_p measured from DPCH 57062 was assumed to apply to all CPT locations, assuming no variation in the depth of full saturation across Section A-A' due to the lack of additional data. The CPTs advanced by Oregon State University indicated that the GWT varied from 1.82 to 2.15 m depth, and concurred with that observed during the earlier CPTs (i.e., 1.95 to 2.07 m). Thus, errors introduced through the assumption that the V_p measured at DPCH 57062 represents the entirety of Section A-A' are likely to be small.

4 Effect of Refinements to Liquefaction Triggering on Liquefaction Severity Indices (LSIs)

This study implements the CPT data presented in Section A-A' to conduct liquefaction triggering analyses using the procedures developed by Boulanger and Idriss (2015) to compute the various LSIs. The factor of safety against liquefaction triggering, FS_{liq}, for two CES events were considered: the M_w 7.1 September 2010 Darfield earthquake (Event 1) and the M_w 6.2 February 2011 Christchurch earthquake (Event 2). Liquefaction severity was computed using five LSIs: the *LPI* (Iwasaki et al. 1978), LPI_{ISH} (Maurer et al. 2015), *LSN* (van Balleygooy et al. 2015), and two procedures to estimate vertical ground settlement including the Zhang et al. (2002) and Yoshimine et al. (2006) procedures. Table 1 presents pertinent details for Events 1 and 2 at the Avondale Playground site including event magnitude, source-to-site distance, *PGA*, and observed liquefaction severity. Aerial imagery and Tonkin+Taylor (2015) suggest the lack of manifestation of liquefaction after Event 1, whereas distinct portions of the site (i.e., from approximately 0 to 6.5 m along Section A-A') exhibited local manifestation of liquefaction as result of Event 2, deemed of moderate severity. Differences between the two events (e.g., source-to-site distance, intensity) help to explain some of the factors producing differing responses to strong ground motion.

Table 1. Characteristics of Event 1 and Event 2 at the Avondale Playground site.

Event parameters & observations	Event 1: September 2010	Event 2: February 2011
Magnitude, M_w	7.1	6.2
Source-to-site distance (km)	47	7.5
Peak ground acceleration, PGA[a] (g)	0.18	0.40
Observed liquefaction severity[b]	None to minor	Moderate

[a] Mean PGA from conditional PGA distributions after Bradley (2014).
[b] Per Tonkin & Taylor (2015); refer to Table 2.

Figure 5 compares the effects of refinements on the *LSN* computed for SCPT 57354 as a representative example of the site. Of the individual refinements, the site-specific *FC* led to the greatest decrease in predicted liquefaction severity, whereas correcting for thin layer effects and partial saturation had the second and third greatest effect, respectively. The *LSN* computed using the measured CPT data (unrefined; Fig. 5) for Event 1 is more than three times larger than that computed using the combination of refinements, to indicate that the majority of the predicted severity is contributed by the depths ranging from approximately 2 m (i.e., the GWT) to 5 m, the depth corresponding to the transition in shear stiffness. The refinement of the liquefaction triggering analysis to include all three adjustments leads to significant decreases in predicted liquefaction severity at this exploration, which is generally consistent with observations based on the other LSIs considered, as described below.

The variation of each LSI for Events 1 and 2 along Section A-A' is presented in Fig. 6 for the unrefined and fully refined liquefaction triggering analyses to identify potential trends between the calculated and observed liquefaction severity and their spatial variation. The comparison of observed and computed ranges in liquefaction severity for indices *LSN*, *LPI*, and *LPI$_{ISH}$* implements the framework suggested by McLaughlin (2017; Table 2) to relate ranges in predicted severity to severity thresholds. The thresholds denote index magnitudes for increasing likelihoods of surficial evidence of liquefaction manifestation. For further details, the reader is referred to Yost et al. (2019), and a discussion of the evolution and adaptation of this framework for assessment of liquefaction severity in the CES.

Considering the *LSN*, *LPI*, and *LPI$_{ISH}$* computed for Event 1 (Figs. 6a–6c), use of the unrefined liquefaction triggering analyses results in a correct prediction of observed liquefaction severity (i.e., none to minor) for the vast majority of CPT profiles. Exceptions

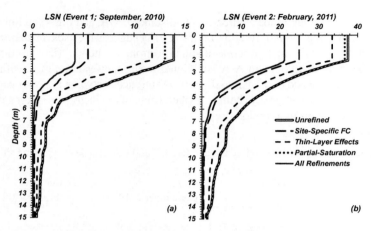

Fig. 5. Comparison of refinement methods for the liquefaction severity index *LSN* for SCPT 57354 for the: (1) September 2010, and (b) February 2011 earthquakes.

Table 2. Tonkin+Taylor (2015) Liquefaction Severity Prediction Categories for LSIs.

Liquefaction severity index	None to marginal	Moderate	Severe
LPI	<8	8 < LPI < 15	>15
LPI_{ISH}	<5	5 < LPI_{ISH} < 11	>11
LSN	<16	16 < LSN < 26	>26

to this observation are identified for *LSN* corresponding to the CPTs located between 14 and 21.2 m along the section, which exhibit *LSN*s ranging from approximately 16 to 20, just above the threshold for moderate severity. However, upon incorporating the site-specific *FC* correlation, accounting for partial saturation below the GWT, and correcting for thin layer effects on CPT data, all of the indices are shown to correctly predict the lack of surficial liquefaction manifestation. These observations for Event 1 extend to the ground surface settlement calculated using the Zhang et al. (2002) and Yoshimine et al. (2006) procedures (Figs. 6d and 6e), which provide average settlements of approximately 70 and 150 mm, respectively, prior to application of the three refinements; adjusting the liquefaction triggering analyses reduces the computed settlement to an average of about 25 and 65 mm, respectively. For reference, the analysis of LIDAR data (NZGD 2012) suggests that 0 to 100 mm of settlement occurred along Section A-A' following Event 1 (the coarse bin size reflects various sources of error). Accordingly, it is suggested that application of the refinements served to improve the accuracy in the computed post-earthquake settlements.

The average ground surface settlement calculated for Event 2 using the Zhang et al. (2002) and Yoshimine et al. (2006) procedures are approximately 160 and 260 mm and 85 and 170 mm, respectively, for the unrefined and fully refined cases (Figs. 6d and 6e), compared to the bin of 100 to 200 mm suggested by the LIDAR data within NZGD (2012). The range in settlement calculated for Event 2 confirm that refining the liquefaction triggering analyses improves the relative accuracy in the prediction for this site. Considering the *LSN*, *LPI*, *LPI$_{ISH}$* computed for Event 2 (Figs. 6a–6c), use of the unrefined liquefaction triggering analyses suggests that severe liquefaction manifestation should have been observed nearly all of the CPTs in Section A-A'. However, the observed severity is considered moderate, with manifestation noted from approximately 0 to 6.5 m along the section. Use of the fully-refined liquefaction triggering analyses reduces the magnitude of each LSI to generally fall within the "moderate" category suggested by Tonkin+Taylor (2015). Figure 6 also illustrates the notable spatial variation in any given LSI across Section A-A', characterized by coefficients of variation ranging from 15 to 23% for the three LSIs. The local reduction in the LSIs from 11 and 12.5 m reflects very soft to soft lens A (Fig. 3), which serves to significantly reduce the liquefiable thickness of soils nearest the ground surface that contribute greatest to the magnitude of the LSI (e.g., Fig. 5). The generally larger magnitude of the LSIs from ~14 to 21 m (relative to 0 to 6.5 m, corresponding to liquefaction manifestation), suggests that use of simplified methods with LSIs is not sufficient for prediction of liquefaction manifestation. Comparison of Figs. 3 and 6 show that beyond ~10 m (along Section A-A'), distinct lenses (A and B) serve to constrain vertical migration of excess pore pressure generated during shaking which can serve to prevent the manifestation of liquefaction (Cubrinovski et al. 2019). The spatially variable stratigraphy at this site leads to different system responses for Event 2 that were not detectable using liquefaction severity indices alone. Such system response-based processes are observed to be repeatable, as demonstrated by Cubrinovski et al. (2019) and illustrated at the Avondale Playground site following the December 2011 event (Fig. 1).

Nonetheless, use of LSIs for routine projects will likely continue where budget for nonlinear effective stress site-response analyses cannot be allocated. Accordingly, global assessments of the accuracy of LSIs remain useful, such as that presented in Fig. 7 for all of the CPTs and LSIs considered herein. The accuracy of the LSIs along Section A-A' were categorized as accurate, overpredicted, or underpredicted relative to the site-wide designation (rather than local, as described above) reported by Tonkin+Taylor (2015), similar to the assessments reported by Cox et al. (2017), McLaughlin (2017), and Yost et al. (2019). Sixty-seven LSIs computed without adjustment to the liquefaction triggering analyses are considered accurate, with 23 resulting in an over-prediction of liquefaction severity for Event 1 (Fig. 7a). Application of refinements for Event 1 increases the correct severity predictions to 86 (of 90 combinations; Fig. 7b). For Event 2 and prior to refinement, 68 cases produced an over-prediction of liquefaction severity with 20 cases accurately predicted (Fig. 7c). Application of all refinements increased the number of accurate cases to 68, but returned 20 cases that under-predicted the severity. Inspection of Fig. 7d shows that 15 of the 20 were associated with the calculation of settlement using the Zhang et al. (2002) procedure. In general, application of the refinements to the liquefaction triggering analyses result in improvements in accuracy of site-wide liquefaction severity; however, the discussions regarding local deviations in observed and predicted accuracy (Fig. 6) suggest that these assessments continue to exhibit distinct shortcomings.

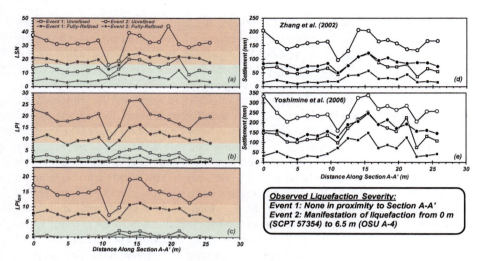

Fig. 6. Variation of fully-refined and unrefined (a) *LSN*, (b) *LPI*, (c) *LPI$_{ISH}$*, and settlement computed using the (d) Zhang et al. (2002) and (e) Yoshimine et al. (2006) procedures with distance across Section A-A' at the Avondale Playground site for the September 2010 (Event 1) and February 2011 (Event 2) earthquakes.

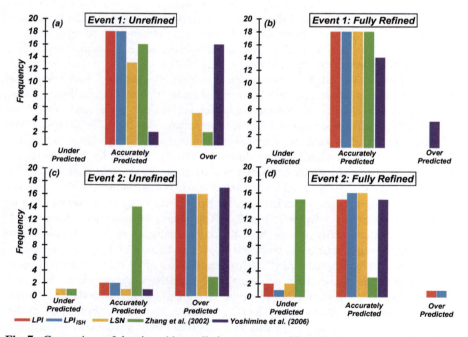

Fig. 7. Comparison of the site-wide prediction accuracy of liquefaction severity indices for the Avondale Playground: (a) Event 1 without refinements, (b) Event 1 with all refinements, (c) Event 2 without refinements, and (d) Event 2 with all refinements.

5 Concluding Remarks

Determining the potential for severe liquefaction during seismic loading continues to present significant challenges to the profession. This study aimed to evaluate the accuracy of selected liquefaction severity indices along a high resolution subsurface cross-section developed for the Avondale Playground case history site to identify the potential for refinements to CPT-based liquefaction triggering procedures to improve the prediction of severity. In general, refinements to account for site-specific fines content correlations, partial saturation, and thin layers served to improve the accuracy of site-wide liquefaction severity. However, the deviation of local liquefaction severity from that observed provides further evidence to support the previously noted role of varying soil stratigraphy with distinct contrasts in hydraulic properties to restrict migration of excess pore pressure during shaking and yield varying system responses across a site on the scale of meters.

Acknowledgments. This work was funded in part by the National Science Foundation (NSF) through Grant CMMI 1931069. Any opinions or findings expressed are those of the authors and do not necessarily reflect the views of NSF.

References

Bastin, S.H., Cubrinovski, M., Van Balleggoy, S., Russell, J.: Geologic and geomorphic influences on the spatial extent of lateral spreading in Christchurch, New Zealand. In: Performance Based Design in Earthquake Geotechnical Engineering, Part III, Vancouver, B.C. (2017)

Beyzaei, C.Z., Bray, J.D., Van Ballegooy, S., Cubrinovski, M., Bastin, S.: Depositional environment effects on observed liquefaction performance in silt swamps during the Canterbury earthquake sequence. Soil Dyn. Earthq. Eng. **107**, 303–321 (2018)

Bradley, B.A.: Site-specific and spatially-distributed ground-motion intensity estimation in the 2010–2011 Canterbury earthquakes. Soil Dyn. Earthq. Eng. **61**, 83–91 (2014)

Boulanger, R.W., Idriss, I.M.: CPT-based liquefaction triggering procedure. J. Geot. Geoenv. Eng. **142**(2), 04015065 (2015)

Boulanger, R.W., Dejong, J.T.: Inverse filtering procedure to correct cone penetration data for thin-layer and transition effects. In: Cone Penetration Testing 2018, pp. 25–44 (2018)

Begg, J.G., Jones, K.E., Barrell, D.J.A.: Geology and geomorphology of urban Christchurch and eastern Canterbury. GNS Science Geological Map 3, Lower Hutt, New Zealand (2015)

Cox, B.R., McLaughlin, K.A., Van Ballegooy, S., Cubrinovski, M., Boulanger, R., Wotherspoon, L.: In-situ investigation of false-positive liquefaction sites in Christchurch, New Zealand: St. Teresa's School case history. In: Performance Based Design in Earthquake Geotechnical Engineering, Part III, Vancouver, B.C. (2017)

Cubrinovski, M., et al.: Geotechnical aspects of the 22 February 2011 Christchurch earthquake. Bull. NZ Soc. Earthq. Eng. **44**(4), 205–226 (2011)

Cubrinovski, M., Rhodes, A., Ntritsos, N., Van Ballegooy, S.: System response of liquefiable deposits. Soil Dyn. Earthq. Eng. **124**, 212–229 (2019)

Hossain, A.M., Andrus, R.D., Camp, W.M., III.: Correcting liquefaction resistance of unsaturated soils using wave velocity. J. Geot. Geoenv. Eng **139**(2), 277–287 (2013)

Iwasaki, T., Tatsuoka, F., Tokida, K. I., Yasuda, S.: A practical method for assessing soil liquefaction potential based on case studies at various sites in Japan. In: 2nd International Conference on Microzonation, Washington, DC, National Science Foundation, pp. 885–896 (1978)

Markham, C.S., Bray, J.D., Macedo, J., Luque, R.: Evaluating nonlinear effective stress site response analysis using records from the Canterbury earthquake sequence. Soil Dyn. Earthq. Eng. **82**, 84–98 (2016)

Maurer, B.W., Green, R.A., Taylor, O.S.: Moving towards an improved index for assessing liquefaction hazard: lessons from historical data. Soils Found. **55**, 778–787 (2015)

Mayne, P.W.: NCHRP Synthesis 368: Cone penetration testing. Transportation Research Board, Washington, DC (2007)

McLaughlin, K.A.: Investigation of False-Positive Liquefaction Case History Sites in Christchurch, New Zealand. M.S. thesis, The University of Texas at Austin, Austin, Texas (2017)

NZGD (New Zealand Geotechnical Database): Vertical Ground Surface Movements. Map Layer CGD0600 (2012). https://canterburygeotechnicaldatabase.projectorbit.com/. Accessed 10 Mar 2020

Robertson, P.K.: Interpretation of cone penetration tests – a unified approach. Can. Geot. J. **46**, 1337–1355 (2009)

Salgado, R., Prezzi, M.: Computation of cavity expansion pressure and penetration resistance in sands. Int. J. Geom. **7**(4), 251–265 (2007)

Tonkin & Taylor Ltd.: Canterbury Earthquake Sequence: Increased Liquefaction Vulnerability Assessment Methodology. Chapman Tripp acting on behalf of the Earthquake Commission (EQC), Tonkin & Taylor ref. 52010.140.v1.0 (2015)

van Ballegooy, S., Green, R.A., Lees, J., Wentz, F., Maurer, B.W.: Assessment of various CPT based liquefaction severity index frameworks relative to the Ishihara (1985) H1-H2 boundary curves. Soil Dyn. Earthq. Eng. **79**(B), 347–364 (2015)

Yoshimine, M., Nishizaki, H., Amano, K., Hosono, Y.: Flow deformation of liquefied sand under constant shear load and its application to analysis of flow slide of infinite slope. Soil Dyn. Earthq. Eng. **26**(2–4), 253–264 (2006)

Yost, K.M., Cox, B.R., Wotherspoon, L., Boulanger, R.W., van Ballegooy, S., Cubrinovski, M.: In situ investigation of false-positive liquefaction sites in Christchurch, New Zealand: Palinurus Road case history. In: Geo-Congress 2019, pp. 436–451 (2019)

Zhang, G., Robertson, P.K., Brachman, R.W.I.: Estimating liquefaction-induced ground settlements from CPT for level ground. Can. Geot. J. **39**(5), 1168–1180 (2002)

Effect of Membrane Penetration on the Undrained Cyclic Behavior of Gravelly Sands in Torsional Shear Tests

Matthew Gapuz Chua[1]($\boxtimes$), Takashi Kiyota[2], Masataka Shiga[2], Muhammad Umar[3], and Toshihiko Katagiri[2]

[1] Department of Civil Engineering, Univeristy of Tokyo, Tokyo, Japan
mattchua@iis.u-tokyo.ac.jp
[2] Institute of Industrial Science, Univeristy of Tokyo, Tokyo, Japan
[3] National University of Computer and Emerging Sciences, Lahore Campus, Pakistan

Abstract. While there is extensive research conducted on liquefaction of sandy soils, only a limited number of studies have been done to understand the liquefaction behavior of gravelly soils. Even with the few case histories that have been documented, several disasters have highlighted the need for further research. To investigate the liquefaction characteristics of gravelly soils in laboratory tests, due consideration must be made on the effect of membrane penetration (MP). MP may be defined as the intrusion/extrusion of membrane into the specimen voids as confining pressure is increased/decreased, respectively. In addition, the effect of membrane force on the measured shear stress must be considered for large strains. This study presents a method for minimizing the effect of MP in undrained torsional shear tests. In this method, a thin layer of sand is used as a MP minimization layer on both the inner and outer surfaces of the specimen. Undrained cyclic torsional shear tests while keeping the specimen height constant were performed on saturated gravelly sand. Based on the results, the undrained cyclic behavior of gravelly soils was investigated.

Keywords: Gravelly soil · Liquefaction · Torsional shear test · Membrane penetration

1 Introduction

In the past, liquefaction was generally considered to occur in sandy soils. Despite this, several cases of gravelly soil liquefaction have been reported during past earthquakes. Some of the more recent case studies include the 2008 Wenchuan earthquake, China [1], the 2014 Cephalonia earthquake, Greece [2], and the 2016 Kaikoura earthquake, New Zealand [3]. It is important to note that the cases varied greatly from natural to artificial fills, flat to sloping ground and small to large deformations. This emphasizes the need for further research on the liquefaction of gravelly soils. However, when the liquefaction characteristics are examined in element tests, the effect of membrane penetration (MP), which hinders pore water pressure generation must be considered.

There are generally two approaches to deal with MP in undrained cyclic loading tests. The first is to apply corrections to the test result. In this approach, the effect of MP on the measured effective stress is corrected after the test [4]. The second approach is to eliminate the MP during the test. Common methods include placing fine sand around the specimen [5, 6], sluicing, or by using an water injection system [7–9]. In this study, MP was eliminated using fine sand placed around the specimen. The results are then compared with available correction methods. It is important to note that, most of the previous studies are based on triaxial tests on uniform soils. As far as the authors know, there have been no study on the use of MP-reducing layer on the hollow cylindrical torsional shear apparatus.

2 Test Apparatus, Material and Procedure

A strain controlled hollow cylindrical torsional shear apparatus was used in this study (Fig. 1). This apparatus was developed at the Institute of Industrial Science, University of Tokyo. By keeping the height constant, it is able to achieve the quasi-simple shear condition which is more representative of actual conditions during an earthquake [10, 11]. Some modifications were made on the system to achieve large torsional shear strain up to 100% [12]. For the specimen dimensions, outer diameter of 200mm, inner diameter of 120 mm, and height of 300mm were used in this study.

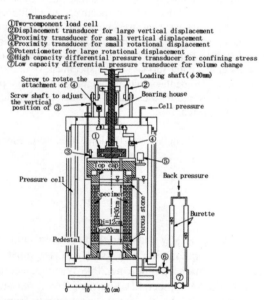

Fig. 1. Schematic diagram of the hollow cylindrical torsional shear apparatus adopted from [13]

Membrane force correction must be considered for torsional shear tests [14]. To calculate the actual stress applied on soils, the shear stress measured by the load cell must be corrected for the apparent shear stress induced by the membrane. This correction is significantly more important at larger shear strains as point out by previous studies [12,

15]. Due to the modification in specimen dimensions, a confirmatory water specimen test was conducted to verify the applicability of the previous corrections. The test was performed by filling both the cell and specimen with water and shearing under the undrained condition. The initial and sheared state of the water specimen are shown in Fig. 2a. and Fig. 2b, respectively. Figure 2c shows the comparison of the current and previous studies [15, 16]. The results show that the previously proposed correction is still valid for this study. This means that the membrane force correction is valid for different specimen size and height/diameter ratios.

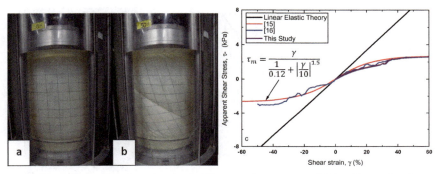

Fig. 2. a. Water specimen before shearing **b.** Water specimen at $\gamma_{sa} = 30\%$ **c.** Comparison between the measured and calculated membrane force.

The material tested in this study was a mixture of silica sand #7 and gravel. Gravel size was restricted to less than 6.4 mm due to sizing requirements for torsional shear tests [17]. Gravel content (GC) is defined as mass ratio between gravel and total dry weight of specimen. For this study, GC = 30% was adopted based on the historical average gravel content of the case histories. The grain size distribution curves of all the materials are shown in Fig. 3.

Table 1. Material properties of tested soil

	Sand	Gravel	GC = 30%
Specific Gravity, G_S	2.64	2.75	2.68
Max density, ρ_{max} (g/cm^3)	1.589	1.853	1.772
Min density, ρ_{min} (g/cm^3)	1.264	1.685	1.513
Max void ratio, e_{max}	1.092	0.632	0.768
Min void ratio, e_{min}	0.664	0.484	0.510

Material properties of the soils are shown in Table 1. Maximum and minimum densities were determined based on [18]. Two test cases were conducted in this study to compare the effect of the sand layer at the specimen surface to the undrained cyclic behavior of gravelly sands.

Dry tamping in 10 layers was used to form the specimens. A relative density (D_r) of 50% was maintained for the tests. Inner and outer split type molds were used to set up the specimen. For the specimen with sand layer, a triple pane mold was used to separate the core material and the outer and inner sand layers (Fig. 4).

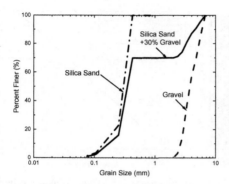

Fig. 3. Grain size distributions

Fig. 4. Specimen preparation to prevent MP

The sand layer's thickness used in this study was set to 4mm. After completing each layer of soil, the mold was carefully removed and a vacuum of −30 kPa was applied to the specimen. The cell was then placed and filled with water to an equal height of the specimen. The specimen was then double vacuumed for 12 h and saturated while maintaining a −30 kPa pressure difference between the specimen and the cell. A back pressure of 200 kPa was applied and a B value not smaller than 0.98 was achieved in all the tests. The specimens were isotropically consolidated to an effective stress of 100 kPa. Undrained cyclic torsional tests with constant shear stress amplitude of 20kPa which is equivalent to a cyclic stress ratio (CSR) of 0.2 were then performed at shear strain rate of 3%/min. The number of cycles to achieve $\gamma_{DA} = 7.5\%$ and excess pore water pressure ratio = 0.95 were used to define liquefaction resistance for this study. The results of tests with sand layer and no sand layer at the specimen surface were compared and MP correction was applied to the test with no sand layer to verify if the use of the sand layer gives reasonable results.

3 Test Results and Discussion

Figure 5 shows effective stress path and torsional shear stress-shear strain relationship.

As shown in Fig. 5, there is a significant difference in the number of cycles (Nc) to reach liquefaction (i.e. γ_{DA} = 7.5%). For the case where no sand layer was used, the specimen liquefied in 29 cycles. In contrast, the specimen with sand layer liquefied in only 4.5 cycles. In Fig. 6, excess pore water pressure ratio (EPWPR) is plotted against number of cycles, liquefaction is judged to have occurred when EPWPR = 0.95. The results show that the specimen with the sand layer liquefied after 3.5 cycles while the specimen with no sand layer liquefied after 22.5 cycles. Regardless of liquefaction criteria, a large disparity in liquefaction resistance was observed, and this may be caused by the MP.

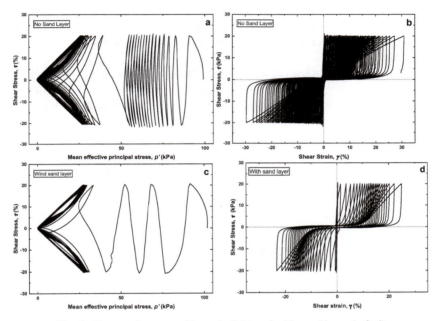

Fig. 5. Test results no sand layer (a & b) and with sand layer (c & d)

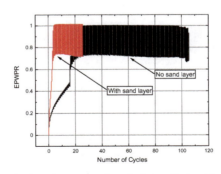

Fig. 6. Pore water pressure generation during undrained torsional loading

To confirm the effectivity of the sand layer in eliminating MP in undrained torsional shear tests, existing MP correction [4, 19] was applied to the results of the no sand layer.

The MP characteristic of any material is usually obtained by doing drained consolidation tests. D_{50} and D_{20} have been shown to have a good correlation with normalized MP. D_{50} is preferred for uniform soils while D_{20} was shown to have a better correlation for non-uniform soils [19]. Shown in Table 2 and Table 3 are the correction factors obtained for the test material using D_{50} and D_{20}, respectively. The variables needed for the correction are rebound factor (C), normalized membrane penetration (S), compliance ratio (Crm) and cyclic ratio (Cn). The number of cycles without membrane penetration (No) is obtained by dividing Nc by Cn.

Regardless of liquefaction criteria, there is significant difference in the number of cycles between the case with the sand layer and that of without the sand layer. The correction using D_{20} is quite near to that of the results from the test with sand layer. This confirms that especially for non-uniform soils, D_{20} is a good indicator of MP characteristics.

As the specimens are sheared, strain accumulates until liquefaction occurs. Figure 7 shows the plot of strain accumulation against the number of cycles. After liquefaction ($\gamma_{DA}= 7.5\%$), the shear strain accumulation of the two tests seems to become parallel.

Table 2. Correction Coefficients [4]

	D_{50}	C	S	Crm	Cn	Nc	No	No'
$\gamma_{DA}=7.5\%$	0.36	0.003	0.002	0.309	1.673	29	17	4.5
EPWPR = 0.95						22.5	13.5	3.5

Table 3. Correction Coefficients [19]

	D_{20}	C	S	Crm	Cn	Nc	No	No'
$\gamma_{DA}= 7.5\%$	0.263	0.003	0.004	0.673	3.181	29	9	4.5
EPWPR = 0.95						22.5	7	3.5

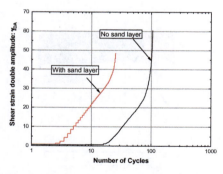

Fig. 7. Large strain development observed during undrained cyclic torsional loading

4 Conclusions

This study investigated the effect of MP to the undrained cyclic behavior of gravelly sands in torsional shear tests. This study makes use of a sand layer to eliminate the MP in torsional shear tests. The results showed that regardless of liquefaction criteria, there is significant effect of MP in undrained torsional shear tests. Although D_{50} has a good correlation with MP for uniform soils, the use of D_{20} correction in non-uniform soils resulted in reasonable estimation compared to that of the sand layer method. Shear strain accumulation seem to become parallel between the two tests. To understand the reasons behind this phenomenon, additional tests will be carried out in the future.

Acknowledgements. This study was supported by JSPS KAKENHI Grant Number 20H02240. The Japanese government (Monbukagakusho: MEXT) is greatly appreciated for the student scholarship provided. Many thanks are also extended to Mr. Mamoru Wakasugi of Kiso-Jiban Consultants Co. For his support in this study.

References

1. Yuan, X., Chen, L., Cao, Z.: The world's largest naturally deposited Gravelly Soils Liquefaction caused by the Wenchuan Ms 8.0 Earthquake. In: IACGE 2018, pp. 245–252, 31 Oct 2019. https://doi.org/10.1061/9780784482049.024
2. Nikolaou, S., Zekkos, D., Assimaki, D., Gilsanz, R.: GEER/EERI/ATC Earthquake Reconnaissance January 26th/ February 2nd 2014 Cephalonia, Greece Events (2014)
3. Cubrinovski, M., et al.: Liquefaction effects and associated damages observed at the Wellington centreport from the 2016 Kaikoura earthquake. Bull. New Zeal. Soc. Earthq. Eng. **50**(2), 152–173 (2017). https://doi.org/10.5459/bnzsee.50.2.152-173
4. Tokimatsu, K., Nakamura, K.: A simplified correction for membrane compliance in liquefaction tests. Soils Found. **27**(4), 111–122 (1987). https://doi.org/10.3208/SANDF1972.27.4_111
5. Miura, S., Kawamura, S.: A procedure minimizing membrane penetration effects in undrained Triaxial Test. Soils Found. **36**(4), 119–126 (1996). https://doi.org/10.3208/SANDF.36.4_119
6. Toyota, H., Takada, S.: Effects of gravel content on liquefaction resistance and its assessment considering deformation characteristics in gravel – mixed sand. Can. Geotech. J. **56**(12), 1743–1755 (2019). https://doi.org/10.1139/cgj-2018-0575
7. Tokimatsu, K., Nakamura, K.: A liquefaction test without membrane penetration effects. Soils Found. **26**(4), 127–138 (1986). https://doi.org/10.3208/SANDF1972.26.4_127
8. Nicholson, P.G., Seed, R.B., Anwar, H.A.: Elimination of membrane compliance in undrained triaxial testing. II. Mitigation by injection compensation, vol. 30(5), pp. 739–746 (1993). https://doi.org/10.1139/T93-066
9. Sivathayalan, S., Vaid, Y.P.: Truly undrained response of granular soils with no membrane-penetration effects, vol. 35(5), pp. 730–739 (1998). https://doi.org/10.1139/T98-048
10. Tatsuoka, F., Sonoda, S., Hara, K., Fukushima, S., Pradhan, T.B.S.: Failure and deformation of sand in torsional shear. Soils Found. **26**(4), 79–97 (1986). https://doi.org/10.3208/SANDF1972.26.4_79
11. Tatsuoka, F., Pradhan, T.B.S., Yoshi-ie, H.: Cyclic undrained simple shear testing method for soils. Geotech. Test. J. **12**(4), 269–280 (1989). https://doi.org/10.1520/gtj10984j

12. Kiyota, T., Sato, T., Koseki, J., Abadimarand, M.: Behavior of liquefied sands under extremely large strain levels in cyclic torsional shear tests. Soils Found. **48**(5), 727–739 (2008). https://doi.org/10.3208/sandf.48.727
13. De Silva, L.I.N.: Deformation characteristics of sand subjected to cyclic drained and undrained torsional loadings and their modelling. University of Tokyo, Tokyo (2008)
14. Koseki, J., Yoshida, T., Sato, T.: Liquefaction properties of Toyoura sand in cyclic Tortional shear tests under low confining stress. Soils Found. **45**(5), 103–113 (2005). https://doi.org/10.3208/SANDF.45.5_103
15. Chiaro, G., Kiyota, T., Miyamoto, H.: Large deformation properties of reconstituted Christchurch sand subjected to undrained cyclic torsional simple shear loadings. In: 2015 NZSEE Conference (2015)
16. Umar, M., Kiyota, T., Chiaro, G., Duttine, A.: Post-liquefaction deformation and strength characteristics of sand in torsional shear tests. Soils Found. **61**(5), 1207–1222 (2021). https://doi.org/10.1016/J.SANDF.2021.06.009
17. JGS, Japanese Geotechnical Society Standard (JGS 0550–2009) Practice for preparing hollow cylindrical specimens of soils for torsional shear test (2009)
18. JGS, Japanese Geotechnical Society Standard (JGS 0162–2009) Test method for minimum and maximum densities of gravels (2009)
19. Nicholson, P.G., Seed, R.B., Anwar, H.A.: Elimination of membrane compliance in undrained triaxial testing. I. Measurement and evaluation, vol. 30(5), pp. 727–738 (1993). https://doi.org/10.1139/T93-065

Implementation and Verification of an Advanced Bounding Surface Constitutive Model

Tony Fierro[1](✉), Stefano Ercolessi[1], Massimina Castiglia[1],
Filippo Santucci de Magistris[1], and Giovanni Fabbrocino[1,2]

[1] University of Molise, via De Sanctis 1, 86100 Campobasso, Italy
tony.fierro@unimol.it
[2] ITC-CNR, L'Aquila Branch, Via G. Carducci, 67100 L'Aquila, Italy

Abstract. The description of the mechanical behavior of loose saturated sandy soils under large cyclic loading represents a complicated task in earthquake geotechnical engineering because of the possible occurrence of liquefaction. The physical and numerical reproduction of this phenomenon is of great support for the protection of the built environment and the assessment of new constructions. To this end, a large number of constitutive models were developed over the years and integrated into computational platforms to describe the soil response to seismic actions. The present paper deals with the implementation and testing of the constitutive model developed by Papadimitriou and Bouckovalas (2002) into the OpenSees environment with the aim of investigating its performance at the element level. The model provides some enhancements with reference to the previous common elastoplastic models, such as a non-linear elastic hysteretic Ramberg-Osgood formulation, and an empirical index that accounts for the fabric evolution. The reliability of the implementation is analyzed through the simulation of drained and undrained, cyclic and monotonic laboratory tests from literature, regarded as a benchmark for the verification and validation of the constitutive model. Overall, the implementation procedure produced satisfactory results and the OpenSees potential in hosting user-defined codes is demonstrated.

Keywords: OpenSees · Constitutive model · Liquefaction · Sand

1 Introduction

The modelling of the response of a soil subjected to liquefaction represents one of the most challenging problems, from a computational point of view, in geotechnical earthquake engineering. Although reliable predictions can dramatically improve the effectiveness of the design procedures in liquefaction-prone areas, the simulations properly describing this phenomenon and its evolution are quite complicated.

It is well-established that the classical elastic perfectly plastic models are not suitable to reproduce the accumulation of high plastic-strain levels in non-cohesive soils, such as liquefaction co-seismic effects (i.e., flow liquefaction or cyclic mobility). To overcome this limitation, specific constitutive frameworks were developed accounting for different initial soil conditions through a single set of parameters (see [5]) using different strategies; in this sense, the bounding surface theory resulted successful. However, these models need to be used in numerical codes with a view to simulating liquefaction-related phenomena. For this reason, a balance between analytical complexity and ease of implementation in numerical tools needs to be achieved.

Based on the above remarks, the present paper deals with the implementation of the constitutive model developed by Papadimitriou and Bouckovalas [8] into the finite element framework OpenSees [6] to increase the set of available constitutive models and guide the interested user towards the solution of soil liquefaction problems. This also highlights the flexibility of the OpenSees numerical framework to support user-defined materials, and the ability of the constitutive platform to interact with finite element codes.

2 Papadimitriou and Bouckovalas (2002) Constitutive Model

The Papadimitriou and Bouckovalas [8] constitutive platform derives from Manzari and Dafalias [5] and represents the multiaxial formulation of the model developed by Papadimitriou et al. [9]. Due to its modularity and flexibility, the Papadimitriou and Bouckovalas [8] model was modified and built in the FDM code FLAC by Andrianopoulos et al. [1], considering a vanished elastic region; Taborda et al. [11] added a low-stress yield surface, included the effect of both void ratio and elastic stiffness on the plastic modulus and slightly modified the distance-to-bounding surface expression. Then, the modified version of the model was implemented in the constitutive framework ICFEP to improve its simulative capabilities. Furthermore, Miriano [7] supplied the original version of the model in Abaqus. The interested reader can refer to [8] for further details since due to space constraints herein only some key points are recalled.

The model is based on the bounding surface theory and is characterized by four different surfaces: yield, dilatancy, critical and bounding. Yield surface is represented by a cone, with the apex in the origin of axes in the p-q plane (Fig. 1a). On the other hand, the intersection of these surfaces with the π-plane (normalized with respect to the mean effective pressure) is reported in Fig. 1b.

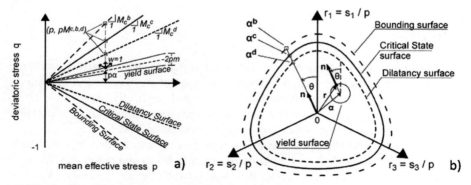

Fig. 1. Constitutive framework in both triaxial (a) and deviatoric (b) planes (from [8]).

Here, the yield surface is represented by a circle with a "diameter" directly related to the constitutive parameter m, while its location in the plane is defined through the back-stress ratio α. The parameter m assumes a constant value, while α can evolve so that only kinematic hardening can be activated and accounted for.

Figure 1b clearly shows that the former three surfaces (dilatancy, critical and bounding surfaces) are wedge-shaped in the π-plane. In particular, the openings of the dilatancy and the bounding surfaces depend on the stress ratio at the critical state, which is kept constant, and on the state parameter ψ, which indicates "how far" the current void ratio is from the void ratio at the critical state.

The loading direction is adopted to map the conjugate critical, bounding and dilatancy back-stress ratios α^i (cfr. Fig. 1b). The signs of the distances d^i between the current back-stress ratio α and its images α^i define whether the soil is looser or denser than critical, hardens or softens, is dilative or contractive.

Another improvement with reference to the Manzari and Dafalias [5] model is a scalar value h_f that directly scales the plastic modulus in order to consider fabric tensor evolution. Finally, the elastic shear modulus recalls the Ramberg-Osgood formulation to reproduce the response at low shear strain levels in a more reliable manner.

3 Implementation in OpenSees

In this section, the implementation of the above-mentioned constitutive model into the OpenSees platform [6] is briefly illustrated. OpenSees is an object-oriented, open-source software framework, mainly developed for research activities, which allows users to develop parallel and serial applications for systems subjected to earthquake loadings, among others. It is primarily written in C++ and the source code is freely available. The implementation of a new multiaxial material requires the addition of a new subclass of the nDMaterial class.

The subclass not only contains all the mandatory methods required by the nDMaterial class, such as those useful for the definition of the current state, but also those aiming at describing the nonlinear response provided by the constitutive framework. For this reason, additional methods for the calculation of state parameter, Lode angle and elastic moduli are implemented, among others.

Different integration scheme performances have been assessed, and Modified Euler with stress correction developed by Sloan et al. [10] was adopted, as in the implementation of the SANISAND [3] model in OpenSees [4]. A relative error is computed and the current state is accepted if this error is lower than a certain tolerance that can be assigned as non-mandatory input parameter.

4 Model Verification

The implementation of the constitutive model proposed by Papadimitriou and Bouckovalas [8] into the OpenSees framework is verified through the simulation of drained and undrained, monotonic and cyclic triaxial tests and direct simple shear tests (DSS) available in the literature. All the analyzed tests have been carried out within the VELACS [2] project on Nevada sand and are summarized in Table 1. The comparative analysis of the simulations performed by various Authors using different versions of the model will be discussed. Constitutive parameters calibrated by [8] for Nevada sand are adopted.

Table 1. Element tests considered for the validation of the model.

Test Type	Test Reference	Simulation Reference	Mean confining Pressure (kPa)	Initial Void Ratio (-)
Undrained Cyclic Triaxial		[8]	80	0.73
Undrained Cyclic Direct Simple Shear	[2]	[8]	160	0.66
Undrained Monotonic		[9]	160/80/40	0.66
Drained Monotonic		[1]	80	0.66/0.73

The performance of different finite elements available in OpenSees is also examined and discussed in the comparisons.

4.1 Undrained Cyclic Direct Simple Shear Tests Simulation

As in the VELACS report [2], the load-control cyclic direct shear test adopted to calibrate and validate the model was performed at the element level on a soil sample of Nevada sand at 60% relative density which corresponds to a 0.66 initial void ratio. Firstly, the consolidation stage with effective consolidation stress equal to 160 kPa was performed and then cyclic shear stress with an amplitude of 13.7 kPa was applied. Furthermore, an initial load offset equal to 5.9 kPa was added before the cyclic load application. Loads are applied as nodal forces.

The τ-σ_y and τ-γ curves resulting from the simulations of the same DSS cyclic undrained test performed by Papadimitriou and Bouckovalas [8] and Miriano [7] are compared to those from the current implementation. In Fig. 2, the tests using quadUP (Fig. 2a and Fig. 2b), SSPquadUP (Fig. 2c and Fig. 2d) and nine-four nodes quadUP (Fig. 2e and Fig. 2f) elements are reported.

The plots in Fig. 2 show a stiffer response exhibited by current simulations in OpenSees (blue curves) than the original ones (black curves). However, better performance is achieved comparing simulations by Miriano [7] and current implementation.

Overall, all the analyzed elements seem to exhibit a good accordance with the stress paths provided by Miriano [7] and Papadimitriou and Bouckovalas [8], while the quadUP shows the shear-stress-shear strain curve closer to that reported in the original model. It is worth noting that the test with nine-four quadUP element seems not to catch stress amplitude accurately.

Furthermore, some multi-element tests were performed by considering the classical plane-strain formulation and meshing the element domain using 2×2 and 4×4 quadUP elements. For the sake of brevity, the comparison is not reported here. However, the response of the single element is confirmed.

The results obtained from the DSS test highlight a stiffer response at low strain levels exhibited by the current implementation and by the Abaqus umat developed by Miriano [7] if compared to that resulting from the simulations provided by Papadimitriou and Bouckovalas [8]. On the contrary, good accordance is reached when the strain level increases. In addition, the classical butterfly-shaped τ-σ_y curve appears properly reproduced. It is worth noting that different numerical platforms are adopted for the simulations.

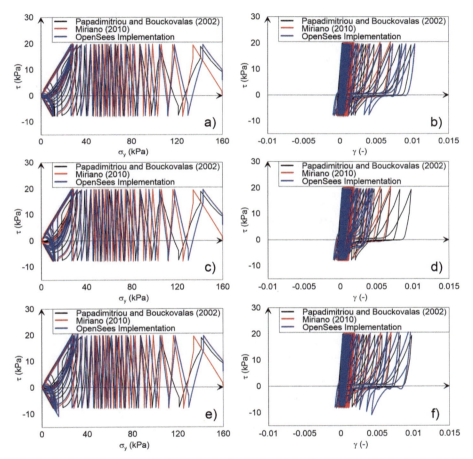

Fig. 2. τ-σ_y and τ-γ curves obtained using plane-strain formulation from [7] (red curves), [8] (black curves) and current implementation (blue curves) using quadUP (a)–(b), SSPquadUP (c)–(d) and nine-four quadUP (e)–(f) elements.

4.2 Undrained Cyclic Triaxial Tests Simulation

The cyclic triaxial test here reproduced was performed on a 2.5 in × 1.0 in sample with 40% relative density, which corresponds to a 0.73 initial void ratio during the VELACS project [2]. In the current simulations, 3D hexahedral elements have been selected. The loading process consists of an anisotropic consolidation of 80 kPa mean effective pressure with 26 kPa offset followed by a cyclic deviatoric load of 43.1 kPa.

Figure 3 and Fig. 4 show the results obtained using brickUP and SSPbrickUP elements.

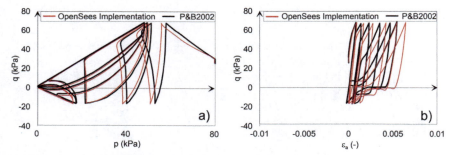

Fig. 3. Deviatoric stress-mean pressure (a) and deviatoric stress-axial strain (b) curves obtained from [8] (black curve) and current implementation adopting a 3D brickUP element (red curve).

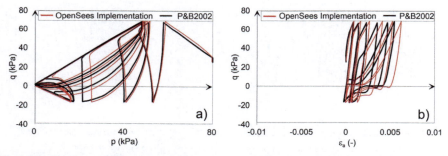

Fig. 4. Deviatoric stress-mean pressure (a) and deviatoric stress-axial strain (b) curves obtained from [8] (black curve) and current implementation adopting a 3D SSPbrickUP element (red curve).

From the simulations of cyclic triaxial tests, a general good accordance is observed. However, the comparison with the original implementation provided by Papadimitriou and Bouckovalas [8] highlights a higher strain reached by the OpenSees implementation. On the other hand, at low strain level, the accordance is satisfactory.

4.3 Undrained Monotonic Triaxial Tests Simulation

Undrained monotonic triaxial tests performed with an initial void ratio equal to 0.66 at different confining pressures (p = 40 kPa, 80 kPa, 160 kPa) are here simulated within the verification procedure. SSPbrickUP hexahedral element is considered, and the resulting deviatoric stress-axial strain responses are shown in Fig. 5.

The undrained monotonic tests show a complete agreement in the comparison between the results obtained in the current study and those resulting from [7].

4.4 Drained Monotonic Triaxial Tests Simulation

Finally, two drained monotonic triaxial tests are investigated. The tests are performed at 80 kPa confining pressure and both 60% and 40% relative density on Nevada Sand. Volumetric vs. axial strain curves simulated in OpenSees are reported in Fig. 6a, while simulations by [1] and test data are shown in Fig. 6b. The SSPbrick element is adopted.

A general agreement is observed; more in detail, both the simulations provided by Andrianopoulos et al. [1] and the current implementation show a lower volumetric strain reached at the end of the test if compared to the real model.

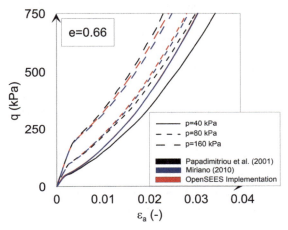

Fig. 5. Simulations of undrained monotonic triaxial tests using the implemented constitutive model (red curve).

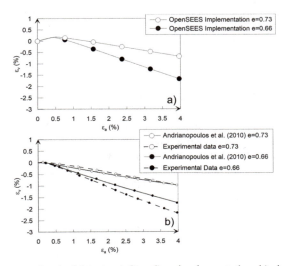

Fig. 6. Drained monotonic triaxial tests: a) OpenSees implementation, b) simulations from [1].

5 Conclusion

Element tests in different conditions have been performed in the verification procedure of the implementation of the constitutive model developed by Papadimitriou and Bouckovalas [9] for liquefaction analyses in OpenSees.

Generally, good accordance between the OpenSees simulations and the results provided by various Authors is observed. The discrepancies can be related either to the different finite element frameworks (e.g., Abaqus and OpenSees) or to the different approaches (e.g., finite elements and finite differences).

The capabilities and the flexibility for the implementation of user-defined materials make OpenSees an extremely valuable tool for geotechnical earthquake engineering applications and for the evaluation of the performances of geotechnical systems.

Finally, it can be stated that the model is successfully verified and its application to boundary value problems is needed.

Acknowledgments. The Authors would like to express their gratitude to Dr. Alborz Ghofrani and Dr. Davide Noè Gorini for their precious support during the implementation of the model.

References

1. Andrianopoulos, K.I., Papadimitriou, A.G., Bouckovalas G.D.: Bounding surface plasticity model for the seismic liquefaction analysis of geostructures. Soil Dyn. Earthq. Eng. **30**, 895–911 (2010)
2. Arulmoli, K., Muraleetharan, K.K., Hossain, M.M., Fruth, L.S.: VELACS verification of liquefaction analyses by centrifuge studies – Laboratory Testing Program – Soil Data Report, Research Report Technology Corporation The Earth (1992)
3. Dafalias, Y.F., Manzari, M.T.: Simple plasticity sand model accounting for fabric change effects. J. Eng. Mech. **130**, 622–634 (2004)
4. Ghofrani, A., Arduino, P.: Prediction of LEAP centrifuge tests results using a pressure dependent bounding surface constitutive model. Soil Dyn. Earthq. Eng. **113**, 758–770 (2018)
5. Manzari, M.T., Dafalias, Y.F.: A critical state two-surface plasticity model for sand. Geotechnique **47**(2), 255–272 (1997)
6. McKenna, F., Fenves, G. L., Scott, M.H., Jeremic, B.: Open system for earthquake engineering simulation (2000). http://opensees.berkeley.edu
7. Miriano, C.: Numerical modelling of the seismic response of flexible retaining structures. Ph.D. thesis, Sapienza University of Rome (in Italian) (2010)
8. Papadimitriou, A.G., Bouckovalas, G.D.: Plasticity model for sand under small and large cyclic strains: a multiaxial formulation. Soil Dyn. Earthq. Eng. **22**, 191–204 (2002)
9. Papadimitriou, A.G., Bouckovalas, G.D., Dafalias, Y.F.: Plasticity model for sand under small and large cyclic strains. J. Geotech. Geoenviron. Eng. ASCE **127**(11), 973–983 (2001)
10. Sloan, S.W., Abbo, A.J., Sheng, D.: Refined explicit integration of elastoplastic models with automatic error control. Eng. Comput. **18**, 121–154 (2001)
11. Taborda, D.M.G., Zdravkovic, L., Kontoe, S., Potts, D.M.: Computational study on the modification of a bounding surface plasticity model for sands. Comput. Geotech. **59**, 145–160 (2014)

Performance of Advanced Constitutive Models in Site Response Analyses of Liquefiable Soils

Tony Fierro(✉), Massimina Castiglia, and Filippo Santucci de Magistris

University of Molise, via De Sanctis 1, 86100 Campobasso, Italy
`tony.fierro@unimol.it`

Abstract. The prediction of liquefiable soils response to earthquake excitations represents one of the most challenging achievements in geotechnical earthquake engineering. For this reason, many constitutive models were formulated to capture the main features of such a complex phenomenon, in the different forms of flow liquefaction and cyclic mobility. These models can catch the soil response at the element level, and nowadays their reliability is increasing with reference to boundary value problems too. To this aim, this paper presents the results of mono-dimensional site response analyses by using an existing advanced bounding surface constitutive model that we implemented in the OpenSees framework. The results are compared to those obtained by adopting other constitutive models already included in the platform. The comparison was possible thanks to the availability of the calibration parameters for the Nevada Sand. The outcomes highlight that this newly implemented constitutive model can be used to address the response of a liquefiable soil profile. However, the discrepancies in the response should be further investigated.

Keywords: OpenSees · Liquefaction · Constitutive modelling · Site response analysis

1 Introduction

A reliable prediction of the response of a soil column subjected to earthquake excitation is a basic although sought goal to numerically address problems of site response analyses. However, when the column is made by liquefiable soil, a reliable prediction becomes extremely hard due to the complexity of the phenomenon.

Indeed, the nonlinear response of the material under earthquake excitations requires the stress and strain distribution in a soil domain to be properly addressed, while the filtering effect exercised by the soil on the input motion should be appropriately reproduced [14]. In this sense, fully-coupled nonlinear dynamic finite element analysis is a precious tool, and the soil non-linearity should be reproduced in a constitutive model able to simulate the shear stress-strain behavior even at higher strain level (e.g., mobilized in liquefaction conditions). To this aim, a lot of constitutive models have been developed over the years and most of them are based on the bounding surface concept.

The current study investigates the response of a 20 m thick soil column of Nevada sand subjected to two different input motions. The analysis is conducted in OpenSees by adopting three bounding surface constitutive models: PM4SAND [2], SANISAND [3], and NTUASand02 [12]. The latter was implemented in the OpenSees platform by the Authors of this paper [4].

2 Finite Element Model

The model of the soil column is built in OpenSees [10] and consists of a 20-m soil column of Nevada sand at about 60% of relative density (corresponding to 0.66 void ratio) modelled using 40 three-dimensional SSPquadUP elements [9]. Quad elements are 1.0×0.5 m^2. The size of the elements was determined to allow the transmission of frequencies up to 20 Hz, a threshold of interest for earthquake engineering problems.

The numerical model was developed so that nodes at the same height are tied together simulating periodic boundaries, while the groundwater table coincides with the top surface of the model. Two Lysmer dashpots [8] were applied at the base nodes in both horizontal and vertical directions to account for the impedance contrast between the soft soil and the underlying bedrock. For the latter, a soil unit weight $\gamma_{bed} = 24.5$ kN/m^3 and a shear wave velocity $V_s = 950$ m/s were assumed. The permeability is assumed to be 1.5×10^{-3} m/s.

Analyses were performed applying kinematic constraints with the Penalty method and a penalty parameter of 1×10^{14}. Finally, the Generalized Alpha method is selected as an integrator with coefficients $\alpha_M = 1.0$ and $\alpha_F = 0.8$. In this way, the unconditional stability $\alpha_M > \alpha_F > 0.5$ is guaranteed. The response is investigated at 1 m, 10 m and 15 m below the ground level (Fig. 1a).

2.1 Constitutive Parameters

In order to avoid the introduction of any bias related to the calibration procedure, we selected for the current study the Nevada Sand, whose calibration with reference to the selected constitutive models is widely available in the literature. Specifically, parameters for NTUASand02 were obtained by Papadimitriou and Bouckovalas [12] on the basis of monotonic and cyclic, drained and undrained triaxial and direct simple shear tests performed within the framework of the VELACS project [1]). The same tests were exploited by Taiebat et al. [13] to calibrate SANISAND. Parameters for PM4SAND were obtained by Kamai and Boulanger [5], based on undrained cyclic triaxial tests, undrained isotropically consolidated torsional hollow cylinder tests, and undrained one-dimensionally consolidated direct simple shear tests. Among others, tests performed by Arulmoli et al. [1] have been exploited. Worth noticing that the constants describing state parameter and peak and dilatancy stress ratios have been modified to match their profiles along the column. A perfect matching cannot be guaranteed due to the different formulations provided by the models.

For the sake of brevity, the parameters adopted in this study are not reported and the interested reader can look at the referenced papers. In addition, the parameter m, which accounts for the "diameter" of the yield surface, was set to 0.0625 for each model: Manzari and Dafalias [11] stated that this value can be assumed equal to 5% of the critical state stress ratio.

The considered constitutive models have three different small-strain stiffness formulations. To allow for the comparison, the parameters defining the shear modulus G for the models are determined based on the shear wave velocity profile shown in Fig. 1b following the relation $V_s = \sqrt{G/\rho}$ where ρ is the soil density.

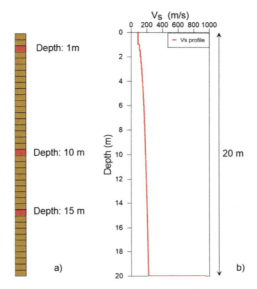

Fig. 1. a) Discretized soil column with control points and b) shear wave velocity profile.

2.2 Input Selection

Two different input motions are applied to investigate the response of the soil column, consisting in recorded acceleration time-histories downloaded from the Engineering Strong Motion Database [7] and PEER Ground Motion Database. These waveforms were selected to examine both cases of liquefaction being triggered or not. The main features are reported in Table 1, while acceleration time-histories and Fourier amplitude spectra are reported in Figs. 2 and 3. The two inputs are referenced in the following recalling the name of the recording station. The sampling interval is 0.0005 s.

Table 1. List of the main features of the selected motions.

Date	Event	Station	Mw (–)	PGA (g)
01/18/2017	Central Italy	Accumoli (ACC)	5.1	0.104
01/16/1995	Kobe	Kakogawa (CUE90)	6.8	0.345

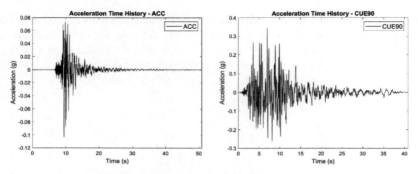

Fig. 2. Input motions adopted for the analyses.

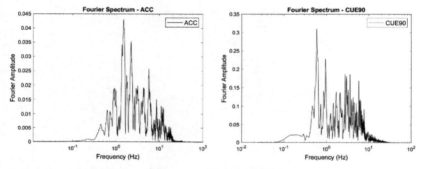

Fig. 3. Fourier spectra of the selected input motions.

2.3 Calibration of Rayleigh Damping

Linear elastic analyses were performed for the calibration of the Rayleigh damping parameters following the procedure suggested by Kwok et al. [6] with two control frequencies. According to this procedure, the first control frequency corresponds to the fundamental frequency of the soil column f_1, while the second control frequency should correspond to f_3, the third natural frequency of the soil column. Consequently, the control frequencies are set as 2.86 Hz and 12.28 Hz respectively.

3 Results

The obtained results are shown at the control points and elements located at 1 m, 10 m and 15 m below the ground level (g.l.), in terms of time-histories of acceleration, excess pore water pressure ratio, and stress-strain loops, in Figs. 4, 5, and 6, respectively. The excess pore water pressure ratio is computed as the ratio between excess pore water pressure and initial vertical effective stress $r_u = \Delta u/\sigma'_{v0}$. Results on the left-hand side are for the ACC input and on the right-hand side for the CUE90 input.

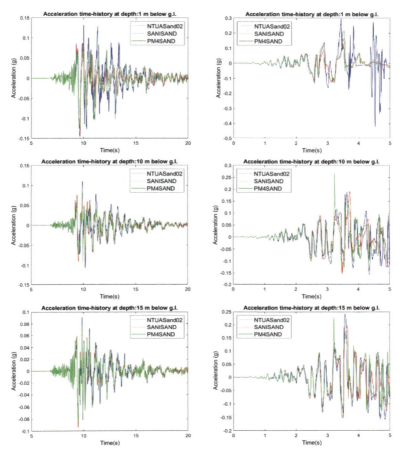

Fig. 4. Acceleration time-histories at 1 m, 10 m and 15 m, below g.l. for ACC (left) and CUE90 (right).

Considering the input named ACC, generally, a good agreement is observed in the acceleration time-histories response at each depth among the different constitutive models (see Fig. 4). Assuming the threshold for liquefaction onset equal to $r_u = 0.95$, the phenomenon is not triggered as it is highlighted by the time-histories of excess pore water pressure ratio reported in Fig. 5. NTUASand02 and PM4SAND show very similar responses while SANISAND generates higher pore water pressure at each depth. Finally, from Fig. 6 emerges that a strong non-linearity is exhibited in the shear stress-shear strain loops for each model.

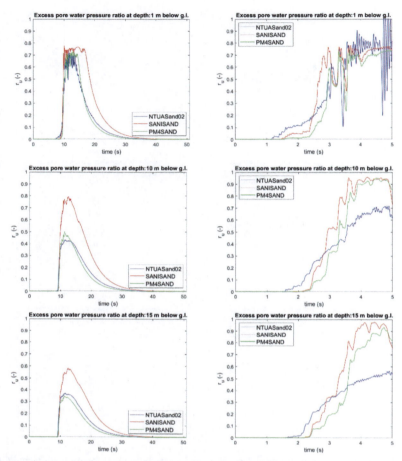

Fig. 5. Excess pore water pressure ratio at 1 m, 10 m and 15 m below g.l. for ACC (left) and CUE90 (right).

More complicated is the response to CUE90, which induces liquefaction since the very early stage of the motion (see Fig. 5). The higher amplitude and the related frequency content make the soil response extremely unstable, with non-significant peaks and a clear limitation in the numerical simulation as it can be observed in Fig. 4 at 1 m below g.l. For this reason, the plots are shown up to 5 s, when the excess pore water pressure ratio is reaching values close to 1 and the models are no longer able to describe the soil patterns. Before the initiation of liquefaction, a general agreement is observed among the models with NTUASand02 showing a more pronounced instability at the top layers (i.e., at about 4 s, at 1 m depth). This is extremely evident in the shear stress-shear strain curves shown in Fig. 6. The same behavior is not observed at higher depths. Finally, at 10 m and 15 m below g.l., only NTUASand02 does not reach liquefaction and this effect is reflected in a stiffer response of shear stress-shear strain curves.

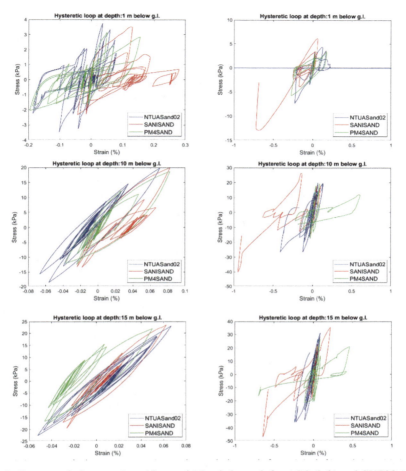

Fig. 6. Stress-strain loops at 1 m, 10 m and 15 m below g.l. for ACC (left) and CUE90 (right).

4 Conclusion

The paper analyzes the response of a 20-m thick soil column of Nevada Sand under two different earthquake motions, using three advanced constitutive models implemented in OpenSees (SANISAND, PM4SAND and NTUASand02). The differences in the response are explained by looking at the acceleration time-histories, the excess pore water pressure ratio time-histories, and the shear stress-shear strain loops at three different depths in the soil profile.

Generally, a good agreement is reached in the simulations. However, the discrepancies should be further investigated with specific evaluations on the formulation of the model, the calibration of the parameters, and the assumptions made for the analysis. These factors should be considered simultaneously, and the parameterization of the analysis would be of great support for the estimation of the sensitivity in the response.

Overall, this preliminary application of the implemented NTUASand02 constitutive model in OpenSees to site response analysis shows a good congruence with the other models and can be considered satisfactory.

Acknowledgments. The Authors would like to thank Dr. Maria Giovanna Durante and Dr. Davide Noè Gorini for their precious suggestions in the development of this study.

References

1. Arulmoli, K., Muraleetharan, K., Hosain, M., Fruth, L.: VELACS laboratory testing program, soil data report Earth Technology Corporation, Irvine, CA Report to the National Science Foundation (1992)
2. Boulanger, R.W., Ziotopoulou, K.: PM4Sand (version 3.1): a sand plasticity model for earthquake engineering applications. Report UCD/CGM-17/01, Center for Geotechnical Modeling, Department of Civil and Environmental Engineering, University of California, Davis, CA (2017)
3. Dafalias, Y.F., Manzari, M.T.: Simple plasticity sand model accounting for fabric change effects. J. Eng. Mech. **130**(6), 622–634 (2004). https://doi.org/10.1061/(ASCE)0733-9399
4. Fierro, T., Ercolessi, S., Castiglia, M., Santucci de Magistris, F., Fabbrocino, G.: Implementation and verification of an advanced bounding surface constitutive model. In: 4th International Conference on Performance- based Design in Earthquake Geotechnical Engineering. Beijing, 15–17 July 2022 (in print)
5. Kamai, R., Boulanger, R.: Simulations of a centrifuge test with lateral spreading and void redistribution effects. J. Geotech. Geoenviron. Eng. **139**, 1250–1261 (2013)
6. Kwok, A.O.L., et al.: Use of exact solutions of wave propagation problems to guide implementation of nonlinear seismic ground response analysis procedures. J. Geotech. Eng. **133**(11), 1385–1398 (2007)
7. Luzi, L., Puglia, R, Russo, E., ORFEUS WG5: Engineering Strong Motion Database, ver. 1.0. Istituto Nazionale di Geofisica e Vulcanologia, Observatories & Research Facilities for European Seismology (2016). https://doi.org/10.13127/ESM
8. Lysmer, J., Kuhlemeyer, R.L.: Finite dynamic model for infinite media. J. Eng. Mech. Div. ASCE **95**(EM4), 859–877 (1969)
9. McGann, C.R., Arduino, P., Mackenzie-Helnwein, P.: Stabilized single-point 4-node quadrilateral element for dynamic analysis of fluid saturated porous media. Acta Geotech. **7**(4), 297–311 (2012)
10. McKenna, F., Fenves, G.L., Scott, M.H., Jeremic, B.: Open system for earthquake engineering simulation (2000). http://opensees.berkeley.edu
11. Manzari, M.T., Dafalias, Y.F.: A critical state two-surface plasticity model for sand. Geotechnique **47**(2), 255–272 (1997)
12. Papadimitriou, A.G., Bouckovalas, G.D.: Plasticity model for sand under small and large cyclic strains: a multiaxial formulation. Soil Dyn. Earthq. Eng. **22**, 191–204 (2002)
13. Taiebat, M., Jeremic, B., Dafalias, Y.F., Kaynia, A.M., Cheng, Z.: Propagation of seismic waves through liquefied soils. Soil Dyn. Earthq. Eng. **30**, 236–257 (2010)
14. Visone, C., Bilotta, E., Santucci de Magistris, F.: One-dimensional ground response as a preliminary tool for dynamic analyses in geotechnical earthquake engineering. J. Earthquake Eng. **14**(1), 131–162 (2010)

A Study on Liquefaction Characteristics of Sandy Soil in Large Strain Levels to Improve the Accuracy of Large Deformation Analysis

Noriyuki Fujii[1,2(✉)], Takashi Kiyota[2], Muhammad Umar[3], and Kyohei Ueda[4]

[1] OYO Corporation, Tokyo Metropolitan, Tokyo, Japan
fj-nr@iis.u-tokyo.ac.jp, fujii-noriyuki@oyonet.oyo.co.jp
[2] Institute of Industrial Science, University of Tokyo, Tokyo, Japan
[3] National University of Computer and Emerging Sciences, Islamabad, Pakistan
[4] Disaster Prevention Research Institute, University of Kyoto, Kyoto, Kyoto, Japan

Abstract. An analytical method to accurately predict the large deformation due to liquefaction is required, especially for the design considering a large earthquake. Therefore, as a basic study of large strain liquefaction behavior for numerical analysis, elemental simulations were carried out based on the results of large strain liquefaction tests for Toyoura sand up to $\gamma_{sa} = 60\%$. The results showed that the normal ϕ_f could not reproduce the test results. However, by using ϕ_f after reduction by cyclic shear stress loading, the results of large strain liquefaction tests up to $\gamma_{sa} = 60\%$ could be generally reproduced. In order to further verify the applicability of using the reduced ϕ_f, a two-dimensional large-deformation analysis was conducted to reproduce the centrifuge model test that obtained large deformation. As a result, compared with the results using the normal ϕ_f, the results using the reduced ϕ_f were closer to the experimental results. Future work includes setting up a method to evaluate the continuous reduction of ϕ_f due to cyclic shear stress, including after reaching the limiting shear strain γ_L, confirming the rate of reduction of ϕ_f in materials other than Toyoura sand, and applying the method to two-dimensional analysis.

Keywords: Liquefaction · Large shear strain · Large deformation analysis

1 Introduction

In the 1995 Hyogoken-Nanbu Earthquake, the Kobe Port was hit by strong motions. The ground motion inflicted significant damage to many port structures, especially those constructed from rigid blocks such as concrete caissons. Historically, Japan has been repeatedly hit by major earthquakes. Other recent earthquakes causing serious damage to social infrastructure facilities include the 2003 Tokachi-Oki, the 2004 Niigata-ken Chuetsu, and the 2011 Great East Japan earthquakes. In the near future, the Nankai Trough earthquake and Tokyo inland earthquake are forecasted to occur in Japan. If these severe earthquakes occur, many soil-structure systems would be seriously damaged. Therefore, it is crucial to predict the structural damage during major earthquakes as accurately as possible and to improve the seismic performance by implementing appropriate remedies as necessary. Generally, FEM analysis based on the infinitesimal deformation theory is used for prediction of displacement caused by liquefaction. However, models have been formulated based on the infinitesimal deformation theory. Strictly speaking, the applications are limited to phenomena with a small deformation and rotation. Consequently, the model has been extended based on large deformation (finite strain) formulations to take the effect of geometrical nonlinearity into account. To more accurately estimate the damage to soil-structure systems during an earthquake, the extended model has been implemented in a large deformation analysis program called "FLIP TULIP [1, 2]". Previous studies have confirmed that FLIP TULIP can generally reproduce the large de-formation behavior of sheet pile structures due to liquefaction. On the other hand, in the analysis to reproduce the results of centrifuge model experiments on inclined ground where the shear strains were several tens of percent or more, the analysis results slightly underestimated the experimental results. One of the rea-sons for this is that the simulation was not conducted targeting up to the strain level targeted in the analysis when setting the liquefaction parameter, which is one of the most important conditions to ensure the accuracy of the analysis. Another reason is that there are almost no examples of liquefaction tests to understand the behavior under liquefaction up to the large strain level.

Therefore, in this study, the previous studies on liquefaction behavior up to the large strain level are analyzed, and a proposal is made to improve the analysis accuracy of FLIP TULIP by utilizing the knowledge obtained from the analysis results.

2 Elemental Simulations Targeting the Large Strain Level

2.1 Elemental Simulation Condition

In this study, elemental simulations were conducted using the results of the large strain liquefaction test obtained from the medium-sized hollow torsion test (specimen size: do = 150 mm, di = 90 mm, h = 300 mm) conducted in the previous study [3] as a target. In this study, the test results using Toyoura sand specimens adjusted to about $D_r = 50\%$ by the air pluviation method were used as the target. After isotropic consolidation to $p' = 100$ kPa, the test specimens were subjected to a prescribed cyclic shear stress ratio under undrained conditions, and the cyclic load was applied at a constant shear strain rate. The cyclic loading was performed under fixed axial strain and constant shear strain rate ($\Delta\gamma/\Delta t = 5.5\%$/min).

On the other hand, FLIP TULIP was used for the elemental simulation. A multi-spring model was adopted as the nonlinear model of the ground, and the following values were used as the ground parameters.

a) Initial shear modulus G_{ma}: The value measured just before the liquefaction test ($p' = 100$ kPa) in the previous study (Kiyota 2007) is used ($G_0 = 71,800$ kN/m^2).
b) Shear resistance angle ϕ_f and phase transformation angle ϕ_p: Estimated from the effective stress path ($\phi_f = 43.8°$, $\phi_p = 30.0°$).
c) Liquefaction parameters: Values that can reproduce the liquefaction test results were set by trial and error (w1 = 0.8, p1 = 0.8, p2 = 0.7, c1 = 1.8, s1 = 0.005).

d) Other soil parameters: General values were adopted (t = 1.9 t/m^3, n = 0.41, h_{max} = 0.24).

2.2 Results of Elemental Simulation

The stress-strain relationship and effective stress path of the elemental simulation results are shown in Fig 1 in comparison with the liquefaction test results. The elemental simulation is terminated when either the number of iterations reaches 50 or the single amplitude shear strain γ_{sa} = 60%. As shown in Fig. 1, the elemental simulation results of CSR = 0.25 did not exceed γ_{sa} = 25%, and the simulation ended at 50 cycles in spite of the laboratory test's getting over γ_{sa} = 60%. In addition, the simulation of CSR = 0.16 was terminated without any decrease in the effective stress.

In the elemental simulation, we focused on ϕ_f as a factor that caused such a result. The critical state line (red solid line), the phase transformation line (red dotted line), and the critical point (red circle and blue circle) are shown on the effective stress path in Fig. 2. Critical points are indicated by red zeros before and blue zeros after reaching the critical shear strain γ_L defined in the previous study [3]. The critical shear strain γ_L is the shear strain at which the specimen becomes localized.

As shown in the effective stress path (see Fig. 2), before reaching γ_L, the critical point tends to be plotted on the left side of the boundary of γ_L. After reaching γ_L, the critical point tends to be plotted on the right side. This means that ϕ_f decreases after reaching γ_L. However, the elemental simulation does not take this trend into account.

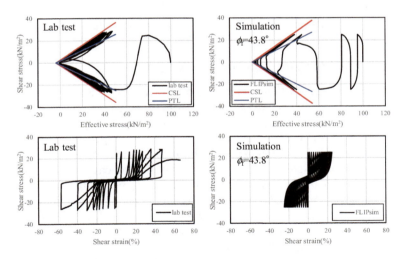

Fig. 1. Results of lab test and elemental simulation (CSR = 0.25)

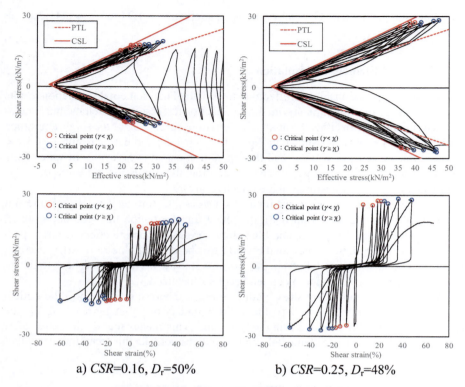

Fig. 2. Zoom up view of liquefaction test results

2.3 Analysis for Getting the Reduction Tendency of ϕ_f

In order to confirm the reduction tendency of ϕ_f in detail, the stress-strain relationship between γ_{sa} and ϕ_f of 10 specimens was read from the results of previous large strain liquefaction tests [3, 4]. Two types of specimens were used: 30 cm and 20 cm in height.

The results of the large strain liquefaction test using each specimen are analyzed in Fig. 3. The vertical axis in this figure is the value of each critical state angle (ϕ) divided by the shear resistance angle ϕ_f, which is the so-called ϕ reduction rate.

As shown in Fig 3, for all the results of ϕ/ϕ_f, after ϕ/ϕ_f reaches the maximum value, the results show that ϕ/ϕ_f tends to decrease as γ_{sa} increases due to the liquefaction development. Therefore, it is found that ϕ_f decreases as the shear strain continues to develop after liquefaction. Here, the behavior in the region of shear strain exceeding γ_L can be interpreted as the shear deformation after the loss of uniformity of the specimen, because the behavior is due to the additional repetitive loading on the localized specimen. Based on this idea, in this study, the reduction of shear resistance angle in the region beyond γ_L is interpreted as the deformation characteristics in the large strain region after liquefaction.

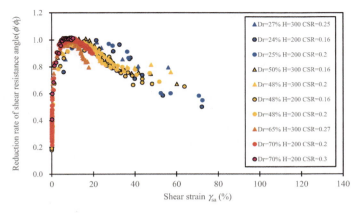

Fig. 3. Relationship between ϕ/ϕ_f and γ_{sa}

2.4 Retry of Elemental Simulation Using the Reduction ϕ_f

As shown in Fig. 3, since ϕ/ϕ_f is 0.7 at approximately $\gamma_{sa} = 60\%$, the elemental simulation was performed again with $\phi_f = 33°$ as the input value. The elemental simulation results of $CSR = 0.16$ are shown in Fig. 4. Focusing on the effective stress path, it can be seen that the path from the elemental simulation has cyclic mobility at an earlier stage than the path from the laboratory test. On the other hand, focusing on the stress-strain relationship, it was confirmed that the maximum shear strain was generally similar to the test results. It is necessary to evaluate the decrease of ϕ_f in order to describe the behavior of the large strain region due to liquefaction by FLIP TULIP.

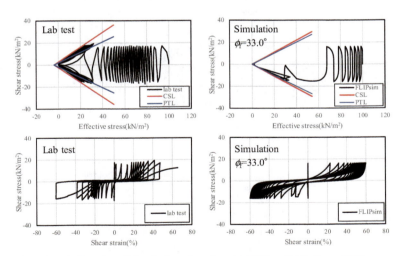

Fig. 4. Results of simulation changing ϕ_f ($CSR = 0.16$, $D_r = 50\%$)

3 Two-Dimensional Numerical Analysis

Using the previous analytical model [5], Two-dimensional large deformation analysis considering the decrease in ϕ_f was carried out. This analytical model was created to reproduce the centrifugal model experiment (see Fig. 5, [6]). This experiment was performed under a centrifugal acceleration of 50 G. After spinning up the centrifuge, the waveform (six-step excitation with sinusoidal waves) was given to the shaking table as an input motion. The analytical conditions were the same as those in the previous literature, and only ϕ_f was changed. Although we wanted to adopt 0.7 as the rate of decrease of ϕ_f in the large strain region, the analysis diverged when the rate was smaller than 33°, so we adopted $\phi_f = 33°$ in this study.

The results of the analysis (residual deformation and excess pore water pressure ratio) using FLIP TULIP are shown in Fig 6. Comparing these results, there was no significant difference in the amount of deformation, but by considering the decrease in ϕ_f, the deformation mode at the bottom of the slope was closer to the experimental results. From this result, it was found that the analysis accuracy in the large strain region could be improved if the reduction rate of ϕ_f is further reduced and the continuous decrease of ϕ_f is taken into account.

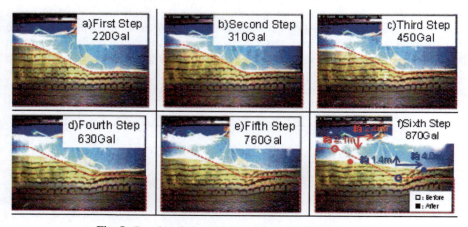

Fig. 5. Results of centrifuge model test (deformation) [6]

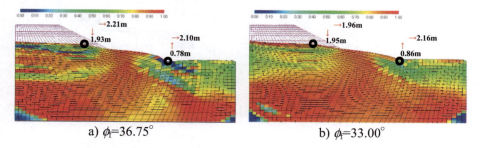

a) $\phi_f = 36.75°$ b) $\phi_f = 33.00°$

Fig. 6. Results of 2D large deformation analysis using FLIP TULIP

4 Conclusions

In this paper, as a basic study for the numerical analysis considering the behavior in the large strain region during liquefaction, the behavior up to $\gamma_{sa} = 60\%$ was attempted to be reproduced by elemental simulation targeting the results of previous large strain liquefaction tests.

As a result, it was found that it is possible to reproduce the behavior up to the large strain region of $\gamma_{sa} = 60\%$ by using ϕ_f after the decrease due to cyclic shear stress loading. However, although the relationship in Fig. 3 is expressed in terms of shear strain on the horizontal axis, we believe that it should be evaluated in terms of accumulated shear strain. For this reason, the relationship with the horizontal axis being the normalized cumulative dissipated energy is shown in Fig. 7. As shown in the figure, by adopting the normalized cumulative dissipated energy as the horizontal axis, the tendency of ϕ/ϕ_f to decrease due to the difference in the D_r of the specimen became more pronounced. Specifically, it was found that the larger the D_r, the more difficult it was for ϕ/ϕ_f to decrease.

Future work includes the introduction of a method for evaluating the continuous decrease in ϕ_f with cyclic shear stress loading using the relationship of Fig. 7, including after the critical shear strain γ_L is reached, confirmation of the rate of decrease in ϕ_f for materials other than the Toyoura sand used in this study, and investigation of its applicability to two-dimensional problems.

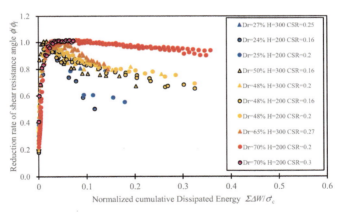

Fig. 7. Relationship between ϕ/ϕ_f and normalized cumulative dissipated energy $\Sigma \Delta W/\sigma'_c$.

References

1. Iai, S., Ueda, K., Tobita, T., Ozutsumi, O.: Finite strain formulation of a strain space multiple mechanism model for granular materials. Int. J. Numer. Anal. Meth. Geomech. **37**(9), 1189–1212 (2013)
2. Ueda, K.: Formulation of large deformation analysis of multiple shear model as sand dynamics model and its applicability, Kyoto University Doctoral thesis (2010). (in Japanese)

3. Kiyota, T.: Liquefaction strength and small strain properties of in-situ frozen and reconstituted sandy soils, Tokyo University Doctoral thesis (2007). (in Japanese)
4. Umar, M.: Degradation of stress-strain properties of sand in undrained torsional shear tests, Tokyo University Doctoral thesis (2019)
5. Fujii, N., Mikami, T., Ueda, K., Tobita, T.: Study on applicability of the strain space multiple mechanism model based on large deformation theory. In: The 52nd JGS National Conference, pp. 1591–1592 (2017). (in Japanese)
6. Fujii, N., Ueda, K., Kuwabara, N., Hyodo, J., Imono, T.: Centrifugal test and analysis for study using the Strain Space Multiple Mechanism Model based on Large Deformation theory, PBDIII, Vancouver, vol. 354 (2017)

A New Biaxial Laminar Shear Box for 1g Shaking Table Tests on Liquefiable Soils

Salvatore Grasso[1(✉)], Valentina Lentini[2], and Maria Stella Vanessa Sammito[1]

[1] Department of Civil Engineering and Architecture, University of Catania, Catania, Italy
sgrasso@dica.unict.it
[2] Faculty of Engineering and Architecture, University "Kore" of Enna, Enna, Italy

Abstract. In this paper, the design of a new laminar shear box at the Laboratory of Earthquake Engineering and Dynamic Analysis (L.E.D.A.) of the University of Enna "Kore" (Sicily, Italy), has been presented. The laminar box has been developed to investigate the liquefaction phenomenon and to validate advanced numerical models and/or the numerical approaches assessed to simulate and prevent related effects.

The paper describes in detail the types of soil container that have been used in the last three decades. Particular attention has been paid to the laminar shear box and liquefaction studies. Moreover, the most important factors that affect the performance of a laminar shear box are reported. The last part of the paper describes components, properties and design advantages of the new laminar shear box for 1g shaking table tests at L.E.D.A.

The new laminar box is rectangular in cross section and consists of 16 layers. Each layer is composed of two frames: an inner frame and an outer frame. The inner frame has an internal dimension of 2570 mm by 2310 mm, while the outer frame has an internal dimension of 2700 mm by 2770 mm. Between the layers, there is a 20 mm gap making the total height of 1.6 m. Aluminum is chosen in order to reduce the inertial effect of the frame on the soil during shaking. Each internal frame is supported independently on a series of linear bearings and rods connected to the external frame, while each external frame is supported independently on a series of linear bearings and rods connected to the surrounding rigid steel walls.

Shaking table tests will be carried out on saturated sandy soils using the 1g 6-DOF 4.0 m × 4.0 m shaking tables at L.E.D.A.

Keywords: Laminar shear box · Shaking table · Liquefaction

1 Introduction

Liquefaction is one of the major reasons for damage during an earthquake [1]. The phenomenon occurs as a result of build-up of pore pressure and hence a reduction of soil strength [2]. A better understanding of this phenomenon is of relevant interest in geotechnical earthquake engineering.

Liquefaction studies involve several approaches and procedures: laboratory cyclic triaxial tests [3–5], stress-based simplified procedures [6–9], numerical models [1, 10,

11], dynamic element tests [12, 13], reduced-scale model tests [10, 14, 15] and full scale field tests [16–18]. Reduced-scale model tests are advantageous for seismic studies thanks to the ability to provide economic and realistic information about the ground amplification, change in water pressure and soil non-linearity [19, 20].

This paper presents the design of a biaxial laminar shear box for reduced-scale model tests. The laminar shear box can be used in liquefaction studies, especially, to validate numerical models assessed to prevent related effects.

2 Background

Geotechnical scaled seismic model tests can be divided into two categories: shaking table test at 1-g and centrifuge test at n-g [19–23]. Shaking table tests have the advantages of well controlled large amplitude, and easier experimental measurements than centrifuge tests. However, high gravitational stresses cannot be produced [15, 20–22, 24]. Most studies of seismic soil behaviours were for one-dimensional shaking [19, 22, 23, 25–27]. However, real earthquake excitations are multiaxial. Zeghal et al. [28] and Ueng et al. [29] investigated the behaviour of saturated sand under two-dimensional earthquake shaking.

A variety of model container configurations have been used in the last three decades. Six type of soil container can be identified: rigid container, rigid container with flexible boundaries, rigid container with hinged end-wall, Equivalent Shear Beam (EBS) container, laminar container, and active boundary container [24].

The simplest way for geotechnical modelling is to use a rigid box. Sadrekarimi & Ghalandarzadeh [2] performed 1g shaking table tests using a transparent Plexiglas container to study liquefaction mitigation methods. However, the rigid box is often not suitable because the waves reach the rigid walls and reflect back to the soil. In order to reduce the reflection of the waves, Saha et al. [30] used a rigid box with absorbent boundaries. Instead, in a rigid container with hinged end-wall, the walls can rotate about the base due to the hinged connection [24, 31].

For increasing the flexibility of the box wall, Equivalent Shear Beam (ESB) containers [32] or laminar box soil containers [33] can be used. The ESB container consists of an alternating stack of aluminium alloy and rubber rings for flexibility [32, 34]. Laminar box soil container consists of a stack of stiff rings supported by ball bearing, linear bearing or rollers. The active boundaries container is very similar to the laminar container but external actuators are connected to each lamina [24]. Laminar box soil container is often used to model liquefaction [14, 15, 20, 21, 28, 29].

Ecemis [21] carried out 1D shaking table tests in order to simulate the liquefaction of loose to medium dense saturated sands. Mohsan et al. [15] designed a laminar soil box for studying the behaviour of saturated soils, especially liquefaction. A 2D laminar shear box was designed by Ueng et al. [29] at the National Centre for Research on Earthquake Engineering (NCREE) in Taiwan for the study of liquefaction and soil-structure interaction. To allow biaxial motion and to ensure non-torsional motion, a special mechanism design was adopted (Fig. 1).

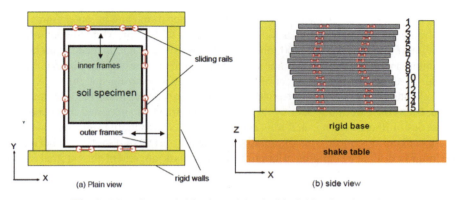

Fig. 1. Plan view and side view of the the biaxial laminar box [29]

3 Designed Laminar Shear Box

Based on the large flexible laminar shear box developed by Ueng et al. [30], a new laminar shear box has been designed at the Laboratory of Earthquake engineering and Dynamic Analysis (L.E.D.A.) of "Kore" University, Enna (Sicily, Italy), for 1g shaking table tests.

The most important facility of the laboratory is an array of two identical 6-DOF 4.0 m × 4.0 m shaking tables. It is possible to use the tables both separately and simultaneously in order to simulate the effects of earthquakes [35]. The 2D laminar box has been developed to monitor liquefaction under two dimensional shaking on a shaking table at L.E.D.A. Figure 2 reports a view of the shaking tables system.

Fig. 2. Shaking tables system at the Laboratory of Earthquake engineering and Dynamic Analysis (L.E.D.A.) [35].

For increasing the flexibility of the box wall, laminar system has been applied. A laminar shear box, during liquefaction, has least undesirable effect in the real behaviour of the model [20]. The laminar box is rectangular in cross section and consists of 16 layers. Each layer is composed of two frames: an inner frame and an outer frame. The inner frame has an internal dimension of 2570 mm by 2310 mm, while the outer frame has an internal dimension of 2700 mm by 2770 mm. Each frame is made from hollow aluminium profiles with 30×80 mm^2 section and 2 mm thickness. Aluminium is chosen in order to reduce the inertial effect of the frame on the soil during shaking.

Each internal frame is supported independently on a series of linear bearings and rods connected to the external frame, while each external frame is supported independently on a series of linear bearings and rods connected to the surrounding rigid steel walls.

Linear bearings allow a maximum displacement of ± 150 mm in the two horizontal directions with almost negligible friction. Thanks to this arrangement, the external frames can move in the x direction and, at the same time, the internal frames can move in the y direction. The 3D view of the laminar box is shown in the Fig. 3.

Fig. 3. 3D view of the laminar shear box.

This design has several advantages: the weight of each frame is transferred to the surrounding rigid steel walls; the effects of inertia and friction do not accumulate along the depth; each frame can move independently without torsion; the horizontal cross section remain horizontal. Between the layers, there is a 20 mm gap making the total height of 1.6 m. The lowest layer is fixed on a steel base with 3274 mm $\times$ 3274 mm $\times$ 20 mm dimensions. To reinforce the base, a steel frame is installed between the base and the shaking table. A drainage system is added on a steel base. It consists of water tubes and four valves. During the shaking, to improve the shear stress transition from the steel base to the soil, a thin layer of gravels will be placed on the base. Moreover, a 2 mm thick rubber membrane with high elasticity will be attached inside the box to provide the water tightness and to protect the external mechanism from soil penetration.

The laminar shear box at the Laboratory of Earthquake engineering and Dynamic Analysis (L.E.D.A.) is shown in Fig. 4. The main components of the laminar shear box are reported in Table 1.

Fig. 4. Laminar shear box at the Laboratory of Earthquake engineering and Dynamic Analysis (L.E.D.A.)

Table 1. Components of the laminar shear box

Component	Property	Value
Inner frame	Mass	12.07 kg
	Internal dimensions	2570 × 2310 mm^2
	Number	16
	Height	80 mm
Outer frame	Mass	12.53 kg
	Internal dimensions	2700 × 2770 mm^2
	Number	16
	Height	80 mm
Rod	Length	370 mm
	Diameter	15 mm
	Number	8

(*continued*)

Table 1. (*continued*)

Component	Property	Value
Linear bearing	Number	8
Gap between frames	Dimension	20 mm
Steel base	Mass	1682.89 kg
	Dimensions	3274 × 3274 mm^2
	Height	20 mm
Steel walls	Total mass	592.94 kg
Steel frame	Total mass	1359.83 kg
Total mass of the laminar box		4033.27 kg

4 Model Preparation and Instrumentation

A liquefiable soil deposit will be placed inside the laminar box using the raining method in order to achieve uniform density. A membrane will be attached inside the box and dry sand will be poured into water in the laminar box. The soil deposit will be made of a commercial sand from Catania area (Italy). The physical properties of sand are given in Table 2.

Table 2. Physical properties of Catania sand

Property	Value
Specific Gravity, G_S	2.65
D_{60} [mm]	0.25
D_{30} [mm]	0.20
D_{10} [mm]	0.15
e_{max}	0.85
e_{min}	0.65
Uniformity Coefficient, U	1.67
Curvature Coefficient, C	1.07

A total of twelve single axis MEMS accelerometers and twelve linear position transducers will be used to monitor the response of the model in term of accelerations and displacements at different depths. They will be placed on the external and internal frames for the x and y directions, respectively.

The instrumentation that will be installed inside the soil includes three submersible biaxial sensors for monitoring the accelerations in both x and y directions and six small-sized pore pressure transducer for measuring the pore pressures during the tests. These sensors will be placed inside the laminar box before the filling.

5 Concluding Remarks

In this paper, the design process of the laminar box system at L.E.D.A. of "Kore" University (Sicily, Italy) has been presented. It has been described along its various components, properties and design advantages. The new laminar shear box will be placed on a 6-DOF 4.0 m × 4.0 m shaking table for use in liquefaction studies.

The laminar box has been realized to investigate the liquefaction phenomenon and to validate advanced numerical models to simulate and prevent related effects.

References

1. Galavi, V., Petalas, A., Brinkgreve, R.B.J.: Finite element modelling of seismic liquefaction in soils. Geotech. Eng. J. SEAGS & AGSSEA **44**(3), 55–64 (2013). ISSN 0046-5828
2. Sadrekarimi, A., Ghalandarzadeh, A.: Evaluation of gravel drains and compacted sand piles in mitigating liquefaction. Ground Improvement **9**(3), 91–104 (2005)
3. Lentini, V., Castelli, F.: Liquefaction resistance of sandy soils from undrained cyclic triaxial tests. Geotech. Geol. Eng. **37**(1), 201–216 (2018). https://doi.org/10.1007/s10706-018-0603-y
4. Castelli, F., Cavallaro, A., Grasso, S., Lentini, V.: Undrained cyclic laboratory behavior of sandy soils. Geosciences **9**(12), 512 (2019). https://doi.org/10.3390/geosciences9120512
5. Ciancimino, A., et al.: Dynamic characterization of fine-grained soils in Central Italy by laboratory testing. Bull. Earthq. Eng. **18**(12), 5503–5531 (2019). https://doi.org/10.1007/s10518-019-00611-6
6. Castelli, F., Cavallaro, A., Grasso, S.: SDMT soil testing for the local site response analysis. In: Proceedings of the 1st IMEKO TC4 International Workshop on Metrology for Geotechnics, Benevento, Italy, pp. 143–148 (2016). ISBN: 978-92-990075-0-1
7. Grasso, S., Maugeri, M.: The Seismic Dilatometer Marchetti Test (SDMT) for evaluating liquefaction potential under cyclic loading. In: Proceedings of IV Geotechnical Earthquake Engineering and Soil Dynamic, Sacramento, USA, 18–22 May 2008. Geotechnical Earthquake Engineering and Soil Dynamics IV GSP 181 © 2008 ASCE, Geo Institute, p. 15 (2008). ISBN: 978-0-7844-0975-6
8. Grasso, S., Massimino, M.R., Sammito, M.S.V.: New stress reduction factor for evaluating soil liquefaction in the Coastal Area of Catania (Italy). Geosciences **11**(1), 12 (2021). https://doi.org/10.3390/geosciences11010012
9. Maugeri, M., Grasso, S.: Liquefaction potential evaluation at Catania Harbour (Italy). WIT Trans. Built Environ. **1**, 69–81 (2013). https://doi.org/10.2495/ERES130061
10. Bhatnagar, S., Kumari, S., Sawant, V.: Numerical analysis of earth embankment resting on liquefiable soil and remedial measures. Int. J. Geom. **16**, 04015029 (2015). https://doi.org/10.1061/(ASCE)GM.1943-5622.0000501
11. Castelli, F., Grasso, S., Lentini, V., Sammito, M.S.V.: Effects of soil-foundation-interaction on the seismic response of a cooling tower by 3D-FEM analysis. Geosciences **11**, 200 (2021). https://doi.org/10.3390/geosciences11050200
12. Ishihara, K., Yamazaki, F.: Cyclic simple shear tests on saturated sand in multi-directional loading. Soils Found. **20**(1), 45–55 (1980)
13. Cavallaro, A., Capilleri, P.P., Grasso, S.: Site characterization by dynamic in situ and laboratory tests for liquefaction potential evaluation during Emilia Romagna earthquake. Geosciences **8**, 242 (2018). https://doi.org/10.3390/geosciences8070242

14. Chen, C.H., Ueng, T.S.: Dynamic responses of model pile in saturated sand in shaking table tests. In: 5th International Conference on Earthquake Geotechnical Engineering, Santiago, Chile (2011)
15. Mohsan, M., Kiyota, T., Munoz, H., Nihaaj, M., Katagiri, T.: Fabrication and performance of laminar soil box with rigid soil box for liquefaction study. Bull. ERS **51**, 1–8 (2018)
16. Ashford, S., Rollins, K., Lane, J.: Blast-induced liquefaction for full-scale foundation testing. J. Geotech. Geoenviron. Eng. **130**(8), 798–806 (2004)
17. Castelli, F., Grasso, S., Lentini, V., Massimino, M.R.: In situ measurements for evaluating liquefaction potential under cyclic loading. In: Proceedings 1st IMEKO TC-4 International Workshop on Metrology for Geotechnics, Benevento, Italy, pp. 79–84 (2016). ISBN: 978-92-990075-0
18. Castelli, F., Cavallaro, A., Ferraro, A., Grasso, S., Lentini, V., Massimino, M.R.: Static and dynamic properties of soils in Catania (Italy). Ann. Geophys. **61**(2), SE221 (2018)
19. Turan, A., Hinchberger, S., El Naggar, H.: Seismic soil–structure interaction in buildings on stiff clay with embedded basement stories. Can. Geotech. J. **50**(3), 858–873 (2013)
20. Bojadjieva, J., Sesov, V., Edip, K., Gjorgiev, I.: Some important aspects in experimental setup for liquefaction studies on shaking table tests. In: 2nd European Conference on Earthquake Engineering and Seismology (2014)
21. Ecemis, N.: Simulation of seismic liquefaction: 1-g model testing system and shaking table tests. Eur. J. Environ. Civ. Eng. **17**(10), 899–919 (2013)
22. Prasad, S.K., Towhata, I., Chandradhara, G.P., Nanjunaswamy, P.: Shaking table tests in earthquake geotechnical engineering. Curr. Sci. **87**(10), 1398–1404 (2004)
23. Zayed, M., Luo, L., Kim, K., McCarteney, J.S., Elgamal, A.: Development and performance of a laminar container for seismic centrifuge modeling. In: Proceedings of 3rd International Conference on Performance-Based Design in Earthquake Geotechnical Engineering, Vancouver, BC, Canada, July 2017
24. Bhattacharya, S., Lombardi, D., Dihoru, L., Dietz, M.S., Crewe, A.J., Taylor. C.A.: Model container design for soil-structure interaction studies. In: Fardis, M.N., Rakicevic, Z.T. (eds.) Role of Seismic Testing Facilities in Performance-Based Earthquake Engineering: SERIES Workshop, Geotechnical, Geological, and Earthquake Engineering, vol. 22. Springer, Cham (2012). https://doi.org/10.1007/978-94-007-1977-4_8
25. Gibson, A.: Physical scale modeling of geotechnical structures at One-G. Ph.D. thesis, California Institute of Technology, Pasadena (1996)
26. Sarlak, A., Saeedmonir, H., Gheyretmand, C.: Numerical and experimental study of soil-structure interaction in structures resting on loose soil using laminar shear box. Int. J. Eng. Trans. B Appl. **30**(11), 1654–1663 (2017)
27. Takahashi, A., Takemura, J., Suzuki, A., Kusakabe, O.: Development and performance of an active type shear box in a centrifuge. Int. J. Phys. Model. Geotech. **1**(2), 1–17 (2001)
28. Zeghal, M., El Shafee, O., Abdoun, T.: Analysis of soil liquefaction using centrifuge tests of a site subjected to biaxial shaking. Soil Dynam. Earthq. Eng. **114**, 229–241 (2018)
29. Ueng, T.S., Wang, M.H., Chen, M.H., Chen, C.H., Peng, L.H.: A large biaxial shear box for shaking table test on saturated sand. Geotech. Test. J. **19**(1), 1–8 (2006)
30. Saha, R., Sumanta, H., Dutta, S.C.: Influence of dynamic soil-pile raft-structure interaction: an experimental approach. Earthq. Eng. Eng. Vib. **14**, 625–645 (2015). https://doi.org/10.1007/s11803-015-0050-1
31. Fishman, K.L., Mander, J.B., Richards, R., Jr.: Laboratory study of seismic free-field response of sand. Soil Dyn. Earthq. Eng. **14**, 33–43 (1995)
32. Zeng, X., Schofield, A.N.: Design and performance of an equivalent-shear-beam container for earthquake centrifuge testing. Geotechnique **46**(1), 83–102 (1996)
33. Hushmand, B., Scott, R.F., Crouse, C.B.: Centrifuge liquefaction tests in a laminar box. Geotechnique **38**(2), 253–262 (1988)

34. Massimino, M.R., Abate, G., Grasso, S., Pitilakis, D.: Some aspects of DSSI in the dynamic response of fully-couplep soil-structure system, Riv. Ital. Geotecnica (2019)
35. Navarra, G., Lo Iacono, F., Oliva, M., Tesoriere, G.: A new research facility: the Laboratory of Earthquake engineering and Dynamic Analysis (L.E.D.A.). In: Proceedings of XXII Congresso - Associazione Italiana di Meccanica Teorica e Applicata - AIMETA 2015, pp. 297–306, Genoa, Italy, 14–17 September 2015. ISBN 978-88-97752-55-4

Assessing the Limitations of Liquefaction Manifestation Severity Index Prediction Models

Russell A. Green[1(✉)], Sneha Upadhyaya[1], Brett W. Maurer[2], and Adrian Rodriguez-Marek[1]

[1] Virginia Tech, Blacksburg, VA 24061, USA
{rugreen,usneha,adrianrm}@vt.edu
[2] University of Washington, Seattle, WA 98195, USA
bwmaurer@uw.edu

Abstract. The severity of surface manifestation of liquefaction is commonly used as a proxy for liquefaction damage potential. As a result, manifestation severity index (MSI) models are more commonly being used in conjunction with simplified stress-based triggering models to predict liquefaction damage potential. This paper assesses the limitations of four MSI models. The different models have differing attributes that account for factors influencing the severity of surficial liquefaction manifestations, with the newest of the proposed models accounting more factors than the others. The efficacies of these MSI models are evaluated using well-documented liquefaction case histories from Canterbury, New Zealand, with the deposits primarily comprising clean to non-plastic silty sands. It is found that the MSI models that explicitly account for the contractive/dilative tendencies of soil did not perform as well as the models that do not account for this tendency, opposite of what would be expected based on the mechanics of liquefaction manifestation. The likely reason for this is the double-counting of the dilative tendencies of medium-dense to dense soils by these MSI models, since the liquefaction triggering model, to some extent, inherently accounts for such effects. This implies that development of mechanistically more rigorous MSI models that are used in conjunction with simplified triggering models will not necessarily result in improved liquefaction damage potential predictions and may result in less accurate predictions.

Keywords: Liquefaction triggering · Liquefaction severity · Liquefaction surficial manifestations

1 Introduction

The objective of this study is to assess the limits of predicting the occurrence and severity of surficial liquefaction manifestation via manifestation severity index (MSI) models that are used in conjunction with simplified stress-based triggering models. An accurate prediction of the severity of surficial liquefaction manifestation is critical for reliably assessing the risk due to liquefaction. This requires a proper understanding of the mechanics of the manifestation of surficial liquefaction features and the controlling factors.

© The Author(s), under exclusive license to Springer Nature Switzerland AG 2022
L. Wang et al. (Eds.): PBD-IV 2022, GGEE 52, pp. 1508–1515, 2022.
https://doi.org/10.1007/978-3-031-11898-2_133

Different models have been proposed in the literature to predict the occurrence/severity of surficial liquefaction manifestation (i.e., MSI models). These models use the results from simplified stress-based liquefaction triggering models and tie the cumulative response of the soil profile to the occurrence/severity of surficial liquefaction manifestation. One of the earliest MSI models is the Liquefaction Potential Index (*LPI*), proposed by Iwasaki et al. [1]. While *LPI* has been widely used to characterize the damage potential of liquefaction throughout the world, it was found to perform inconsistently during the 2010–2011 Canterbury earthquakes in New Zealand (e.g., [2–4]). This inconsistency can be attributed to limitations in the *LPI* formulation to appropriately account for some of the factors influencing surficial manifestation of liquefaction. Specifically, the *LPI* formulation may not adequately account for the contractive/dilative tendencies of the soil on the potential consequences of liquefaction. Additionally, the *LPI* formulation does not account for the limiting thickness of the non-liquefied crust and/or the effects of non-liquefiable, high fines-content (*FC*), high-plasticity strata on the severity of surficial liquefaction manifestations. Although the influence of these effects could be accounted for by using different *LPI* manifestation severity thresholds (i.e., *LPI* values distinguishing between different manifestation severity classes, e.g., [3, 5]), it is preferred to have a model that can explicitly account for these conditions in a less ad hoc manner.

In efforts to address some of the shortcomings in the *LPI* formulation, alternative MSI models have been proposed, such as the Ishihara-inspired *LPI* (*LPI$_{ish}$*) [6] and Liquefaction Severity Number (*LSN*) [7]. A major improvement of *LPI$_{ish}$* over *LPI* is that it explicitly accounts for the phenomenon of limiting-crust-thickness, where a non-liquefied capping stratum having an equal or greater thickness than the limiting crust thickness inhibits any surficial liquefaction manifestations regardless of the liquefaction response of the underlying strata. This attribute of the *LPI$_{ish}$* model is derived from Ishihara's [8] empirical relationship that relates the thicknesses of the non-liquefied crust (H_1) and of the liquefied stratum (H_2) to the occurrence of surficial liquefaction manifestations. However, as with *LPI*, *LPI$_{ish}$* does not explicitly account for the contractive/dilative tendencies of the soil on the severity of manifestations. The *LSN* formulation conceptually overcomes this limitation of *LPI*, as well as *LPI$_{ish}$*, in that it explicitly accounts for the additional influence of contractive/dilative tendencies of the soil via a relationship among *FS*, D_r, and the post-liquefaction volumetric strain potential (ε_v) [9]. However, *LSN* does not account for the phenomenon of limiting-crust-thickness, as *LPI$_{ish}$* does.

Based on the identified limitations of previously proposed MSI models, Upadhyaya et al. [10] proposed a new MSI model that accounts for the limiting-crust-thickness phenomenon and the effects of contractive/dilative tendencies of the soil on the severity of surficial liquefaction manifestations is proposed. The new model, termed *LSN$_{ish}$*, combines the positive attributes of *LPI$_{ish}$* and *LSN* in a single formulation that mechanistically accounts for the limiting-crust-thickness phenomenon based on Ishihara's H_1-H_2 boundary curves and the contractive/dilative tendencies of the soil on the severity of surficial liquefaction manifestation via an *FS*-D_r-ε_v relationship [9]. Similar to the derivation of *LPI$_{ish}$* [6], the new index is a conceptual and mathematical merger of the Ishihara [8] H_1-H_2 relationships and the *LSN* formulation. In the following, overviews of *LPI*, *LPI*$_{ish}$, *LSN*, and *LSN*$_{ish}$ models are presented first. Next, these four MSI models are evaluated using a large dataset of liquefaction case histories from the 2010–2011 Canterbury earthquake sequence and the 2016 Valentine's Day earthquake that impacted Christchurch, New Zealand, and the MSI models' predictive efficiencies are assessed.

2 Overview of MSI Models

2.1 Liquefaction Potential Index (*LPI*)

The liquefaction potential index (*LPI*) is defined as [1]:

$$LPI = \int_0^{z_{max}} F_{LPI}(FS) \cdot w_{LPI}(z) \, dz \tag{1}$$

where: *FS* is the factor of safety against liquefaction triggering, computed using a liquefaction triggering model; z is depth below the ground surface in meters; z_{max} is the maximum depth considered, generally 20 m; and $F_{LPI}(FS)$ and $w_{LPI}(z)$ are functions that account for the weighted contributions of *FS* and z on surface manifestation. Specifically, $F_{LPI}(FS) = 1 - FS$ for $FS \leq 1$ and $F_{LPI}(FS) = 0$ otherwise; and $w_{LPI}(z) = 10 - 0.5z$. Thus, *LPI* assumes that the severity of surface manifestation depends on the cumulative thickness of liquefied soil layers, the proximity of those layers to the ground surface, and the amount by which *FS* in each layer is less than 1.0.

2.2 Ishihara-Inspired Liquefaction Potential Index (*LPI*$_{ish}$)

Using the data from the 1983, M_w7.7 Nihonkai-Chubu and the 1976, M_w7.8 Tangshan earthquakes, along with considerable judgement, Ishihara [8] proposed a generalized relationship relating the thicknesses of the non-liquefiable crust (H_1) and of the underlying liquefied strata (H_2) to the occurrence of liquefaction-induced damage at the ground surface. This relationship is presented in the form of boundary curves that separate cases with and without surficial liquefaction manifestation as a function of peak ground acceleration (a_{max}). The H_1-H_2 boundary curves imply that, for a given a_{max}, there exists a limiting H_1, thicker than which surficial liquefaction manifestations will not occur regardless of the value of H_2 (i.e., the limiting-crust-thickness phenomenon mentioned in the Introduction). While Ishihara's H_1-H_2 curves have been shown to be conceptually correct, they are not easily implementable for more complex soil profiles that have multiple interbedded non-liquefied/non-liquefiable soil strata, such as those in Christchurch, New Zealand (e.g., [7, 11]). Additionally, the curves were derived from earthquakes that have a narrow magnitude range (i.e., M_w7.7–7.8).

To account for the limiting-crust-thickness phenomenon on the severity of surficial liquefaction manifestations using a more quantitative approach, Maurer et al. [6] utilized Ishihara's boundary curves to derive an alternative MSI model, *LPI*$_{ish}$:

$$LPI_{ish} = \int_{H_1}^{z_{max}} F_{LPI_{ish}}(FS) \cdot \frac{25.56}{z} dz \tag{2a}$$

where

$$F_{LPI_{ish}}(FS) = \begin{cases} 1 - FS & \text{if } FS \leq 1 \cap H_1 \cdot m(FS) \leq 3 \text{ m} \\ 0 & \text{otherwise} \end{cases} \quad (2b)$$

and

$$m(FS) = exp\left[\frac{5}{25.56 \cdot (1 - FS)}\right] - 1; \quad m(FS > 0.95) = 100 \quad (2c)$$

where FS and z_{max} are defined the same as they are for LPI. The LPI_{ish} framework explicitly accounts for the limiting thickness of the non-liquefied crust by imposing a constraint on $F_{LPIish}(FS)$ and uses a power-law depth weighting function, which is consistent with Ishihara's H_1-H_2 boundary curves. The power-law depth weighting function results in LPI_{ish} model giving a higher weight to shallower layers than the LPI model in predicting the severity of surficial liquefaction manifestations.

2.3 Liquefaction Severity Number (LSN)

As stated in the Introduction, LSN was proposed by van Ballegooy et al. [7] and uses a relationship relating FS, D_r, and ε_v to account for the contractive/dilative tendencies of the soil on the severity of surficial liquefaction manifestations [9]. LSN is given by:

$$LSN = \int_0^{z_{max}} 1000 \cdot \frac{\varepsilon_v}{z} dz \quad (3)$$

where z_{max} is the maximum depth considered, generally 10 m, and ε_v is estimated by using the relationship proposed by Zhang et al. [12] (entered as a decimal in Eq. 3), which is based on the FS-D_r-ε_v relationship proposed by Ishihara and Yoshimine [9]. Thus, unlike the LPI and LPI_{ish} models, which only consider the influence of soil strata with $FS < 1$ on the severity of surficial liquefaction manifestations, the LSN model considers the contribution of layers with $FS \leq 2$ via the FS-D_r-ε_v relationship [9].

2.4 Ishihara-Inspired LSN (LSN_{ish})

As mentioned previously, the LSN_{ish} model merges the positive attributes of the LPI_{ish} and LSN models. The derivation of the LSN_{ish} model follows a procedure similar to the derivation of the LPI_{ish} model [6] and is detailed in Upadhyaya et al. [10]. LSN_{ish} is given by:

$$LSN_{ish} = \int_{H_1}^{z_{max}} F_{LSN_{ish}}(\varepsilon_v) \cdot \frac{36.929}{z} \cdot dz \quad (4a)$$

where

$$F_{LSN_{ish}}(\varepsilon_v) = \begin{cases} \frac{\varepsilon_v}{5.5} & \text{if } FS \leq 2 \text{ and } H_1 \cdot m(\varepsilon_v) \leq 3m \\ 0 & \text{otherwise} \end{cases} \quad (4b)$$

and

$$m(\varepsilon_v) = \exp\left(\frac{0.7447}{\varepsilon_v}\right) - 1; m(\varepsilon_v < 0.16) = 100 \quad (4c)$$

where ε_v is expressed in percent. The LSN_{ish} model explicitly accounts for: (1) the influence of ε_v on the severity of surficial liquefaction manifestations; (2) the limiting-crust-thickness phenomenon; and (3) the contribution of liquefiable layers with $FS \leq 2$ to the severity of surficial liquefaction manifestations.

Specific to item (2), the limiting crust thickness is accounted for in the LSN_{ish} model via the requirement that $H_1 \cdot m(\varepsilon_v) \leq 3$ m in Eq. 4b. Since m is a function of ε_v (which in turn is a function of normalized penetration resistance and FS), it is implied that as ε_v increases, the thickness of the non-liquefiable crust required to suppress manifestations increases. The limiting crust thickness is equal to 3 m/m, where m is a function of the penetration resistance of the soil (e.g., normalized cone penetration tip resistance, q_{c1Ncs}) and FS against liquefaction triggering.

3 Evaluation of MSI Models

3.1 Canterbury Earthquake Liquefaction Case-History Dataset

The LPI, LPI_{ish}, LSN, and LSN_{ish} models were evaluated using 7167 Cone Penetration Test (CPT) liquefaction case histories from the M_w7.1 September 2010 Darfield (2574 cases), the M_w6.2 February 2011 Christchurch (2582 cases), and the M_w5.7 February 2016 Valentine's Day (2011 cases) earthquakes in Canterbury, New Zealand, largely assembled by Maurer et al. [2–4, 13] and Geyin et al. [14]. Collectively these earthquake case histories are referred to as the Canterbury earthquakes (CE) case histories. The case histories consist of classifications of liquefaction manifestations, geotechnical and hydrological data, and ground-motion intensity measures. The severity of the liquefaction manifestations was based on post-event observations and high resolution aerial photographs and satellite imagery taken within a few days after the earthquakes. It should be noted that none of the MSI models being evaluated account for the influence of non-liquefiable, high fines content, high plasticity interbedded soil strata on the occurrence/severity of surficial liquefaction manifestations. Therefore, the MSI models can be best evaluated using case histories comprised predominantly clean to non-plastic silty sand profiles. Maurer et al. [3] found that sites in the region that have an average soil behavior type index (I_c) for the upper 10 m of the soil profile (I_{c10}) less than 2.05 generally correspond to sites having predominantly clean to non-plastic silty sands. Accordingly, the 7167 liquefaction case histories used in this study only comprised CPT soundings that have $I_{c10} < 2.05$. Of the 7167 case histories, 38% of the case histories were categorized as "no manifestation" and the remaining 62% were categorized as either "marginal," "moderate," or "severe" manifestations following the Green et al. [15] classification.

3.2 Evaluation of Liquefaction Triggering and Severity of Surficial Liquefaction Manifestation

In evaluating the MSI models, *FS* is used as an input parameter. In the present study, *FS* was computed using the deterministic BI14 CPT-based liquefaction triggering model. Inherent to this process, soils with $I_c > 2.5$ were considered to be non-liquefiable [13]. Additionally, the *FC* required to compute q_{c1Ncs} was estimated using the Christchurch-specific I_c - *FC* correlation proposed by Maurer et al. [13].

For each CE case history, the predictive efficacies of the *LPI*, *LPI*$_{ish}$, *LSN*, and *LSN*$_{ish}$ models were compared by performing receiver operating characteristic (ROC) analyses on the CE dataset. In ROC analyses, the area under the ROC curve (*AUC*) can be used as a metric to evaluate the predictive performance of a diagnostic model (e.g., MSI model), where a higher *AUC* value indicates better predictive capabilities (e.g., [16]), e.g., a random guess returns an *AUC* of 0.5 and a perfect model returns an *AUC* of 1.

3.3 Results and Discussion

The results from ROC analyses show that the *AUC* values returned by the four different MSI models follow the order: $LPI \approx LPI_{ish} > LSN \approx LSN_{ish}$. As such, two main observations can be made. *First*, despite accounting for the limiting-crust-thickness phenomenon, *LPI*$_{ish}$ and *LSN*$_{ish}$ did not show improvements over *LPI* and *LSN*, respectively. This is likely due to the fact that the majority of case histories are located in eastern Christchurch where the groundwater table is shallow (usually ranging between ~1–2 m). As a result, the limiting-crust-thickness phenomenon may not have much of an influence on the severity of surficial liquefaction manifestations for the cases analyzed. *Second*, the higher *AUC*s for the *LPI* and *LPI*$_{ish}$ models than for the *LSN* and *LSN*$_{ish}$ models indicate that the latter group performs more poorly despite accounting for the influence of soil density on the occurrence/severity of surficial liquefaction manifestation via the FS-D_r-ε_v relationship, contrary to what would be expected. The most likely reason for the poorer performance of the *LSN* and *LSN*$_{ish}$ models is that the influence of post-triggering volumetric strain potential of medium-dense to dense soils on the severity of surficial liquefaction manifestations is being double-counted by these models. This is because *FS*, which is used as an input to compute ε_v, inherently accounts for such effects via the shape of the cyclic resistance ratio curve ($CRR_{M7.5}$ curve).

Specifically, the $CRR_{M7.5}$ curves likely tend towards vertical at medium to high penetration resistance due to dilative tendencies of medium-dense to dense soils that inhibit the surficial liquefaction manifestation, even if liquefaction is triggered at depth. Accordingly, while the existing triggering curves are often thought of as "actual" or "true" triggering curves in current practice, in reality they are combined "triggering" and "manifestation" curves. This is mainly because the $CRR_{M7.5}$ curves are based on the liquefaction response of profiles inferred from post-earthquake surface observations at sites. Sites without surficial evidence of liquefaction are classified, by default, as "no liquefaction," despite the possibility of liquefaction having been triggered at depth, but not manifesting at the ground surface. Consequently, inherent to the resulting triggering curve are factors that relate not only to triggering, but also to post-triggering surface manifestation.

4 Conclusions

- The predictive efficacies of the four MSI models were evaluated using 7167 well-documented CPT liquefaction case histories from the 2010–2011 Canterbury earthquake sequence and the 2016 Valentine's Day earthquake; the case histories comprised predominantly clean to non-plastic silty sand profiles. These models were evaluated in conjunction with the deterministic BI14 triggering model to compute FS.
- The predictive efficacies of LSN_{ish} and LSN models were lower than those of LPI and LPI_{ish}, despite the former two MSI models accounting for the additional influence of soil density on the severity of surficial liquefaction manifestation via the FS-D_r-ε_v relationship. The likely reason for this is that the influence of post-triggering volumetric strain potential on the severity of surficial liquefaction manifestation is being "double-counted" by the LSN and LSN_{ish} models, since the shape of the $CRR_{M7.5}$ curve inherently accounts for the dilative tendencies of medium-dense to dense soils, which inhibit surficial liquefaction manifestations even when liquefaction is triggered at depth.
- These findings suggest that current frameworks for predicting the occurrence/severity of surficial liquefaction manifestation do not account for the mechanics of triggering and manifestation in a proper and sufficient manner. While the triggering curves are assumed to be "true" (i.e., free of factors influencing manifestation), in reality they inherently account for some of the factors controlling surficial manifestation of liquefaction, particularly for denser soils. This implies that development of mechanistically more rigorous MSI models that are used in conjunction with simplified triggering models will not necessarily result in improved liquefaction damage potential predictions and may result in less accurate predictions.

Acknowledgements. This research was funded by National Science Foundation (NSF) grants CMMI-1751216, CMMI-1825189, and CMMI-1937984, as well as Pacific Earthquake Engineering Research Center (PEER) grant 1132-NCTRBM and U.S. Geological Survey (USGS) award G18AP-00006. This support is gratefully acknowledged, as well as access to the NZGD. However, any opinions, findings, and conclusions or recommendations expressed in this paper are those of the authors and do not necessarily reflect the views of NSF, PEER, USGS, or the NZGD.

References

1. Iwasaki, T., Tatsuoka, F., Tokida, K., Yasuda, S.: A practical method for assessing soil liquefaction potential based on case studies at various sites in Japan. In: 2nd International Conference on Microzonation, 26 November–1 December, San Francisco, CA, USA, pp. 885–896 (1978)
2. Maurer, B.W., Green, R.A., Cubrinovski, M., Bradley, B.A.: Evaluation of the liquefaction potential index for assessing liquefaction hazard in Christchurch, New Zealand. J. Geotech. Geoenviron. Eng. **140**(7), 04014032 (2014)
3. Maurer, B.W., Green, R.A., Cubrinovski, M., Bradley, B.: Fines-content effects on liquefaction hazard evaluation for infrastructure during the 2010–2011 Canterbury, New Zealand earthquake sequence. Soil Dyn. Earthq. Eng. **76**, 58–68 (2015)

4. Maurer, B.W., Green, R.A., Cubrinovski, M., Bradley, B.: Assessment of CPT-based methods for liquefaction evaluation in a liquefaction potential index framework. Géotechnique **65**(5), 328–336 (2015)
5. Upadhyaya, S.: Development of an improved and internally-consistent framework for evaluating liquefaction damage potential. Doctoral dissertation, Virginia Tech, Blacksburg, VA (2019)
6. Maurer, B.W., Green, R.A., Taylor, O.S.: Moving towards an improved index for assessing liquefaction hazard: lessons from historical data. Soils Found. **55**(4), 778–787 (2015)
7. van Ballegooy, S., et al.: Assessment of liquefaction-induced land damage for residential Christchurch. Earthq. Spectra **30**(1), 31–55 (2014)
8. Ishihara, K.: Stability of natural deposits during earthquakes. In: 11th International Conference on Soil Mechanics and Foundation Engineering, San Francisco, CA, USA, vol. 1, pp. 321–376 (1985)
9. Ishihara, K., Yoshimine, M.: Evaluation of settlements in sand deposits following liquefaction during earthquakes. Soils Found. **32**(1), 173–188 (1992)
10. Upadhyaya, S., Green, R.A., Maurer, B.W., Rodriguez-Marek, A., van Ballegooy, S.: Limitations of surface liquefaction manifestation severity index models used in conjunction with simplified stress-based triggering models. J. Geotech. Geoenviron. Eng. **148**(3), 04021194 (2022)
11. van Ballegooy, S., Green, R.A., Lees, J., Wentz, F., Maurer, B.W.: Assessment of various CPT based liquefaction severity index frameworks relative to the Ishihara (1985) H_1-H_2 boundary curves. Soil Dyn. Earthq. Eng. **79**(Part B), 347–364 (2015)
12. Zhang, G., Robertson, P.K., Brachman, R.W.I.: Estimating liquefaction-induced ground settlements from CPT for level ground. Can. Geotech. J. **39**(5), 1168–1180 (2002)
13. Maurer, B.W., Green, R.A., van Ballegooy, S., Wotherspoon, L.: Development of region-specific soil behavior type index correlations for evaluating liquefaction hazard in Christchurch, New Zealand. Soil Dyn. Earthq. Eng. **117**, 96–105 (2019)
14. Geyin, M., Maurer, B.W., Bradley, B.A., Green, R.A., van Ballegooy, S.: CPT-based liquefaction case histories compiled from three earthquakes in Canterbury, New Zealand. Earthq. Spectra (2021) https://doi.org/10.1177/8755293021996367
15. Green, R.A., et al.: Select liquefaction case histories from the 2010–2011 Canterbury earthquake sequence. Earthq. Spectra **30**(1), 131–153 (2014)
16. Fawcett, T.: An introduction to ROC analysis. Pattern Recogn. Lett. **27**(8), 861–874 (2005)

Assessment of Stone Column Technique as a Mitigation Method Against Liquefaction-Induced Lateral Spreading Effects on 2 × 2 Pile Groups

S. Mohsen Haeri[(✉)], Morteza Rajabigol, Milad Zangeneh, and Mohammad Moradi

Sharif University of Technology, Tehran, Iran
smhaeri@sharif.edu

Abstract. Two shake table experiments were conducted on two 2 × 2 pile groups to investigate the efficacy of stone column technique as a mitigation method against liquefaction-induced lateral spreading. The experiments were performed employing a designed and manufactured laminar shear box by Sharif University of Technology. Two separate 2 × 2 pile groups (with and without superstructure weight) were installed in the models to investigate the effects of superstructure weight on the behavior of pile groups in treated and untreated liquefiable layers as well. The models were shaken with a sinusoidal base acceleration having a frequency of 3 Hz and amplitude of 0.3 g. The piles and the soil at far field were instrumented to measure various parameters during and after shaking. The results including acceleration, pore water pressure, displacement at free field and bending moment of the piles are briefly presented and discussed in this paper. The results illustrate that stone column technique significantly decreased lateral displacemnts in the free field and bending moments in the piles, while increased the accelearions in the pile caps and rate of dissipation of excess pore water pressure in the liquefiable layer.

Keywords: Physical modeling · Stone column · Lateral spreading

1 Introduction

Liquefaction-induced lateral spreading generaly occurs in midely sloping grounds or grounds ending in a free face of water. In some cases, lateral movement of the liquefied and crust layers may reach to several meters and generate significant bending moment in deep foundations of buildings and bridges. Considarable damages of pile-supported structures founded in liquefiable soils have been observed due to lateral spreading, e.g., 1964 Niigata and 1995 Hyogoken-Nambu earthquakes [1, 2]. Ground improvement using stone columns (gravel drains) as a mitigation method against liquefaction-induced lateral spreading has been employed by several researches. Seed and Booker [3] were the first researchers who investigated the effectiveness of stone solumns for mitigation of liquefaction [4]. Elgamal et al. [4] employed Opensees (3D finite element software) to study the effects of stone columns on lateral deformation in a sloping ground. Their results showed that stone columns were generally effective in reducing lateral deformation of a sand stratum, but highly ineffective in a silt stratum. Kavand et al. [5] conducted a series of shake table experiments to study the behavior of 3 × 3 flexible piles in a gently sloped ground in presence of stone columns. They found that stone columns effectively reduced the kinematic bending moments due to lateral spreading in the piles up to a particular level of earthquake magnitude. Forcellini and Tarantino [6] performed a 3D finite element study to assess effectiveness of stone column against lateral spreading. Their results illustrate that the method was effective in reduction of ground lateral movement. Tang and Orense [7] used $FLAC^{2D}$ (a finite difference software) to study improvement mechanism of stone column against liquefaction-induced lateral spreading. They found that stone columns were effective in reduction of lateral displacement of liquefiable layer. Lu et al. [8] employed Opensees (a 3D finite element software) to evaluate three different configurations of stone columns remediation method against lateral spreading.

The results of aforementioned studies showed that further experiments are required to reveal the various aspects of performance and effectiveness of stone columns in different situations. In the present study, two shake table experiments (with and without stone columns) were conducted on two sets of 2 × 2 pile groups. In order to investigate the efficacy of stone columns, the main results of these experiments are presented and compared in this paper.

2 Physical Modeling

To achieve purposes of this research, Sharif University of Technology (SUT) shake table (4 m × 4 m, 3DOFS facility, capable of taking models of up to 300 kN) and a newly designed and built large laminar shear box were employed. Figures 1 and 2 shows the schematic plan and cross section views of the physical models without and with the mitigation method, respectively. Constructed models without and with the mitigation method, prior to the tests loadings, are also exhibited in Figs. 1 and 2. Both physical models were fully instrumented with different types of transducers (e.g. displacement, pore water pressure and acceleration). In order to determine mechanical and geometrical properties of the piles in the physical model, the generalized similitude law proposed by Iai [9, 10] were used, considering a geometrical scale (prototype/model) of $\lambda = 8$. Two 2 × 2 piles made of aluminum pipes (T6061 alloy) with 150 cm length, 5 cm outer diameter and 0.15 cm thickness were implemented in each model. A lumped mass of 500 N was attached to the cap of one of the pile groups to model the weight of a bridge pier and study the effect of inertial loading due to the presence of a superstructure. According to the similitude law, this mass can represent the dead load (bridge deck and pier) and live load (a 20 tons truck) of a bridge section. In each model, six piles (three piles in each group) named by numbers 1 to 6 (Figs. 1 and 2) were instrumented with pair strain gauges to obtain profiles of bending moment and lateral pressure on the piles due to lateral spreading. This paper does not include the results of the single pile because the focus of this paper is on the behavior of 2 × 2 pile groups. According to Figs. 1 and 2, the models consisted of three soil layers: 1) a non-liquefiable crust layer made of dry clayey sand, with a thickness 0f 20 cm 2) a liquefiable layer of 1 m thick with a relative density of

15%, 3) a very dense non-liquefiable sand with a relative density of 80%. The crust layer was constructed by method of dry deposition in air, while the very dense non-liquefiable layer was constructed using wet tamping method. Water sedimentation method was used to construct the very loose liquefiable soil layer. The stratum was built with standard Firuzkuh silica sand no. 161 which is a crushed sand with a uniform gradation of mean grain size of about 0.24 mm. This sand is widely used in geotechnical experimental researches in Iran, especially for those related to liquefaction. All layers in the models had a ground slope of 4° in longitudinal direction. In the second model, stone columns were employed as a remediation method against lateral spreading. 29 polyethylene pipes with a diameter of 60 cm at prototype scale and triangular configurations were installed in this model. A coarse sand as the stone column material was purred into the pipes slowly. The pipes were removed from the model after construction of the stratum and before the input loading. The stone columns technique can combine the beneficial effects of densification, reinforcement, and increased drainage [11]. In this study, the stone

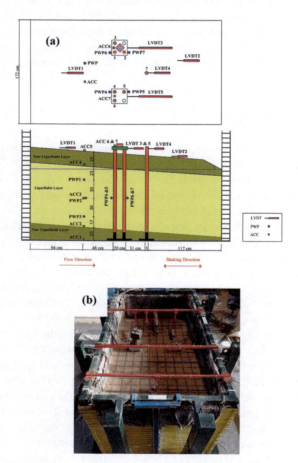

Fig. 1. Model without stone column: a) schematic plan and section views, b) after construction

columns were constructed at minimum density and densification of the surrounding soil, to observe mainly the drainage and some reinforcement effects of stone columns on liquefaction induced lateral spreading and its effects on studied pile groups.

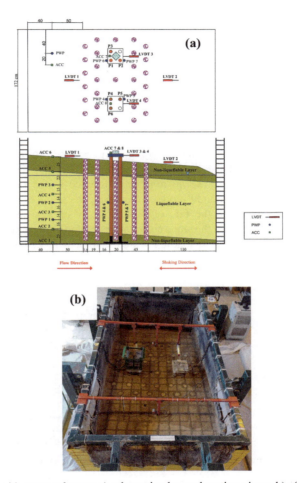

Fig. 2. Model with stone columns: a) schematic plan and section views, b) after construction

3 Results

Similar loadings with amplitude of 0.3 g and frequency of 3 Hz and different numbers of load cycles (30 cycles for the 1st model and 15 cycles for the 2nd model) were applied to the base of the models (Fig. 3). As the liquefaction trigerring, and maximum amounts of displacement, accelerations and bending moments occurred at first cycles of the loadings, the effects of difference between the input motions on the main results were almost negligible.

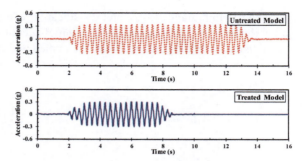

Fig. 3. Base input motion in treated and untreated models

Time histories of r_u (ratio of excess pore water pressure to initial vertical effective stress) values at free field and near the piles are exhibited in Fig. 4 and Fig. 5, respectively. As seen in these figures, there are no notable difference in the time of liquefaction triggering for treated and untreated models. However, the stone columns significantly changed the starting time and the rate of dissipation of excess pore water pressure. Another remarkable point is that the stone columns generated high spikes and fluctuations of pore water pressure during test loading which is indicative of fast dissipation after generation of pore water pressure in liquefiable soil.

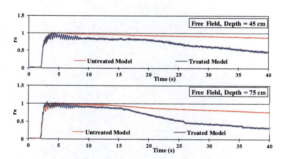

Fig. 4. Comparison of time histories of excess pore pressure at depths of free field in treated and untreated models

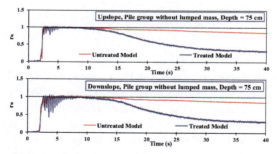

Fig. 5. Comparison of time histories of excess water pore pressure near the piles in treated and untreated models

Time histories of soil displacement at free field for models with and without mitigation with stone columns are exhibited in Fig. 6. The maximum displacement for the treated model was 26 mm which is significantly less than that for the model without mitigation (45 mm). It should be noticed that the maximum displacement for the treated model occurred at 8th cycle of the loading, hence the difference between the numbers of the cycles of input motions did not influence the maximum values of displacements.

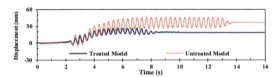

Fig. 6. Comparison of time histories of soil displacement at free field in treated and untreated models

Bar graph of the maximum acceleration of the pile caps for both models are depicted in Fig. 7. As seen in this figure, acceleration amplitude of the pile caps of the model using stone columns was amplified due to both pore water pressure dissipation and reinforcement effects of stone columns. The ratio of the maximum acceleration of the pile caps with and without the lumped mass in the model using stone columns to those in the model without stone column were approximately 2.55 and 2.79, respectively.

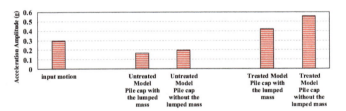

Fig. 7. Comparison of maximum acceleration of the piles in treated and untreated models

Time histories of bending moments in the piles of the models are compated in Fig. 8. According to this figure, two remarkable effects of stone columns are detectable: 1) The maximum bending moment of the piles in the model without stone column occurred prior to liquefaction triggering, however that occured after liquefaction trigering in the treated model using stone columns. 2) The maximum bending moments in the piles significantly reduced due to presence of stone columns. Bar graphs of the maximum bending moments in two models are compared in Fig. 9. As seem in this figure, stone columns reduced the maximum bending moments of the piles between 51% to 77% in this study.

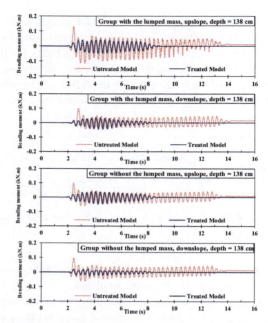

Fig. 8. Comparison of the time histories of bending moments of the pile groups with and without the lumped mass, in treated and untreated models

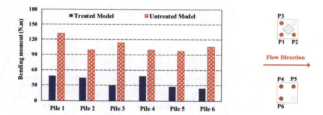

Fig. 9. Bar graph of maximum bending moment of the piles in treated and untreated models

4 Conclusions

Two shake table experiments were conducted to study the effects of stone columns (gravel drains), especially the effects of drainage and partial reinforcement on the responses of two 2 × 2 pile groups (with and without the superstructure weight) to lateral spreading. The results reveal that stone columns significantly reduce the lateral deformation of the ground and bending moments in the piles, whereas those increase the pile caps accelerations and the rate of pore water pressure dissipation. Further experiments are still required to investigate the efficacy of this mitigation method in different circumstances.

Acknowledgements. The partial financial supported by Construction and Development of Transportation Infrastructures Company, Ministry of Roads & Urban Development of Iran and also the partial financial support granted by Research Deputy of the Sharif University of Technology are acknowledged. The experiments were conducted at Shake Table Facilities of Civil Engineering Department, Sharif University of Technology. The contributions by all faculty, students and technicians in performing the experiments are acknowledged as well.

References

1. Hamada, M., Yasuda, S., Isoyama, R., Emoto, K.: Study on Liquefaction Induced Permanent Ground Displacements. Association for the Development of Earthquake Prediction, Japan (1986)
2. Tokimatsu, K., Asaka, Y.: Effects of liquefaction-induced ground displacements on pile performance in the 1995 Hyogoken–Nambu earthquake. Spec. Issue Soils Found. **38**, 163–177 (1988). https://doi.org/10.3208/sandf.38.Special_163
3. Seed, H.B., Booker, J.R.: Stabilization of potentially liquefiable sand deposits using gravel drains. J. Geotech. Geoenviron. Eng. **103**(7), 757–768 (1977)
4. Elgamal, A., Lu, J., Forcellini, D.: Mitigation of liquefaction-induced lateral deformation in a sloping stratum: three-dimensional numerical simulation. J. Geotech. Geoenviron. Eng. **135**(11), 1672–1682 (2009). https://doi.org/10.1061/(ASCE)GT.1943-5606.0000137
5. Kavand, A., Haeri, S.M., Raisianzadeh, J., Padash, H., Ghalandarzadeh, A.: Performance evaluation of stone columns as mitigation measure against lateral spreading in pile groups using shake table tests. In: International Conference on Ground Improvement and Ground Control (ICGI 2012), Paris, France (2012)
6. Foellini, D., Tarantino, A.M.: Assessment of stone columns as a mitigation technique of liquefaction-induced effects during Italian earthquakes (May 2012). Sci. World J. (2014). https://doi.org/10.1155/2014/216278. Article ID 216278
7. Tang, E., Orense, R.P.: Improvement mechanisms of stone columns as a mitigation measure against liquefaction-induced lateral spreading. Paper Number 082. New Zealand Society for Earthquake Engineering, Aukland (2014)
8. Lu, J., Kamatchi, P., Elgamal, A.: Using stone columns to mitigate lateral deformation in uniform and stratified liquefiable soil strata. Int. J. Geomech. **19**(5), 04019026 (2019). https://doi.org/10.1061/(ASCE)GM.1943-5622.0001397
9. Iai, S.: Similitude for shaking table tests on soil–structure–fluid model in 1g gravitational field. Soils Found. **29**(1), 105–118 (1989). https://doi.org/10.3208/sandf1972.29.105
10. Iai, S., Tobita, T., Nakahara, T.: Generalized scaling relations for dynamic centrifuge tests. Géotechnique **55**(5), 355–362 (2005). https://doi.org/10.1680/geot.2005.55.5.355
11. Shenthan, T., Nashed, R., Thevanayagam, S., Martin, G.R.: Liquefaction mitigation in silty soils using composite stone columns and dynamic compaction. Earthq. Eng. Eng. Vib. **3**(1), 39–50 (2004). https://doi.org/10.1007/BF02668849

Influence of Lateral Stress Ratio on N-value and Cyclic Strength of Sands Containing Fines

Kenji Harada[1(✉)], Kenji Ishihara[2], and Hiroshi Yabe[1]

[1] Fodo Tetra Corporation, Tokyo, Japan
kenji.harada@fudotetra.co.jp
[2] Chuo University, Tokyo, Japan

Abstract. The compaction method, such as the sand compaction pile (SCP) method, is a method to increase the strength of the ground by installing the material into the ground to be improved. However, in the standard penetration test (SPT) and cone penetration test (CPT) of the in-situ sounding test to confirm the improvement effect, the effect of density increase has been sufficiently evaluated, but the effect of lateral stress has not been evaluated. The authors investigated the effect of the lateral stress ratio, Kc-value and presented a relationship between the normalized N-value, N_1 of the standard penetration test and cyclic strength based on the results of soil chamber tests and hollow torsional shear tests under anisotropic condition of clean sand. This relationship was analyzed using the same method that led to this relationship against sands containing fines. As a result, a design chart of the ground (recommended chart), which is the relationship between the two considering the fine content, was presented.

Keywords: Lateral stress ratio · Cyclic strength · N-value · Recommended chart

1 Introduction

In a previous study, the authors investigated the influence of the effective lateral stress ratio, Kc (= σ_h'/σ_v': effective lateral stress/effective vertical stress), on the penetration resistance and cyclic strength of clean sand, and presented a recommended chart showing the relationship between the normalized N-value (SPT blow count), N_1, obtained from the standard penetration test (SPT) and cyclic strength R based on the results of soil chamber tests and hollow torsional shear tests [1]. Increases in the cyclic strength of ground improved by compaction using the sand compaction pile (SCP) method and other similar methods include not only increases in density but also increases in lateral stress caused by the

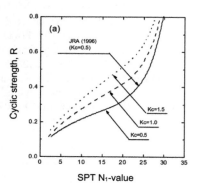

Fig. 1. Relationship between cyclic strength and N_1-value [1]

casing pressing in and sand pile construction, so the recommended chart is useful when trying to quantitatively take into account the influence of those increases at the design stage. However, in reality, sand contains fines, so it is necessary to analytically determine the practical relationship between penetration resistance and cyclic strength and present a design chart applicable to sandy soil that takes fines into consideration. With this objective in mind, the authors investigated the influence of lateral stress on the penetration resistance and cyclic strength of sand that contains fines (Fig. 1).

2 Basic Methodology

The methodology adopted in this study is basically the same as the concept applicable to clean sand [1]. The approach in this study, therefore, consists of five steps:

(1) Definition of Dr^* for relative density correction that takes the fines content, Fc, into consideration

   A number of standards and specifications, including the Highway Bridge code, indicates *the relationship between N_1 and cyclic strength based on site-specific data during past* earthquakes and laboratory test data under the condition of $Kc = 0.5$. No relationship is indicated, however, with relative density Dr (Eq. 1) associated with those factors. The definition of Dr is applicable to clean sand with a fines content Fc of 5% or less. Therefore, for sand with a higher fines content, it is necessary to consider a method for determining the minimum void ratio, e_{min}. Ishihara et al. [2] proposed a modified relative density, Dr^*, that uses the minimum void ratio converted from the maximum dry density ρ_{dmax} obtained from a soil compaction test method that uses a rammer (JIS A 1210:2009). Figure 2 shows the relationship between $e_{min}{}^*/e_{min}$ and the fines content Fc. As shown, when Fc is 5% or more, $e_{min}{}^*$ tends to be smaller than e_{min}, and $e_{min}{}^*/e_{min}$ tends to be smaller than 1.0. An equation expressing the relationship between the void ratio range e_{max}-$e_{min}{}^*$ in Eq. 2 and Fc, Eq. 3, has also been proposed.

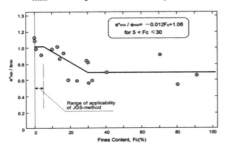

$$D_r = \frac{e_{max} - e}{e_{max} - e_{min}} \quad (1)$$

$$D_r{}^* = \frac{e_{max} - e}{e_{max} - e_{min}{}^*} \quad (2)$$

e_{max}: maximum void ratio

e_{min}: minimum void ratio (JIS A 1224:2009)

Fig. 2. Relationship between $e^*{}_{min}/e_{min}$ and fines content [2]

$$e_{max} - e^*_{min} = \begin{cases} 0.43 & (Fc \leq 5\%) \\ 0.394 + 0.0154Fc + 0.000096Fc^2 \\ & (5 < Fc \leq 30\%) \\ 0.75 + 0.0064Fc & (30\% < Fc) \end{cases} \quad (3)$$

(2) Relationship between the penetration resistance and modified relative density of normally consolidated soil at different Fc values (Sect. 4.1)

If the influence of a given value of Kc on N_1 is to be examined for sand containing fines, it is necessary, as the first step, to formulate the relationship between the penetration resistance and Dr^* at $Kc = 0.5$ where Fc is taken into consideration. This formula is derived based on the results of a soil chamber test (described in detail in Sect. 3) in which sand specimens containing fines are used and Kc can be controlled.

(3) Relationship between the cyclic strength and modified relative density of normally consolidated soil at different Fc values (Sect. 4.2)

The relationship between the cyclic strength and N_1 of sand containing fines is shown in various standards. The relationship between the cyclic strength and th modified relative density at $Kc = 0.5$ is derived from the relation between the N_1-value and Dr^* derived in Step 2.

(4) The influence of lateral stress ratio on penetration resistance and cyclic strength (Sect. 5)

To investigate the influence of Kc on the N-value, the effects of Dr^* and Fc are corrected by using soil chamber test results obtained by parameterizing Dr^* and using apparatus capable of controlling the Kc of the same material as the one used in the test in Step 2. To investigate the influence of Kc on cyclic strength, correction is made based on the results of a hollow torsional shear test conducted on anisotropically consolidated soil samples.

(5) Relationship between the penetration resistance and cyclic strength at a given value of Kc (Sect. 6)

The relationship between N_1 and the cyclic strength at a given value of Kc is derived for different values of Fc from the N_1–Dr^* relationship derived at Steps 2 and 4 and the relationship between the cyclic strength and Dr^* derived at Steps 3 and 4.

3 Standard Penetration Test Using Pressurized Soil Chamber

Figure 3 shows the standard penetration test apparatus and soil chamber used for the testing [3]. The sampler and the knocking head of the standard penetration test apparatus are directly connected by a rod coupling (50 mm long). The 63.5 kg hammer is allowed to fall freely from a height of 760 mm. The soil chamber is a cylindrical steel chamber with an inside diameter of 775 mm and a height of 916 mm. To simulate lateral and vertical stresses acting on the model ground, the soil chamber is designed so that Kc can be freely adjusted by independently applying water pressure laterally and vertically through the membrane of the circumferential surface and the bottom of the chamber. The Kc value is adjusted by adjusting the effective lateral stress while keeping the effective vertical stress constant at 98 kPa. The standard penetration test was conducted after the model ground was fabricated and then consolidation was completed by applying lateral and vertical stresses. To obtain N-value measurements in the central region of the soil chamber, preliminary driving was done to a depth of 300 mm. During the main driving, confining pressure was adjusted for each hammer impact so that Kc remained constant. The weight was dropped using the trip monkey method.

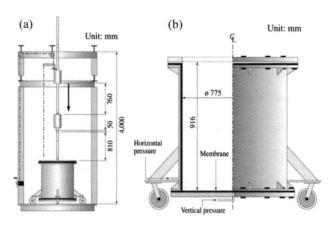

Fig. 3. Standard penetration test equipment ((a): whole device, (b): soil chamber)

For the purpose of testing, the model ground was designed to simulate saturated soil and was converted into three types in terms of density and Kc by using the three types of samples shown in Table 1. The Fc values of the three samples were about 0, 20, and 70%. The $Dr*$ of the model ground was converted into three ranges, namely, loose (55–70%), moderately dense (75–90%), and dense (85–100%), and Kc was adjusted to three levels: 0.5, 1.0, and 1.5. Each sample was prepared by either "dropping in water", "consolidation", or "tamping", depending on

Table 1. Used materials

Material	ρ_s (g/cm³)	F_C (%)	U_C	e_{max}	e_{min}	e_{min}*
Toyoura	2.650	0.0	1.73	0.985	0.611	–
Takahama	2.688	17.7	5.00	1.207	0.677	0.616
Ogu	2.713	70.0	–	1.605	0.685	0.624

ρ_s: soil particle density Uc: uniformity coefficient

the fine grain content, to achieve a predetermined density. All tests were conducted under an effective overburden pressure, σ_v', of 98 kPa, so the normalized N-value, $N_1 = 170\ N/(70 + \sigma_v')$, is actually the same as the N-value.

4 Relationship of N_1 and the Cyclic Strength with the Modified Relative Density of Normally Consolidated Soil

4.1 Relationship Between N_1 and the Modified Relative Density

In a previous study, the relationship between the N-value and relative density for clean fine sand and clean coarse sand was expressed by using the results obtained from a standard penetration test soil chamber as shown in Fig. 4 (a) [1]. The same figure also shows plots indicating the results obtained by using two types of samples containing fines (Takahama and Ogu) shown in Table 1. Although the data varies to a certain degree, the results tend to be plotted under the clean sand results as Fc increases, while

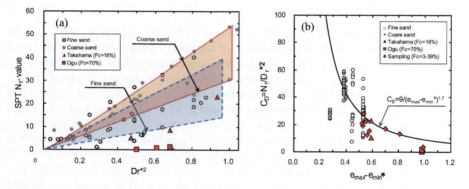

Fig. 4. Relationship (a) between N_1-value and modified relative density, and (b) between N_1/Dr^{*2} and void ratio

the slope N_1/Dr^{*2} (= C_D) decreases.

Cubrinovski and Ishihara [4] proposed an equation (Eq. 4) expressing the relationship between N_1 and Dr based on in-situ sampling data. Figure 4 (b) shows the plots of the results for the two samples containing fines mentioned above (▲, ■) and the results for the sampled material [5] (◆). The plots almost follow the curve, indicating that Eq. 5, which can be derived by replacing Dr and e_{min} with Dr^* and e_{min}^* without changing C_D in Eq. 4, holds true [6].

$$N_1 = C_D \cdot D_r^2 = \frac{9}{(e_{max} - e_{min})^{1.7}} D_r^2 \qquad (4)$$

$$N_1 = C_D \cdot D_r^{*2} = \frac{9}{(e_{max} - e_{min}^*)^{1.7}} D_r^{*2} \tag{5}$$

4.2 Relationship Between the Cyclic Strength and Modified Relative Density

For clean sand, the equation shown in Specifications for Highway Bridges (1996) [7(a)] was used to express the relationship between Na, which is an N-value corrected using Fc, and cyclic strength R. In this study, the authors used the relation (Eqs. 6, 7) between Na and R at Kc = 0.5 derived using the Na-value converted from the N-value correction factor, C_{Fc}, based on Fc, shown in the Specifications for Highway Bridges revised in 2017 [7(b)].

$$N_a = C_{Fc}(N_1 + 2.47) - 2.47 \qquad C_{Fc} = \begin{cases} 1 & (0 \leq F_c < 10) \\ (F_c + 20)/30 & (10 \leq F_c < 40) \\ (F_c/16)/12 & (F_c \geq 40) \end{cases} \tag{6}$$

$$R = \begin{cases} 0.0882\sqrt{(0.85 N_a + 2.1)/1.7} & (N_a < 14) \\ 0.0882\sqrt{N_a/1.7} + 1.6 \times 10^{-6} \cdot ((N_1)_{80} - 14)^{4.5} & (N_a \geq 14) \end{cases} \tag{7}$$

5 Influence of *Kc* on N_1-value and Cyclic Strength

5.1 Influence of *Kc* on Penetration Resistance

Figure 5 shows the standard penetration test results shown in Sect. 3 in the form of the Kc–N_1 relationship for each sample. Harada et al. [1] defined a coefficient for the impact of an increase in Kc on an increase in N_1 applicable to clean sand as $C_{SPH} = (N_1)_{Kc}/(N_1)_{Kc,NC}$ (where $(N_1)_{Kc}$: N_1-value in a given Kc condition, $(N_1)_{Kc,NC}$: N_1-value in a normally consolidated condition) in expressing this relationship. In this study, the fitting curve was modified by adding data obtained by using the samples containing fines, and the result was formulated, so as to maintain overall consistency, as shown in Eq. 8.

$$C_{SPH} = \left(\frac{K_C}{K_{C,NC}}\right)^{0.90 - 0.75 Dr^*} \tag{8}$$

For sand containing fines, a coefficient for the impact of an increase in Kc on Na was defined as $C_{SPH}{}^* = (Na)_{Kc}/(Na)_{Kc,NC}$ ($(Na)_{Kc}$: Na-value in a given Kc condition, $(Na)_{Kc,NC}$: Na-value in a normally consolidated condition). Figure 6 shows plots relative to Fc, obtained by using the correction factor $F_{Na}(Fc)$ shown in Eq. 9, in order to investigate the influence of Fc on C_{SPH}. The plots show good approximations to Eq. 10. Figure 7 plots the relationship between $C_{SPH}{}^*$ and Dr^* obtained from the test

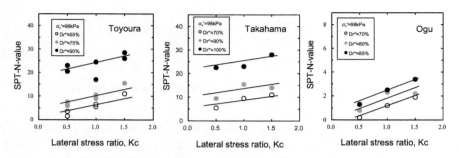

Fig. 5. Relationship between lateral stress ratio and normalized N-value, N_1

results for the three samples shown in Table 1. Figure 7 also shows the curves formulated by substituting the density correction coefficient C_{SPH} and the fines content correction factor $F_{Na}(Fc)$ expressed as Eq. 8 and Eq. 10 in Eq. 9. As shown, the plots and the curves show strong correlations.

$$C^*_{SPH} = F_{Na}(F_C) \cdot C_{SPH} \quad (9)$$

$$F_{Na}(F_C) = \left(\frac{K_C}{K_{C,NC}}\right)^{-0.10\left(\frac{FC}{100}\right)^{0.40}} \quad (10)$$

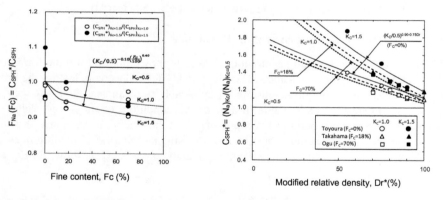

Fig. 6. Relationship between $F_{Na}(Fc)$ and Fc

Fig. 7. Relationship between $C_{SPH}*$ and $Dr*$

5.2 Effect of Kc on Cyclic Strength

Figure 8 shows the hollow torsional shear test results in the form of the relationship between cyclic strength R and Kc as in the case of the N-value test results. As shown, it can be said that as in the case of the N-value, the increase effect of Kc on R remains nearly constant, regardless of the relative density, for all samples. Ishihara et al. [8]

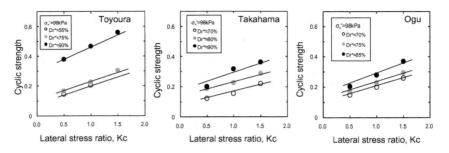

Fig. 8. Relationship between lateral stress ratio and cyclic strength

presented Eq. 11 for the impact of Kc on cyclic strength. If the test results shown in Figure 8 are plotted on the cyclic strength graphs for isotropic ($Kc = 0.5$) and anisotropic ($Kc = 0.5, 1.5$) conditions shown in Fig. 9, it can be seen that this relation also roughly holds true for sand containing fines, regardless of the density.

$$(R)_{K_C} = \frac{1+2K_C}{1+2K_{C,NC}} \cdot (R)_{K_{C,NC}} \qquad (11)$$

$(R)_{Kc}$: R in a given Kc condition
$(R)_{Kc,NC}$: R in a normally consolidated condition

Comparing the cyclic strength based on Eq. 7 (estimated cyclic strengths) with the experimental results (measured cyclic strengths) at Kc = 0.5, it can be seen that the experimental results tend to be smaller as the fines content becomes larger, as shown in Fig. 10.

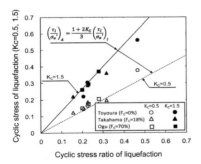

Fig. 9. Experimental results related influence of lateral stress ratio

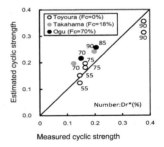

Fig. 10. Comparison with measured and estimated cyclic strengths

6 Relationship Between N_1-value and Cyclic Strength

If the modified relative density Dr^* defined by Eq. 2 is introduced, $(Na)_{Kc}$ and $(R)_{Kc}$ for a given Kc value can be estimated by density correction based on Eq. 8 and fines content correction based on Eq. 10 for the effect of Kc on Na obtained by correcting N_1 with respect to the fines content, and they can be estimated by using Eq. 11 for R. Since the N_1-value can be converted by Eq. 6 to the Fc-dependent Na-value, the relationship between N_1 and $(R)_{Kc}$ at each Fc value can be expressed by combining the $(R)_{Kc}$–Dr^* relation and the N_1–Dr^* relation at each Kc value through Dr^*. Figure 11 shows this relationship at Fc values of

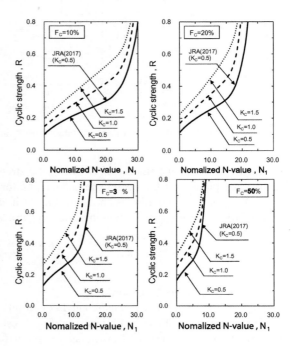

Fig. 11. Relationship between N_1-value and cyclic strength of each fines content

10, 20, 35, and 50%. From Fig. 11, it can be seen that as the Kc value increases, R increases at all values of Fc even if the same measured value of N_1 is obtained.

7 Conclusion

To investigate the effects of increases in the lateral stress ratio Kc on the N_1 value and cyclic strength, the authors examined and analyzed the results of soil chamber tests and laboratory element tests capable of controlling the value of Kc for both clean sand and sand containing fines. By integrating the results thus obtained, the authors then presented charts showing the relationship between the N_1-value and cyclic strength for a given value of Kc at different values of Fc. It is believed that these charts make it possible to quantitatively evaluate the effect of the lateral stress ratio Kc on the cyclic strength of ground improved by the sand compaction pile (SCP) method.

References

1. Harada, K., Ishihara, K., Orense, R.P., Mukai, J.: Relations between penetration resistance and cyclic strength to liquefaction as affected by Kc–conditions. In: Proceedings of GEESD IV Paper 111 (2008)
2. Ishihara, K., Harada, K., Lee, W.F., Chan, C.C., Safiullah, A.M.M.: Post-liquefaction settlement analyses based on the volume change characteristics of undisturbed and reconstituted samples. Soils Found. **56**(3), 545–558 (2016)
3. Harada, K., Yasuda, S., Niwa, T., Shinkawa, N., Ideno, T.: Evaluation of improved ground by compaction containing fines. JSCE J. Earthq. Eng. **27** (2003). (in Japanese)
4. Cubrinovski, M., Ishihara, K.: Empirical correlation between SPT N-value and relative density for sandy soils. Soils Found. **39**(5), 61–71 (1999)
5. Harada, K., Ishihara, K.: Study on minimum void ratio and relative density of soils containing fines. In: Annual Meeting of JEE, pp. 276–277 (2012). (in Japanese)
6. Harada, K., Ishihara, K.: Relation between penetration resistance and relative density of soils containing fines. In: The 67th Annual Meeting of JSCE, pp. 411–412 (2012). (in Japanese)
7. Japan Road Association: ((a): 1996, (b): 2017) Specifications for Highway Bridges, Part V, Seismic Design. (in Japanese)
8. Ishihara, K., Iwamoto, S., Yasuda, S., Takatsu, H.: Liquefaction of anisotropically consolidated sand. In: Proceedings of the 9th International Conference on SMFE, pp. 11–15 (1977)

Effect of Water Flow Rate and Insertion Velocity on Soil Disturbance Due to Insertion of Small-Scale Self-boring Tube

Pei-Chen Hsieh[1(✉)], Takashi Kiyota[1], Toshihiko Katagiri[1], Masataka Shiga[1], and Manabu Takemasa[2]

[1] Institute of Industrial Science, The University of Tokyo, Tokyo, Japan
hsieh@iis.u-tokyo.ac.jp
[2] Kiso-Jiban Consultants Co., Ltd., Tokyo, Japan

Abstract. The freezing sampling is considered as a method for high quality undisturbed soil sample especially for evaluating the liquefaction strength on cohesionless and saturated sandy soils accurately. However, since the conventional freezing sampling method takes much coolant consumption and time, a newly "small-scale" freezing sampling method was developed recently. Due to the procedure of inserting the freezing tube into the ground, some disturbances around the freezing tube might be occurred before the soil has been frozen. In this study, in order to obtain undisturbed soil samples for the laboratory test, attempts were made to evaluate the area of disturbance in the collected sample by the "small-scale" freezing sampling. A 750 mm-height, 186 mm-diamater sand box with an insert machine is adopted in this research to observe the disturbance after insert the freezing tube. From the changes in density, soil hardness and other factors after the test, it was found that flow rate of water and the insert velocity of freezing tube affected the area of sample disturbance around the freezing tube.

Keywords: Freezing sampling · Density · Soil box test · Soil disturbance · Undisturbed sample

1 Introduction

1.1 Small-Scale Freezing Sampling

In order to obtain the in-situ liquefaction strength of the ground by the laboratory test, it is important to avoid soil disturbance by sampling. In common sampling method, the soil skeleton of the specimen is easily disturbed due to the collection and transportation procedure. The freezing sampling is recognized as a method that it could solve the disturbance issue and be applied for high quality (undisturbed) in-situ sandy soil sample. However, it is rarely used in common construction works because of its high cost and large scale of construction works.

In Japan, the first case of freezing sampling was reported in 1973 [1], and the method has been improved in the past decades. In typical cases, it usually takes one day or more for freezing the ground and 10 tons to 100 tons of coolant, in the case of using liquid nitrogen (LN_2), are consumed. In addition, it is estimated that the volume which used for collecting specimen is only 6% of the total frozen area [2].

In order to improve the above disadvantages, a new method called "small-scale freezing sampling" was proposed in 2018, and its feasibility is also demonstrated in 2019 [2, 3]. This new method can reduce the required time and the coolant consumption to about 1% of that of the conventional freezing sampling method, and it can be applied to common borehole machine which widely used in site investigations. In the conventional method, a borehole with around 76 mm in diameter is required to insert a freezing tube. On the other hand, in the small-scale freezing sampling, a tube with double layer structure with 20 mm in outer diameter, which integrates drilling and cooling functions, is inserted by self-boring method.

1.2 Disturbance Due to Insert the Freezing Tube

Although freezing sampling can prevent the disturbance induced by sample collection and transportation, the installation of the freezing tube still may cause some disturbance to the ground around the freezing tube. It is necessary for both conventional freezing sampling and small-scale freezing sampling to collect samples from out of the disturbed area. Therefore, to clarify the size of the disturbed area around the freezing tube is an important issue.

Previous studies, such as Yoshimi et al. [4], measured the change of soil density around a freezing tube with 73 mm in diameter, which is used in the conventional freezing sampling method, and the disturbed area was determined by the significant change of soil density. The result of previous studies reveals that the diameter of disturbed area around freezing tube, which is unsuitable for the experiment, is about 3–5 times of the diameter of freezing tube.

In this study, small-scale freezing sampling using a double layer structure freezing tube is considered, and the disturbance of the surrounding ground due to the insertion of freezing tube by self-boring was investigated. A block sampling method [5] was used in this study to measure the dry density at several locations around the freezing tube before and after the insertion. In addition, the soil hardness test (using Yamanaka type) [6], and the insertion resistance of the freezing tube was also measured.

2 Methodology

A 750 mm-height and 186 mm-diameter cylindrical sand box with an insert machine shown in Fig. 1 was adopted in this research. The soil material is silica sand No. 5 ($D_{50} = 0.4$ mm, $\rho_s = 2.643$ g/cm^3, $e_{min} = 0.683$, $e_{max} = 1.098$) and the average relative density (Dr) of the sand was 66%. The specimens were prepared by air pluviation method by multiple sieves to make sure the distribution of density is uniform. After the specimens reached the specified height, it was saturated through the tube at the bottom of the sand box and the freezing tube was inserted at the constant velocity.

The freezing tube used in this study has a double structure, the inner tube is 40 mm shorter than the outer tube, and the angle of shoe at the tip of the outer tube is 30°. A strain gage was also installed on the freezing tube to measure the resistance force during the insertion. During insertion, water is pumped to the bottom between the inner and outer tubes, excavates the sand at the tip of the freezing tube, and water is discharged with the excavated soil through the inner tube.

After the insertion of the freezing tube was completed, the water in the sand box was drained. Then the sand box was laid into horizontally, the split mold of the column was removed, and the cross section of the sample was shaped to observe the deformation around the freezing tube. The soil density was measured by block sampling method at several locations using a cylindrical tube with a diameter of 50 mm and a height of 25 mm, and the soil hardness test was also conducted.

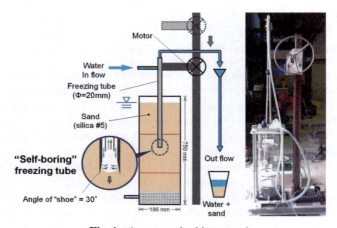

Fig. 1. Apparatus in this research

As shown in Table 1, six tests were conducted by varying the insertion velocity of the freezing tube and the water flow rate: a blank test without inserting a freezing tube

Table 1. The condition of each test case.

Test #	Insertion velocity (cm/s)	Water flow rate (cm^3/s)
0	No insertion	No water flow
1	1.5	10.1
2	1.5	14.7
3	1.5	25.1
4	1.5	56.5
5	0.5	15.6
N	1.5	No water flow

(#0), a series of tests with a freezing tube insertion velocity of 1.5 cm/s and varying the water flow rate (#1 to #4), and a test with a slower insertion velocity of 0.5 cm/s with an insertion flow rate of about 15 cm^3/s (#5). In addition, a test which the freezing tube was inserted without water supply was conducted (#N) to compare with the result without self-boring. The target insertion depth of the freezing tube is 50 cm.

3 Results and Discussion

As a result of each test, the central cross section is shown in Fig. 2. In test #1, a very large deformation was observed in the lower part. This is due to the drainage tube was extended to the floor level and a siphon phenomenon occurred, causing a large amount of sand to be sucked out. In order to prevent siphon phenomenon, the drainage tube was placed above the water level of the sand box in other experiments. For the test #3, the test was failed due to the insertion resistance force of the freezing tube was too large and it slipped between the holder and the tube, so the insertion was stopped at 35 cm. Since the siphon phenomenon in #1 is an impossible condition in the field, and #3 is unreliable, the results of both tests were excluded from the following discussion. While in the test #5, the motor stopped when the insertion depth reached 38 cm because the maximum insertion resistance became smaller as the insertion velocity became slower.

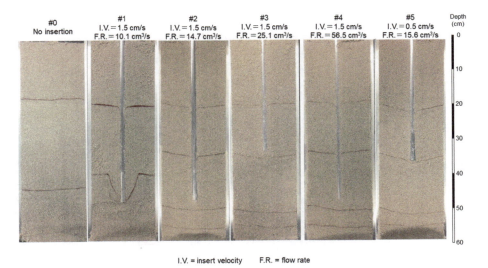

Fig. 2. Central cross section of each case

The change in the dry density and the soil hardness of the surrounding ground in each test were compared with the dry density and soil hardness of test #0, which is assumed as the reference values before inserting the tube. The differences were plotted as density and soil hardness change distribution diagram shown in Fig. 3 and Fig. 4, respectively.

3.1 Change in Density

The dry density was measured in four different depths with three different distances from the freezing tube, and the change in dry density was converted to relative density. According to the result from Tatsuoka et al. [7], it was shown that for the specimens with Dr of 50–70%, the change in liquefaction strength ratio was within ±10% if the change in Dr was ±6% (±1.3% in terms of dry density). In Fig. 3, the area where the Dr changes by more than ±6% is marked with a red frame to indicate the disturbance area.

Figure 3 shows that in all cases, there is a significant density decreasing area around the freezing tube at the top of the specimen. It is assumed that this is due to the vibration generated during the insertion of the freezing tube and the piping phenomenon occurred in the surface layer. In case #2, the disturbed area was located within 6 cm from the center of the specimen (5 cm from the freezing tube). The tendency that the upper part became looser and the lower part became denser is considered to be the same mechanism as pile penetration which shown in a previous study [8]. The disturbed area narrowed in case #4, which the flow rate and insertion velocity were both higher, the density in the lower part remained almost unchanged. Since the outer diameter of the sample by small-scale freezing sampling method that the authors intend to develop is about 10–12 cm, there is a high possibility of collecting undisturbed samples with less disturbance in the middle to lower part of #4 in terms of density change.

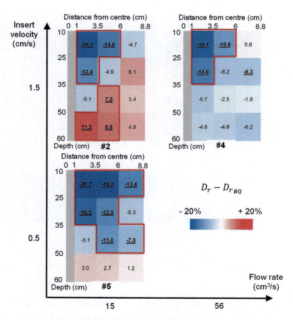

Fig. 3. Distribution of density change in each case

3.2 Change in Soil Hardness

The soil hardness test was conducted after the block sampling was completed and the cross section was shaped so that the disturbance induced by block sampling would be prevented. The results in Fig. 4 show that the value of soil hardness decreased in both tests #2, #4 and #5. The values within 2–3 cm around the freezing tube tended to decrease more than the outside. The change in the value of soil hardness may indicate the disturbance of the soil particle structure. Further investigations are necessary to determine the allowable value in future works.

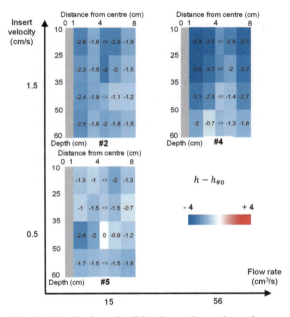

Fig. 4. Distribution of soil hardness change in each case

3.3 Discharged Soil

The ideal insertion condition can be assumed that there is neither compaction nor erosion of the surrounding ground during inserting the freezing tube, which means that the volume of discharged sand through inner tube should be equal to the volume replaced by the freezing tube. However, the amount of sand discharged was about 20 to 40% of the ideal value, as shown in Table 2. From this result, it is implied that the ground around the freezing tube become denser of the insufficiency of discharge sand, which match the result in Fig. 3 (#2 and #5). Although the denser part of test #4 is not showing in Fig. 3, it is assumed that it appeared in the area deeper than 60 cm in the sand box, which the block sampling cannot be conducted.

Table 2. Amount of out flow sand in each test case

Test #	w_i Ideal amount of discharged soil (g)	w_m Measured amount of discharged soil (g)	w_m/w_i (%)
2	202.7	78.8	38.9
4	255.2	62.6	27.8
5	166.6	32.1	19.3

3.4 Insertion Resistance Force

Figure 5 shows the results of insertion resistance force measured by strain gauges installed on the freezing tube. In test #N, since there was no water supply, the resistance force after 10 cm depth was significantly higher than the other tests. The tests #2, #4 and #5 show a similar trend of increase in resistance force with depth. The results also show that the effect of water flow rate and insertion velocity are insignificant on the resistance force. In addition, the increase rate of the resistance force was gentle up to the depth of 20 cm, while it suddenly increased after that.

The behavior of the freezing tube used in this study may be considered to be the same structure as an open-end pile. It is known that the tip of the open-end pile is clogged with soil as the pile penetrated into the ground, called "plugging effect". According to Mizutani et al. [9], the initial increase rate of penetration resistance force of an open-end pile is gentle, however, when the plugging effect starts to occur, the increase rate in

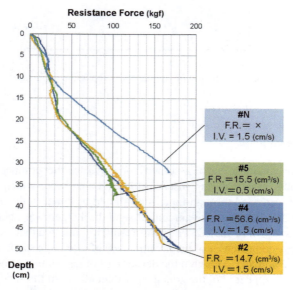

Fig. 5. Relationship between insertion depth and insertion resistance

penetration resistance force increases and becomes almost the same as that of closed-end piles. This mechanism may be the reason for the change in the increase rate of the resistance force in Fig. 5.

Therefore, before the freezing tube reached 20 cm, the sand near the tip of freezing tube was able to enter the tube and some of it was lifted and discharged with the water, which means self-boring was working. While when the freezing tube was deeper than 20 cm, the plugging effect occurred, and it was considered that the sand could not be discharged anymore. The plugging effect of the freezing tube can also be a possible reason to explain that the discharge sand is insufficient which mentioned in Sect. 3.3, and it might cause a large disturbance in the surrounding ground.

4 Conclusion

This study conducted cylindrical sand box test to investigate the disturbed area of the surrounding ground due to the insertion of the freezing tube for small-scale freezing sampling method. The changes in dry density and soil hardness around the freezing tube, the amount of discharged sand through inner tube, and the insertion resistance during the test were focused.

The result shows that it is possible to insert the small-scale freezing tube with relatively less disturbance by using self-boring method if the siphon phenomenon could be avoided and the apparatus could provide enough resistance force. In the case of sandy ground with Dr of 66% and using double layer structure self-boring tube with an outer diameter of 20 mm, the disturbance area decreased as the higher water flow rate and higher insertion velocity condition. The results showed that it may be possible to collect samples with less disturbance if the allowable difference of the liquefaction strength ratio is set to $\pm 10\%$.

On the other hand, although the water flow rate and insertion velocity influenced the change in density and soil hardness of the ground around the freezing tube, the amount of discharged sand was not affected by the insertion conditions and was only 20–40% of the ideal value. It was observed that the increasing rate of the insertion resistance force suddenly increased during the test. The reason for this phenomenon was that the tip of the self-boring freezing tube was clogged by sand, which can be considered as the same as the plugging effect of the open-end pile penetration.

References

1. Yoshimi, Y., Hatanaka, M.: In-situ density measurement of a loose saturated sand deposit. In: Proceedings of 10th Symposium of Disaster Science, pp. 341–342 (1973). (in Japanese)
2. Sakai, K., Yukawa, N.: Ground investigation beginning the future III Sampling. In: 53rd Japanese Geotechnical Society Annual Conference, pp. 217–218 (2018). (in Japanese)
3. Kobayashi, R., Yukawa, H., Kiyota T.: Basic experiment for practicability of small-scale freezing sampling. In: 54th Japanese Geotechnical Society Annual Conference, pp. 175–176 (2019). (in Japanese)
4. Yoshimi, Y., Hatanaka, M., Oh-oka, H.: Undisturbed sampling of saturated sands by freezing. Soils Found. **18**(3), 59–73 (1978)

5. Japanese Geotechnical Society: JGS1231-2003. In: Japanese standard for geotechnical and geoenvironmental investigation methods –standards and explanations– (2013). (in Japanese)
6. Japanese Geotechnical Society: JGS1441-2012. In: Japanese standard for geotechnical and geoenvironmental investigation methods –standards and explanations– (2013). (in Japanese)
7. Tatsuoka, F., Muramatsu, M., Sasaki, T.: Cyclic undrained stress strain behavior of dense sands by torsional simple shear test. Soils Found. **22**(2), 55–70 (1982)
8. Hayashi, K.: The influence of the pile-point angle on the bearing capacity, point resistance and pulling resistance of model piles. Trans. Jpn. Soc. Civil Eng. **105**, 19–26 (1964). (in Japanese)
9. Mizutani, T., Kikuchi, Y., Taguchi, H.: Model test on change of characteristics of ground induced by plugging effects of open-ended piles. Technical Note of the Port and Airport Research Institute No. 1053 (2003). (in Japanese)

Liquefaction Countermeasure for Existing Structures Using Sustainable Materials

Yutao Hu[1(✉)], Hemanta Hazarika[1], Gopal Santana Phani Madabhushi[2], and Stuart Kenneth Haigh[2]

[1] Kyushu University, Fukuoka 819-0382, Japan
hu.yutao.594@s.kyushu-u.ac.jp
[2] University of Cambridge, Cambridge CB2 1PZ, UK

Abstract. Many liquefaction countermeasures have been developed by researchers all over the world. The drainage approach is considered to be one of the most successful among those. However, most research focused on utilizing this method for newly constructed structures. In this research, a sustainable and low-cost technique for existing infrastructures is proposed. Gravel-tire chips mixture (GTCM) as an alternative drainage enhancing geomaterial has been adopted here. A series of model tests were conducted using shaking table. The results indicated that excess pore water pressure beneath the building quickly dissipated through GTCM drains during the shaking. The drainage system successfully prevented the liquefaction leading to the significant reduction of the settlement of the structure.

Keywords: Liquefaction · Earthquake · Vertical drain · Shaking table · GTCM

1 Introduction

In Japan, the total number of end-of-life tires produced was 86 million, which was near 1 million tons by weight in 2020, according to the report by JATMA (Japan Automobile Tyre Manufacturers Association 2021). To recycle such industrial waste, scrap tire-derived materials (STDMs) have been utilized as geomaterials in recent years. In addition to its low-carbon-release characteristics when used as geomaterials, other advantageous material characteristics of STDM include lightweight, excellent vibration absorption capability, and high permeability. Furthermore, unlike other granular geomaterials, these materials are non-dilatant in nature (Hazarika 2013). Gravel-tire chips mixture (GTCM), as an alternative geomaterial, has been introduced by Hazarika and Abdullah (2016).

On the other hand, large-scale earthquake-induced hazards have caused huge damage to infrastructures worldwide. One of such hazards, liquefaction, has become much frequent in recent years, especially in Japan. The Ministry of Land, Infrastructure, Transport and Tourism reported that, during the 2011 off the Pacific coast of Tohoku earthquake, about 27,000 houses were damaged due to liquefaction. Liquefaction-induced damage was also observed over a wide area following the 2016 Kumamoto Earthquake (Hazarika et al. 2017) as well. The severe damage caused by liquefaction has underlined the importance of adopting precautions to preserve buildings and facilities by limiting ground settlement and lateral spreading. Furthermore, in most developing countries, where infrastructure growth is still in its infancy, suitable and cost-effective mitigation measures are urgently needed. The main problem, however, is striking a balance between the cost and the increased environmental impact of any infrastructure project.

One such low-cost technique was developed by Hazarika et al. (2009), which utilizes a layer of tire chips as the horizontal inclusion under the foundation of residential housing. Horizontal reinforcing inclusion refers to a layer of tire chips which is placed horizontally. GTCM was then used to replace the pure tire chips. Since GTCM can provide sufficient bearing capacity to the foundation that otherwise has to rest on a highly compressible layer of tire chips, the improved technique is more practical (Hazarika et al. 2020). However, the above-mentioned techniques, as well as most of other liquefaction countermeasures, were developed for new constructions other than existing infrastructures.

Vertical drain method is one of effective liquefaction countermeasures, that has been widely used in practice. The principal objective of using vertical drains is to relieve the excess pore water pressure generated during the earthquake before they reach high values that can finally cause liquefaction (Brennan and Madabhushi 2006). More recently, Garcia-Torres and Madabhushi (2019) have investigated performance of the drains underneath structures. Considering the performance of GTCM in horizontal reinforcing inclusion mentioned before, it could be possible to utilize this material in vertical drain method as well.

In this paper, at first a brief description on previous research on vertical drain using pure tire chips is given. In reference to the developed technique, a series of model tests using shaking table were conducted to evaluate the performance of GTCM drains during earthquake. The prefabricated GTCM drains were installed around existing model structures. With dynamic loading applied, the development of excess pore water pressure in the foundation and progress of settlement were observed to evaluate the effectiveness of this drainage technique in mitigating liquefaction.

2 Behavior of Tire Chips Drains in Free Field Condition

An innovation cost-effective earthquake induced hazards mitigation technique has been developed by Hazarika et al. (2008) using tire chips. This such novel technique involves placing cushion layer made out of tire chips as a vibration absorber immediately behind the structures (Hazarika 2007). In addition, vertical drains made out of tire chips are installed in the backfill as a preventive measure against soil liquefaction.

To evaluate the performance of such technique, two cases of 1-g shaking table tests were conducted. The cross section of the soil box is shown in Fig. 1, in which the model caisson and the location of the various measuring devices (load cells, earth pressure cells, pore water pressure cells, accelerometers and displacement gauges) can also be seen. The model caisson was made of steel plates filled with dry sand and sinker. Behind the caisson, a cushion layer of tire chips was placed vertically down. Vertical drains using tire chips were installed in the backfill as well. The top of the entire drains was covered with a 50 mm thick gravel layer underlying a 50 mm thick soil cover. The purpose of such cover layer is twofold: one is to allow the free drainage of the water and other is to prevent the likely uplifting of the tire chips during shaking due to its lightweight nature.

In order to evaluate whether the technique can mitigate the liquefaction related damages, the time histories of the excess pore water pressure during the loading are compared. The comparison at the particular point (B in Fig. 1) in with and without improvement tests is shown in Fig. 2, which reveal that the pore water pressure build up is restricted due to dissipation by the permeable backfill condition. In the case of without any improvement, the pore water pressure built up and it took considerable time (about 25 s) to dissipate. Meanwhile, in the case of with tire chips drains installed, the built-up pore water pressure dissipated within a very short interval (2.5 s), preventing any chance for the backfill to liquefy.

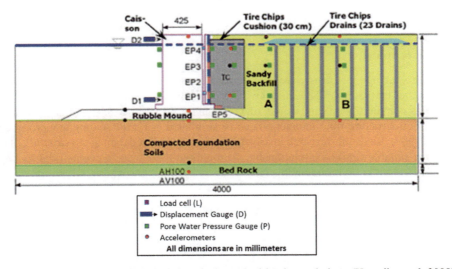

Fig. 1. Outline of the developed earthquake hazard mitigation technique (Hazarika et al. 2008)

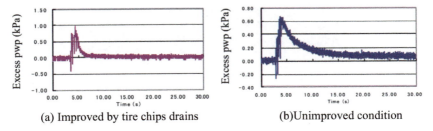

(a) Improved by tire chips drains   (b) Unimproved condition

Fig. 2. Time history of excess pore water pressure during the loading

3 Design of GTCM Drainage System

A new liquefaction mitigation technique, which is called GTCM drains system, is developed as a modification of pure tire chips drains developed in the past. This technique can be applied for existing infrastructures to reduce liquefaction induced damage (Fig. 3).

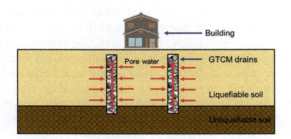

Fig. 3. Principle of GTCM drains technique

3.1 Tests Model

Series of shaking table tests were conducted using the 1-g shaking table facility at the geo-disaster laboratory of Kyushu University. Models were constructed in a transparent Plexiglas container with dimensions of 1800 mm × 400 mm × 850 mm, as shown in Fig. 4. Soil-structure-fluid interaction can be simulated using the scaling law proposed by Iai (1989). Since this research involves liquefaction induced damage to structure, it is the most suitable similitude relationship. Throughout these tests, a geometrical scaling factor of 1:32 was set based on this law.

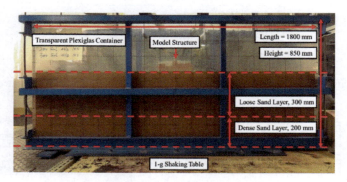

Fig. 4. Test model on 1-g shaking table

3.2 Tests Conditions

Toyoura sand was used as foundation soil in these tests. A dense layer of such sand (Dr = 90%, 200 mm in depth) representing non-liquefiable ground was constructed using both dry deposition and tamping techniques. The upper liquefiable layer (Dr = 50%, 300 mm in depth) was constructed only using dry deposition technique. The saturation process could be performed by percolating water gradually and uniformly through 3 water inlets from the bottom of the container.

GTCM with volumetric gravel fraction of 50% was used to make drains. The properties of GTCM are shown in Table 1. Depending on GTCM drains or not, two cases of 1-g shaking table tests were performed. As a compassion, Case 1 was set as default condition, with no drains installed. While in Case 2, 4 parallel arrays of prefabricated GTCM drains with diameter of 50 mm and height of 300 mm were installed vertically around the four sides of residential building from the surface level up to the bottom of the loose sandy layer and extended into hard layer. The distance between the center of two neighboring drains is 100 mm. A shallow foundation of a structure with a bearing pressure of 3 kPa, represented by a rectangular block of brass material, with cross-sectional area of 230 mm × 100 mm in model scale, was set upon the soil. Cross and top sectional views of the tests performed are shown in Fig. 5. Furthermore, another model with neither GTCM drains nor model structure was tested, named Case 0.

A sinusoidal acceleration of 200 Gal with frequency of 4 Hz and duration of 10s was applied to the model in both three cases. Different types of transducer were employed to measure acceleration, pore water pressure and dis-placement at different positions, as shown in this figure. The pore pressure transducers (PPTs) were fixed in place to monitor the pore water pressure in the exact locations. The laser micro displacement transducers (LMDTs) were set at the middle of short sides on the top of the model buildings. The locations of those transducers are shown in Fig. 5.

Table 1. Properties of GTCM in used

Property	Value
G_s	1.910
Dry density (g/cm^3)	1.022
ρ_{max} (g/cm^3)	1.036
ρ_{min} (g/cm^3)	0.842
D_{50} (mm)	4

3.3 Results and Discussion

Settlement. Figure 6(a) shows the position of the model structure before and after shaking in Case 1. Located on the unreinforced soil, the structure suffered significant settlement during the earthquake, as it sank until the original ground level and disappeared. The record of the LMDTs indicates the maximum displacement of 40.13 mm at D1 and 32.02 mm at D2. The rotation due to the uneven settlement was about 2°. In contrast, the settlement of the structure in Case 2 was less obvious. As shown in Fig. 6(b), the structure was nearly at the same position as before the earthquake. The LMDTs, as

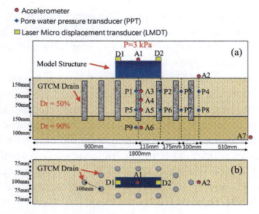

Fig. 5. Cross (a) and top (b) section views of the model

expected, captured the maximum displacement of only 1.02 mm at D1 and 0.88 mm at D2. This may be due to the excess pore water pressures not reaching the full liquefaction levels below the structure in Case 2. Since the only variable is the installation of GTCM drains, the difference of almost 40 times implies the effect of such technique.

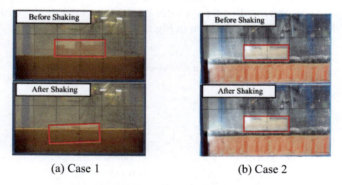

(a) Case 1   (b) Case 2

Fig. 6. Displacement of model structure

Excess Pore Water Pressure. The settlements of the buildings discussed above were induced by generation of excess pore water pressures and consequent liquefaction of the loose sand ($Dr = 50\%$). Figure 7(a) shows the time history of excess pore water pressure ratio, defined as $R_u = u_{expp}/\sigma'_{vo}$, recorded by PPTs at P1, P2, P3 and P4 in Case 0 (shown in Fig. 5). As a benchmark, with no influence from any drains, only P1 does not reach the full liquefaction value of 1. The Ru initiated a speedy increase as the earthquake started and remained in high value until the end of shaking. The similar phenomenon was observed at same locations in Case 1 as well. As shown in Fig. 7(b), the maximum Ru at P1 was smaller than that in Case 0, which is considered due to the compaction from the model structure. The generation of excess pore water pressure

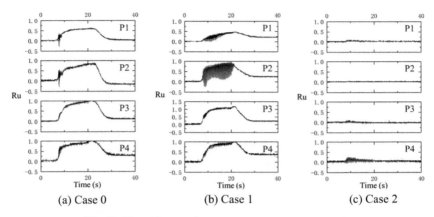

Fig. 7. Time history of excess pore water pressure ratio

during the earthquake resulted in liquefaction, which is hypothesized as the main reason for the considerable settlement of the structure in this case. On the contrary, the Ru at each place remains at an extremely low level in Case 2, as shown in Fig. 7(c). The GTCM drains, therefore, did effectively dissipate the excess pore water and make a great effort in preventing liquefaction.

Acceleration. Figure 8 shows the time history of acceleration recorded by accelerometers A1, A2, A3, A4, A5 and A6 (shown in Fig. 5) in both three cases. During the earthquake with input motion of 200 Gal, the acceleration enlarged from the bottom to the top of the model. The amplification was significant, especially in the shallow layer. A1 and A2, located on the surface of the soil, in particular, were overturned by an instant strong acceleration. Larger acceleration was also observed in Case 1. However, the amplification, in comparison to Case 0, was limited by the 3 kPa bearing pressure from the model structure. The maximum acceleration inside the soil was smaller than in Case 0. The amplification was more significant where liquefaction occurred due to the

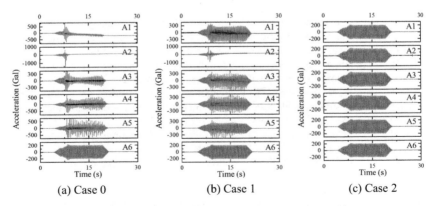

Fig. 8. Time history of excess pore water pressure ratio

softening of liquefied soil. Under the same input excitation with amplitude of 200 Gal, as shown in Fig. 8(c), the amplification did not take place in Case 2, which subsequences from the effective installation of GTCM drains.

4 Conclusions

When GTCM drainage system is applied, excess pore water can quickly dissipate through the GTCM drains from the foundation to the soil surface. As a result, the risk of liquefaction beneath the existing structures would be reduced. Liquefaction induced settlement of the building can be controlled to a low level during the earthquake since the liquefaction would be prevented through GTCM drains. The technique, thus, is expected to have a great potential in the cost-effective seismic design and retrofitting of structures. If adequate design guideline is established, the developed liquefaction mitigation technique could be applied for upgrading (retrofitting) of the existing structures that run the risk of damages during devastating future earthquakes.

Acknowledgement. The authors would like to acknowledge the financial support provided by Kyushu University under Progress 100 project. Thanks to Mr. Yuichi Yahiro, technical assistant of Geo-disaster Prevention Laboratory of Kyushu University, for his help and support while conducting experiments.

References

Brennan, A.J., Madabhushi, S.P.G.: Liquefaction remediation by vertical drains with varying penetration depths. Soil Dyn. Earthq. Eng. **26**(5), 469–475 (2006)

García-Torres, S., Madabhushi, G.S.P.: Performance of vertical drains in liquefaction mitigation under structures. Bull. Earthq. Eng. **17**(11), 5849–5866 (2019). https://doi.org/10.1007/s10518-019-00717-x

Hazarika, H.: Paradigm shift in earthquake induced geohazards mitigation -emergence of nondilatant geomaterials-. In: Keynote Lecture for the Annual Conference of Indian Geotechnical Society, Roorkee, India (2013)

Hazarika, H.: Structural Stability and Flexibility during Earthquakes using Tyres (SAFETY) - a novel application for seismic disaster mitigation-. In: Hazarika, H., Yasuhara, K. (eds.) Scrap Tire Derived Geomaterials - Opportunities and Challenges, pp. 115–125 (2007)

Hazarika, H., Abdullah, A.: Improvement effects of two and three dimensional geosynthetics used in liquefaction countermeasures. Jpn. Geotech. Soc. Spec. Publ. **2**(68), 2336–2341 (2016)

Hazarika, H., Igarashi, N., Yamagami, T.: Evaluation of ground improvement effect of tire recycle materials using shaking table test. In: 64th Annual Conference of Japan Society of Civil Engineers on Proceedings, pp. 931–932 (2009). (in Japanese)

Hazarika, H., et al.: Geotechnical damage due to the 2016 Kumamoto earthquake and future challenges. Lowland Technol. Int. **19**(3), 189–204 (2017)

Hazarika, H., et al.: Tire chip reinforced foundation as liquefaction countermeasure for residential buildings. Soils Found. **60**(2), 315–326 (2020)

Hazarika, H., Yasuhara, K., Hyodo, M., Karmokar, A.K., Mitarai, Y.: Mitigation of earthquake induced geotechnical disasters using a smart and novel geomaterial. In: 14th World Conference on Earthquake Engineering on Proceedings, Beijing, China (2008)

Iai, S.: Similitude for shaking table tests on soil-structure-fluid model in 1g gravitation field. Soils Found. **29**(1), 105–118 (1989)

Japan Automobile Tyre Manufacturers Association. Tyre industry of Japan (2021). http://www.jatma.or.jp/media/pdf/tyre_industry_2020.pdf

Undrained Monotonic Compression, Cyclic Triaxial and Cyclic Simple Shear Response of Natural Soils: Strength and Excess Pore Water Pressure Response

Majid Hussain[1(✉)] and Ajanta Sachan[2]

[1] Department of Civil Engineering, NIT Srinagar, Srinagar, India
majid.h@nitsri.ac.in
[2] Civil Engineering, IIT Gandhinagar, Gandhinagar, India

Abstract. The mechanical response of soils is governed by several factors including loading and boundary conditions. Under undrained boundary conditions, the nature and magnitude of excess pore water pressure (PWP) control the evolution of effective confining pressure (p') which in turn controls the evolution of shear stress. In this study, we investigate the shear strength and excess PWP response of natural soils under monotonic triaxial compression (TX), cyclic triaxial (CTX) and cyclic simple shear (CSS) testing conditions. The experimental study consisted of evaluating the undrained response of 31 natural soils collected from 10 locations (including 5 dams) in the Kutch region of India. The significance of the investigation lies in the fact that the region is seismically active with a proven history of devastating earthquakes. The most recent earthquake, the 2001 Bhuj earthquake, created large scale destruction with incidences of widespread earthquake liquefaction. The experimental investigation revealed that the undrained response of the soils at the in-situ density is controlled by both the fines content (FC) and plasticity index (PI). For cohesionless soils, FC governed the soil behaviour whereas for cohesive soils PI dominated the soil behaviour. Cohesionless soils exhibited intense strain softening (SS) under monotonic triaxial compression whereas cohesive soils displayed limited strain softening (LSS). Under CTX and CSS testing conditions, cohesionls soils exhibited very low liquefaction resistance (less than 10 cycles) whereas cohesive soils did not liquefy in 50 cycles. However, cohesive soils did exhibit significant degradation in cyclic strength, which was controlled by PI. The excess PWP was found to be contractive for all three conditions. For cyclic loading, PWP was found to be 30% higher for CSS conditions compared to the CTX conditions. Cyclic simple shear simulates the earthquake conditions better and should be considered for seismic and liquefaction analysis.

Keywords: Liquefaction · Cyclic triaxial · Cyclic simple shear · Earthquake · Excess pore water pressure

1 Introduction

Soil behavior is governed by several factors including nature and type of loading, boundary conditions and soil properties [1, 2]. Soil behavior under static and dynamic loading has been explored by many researchers [3–5]. The effect of strain rate, relative density and fines content on the soil behavior has been explored. Increased stiffness, increased peak shear strength and reduced strain at peak shear strength were observed with increased strain rate [6]. The influence of loading history on the undrained cyclic response of granular soils was explored under both the isotropic and anisotropic conditions [7, 8]. Soil behavior under cyclic loading is affected by characteristics of the input loading such as stress amplitude (γ), frequency (f) and the number of loading cycles (N) [9–12]. Loading under undrained conditions static as well as cyclic may lead to static and cyclic liquefaction respectively. This is characterized by lower mobilized shear strength and large excess pore water pressure. Fewer incidences of static liquefaction (Nerleck Berm, Aberfan Landslide, Merriespruit tailings dam failure) as compared to cyclic liquefaction are documented [13]. In the knowledge of the authors, very few studies have explored the link between static and cyclic liquefaction [14]. In these studies, a strong correlation between the state of the stress at the onset of instability under static and cyclic loading has been observed. The impending cyclic liquefaction could very well be predicted from the monotonic response. In this study, undrained soil behavior under monotonic triaxial compression, cyclic triaxial and cyclic simple shear conditions from the perspective of shear strength and excess pore water pressure is studied. Soil behavior of 31 natural soils from Kutch, a region in western India is investigated. Hussain and Sachan [15] investigated the link between the conditions triggering the static and cyclic liquefaction for a typical silty-sand of the region. The results were found to be in agreement with the those reported by [14].

2 Materials and Methods

In this study, 32 disturbed representative soil specimens were collected from 10 locations, including 5 major dams, at depths ranging from 0.5 m to 2.5 m from the Kutch region (Fig. 1). The locations were primarily selected on the basis of evidence of earthquake liquefaction during the 2001 Bhuj earthquake. During this seismic event, widespread liquefaction over an extent of greater than 15,000 m^2 was observed, thereby aggravating the earthquake damage. The basic geotechnical properties of the soils of the region can be found elsewhere [15] and are presented in Table 1. The soils in the region are mostly sandy soils with fines content varying from 11% to 83%. The in-situ dry density of the soils, in the present investigation, varied from 13.37 kN/m^3 to 17.71 kN/m^3. The plasticity index of the soils varied from non-plastic to 23%. From the total of 32 soil samples, 17 classified as silty-sands, 9 as clayey sands, 1 as low plasticity silt, 3 as clay with low plasticity, and 2 clay with high plasticity. Thus the soils explored in the current study, represent natural soil deposits with varying characteristics including in-situ density, particle characteristics (size, shape and texture), fines content, nature of fines and plasticity index.

Three series of tests including isotropically consolidated undrained compression triaxial (TX) tests, isotropically consolidated undrained cyclic triaxial (CTX) tests, constant volume cyclic simple shear (CSS) tests were performed on the 31 soil samples. The specimens for the three series of tests were prepared by the moist tamping technique at the in-situ density and moisture content. The specimens for TX and CTX were consolidated to an effective confining pressure of 100 kPa and then subjected to monotonic

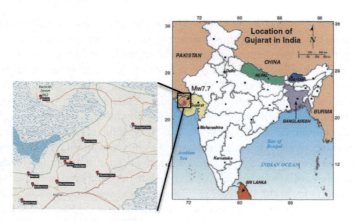

Fig. 1. Map of India showing sample collection locations and location of the 2001 Bhuj earthquake

Table 1. Geotechnical properties of soils of the Kutch region

Soil name	Depth m	γ_{di} kN/m^3	G_S	GSD				FC %	Atterberg limits			Soil class
				G %	S %	M %	C %		LL %	PL %	PI %	
Chang Dam		23°27.591′ N		70°24.408′ E								
S1 (L1)	0.5	15.00	2.67	6	78	11	5	16	–	–	–	SM
S2 (L2)	0.5	15.69	2.66	0	82	15	3	18	15.5	NP	NP	SM
S3 (L2)	1.5	15.70	2.68	5	76	17	2	19	20.0	NP	NP	SM
Kharoi		23°28.367′ N		70°23.330′ E								
S4	0.5	16.01	2.67	0	82	13	5	18	15.7	NP	NP	SM
S5	1.5	16.90	2.67	5	84	9	2	11	13.8	NP	NP	SP-SM
S6	2.5	16.00	2.67	1	86	11	2	13	12.7	NP	NP	SM
Suvai Dam		23°36.428′ N		70°29.821′ E								
S7	0.5	17.03	2.67	0	72	21	7	28	15.1	NP	NP	SM
S8	1.0	14.37	2.66	2	74	19	5	24	14.6	NP	NP	SM
S9	1.5	13.55	2.66	1	82	14	3	17	14.8	NP	NP	SM
Fatehgarh Dam		23°41.369′ N		70°48.057′ E								
S10	0.5	17.17	2.72	0	1	62	37	99	54.0	19.0	35	CH
S11	1.5	15.53	2.67	1	54	42	3	45	19.9	NP	NP	SM
S12	2.5	15.45	2.69	0	78	21	1	22	16.3	NP	NP	SM

(*continued*)

Table 1. (*continued*)

Soil name	Depth m	γ_{di} kN/m³	G_S	GSD				FC %	Atterberg limits			Soil class
				G %	S %	M %	C %		LL %	PL %	PI %	
Chobari		23°30.722' N		70°20.881' E								
S13	0.5	17.51	2.70	0	56	42	2	44	24.2	14.2	10.0	SC
S14	1.5	16.96	2.71	0	51	42	7	49	26.2	14.8	11.4	SC
S15	2.5	17.57	2.70	0	59	37	4	41	24.6	16.2	8.4	SC
Khadir		23°50.82' N		70°14.39' E								
S16	0.5	15.94	2.66	2	79	17	2	19	16.9	NP	NP	SM
S17	1.5	16.82	2.66	1	74	22	3	25	15.6	NP	NP	SM
S18	2.5	16.96	2.66	2	88	9	1	10	13.7	NP	NP	SP-SM
Tappar Dam		23°15.017' N		70°07.586' E								
S19	0.5	17.36	2.67	0	58	24	18	42	34.1	11.2	22.9	SC
S20	1.5	16.39	2.66	5	66	14	15	29	31.4	10.1	21.3	SC
S21	2.5	17.67	2.68	4	72	14	10	24	22.2	10.5	11.7	SC
Budharmora		23°20.634' N		70°11.501' E								
S22	0.5	17.71	2.68	2	69	21	8	29	23.2	14.6	8.6	SC
S23	1.5	14.27	2.71	1	34	46	19	65	44.3	15.7	28.6	CL
S24	2.5	12.26	2.70	2	18	57	23	80	65.8	26.9	38.9	CH
Banniari		23°24.299' N		70°09.910' E								
S25	0.5	13.37	2.74	0	17	81	2	83	26.4	NP	NP	ML
S26	1.5	14.59	2.75	0	5	68	27	95	47.2	18.6	28.6	CL
S27	2.0	16.26	2.68	0	68	26	6	32	24.6	11.6	13.0	SC
S28	2.5	17.60	2.69	1	78	13	8	21	28.0	11.7	16.3	SC
Shivlakha Dam		23°24.659' N		70°35.128' E								
S29	0.5	14.43	2.69	0	71	25	4	29	16.8	NP	NP	SM
S30	1.5	14.88	2.70	1	88	9	2	11	17.4	NP	NP	SP-SM
S31	2.0	16.37	2.69	1	74	18	7	25	15.0	NP	NP	SM
S32	2.5	13.40	2.68	0	28	50	22	72	39.0	15.5	23.5	CL

compression and axial cyclic loading respectively. Specimens for CSS were consolidated to vertical effective stress of 100 kPa under K_0 conditions and subjected to horizontal cyclic loading. The details of the experimental programme for TX, CTX and CSS tests can be found elsewhere [16–18]. Specimen S1 was not included in any of the test series whereas specimens S10, S28, and S32 were not included in the CTX and CSS test series.

These samples could not be tested as they could not be saturated due to the presence of large amount of high plasticity fines. Soils samples S1 and L2 were collected from very close locations and had similar properties.

2.1 Loading Conditions

The loading conditions for three series of tests were different. For the CIUC tests, the nature of the loading was static whereas for CTX and CSS the loading was cyclic in nature. The shearing of the specimens for all three series was deformation controlled. The deformation rate for CIUC tests was 0.1 mm/min, for CTX and CSS tests the frequency of cyclic loading was the same 0.1 Hz, however, the deformation amplitude was 0.4 mm and 0.12 mm respectively resulting in the strain amplitude of 0.6 % for both series of the tests. Figure 2 shows the pictorial representation of the loading for the three conditions along with the specimen size.

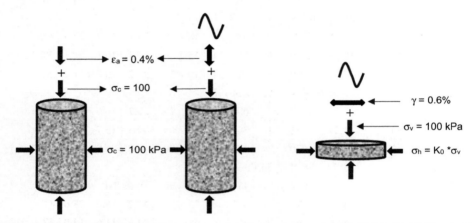

Fig. 2. Loading conditions and specimens size for the three series (a) CIUC, (b) CTX and (c) CSS.

3 Results and Discussion

The strength and excess pore pressure response of the Kutch soils as observed from the carefully conducted experiments including three series CIUC triaxial, CTX, and CSS tests are analyzed. The results are generalized for typical silty-sand and clayey-sand behaviour under the three loading conditions. The effect of fines content and plasticity index on the observed behavior is discussed.

3.1 Response Under CIUC Conditions

Figure 3 shows the stress-strain and pore pressure response of a typical silty-sand (S2) from the Kutch region under monotonic compression (CIUC) loading. The specimen mobilized deviatoric stress (σ_d) rapidly and attained peak value at axial strain (ε_p) value of 0.47%. After attaining the peak value, intense strain-softening behavior was observed. The specimen displayed a very low deviatoric stress value at large strains (residual strength) (Fig. 3a). The specimen developed rapid and large excess pore pressure (u) response in agreement with the stress-strain response (Fig. 3a). The intense strain-softening could be attributed to the extremely large pore pressure values which nearly reached to initial effective confining pressure (p'_i) of 100 kPa. Figure 4 shows the effective stress path (ESP) response of the typical silty-sand (S2) from the Kutch region. The ESP captures the signature of both the stress-strain and pore pressure response. Right from the onset of shearing, the effective confining pressure (p') reduced and attained very low values. The stress-path after attaining peak deviatoric stress (σ_{dp}) moved towards the stress origin and mobilized very low shear strength nearly manifesting static liquefaction. Similar behavior was displayed by the reaming silty-sands. However, σ_{dp}, rate of development of excess pore water pressure, and residual strength (σ_{dr}) was significantly affected by the amount of FC and PI (Table 2).

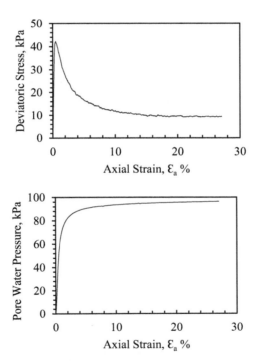

Fig. 3. Behavior of typical silty-sand (S2) under CIUC conditions. (a) Stress-strain and (b) Excess pore pressure

Table 2. Summary of CIUC triaxial tests. Parameters at peak deviatoric stress, and large strains

Soil name	M (%)	C (%)	FC (%)	σ_{dp}	ε_p	u_p	σ_{dr}	ε_r	u_r
Chang Dam									
S2	15	3	18	42	0.45	46	9.3	25	96
S3	17	2	19	38	0.42	48	4.6	25	96
Kharoi									
S4	13	5	18	34	0.6	57	16	25	92
S5	9	2	11	40	0.5	53	13	25	96
S6	11	2	13	42	0.48	52	9.2	25	98
Suvai Dam									
S7	21	7	28	38	0.54	51	14	23	95
S8	19	5	24	38	0.48	49	18	22	89
S9	14	3	17	36	0.42	45	5	25	97
Fatehgarh Dam									
S10	62	37	99	–	–	–	–	–	–
S11	42	3	45	44	0.55	51	18	25	95
S12	21	1	22	67	0.56	46	1.2	19	95
Chobari									
S13	42	2	44	48	0.76	53	42	25	85
S14	42	7	49	49	0.5	34	52	25	75
S15	37	4	41	37	0.5	28	47	25	76
Khadir									
S16	17	2	19	33	0.46	50	5.8	25	96
S17	22	3	25	38	0.54	53	13	25	93
S18	9	1	10	42	0.38	44	29	25	90
Tappar Dam									
S19	24	18	42	45	0.5	35	60	25	73
S20	14	15	29	49	0.5	41	54	25	72
S21	14	10	24	49	0.5	41	53	25	80
Budharmora									
S22	21	8	29	46	0.5	44	29	25	90
S23	46	19	65	27	0.5	20	72	25	76
S24	57	23	80	31	0.5	15	61	25	66
Banniari									

(*continued*)

Table 2. (*continued*)

Soil name	M (%)	C (%)	FC (%)	σ_{dp}	ε_p	u_p	σ_{dr}	ε_r	u_r
S25	81	2	83	49	0.56	46	6	25	97
S26	68	27	95	37	0.5	21	77	25	70
S27	26	6	32	45	0.76	52	43	25	85
S28	13	8	21	42	0.7	46	37	25	85
Shivlakha Dam									
S29	25	4	29	40	0.47	44	8	25	96
S30	9	2	11	53	0.45	48	5.6	25	98
S31	18	7	25	50	0.5	37	7.6	25	96
S32	50	22	72	38	0.5	30	50	25	75

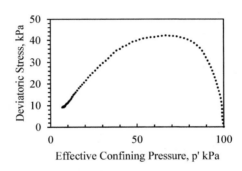

Fig. 4. Effective stress path (ESP) response of typical silty-sand (S2)

Figure 5 shows the stress-strain and pore pressure response of a typical clayey-sand (S19) of the Kutch region. It is evident that the response depicted by S19 is significantly different from the S2. Specimen S19 mobilized higher deviatoric stress as compared to S19 and did not show any strain softening. The mobilized deviatoric stress continued to increase slowly (Fig. 5b).

Specimen S19 exhibited gradual development of pore water pressure and reached a value of 73 kPa, significantly lower as compared to S2. The presence of higher and plastic fines led to the behavior observed in S19. Figure 6 shows the ESP response of S19. After an initial decrease, p' increased displaying strain hardening behavior. This could be attributed to the nature of fines. While S11 was non-plastic (silty-sand), the nature of fines was plastic in case of S19. Specimen S11 and S19 had nearly same FC (Table 2). However, the response of the two specimens was observed to be significantly different (Figs. 5 and 6).

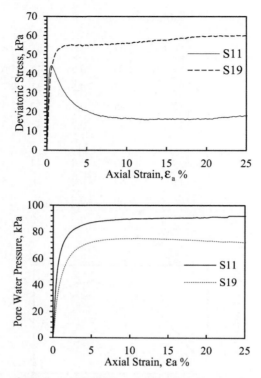

Fig. 5. Behavior of typical silty-sand (S19) and silty-sand (S11) under CIUC conditions. (a) Stress-strain and (b) Excess pore water pressure

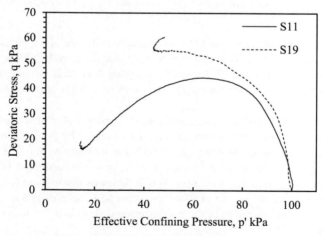

Fig. 6. Effective stress path (ESP) response of typical clayey-sand (S19) and silty-sand (S11)

3.2 Response Under CTX Conditions

Figure 7 shows the response of a typical silty-sand specimen S2 subjected to CTX loading after isotropic consolidation at 100 kPa. The observations reveal that after four loading cycles the hysteresis loops flattened and mobilized very low cyclic stresses (Fig. 7a) while large pore pressures developed rapidly (Fig. 7a). Large and rapid pore pressure developed leading to reduced mean effective pressures which subsequently led to very low cyclic stresses at constant strain. After four loading cycles, pore pressure ratio ($r_u = \frac{excess pore water pressure, \Delta u}{initial effective confng pressure, p_i}$) attained values nearly equal to unity signifying initial liquefaction (Fig. 7b). Figure 8 shows the effective stress of the specimen S2 subject CTX loading. It is evident that after four cycles of loading mean effective confining pressure reduced nearly to zero. At reduced p', the specimen exhibited very low cyclic resistance leading to liquefaction. The cyclic stress ratio ($CSR = \frac{\sigma_d}{p_i}$) at the end of the 5th cycle was observed to be very low, 0.03, signifying σ_d of 3 kPa (Table 3). Soils classified as silty-sand showed similar behavior with large and rapid generation of Δu leading to huge reduction in p'. Reduced p' values mean lower shear strength for cohesionless soils which is evident from the stress-strain response of the silty-sand specimens. With reduced mobilized shear strength (σ_d), CSR values also decreased.

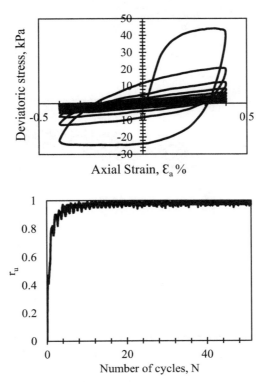

Fig. 7. Behavior of typical silty-sand (S11) under CTX conditions. (a) Stress-strain and (b) Excess pore water pressure

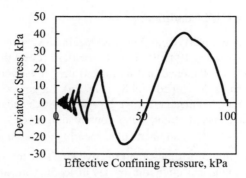

Fig. 8. Effective stress path (ESP) response of typical silty-sand (S11)

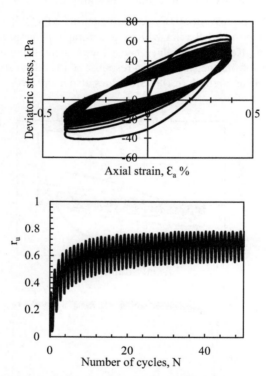

Fig. 9. Behavior of typical clayey-sand (S19) under CTX conditions. (a) Stress-strain and (b) Excess pore water pressure

Figure 9 shows the cyclic behavior of a typical clayey-sand (S19) of Kutch region. It is evident that the specimen exhibited higher resistance to cyclic loading as compared to S2. The presence of plastic fines resulted in mobilization of higher cyclic stresses as well as reduced and delayed excess pore water pressure. The hysteresis loops instead of flattening stabilized after five cycles (Fig. 9a). Excess pore pressure values also stabilized, however, the difference between the minimum and maximum r_u was higher as compared to S2. Even after 50 cycles of loading the r_u value was observed to be 0.75. Figure 10 shows the evolution of effective stress path for specimen S19. It could be observed during the 50 cycles not much degradation in cyclic strength occurred and p' values did not go below 30 kPa which lead to specimen not liquefying in 50 cycles. After 5 loading cycles, CSR was observed to be 0.28 much higher as compared to S2 (0.03) (Table 3). Specimen S11 with similar fines content as that of S19 displayed significantly different behavior and liquefied in 27 cycles with CSR at the end of 5th cycle equal to 0.11. The results from the CTX tests are presented in Table 3. Silty-sands liquefied within 5 cycles whereas clayey-sand did not liquefy. The amount of fines as well as their nature controlled the cyclic behavior of Kutch soils.

Table 3. Summary of CTX tests

Soil name	FC (%)	Clay (%)	Peak parameters at the end of								Number of cycles to liquefaction, N_L
			1st Cycle				5th Cycle				
			σ_{dmax} (kPa)	p' (kPa)	Δu (kPa)	CSR	σ_{dmax} (kPa)	p' (kPa)	Δu (kPa)	CSR	
Chang Dam											
S2	18	3	44.8	27.0	75	0.22	6.6	9.0	96	0.03	4
S3	19	2	31.3	35.0	67	0.15	8.0	11.0	89	0.04	9
Kharoi											
S4	18	5	36.2	32.0	70	0.18	10.9	12.0	90	0.05	31
S5	11	2	43.2	27.0	79	0.21	5.7	9.0	96	0.03	4
S6	13	2	47.7	25.0	80	0.24	3.4	4.0	95	0.02	5
Suvai Dam											
S7	28	7	38.6	34.0	67	0.19	9.4	10.0	90	0.05	23
S8	24	5	35.3	42.0	69	0.17	9.8	15.0	90	0.05	18
S9	17	3	41.8	28.0	74	0.21	7.0	5.0	95	0.03	5
Fatehgarh Dam											
S11	45	3	43.8	46.0	59	0.22	21.1	20.0	84	0.11	27
S12	22	1	52.2	25.0	78	0.26	4.0	4.0	97	0.02	3
Chobari											
S13	44	2	58.1	44.0	61	0.29	29.1	20.0	82	0.15	>50
S14	49	7	46.7	86.0	16	0.23	21.9	67.0	35	0.11	>50
S15	41	4	47.7	68.0	35	0.24	12.0	42.0	60	0.06	>50
Khadir											

(continued)

Table 3. (*continued*)

Soil name	FC (%)	Clay (%)	Peak parameters at the end of								Number of cycles to liquefaction, N_L
			1st Cycle				5th Cycle				
			σ_{dmax} (kPa)	p' (kPa)	Δu (kPa)	CSR	σ_{dmax} (kPa)	p' (kPa)	Δu (kPa)	CSR	
S16	19	2	39.5	33.0	71	0.20	9.5	12.0	93	0.05	7
S17	25	3	41.7	26.0	76	0.21	7.4	4.0	97	0.04	5
S18	10	1	67.9	28.0	75	0.34	3.6	3.0	97	0.02	3
Tappar Dam											
S19	42	18	66.4	73.0	36	0.33	55.4	51.0	64	0.28	>50
S20	29	15	55.4	66.0	38	0.27	38.5	48.0	62	0.19	>50
S21	24	10	45.2	64.0	38	0.23	21.2	41.0	60	0.11	>50
Budharmora											
S22	29	8	51.7	58.0	44	0.26	20.1	11.0	87	0.1	10
Banniari											
S25	83	2	53.3	41.0	63	0.27	7.7	5.0	95	0.04	5
S26	32	6	39.3	60.0	42	0.20	18.0	31.0	71	0.09	>50
S27	21	8	45.1	53.0	36	0.27	23.6	24.0	80	0.12	>50
Shivlakha Dam											
S29	29	4	39.0	30.0	73	0.20	4.5	7.0	97	0.02	5
S30	11	2	59.6	32.0	72	0.29	4.0	5.0	96	0.02	5
S31	25	7	62.2	33.0	73	0.31	13.5	6.0	95	0.07	5

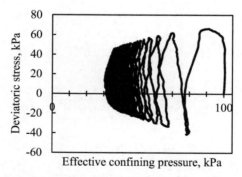

Fig. 10. Effective stress path (ESP) response of typical clayey-sand (S19)

3.3 Response Under CSS Conditions

Figure 11 shows the stress-strain and pore pressure response of a typical silty-sand (S2) of Kutch region. The hysteresis loop flattened after second cycle of loading indication large and rapid reduction in cyclic resistance (Fig. 11a).

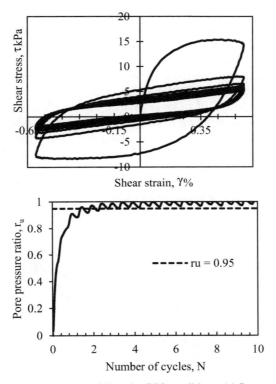

Fig. 11. Behavior of typical silty-sand (S2) under CSS conditions. (a) Stress-strain and (b) Excess pore water pressure

Shear stresses of magnitude 15.6 kPa were mobilized during the first cycle; much lower as compared to the deviatoric stress under the CTX conditions. Pore pressure ratio exceeded 0.95 just after two cycles of loading resulting to negligible confinement; a state of initial liquefaction (Fig. 11b). The specimen reached the state of initial liquefaction in two cycles.

Figure 12 shows the response of a typical clayey-sand (S19) of Kutch region. The hysteresis loops stabilized after a small reduction in mobilized shear stress (Fig. 12a). With further loading, the reduction in cyclic strength was lower and slower. Pore pressure ratio response also showed a similar trend (Fig. 12b). More than 50% (43 kPa) of the ultimate (80 kPa) pore water pressure developed during the first cycle. After the first loading cycle, the development of pore pressure was lower and slower. The specimen did not liquefy in 100 cycles of loading indicating large liquefaction to resistance. Specimen S11 with similar fines content, however, non-plastic attained liquefaction ($r_u > 0.95$) in 40 cycles under CSS conditions. The results from the CSS tests on Kutch soils are summarized in Table 4. Maximum shear stress (τ_{max}) and pore pressure water ratio during the 1st and the 5th cycle are shown. It is evident from the data that the mobilization of shear stresses and development of pore pressure varies over a wide range and is controlled by the FC in silty-sands and PI in clayey-sands. Silty-sand specimens did attain the state of liquefaction whereas clayey-sand specimens did not reach liquefy in 100 cycles. However, the clayey-sand specimens showed cyclic degradation the magnitude of which was controlled by the PI. At the end of the 5th cycle, the r_u value was found to be higher than 0.62 for all the soils signifying large reduction in the effective confining pressure. The reduced confinement resulted in lower mobilized cyclic strength which was evident from the stress-strain response (Figs. 10 and 12).

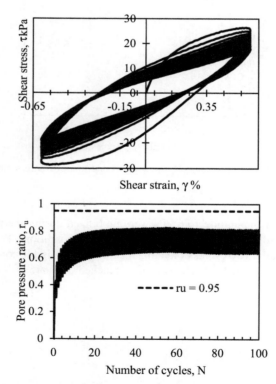

Fig. 12. Behavior of typical clayey-sand (S19) under CSS conditions. (a) Stress-strain and (b) Excess pore water pressure

Table 4. Summary of CSS tests

Soil name	FC (%)	C (%)	Parameters at the end of				Number of cycles to liquefaction
			1st Cycle		5th Cycle		
			τ_{max} (kPa)	r_u	τ_{max} (kPa)	r_u	N_L
Chang Dam							
S2	18	3	15.6	0.91	1.0	0.99	2
S3	19	2	20.1	0.87	2.0	0.98	3
Kharoi							
S4	18	5	18.4	0.81	6.0	0.93	7
S5	11	2	18.3	0.81	5.0	0.95	5
S6	13	2	15.3	0.87	1.1	0.97	3
Suvai Dam							
S7	28	7	16.3	0.85	4.1	0.97	3
S8	24	5	15.2	0.84	2.8	0.95	5
S9	17	3	14.0	0.88	1.7	0.96	4
Fatehgarh Dam							
S11	45	3	22.5	0.62	13.9	0.84	40
S12	22	1	18.4	0.87	3.8	0.98	2
Chobari							
S13	44	2	30.5	0.61	23.7	0.77	>100
S14	49	7	26.5	0.16	16.4	0.76	>100
S15	41	4	23.2	0.35	12.0	0.85	>100
Khadir							
S16	19	2	17.0	0.71	4.5	0.94	10
S17	25	3	17.7	0.76	4.6	0.93	20
S18	10	1	15.6	0.75	2.6	0.95	5
Tappar Dam							
S19	42	18	26.4	0.43	22.5	0.69	>100
S20	29	15	21.6	0.41	18.0	0.62	>100
S21	24	10	21.6	0.37	17.6	0.57	>100
Budharmora							
S22	29	8	20.2	0.78	7.4	0.94	6
Banniari							
S26	83	2	18.4	0.75	4.8	0.96	5

(*continued*)

Table 4. (*continued*)

Soil name	FC (%)	C (%)	Parameters at the end of				Number of cycles to liquefaction
			1st Cycle		5th Cycle		
			τ_{max} (kPa)	r_u	τ_{max} (kPa)	r_u	N_L
S27	32	6	19.9	0.79	6.4	0.93	6
S28	21	8	18.0	0.79	8.5	0.95	5
Shivlakha Dam							
S29	29	4	18.0	0.79	4.4	0.95	5
S30	11	2	19.2	0.88	3.1	0.99	2
S31	25	7	19.9	0.81	6.1	0.95	5

3.4 Discussion

The experimental results presented in Sects. 3.1, 3.2 and 3.3 reveal that the monotonic and cyclic strength under undrained conditions of Kutch soils is controlled by the amount and nature of fines. The response of the silty-sands at in-situ density resembled that of static and cyclic liquefaction. The observations from CIUC triaxial tests revealed intense strain softening (static liquefaction) in silty-sands as a common feature, intensity of which was controlled by the fines content (Table 2). As the nature of the fines changes from non-plastic (silty-sands) to plastic (clayey-sands), the response changes from intense strain softening strain hardening. However, for clayey-sands it is not the fines content rather the plasticity index of the fines that governs the behavior. The mechanism governing the effect of plastic and non-plastic fines at particle level can be found elsewhere [19]. The behavior of typical silty-sand from the Kutch under drained, undrained and mix-drained conditions can be found elsewhere [20]. The authors have discussed the effect of stress-ratio on the mix-drained behavior of a typical silty-sand.

Under cyclic conditions, both CTX and CSS, observations regarding the general behavior of Kutch soils were similar as that under CIUC conditions. However, the cyclic resistance of the soils under CSS was observed to lower as compared to CTX conditions. The nature of loading under the two conditions is quite different with CSS conditions more closely simulating the earthquake loading. The magnitude and rate of pore pressure development was higher and faster respectively under CSS conditions. This could be attributed to the nature of the cyclic loading and the mean effective confining pressure. The value of mean effective confining pressure at the start of the test for CTX and CSS is 100 kPa and 67 kPa (assuming at rest conditions with $K_0 = 0.5$). The lower value of p'_i leads to reduced cyclic strength under CSS conditions.

4 Conclusions

Deformation controlled triaxial (CIUC), cyclic triaxial (CTX), and cyclic simple shear (CSS) tests were performed on 31 soil samples from Kutch region at in-situ density. The results were analyzed in the context of shear strength and excess pore water pressure response. Following conclusions could be drawn from the study:

1. Soil response under the conditions is controlled by both fines content as well as nature of fines. While fines content dominates cohesionless soil behavior, plasticity index dominates the cohesive soil behavior.
2. Silty-sands of Kutch region have inherent tendency to liquefy under static as well as cyclic conditions. This could be attributed to many factors including lower density.
3. Kutch soils showed lower cyclic resistance under CSS conditions. Thus, CTX overestimates the liquefaction resistance and needs a careful evaluation

References

1. Wood, D.M.: Soil Behaviour and Critical State Soil Mechanics. Cambridge University Press, Cambridge (1990)
2. Mitchell, J.K., Soga, K.: Fundamentals of Soil Behavior, vol. 3. Wiley, New York (2005)
3. Schimming, B.B., Haas, H.J., Saxe, H.C.: Study of dynamic and static failure envelopes. J. Soil Mech. Found. Div. **92**(2), 105–124 (1966)
4. Ibsen, L.B.: The Static and Dynamic Strength of Sand. Aalborg Univ. (1995)
5. Inci, G., Yesiller, N., Kagawa, T.: Experimental investigation of dynamic response of compacted clayey soils. Geotech. Test. J. **26**(2), 125–141 (2003)
6. Yamamuro, J.A., Abrantes, A.E., Lade, P.V.: Effect of strain rate on the stress-strain behavior of sand. J. Geotech. Geoenviron. Eng. **137**(12), 1169–1178 (2011)
7. Ishihara, K., Sodekawa, M., Tanaka, Y.: Effects of overconsolidation on liquefaction characteristics of sands containing fines. In: Dynamic Geotechnical Testing, pp. 246–264. ASTM, West Conshohocken, PA (1978)
8. Ishihara, K., Takatsu, H.: Effects of overconsolidation and K0 conditions on the liquefaction characteristics of sands. Soils Found. **19**(4), 59–68 (1979)
9. Tatsuoka, F., et al.: Some factors affecting cyclic undrained triaxial strength of sand. Soils Found. **26**(3), 99–116 (1986)
10. Tatsuoka, F., Kato, H., Kimura, M., Pradhan, T.B.: Liquefaction strength of sands subjected to sustained pressure. Soils Found. **28**(1), 119–131 (1988)
11. Ishihara, K.: Soil Behavior in Earthquake Geotechnics. The Oxford Engineering Science Series, No. 46, Oxford, England (1996)
12. Kramer, S.L.: Geotechnical Earthquake Engineering. Prentice–Hall International Series in Civil Engineering and Engineering Mechanics. Prentice-Hall, New Jersey (1996)
13. Jefferies, M., Been, K.: Soil Liquefaction: A Critical State Approach. CRC Press, Boca Raton (2019)
14. Baki, M.A., Rahman, M.M., Lo, S.R., Gnanendran, C.T.: Linkage between static and cyclic liquefaction of loose sand with a range of fines contents. Can. Geotech. J. **49**(8), 891–906 (2012)
15. Hussain, M., Sachan, A.: Post-liquefaction reconsolidation and undrained cyclic behaviour of Chang Dam soil. In: Prashant, A., Sachan, A., Desai, C.S. (eds.) Advances in Computer Methods and Geomechanics. LNCE, vol. 55, pp. 77–90. Springer, Singapore (2020). https://doi.org/10.1007/978-981-15-0886-8_7
16. Hussain, M., Sachan, A.: Static liquefaction and effective stress path response of Kutch soils. Soils Found. **59**(6), 2036–2055 (2019)

17. Hussain, M., Sachan, A.: Dynamic characteristics of natural Kutch sandy soils. Soil Dyn. Earthq. Eng. **125**, 105717 (2019)
18. Hussain, M., Sachan, A.: Cyclic simple shear behaviour of saturated and moist sandy soils. Geomech. Geoeng., 1–24 (2021). https://doi.org/10.1080/17486025.2021.1975045
19. Hussain, M., Sachan, A.: Dynamic behaviour of Kutch soils under cyclic triaxial and cyclic simple shear testing conditions. Int. J. Geotech. Eng. **14**(8), 902–918 (2020). https://doi.org/10.1080/19386362.2019.1608715
20. Gujrati, S., Hussain, M., Sachan, A.: Liquefaction susceptibility of cohesionless soils under monotonic compression and cyclic simple shear loading at drained/undrained/partially drained modes. Transp. Infrastruct. Geotechnol., 1–33 (2022). https://doi.org/10.1007/s40515-022-00226-6

Simple Countermeasure Method to Mitigate the Settlement and Tilting of Existing Detached Houses Owing to Liquefaction

Keisuke Ishikawa[1(✉)], Susumu Yasuda[1], Motomu Matsuhashi[2], and Toshifumi Fukaya[3]

[1] Tokyo Denki University, Hatoyama, Saitama, Japan
ishikawa@g.dendai.ac.jp
[2] Graduate School of Tokyo Denki University, Hatoyama, Saitama, Japan
[3] HyAS & Co., Inc., Shinagawa, Tokyo, Japan

Abstract. In this study, we propose a new liquefaction countermeasure to suppress the settlement and tilting of detached houses that can satisfy the three conditions of "narrow construction space," "cost," and "impact on neighboring houses and life." The concept of this liquefaction countermeasure is to compensate for the reduced bearing capacity of a spread foundation when the bearing layer liquefies using the axial force of the pile and tension of the wire. In this paper, the mechanism of the countermeasure effect of this method is discussed using 1/25 scale 1G shaking-table tests under the experimental conditions of pile arrangement and wire arrangement between piles. From a series of shaking table experiments, it was observed that the piles were placed in pairs near the corners of the building model, and the horizontal displacement of the pile heads and pile tips was constrained, which significantly reduced the amount of settlement. However, the deformation did not cause the pile to yield.

Keywords: Existing detached house · Liquefaction countermeasure · Shaking table test · Penetration settlement

1 Introduction

The 2011 Great East Japan Earthquake caused severe damage due to liquefaction over a wide area from the Tohoku to the Kanto regions of Japan. Damage from, the settlement and tilting of detached houses was particularly noticeable. In eastern Japan, the earthquake caused about 27,000 cases of such damage, half of which in the urban residential areas in the Tokyo Bay area [1]. This was the first time that so many houses had been damaged in this way, not only in Japan but anywhere in the world. On the other hand, no damage was observed in various public structures, had been taken in advance against liquefaction of the ground.

In Japan, the danger posed by ground liquefaction to small buildings, such as detached houses, has not been fully recognized by owners and designers. Prediction methods and countermeasure systems for such structures have not been sufficiently developed.

Liquefaction countermeasures for existing detached houses and other small buildings in Japan must take into account the following factors: [2].

- Residential areas in Japan's urban areas are narrow and require small construction equipment:
- Houses are adjacent to each other, so adverse effects on neighboring houses must be avoided:
- Because they are individual properties, only inexpensive measures are desired compared to public structures:
- The cost of countermeasures differs significantly for new and existing structures.

Thus, liquefaction countermeasures for existing detached houses are strongly constrained by costs and construction methods. For this reason, the development of new liquefaction countermeasures specific to detached houses is desirable. Since the Great East Japan Earthquake, various countermeasures have been proposed to mitigate the settlement of existing detached houses and their surroundings. Yasuda et al. [3] showed experimentally that enclosing a building with thin steel sheet piles can suppress uneven settlement. Miwa et al. [4] used model experiments and numerical analysis to explore the effectiveness of piling logs around houses to suppress uneven settlement. Mano et al. [5] experimentally demonstrated that the replacement of the surface layer around the building perimeter with gravel can significantly mitigate the tilting of a building due to liquefaction. Nevertheless, there are still problems that need to be solved, such as the difficulties of construction in a narrow space, the effect on neighboring houses, and the influence of discontinuity caused by buried pipes.

In this study, a new and simple liquefaction-countermeasure method using piles and wires is proposed for existing detached houses. Shaking-table tests under a gravity field were conducted to investigate the mechanism of settlement prevention in this method and to evaluate its effectiveness for the prevention of penetration settlement.

2 Shaking-Model Test Method

Figure 1 shows the layout of the model and the instruments. The soil container was 1200 mm × 800 mm wide and 1000 mm high, with 15 laminar layers. The scale of the model was 1/25. The model ground was made of Toyoura sand ($\rho_s = 2.659$ Mg/m^3, $e_{max} = 0.993$, $e_{min} = 0.613$, FC = 0.0%) and two layers (liquefied and non-liquefied). The liquefied layer was a loose sand layer with a relative density of 50% and a layer thickness of 400 mm. A non-liquefied layer of the same thickness and relative density of 90% was used as the lower layer. An anti-extraction filter (10 mm thick) was installed at the boundary between the liquefied and non-liquefied layers. This was done to prevent sand movement between the two layers during the repeated model tests. In both layers, the density was controlled by the layer thickness. A hydraulic jet nozzle penetrating the ground from the surface agitated the liquefied layer to create a loose state. The non-liquefied layer was compacted by micro-shaking. The groundwater level was at the ground surface in all the cases. The house model was a two-story wooden structure with a ground load of 0.48 kN/m^2 (12 kN/m^2 in real scale). Acrylic and aluminum pipes were selected based on the similarity rule of bending stiffness proposed by Iai [6], assuming that general steel pipes (outside diameter 60.5 mm, wall thickness 4 mm) were used as the pile material for countermeasures. The wire used was a non-stretch polyethylene rope (1.5 mm thick) that was rigidly tied to the pile head with screws.

The accelerometers (A1–A9) and pore water pressure gauges (P2–P8) were placed in pairs at the same depth, as shown in Fig. 1. The sensors were glued to a thin rubber plate fixed to the bottom of soil container, which restrained the movement of the sensors underground. The displacement of the house model was measured using potentiometers (D1–D4) installed at the four corners of the model. The amount of ground settlement due to liquefaction was determined from the surface changes before and after shaking. Measurements were made at ten points at a pitch of 25 mm in the longitudinal direction of the soil container from the center line of the house model.

The shaking condition was a sine wave (2 Hz, 20 s); the table acceleration was set to 1.40 m/s^2. The shaking was unidirectional, in the longitudinal direction of the soil container. The measurements were completed after confirming that the excess pore-water pressure dissipated completely after the end of the shaking.

As shown in Table 1, the conditions of the pile-head and pile-tip restraints, pile diameter, and number of piles were varied in a total of seven cases. Case 1 was the no-countermeasure condition. Cases 2–4 represented a countermeasure model using four acrylic-pipe piles and wires, with different restraint conditions and placement methods at the pile edges. In Cases 5–7, the number of piles of aluminum pipes was varied. In Cases 2–3, the pile head was free, and the pile tip was grounded in the non-liquefied layer. In Cases 4–7, the pile end was constrained by a screw eye placed on the house model to restrain the horizontal displacement between the house model and the pile head; the vertical displacement of the house model was not constrained by the screw eye. A 30 mm conical round bar was connected to the pile tip, and pile tip was embedded in the non-liquefied layer to restrain the pile. The wires were orthogonal to the shaking direction in Case 2, and parallel to it in Cases 3–7.

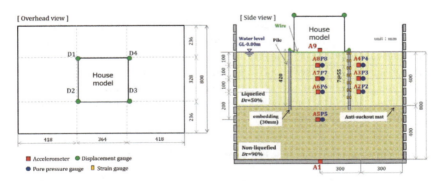

Fig. 1. Model ground and location of instruments

Table 1. List of experimental cases

Case	Piles Material	Piles Number	Pile diameter (mm)	Pile edge constraints Head	Pile edge constraints Tip	Wires Number	Wires Direction
Case 1	–	–	–	–	–	–	–
Case 2	Acrylic	4	8	Free	Fixed	2	Orthogonal
Case 3							Paralel
Case 4	Aluminum		4	Horizontal restraints	Embedding		
Case 5		4					
Case 6		8				4	
Case 7		12				6	

- Fixed: The pile tip is in contact with upper mat of the non-liquefied layer.
- The direction of the wire is relative to the direction of the shaking table motion.

3 Verification of Countermeasure Effects by Shaking-Table Model Tests

3.1 Settlement Behavior of a House Model Without Countermeasures

Figure 2 shows the output of each instrument in Case 1 over time. The excess pore-water pressure ratio in the liquefied layer increased when the shaking began and reached the effective overburden pressure at each depth in approximately 10 s. After 20 s, the excess pore-water pressure from the lower layer dissipated, and the effective overburden pressure near the ground surface continued to disappear until 30 s. On the other hand, as seen in the bottom row of the figure, the excess pore-water pressure ratio of the non-liquefied layer reached an upper limit of approximately 0.6, and liquefaction did not occur.

The settlement behavior of the house model exhibited three phases. In the first phase, the excess pore-water pressure ratio of P8 reached 0.7 for 8 s after shaking, and the house model underwent minor settlement due to shaking. The second phase was between 8 s and 30 s; major settlement occurred due to penetration settlement and volume compression caused by the dissipation of excess pore-water pressure in the lower liquefaction layer. The final phase, after 30 s, featured volume-compression behavior due to the dissipation of excess pore-water pressure.

3.2 Effect of Countermeasures Using Piles and Wires

Figure 3 shows the results of the countermeasure model for Cases 2–4. The changes with time of the excess pore-water pressure ratio and the displacement of the house model at the near the top of the liquefied layer (100 mm below water level) are shown in this figure.

In case 2, the piles and wires ware placed orthogonal to the shaking direction. The excess pore-water pressure at the liquefied layer and settlement of the house model behaved much as they did without countermeasures. Also, the house model settled after 29 s, because of the dissipation of the excess pore-water pressure. The reason for the ineffectiveness of the countermeasures was that the shear deformation of the liquefied layer caused the pile head to move in the shaking direction and the wires under the house to loosen.

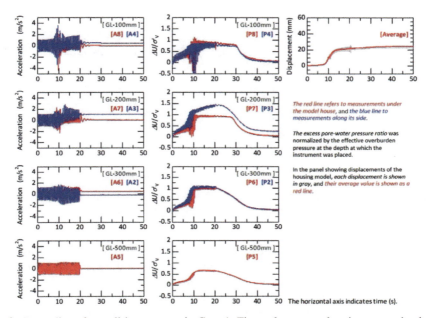

Fig. 2. Recordings from all instruments in Case 1. First column: acceleration; second column: excess pore-water pressure ratio; third column: displacement of housing model.

In Case 3, the piles and wires ware placed parallel to the shaking direction. The excess pore-water pressure reached the effective overburden pressure 4 s after the shaking started, and the settlement of the house model began near this time. The settlement of the house model reached its upper limit at approximately 10 s; the settlement behavior was unlike that of Case 1 or Case 2 during the subsequent shaking and volume compression because of the dissipation of excess pore-water pressure. The pile head contacted the house model 7 s after the start of shaking. This suggests that the house model settled because of the decrease in bearing capacity due to liquefaction, and the resultant tension of the wire caused the pile head to contact the house model.

In Case 4, the pile and wire were placed parallel to the shaking direction, the horizontal displacement of the pile head was constrained, and the pile tip was embedded in the non-liquefied layer. The excess pore-water pressure in the liquefied layer increased immediately after shaking, and reached an effective top pressure in approximately 1 s. The monotonic increase in the displacement of the house model during the shaking in the liquefied state between 1 s and 20 s was unlike the behavior seen in Cases 2 and 3. We suppose that the shear deformation of the liquefied layer increased during shaking,

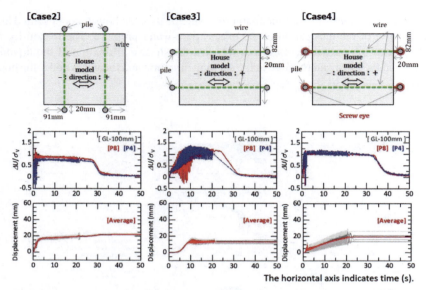

Fig. 3. Recording from all instruments in Case 2–4. Top row: schematic illustration of cases; middle row: excess pore-water pressure ratio; third row: displacement of housing model

and the piles installed in the liquefied ground were affected by the shear deformation of the ground, causing horizontal movement of the house model. However, after the excess pore-water pressure dissipated completely, no penetration settlement of the house model was observed, and the effectiveness of the countermeasure was confirmed.

These results indicate that horizontal displacement of the pile head and embedding of the pile tip can be used to control the penetration settlement of a spread-foundation structure during liquefaction.

4 Evaluation of a Simple Liquefaction Countermeasure Using Piles and Wire

Reports on cases of damage to detached houses indicate that, when the slope reaches 10/1,000, inclination-repair work is necessary to avoid dizziness, headaches, and other problems in daily life [7]. This inclination is classified as a half collapse in the Cabinet Office's damage assessment for affliction certification [8]. In houses that were restored after the Great East Japan Earthquake, the amount of uneven settlement was approximately 80 mm and the inclination was 6/1,000–10/1,000. Accordingly, in the present study the maximum inclination was set to 10/1,000 to avoid harmful settlement deformation of spreading-foundation structures.

However, because no inclination was observed in a single house model with a uniformly distributed load, a different but related quantity was used to evaluate the effectiveness of the countermeasures: penetration settlement (i.e. the difference between the average settlement of the house model and that of the surrounding ground). In the Great

East Japan Earthquake, the average penetration settlement of detached houses at an inclination of 10/1,000 was 70–150 mm, depending on the densification of the houses [9]. In this study, the average penetration settlement of a house with an inclination of 10/1,000 was therefore defined as 100 mm (4 mm in the scale model), and whether this value was satisfied was chosen as the performance-effectiveness criterion of the countermeasures.

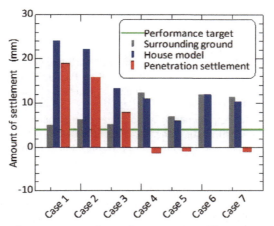

Fig. 4. Penetration settlement compared to performance target. The settlement of the house model and surrounding ground are also shown.

A positive value of the penetration settlement indicates that the house model has settled below the ground surface, while a negative value indicates that the house model has settled above the ground surface because of the countermeasures. Figure 4 shows the house-model, surrounding ground, and penetration settlements for each case studied. The average settlement of the house models in Cases 4–7, where horizontal movement of the models due to shaking during liquefaction was confirmed, was calculated by determining the positions of the models before and after shaking, and dividing the displacement from the displacement meter into vertical and horizontal components. In Cases 2 and 3, where pile edges were not subjected to displacement restraint, some effect of the countermeasure was observed, but the performance target was not reached. By contrast, in Cases 4–7, when the horizontal displacement of the pile head from the house model was constrained and the pile tip was embedded in the non-liquefied layer, the settlement of the surrounding ground and that of the house model were almost the same, and there was no significant penetration settlement. In these cases, the performance target of the countermeasure was satisfied.

5 Conclusion

In this study, a simple ground-liquefaction countermeasure using piles and wires was proposed for existing detached houses. Its effectiveness was verified by 1G shaking-model tests. The main conclusions are as follows:

(1) The effectiveness of the countermeasure varied significantly, depending on the restraining condition at the pile edge.
(2) The countermeasure was most effective when the horizontal movement of the pile head was restrained, and the pile tip was embedded in the non-liquefied layer to control the penetration settlement during liquefaction.

References

1. Yasuda, S., Harada, K., Ishikawa, K., Kanemaru, Y.: Characteristics of liquefaction in Tokyo Bay area by the 2011 Great East Japan Earthquake. Soils Found. **52**(5), 793–810 (2021)
2. The Japanese Geotechnical Society, Kanto Branch: A Guide for Protecting Detached Houses from Liquefaction (2013). (in Japanese)
3. Yasuda, S., Ishikawa, K.: Appropriate measure to prevent the liquefaction-induced inclination of existing houses. Soil Dyn. Earthq. Eng. **15**, 652–662 (2018)
4. Miwa, S., Yoshida, M., Murata, T., Numata A.: Shaking table tests and numerical analysis on log piling as a liquefaction countermeasure for existing houses. J. Jpn. Soc. Civil Eng. Ser. A1 **72**(4), I_117–I_128 (2016). (in Japanese)
5. Mano, H., Shamoto, Y., Ishikawa, A., Yoshinari, K.: Liquefaction damage reduction measures for small-scale structures by gravel replacement under the outer circumference of the structure. J. Jpn. Assoc. Earthq. Eng. **18**(3), 3_40–3_51 (2018). (in Japanese)
6. Iai, S.: Similitude for shaking table tests on soil-structure model in 1G gravitational field. Soils Found. **29**(1), 105–118 (1989)
7. Yasuda, S., Ariyama, Y.: Study on the mechanism of the liquefaction-induced differential settlement of timber houses occurred during the 2000 Tottoriken-seibu earthquake. In: Proceedings of 14th World Conference on Earthquake Engineering (2008)
8. Cabinet Office Japan, Disaster Management in Japan. http://www.bousai.go.jp/taisaku/unyou.html. Accessed 14 Oct 2021
9. Hashimoto, T., Yasuda, S., Yamaguchi, M.: Relationship between the slope of the house and the subsidence of grounds by liquefaction during the 2011 Tohoku-Pacific ocean earthquake. In: Proceedings of the 47th Japan National Conference on Geotechnical Engineering, No. 748 (2012). (in Japanese)

Effect of Artesian Pressure on Liquefaction-Induced Flow-Slide: A Case Study of the 2018 Sulawesi Earthquake, Indonesia

Takashi Kiyota[1]([✉]) [iD], Masataka Shiga[1], Toshihiko Katagiri[1], Hisashi Furuichi[2], and Hasbullah Nawir[3]

[1] Institute of Industrial Science, The University of Tokyo, Tokyo, Japan
kiyota@iis.u-tokyo.ac.jp
[2] International Division, Yachiyo Engineering Co., Ltd., Tokyo, Japan
[3] Faculty of Civil and Environmental Engineering, Bandung Institute of Technology, Bandung, Indonesia

Abstract. This paper discusses the trigger of the long-distance flow-slide that occurred in the 2018 Sulawesi earthquake, Indonesia. As a result of post-earthquake field surveys and interviews with local residents, it was confirmed that artesian groundwater existed before the earthquake in the area where the long-distance flow-slide occurred. This fact may be due to the liquefaction of the deep ground by the earthquake and the inflow of a large amount of groundwater from the aquifer below the liquefaction to the surface layer, which may have caused the long-distance flow-slide. In this study, the possibility of the presence of artesian pressure was examined by preparing a soil cross section of the affected areas estimated from the results of borehole surveys conducted after the earthquake and grain size analysis. In addition, the cyclic resistance ratio of the supposed liquefied soil was obtained by undrained triaxial cyclic loading test using in-situ undisturbed samples, and the effect of artesian pressure on the occurrence of liquefaction was examined by simplified liquefaction analysis.

Keywords: Sulawesi earthquake · Flow-slide · Liquefaction

1 Introduction

On 28th September 2018, an earthquake with Mw 7.5 hit Palu City in Sulawesi Island, Indonesia. The earthquake triggered long-distance flow-slides on very gentle slopes with slope of 1–5%, and led to the deaths of more than 2,000 people (BNPB [1]). Figure 1 shows the locations of the long-distance flow-slide and seismometer. An acceleration of about 200–300 gal was recorded by this seismometer during the earthquake (BMKG [2]). The topography is characterized by 10 km-wide lowland and mountains on both east and west banks and Palu-Koro Fault which caused the earthquake lies along the west side of Palu River.

Since the earthquake, the mechanism of the long-distance flow-slide has been studied by many researchers (e.g., Okamura et al. [3]; Hazarika et al. [4]). Cummins [5], Watkinson and Hall [6] and Bradley et al. [7] reported that the presence of irrigation channel affected the shallow groundwater condition and played a role in the liquefaction and flow-slides. However, there was no irrigation channel at the west side of Palu Valley, so that it cannot be deemed that only the irrigation channel was the cause of the flow-slides. On the other hand, Kiyota et al. [8] reported that the affected areas before the earthquake have a shallow groundwater table and it was under pressure (artesian pressure). They also reported that the shear strength of the surface soils significantly reduced by a large amount of groundwater supplied from the aquifer causing mudflows as shown in Fig. 2.

In this study, a soil cross section of the affected areas was prepared based on the results of borehole surveys conducted after the earthquake and grain size analysis. In addition, as part of the verification of the hypothesis that a large amount of groundwater flows from the aquifer to the ground surface after the earthquake, laboratory liquefaction tests using in-situ undisturbed samples and simplified liquefaction analysis considering the artesian pressure were conducted.

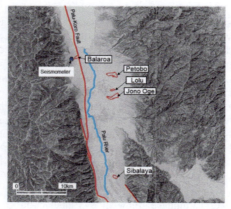

Fig. 1. Locations of long-distance flow-slide and seismometer in Palu City (after Kiyota et al. [8])

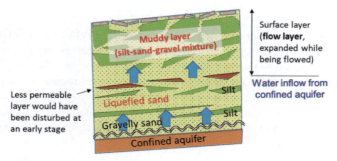

Fig. 2. Schematic illustration of long-distance flow-slide based on water inflow from confined aquifer (Kiyota et al. [8])

2 Characteristic Geotechnical Structure of the Affected Area

As shown in Fig. 1, the Sulawesi earthquake caused the long-distance flow-slide at five locations. Figures 3, 4 and 5 show a google earth image and the estimated soil cross section of Balaroa, Petobo and Jono Oge, respectively. The cross sections were set at locations in the damaged area where flow was significant. The cross section shows the post-earthquake situation, and the ground surface before the earthquake is indicated by the dashed line. The borehole data used was given by JICA [9] and the soil type was revised based on the results of grain size analysis.

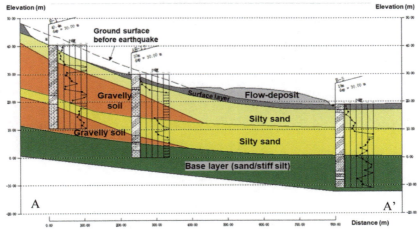

Fig. 3. Balaroa area after the earthquake from google earth and estimated soil cross section

One of the common features in the cross sections of each place is that the gravelly soil layer becomes extremely thin in the flow area. The gravelly soil layer is a continuous aquifer from upstream region, and is covered by a silty sand or sandy silt layer with relatively low permeability. The locations where the gravelly soil layer becomes thinner are directly beneath the area where the flow-slide occurs. The surface ground also consists of clayey soil with low permeability. Therefore, it is considered that the groundwater in

1582 T. Kiyota et al.

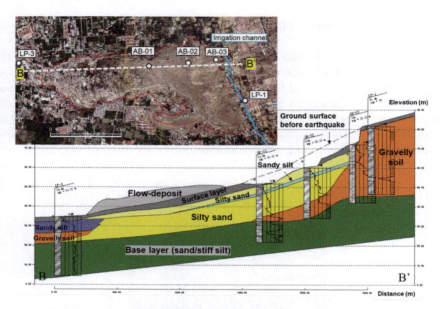

Fig. 4. Petobo area after the earthquake from google earth and estimated soil cross section

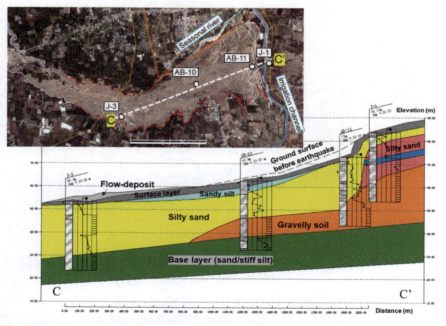

Fig. 5. Jono Oge area after the earthquake from google earth and estimated soil cross section

the gravelly soil layer and the silty sand/sandy silt layer above it may have been under high pressure before the earthquake.

The SPT-N values of the silty sand layer above the gravelly layer were relatively large, ranging from 10 to 20, however, they were measured after the earthquake, and the artesian pressure had already been almost released. On the other hand, assuming that the ground was under the artesian pressure before the earthquake, the effective stress acting on the silty sand layer would have been quite small. Therefore, the shear strength and liquefaction resistance of the silty sand layer before the earthquake may have been considerably lower than those estimated from the present N value. If the silty sand layer was liquefied during the earthquake, its permeability would have increased significantly, and a large amount of groundwater would have been supplied to the ground surface not only from the liquefied layer but also from the gravelly soil layer, which may have induced long-distance flow-slide.

Figure 6 shows a cross-section of the area where no flow-slide occurred (about 2 km south of Jono Oge). In this place, the gravelly layer is thickly deposited from upstream to downstream. Therefore, the groundwater from the upstream flowed smoothly downstream, and it is considered that no artesian pressure was produced even before the earthquake.

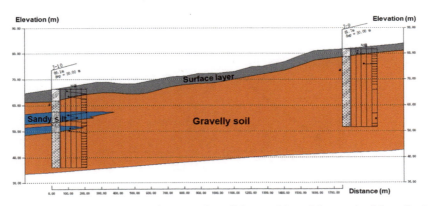

Fig. 6. Estimated soil cross section at no flow-slide area (about 2 km south of Jono Oge)

3 Effect of Artesian Pressure on Occurrence of Liquefaction

Triaxial liquefaction tests using undisturbed samples were conducted to investigate the occurrence of liquefaction in the silty sand layer with relatively high SPT-N value above the gravelly soil layer. The simplified liquefaction analysis was conducted at AB-01 in Petobo. However, since sampling was not conducted in the target layer at AB-01, samples were collected from the same layer (SPT-N = 10) at a borehole point 1.5 km south of Pedobo. Figure 7 shows the particle size distribution of tested sample ($FC = 53\%$, $D_{50} = 0.066$ mm, $G_s = 2.720$, $I_p = 10.6\%$, $e = 0.537$).

Figure 8 shows the relationship between the cyclic stress ratio, CSR, and the number of cycles to cause 5% of double amplitude of vertical strain. The cyclic resistance ratio, CRR, defined at 20 cycles for the sandy silt layer is estimated to be 0.39, which is relatively high. Using the obtained CRR, the results of a simplified liquefaction analysis (method of Japan Road Association [10]) conducted on the silty sand layer at site AB01 are shown in Fig. 9. In the analysis, the silty sand layer above the gravelly soil layer was subjected to different degrees of initial excess pore water pressure considering artesian pressure, and its effect on the liquefaction safety factor, F_L, was investigated. The surface accelerations that we set are 200 gal and 423 gal. The latter value is the surface acceleration in the flow-slide area estimated using the measured seismic motions (JICA [9]).

As a result, it was found that the silty sand layer may not liquefy even if the surface acceleration was 423 gal without considering the artesian pressure. However, when the artesian pressure of about 30–40% of the effective vertical stress was considered, most of the layer liquefied. It was also shown that liquefaction may occur even when the surface acceleration was 200 gal if the artesian pressure was about 70% of the effective vertical stress.

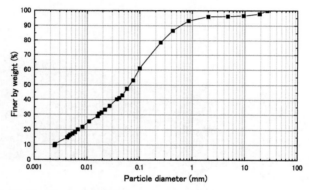

Fig. 7. Grain size distribution of tested sample (undisturbed sample)

Although this study is very rough, it shows that artesian pressure has a significant effect on the occurrence of liquefaction of silty sands above the gravelly layer. Considering that the permeability of liquefied soil becomes very large, this result suggests that a large amount of groundwater may have been supplied to the ground surface from the liquefied layer and the underlying aquifer (gravelly soil layer) after the earthquake. In order to quantitatively estimate the degree of artesian pressure, it is necessary to model the soil cross sections obtained in this study and conduct seepage flow analysis in the future.

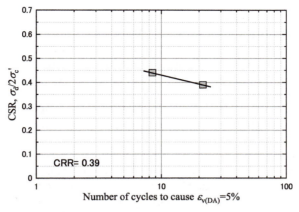

Fig. 8. Liquefaction resistance curve of undisturbed sample extracted from the silty sand layer above the gravelly soil layer

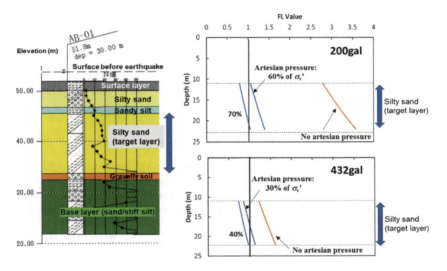

Fig. 9. Soil profile at AB-01 in Petobo area and the effect of artesian pressure on FL of the silty sand layer above the gravelly soil layer

4 Conclusions

In this study, the trigger of the long-distance flow-slide occurred on a very loose slope due to the 2018 Sulawesi earthquake are investigated. The conclusions obtained are as follows.

1) From the soil cross-section obtained from the borehole data and revised soil type based on the particle size test, it was confirmed that the gravelly soil layer (aquifer) becomes extremely thin in the flow-slide area. On the other hand, in the area where

no flow-slide occurred, it was confirmed that a thick layer of gravelly layer was formed continuously from upstream to downstream.
2) The gravelly soil layer, which becomes thinner within the flow-slide area, is covered by silty sand layer and surface layer that is considered to have low permeability. This indicates that the groundwater in the area may have been under artesian pressure before the earthquake.
3) From the triaxial liquefaction tests using undisturbed samples, the cyclic resistance ratio, CRR, of the silty sand layer overlying the gravelly layer was estimated to be 0.39.
4) The effect of artesian pressure on the occurrence of liquefaction of silty sand layer overlying gravelly soil layer was investigated by simplified liquefaction analysis. The results showed that the silty sand layer may be susceptible to liquefaction depending on the degree of artesian pressure. This suggests that a large amount of groundwater may have been supplied to the ground surface from the liquefied layer and gravelly soil layer after the earthquake, which may have triggered the occurrence of long-distance flow-slide.

Acknowledgement. The authors would like to express their sincere gratitude to JICA for providing borehole and grain size analysis data. This work was supported by grant funding of the JSCE committee of Promoting Innovation in Infrastructure Management and JSPS KAKENHI Grant Number JP19KK0108 and JP20H02244.

References

1. BNPB: Situation report of Central Sulawesi earthquake M7.4 and tsunami (2019) https://bnpb.go.id/infografis/infografis-gempabumi-m74-tsunami-sulawesi-tengah. (in Indonesian)
2. Sahadewa, A., et al.: Overview of the 2018 Palu Earthquake, Earthquake Geotech-nical Engineering for Protection and Development of Environment and Constructions, pp. 857–870 (2019)
3. Okamura, M., Ono, K., Arsyad, A., Minaka, U.S., Nurdin, S.: Large-scale flowslide in Sibalaya caused by the 2018 Sulawesi Earthquake. Soils Found. **60**(4), 1050–1063 (2020)
4. Hazarika, M., Pasha, S.M.K., Rohit, D., Masyhur, I., Arsyad, A., Nurdin, S.: Large distance flow-slide at Jono-Oge due to the 2018 Sulawesi Earthquake. Indonesia Soils Found. **61**(1), 239–255 (2021)
5. Cummins, P.R.: Irrigation and the Palu landslides. Nat. Geosci. **12**, 881–882 (2019)
6. Watkinson, I.M., Hall, R.: Impact of communal irrigation on the 2018 Palu earthquake-triggered landslides. Nat. Geosci. **12**, 940–947 (2019)
7. Bradley, K., et al.: Earthquake-triggered 2018 Palu valley landslides enabled by wet rice cultivation. Nat. Geosci. **12**, 935–939 (2019)
8. Kiyota, T., Furuichi, H., Hidayat, R.F., Tada, N., Nawir, H.: Overview of long-distance flow-slide caused by the 2018 Sulawesi earthquake. Indonesia Soils Found. **60**(3), 722–735 (2020)
9. JICA: Technical report by Support Committee on liquefaction-induced landslides (Inland area) (2021). (in Japanese)
10. Japan Road Association: Specification for Highway Bridges (2017). (in Japanese)

Physical Modeling and Reliability Assessment of Effectiveness of Granular Columns in the Nonuniform Liquefiable Ground to Mitigate the Liquefaction-Induced Ground Deformation

Ritesh Kumar[1] and Akihiro Takahashi[2(✉)]

[1] Indian Institutes of Technology Roorkee, Roorkee, India
`ritesh.kumar@eq.iitr.ac.in`
[2] Tokyo Institute of Technology, Tokyo, Japan
`takahashi.a.al@m.titech.ac.jp`

Abstract. Granular columns have been widely used to mitigate the liquefaction-induced effects on the built environment. Previous studies based on physical and numerical modeling and post-earthquake site investigations consolidate the efficacy of granular columns to mitigate the liquefaction-induced effects under small earthquakes. The increment in lateral stress due to densification, shear reinforcement, and drainage capacity of granular columns are believed to increase the liquefaction resistance of the ground. However, several case histories and recent research development exhibited the limitations of the effectiveness of granular columns under strong earthquakes. Therefore, a series of dynamic centrifuge experiments are carried out to investigate the effectiveness of granular columns in the liquefiable ground under strong ground motion recorded at Hachinohe Port during the 1968 Tokachi-Oki Earthquake. The performance of granular columns is evaluated by examining the evolution of excess pore water pressure, evolution of co-shaking and post-shaking settlement of foundation-structure system. A series of three-dimensional nonlinear stochastic analyses are also carried out using the OpenSees framework with PDMY02 elasto-plastic soil constitutive model to map the reliability of the performance of equally-spaced granular columns. The spatial nonuniformity of the ground should be considered for a reliable engineering assessment of the performance of granular columns which is implemented with stochastic realizations of overburden and energy-corrected, equivalent clean sand, $(N1)_{60cs}$ values using spatially correlated Gaussian random field. The reliability of the performance of the granular column is assessed based on the stochastic distributions of average surface settlement and horizontal ground displacement associated with the degree of confidence.

Keywords: Centrifuge modeling · Granular column · Liquefaction · Spatial nonuniformity · Stochastic analyses

1 Introduction

Liquefaction has caused severe damage to the built environment, for instance, settlement, tilting, and sinking of the foundation-structure system all over the world during many past earthquakes. The construction on the liquefiable ground is not recommended unless the appropriate liquefaction mitigation measures are taken at such sites. Soil remediation measures are requisite for liquefaction prone sites. Over the past few decades, extensive efforts and contributions have been made by the geotechnical earthquake engineering society to grasp the physics behind the liquefaction and for the development of remedial measures for the liquefaction-induced effects. Liquefaction mitigation by granular columns is one of the well-established techniques, which is used to facilitate quick dissipation of excess pore water pressure generated during the earthquake. Besides, the granular columns densify the surrounding soil during installation and believe to re-distribute the earthquake-induced or pre-existing stresses [1–3].

Many researchers have found that the pioneering design charts for granular columns developed by Seed and Booker [1] overestimate their performance [3–7]. Brennan and Madabhushi [8] performed centrifuge experiments to investigate the effectiveness of vertical drains in the mitigation of liquefaction-induced effects. They reported that the flow front (zone of adequate drainage at any time) play a vital role in the performance of gravel drains. The flow front slows down with distance from the gravel drain, and hence it is highly relevant to consider the effective radius and adequate spacing between the gravel drains. Adalier and Elgamal [3] have performed centrifuge experiments to understand the liquefaction mitigation capabilities of granular columns and associated ground deformations. They concluded that the performance of granular columns depends on their drainage capacity, and the densification of the ground during the installation of granular columns is inevitable. The ancillary benefits of treating the ground with granular columns are the restriction of shear deformation, offering the containment of the encapsulated soil, and providing stiffening-matrix effects (reducing the stress in adjacent soil) [3–7]. However, these effects are not well established yet, and more research is needed in this direction. Raymajhi et al. [9] and Kumar and Takahashi [10] investigated the contribution mechanism of shear reinforcement, increment in lateral stress, and drainage effects with the help of three-dimensional finite-element analyses. They reported that the granular columns undergo a shear strain deformation pattern, which is noncompatible with the surrounding soil contrasting with the conventional design assumption of shear strain compatibility. Many researchers [11–14] also suggested that the granular columns may deform in both flexure and shear modes which are not considered in the conventional design charts.

The ground is also prone to spatial nonuniformity and needs to be taken into account for a reliable engineering assessment of the performance of granular columns. The modeling of inherent soil variability can be achieved utilizing advanced nonlinear finite element analyses and well-calibrated sophisticated elasto-plastic soil constitutive models. Reliability analyses provide a means of evaluating the combined effects of uncertainties in the parameters involved in the calculations, and they offer a useful supplement to traditional engineering judgment. In this paper, test results of centrifuge modeling are discussed to better understand the performance of granular columns. Three types of model tests are performed (using Models 1–3) in which two model tests consist of the gravel drainage system of different drainage capacities, and one model test consists of untreated ground without any granular columns. Further, a series of nonlinear stochastic analyses are carried out using the OpenSees framework with PDMY02 elasto-plastic soil constitutive model. The soil variability is implemented with stochastic realizations of overburden and energy-corrected, equivalent clean sand, $(N1)_{60cs}$ values using spatially correlated Gaussian random field. Three-dimensional finite element simulations are performed for a sufficient number of realizations to map the reliability of the effectiveness of equally-spaced granular columns to mitigate the liquefaction-induced ground deformation.

2 Physical Modeling

A series of dynamic centrifuge experiments are carried out to investigate the liquefaction-induced effects on shallow foundations and the performance of granular columns resting on a level deposit of liquefiable Toyoura sand (properties are tabulated in Table 1). A flexible laminar container with inner dimensions of 600 × 250 × 438 mm (model scale) in length, width and height respectively, is used to frame the centrifuge model. The laminar box is composed of many aluminum rectangular alloy rings which allow its movement along with soil mass, creating a flexible boundary and establishing the uniform dynamic shear stresses within the model ground during the dynamic excitation. The centrifuge model contains two shallow foundations and associated superstructures, namely, buffer tank (BT) and flare stack (FS), imposing average bearing pressures of 51.2 kPa and 71.2 kPa, respectively, at 0.8 m below the surface of the model ground at the prototype scale, as shown in Fig. 1. The details of model scaling for BT and FS and physical properties can be found in Kumar et al. [7].

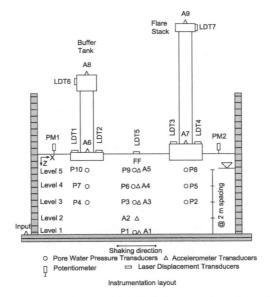

Fig. 1. Centrifuge model layout for Model 1: LG (no granular columns) and positioning of different transducers

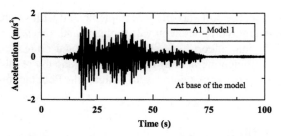

Fig. 2. Tokachi-Oki ground motion used as dynamic excitation

Table 1. Index properties of Toyoura sand and Silica no. 3

Description	Toyoura sand	Silica no. 3
Specific gravity, G_s	2.65	2.63
D_{50} (mm)	0.19	1.72
D_{10} (mm)	0.14	1.37
Maximum void ratio, e_{max}	0.973	1.009
Minimum void ratio, e_{min}	0.609	0.697
Permeability, k (m/s)	2×10^{-4}	6.6×10^{-3}
Relative density, D_R	~50%	30%

Dynamic centrifuge model tests are carried out utilizing the Tokyo Tech Mark III centrifuge facility with a radius of 2.45 m, at a centrifugal acceleration of 40g (N = 40; scaling law is tabulated in Table 2). The centrifuge model tests simulate a prototype saturated soil deposit at a depth of 10 m, with a water table located 1.8 m below the top surface. Multiple transducers (e.g., pore pressure transducers, accelerometers, laser displacement transducers, and potentiometers) are carefully placed at desirable locations (see Table 3) during the model ground preparation. The model ground is saturated with the viscous fluid, i.e., a mixture of water and 2.0% Hydroxypropylmethyl cellulose solution (Metolose by Shin-Etsu Chemical Co., Ltd.; grade 60SH-50) by weight of water, to achieve a viscosity about 50 times that of water. This solution is used to ensure the compatibility of prototype permeability of the soil to set up the affinity between dynamic and diffusion scaling laws [15]. Centrifuge model tests are performed under the earthquake ground motion recorded at the Hachinohe Port in the 1968 Tokachi-Oki Earthquake (NS component, see Fig. 2).

Table 2. Scaling law for centrifuge test (Schofield, 1981)

Parameters	Ratio of model to prototype
Length	$1/N$
Area	$1/N^2$
Volume	$1/N^3$
Acceleration	N
Stress	1
Strain	1
Time (dynamic)	$1/N$
Force	$1/N^2$
Bending moment	$1/N^3$

Table 3. Locations of different transducers within the model grounds

Level	Transducers*	Location (prototype scale)		Initial vertical effective stress (σ'_{vo}) at different levels**	
		X m	Z (depth) m	Magnitude, kPa (prototype scale)	Description
Level 1	P1, A1	12	10	102.40	Model centerline
Level 2	A2	12	8	73.92	Model centerline
Level 3	P2	18	6	61.22	Below FS footing
	P3, A3	12	6	65.44	Model centerline
	P4	6	6	57.52	Below BT footing
Level 4	P5	18	4	54.49	Below FS footing
	P6, A4	12	4	41.96	Model centerline
	P7	6	4	47.49	Below BT footing
Level 5	P8	18	2	53.80	Below FS footing
	P9, A5	12	2	26.16	Model centerline
	P10	6	2	41.80	Below BT footing

* P: Pore water pressure transducers, A: Accelerometers.
** Including vertical stress induced by the foundation-structure systems.

2.1 Design of Gravel Drainage System

The Design charts reported by Seed and Booker [1] in their seminal work and the revised guidelines presented by Bouckovalas et al. [16] are used to design the gravel drainage system for Models 2 and 3 (as shown in Fig. 3). Many parameters, e.g., replacement area, target excess pore water pressure ratio, earthquake intensity, reported case histories, and installation methodology of gravel drains, are considered while designing the gravel drainage system. Design specifications of both gravel drainage Types 1 and 2 are tabulated in Table 4. Liquefaction resistance curves for saturated Toyoura sand with a relative density of 50% ± 5% for different confining pressure are obtained using the laboratory test results from Chiaro et al. [17]. Then, based on Seed and Booker [1] it is found that Tokachi-Oki ground motion can be considered as a medium EQ to strong EQ (as specified in Table 4) for Models 1–3. It is evident that the gravel drainage Type 1 does not satisfy the design guidelines as the clear spacing between the drains are more than the maximum allowable spacing for medium EQ to strong EQ. Initially, it is hypothesized that a gravel drainage system (Type 1) could render the targeted performance of hybrid foundation along with friction piles having spiral blades [7]. However, centrifuge test results of Model 2 demonstrated the inefficiency of gravel drainage Type 1 in terms of both generation and dissipation of excess pore water pressure (EPWP). Based on the performance of gravel drainage Type 1, redesign of the gravel drainage system is done, and new gravel drainage Type 2 is tested in Model 3 which satisfies the design guidelines as tabulated in Table 4. In addition, to provide the significant drainage for developed EPWP to dissipate, the focus is put on shear-induced and post-liquefaction/shaking settlement due to the presence of gravel drains which alter the stresses and strains applied to the improved ground as highlighted by Priebe [18], Baez and Martin [19], and Adalier et al. [5].

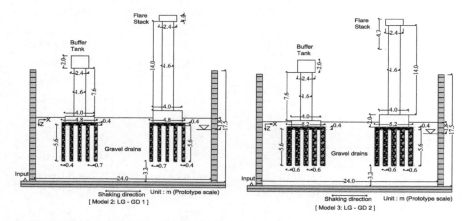

Fig. 3. Models configuration with gravel drainage systems (types 1 and 2) in prototype scale

Table 4. Design specifications of gravel drainage Types 1 and 2

Specifications*	Gravel drainage Type 1	Gravel drainage Type 2
Drain diameter (m)	0.40	0.60
Length of gravel drain (m)	5.6	5.6
Clear spacing (m)	0.70	0.55
Treated Plan	4.8 m × 4.8 m	5.2 m × 5.2 m
Replacement area (%)	13.63	26.14
Maximum allowable clear spacing between gravel drains for target $r_u = 0.7$		
Earthquake intensity Small EQ (Neq/NL = 1) **	For gravel drainage Type 1	For gravel drainage Type 2
Medium EQ (Neq/NL = 2) **	0.74 m	1.12 m
	0.44 m	0.80 m
Strong EQ (Neq/NL = 3) **	0.28 m	0.64 m

*The layouts for both gravel drainage types are shown in Fig. 3.
**For more details, it is recommended that readers refer to Seed and Booker [1] and Bouckovalas et al. [16].

2.2 Making of Granular Columns

The guide frames are prepared to make the gravel drainage system as shown in Fig. 4. The guide frames are kept at desirable locations (under both BT and FS) while preparing the model ground. Initially, the gravel drain casings are covered with tape while pouring the Toyoura sand. The guide frame has enough opening for Toyoura sand to pass through and to form uniform model ground in the vicinity of gravel drains. After achieving the required level of the model ground with Toyoura sand, Silica no. 3 (see Table 1) is poured inside all the gravel drain casings carefully. Then, guide frames are taken out from the model ground with due care to avoid any possible disturbance and densification within the model ground. It is to be noted that the length of gravel drain casings attached to guide plates are kept 10 mm (in model scale) longer than the required length of gravel drains in the model ground. The reason for this is to form a 10 mm deep (0.4 m in prototype scale) gravel mat over the group of gravel drain piles. From Fig. 4, a little

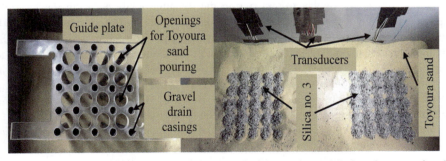

Fig. 4. Guide plate to form the gravel drains and typical array of gravel drains after removing the guide plates (for both BT and FS)

overflow of Silica no. 3 is evident which in turn is used to form the gravel mat of 10 mm (in model scale) thickness. By adding Silica no.3, the gravel mat is formed with due care to avoid any densification around the gravel drains.

3 Numerical Modeling

There are a few parameters that need to be considered to ensure the reliability of the performance of the granular column. For instance, the ground is prone to spatial nonuniformity, which was not considered in the centrifuge experiments. Besides, the granular columns only provided additional drainage to rapidly dissipate the excess pore water pressure, and the contribution in the shear reinforcement was ignored in the centrifuge experiments. Moreover, the density of granular columns in the centrifuge experiments was ~30%, which is significantly less than the density of constructed granular columns at the site (which is usually in the range of 75–85%). Considering these site-specific conditions, a half of the single granular column (with D_R ~ 80%) in the middle of the gravel drainage system (due to symmetry) under the buffer tank (BT) and associated model ground (effective drainage zone of granular column) is considered for the numerical simulations, as shown in Fig. 5 by assuming that many equally-spaced granular columns are constructed in a large area. The reason for this assumption is that the modeling of the whole centrifuge model and gravel drainage system is computationally expensive and not feasible for stochastic analyses as the reliability assessment requires thousands of analyses. Similar simplifications have been well-adopted by many researchers [14, 20]. This approach does not account for the distinct stress distribution to the individual granular column (in the gravel drainage system) coming from the foundation-structure system during the dynamic event. Instead, the intent is to explore the reliability at a single granular column to get an insight into the overall performance of the whole gravel drainage system.

Numerical simulations are carried with Rayleigh damping of 1% at a frequency of 1 Hz corresponding to the first-mode of a typical nonlinear ground response is used in the analyses [21]. The ground is modeled using brick u-p (8-node brickUP) elements. The load from the foundation-structure system is modeled as surface pressure for simplicity. The effects of the superstructure inertia are ignored in this study. The bottom nodes of the ground are kept fixed in all degrees of freedom. Tokachi-Oki ground motion (NS component of recorded shaking at the Hachinohe Port in 1968, see Fig. 2) is imposed on the bottom nodes of the ground during the dynamic analyses using the multiple support excitation technique in OpenSees. All the nodes on the side boundary with the same elevation are tied to move together (in X and Y direction) using equalDOF command in OpenSees. The vertical movement of side boundary nodes is kept free. The nonuniformity of the liquefiable ground is considered in the presented study. Based on the random realization of the nonuniformity of the ground, the relative density of the elements would fall into a wide range (D_R = 30–75%). The dynamic behavior of the liquefiable element significantly depends on the relative density and its corresponding calibrated parameters. In this case, tie the vertical movement of side nodes with periodic boundary (as adopted by Law and Lam [22] and Rayamajhi et al. [14], for a uniform ground) would enforce the side boundary elements to have the same settlement which is not reasonable for the nonuniform ground even though the extent of the model in the X and Y directions (see Fig. 5) are small compared to the size of the granular column. All the nodes above the water table are assigned zero pore water pressure. The nodes of the planes of Y = 0 and 0.7 are kept fixed against the out-of-plane displacement.

PDMY02 soil constitutive model is used to model the dynamic behavior of the ground. The PDMY02 Model is an elastoplastic soil-liquefaction constitutive model originally developed to simulate the cyclic liquefaction response and the associated accumulation of cyclic shear deformation in clean sand and silt [23]. Within a stress–space plasticity framework, PDMY02 Model employs a new flow rule and strain–space parameters to simulate the cyclic development and evolution of plastic shear strain. PDMY02 does not include a critical state soil mechanics framework.

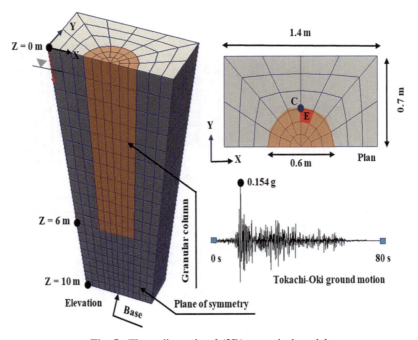

Fig. 5. Three-dimensional (3D) numerical model

The parameters of the PDMY02 Model are calibrated to achieve the single-amplitude shear strain of 3% in cyclic undrained simple shear loading with zero initial static shear stress ratio on a horizontal plane at a single element level. Laboratory test results from Chiaro et al. [17] are considered as the dynamic behavior of saturated Toyoura sand with a relative density of 50% at a single element level for the calibration purpose. Figure 6(a) shows a typical response of calibrated PDMY02 Model for cyclic stress ratio (CSR) = 0.171, D_R = 50%, and σ'_{vc} = 100 kPa in cyclic undrained simple shear loading with zero initial static shear stress ratio on a horizontal plane. The PDMY02 Model exhibits the ability of shear strain accumulation, commonly referred to as cyclic mobility, which is evident from the stress-strain behavior. The stress path is shown in

Fig. 6(b). The vertical effective stress ratio drops down to nearly zero within 15 cycles and triggers large shear strains afterward. Numerically simulated cyclic response at the single element level is obtained after calibrating the parameters of the PDMY02 Model to achieve a similar response as observed in the experiment in terms of cyclic mobility, initial shear modulus, and the accumulation rate of shear strain. Figure 6(c) shows the shear strain accumulation with the drop in vertical effective stress ratio. Figure 6(d) shows the CSR curves corresponding to single-amplitude shear strains of 3% with zero initial static shear stress ratio. The calibrated values of the PDMY02 Model for Toyoura sand (D_R ~ 50%) and granular column (D_R ~ 80%) are shown in Table 5. Besides, the calibrated parameters for 17 different individual relative densities of Toyoura sand ranging from $D_R = 30-75\%$ are tabulated in Table 6. For intermediate relative densities, linear interpolation is used to get the calibrated parameters.

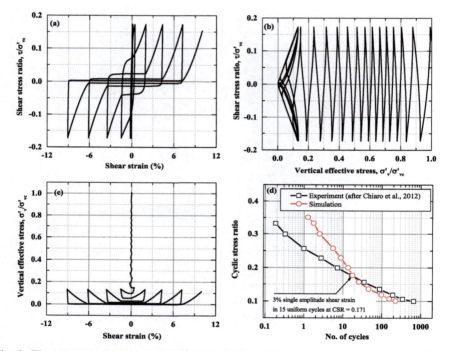

Fig. 6. The response of the calibrated PDMY02 Model at the element level Toyoura sand ($D_R = 50\%$)

Table 5. Calibrated parameters for Toyoura sand and granular column

Material/Parameters*	ρ (ton/m^3)	G_{max} (kPa)	B (kPa)	φ	PT_{ang}	C1	C3	D1	D3
Toyoura sand (D_R = 50%)	1.94	3.54E4	7.50E4	33.5	25.5	0.07	0.20	0.06	0.20
Granular column (D_R = 80%)**	2.14	10.4E4	26.0E4	48	30	0.006	0.0	0.42	0.0

*Remaining parameters (total number of parameters are 22) received default values as reported by Khosravifar et al. [20].
**The parameters for the granular column are selected per Rayamajhi et al. [9] and Khosravifar et al. [20]

Table 6. Calibrated parameters for Toyoura sand with different relative densities (D_R = 30–75%)

D_R (%)	G_{max} (kPa)	B (kPa)	φ	PT_{ang}	C3	D1	D3
30	2.5E4	5.30E4	31.0	31.0	0.52	0.00	0.00
33	2.6E4	5.51E4	31.4	29.0	0.48	0.00	0.00
36	2.7E4	5.72E4	31.8	27.0	0.40	0.01	0.10
39	2.8E4	5.93E4	32.0	26.0	0.30	0.04	0.10
42	2.9E4	6.14E4	32.2	25.9	0.27	0.06	0.25
45	3.0E4	6.36E4	32.5	25.8	0.20	0.06	0.20
48	3.1E4	6.57E4	32.9	25.7	0.20	0.06	0.20
50	3.5E4	7.50E4	33.5	25.5	0.20	0.06	0.20
52	3.7E4	7.73E4	33.8	25.6	0.15	0.06	0.15
55	3.8E4	8.05E4	34.1	25.7	0.10	0.09	0.10
58	4.0E4	8.47E4	34.5	25.8	0.06	0.09	0.10
61	4.5E4	9.53E4	35.0	26.0	0.06	0.09	0.10
64	5.2E4	1.09E5	35.3	26.0	0.06	0.10	0.10
67	5.9E4	1.24E5	35.6	26.0	0.04	0.13	0.10
71	6.3E4	1.32E5	35.7	26.0	0.04	0.14	0.00
73	7.1E4	1.49E5	35.9	26.0	0.00	0.17	0.00
75	8.0E4	1.68E5	36.0	26.0	0.00	0.20	0.00

Note: C1 is 0.07 for all values of D_R. The remaining parameters (total number of parameters are 22) received default values as reported by Khosravifar et al. [20].

4 Physical Modeling Results and Discussion

The performance of gravel drainage systems (Types 1 and 2 as described in Table 4) are investigated in Models 2 and 3. Excess pore water pressure (EPWP) time histories of different pore pressure transducers (PPTs) for Models 2 and 3 are presented and compared with EPWP time histories of respective PPTs of Model 1 in Fig. 7. It is evident from Fig. 7 that the dissipation rate of EPWP increases in accordance with gravel drainage capacity. It is to be noted that the drainage capacity of the gravel drainage system in Model 3 is more than the drainage capacity in Model 2. The larger the drainage capacity, the quicker is the dissipation of EPWP as designed. For instance, at Level 3 (at P2, P3, and P4), the EPWP takes approximately 1000s to dissipate in the case of Model 1 (no gravel drains), whereas in the case of Model 3 the EPWP dissipates within 400 s. The foremost reason for the inefficiency of gravel drainage Type 1 in the case of Model 2 is the less drainage capacity. The targeted excess pore water pressure ratio (as described in Table 4) at the liquefied zone within the ground could not be achieved even in the case of Model 3 as the maximum magnitude of EPWP at all PPTs (P1–P10) for Models 2 and 3, are almost similar to the one observed in Model 1 for respective PPTs. Surprisingly, the maximum EPWP at P9 and P10 in the case of Model 2 is more than the one observed in the case of Model 1. This might be associated with the nonuniformity of the model ground and possible densification because of the presence of gravel drains or relatively deeper positioning of PPTs. Although the capacity of the gravel drainage Type 1 is not sufficient to dissipate the EPWP; however, relatively larger permeability of gravel drains might have led to the flow of water from the deeper portion to the shallower portion with more ease, resulting in large EPWP at the shallower portion in case of Model 2.

The settlement time histories of both BT and FS foundation in Models 2 and 3 are presented and compared with the respective settlement time histories of Model 1 in Fig. 8. The influence of gravel drainage Type 1 (Model 2) on settlement mitigation is not significant. However, the overall settlement in the case of Model 3 is less in comparison with Models 1 and 2 due to the presence of sufficient gravel drainage capacity. Considerable differential settlement for BT in the case of Model 3 is occurred because of unusual change in the initial condition before dynamic excitation. While spinning up the centrifuge up to 40 g; the BT foundation experienced a significant amount of differential settlement probably because of the nonuniformity of the model ground. This uneven settlement of BT before shaking exaggerated the differential settlement during Tokachi-Oki ground motion as shown in Fig. 8.

Figure 9 depicts the average settlement of the foundations during both the co-shaking and post-shaking phases. In the case of Model 1, the foundations undergo excessive settlement during both the co-shaking and post-shaking phase of the dynamic event. However, in the case of Models 2 and 3, the significant settlement of both BT and FS foundation occurred during the co-shaking phase. Figure 9 depicts that the effect of the post-liquefaction/shaking reconsolidation mechanism seems to be overshadowed by the presence of a gravel drainage system which helped to reduce the total settlement of the shallow foundation.

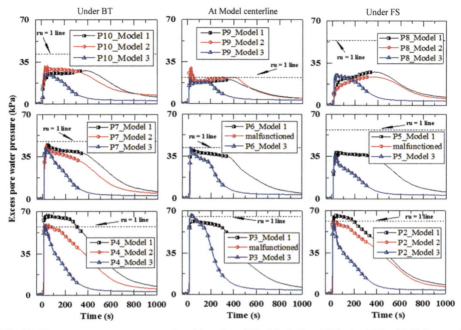

Fig. 7. Excess pore water pressure time histories of Models 1, 2 and 3 during Tokachi Oki ground motion

5 Reliability Analyses Results and Discussion

The modeling of inherent soil variability can be achieved utilizing advanced nonlinear finite element analyses and well-calibrated sophisticated elasto-plastic soil constitutive models. The nonuniformity of the ground is mapped using the overburden and energy-corrected, equivalent clean sand, SPT $(N1)_{60cs}$. The coefficient of variation (COV = 40%) and scale of fluctuation (θx = 5.0 m and θz = 0.5 m) are considered to model the nonuniformity of the ground according to Montgomery and Boulanger [24] and Phoon and Kulhawy [25]. The nonuniformity of the ground is modeled with a mean $(N1)_{60cs}$ = 12 (D_R ~ 50%), as shown in Fig. 10. A series of three-dimensional stochastic dynamic analyses are performed considering the nonuniformity of the ground using anisotropic, spatially correlated Gaussian random fields of $(N1)_{60cs}$ values. The parameters of PDMY02 are calibrated for a wide range of relative densities corresponding to $(N1)_{60cs}$ of 5 (D_R ~ 32%) to $(N1)_{60cs}$ of 26 (D_R ~ 75%). All the parameters of PDMY02 Model for different relative densities are calibrated as described earlier. Numerically simulated CSR curves for loose sand (D_R = 30%) and dense sand (D_R = 75%) are compared with the CSR curves obtained in the experiment, as shown in Fig. 11. Figure 11 exhibits that the calibrated parameters reasonably approximate the behavior of Toyoura sand for a wide range of relative densities of interest (D_R = 30–75%). The granular column and associated ground (Toyoura Sand) is assigned a uniform permeability value of 0.0066 and 0.0002 m/s, respectively. The assigned uniform properties for the granular column is corresponding to $(N1)_{60cs}$ of 30 (D_R ~ 80%), per Raymajhi et al. [9] and

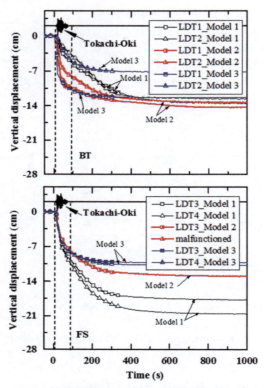

Fig. 8. Settlement time histories of BT (LDT1 & 2) and FS (LDT3 & 4) during Tokachi-Oki ground motion

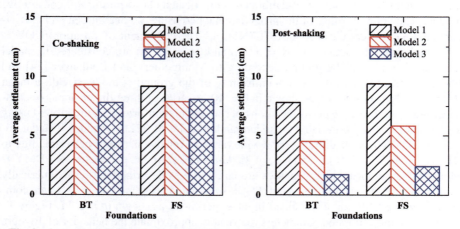

Fig. 9. Average settlement during co-shaking and post-shaking phase for Models 1, 2, and 3

Khosravifar et al. [20]. The random field of $(N1)_{60cs}$ values with calibrated parameters of the PDMY02 Model are implemented into the OpenSees numerical model with the help of Matlab code.

The results of three-dimensional stochastic analyses are presented and compared with the deterministic analysis results [10]. Figure 12 illustrates the stochastic distribution of the average settlement and horizontal displacement of the top surface of the ground (Z = 0 plane, see Fig. 5) for the grounds with and without a granular column.

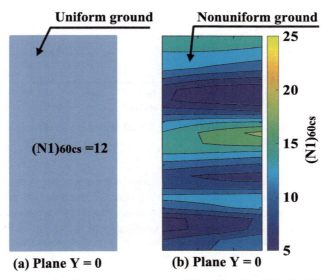

Fig. 10. A typical scenario of the ground condition at Plane Y = 0 (see Fig. 5): (a) uniform ground and (b) nonuniform ground

Figure 12(a) depicts that the mean (μ) and the standard deviation (σ) of the average surface settlement is 4.31 cm and 0.23 cm, respectively, for the ground with the granular column. It is observed that the mean of stochastic average surface settlement is significantly larger than the respective deterministic value. Whereas, the mean (μ) and the standard deviation (σ) of the average surface settlement is 4.80 cm and 0.10 cm, respectively, and the mean value is comparable to the deterministic value for the ground without the granular column. A relatively wider stochastic distribution and considerable standard deviation in the average surface settlement is evident in the case of the ground with the granular column in comparison with the ground without a granular column. This emphasizes that the presence of a granular column may adversely affect the uncertainty in the prediction of the average surface settlement due to the inherent ground nonuniformity. Figure 12(b) depicts that the mean (μ) and the standard deviation (σ) of the horizontal surface displacement is 0.31 cm (distribution in the range of −1.0 cm to 1.5 cm) and 0.60 cm, respectively, for the ground with the granular column. Whereas, the mean (μ) and the standard deviation (σ) of t surface horizontal displacement is − 3.17 cm (distribution in the range of −6.0 cm to −1.0 cm) and 1.34 cm, respectively,

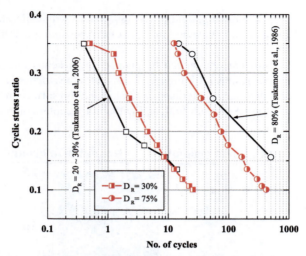

Fig. 11. The response of the calibrated PDMY02 Model at element level for loose ($D_R = 30\%$) and dense sand ($D_R = 75\%$)

for the ground without granular column. The mean of the stochastic distribution is comparable with the deterministic values for both the cases of grounds with and without the granular column. However, a relatively wider stochastic distribution and considerable standard deviation in the horizontal surface displacement is evident in the case of the ground without the granular column in comparison with the ground with the granular column. This emphasizes that the presence of a granular column may favorably affect the uncertainty in the prediction of horizontal surface displacement. The reason for this is the shear reinforcement of the ground due to the stiffness of the granular column, which is the governing factor for the residual amount of horizontal surface displacement, as discussed earlier. Besides, the granular column is considered with uniform properties, which facilitated relatively less uncertainty in the prediction of surface horizontal displacement in the case of the ground with the granular column in comparison with the ground without a granular column.

The probability of deviation of the stochastic average settlement and horizontal displacement of the top surface of the ground from their deterministic values are evaluated and presented in Fig. 13 for the grounds with and without the granular column. The deviations of the average settlement and horizontal displacement of the top surface of the ground are considered on the positive side (more than the deterministic value) and the negative side (less than the deterministic value). Figure 13(a) shows that total probabilities of the stochastic average surface settlement being deviated on the negative side from the deterministic value are 13.07 and 40.03%, respectively, and on the positive side from the deterministic value are 86.45 and 56.65%, respectively, for the grounds with and without granular column. The maximum deviation of the average surface settlement on the negative side is 0.20 cm (with a 2.32% probability of occurrence) and 0.29 cm (with a 0.09% probability of occurrence), respectively, for the grounds with and without granular column. The maximum deviation of the average surface settlement on the

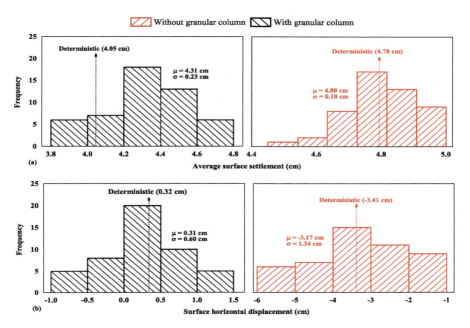

Fig. 12. Stochastic distribution of model ground deformation for the grounds with and without granular column: (a) average surface settlement and (b) surface horizontal displacement

positive side is 0.62 cm (with a 5.97% probability of occurrence) and 0.19 cm (with a 4.43% probability of occurrence), respectively, for the grounds with and without granular column. A relatively larger deviation from the deterministic value associated with a significant probability of occurrence in the case of the ground with granular column signifies that the presence of granular column adversely affects the uncertainty in the prediction of the average surface settlement due to the inherent ground nonuniformity as discussed earlier. Figure 13(b) shows that total probabilities of stochastic surface horizontal displacement being deviated on the negative side from the deterministic value are 57.05 and 42.68%, respectively, and on the positive side from the deterministic value are 41.92 and 51.69%, respectively, for the grounds with and without granular column. The maximum deviation of the horizontal surface displacement on the negative side is 1.48 cm (with a 1.12% probability of occurrence) and 2.32 cm (with a 2.96% probability of occurrence), respectively, for the ground with and without granular column. The maximum deviation of the horizontal surface displacement on the positive side is 1.04 cm (with a 2.74% probability of occurrence) and 3.30 cm (with a 1.05% probability of occurrence), respectively, for the grounds with and without granular column. A relatively larger deviation from the deterministic value associated with a significant probability of occurrence in the case of the ground without granular column signifies that the presence of granular column favorably affects the uncertainty in the prediction of the horizontal surface displacement due to the inherent ground nonuniformity as discussed earlier.

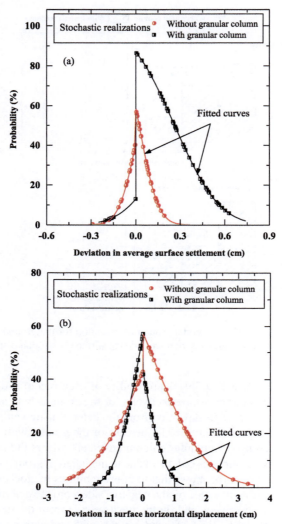

Fig. 13. Probability of deviation from the deterministic values for the grounds with and without granular column: (a) for average surface settlement and (b) for surface horizontal displacement

6 Summary

This paper presents the physical modeling of granular columns and assess their reliability of effectiveness in the nonuniform liquefiable ground to mitigate the liquefaction-induced ground deformation. The intended purpose of providing the gravel drainage is to mitigate the liquefaction-induced effects through rapid dissipation of excess pore water pressure (EPWP). Three types of model tests are performed for this purpose in which two model tests consist of the gravel drainage system of different drainage capacities, and one model test consists of only liquefiable ground. Treating the ground with a gravel drainage system was found to be useful to mitigate the liquefaction-induced deformation. The presence of granular columns increased the dissipation rate (through radial flow towards the gravel drainage zone) of generated EPWP and reduced the post-shaking settlement. Besides, a reliability assessment of the performance of equally-spaced granular columns in a nonuniform liquefiable ground is carried out to mitigate the liquefaction-induced ground deformation using three-dimensional (3D) stochastic numerical analyses. The PDMY02 elastoplastic soil constitutive model is used to simulate the dynamic behavior of the liquefiable ground treated with the granular column. The nonuniformity in the ground is mapped with the stochastic realizations of the overburden and energy-corrected, equivalent clean sand, SPT $(N1)_{60cs}$ values using a spatially correlated Gaussian random field. The spatial nonuniformity in the ground is found to affect the liquefaction-induced ground deformation. Stochastic results depicted that the presence of the granular column reduces the uncertainty in the estimation of horizontal displacement; however, it adversely affects the uncertainty in the prediction of the average surface settlement of the ground. The reliability assessment of the performance of the granular column is essential for better engineering judgment associated with a desired level of confidence. The presented probabilistic estimation of liquefaction-induced ground deformation possesses significant practical importance and provides useful information to assess the reliability of the performance of the granular column.

References

1. Seed, H.B., Booker, J.R.: Stabilization of potentially liquefiable sand deposits using gravel drains. J. Geotech. Geo-Environ. Eng. **103**, 757–768 (1977)
2. Tokimatsu, K., Yoshimi, Y., Ariizumi, K.: Evaluation of liquefaction resistance of sand improved by deep vibratory compactions. Soils Found. **30**(3), 153–158 (1990). https://doi.org/10.3208/sandf1972.30.3_153
3. Adalier, K., Elgamal, A.: Mitigation of liquefaction and associated ground deformations by stone columns. Eng. Geol. **72**(3), 275–291 (2004). https://doi.org/10.1016/j.enggeo.2003.11.001
4. Boulanger, R.W., Idriss, I.M., Stewart, D.P., Hashash, Y., Schmidt, B.: Drainage capacity of stone columns or gravel drains for mitigating liquefaction. In: Geotechnical Earthquake Engineering and Soil Dynamics, vol. III, 678–690 (1998)
5. Adalier, K., Elgamal, A., Meneses, J., Baez, J.I.: Stone columns as liquefaction countermeasure in non-plastic silty soils. Soil Dyn. Earthq. Eng. **23**(7), 571–584 (2003). https://doi.org/10.1016/S0267-7261(03)00070-8
6. Olarte, J., Paramasivam, B., Dashti, S., Liel, A., Zannin, J.: Centrifuge modeling of mitigation-soil-foundation-structure interaction on liquefiable ground. Soil Dyn. Earthq. Eng. **97**, 304–323 (2017). https://doi.org/10.1016/j.soildyn.2017.03.014
7. Kumar, R., Sawaishi, M., Horikoshi, K., Takahashi, A.: Centrifuge modeling of hybrid foundation to mitigate liquefaction-induced effects on shallow foundation resting on liquefiable ground. Soils Found. **59**(6), 2083–2098 (2019)
8. Brennan, A.J., Madabhushi, S.P.G.: Effectiveness of vertical drains in mitigation of liquefaction. Soil Dyn. Earthq. Eng. **22**(9), 1059–1065 (2002). https://doi.org/10.1016/S0267-7261(02)00131-8

9. Rayamajhi, D., Ashford, S.A., Boulanger, R.W., Elgamal, A.: Dense granular columns in liquefiable ground. I: shear reinforcement and cyclic stress ratio reduction. J. Geotech. Geoenviron. Eng. **142**(7), 04016023 (2016). https://doi.org/10.1061/(ASCE)GT.1943-5606.0001474
10. Kumar, R., Takahashi, A.: Reliability assessment of the performance of granular column in the nonuniform liquefiable ground to mitigate the liquefaction-induced ground deformation. Georisk: Assess. Manag. Risk Eng. Syst. Geohazards **16**, 376–395 (2020). https://doi.org/10.1080/17499518.2020.1836378
11. Goughnour, R.R., Pestana, J.M.: Mechanical behavior of stone columns under seismic loading. In: Proceedings of 2nd International Conference on Ground Improvement Techniques, Singapore, pp. 157–162 (1998)
12. Green, R.A., Olgun, C.G., Wissmann, K.J.: Shear stress redistribution as a mechanism to mitigate the risk of liquefaction. In: Geotechnical Earthquake Engineering and Soil Dynamics, vol. IV, pp. 1–10 (2008)
13. Olgun, C.G., Martin, J.R., II: Numerical modeling of the seismic response of columnar reinforced ground. In: Geotechnical Earthquake Engineering and Soil Dynamics, vol. IV, pp. 1–11 (2008)
14. Rayamajhi, D., et al.: Numerical study of shear stress distribution for discrete columns in liquefiable soils. J. Geotech. Geoenviron. Eng. **140**(3), 04013034 (2014). https://doi.org/10.1061/(ASCE)GT.1943-5606.0000970
15. Schofield, A.N.: Dynamic and earthquake geotechnical centrifuge modelling (1981)
16. Bouckovalas, G.D., Papadimitriou, A.G., Kondis, A., Bakas, G.J.: Equivalent-uniform soil model for the seismic response analysis of sites improved with inclusions. In: Proceedings of 6th European Conference on Numerical Methods in Geotechnical Engineering, pp. 801–807. Taylor & Francis, London (2006)
17. Chiaro, G., Koseki, J., Sato, T.: Effects of initial static shear on liquefaction and large deformation properties of loose saturated Toyoura sand in undrained cyclic torsional shear tests. Soils Found. **52**(3), 498–510 (2012). https://doi.org/10.1016/j.sandf.2012.05.008
18. Priebe, H.J.: The prevention of liquefaction by vibro replacement. In: Proceedings of the 2nd International Conference on Earthquake Resistant Construction and Design, pp. 211–219 (1989)
19. Baez, J.I., Martin, G.R.: Advances in the design of vibro systems for the improvement of liquefaction resistance. In: International Symposium for Ground Improvement, pp. 1–16 (1993)
20. Khosravifar, A., Elgamal, A., Lu, J., Li, J.: A 3D model for earthquake-induced liquefaction triggering and post-liquefaction response. Soil Dyn. Earthq. Eng. **110**, 43–52 (2018). https://doi.org/10.1016/j.soildyn.2018.04.008
21. Stewart, J.P.: Benchmarking of nonlinear geotechnical ground response analysis procedures. Pacific Earthquake Engineering Research Center (2008)
22. Law, H.K., Lam, I.P.: Application of periodic boundary for large pile group. J. Geotech. Geoenviron. Eng. **127**(10), 889–892 (2001). https://doi.org/10.1061/(ASCE)1090-0241(2001)127:10(889)
23. Yang, Z., Elgamal, A., Parra, E.J.: Computational model for cyclic mobility and associated shear deformation. Geotech. Geoenviron. Eng. **129**(12), 1119–1127 (2003). https://doi.org/10.1061/(ASCE)1090-0241(2003)129:12(1119)
24. Montgomery, J., Boulanger, R.W.: Effects of spatial variability on liquefaction-induced settlement and lateral spreading. J. Geotech. Geoenviron. Eng. **143**(1), 04016086 (2016). https://doi.org/10.1061/(ASCE)GT.1943-5606.0001584
25. Phoon, K.K., Kulhawy, F.H.: Characterization of geotechnical variability. Can. Geotech. J. **36**(4), 612–624 (1999). https://doi.org/10.1139/t99-038

Experimental Study on the Effect of Coexistence of Clay and Silt on the Dynamic Liquefaction of Sand

Tao Li[1(✉)] and Xiao-Wei Tang[2]

[1] Changsha University of Science and Technology, Changsha 410114, China
litaodut@163.com
[2] Dalian University of Technology, Dalian 1160224, China

Abstract. In order to study the effect of coexistence of clay and silt on the dynamic liquefaction of sand under different void ratios, the characteristics of sand with two different fines contents (FC = 5% and 10%), three different clay silt ratios (CS = 0.25, 1 and 4) under two different void ratios were investigated by the cyclic undrained triaxial test. The test results show that dynamic test results of specimens with different FC, CS and void ratios are different. Liquefaction happens for all specimens and there are two liquefaction types: flow liquefaction (brittle failure) and cyclic mobility (ductility failure). With the increase of FC or CS, two liquefaction types change mutually. When FC is same, the liquefaction resistance of specimens (with different CS) is different. When FC is different, the liquefaction resistance is different with the monotonous change of clay or silt content in fines. When the void ratio is different, the dynamic strength of specimens is different with same FC and CS. The clay and silt play roles of filling, lubrication, bonding and skeleton effect on the sand particles. With the change of clay and silt content in fines, the proportion of four roles is different under different fines content.

Keywords: Fines content · Clay silt ratio · Sand-silt-clay mixture · Liquefaction · Particle contact

1 Introduction

According to the geotechnical survey data, it shows that the pure sand is less distribution in nature, also the single silt or clay mixed with the sand is rare. Most sand contains a small amount of silt and clay at the same time (Yang and Wei 2012). Many researchers have investigated the effect of fines content (FC) on soil liquefaction. And they pointed out that with the increase of fines, the liquefaction resistance of sand showed the characteristic with increase, decrease and non-monotonous change (Amini and Qi 2000; Lade and Yamamuro 1997; Xenaki and Athanasopoulos 2003). Some scholars also observed that in the mixed soil of sand and fines, there was a turning fines content, when the fines content was less than this content, the soil was similar as sand; when the fines content exceeded this content, the soil has the similar properties as fines (Yang et al. 2005; Polito and Martin II 2001). So far, the main research work on sand liquefaction can be divided into three parts: sand mixes with single fines (silt or clay); sand mixes with clay and silt (no difference between clay and silt); sand mixes with clay and silt at same time (silt and sand are believed to have similar properties), so the effect of fines on the liquefaction mainly depends on the clay. However, as we know, silt and clay are two different types of fines. According to other research results, it shows that silt and clay have different effects on the sand liquefaction (Gratchev et al. 2006; Cabalar and Mustafa 2017; Kim et al. 2016; Porcino et al. 2018; Prashant et al. 2018; Porcino and Diano 2017; Carrera et al. 2011; Chang and Hong 2008). On the whole, the research on the effect of clay or silt content on sand liquefaction and its influence mechanism are relatively mature. While when different contents of clay and silt are added to the sand at the same time, what will happen to the dynamic property of sand-silt-clay mixture? So far, there is few research on the liquefaction of sand-silt-clay mixture. As we know, the composition of soil is the most basic factor affecting the shear strength, which includes the mineral composition, particle size, particle gradation and particle shape. Explaining the macro mechanical behavior of soil based on the meso perspective of particle contact has become the focus of research in recent years.

In order to study the influence of coexistence of clay and silt on the dynamic liquefaction of sand, fines content $FC = 0\%$, 5% and 10% were added to the sand, and through the CKC triaxial testing system, the consolidated-undrained test was conducted. Here the fines included clay and silt. When FC was same, change the clay silt ratio (CS) in the fines. For example, when $FC = 5\%$, the mass percent of clay in fines were 20%, 50%, and 80%, so the corresponding percent of silt were 80%, 50%, and 20%, respectively. The soil with this fines content and clay silt ratio can be abbreviated as $FC = 5\%\text{-}CS = 0.25$, $FC = 5\%\text{-}CS = 1$ and $FC = 5\%\text{-}CS = 4$. Same method was used for specimen with $FC = 10\%$. According to dynamic test results, the influence of coexistence of clay and silt on the liquefaction of sand had been studied. At the same time, based on the mesostructure observation results of sand-silt-clay mixture, the effect of particle contact on the macroscopic mechanical behavior of mixed soil was explained.

2 Test Overview

The sand is Fujian standard sand with the particle size of 0.1–0.25 mm. The silt is come from the construction site and the particle size is 0.02–0.075 mm. The clay is the bentonite (Ca-based). Figure 1 shows the grading curve and meso morphology of sand and silt.

Fig. 1. Grain size distributions and meso morphology of sand and silt.

3 Test Program

The size of specimen is 39.1 × 8 mm (diameter × height). In order to minimize the impact of the non-uniformity on the test results, wet compaction method was used to obtain the uniform specimen (Amini and Qi 2000; Kim et al. 2017). The specimen was divided into 4 layers to compact, after each compaction, the surface of the layer was shaved to ensure good contact between the upper and lower. After preliminary test, it is indicated the specific gravity of sand, silt and clay is not so big. For convenience, it can be assumed that the sand, silt and clay have the same specific gravity $G_s = 2.65$. This method can be also found in other studies (Thevanayagam 1998; Thevanayagam and Mohan 2000).

Table 1 is the program of test. The fines content (FC) and clay silt ratio (CS) are defined as follows:

$$FC = m_f/m_s \tag{1}$$

$$CS = CC/SC \tag{2}$$

$$CC = m_c/m_s \tag{3}$$

Table 1. The program of test

Fines content $FC/\%$	Clay silt ratio $CS/\%$		Sand skeleton void ratio e_s	Void ratio e	Specimen
	Clay	Silt			
5	20	80	0.9800	0.8823	$FC = 5\%$-$CS = 0.25$ (constant e_s/e)
	50	50			$FC = 5\%$-$CS = 1$ (constant e_s/e)
	80	20			$FC = 5\%$-$CS = 4$ (constant e_s/e)
10	20	80		0.7967	$FC = 10\%$-$CS = 0.25$ (constant e_s)
	50	50			$FC = 10\%$-$CS = 1$ (constant e_s)
	80	20			$FC = 10\%$-$CS = 4$ (constant e_s)
10	20	80	1.0704	0.8823	$FC = 10\%$-$CS = 0.25$ (constant e)
	50	50			$FC = 10\%$-$CS = 1$ (constant e)
	80	20			$FC = 10\%$-$CS = 4$ (constant e_s)

$$SC = m_{ss}/m_s \qquad (4)$$

where:

m_c—mass of clay, g
m_{ss}—mass of silt, g
m_s—mass of sand, g
m_f—mass of fines (clay and silt), g
CC—clay content, %
SC—silt content, %

Sand skeleton void ratio $e_s = 0.9800$ and void ratio $e = 0.8823$ were used as the control parameter when making specimen. When making the specimen with constant e_s, first ensure that total mass of the sand in each specimen was same, and then added different content of silt and clay into the sand. When making the specimen with constant e, ensure that total mass of each specimen was same, and then added different content of silt and clay into the sand. Hydraulic saturation was used in the saturation process, which was completed on the dynamic triaxial apparatus. After saturation, Skempton's parameter (B-value) needed to be checked, and the specimen is considered to be saturated when $B > 0.95$. The fully digital closed-loop control pneumatic triaxial system (CKC) was used in the undrained cyclic triaxial test, as shown in the Fig. 2.

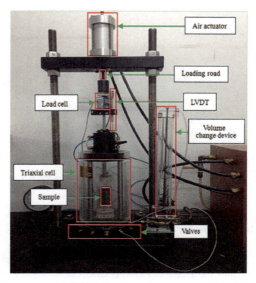

Fig. 2. Photograph of CKC triaxial testing system (Li and Tang 2019).

Specimen was subjected to the consolidation test under the confining pressure $\sigma_3 = 200$ kPa first, and then a constant-amplitude half-sine wave cyclic loading was applied to the specimen. The loading frequency $f = 0.5$ Hz. The cyclic stress ratio CSR (Kim et al. 2016; Huang and Zhao 2018) was used to evaluate the dynamic behavior of the specimen. The CSR is defined as follows:

$$CSR = \sigma_d/2\sigma_3 \tag{5}$$

where:

σ_d—dynamic stress, kPa

σ_3—confining pressure, kPa

Axial strain $\varepsilon_a = 2.5\%$ was considered as the failure criteria (Kim et al. 2016; Wang et al. 2018) and $\varepsilon_a = 30\%$ was used as the sign of ending in the tests. Specimen preparation, saturation, consolidation and loading were carried out according to ***Standard for geotechnical testing method*** (GB/T50123-2019).

4 Repeatability of Test Results

Figure 3 shows the ε_a-N/N_f (axial strain versus normalized cycle numbers) and u-N/N_f (pore water pressure versus normalized cycle numbers) curves of specimen with $FC = 5\%$-$CS = 4$ under two parallel tests. Here the N_f indicted the failure cycle number. It was shown that ε_a-N/N_f and u-N/N_f curves of two parallel tests were highly consistent, which indicated good repeatability for dynamic test results of sand-silt-clay mixtures in this study.

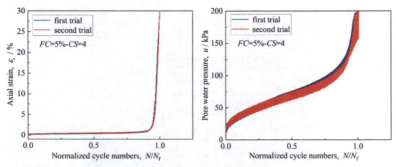

Fig. 3. Dyamic test results of specimen with $FC = 5\%$-$CS = 4$ showing the repeatability of the test results.

5   Test Results and Discussion

Figure 4 and Fig. 5 give the typical dynamic test results of specimens with same FC, CS and different void ratio, include ε_a-N (axial strain versus cycle numbers), u-N (pore water pressure versus cycle numbers), σ_d-ε_a (dynamic stress versus axial strain) and σ_d-p' (dynamic stress versus average effective stress).

From figures it can be seen that all specimens liquefied under the dynamic loading. The failure cycle numbers or failure instability point of specimens showed big differences under different loading amplitudes. The axial strain of the specimen increased a little during the early period of vibration, and it changed abruptly until liquefaction occurred in the later stage. In contrast, the pore water pressure of the specimen experienced a short period of rapid growth in the early stage, then the growth rate decreased. The pore water pressure increased rapidly until it reached the confining pressure at the later stage of vibration, and the specimen liquefied and destroyed.

As can be seen from these figures, the dynamic load amplitudes of specimens with $FC = 10\%$-$CS = 1$ (constant e_s) were generally higher than those of specimens with $FC = 10\%$-$CS = 1$ (constant e). For specimens with two different void ratios, with increasing CSR, the tangent slope of ε_a-N curve decreased (Fig. 4(a) and Fig. 5(a)), and σ_d-ε_a, σ_d-p' curves became more and more sparse. For specimen with same FC, CS but different void ratio, the development model of dynamic results was quite different. For the specimen with $FC = 10\%$-$CS = 1$ (constant e_s), the turning point of axial strain (failure instability point) appeared gently, and the pore water pressure reached the confining pressure alternately at the end of cycle numbers. The σ_d-ε_a and σ_d-p' curves were dense at the initial vibration stage, and these curves became loose at the final vibration stage. The σ_d-ε_a and σ_d-p' curves were relatively dense on the whole. Overall, the specimen with $FC = 10\%$-$CS = 1$ (constant e_s) showed similar phenomenon of alternating mobility or cyclic mobility (Fang 2002) of soil under alternating load.

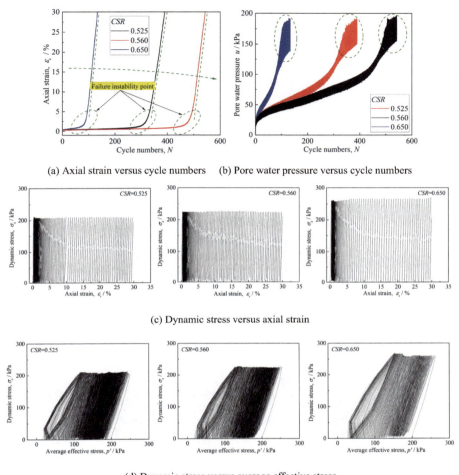

Fig. 4. Results of tests under different loading ($FC = 10\%$-$CS = 1$ (constant e_s)).

For the specimen with $FC = 10\%$-$CS = 1$ (constant e), the turning point of axial strain (failure instability point) appeared suddenly, and the pore water pressure reached the confining pressure directly at the end of cycle numbers. The σ_d-ε_a curves were dense at the initial vibration stage, and these curves became very loose at the final vibration stage. The σ_d-ε_a and σ_d-p' curves of the specimen with $FC = 10\%$-$CS = 1$ (constant e) were relatively loose generally. Compared with the specimens with $FC = 10\%$-$CS = 1$ (constant e_s), the σ_d-p' curves of specimens with $FC = 10\%$-$CS = 1$ (constant e) were relatively loose at the initial vibration stage, and these curves became more looser at the final vibration stage. Generally speaking, specimens with $FC = 10\%$-$CS = 1$ (constant e) showed phenomenon of flow liquefaction (Zhang et al. 2007).

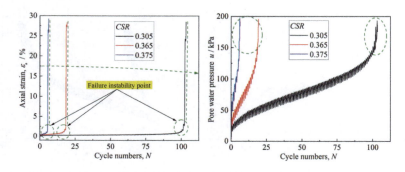

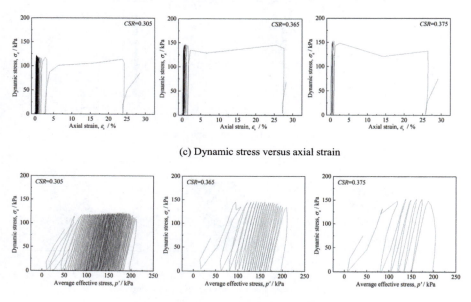

(a) Axial strain versus cycle numbers (b) Pore water pressure versus cycle numbers

(c) Dynamic stress versus axial strain

(d) Dynamic stress versus average effective stress

Fig. 5. Results of tests under different loading ($FC = 10\%$-$CS = 1$ (constant e)).

Figure 6 presents the relationship between CSR and N_f (cycle numbers to failure) of specimens with different FC and CS under different void ratios.

Comparing the results of the specimens, it can be found that for specimens with same FC and different CS, liquefaction resistance were different. This result was different from the conclusion obtained by other scholars, in which the silt and the clay were treated as same fines particles to add into the sand. In Fig. 6, N_f increased with the decrease of CSR. The $CSR \sim N_f$ curves of specimens with $FC = 10\%$ were all higher than those of the specimens with $FC = 5\%$ under constant e_s. The $CSR \sim N_f$ curves of specimens with $FC = 10\%$-$CS = 1$ (constant e_s) were all higher than those of the specimens with $FC = 10\%$-$CS = 1$ (constant e). When $FC = 10\%$, the liquefaction resistance decreased with

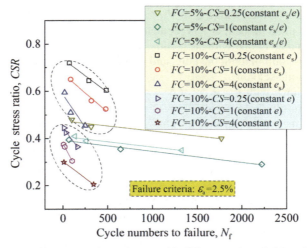

Fig. 6. Dynamic strength curves of specimens with different FC and CS under different void ratios.

the increase of clay contents (decrease of silt contents) in the case of constant e and e_s. Whereas when $FC = 5\%$, the liquefaction resistance did not change monotonously with monotonic change of clay or silt contents, and the liquefaction resistance of the specimen was the lowest when the clay content equalled to the silt content ($FC = 5\%$-$CS = 1$). The slope of $CSR \sim N_f$ curves of specimens with $FC = 10\%$ decreased more quickly than those of specimens wisth $FC = 5\%$ with increasing N_f.

If the test results are analyzed solely from the filling effect of fines on sand particles (Chang et al. 2016; Zhou et al. 2018; Chang and Deng 2017; Monkul et al. 2016; Cabalar and Hasan 2013), when $FC = 5\%$, the fines will not or rarely participate in the construction of the soil skeleton, and test results of specimens with same FC should be same. However, the results of specimens with $FC = 5\%$ showed great differences under different CS. When $FC = 10\%$, development laws of the dynamic strength changed more obviously with the monotonous increasing of clay or silt content. So it can be seen that when $FC = 5\%$, fines particles are not just filled the pore of sand particles. Similarly, for the role of fines as the soil skeleton, there was no reasonable explanations for the test results of specimens with $FC = 5\%$ and 10% in both constant e_s and e..Considering results difference between specimens with $FC = 5\%$ and 10% under differetn void ratios, it was inferred that in sand-silt-clay mixtures, clay and silt particles cannot fully fill in the space between sand or act as the soil skeleton, regardless of the FC, CS and void ratio.

From the results of specimens with different FC and CS, it can be inferred that in the sand-silt-clay mixtures, the filling, bonding and soil skeleton effects of fines on sand particles exist simultaneously. Proportion of these effects are different with the change of clay and silt content in the mixed soil, specific performances: for the specimen with $FC = 5\%$-$CS = 1$, "equal" content of clay and silt fill in the pore of sand particles and bond to each other, and only few fines particles participat in the construction of soil skeleton. For the specimen with $FC = 5\%$-$CS = 0.25$ and $FC = 5\%$-$CS = 4$, the clay or silt content is predominant, at this moment, the "equal" content of silt and clay fill in the pore of sand and bond with each other, while in the "remaining" clay or silt, one part of these fines continu to fill the pore, and the other part of fines were involved in the construction of soil skeleton.

For the specimen with $FC = 5\%$-$CS = 4$ and $FC = 5\%$-$CS = 0.25$, clay and silt particles were "surplus". As we all know, the average particle size of silt is larger than the clay, which means that in the "surplus" clay or silt, more silt particles, together with sand particles will play the role of soil skeleton under the same conditions. These silt particles may have higher chances to contact the sand particles and can enhance the friction between the sand particles, and finally the specimen with $FC = 5\%$-$CS = 0.25$ (more silt particles) showed the maximum liquefaction resistance due to the large intergranular friction; for the specimen with $FC = 5\%$-$CS = 4$, the "surplus" clay particles are attracted to each other (forming the clay aggregate), distribute around the sand and adhered to the sand particles to some extent, these clay aggregates can bond the sand to a certain extent at the same time. Ultimately, the liquefaction resistance of the specimen of $FC = 5\%$-$CS = 4$ was higher than that of the specimen with $FC = 5\%$-$CS = 1$ due to the larger cohesive force between particles.

When $FC = 10\%$ with conatant e_s, the fines content in specimens increased as a whole, except some fines used to fill the pores of sand, more fines particles have been involved in the construction of soil skeleton. The friction effect of silt on soil particles is stronger than the bonding effect of clay, at this time, friction is dominant. Therefore, the liquefaction resistance of specimen increased monotonically with the increase of silt content in fines. Compared with specimens of $FC = 10\%$ with conatant e_s, when $FC = 10\%$ with conatant e, it was equivalent to increasing the gap spacing between soil particles. So the interaction mechanism between different particles is consistent with specimens of $FC = 10\%$ with conatant e_s, and here we will not repeat.

6 Meso Particle Contact Analysis

For a better and direct comparison of the soil particle contact state in different sand-silt-clay mixtures, the meso mechanical analysis was conducted by Li and Tang (2020), which based on the combination of particle dyeing and stereomicroscope. The specific experimental steps were as follows:

Firstly, the clay and silt were dyed with red and blue dyes. And after repeated attempts, it can be found that the properties of the dye meet the experimental requirements.

Secondly, the dyed clay, dyed silt and undyed sand particles were dried and then mixed according to the given FC and CS. Here the mixing steps were completely consistent with the triaxial specimens preparation process.

Thirdly, putting the dyed sand-silt-clay mixtures under the stereomicroscope, and then selecting the same field of view to observe and photograph with SLR camera for many times.

Finally, multiple photos were synthesized by image processing software and then we could get the meso particle contact state of specimen with different FC and CS.

Figure 7 shows the process of particle dyeing and meso photography. Figure 8 gives the microscope images of different sand-silt-clay mixtures. Here the specimen was loose mixed state without water.

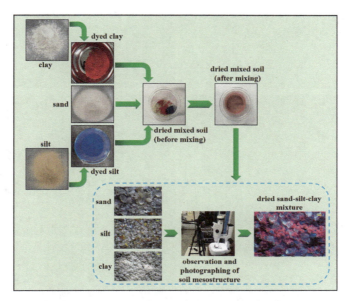

Fig. 7. Process of particle dyeing and meso photography.

In microscope images, the red particles was clay, the blue particles was silt, and the transparent particles was sand. Based on the microscope test, the discussion was conducted to demonstrate the mechanism of soil particle contact. From Fig. 8 it could be seen that clay, silt and sand particles were mixed uniformly. For the specimen of $FC = 5\%\text{-}CS = 0.25$, silt was mainly filled in the pores between sand, and a small amount of silt distributed on the sand surface. Besides that, some silt played the role of soil skeleton. As for clay, some clay distributed on the sand surface uniformly, and there are still some clay concentrated around the silt or distributed on the silt surface (Fig. 8(a)). In the microscope image of $FC = 5\%\text{-}CS = 1$, with increasing clay content, more and more clay gathered around the silt and formed the "clay-silt aggregate" which can fill pores between sand (Fig. 8(b)). In the microscope image of $FC = 5\%\text{-}CS = 4$ (Fig. 8(c)), large portions of silt were covered by clay particles. Futhermore, some clay attracted to each other and then forming "clay aggregates" (Gratchev et al. 2006), whcih could bond sand particles. In microscope images of $FC = 10\%$ (Fig. 8(d)–(f)), the content of fines (clay and silt) in specimens increased obviously. In the specimen of $FC = 10\%\text{-}CS = 0.25$ (Fig. 8(d)), more clay distributed on the silt and sand surface, and compared with $FC = 5\%\text{-}CS = 0.25$, more silt filled in the pores and played the role of soil skeleton. In the specimen of $FC = 10\%\text{-}CS = 4$ (Fig. 8(f)), lots of clay distributed on the sand surfaces, and there were still a large part of clay congregated. It can be also found that silt were almost submerged by clay particles at this time. In the specimen

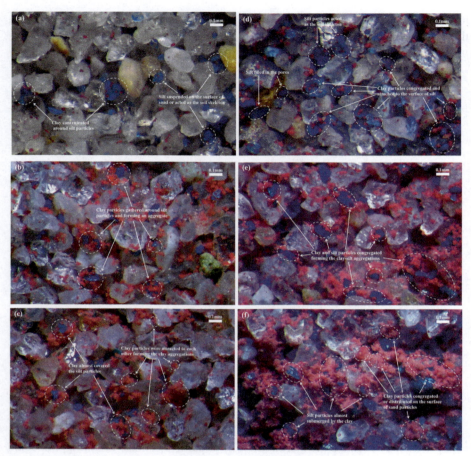

Fig. 8. Microscope images of sand with different fines contents and clay silt ratios: (a) $FC = 5\%$-$CS = 0.25$, (b) $FC = 5\%$-$CS = 1$, (c) $FC = 5\%$-$CS = 4$, (d) $FC = 10\%$-$CS = 0.25$, (e) $FC = 10\%$-$CS = 1$, (f) $FC = 10\%$-$CS = 4$ (Li and Tang 2020)

of $FC = 10\%$-$CS = 1$ (Fig. 8(e)), quite a number of clay and silt congregated, which formed the "clay-silt aggregate" (silt were wrapped by clay particles). These "clay-silt aggregate" played the role of soil skeleton or bonded sand.

It is widely known that soil is a friction material, and it can be inferred that in the "surplus" clay or silt, the friction effect of silt was stronger than the bonding effect of clay. Therefore, when $FC = 5\%$, the liquefaction resistance of the specimen with $FC = 5\%$-$CS = 0.25$ was the highest, followed by the specimen with $FC = 5\%$-$CS = 4$, and the liquefaction resistance of the specimen with $FC = 5\%$-$CS = 1$ was the lowest. When $FC = 10\%$, the fines content in the specimen increased as a whole. Some fines filled the pores of sand, and more fines had participated in the construction of soil skeleton. Similarly, the friction effect of silt was stronger than the bonding effect of clay, therefore, the frictional effect was dominant at this time. As a consequence, the liquefaction resistance of specimens with $FC = 10\%$ increased monotonously with the increase of silt content.

7 Conclusion

In this paper, the sand with two different fines content (FC) and three different clay silt ratios (CS) under two different void ratios were taken as the research object. Both cyclic triaxial test and meso photography test were conducted in the research. The influence of coexistence of clay and silt on the liquefaction resistance of sand was studied. The following conclusions can be obtained:

(1) The effects of coexistence of clay and silt on the dynamic behaviour of sand were complex. The major factors included the overall fines content, the relative content between clay and silt (clay silt ratio) and void ratio of mixtures. In the sand-silt-clay mixture, the filling, bonding and skeleton effects of clay and silt on the sand existed simultaneously.
(2) Liquefaction resistance of specimens with same FC and different CS were different. The $CSR \sim N_f$ curves of specimens with $FC = 10\%$ were all higher than those of the specimens with $FC = 5\%$ under constant e_s. The $CSR \sim N_f$ curves of specimens with $FC = 10\%$ (constant e_s) were all higher than those of specimens with $FC = 10\%$ (constant e).
(3) With increasing N_f, the slope of $CSR \sim N_f$ curve of specimens with $FC = 10\%$ decreased more quickly than that of specimens with $FC = 5\%$ under constant e_s and e. With the increase of clay content (decrease of silt content), the liquefaction resistance of specimens with $FC = 5\%$ changed non-monotonously. While the liquefaction resistance of the specimen with $FC = 10\%$ (both constant e_s and e) decreased monotonically.
(4) When e_s was constant, most of the fines filled in the pores of sand and bonded to each other in specimens of $FC = 5\%$, and small part of fines acted as the soil skeleton. Compared with specimens of $FC = 5\%$, when $FC = 10\%$, part of fines filled the pores of sand, and more fines particles involved in the construction of soil skeleton. For specimens of $FC = 10\%$ with constant e, the contact state of soil particles was same as specimens of $FC = 10\%$ with constant e_s.
(5) The stereomicroscope images revealed the particle contact meso-mechanism and the interaction between sand and fines particles in the sand-silt-clay mixtures, once again confirmed that the filling, bonding and soil skeleton effects of fines on sand existed at the same time.

References

Amini, F., Qi, G.Z.: Liquefaction testing of stratified silty sands. J. Geotech. Geoenviron. Eng. **126**(3), 208–217 (2000)

Cabalar, A.F., Mustafa, W.S.: Behaviour of sand-clay mixtures for road pavement subgrade. Int. J. Pavement Eng. **18**(8), 714–726 (2017)

Cabalar, A.F., Hasan, R.A.: Compressional behaviour of various size/shape sand-clay mixtures with different pore fluids. Eng. Geol. **164**(Complete), 36–49 (2013)

Carrera, A., Coop, M.R., Lancellotta, R.: Influence of grading on the mechanical behaviour of stava tailings. Géotechnique **61**(11), 935–946 (2011)

Chang, C.S., Deng, Y.: A particle packing model for sand–silt mixtures with the effect of dual-skeleton. Granul. Matter **19**(4), 1–15 (2017). https://doi.org/10.1007/s10035-017-0762-1

Chang, C.S., Wang, J.Y., Ge, L.: Maximum and minimum void ratios for sand-silt mixtures. Eng. Geol. **211**, 7–18 (2016)

Chang, W.J., Hong, M.L.: Effects of clay content on liquefaction characteristics of gap-graded clayey sands. Soils Found. **48**(1), 101–114 (2008)

Fang, H.L.: 3D multi-mechanism model for cyclic mobility of saturated sands. Chin. J. Geotech. Eng. **24**(3), 376–381 (2002)

Gratchev, I.B., Sassa, K., Osipov, V.I., Sokolov, V.N.: The liquefaction of clayey soils under cyclic loading. Eng. Geol. **86**(1), 70–84 (2006)

Huang, Y., Zhao, L.: The effects of small particles on soil seismic liquefaction resistance: current findings and future challenges. Nat. Hazards **92**(1), 567–579 (2018). https://doi.org/10.1007/s11069-018-3212-4

Kim, U., Kim, D., Zhuang, L.: Influence of fines content on the undrained cyclic shear strength of sand-clay mixtures. Soil Dyn. Earthq. Eng. **83**, 124–134 (2016)

Kim, U.G., Zhuang, L., Kim, D., Lee, J.: Evaluation of cyclic shear strength of mixtures with sand and different types of fines. Mar. Georesour. Geotechnol. **35**(4), 447–455 (2017)

Lade, P.V., Yamamuro, J.A.: Effects of nonplastic fines on static liquefaction of sands. Can. Geotech. J. **34**(6), 918–928 (1997)

Li, T., Tang, X.W.: Influences of low fines content and fines mixing ratio on the undrained static shear strength of sand-silt-clay mixtures. Eur. J. Environ. Civil Eng. **26**, 3706–3728 (2020)

Li, T., Tang, X.W., Wang, Z.T.: Experimental study on unconfined compressive and cyclic behaviors of mucky silty clay with different clay contents. Int. J. Civil Eng. **17**(6B), 841–857 (2019)

Monkul, M.M., Etminan, E., Şenol, A.: Influence of coefficient of uniformity and base sand gradation on static liquefaction of loose sands with silt. Soil Dyn. Earthq. Eng. **89**, 185–197 (2016)

National Standards Compilation Group of People's Republic of China: Standard for geotechnical testing method (GB/T 50123-2019). China Water & Power Press, Beijing (2019)

Wang, J., et al.: Liquefaction behavior of dredged silty-fine sands under cyclic loading for land reclamation: laboratory experiment and numerical simulation. Environ. Earth Sci. **77**(12), 1–15 (2018). https://doi.org/10.1007/s12665-018-7631-z

Polito, C.P., Martin II, J.R.: Effects of nonplastic fines on the liquefaction resistance of sands. J. Geotech. Geoenviron. Eng. **127**(5), 408–415 (2001)

Porcino, D.D., Diano, V.: The influence of non-plastic fines on pore water pressure generation and undrained shear strength of sand-silt mixtures. Soil Dyn. Earthq. Eng. **101**, 311–321 (2017)

Porcino, D.D., Tomasello, G., Diano, V.: Key factors affecting prediction of seismic pore water pressures in silty sands based on damage parameter. Bull. Earthq. Eng. **16**(12), 5801–5819 (2018). https://doi.org/10.1007/s10518-018-0411-z

Prashant, A., Bhattacharya, D., Gundlapalli, S.: Stress-state dependency of small-strain shear modulus in silty sand and sandy silt of Ganga. Géotechnique **69**(1), 42–56 (2019)

Thevanayagam, S.: Effect of fines and confining stress on undrained shear strength of silty sands. J. Geotech. Geoenviron. Eng. **124**(6), 479–491 (1998)

Thevanayagam, S., Mohan, S.: Intergranular state variables and stress–strain behaviour of silty sands. Geotechnique **50**(1), 1–23 (2000)

Xenaki, V.C., Athanasopoulos, G.A.: Liquefaction resistance of sand-silt mixtures: an experimental investigation of the effect of fines. Soil Dyn. Earthq. Eng. **23**(3), 1–12 (2003)

Yang, J., Wei, L.: Collapse of loose sand with the addition of fines: the role of particle shape. Géotechnique **62**(12), 1111–1125 (2012)

Yang, S., Lacasse, S., Sandven, R.: Determination of the transitional fines content of mixtures of sand and non-plastic fines. Geotech. Test. J. **29**(2), 102–107 (2005)

Zhang, F., Ye, B., Noda, T.: Explanation of cyclic mobility of soils: approach by stress-induced anisotropy. Soils Found. **47**(4), 635–648 (2007)

Zhou, W., Wu, W., Ma, G., Ng, T.T., Chang, X.: Undrained behavior of binary granular mixtures with different fines contents. Powder Technol. **340**, 139–153 (2018)

The Prediction of Pore Pressure Build-Up by an Energy-Based Model Calibrated from the Results of In-Situ Tests

Lucia Mele[1(✉)], Stefania Lirer[2], Alessandro Flora[1], Alfredo Ponzo[1], and Antonio Cammarota[1]

[1] University of Napoli Federico II, Naples, Italy
lucia.mele@unina.it
[2] University of Roma Guglielmo Marconi, Rome, Italy

Abstract. The excess pore water pressure generation induced by rapid forms of loading can lead to a significant reduction in soil stiffness and strength until reaching liquefaction. The estimate of earthquake-induced pore pressure is important to predict accurately the response of soil deposits and consequently earthquake effects on built environment. Traditionally, the excess pore water pressure generation is linked to stress and strain, however, recently, more innovative and promising energy-based models are developing. Although they allow avoiding the conversion of irregular earthquake load in an equivalent number of cycles – necessary for traditional models – their calibration is often complex, limiting considerably their application. In this paper, the calibration procedure proposed by Mele et al. [10] for the model of Berrill and Davis [5] is used. It consists of linking the two parameters of the model to the well-known equivalent cone tip resistance (q_{c1Ncs}) or the corrected SPT blowcount (($N_1)_{60cs}$). This procedure has been used to predict the excess pore water pressure of the case history of Scortichino (Italy), affected by liquefaction phenomena during the 2012 Italian earthquake. The results of 1D site response analysis of Scortichino dyke, performed with a non-linear code, in which the energy-based pore pressure generation model is implemented, show that the results are in good agreement with those deriving from susceptibility analysis and with the experimental evidence, confirming the effectiveness of the calibration procedure and the reliability of the energy-based models in the prediction of pore pressure build-up.

Keywords: Liquefaction · Pore pressure generation model · Energy-based models · 1-D site response analysis

1 Introduction

The soil response of saturated soils under strong seismic motion is ruled by complex mechanisms, which may be basically ascribed to hysteretic behaviour and volumetric-distorsional coupling. The latter induces volumetric strains under drained conditions or pore pressure changes under undrained conditions. In loose and saturated sandy soil deposits these conditions may lead to liquefaction, which may cause serious damage to structures and infrastructures. The correct estimate of the liquefaction effects on built environment is strongly linked to the knowledge of the excess pore pressure induced by shaking within the soil. Several relationships have been proposed in literature to predict the pore pressure build-up induced by cyclic loadings, traditionally divided into three groups: stress-based, strain-based and energy-based models. The stress and strain-based models are calibrated on the results of cyclic stress and strain-controlled tests, respectively. The first ones link the excess pore pressure ratio r_u (defined as the ratio between the pore pressure increment Δu and the initial effective overburden stress, σ'_{v0}) to the cycle ratio N/N_{liq}, where N is the number of loading cycles and N_{liq} is the number of cycles required to attain liquefaction (i.e., $r_u = 0.9$), while in the strain-based models the pore pressure build-up is controlled by the amplitude of cyclic shear strains and number of loading cycles. More recently, Chiaradonna & Flora [1] proposed an empirical relationship between r_u and the factor of safety (FS):

$$r_u = \frac{2}{\pi} \cdot arcsin\left(FS^{-\frac{1}{2b\beta'}}\right) with \quad FS \geq 1 \tag{1}$$

where b rules the slope of the cyclic resistance curve and β' is defined by Polito et al. [2]. FS is defined as the ratio between the seismic loading required to trigger liquefaction and the one expected from the earthquake in free field conditions (Seed and Idriss, [3]). Typically, both the liquefaction resistance (capacity) and the seismic demand are written as cyclic stress ratios (respectively CRR and CSR), and the factor of safety is expressed as:

$$FS = \frac{CRR}{CSR} = \frac{CRR_{M=7.5,\sigma'v=1}}{CSR} \cdot MSF \cdot K_\sigma \cdot K_\alpha \tag{2}$$

where $CRR_{M=7.5,\sigma'v=1}$ is the resistance referred to a magnitude $M = 7.5$ and to $\sigma'_v = 103$ kPa, MSF is the magnitude scaling factor, introduced to account for the effect of the duration of the seismic event, K_σ and K_α are correcting factors to account respectively for the effective overburden stress and for an initial static shear stress on the horizontal plane. In the simplified stress-based procedure, CSR is usually expressed as:

$$CSR = 0.65 \cdot \frac{\tau_{max}}{\sigma'_v} = 0.65 \cdot \frac{a_{max}}{g} \cdot \frac{\sigma_v}{\sigma'_v} \cdot r_d \tag{3}$$

where σ_v and σ'_v are the vertical total and effective stresses at a depth z, a_{max} is the maximum horizontal acceleration, g is the gravity acceleration and r_d is a shear stress reduction factor accounting for soil deformability. For the sake of brevity, the expressions of all the factors of Eqs. (2–3) are not reported, since they can be easily found in literature (Boulanger and Idriss [4]). The coefficient 0.65 (Eq. 3) is introduced to transform the irregular shear stress history (represented by τ_{max}) in one having an equivalent constant shear stress amplitude. Indeed, the aforementioned models, although simple to use, need to convert the earthquake shaking into an equivalent number of cycles of uniform shear stress. Such conversion is often complex and not always reliable, depending on the conversion curve adopted and on the techniques for choosing and counting the stress cycles. This drawback may be overcome by the energy-based pore pressure generation methods [5]. These models relate the pore pressure generation to the energy dissipated per unit volume of soil or specific deviatoric energy (E_s). The specific deviatoric energy $E_{s,i}$ at a generic load cycle i (N_i) is computed as the sum of the areas bounded by stress-strain hysteresis loops [6]. Several experimental studies demonstrated that the relationship r_u - E_s is unique, regardless of the performed tests and adopted loading path [7, 8]. In 1985, Berrill & Davis [5] proposed the following simple empirical formulation:

$$r_u = \alpha \cdot W_s^\beta \quad (4)$$

where W_s is the energy dissipated per unit volume of the soil normalized to the initial mean effective stress ($W_s = E_s/\sigma'_0$), and α and β are parameters to be calibrated through cyclic laboratory tests. E_s is computed as the sum of the areas bounded by stress-strain hysteresis loops. Due to the simplicity of Eq. (4), the model of Berrill & Davis [5] has been implemented in a 1D non-linear computer code, as DEEPSOIL (v. 7.0; [9]). Although the energy-based models are very simple, the calibration of the parameters from laboratory tests, processed with an energetic perspective is not easy, limiting, therefore their applications. In order to expand the use of such models, Mele et al. [10] proposed a simple calibration procedure deriving from the results of in situ tests, which is briefly described in the following section (Sect. 2). The proposed procedure is validated through the case study of Scortichino (Italy), performing 1D site response analyses by means of DEEPSOIL. The good agreement with the results of the analysis with the experimental evidence confirms the effectiveness of the proposed calibration approach and the reliability of energy-based models to predict the pore pressure build-up.

2 Calibration Procedure of the Energy-Based Model of Berrill & Davis (1985)

Starting from the experimental evidences collected through an extensive experimental study performed on different sandy soils, Mele et al. [10] proposed a useful and simple procedure to practitioners to calibrate the parameters α and β of Berrill & Davis model (Eq. 4) by means of the results of CPT and SPT in-situ tests. The authors showed that not only the parameters α and β (Eq. (4)) seem to be dependent on each other according to the following equation:

$$\alpha = 0.75 \cdot exp(5.29 \cdot \beta) \quad (5)$$

but additionally, the exponent of Eq. (4) may be related to the well-known equivalent cone tip resistance (q_{c1Ncs}) or the corrected SPT blowcount (($N_1)_{60cs}$) parameters. For brevity only the relationship for CPT tests has been reported (Eq. (6)), that relative to SPT tests may be found in Mele et al. [10].

$$\beta = -0.073 \cdot ln(q_{c1Ncs}) + 0.74 \, for \, CPT \, tests \quad (6)$$

In other words, known the results of in-situ tests such as the profile of cone tip resistance (q_c) - or blowcount (N) - it is possible to evaluate q_{c1Ncs} - or ($N_1)_{60cs}$ -via equations proposed by Boulanger and Idriss [7]. From the profile of q_{c1Ncs} that of β may be obtained by Eq. (6). Known the values of β at each depth, α may be estimated by Eq. (5), as well. Further details regarding the calibration procedure may be found in Mele et al. [10]. Once the model of Berrill & Davis [5] has been calibrated, the 1D site response analyses may be performed through DEEPSOIL, enabling the option of the excess pore pressure generation and choosing the energy-based model of Berrill & Davis [5]. The described procedure has been applied for the case study of Scortichino (Italy).

3 Case Study

In May 2012 a seismic sequence hit a large area of the river Po valley (Emilia Romagna region, Northern Italy). The consequences were damaged structures turned out to be the banks of an irrigation canal known as Diversion channel of Burana, flowing through the small village of Scortichino (Municipality of Bondeno), near the historic town of Ferrara [11]. The damage survey carried out along and around the levee after the event revealed liquefaction evidence, such as sand boils and lateral spreading mechanisms. The seismic response of the dyke has been studied in this research work in order to confirm the effectiveness of energy-based models in the prediction of excess pore water pressure build-up.

3.1 Stratigraphy, Field and Laboratory Tests

The site of Scortichino was studied in depth performing an extensive experimental programme, composed by laboratory and in-situ tests, in order to define accurately the geotechnical model. The soil investigation was concentrated around four cross sections of the dyke. In this work the section c-c' was considered, as a matter of the fact that it was the most damaged area [12]. The field investigation consists of 5 boreholes, during with were retrieved 29 undisturbed samples, 12 static penetration tests with piezocone (CPTU) and 4 seismic dilatometer tests (SDMT). The laboratory tests allowed to determinate both static and dynamic properties. Further details regarding laboratory and field tests may be found in Tonni et al. [12]. The stratigraphic model is composed mainly by four deposits: dyke, composed by subsequent layers of sand, sandy silt and silty sand (AR); silty sand (B); sand (A) and clay (C) (Fig. 1a; [11]), while the ground water table is located at 7 m from ground surface. Before performing dynamic analyses, the depth of bedrock and shear wave velocity profile (V_s) need to be determined. As reported by Tonni et al. [12], V_s profile has been defined by integrating the results of the SDMTs with measurements of two Cross-Hole tests carried out in the towns of Mirandola and Medolla (about 20 km distant from the area of study) which reach 130 m depth. Observing V_s profile (Fig. 1b) it can be deduced that the bedrock is placed at a depth of 120 m from ground surface ($V_s > 800$ m/s). The variation of the normalized shear modulus (G/G_0) and the damping ratio (D) with shear strain (Υ), necessary to simulate the non-linear and dissipative behaviour of the soil has been achieved by resonant column and cyclic simple shear tests [12] and reported in Fig. 2.

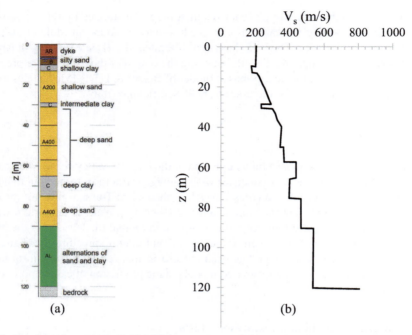

Fig. 1. Stratigraphy sequence of Scorticino, section c-c' ([11]) (a) and shear wave velocity profile (b).

3.2 Susceptibility Analysis

In order to identify the liquefiable layers a susceptibility analysis has been performed according to the procedure of Boulanger & Idriss [4], by using empirical charts, linking CRR with q_{c1Ncs} or $(N_1)_{60cs}$. In Fig. 3a, the q_c profile of CPTU7 has been plotted [12]. The q_c profile has been transformed in q_{c1Ncs} profile (Fig. 3b) according to the equations of Boulanger & Idriss [4]. Known q_{c1Ncs}, $CRR_{M=7.5,\sigma'v=1}$ is obtained. CRR ($M_w = 6.1$) has been finally evaluated taking into account the coefficients MSF, K_σ e K_α. On the other hand, CSR has been evaluated via Eq. (3), where a_{max} is assumed equal to 0.27 g, taking into account the effect of stratigraphy and topography. Finally, the factor of safety

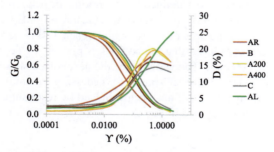

Fig. 2. Normalized shear modulus (G/G$_0$) and damping ratio (D) with strain.

is obtained by Eq. (2) and reported in Fig. 3c. It shows that sandy shallow layers, from 7 to 25 m, are susceptible to liquefaction, even though the fully liquefaction ($r_u = 1$) seems to be reached clearly between 7 and 18 m, and punctually at 23 and 25 m, as confirmed by the r_u profile, achieved via Eq. (1).

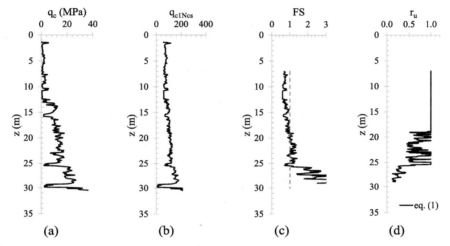

Fig. 3. Cone tip resistance (CPTU7) profile (a); q_{c1Ncs} profile (b); factor of safety (c) and excess pore pressure profile evaluated according to Eq. (1).

4 1D Site Response Analysis of Scortichino Dyke

In order to predict the pore pressure generation of the case study of Scortichino, described in the previous section, a 1D site response analysis has been performed by means of DEEPSOIL v7.0, performing a non-linear analysis in effective stress conditions. The non-linear soil response is modelled with the hysteretic constitutive model in time domain.

4.1 Input Motion

Firstly, the input motion has been defined. As reported by Chiaradonna [11], since no acceleration records were available at the site of Scortichino, a selection of recorded accelerograms was adopted in the simplified analysis in order to simulate the 20th May event. The input motion was defined through a selection of records within the Italian database ITACA, based on the magnitude (5.5–6.5) and distance (5 – 10 km) bins approach. The NS component of the mainshock of Irpinia earthquake (11/23/1980), recorded at the Lauria station was finally selected and adopted as input motion in the analyses. The selected time history was then scaled to the PGA estimated at the site through the hazard map (PGA = 0.192 g).

4.2 Calibration of α and β and Prediction of Pore Water Pressure Generation

The prediction of the excess pore pressure ratio profile for Scortichino site has been obtained by using the energy-based model of Berrill & Davis (1985), calibrated according to the procedure proposed by Mele et al. [10] and briefly summarized in paragraph 2. The results have been plotted in Fig. 4. Both β (Fig. 4b) and α (Fig. 4c) profiles seem to be roughly homogeneous. Indeed, β ranges between 0.37–0.43, while α between 5.51 and 7.47. The non-linear hysteretic response of the soil is simulated adopting the curves shown in Fig. 2, which were fitted by the MKZ model [13], along with the modified Masing rules, according to Philips & Hashash [14]. The result in terms of r_u has been plotted in Fig. 4d (red dots). It shows as fully liquefaction occurs between 7 and 18 m from ground surface, on the contrary the lower layers do not reach the complete loss of strength and stiffness, induced by liquefaction. Between 18 and 28 m r_u does not exceed the value of 0.85. The energy-based model returns a result congruent with those obtained by means of a simplified analysis (Fig. 3d; Eq. 1), confirming the reliability of the proposed calibration procedure for the energy-based model.

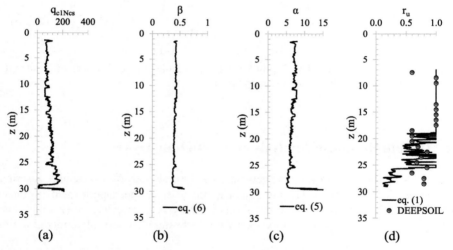

Fig. 4. q_{c1Ncs} profile (a); β (b) and α (c) and r_u profile from DEEPSOIL analysis (d).

5 Final Remarks

In order to confirm the potentialities of the energy-based pore pressure generation model proposed by Berrill & Davis [5], the case study of Scortichino (Italy) has been considered. The energy-based model has been calibrated following the procedure proposed by Mele et al. [10], known the results of CPT in situ tests. Calibrated the pore pressure model, a 1D site response analysis of Scortichino dyke has been performed in effective stress conditions. The results show that liquefaction occurs in the shallowest sandy layers, congruently with the results derived from the susceptibility analysis and the experimental evidence, confirming the reliability of the energy-based model to predict the excess pore water pressure build-up.

References

1. Chiaradonna, A., Flora, A.: On the estimate of seismically induced pore-water pressure increments before liquefaction. Géotechnique Lett. **10**(2), 128–134 (2020)
2. Polito, C.P., Green, R.A., Lee, J.: Pore pressure generation models for sands and silty soils subjected to cyclic loading. J. Geotech. Geoenviron. Eng. **134**(10), 1490–1500 (2008)
3. Seed, H.B., Idriss, I.M.: Simplified procedure for evaluating soil liquefaction potential. J. Soil Mech. Found. Div. **97**(SM9), 1249–1273 (1971)
4. Boulanger, R.W., Idriss, I.M.: CPT and SPT liquefaction triggering procedures. Report No UCD/GCM14/01, University of California at Davis, California (2014)
5. Berrill, J.B., Davis, R.O.: Energy dissipation and seismic liquefaction of sands: revised model. Soils Found. **25**(2), 106–118 (1985)
6. Mele, L., Flora, A.: On the prediction of liquefaction resistance of unsaturated sands. Soil Dyn. Earthq. Eng. **125**, 105689 (2019)
7. Polito, C., Green, R.A., Dillon, E., Sohn, C.: Effect of load shape on relationship between dissipated energy and residual excess pore pressure generation in cyclic triaxial tests. Can. Geotech. J. **50**(11), 1118–1128 (2013)
8. Mele, L.: Experimental and theoretical investigation on cyclic liquefaction and on the effects of some mitigation techniques. Ph.D. thesis, Università degli Studi di Napoli Federico II, Napoli, Italy (2020)
9. Hashash, Y.M.A., et al.: DEEPSOIL 7.0, User Manual. Board of Trustees of University of Illinois at Urbana-Champaign, Urbana (2020)
10. Mele, L., Lirer, S., Flora, A.: A simple procedure to calibrate a pore pressure energy-based model from in situ tests. Acta Geotech. (2022). https://doi.org/10.1007/s11440-022-01650-1
11. Chiaradonna, A.: Development and assessment of a numerical model for non-linear coupled analysis on seismic response of liquefiable soils. University of Napoli Federico II, Naples (2016)
12. Tonni, L., Gottardi, G., Amoroso, S., …, Vannucchi, G., Aversa, S.: Interpreting the deformation phenomena triggered by the 2012 Emilia seismic sequence on the Canale Diversivo di Burana banks. Ital. Geotech. J. **2**, 28–58 (2015). (in Italian)
13. Matasović, N., Vucetic, M.: Cyclic characterization of liquefiable sands. J. Geotech. Eng. **119**(11), 1805–1822 (1993)
14. Phillips, C., Hashash, Y.M.: Damping formulation for nonlinear 1D site response analyses. Soil Dyn. Earthq. Eng. **29**(7), 1143–1158 (2009)

CDSS Tests for Evaluation of Vibration Frequency in Liquefaction Resistance of Silica Sand

Zhen-Zhen Nong[1(✉)], Sung-Sik Park[2], and Peng-Ming Jiang[1]

[1] Jiangsu University of Science and Technology, Zhenjiang 212100, Jiangsu, China
zznong@foxmail.com
[2] Kyungpook National University, Daegu 702-701, Republic of Korea

Abstract. The frequency of ground motions during earthquakes is typically on the order of a few hertz. In the evaluation of the liquefaction resistance of soil in laboratory tests, it is necessary to consider various vibration frequencies generated by real earthquakes. The effect of vibration frequency has been studied by cyclic triaxial tests; however, it has rarely been investigated by cyclic direct simple shear (CDSS) tests, which are more similar to the cyclic loading conditions associated with earthquakes. In this study, a series of CDSS tests was performed on relative density of 40% of sand obtained from Nakdong River. Two different initial vertical effective stresses (σ'_{v0}, 100 and 200 kPa) and four different frequencies (f, 0.05, 0.1, 0.5, and 1 Hz) were applied to evaluate the effect of the vibration frequency on the liquefaction resistance of clean sand for both the undrained and drained conditions. For the undrained CDSS tests, the liquefaction resistance of the sand was observed to increase with f, regardless of σ'_{v0}. The maximum increase in the cyclic resistance was 15% when f was increased from 0.1 to 1 Hz. For the drained CDSS tests, with an increase in f, the rate of volumetric strain accumulation decreased and the shear modulus ratio increased.

Keywords: Sand liquefaction · Vibration frequency · Undrained CDSS test · Drained CDSS test

1 Introduction

The frequency of a seismic excitation, varying mostly from 0 to 15 Hz, indicates that vibration frequency (f) is an important factor influencing the liquefaction behavior of soil layers. The evaluation of the dynamic properties of soil in laboratory tests requires the consideration of various vibration frequencies. The effect of f for the liquefaction resistance of sand has been studied mostly by conducting cyclic triaxial tests on medium saturated sand. Nevertheless, conclusions about the effect of the vibration frequency on the liquefaction resistance of sand are contradictory. The previous study results were concluded that the undrained cyclic strength of saturated sand can remain uninfluenced (Sze 2010), can be influenced only slightly (Wang and Zhou 2003), or can increase (Zhang et al. 2015) with f. These results were obtained from the cyclic triaxial tests. For the cyclic direct simple shear (CDSS) tests, Peacock and Seed (1968) investigated the effect of f in the range 0.17–4 Hz on the liquefaction resistance of saturated Monterey sand by using a Roscoe-type simple shear apparatus. They noted that the effect of f was small since their data for medium sand with relative density (D_r) of 50% were scattered within ±10% of the mean shear stress, causing failure in 10 cycles, with no special order prevailing. Cyclic direct simple shear loading causes the continuous rotation of principle stress axes, which can be better to simulate the rotation of principal stress caused by earthquakes. Due to the contradictory conclusions and limited data on CDSS tests on the effect of f on the liquefaction prediction of sand, the effect of f on the liquefaction prediction of sand requires further attention.

The volumetric strain–induced settlement of soils because of drainage immediately after an earthquake has been recognized as an important phenomenon in soil dynamics (Wartman et al. 2003; Whang et al. 2004). Whereas the effect of factors such as relative density, vertical effective stress (σ'_{v0}), and cyclic amplitude (γ_c) on the volume change of soils has been investigated widely (Sriskandakumar 2000; Wu et al. 2020), studies that examined the effect of the vibration frequency on the cyclic behavior of sand on the basis of the drained cyclic test have been rare. Duku et al. (2008) observed that the effect of f on the cyclic volume change behavior of clean dry silica sand was insignificant on the basis of CDSS tests for f range of 0.1–10 Hz. However, the rare study results on the cyclic volume change behavior of sand indicate the need for more research to investigate the effect of f on the drained cyclic behavior of soils.

In this study, a series of CDSS tests were performed on clean dry silica sand to determine the effect of vibration frequency on both undrained and drained conditions. Frequencies of 0.05, 0.1, 0.5, and 1 Hz were considered. In undrained tests, the influence of f on the liquefaction resistance of sand was determined. In drained tests, the effect of f on the volumetric strain accumulation (ε_v) and shear modulus ratio were investigated.

2 Cyclic Direct Simple Shear Tests

2.1 Test Material

This study used sand obtained from the Nakdong River. This river flows in the center lowlands of the Yongnam district and stretches to Elsuk Island in Busan, South Korea. The sand was washed and sieved to obtain grain sizes in the range 0.85–0.075 mm. The sand was siliceous–medium nonplastic sand with subangular grains. Its specific gravity was 2.64, the value of D_{50} was 0.32 mm, and the minimum and maximum void ratios were 0.65 and 1.18, respectively. It was classified as poorly graded sand (SP) according to the Unified Soil Classification System.

2.2 Laboratory Device, Sample Preparation, and Test Conditions

The automatic control CDSS device used in this study was developed by the Norwegian Geotechnical Institute (NGI) and manufactured by Geocomp. Details of the design and capabilities of this device were presented by Nong et al. (2021).

All samples were prepared inside a wire-reinforced membrane with an initial diameter of 63.5 mm and an initial height of 30 mm. A static compaction method that applied less energy to the lower layer of the sample was used to achieve a uniform density in the height direction, and the target relative density after consolidation was approximately 40%. After preparation, samples were consolidated at initial σ'_{v0} of 100 and 200 kPa, and then they were sheared with a sinusoidal wave of at f values of 0.05, 0.1, 0.5, and 1 Hz, respectively. For a given f and σ'_{v0}, in undrained tests, three distinct tests with different levels of the cyclic stress ratio (CSR) were conducted to obtain a cyclic resistance curve (number of cycles to liquefaction versus CSR); and in drained tests, samples were sheared at a γ_c of 1.0%. The test conditions for undrained and drained CDSS tests are listed in Table 1.

Table 1. Test conditions for undrained and drained CDSS tests.

D_r (%)	σ'_{v0} (kPa)	f (Hz)	Undrained tests	Drained tests
			CSR	γ_c (%)
40	100	0.05	Appropriate values	1.0
		0.1		
		0.5		
		1		
	200	0.05		
		0.1		
		0.5		
		1		

3 Results

3.1 Results of Undrained CDSS Tests

In undrained CDSS tests, a sample was deemed to have liquefied when a double amplitude of the shear strain exceeds 7.5%. The cyclic resistance ratio (CRR_{15}) reported herein corresponds to liquefaction in 15 cycles.

The undrained CDSS response at $f = 0.05$ Hz and 0.5 Hz for $\sigma'_{v0} = 100$ kPa and $CSR = 0.18$ was showed in Fig. 1. As loading proceeded, shear strain was accumulated continuously, and initially, the excess pore pressure rose rapidly. Subsequently, the increase in the excess pore pressure became gradual and was accompanied by a decrease in the initial σ'_{v0}. Once the liquefaction criterion was satisfied, the excess pore pressure at the end of the test exceeded 90% of the initial σ'_{v0}. A comparison of Fig. 1c and 1f shows that the magnitude of initial σ'_{v0} decrease was influenced by the vibration frequency. At $f = 0.05$ Hz, the value of σ'_{v0} decreased approximately 45%, from 100 kPa to 55 kPa in the first cycle. However, at $f = 0.5$ Hz, the value of σ'_{v0} decreased only 35%, from

100 kPa to 65 kPa in the first cycle. The faster decrease in initial σ'_{v0} in the first cycle is resulted in a less numbers of cycles to liquefy. The different changes in sample fabric under different f contribute to observed differences in cyclic response.

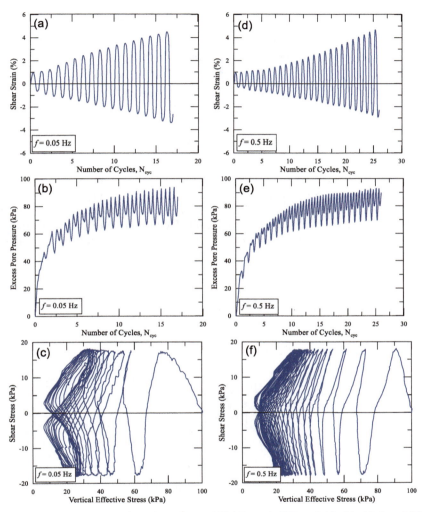

Fig. 1. Undrained response of sand at $\sigma'_{v0} = 100$ kPa and $CSR = 0.18$: (a)–(c) $f = 0.05$ Hz; (d)–(f) $f = 0.5$ Hz.

The cyclic resistance ratios at 15 cycles (CRR_{15}) were used to estimate the sand liquefaction. Figure 2 shows the relationship between f and CRR_{15} variation with σ'_{v0}. At both $\sigma'_{v0} = 100$ and 200 kPa, the value of CRR_{15} increased linearly with the logarithmic value of f. For a given f, the value of CRR_{15} decreased when the initial σ'_{v0} increased.

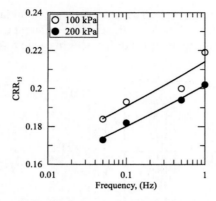

Fig. 2. The relationship between CRR_{15} and f variation with σ'_{v0}

The percentage increase in the cyclic resistance at 0.1, 0.5, and 1 Hz relative to the cyclic resistance at 0.05 Hz is presented in Fig. 3. At $\sigma'_{v0} = 100$ kPa, an increase in f from 0.05 Hz to 0.1, 0.5, and 1 Hz led to the cyclic resistance increasing by 3.7%, 7.9%, and 12%, respectively. Similarly, at $\sigma'_{v0} = 200$ kPa, the values of CRR_{15} increased by 6.8%, 9%, and 15% when the values of f increased from 0.05 Hz to 0.1, 0.5, and 1 Hz.

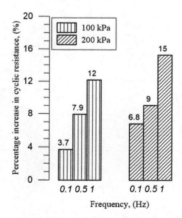

Fig. 3. Increase in CRR_{15} at 0.1, 0.5 and 1 Hz relative to the cyclic resistance at 0.05 Hz

3.2 Results of Drained CDSS Tests

The drained cyclic loading responses in the first three cycles for Nakdong river sand sheared at low and high frequencies are compared at Fig. 4. The dashed and solid lines correspond to results for samples sheared at $f = 0.05$ and 1 Hz, respectively. For a given γ_c, the cyclic shear stress in the sample sheared at $f = 1$ Hz was lower than that in the sample sheared at $f = 0.05$ Hz from the beginning of the test. Furthermore, with an increase in the number of cycles, the increase in cyclic shear stress in the sample sheared at $f = 0.05$ Hz was more obvious than that in the sample sheared at $f = 1$ Hz. The shear-induced volumetric strain during both loading and unloading phases of the first half cycle increased with a decrease in f. During the cyclic process, ε_v accumulation for the sample sheared at the lower f was always more significant than that for the sample sheared at the higher f. At the end of the third cycle, ε_v accumulation for the sample sheared at $f = 1$ Hz was 0.529%, which was only 64% that of the sample sheared at $f = 0.05$ Hz.

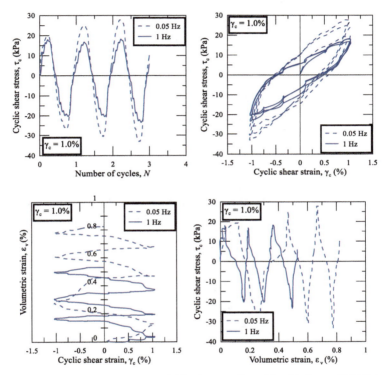

Fig. 4. Typical results in drained CDSS tests affected by f ($\sigma'_{v0} = 100$ kPa)

The effect of the vibration frequency on the volumetric strain accumulation of Nakdong river sand is shown in Fig. 5. The results show that at the same γ_c and σ'_{v0}, the value of ε_v consistently increased with a decrease in the frequency. In drained tests, volumetric strain accumulation was related to the drainage. The sand sample sheared at a lower f had more time to drain in a cycle, resulting in a higher volumetric strain accumulation. On the other hand, the higher the f value, the greater was the number of cycles required to reach a specified ε_v value. For example, for the case of $\sigma'_{v0} = 200$ kPa, when f increased from 0.05 to 1 Hz, the number of cycles required to reach a volumetric strain of 1.0% more than doubled (from 3.5 to 7.5 cycles). For engineering practice, this implies that for given shear strain and stress conditions, the volumetric strain accumulation for a loose sand sheared at a low vibration frequency with a small number of cycles is similar to that for sand sheared at a high vibration frequency with a larger number of cycles.

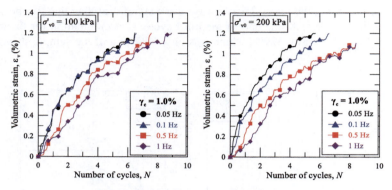

Fig. 5. Effect of f on the ε_v accumulation

Figure 6 shows the relationship between the shear modulus ratio (G_N/G_0) and the number of cycles for different f values, where G_0 and G_N denote the shear secant moduli in the first and Nth cycles, respectively. The result shows that the value of G_N/G_0 increased rapidly in the first few cycles, and then increased slowly, indicating that the soil's shear modulus had stabilized. For given values of γ_c, σ'_{v0}, a higher f always resulted in a larger G_N/G_0. However, the difference of G_N/G_0 value for the different values of f became less significant with the increase of σ'_{v0}.

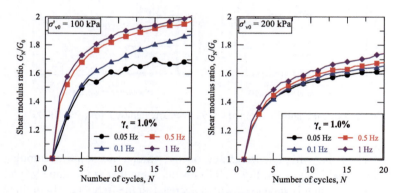

Fig. 6. Effect of f on the shear modulus ratio

4 Discussion and Conclusions

This study investigated the undrained and drained cyclic behavior of silica sand that was affected by vibration frequency by conducting CDSS tests. Both the undrained and drained tests were affected by vibration frequency. In the undrained tests, the liquefaction resistance of silica sand increased with an increase of f. The maximum increase of liquefaction resistance was approximately 15% when f increased from 0.05 to 1 Hz. In drained tests, the volumetric strain accumulation can be affected by vibration frequency. The higher f, the larger was the number of cycles required to reach a constant volumetric strain. Moreover, the shear modulus ratio increased with the increase in f.

Although the effect of the vibration frequency determined from the undrained test cannot be compared with that obtained from the drained test directly because of the difference in the test modes, we can examine whether the effect is identical in the two types of tests by comparing the cyclic behavior in drained and undrained tests. In undrained tests, an increase in f from 0.05 to 1 Hz resulted in an increase in liquefaction resistance. It is indicated that a larger number of cycles being required to reach liquefaction when a higher f was applied. In drained tests, an increase in f led to a decrease in the accumulated volumetric strain and an increase in the shear modulus ratio, which resulted in a larger number of cycles being required to reach a given volumetric strain. This requirement of a larger number of cycles indicated an increase in the stiffness of sand. Thus, the effect of the vibration frequency on sand appears to be similar in the undrained and drained CDSS tests. This conclusion is consistent with that obtained by Normandeau and Zimmie (1991).

References

Duku, P.M., Stewart, J.P., Whang, D.H., Tee, E.: Volumetric strains of clean sands subjected to cyclic loads. J. Geotech. Geoenviron. Eng. **138**(8), 1073–1085 (2008)

Nong, Z.Z., Park, S.S., Lee, D.E.: Comparison of sand liquefaction in cyclic triaxial and simple shear tests. Soils Found. (2021). https://doi.org/10.1016/j.sandf.2021.05.002

Normandeau, D.E., Zimmie, T.F.: The effect of frequency of cyclic loading on earth structures and foundation soils. In: 2nd International Proceedings on Recent Advances in Geotechnical Earthquake Engineering and Soil Dynamics, p. 39. Scholars' Mine, State Louis, USA (1991)

Peacock, W.H., Seed, H.B.: Sand liquefaction under cyclic loading simple shear conditions. J. Soil Mech. Found. Div. **94**(SM 3), 689–708 (1968)

Sriskandakumar, S.: Cyclic loading response of Fraser River sand for validation of numerical models simulating centrifuge tests. Thesis, The University of British Columbia, Canada (2000)

Sze, H.Y.: Initial shear and confining stress effects on cyclic behaviour and liquefaction resistance of sands. Ph.D. thesis, Hong Kong University, Hongkong, China (2010)

Wang, X.H., Zhou, H.L.: Study on dynamic steady state strength of sand liquefaction. J. Rock Mech. Eng. **22**(1), 96–102 (2003)

Wartman, J., et al.: Ground failure. Earthq. Spectra **19**(S1), 35–56 (2003)

Whang, D.H., Stewart, J.P., Bray, J.D.: Effect of compaction conditions on the seismic compression of compacted fill soils. Geotech. Test. J. **27**(4), GTJ11810 (2004)

Wu, Z.X., Yin, Z.Y., Dano, C., Hicher, P.Y.: Cyclic volumetric strain accumulation for sand under drained simple shear condition. Appl. Ocean Res. **101**, 102200 (2020)

Zhang, S., Zhang, Y.F., Zhang, L.K., Liu, C.J.: Influence of confining pressure and vibration frequency on the liquefaction strength of the saturated gravel sand. J. Xinjiang Agric. Univ. **38**(1), 68–71 (2015)

Some Important Limitations of Simplified Liquefaction Assessment Procedures

Nikolaos Ntritsos(✉) and Misko Cubrinovski

University of Canterbury, Christchurch, New Zealand
nikolaos.ntritsos@canterbury.ac.nz

Abstract. A subset of a recently compiled database of CPT-based liquefaction case histories from Canterbury, New Zealand, is used to scrutinize the performance of simplified liquefaction assessment procedures, for different soil types and ground conditions. In general terms, simplified procedures are shown to perform well (i.e. correctly predict severe manifestation of liquefaction at the ground surface) for deposits that have a critical zone of vertically continuous low-resistance liquefiable soils at shallow depth (true-positive sites), and also (correctly predict the absence of liquefaction manifestation) for relatively uniform high-resistance clean sand or fine sand deposits (true-negative sites). In contrast, the severity of liquefaction manifestation in intermediate-resistance clean sand to silty sand deposits is generally underpredicted by the simplified procedures (false-negative sites). Lastly, systematic overprediction of liquefaction manifestation is observed in deposits with interbedded non-liquefiable soils and liquefiable soils of low-resistance (false-positive sites). The poor performance of the simplified procedures for the false-negative and the false-positive sites can be attributed to the neglect in the evaluation of important system response effects which, on the one hand, intensify the severity of liquefaction manifestation for the false-negative sites and, on the other hand, mitigate liquefaction manifestation for the false-positive sites.

Keywords: Case history · Liquefaction · Simplified method · System response

1 Introduction

In current engineering practice, liquefaction assessment (i.e. evaluation of triggering and consequences of liquefaction) is typically performed using semi-empirical simplified procedures that were originally developed based on observations from liquefaction case histories on relatively uniform clean sands or sands with small amounts of fines. In these procedures, the triggering potential and resulting strains are estimated separately for each individual layer in the deposit and, subsequently, liquefaction damage indices, such as *LSN* [1] and *LPI* [2], are calculated using specific weighting functions to quantify the damage potential of liquefying layers depending on their proximity to the ground surface. Increasing values for the calculated damage indices are generally associated with increasing severity of surface liquefaction manifestation and associated damage (e.g. severe liquefaction manifestation would be expected for $LSN > 30$ and $LPI > 15$).

Since their initial development in the early 70's [3], simplified procedures have undergone a number of updates and have nowadays gained wide acceptance thanks to their simplicity and accumulated experience over their application. However, recent studies have questioned the predictive performance and generalized applicability of simplified procedures to soils and ground conditions other than uniform clean sands (e.g. [4]). The significant inconsistencies between the predictions of simplified procedures and the actual observations of liquefaction manifestation that were noted in several studies of the 2010–2011 Canterbury, New Zealand (NZ), earthquakes (e.g. [4–7]) corroborate the above concerns.

The limitations of simplified procedures may be related to, among other factors, the material and behavioural characterization of liquefiable soils at the *element level* (e.g. uncertainty in the estimation of liquefaction resistance) or to the overall *system-level response* of liquefying deposits which may involve important dynamic cross-interactions between different layers in the deposit that are not accounted for in the simplified procedures. In fact, previous work on the Canterbury case histories has demonstrated that such interactions between different layers in the dynamic response and through excess pore water pressure redistribution and water flow have had a governing influence on the development of liquefaction and its manifestation at the ground surface [7]. Depending on the overall configuration of the soil profile, these system response effects can either intensify or mitigate the severity of liquefaction manifestation at the ground surface [7].

This paper uses a recently compiled liquefaction database from the Canterbury earthquakes [8] to discuss key soil profile characteristics of case-history sites where the simplified procedures perform well (i.e. their predictions are consistent with field observations) and those of sites where the simplified procedures overpredict or underpredict the severity of liquefaction manifestation at the ground surface. Results from simplified triggering analysis of selected representative sites are used to discuss system response processes that are potentially activated in each case and which can explain why simplified procedures may not be as effective for certain types of soil profile configurations. The paper serves to elucidate some key limitations of simplified procedures and emphasizes the need to consider system response effects in the engineering assessment of liquefaction and associated damage.

2 The Canterbury Liquefaction Database

The series of earthquakes that occurred between 2010 and 2016 in Canterbury (NZ) included several strong events ($M_w > 5$) in the proximity, or within the boundaries, of the city of Christchurch that caused widespread and damaging liquefaction. The unprecedented extent and severity of liquefaction in a major urban centre motivated the collection of vast amounts of data, including field observations of liquefaction manifestation after each earthquake, as well as seismologic, hydrologic, geospatial, and geotechnical measurements. Geyin et al. [8] recently compiled some of these data into a curated digital dataset of approximately 15,000 CPT-based case histories.

The present paper examines a subset of 712 case histories from the Canterbury database which are classified into four different groups, as outlined in Table 1. The sites are grouped in accordance with the severity of liquefaction manifestation at the ground surface ('positive' for severe manifestation and 'negative' for no manifestation) documented after the $M_w6.2$ 22 February 2011 earthquake (i.e. the most devastating earthquake in the Canterbury earthquake sequence) and the associated performance of the simplified procedures for this earthquake ('true' for correct prediction of the severity of liquefaction manifestation and 'false' for incorrect prediction). The performance of the simplified procedures (i.e. whether they provide 'true' or 'false' predictions) is assessed based on the computed values for the liquefaction severity number, LSN [1]. In the computation of LSN, factors of safety against liquefaction triggering (FS) were computed using the procedure of [9] and post-liquefaction volumetric strains were computed in accordance with [10]. Yet, it should be noted that the general conclusions of this study hold true irrespective of the specific liquefaction damage index and the triggering procedure used in the evaluation.

Table 1. Classification of examined sites

Classification (# sites, N)	LSN	Observed manifestation
True-positive (196)	>30	Severe
True-negative (243)	<10	None
False-negative (84)	<20	Severe
False-positive (189)	>20	None

3 Analysis of Case-Histories

Figures 1, 2, 3 and 4 show, on the left hand side, the I_c and q_{c1Ncs} profiles for the top 10 m of all of the CPT records in each of the above groups (gray lines), along with the median (thick solid lines) and the 16th–84th percentile values (dashed lines) at each depth. The number of CPT records at each depth (N) is also indicated in a separate plot, for reference. The right hand side of each figure illustrates the characteristics (soil type and q_{c1Ncs}) of a selected representative soil profile from each group, together with the results from simplified triggering analyses (FS throughout the deposit) and a description of key processes anticipated in the actual field response. The latter are substantiated by comprehensive effective stress analyses of a large number of case-history sites which are not reported in this paper but can be found in [7] and [11].

3.1 True-Positive Sites

True-positive sites (Fig. 1) with severe liquefaction manifestation are generally composed of low-resistance ($q_{c1Ncs} \approx 75\text{--}85$, on average) silty sand to sandy silt soils ($I_c \approx$ 1.85–2.3, on average) in the top 4 m, overlying clean sands or fine sands of progressively higher resistance with depth ($q_{c1Ncs} \approx 100$, from 4 to 6 m and $q_{c1Ncs} \approx 105\text{--}125$, from 6 to 10 m, on average).

Simplified liquefaction analysis predicts triggering of liquefaction for the majority of soils below the water table as illustrated for the example site shown in Fig. 1. In reality, because the dynamic characteristics of the site and its capacity for energy dissipation change rapidly upon the first occurrence of liquefaction [7], liquefaction triggering may be limited to only the weakest parts of the deposit which, in this case, are the soils in the critical zone from about 2 to 6 m depth. Some of the deeper layers that were predicted to have low FS may liquefy as well, or develop excess pore water pressures (EPWPs) equal or higher than those in the critical zone, therefore supplying additional water and sustaining the upward flow of water towards the ground surface. The soils above the water table are also subject to inflow of water from the heavily liquefied critical zone and may hence lose their effective stress and liquefy (due to seepage-induced liquefaction). Thus, true-positive sites are characterized by strong vertical communication of EPWPs with contributions from the entire deposit and large volumes of water discharged at the ground surface.

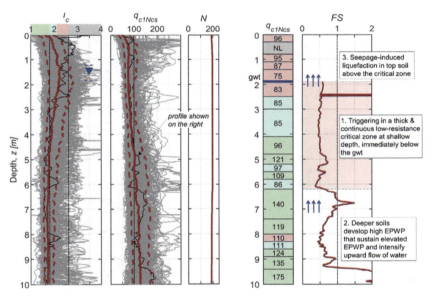

Fig. 1. True-positive sites: I_c and q_{c1Ncs} profiles (left); triggering analysis and key anticipated response processes for a severe manifestation site with $LSN = 35$ and $LPI = 21$ (right).

As a key feature of the true-positive sites is the presence of thick critical zone of low-resistance soils at shallow depth, liquefaction damage indices that combine thickness, depth, and severity of liquefaction in the assessment (e.g. LSN and LPI) can correctly predict the severe liquefaction manifestation at the ground surface, even though water flow and seepage effects are not explicitly considered in the assessment.

3.2 True-Negative Sites

True-negative sites (Fig. 2) are characterized by higher variability compared to the true-positive sites, especially with respect to q_{c1Ncs}. On average, they consist of relatively uniform clean sands or fine sands with high penetration resistance ($q_{c1Ncs} > 150$ for depths greater than 3 m). However, thin layers of lower resistance are occasionally present in these sites as shown for the example soil profile in Fig. 2. Under the severe shaking of the February 2011 earthquake, liquefaction may be triggered (marginally) in these layers. Yet, the liquefaction in thin, isolated layers of relatively high penetration resistance can produce only small amounts of excess water which can be quickly dissipated throughout the deposit without causing sufficiently high hydraulic gradients to transport the liquefied soil to the ground surface. Similarly, because of their small thickness, relatively high resistance and large depth (on average), the liquefied layers do not contribute significantly to *LSN* or *LPI*, which have low values, consistent with the absence of surface liquefaction manifestation at the true-negative sites.

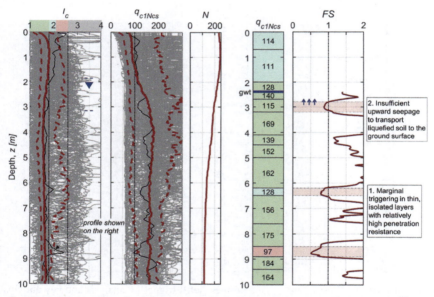

Fig. 2. True-negative sites: I_c and q_{c1Ncs} profiles (left); triggering analysis and key anticipated response processes for a no-manifestation site with $LSN = 5$ and $LPI = 2$ (right).

3.3 False-Negative Sites

False-negative sites (Fig. 3) have characteristics in between those of the true-positive (Fig. 1) and true-negative sites (Fig. 2). They predominantly consist of clean sands to silty sands of intermediate penetration resistance, below the water table. In these deposits, as shown for the example site in Fig. 3, liquefaction is predicted to be more extensive (i.e. it involves larger volumes of soil) and more severe (lower FS) than it is for the true-negative sites, meaning that larger volumes of water can be expelled from the liquefied soils. In addition, the high-permeability soils that connect the liquefied layers allow the expelled pore water to flow through and between layers. Seepage action and upward flow of water is expected to exacerbate the fluidization and instability of the shallow soils that liquefied during shaking and even lead to seepage-induced liquefaction of initially stable soils, both within the deposit and above the water table.

Simplified procedures ignore these important seepage effects which are expected to intensify the spread and severity of liquefaction throughout the deposit including the severity of its manifestation at the ground surface. Therefore, they tend to somewhat underestimate the severity of manifestation for sites with the above characteristics.

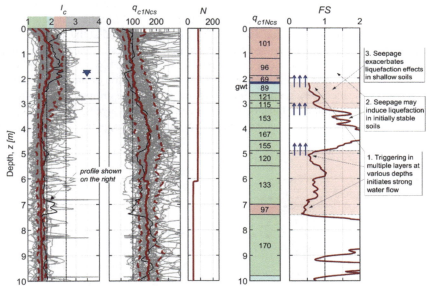

Fig. 3. <u>False-negative sites</u>: I_c and q_{c1Ncs} profiles (left); triggering analysis and key anticipated response processes for a severe manifestation site with $LSN = 18$ and $LPI = 11$ (right).

3.4 False-Positive Sites

False-positive sites (Fig. 4), for which severe liquefaction manifestation is predicted but there was no evidence of liquefaction at the ground surface after the earthquake, have CPT characteristics that, on average, are not very different from those of the true-positive sites (Fig. 1) where severe manifestation was both predicted and observed. In the top 6 m, for instance, the median I_c is between 1.9 and 2.2 and the median q_{c1Ncs} between 80 and 90, similar to the respective values for the true-positive sites. Yet, the false-positive sites show a significant variation in the I_c values compared to the true-positive sites. A more detailed scrutiny of the characteristics of the false-positive sites unveils that, in contrast to the previous sites, in this case the liquefiable soils in the deposit are rarely continuous, instead they are most often interrupted by non-liquefiable layers, as shown for the example site in Fig. 4. False-positive sites are therefore of very different nature as they typically consist of highly stratified deposits with interbedded liquefiable soils of low penetration resistance (silty sands, non-plastic or low-plasticity silts) and non-liquefiable plastic silts.

The above characteristics of the false-positive sites give rise to a series of mechanisms that can work together to effectively mitigate the development of liquefaction and its manifestation at the ground surface. The existence of multiple low resistance layers at depth (e.g. >5 m) can be beneficial from a liquefaction manifestation viewpoint, as liquefaction of these loose deeper layers is expected to cause a substantial reduction of acceleration and seismic demand for the overlying soils. Therefore, the shallower layers, which are closer to the ground surface and therefore more critical for liquefaction manifestation, will be subject to reduced demand (reduced CSR) which in combination with partial saturation effects (increased CRR), often encountered at the shallow depths of interbedded deposits, can increase FS and even prevent the occurrence of liquefaction at shallow depth. In addition, contrary to the previous cases, the high-proportion of non-liquefiable soils in the deposit impede the vertical communication of the liquefied layers through water flow including seepage-induced flow towards the ground surface. In other words, while liquefaction may have been triggered in multiple layers, these are isolated and capped by non-liquefiable layers at sufficient depths to suppress water and ejecta reaching the ground surface.

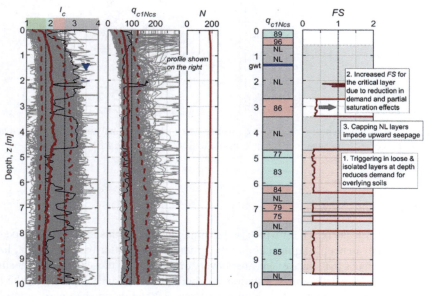

Fig. 4. False-positive sites: I_c and q_{c1Ncs} profiles (left); triggering analysis and key anticipated response processes for a no-manifestation site with $LSN = 24$ and $LPI = 24$ (right).

Simplified analysis predicts triggering of liquefaction with low *FS* for all the liquefiable layers in the deposit and relatively high values for *LPI* and *LSN*, despite the fact that liquefaction was not manifested at the ground surface. The fact that the above mitigating mechanisms of system response are not considered in the simplified procedures explains the discrepancy between predictions and actual observations for the false-positive sites.

4 Conclusions

The wealth of high-quality data collected from the Canterbury earthquakes provide an exceptional opportunity to extract new knowledge and advance the state-of-practice in liquefaction assessment. The present study used a subset of these data to scrutinize the performance of current state-of-practice simplified procedures for different types of soil profiles. In general terms, simplified procedures were shown to perform well (i.e. correctly predict severe liquefaction manifestation at the ground surface) for deposits which have a critical zone of vertically continuous liquefiable layers of low-*FS* at shallow depth (true-positive sites), and also (correctly predict no manifestation) for relatively uniform clean sand or fine sand deposits with high-*FS* values where only few thin layers are predicted to marginally liquefy (true-negative sites). In contrast, liquefaction manifestation of clean sand to silty sand deposits of intermediate-*FS* (false-negative sites) are generally underpredicted by the simplified procedures. Lastly, in deposits with interbedded non-liquefiable soils and liquefiable soils with low-*FS* values (false-positive sites), simplified procedures systematically and substantially overpredict liquefaction manifestation.

The poor performance of the simplified procedures for the false-negative and the false-positive sites can be largely attributed to the neglect of important system response effects which, on the one hand, intensify the severity of liquefaction manifestation for the false-negative sites and, on the other hand, mitigate liquefaction manifestation for the false-positive sites. Intensification of liquefaction manifestation is caused in deposits with high overall permeability (or vertical continuity of liquefiable soils) by the vertically unconstrained water flow and associated seepage action. These water-flow related mechanisms result in additional disturbance to that caused by the ground shaking, leading to further fluidization of liquefied soils or seepage-induced liquefaction and hydraulic fracturing of initially stable soils. Mitigation of liquefaction manifestation occurs in interbedded deposits as a result of a deep liquefaction-induced 'base isolation' effect, partial saturation effects, isolation or vertical confinement of liquefied layers and at-depth suppression of ejecta movement by capping non-liquefiable layers. The above hypotheses have been confirmed by advanced effective stress analyses of a large number of Canterbury case-history sites [11].

In conclusion, it can be argued that without adequate consideration of fundamental mechanisms of system response of liquefying deposits, it would be difficult to achieve the required level of accuracy for effective assessment and mitigation of liquefaction.

References

1. van Ballegooy, S., et al.: Assessment of liquefaction-induced land damage for residential Christchurch. Earthq. Spectra **30**(1), 31–55 (2014)

2. Iwasaki, T., Tatsuoka, F., Tokida, K., Yasuda, S.: A practical method for assessing soil liquefaction potential based on case studies at various sites in Japan. In: 2nd International Conference on Microzonation for Safer Construction, San Francisco, CA, USA, pp. 885–896 (1978)
3. Seed, H.B., Idriss, I.M.: Simplified procedure for evaluating soil liquefaction potential. J. Soil Mech. Found. Div. **92**(SM2), 53–78 (1971)
4. Cubrinovski, M., Ntritsos, N., Dhakal, R., Rhodes, A.: Key aspects in the engineering assessment of soil liquefaction. In: Silvestri, F., Moraci, N. (eds.) Earthquake Geotechnical Engineering for Protection and Development of Environment and Constructions, pp. 189–208. Taylor & Francis Group, London, UK (2019)
5. Maurer, B.W., Green, R.A., Cubrinovski, M., Bradley, B.A.: Assessment of CPT-based methods for liquefaction evaluation in a liquefaction potential index framework. Geotechnique **65**(5), 328–336 (2015)
6. Beyzaei, C.Z., Bray, J.D., van Ballegooy, S., Cubrinovski, M., Bastin, S.: Depositional environment effects on observed liquefaction performance in silt swamps during the Canterbury earthquake sequence. Soil Dyn. Earthq. Eng. **107**, 303–321 (2018)
7. Cubrinovski, M., Rhodes, A., Ntritsos, N., van Ballegooy, S.: System response of liquefiable deposits. Soil Dyn. Earthq. Eng. **124**, 212–229 (2019)
8. Geyin, M., Maurer, B.W., Bradley, B.A., Green, R.A., van Ballegooy, S.: CPT-based liquefaction case histories compiled from three earthquakes in Canterbury, New Zealand. Earthq. Spectra **37**, 2920–2945 (2021)
9. Boulanger, R.W., Idriss, I.M.: CPT-based liquefaction triggering procedure. J. Geotech. Geoenviron. Eng. **142**(2), 1–11 (2016)
10. Zhang, G., Robertson, P.K., Brachman, R.W.I.: Estimating liquefaction-induced ground settlements from CPT for level ground. Can. Geotech. J. **39**, 1168–1180 (2002)
11. Ntritsos, N.: System response of liquefiable deposits: insights from advanced analyses of case-histories from the 2010–2011 Canterbury earthquakes. Ph.D. thesis, University of Canterbury, Christchurch, New Zealand (2021)

Dynamic Behavior of Pipe Bend Subjected to Thrust Force Buried in Liquefiable Sand

Kohei Ono[✉] and Mitsu Okamura

Ehime University, Matsuyama, Japan
`ono.kohei.vb@ehime-u.ac.jp`

Abstract. At pressure pipe bends, thrust forces cause the lateral displacement of pipes and subsequent detachment of joints. The current design requires the calculation of the safety factor for pipe sliding based on the passive earth pressure behind the pipe bends; however, the resistance force of the ground and the dynamic behavior of the pipe during liquefaction are not elucidated. Therefore, efficient countermeasures and their design have not yet been established. The objective of this study is to understand the dynamic behavior of buried pipes subjected to thrust forces during liquefaction through a series of dynamic centrifuge tests. One-dimensional horizontal shaking is applied to the model at 40g whereas the lateral load, which simulates the thrust force, is applied to the model pipe. The lateral movement mechanism of the pipe is clarified by measuring the excess pore water pressure, acceleration response, and horizontal displacement of the pipe. Based on the test results, a countermeasure using a gravel layer is devised and its effectiveness is examined.

Keywords: Liquefaction · Centrifuge · Pipeline

1 Introduction

Agricultural pipelines involving various branches and bends are often arranged along complex terrains, from water sources to farmlands. Agricultural water is generally pressurized to several hundred kilopascals to ensure its efficient distribution. Therefore, an unbalanced force, known as the thrust force, is exerted toward the outside of the pipe bend owing to the imbalance of internal water pressure and centrifugal force by the flow. The thrust force is expressed as follows:

$$P = 2Ha_c \sin\frac{\theta}{2} \qquad (1)$$

where H is the internal water pressure (kPa), a_c the cross-sectional area of the pipe (m^2), and θ the bending angle of the pipe (°). Agricultural pipelines generally experience a greater thrust force than waterworks because of their larger diameter and higher internal water pressure. Consequently, the thrust force may cause the pipe to displace and detach. According to the current Japanese design guidelines for agricultural pipelines (MAFF

2009), the combined force of the passive earth pressure in the ground behind pipe bends, calculated using Eq. (2), is expected to be resistant to the thrust force.

$$R_h = 0.65 \frac{1}{2} K_p B_b \gamma' \left(H_2^2 - H_1^2 \right) \quad (2)$$

where K_p is the passive earth pressure coefficient, B_b the pipe length (m), γ' the effective unit weight of soil (kN/m^3), H_1 the depth of the soil cover (m), H_2 the depth from the ground surface to the bottom of the pipe (m), and 0.65 the correction factor for the passive earth pressure exerting on the curved surface of the pipe. The ratio of the thrust force in Eq. (1) into the lateral resistance force in Eq. (2) is the safety factor pertaining to the possibility of pipe sliding. When the safety factor is less than 1.5, additional thrust restraint must be provided to prevent pipe movement. Typically, a concrete block is enveloped around a pipe bend to increase the area where the passive earth pressure is exerted and to enhance the frictional resistance at the bottom. However, these design methods only guarantee safety against the displacement of pipe bends under normal conditions and do not consider seismic behaviors such as inertial forces and backfill deformation due to liquefaction. In a field survey of the 1993 earthquake southwest off Hokkaido, Mohri et al. (1995) reported that the liquefaction of a backfill caused a pipe bend to shift 0.6–0.8 m backward and subsequently detach. Reduced passive earth pressure behind a pipe bend due to liquefaction was not considered in the current design. Therefore, the seismic resistance of pipe bends must be improved.

Two possible scenarios might destabilize a buried pipe bend subjected to thrust forces: (1) liquefaction of the backfill material for pipe bends, and (2) liquefaction of natural ground. Although seismic damage has been observed in both cases, backfill liquefaction can be prevented relatively easily via the appropriate compaction of high-quality materials with high liquefaction strength. Ohta et al. (2021) conducted centrifuge model tests and reported that the displacement of a pipe can be maintained below the allowable value by backfilling the pipe with gravel on the passive side of the pipe when the natural ground is not liquefiable. Meanwhile, several thrust restraint methods have been devised for liquefiable natural grounds, e.g., wrapping a geogrid around a pipe bend (Kawabata et al. 2011), using a pipe with detachment prevention joints (Itani et al. 2016), and installing a gabion filled with gravel on the passive side (Araki and Hirakawa 2018). However, these design methods are yet to be well established.

This study examines thrust restraint using gravel as a backfill material for pipe bends buried in liquefiable natural ground. When the pipe is backfilled with gravel, the excess pore water pressure around the pipe is suppressed by the higher coefficient of permeability of gravel compared with that of sand. Moreover, gravel is expected to dissipate the excess pore water pressure of natural ground, similarly in gravel drains, as a liquefaction countermeasure. It is noteworthy that the dissipation effect of gravel drains depends significantly on the drain diameter and drain spacing (Minaka et al. 2021). However, a prominent dissipation effect is expected from this method because the backfilled area of the pipe is generally several times larger than the pipe diameter. This paper presents the results of a series of centrifuge model tests that evaluate both the effect of the thrust force on the dynamic displacement behavior of the pipe and the effect of backfilling the pipe with gravel.

2 Centrifuge Model

2.1 Model Configuration

Figure 1 shows the cross-section of the test model. A rigid container with internal dimensions of 510 mm × 230 mm × 120 mm was used. The front of the container was made of a transparent acrylic plate, which allowed the soil inside to be observed.

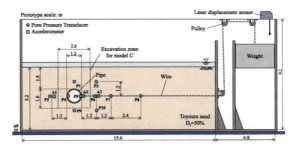

Fig. 1. Cross-section of test model.

Model Pipe. The model pipe was assumed to be rigid with no deformation in the cross-sectional direction. In practice, the thrust force exerts on the pipe bends; however, the model pipe was simplified to a plane strain condition, such that a straight pipe was modeled to reduce the factors affecting the displacement of the pipe, such as the bending angle or pipe length. A hollow aluminum rod with a diameter of 30 mm and length of 120 mm was used. The specific gravity of the pipe was adjusted to be approximately equal to that of saturated sand such that the effect of buoyancy during liquefaction was negligible. Both ends of the pipe were closed and covered with non-woven fabric to mitigate friction between the pipe end and container. Two methods can be used to apply a constant lateral load to the pipe for simulating the thrust force: one is to generate the thrust force by applying an internal water pressure to a pipe bend model (e.g., Kawabata et al. 2011), and the other is to apply a lateral load by converting the vertical load of a weight with pulleys (e.g., Ono et al. 2018). In the present experiments, the latter method, using a weight, was used to apply the lateral load because the model pipe was straight. A wire with a diameter of 1 mm and three pulleys were combined, and a weight made of a lead shot was attached to the end. The wire was passed through a polyethylene tube with an inner diameter of 2 mm to reduce the friction with the soil.

Soil Properties and Model Preparation. Toyoura sand whose properties are listed in Table 1, was used as the natural ground and backfill material. Ground with a target relative density of 50% was prepared via the wet tamping method. Two types of backfill conditions were modeled: a uniform condition with no distinction between the natural ground and backfill, and a condition where gravel was used as the backfill material. The pipe was embedded 1.8 m deep in the prototype scale, which was 1.5 times the pipe diameter. The model pipe, pore pressure transducers, and accelerometers were buried at the positions shown in Fig. 1. The pore pressure transducers (P4 and P5) and accelerometer (A2) were attached directly to the spring line of the pipe. The vertical displacement of the weight, as using the laser displacement sensor, was regarded as the displacement of the pipe. After the model was prepared completely, the air in the model was fully replaced with CO_2 in a vacuum chamber. The model was saturated by introducing a de-aired viscous fluid, whereas the pressure in the chamber was maintained at −95 kPa (Okamura and Inoue 2012). The viscous fluid was prepared by dissolving 1.8% methylcellulose by weight in water to achieve a kinematic viscosity of 40 cSt.

Table 1. Soi properties.

Specific gravity	2.66
Maximum void ratio	0.97
Minimum void ratio	0.61
Saturated unit weight (kN/m^3)	18.9

2.2 Test Procedure

Shaking tests were conducted under a centrifugal field of 40g using a geotechnical beam centrifuge at Ehime University. The mass of the weight for the lateral loading was adjusted such that the weight at 40g corresponded to the target thrust force. The combined force of the passive earth pressure under the present experimental condition, as calculated using Eq. (1), was 350 kN. Based on Eq. (2), a lateral load of 350 kN corresponds to the thrust force when an internal water pressure of 400 kPa is exerted on a 45° pipe bend with a diameter of 1.2 m.

After the centrifugal acceleration was gradually increased to 40g, one-dimensional horizontal shaking was applied to the model using an electronic mechanical shaker. The input base motion for the shaking event was a tapered sine wave with a maximum acceleration amplitude of 2.3 m/s^2 and dominant frequency of 1.3 Hz. The input motions were the same for all the tests. The directions of the shaking and lateral loading were parallel, which implies that inertial force due to shaking was applied to the pipe. Under the present experimental conditions, the maximum inertia force exerting on the pipe was approximately 20 kN, which was approximately 5% or 10% of the thrust force.

Table 2. Parameters of test models.

Model	Lateral load (prototype)	Backfill material
A	200 kN	Sand (Toyoura sand)
B	400 kN	Sand (Toyoura sand)
C	200 kN	Gravel (Ube-Keisa No. 1)

The test models are presented in Table 2. Models A and B involved the same backfill conditions, but different magnitudes of the lateral load. Meanwhile, in Model C, an area 2.6 m wide and 3.4 m deep, as shown in Fig. 1, was excavated after uniform sand layer was made and backfilled with Ube-Keisa No.1 gravel. The coefficient of permeability of the gravel was approximately 100 times higher than that of Toyoura sand.

All the test results are shown in the prototype scale, unless otherwise mentioned.

3 Results and Discussion

3.1 Response of Ground and Lateral Displacement of Pipe

Pore Water Pressure and Acceleration Response. Figure 2 shows the time histories of the excess pore water pressure ratio r_u of Model A. The vertical effective stress for calculating r_u was the effective overburden pressure in the initial state. The shaking event was initiated at $t = 10$ s, and excess pore water pressure began to generate at approximately $t = 12$ s. Meanwhile, r_u reached unity at approximately $t = 18$ s, except for P3 and P4, which were embedded on the active side of the pipe. The sand on the active side did not liquefy because of the dilative response of sand caused by the movement of the pipe, as will be described later. The r_u of P5 and P6 at the passive side of the pipe exceeded unity, owing to sand compression or sensor settlement caused by the pipe movement.

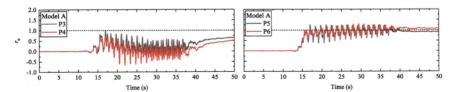

Fig. 2. Time history of excess pore water pressure ratio in Model A.

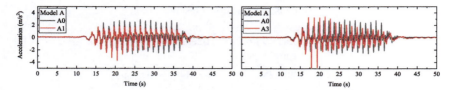

Fig. 3. Time history of acceleration response in Model A.

Figure 3 shows a comparison between the input acceleration measured using the accelerometer attached on the shaker and the acceleration response of Model A. The accelerations at A1 on the active side indicated the same phase with the input acceleration. The negative response was significant immediately after shaking commenced, indicating the lateral movement of the sand. At A3, typical dilation spikes were observed at approximately $t = 18$ s, which is consistent with the time when the r_u of P6 at the same location decreased. Subsequently, the phase difference with the input acceleration increased and the response acceleration declined, indicating the typical softening of the ground due to liquefaction.

Lateral Displacement of Pipe. Figure 4 shows a comparison of the lateral displacements of the pipe. The displacement was defined as positive (forward) in the direction where the lateral load was exerted and negative (backward) in the opposite direction. The displacements of the pipe accumulated on the forward side when the forward and backward movements were repeated, whereas the average velocity was almost constant. The displacement of Model B, which was subjected to a greater lateral load, was larger than that of Model A, indicating that the displacement depended on the magnitude of the thrust force. The pipes almost stopped moving at approximately $t = 38$ s, i.e., when shaking ceased.

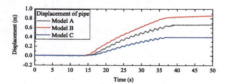

Fig. 4. Time history of lateral displacement of pipes.

Figure 5 shows the relationships between the displacement of the pipe and r_u of P5 at the passive side. Since the soil deformation of Model C differed completely to that of Model A and B, the result of Model C is not discussed herein. The relationships reveal the following three basic dynamic behaviors of the pipe: First, the displacement of the pipe increased after r_u increased to approximately unity. Second, whereas r_u repeatedly increased and decreased because of the shaking, the pipe did not displace when r_u decreased below a certain level. Third, r_u remained high for a certain duration after shaking ended, but the pipe did not displace further. The first and second findings indicate the possibility of reducing the displacement of the pipe by suppressing the increase in the excess pore water pressure. The third finding indicates that the displacement of the pipe may be affected not only by the decrease in the soil strength, but also by the inertia force due to seismic waves.

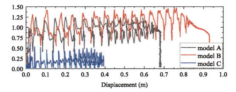

Fig. 5. Relationship between displacement of pipe and r_u at P5.

3.2 Thrust Restraint with Gravel

As shown in Fig. 4, the ultimate lateral displacement of the pipe in Model C, which was backfilled with gravel, was suppressed to approximately 60% that in Model A when subjected to the same lateral load.

Figure 6 shows a comparison of the distribution of the maximum r_u for Model A and C at the same pipe level. The distance in the longitudinal direction was the initial configuration, and $r_{u\,max}$ was expressed as unity when it exceeded unity. In Model C, the r_u inside the gravel was predictably low and maintained low up to a distance of approximately 2 m from the pipe. Even when r_u reached unity, the rate of increase in the excess pore water pressure was lower than that of Model A, indicating the dissipation effect of the excess pore water pressure.

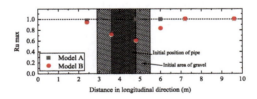

Fig. 6. Distribution of maximum r_u in longitudinal direction.

Figure 7 shows the images of Models A and C after shaking ceased. The colored sands clearly show the difference in deformation between the two cases. The vertical-colored lines on the right side of the pipe shows that the uniform sand of Model A deformed into a convex shape to the right, whereas the sand at shallower depth shifted in the direction opposite to that of the pipe movement. On the other hand, no significant deformation of the gravel and the surrounding sand was observed in Model C. The gravel and pipe shifted simultaneously, and the sand distant from the pipe deformed because the shear strength of the gravel was higher than that of the liquefied sand.

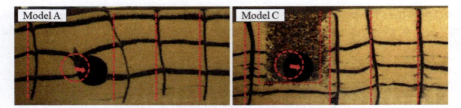

Fig. 7. Images of model after shaking ceased.

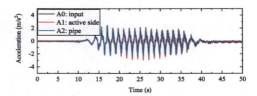

Fig. 8. Time history of acceleration response of pipe.

The displacement of Model C was suppressed owing to the increase in the lateral resistance force due to the expansion of the vertical plane, which maintained the effective stress. The effect of this countermeasure is comparable to that of a general thrust restraint enveloped by concrete blocks. Kawabata et al. (2011) reported that thrust restraints using concrete blocks increased the lateral displacement of a pipe owing to the phase difference between the pipe and ground on the active side, which was caused by the increased inertia force during liquefaction. In the proposed method, however, no phase difference was observed between the accelerations of the pipe and ground on the active side, as shown in Fig. 8, because the gravel and pipe were not bonded. The experimental results showed that backfilling the pipe with gravel reduced the displacement of the pipe including when the natural ground liquefied. Hence, this method is potentially better than using thrust restraints based on concrete blocks.

4 Conclusions

A series of centrifuge test results for investigating the dynamic behavior of a buried pipe subjected to thrust force was reported herein. The pipe subjected to a lateral load accumulated lateral displacement during shaking under the effects of the increase and decrease in the excess pore water pressure on the passive side and the inertia force. When the pipe was backfilled with gravel, the ultimate displacement reduced to 60% because of the expansion of the area subjected to the resistance force by maintaining the effective stress around the pipe. Further studies are required to model the proposed thrust restraints.

References

Mohri, Y., Yasunaka, M., Tani, S.: Damage to buried pipeline due to liquefaction induced performance at the ground by the Hokkaido-Nansei-Oki Earthquake in 1993. In: Proceedings of 1st International Conference on Earthquake Geotechnical Engineering, pp. 31–36 (1995)

Ministry of Agriculture, Forestry and Fisheries of Japan (MAFF): Planning and design criteria of land improvement project (pipeline), JSIDRE, 321-409 (2009). (in Japanese)

Ohta, Y., Sawada, Y., Ariyoshi, M., Mohri, Y., Kawabata, T.: Effects of gravel layer as thrust restraint for pipe bends subjected to earthquake loading. Int. J. Phys. Model. Geotech. **22**(2), 99–110 (2022)

Kawabata, T., Sawada, Y., Mohri, Y., Ling, H.I.: Dynamic behavior of buried bend with thrust restraint in liquefying ground. J. Jpn. Soc. Civ. Eng. Ser. C **67**(3), 399–406 (2011). (in Japanese)

Itani, Y., Fujita, N., Ariyoshi, M., Mohri, Y., Kawabata, T.: Dynamic behavior of flexibly jointed pipeline with a bend in liquefied ground. IDRE J. **301**, 1–8 (2016). (in Japanese)

Araki, H., Hirakawa, D.: Model tests on thrust protecting method for buried pipe by using geogrid cabion. J. Jpn. Soc. Civ. Eng. Ser. C **74**(1), 106–117 (2018). (in Japanese)

Minaka, U., Okamura, M., Ono, K.: Verification of effectiveness and design procedure of gravel drains for liquefaction remediation. Soils Found. **61**(5), 1191–1206 (2021)

Ono, K., Yokota, Y., Sawada, Y., Kawabata, T.: Lateral force-displacement prediction for buried pipe under different effective stress condition. Int. J. Geotech. Eng. **12**(4), 420–428 (2018)

Okamura, M., Inoue, T.: Preparation of fully saturated model for liquefaction study. Int. J. Phys. Model. Geotech. **12**(1), 39–46 (2012)

Liquefaction Resistance of Solani Sand Under Normal and Sequential Shaking Events

Gowtham Padmanabhan and B. K. Maheshwari[✉]

Department of Earthquake Engineering, Indian Institute of Technology Roorkee, Roorkee, Uttarakhand, India
bk.maheshwari@eq.iitr.ac.in

Abstract. Earthquake damage caused by liquefaction is intense and innumerable. This led to the research worldwide on understanding the mechanism and designing suitable ground improvement systems. In the present study, 1g shake table experiments were conducted to investigate the liquefaction resistance and the intrinsic mechanism associated with the sequential shaking events compared to normal shaking events. A tank of dimension 1 × 0.6 × 0.6 m was used for preparing the saturated sand-bed of 25% relative density. Poorly graded liquefiable Solani sand collected from the bed of Solani river near Roorkee, India was used for testing. A total of 8 shaking events were conducted with varying acceleration amplitudes from 0.1g to 0.4g under constant frequency 2 Hz and 1-min shaking duration. In sequential shaking events, the prepared sand-bed was shaken four times in succession with increment in acceleration after the dissipation of pore water pressure. The variation in the generated excess pore-water pressure was monitored continuously using pore-pressure transducers placed at different depths. Sand-bed subjected to sequential events were found susceptible to reliquefaction despite improved sand density under incremental acceleration loading. The experimental data shows that liquefaction resistance increased under the sequential shaking compared to normal events even at higher acceleration amplitude.

Keywords: Shaking table · Liquefaction resistance · Sequential acceleration

1 Introduction

Soil liquefaction is a common geotechnical hazard during earthquakes, which occurs in saturated, loose sand deposits. Lateral spreading, structural and foundation damages and ground subsidence are the common damages observed during liquefaction. In recent years, the occurrence of repeated seismic events resulted in the reliquefaction phenomenon. Damages that occurred during the 2010–2011 Christchurch, 2011 Tohoku, and 2016 Kumamoto earthquakes bear testimony to reliquefaction-induced damages [1]. In India, the recent 2021 Assam earthquake series witnessed repeated liquefaction and associated damages. However, limited research works have been reported in understanding the mechanism of reliquefaction subjected to repeated shaking events and difference in liquefaction potential of sand deposits subjected to normal and repeated shaking events.

© The Author(s), under exclusive license to Springer Nature Switzerland AG 2022
L. Wang et al. (Eds.): PBD-IV 2022, GGEE 52, pp. 1656–1663, 2022.
https://doi.org/10.1007/978-3-031-11898-2_147

Some researchers focused on understanding the induced reliquefaction resistance through repeated seismic shaking events. Ha et al. [2] reported liquefaction/reliquefaction resistance of sand with different gradation characteristics using shaking table experiments. Reliquefaction resistance trend was found different from the past published works. Through field observations and centrifuge experiments, El-Sekelly et al. [3] discussed the effect of Preshaking followed by extensive liquefaction. Authors stated that, liquefaction resistance of silty sand deposit increased significantly after each earthquake, which predominantly of lower magnitude and do not extensively liquefy. With help of shaking table experiments Ecemis et al. [4] assessed the reliquefaction potential of sands with varying relative density of sand deposits and concluded that apart from density and initial sand properties, consolidation characteristics also influence the reliquefaction resistance of sand deposits. Ye et al. [5] investigated the macroscopic and mesoscopic mechanism of induced sand liquefaction resistance through seismic Pre-shaking. The authors concluded that the induced anisotropy and sand fabric caused by seismic pre-shaking events are responsible for sand resistance. The seismic Preshaking is defined here by the total number of earthquakes shaking applied to the sand deposit in a specified time period. With help of field evidences El-Sekelly et al. [6] substantiated that the lower bound of the acceleration amplitude should be a minimum of 0.10 g to affect the critical layer of the soil deposit and to generate sufficient excess pore water pressure to induce liquefaction. Padmanabhan and Shanmugam [7] examined the behavior of Solani sand under repeated shaking events and reported decrease in reliquefaction potential when subjected to increased acceleration amplitude despite a significant increase in sand density due to shaking.

The objective of this study is to investigate the liquefaction behavior under normal and repeated shaking events. Shaking table experiments were performed to demonstrate the effect of sequential acceleration during repeated shaking events and its influence on reliquefaction potential. The measured excess pore water pressure and soil displacement are presented and discussed the critical insights behind the varied liquefaction/reliquefaction resistance subjected to normal and repeated shaking events.

2 Experimental Setup

Experimental works were conducted by 1-g uniaxial shake table available in the Soil dynamics laboratory, IIT Roorkee, India. Figure 1 shows the experimental test setup used in the study. The shake table can operate in a frequency range of 0 to 5 Hz and an acceleration amplitude (0.05 to 1g). The shake table consists of a tank of dimension $1 \times 0.6 \times 0.6$ m which was used to prepare the saturated sand-bed of 25% relative density. The pressure transducers were installed at different depths 40, 200, and 360 mm from the bottom of the tank representing the bottom (B), middle (M), and top point (T) respectively as shown in Fig. 1b.

To counter the rigid boundary effects and constraint on horizontal movement, the pore pressure response was measured at the center of the tank. Sinusoidal harmonic motion was applied at the base of the tank to impart seismic motion under constant excitation frequency. Considering the capacity of the shaking table and the dimension of the prototype, similarity ratio of the geometry type was determined to be 1:50 [8]. The

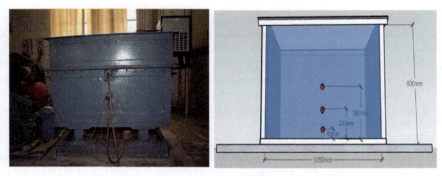

Fig. 1. Experimental setup (a) 1-g shake table; (b) instrumentation details

similarity law was proposed based on Buckingham π theory with the assumption that the soil obeys equivalent viscoelastic criteria under dynamic conditions [9]. In the shaking table experiments, acceleration amplitude and shaking duration were set as input control variables.

Test sand was procured from Solani riverbed near Roorkee, Uttarakhand, India and to be referred to as Solani sand hereinafter. From grain size analysis, it is found to be poorly graded sand and lies in the range of liquefiable soils (Fig. 2). The index properties of the sand were estimated in the laboratory and listed in Table 1. The amount of sand and water required to fill the tank to the desired depth was estimated using the procedure adopted by Maheshwari et al. [10]. Then, the estimated total quantity of sand and water was divided into three layers and the ground bed was formed layer by layer to achieve more uniformity in sample preparation. First, the container was filled with the calculated quantity of water to the pre-calculated depth. Then, sand was rained down into the container from a calculated height through a conical hopper arrangement for achieving uniform distribution of loose soil. The height at which the sand is to be poured for achieving respective density was pre-calculated by performing repeated relative density tests. The sample preparation procedure was repeated up to a total depth of 570 mm. The repeated acceleration loading was applied only after the complete dissipation of excess pore water pressure generated from previous acceleration loading.

A total of 8 shaking events (4 events of normal shaking and 4 events of repeated shaking) were conducted with an acceleration amplitude varying from 0.1g to 0.4g under constant frequency 2 Hz and 1-min shaking duration. Repeated shaking test series was proposed for stimulating foreshocks and its associated main-shock event. The acceleration loading was applied from 0.1g to 0.4g with an incremental interval of 0.1g. Next incremental acceleration loading of 0.2g was applied only after 24 h to allow complete dissipation of pore water pressure. Similar procedure was repeated for 0.3g, and then to 0.4g acceleration loading. Soil settlement was measured after leveling the ground surface and dissipation of pore pressure.

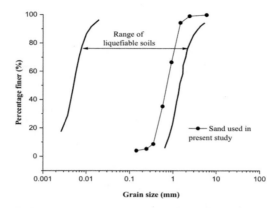

Fig. 2. Grain size distribution curve of Solani sand

Table 1. Index properties of Solani sand

S. No	Property		Value
1	Soil type (SP)		Poorly graded
2	Specific Gravity (G)		2.64
3	Uniformity coefficient (C_u)		2.26
4	Coefficient of curvature (C_c)		1.00
5	Grain size	D_{60}	0.262 mm
		D_{10}	0.108 mm
6	Maximum void ratio (e_{max})		0.923
7	Minimum void ratio (e_{min})		0.580

3 Results and Discussions

The soil is said to be liquefiable based on the generated pore pressure ratio (r_u) defined as ratio of excess pore water pressure (U_{excess}) to the effective overburden pressure (σ'_{vo}) at a particular depth defined as follows:

$$r_u = \frac{U_{excess}}{\sigma'_{vo}}$$

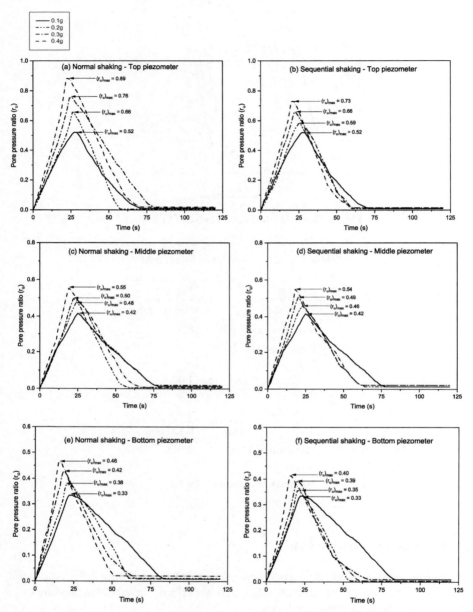

Fig. 3. Time history of generated pore pressure ratio for normal and sequential shaking

The shear stress will not be uniform throughout the depth; however, it is compensated by the different effective overburden pressure obtained at different depths. The time history of pore pressure ratio subjected to normal and repeated shaking events were presented in Fig. 3. The generation of pore water pressure during sequential shaking induces non-uniformity in sand deposition and anisotropy, which in turn increases the potential to reliquefaction. The increase/decrease in liquefaction resistance was decided based on the extent of pore water pressure ratio. The liquefaction occurred due to preshaking/foreshocks reduced the resistance to liquefaction in the further shaking events. Here, only the partial liquefaction was observed during the incremental shaking events (0.1g to 0.4g). The effect of preshaking increased the resistance to liquefaction under sequential shaking events. The time taken to attain maximum pore pressure ratio decreased with the increase in acceleration for both normal and sequential shaking events. Similar observation was also reported by Padmanabhan and Shanmugam [11].

The possibility of liquefaction was found to be higher under normal shaking compared to sequential shaking events. For instance, the pore pressure ratio for 0.4g loading at top piezometer was found to be 0.89 and 0.73 for normal and sequential shaking respectively. In addition, the time taken to attain maximum pore pressure ratio was bit higher in case of sequential shaking. It infers that the preshaking is beneficial in increasing the resistance to liquefaction, when subjected to repeated seismic events.

Figure 4 shows the difference in the generated pore pressure ratio at top piezometer subjected to repeated and normal shaking events. The 0.1g acceleration was not presented; as loading was similar in both the normal and repeated shaking events. Top piezometer was selected for comparison, because it was found to be more vulnerable to liquefaction compared to middle and bottom piezometers due to low overburden pressure at the top surface. Another reason is that, the liquefaction at shallow depth results in failure of built environment located with shallow foundations. The generated pore pressure ratio was smaller for the sequentially shaken sand-bed compared to that for normally shaken irrespective of acceleration amplitude. The percentage increment in sand reliquefaction resistance was found to be 11%, 13% and 18% at top piezometer subjected to 0.2g, 0.3g and 0.4g acceleration respectively. In case of middle and bottom piezometers, the resistance to generation of pore pressure ratio was in the range of (2 to 4%) and (8 to 13%) respectively as evident from Fig. 3. Increase in resistance to liquefaction was observed higher at top piezometer. Resistance was contributed due to rearrangement of soil grains and sand densification due to preshaking. The variation in soil settlement after each shaking event (Fig. 5) confirms the reconsolidation and increased sand density.

From the experimental results, it can be concluded that the acceleration amplitude was critical in influencing the excess pore water pressure generation compared to increased sand density due to sequential shaking and the effect of seismic pre-shaking. Experimental evidences obtained from the present study finds good correlation with the case study findings reported by Padmanabhan and Maheshwari [1]. The case study reports that there was a significant increase in the liquefaction and reliquefaction resistance of sand deposits to both mainshock and subsequent aftershock events; as the Kumamoto deposit was exposed to an increasing number of earthquakes.

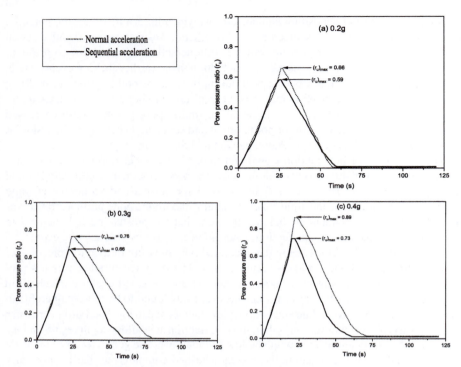

Fig. 4. Comparison of pore pressure ratio for normal and sequential shaking for top piezometer (a) 0.2g; (b) 0.3g; and (c) 0.4g.

Fig. 5. Variation of soil settlement after shaking (a) first event; (b) second event.

4 Conclusions

1. Acceleration amplitude is an important parameter in controlling the generation of excess pore water pressure even under repeated shaking events. In this case also, the magnitude of excess pore water pressure ratio increases with the increase in acceleration amplitude.

2. Due to the effect of seismic pre-shaking and achieved sand densification, the generated pore water pressure is less in the repeated shaking compared to the normal shaking events. Preshaking is the major difference between the normal and sequentially accelerated sand deposits. The increased reliquefaction resistance was contributed due to rearrangement of soil grains and resulting sand densification.

References

1. Padmanabhan, G., Maheshwari, B.K.: Case studies on preshaking and reliquefaction potential for different earthquakes in Japan. In: Sitharam, T.G., Jakka, R., Govindaraju, L. (eds.) Local Site Effects and Ground Failures. Lecture Notes in Civil Engineering, vol. 117. Springer, Singapore (2021). https://doi.org/10.1007/978-981-15-9984-2_12
2. Ha, I.S., Olson, S.M., Seo, M.W., Kim, M.M.: Evaluation of reliquefaction resistance using shaking table tests. Soil Dyn. Earthq. Eng. **31**(4), 682–691 (2011)
3. El-Sekelly, W., Abdoun, T., Dobry, R.: Liquefaction resistance of a silty sand deposit subjected to preshaking followed by extensive liquefaction. J. Geotech. Geoenviron. Eng. **142**(4), 04015101 (2016)
4. Ecemis, N., Demirci, H.E., Karaman, M.: Influence of consolidation properties on the cyclic re-liquefaction potential of sands. Bull. Earthq. Eng. **13**(6), 1655–1673 (2014)
5. Ye, B., Hu, H., Bao, X., Lu, P.: Reliquefaction behavior of sand and its mesoscopic mechanism. Soil Dyn. Earthq. Eng. **114**, 12–21 (2018)
6. El-Sekelly, W., Dobry, R., Abdoun, T., Steidl, J.H.: Centrifuge modeling of the effect of preshaking on the liquefaction resistance of silty sand deposits. J. Geotech. Geoenviron. Eng. **142**(6), 04016012 (2016)
7. Padmanabhan, G., Shanmugam, G.K.: Reliquefaction assessment studies on saturated sand deposits under repeated acceleration loading using 1-g shaking table experiments. J. Earthq. Eng. **26**(6), 2888–2910 (2020)
8. Zhou, Z., Lei, J., Shi, S., Liu, T.: Seismic response of aeolian sand high embankment slopes in shaking table tests. Appl. Sci. **9**(8), 1677 (2019)
9. Wang, J., Yao, L., Hussain, A.: Analysis of earthquake-triggered failure mechanisms of slopes and sliding surfaces. J. Mt. Sci. **7**(3), 282–290 (2010)
10. Maheshwari, B.K., Singh, H.P., Saran, S.: Effects of reinforcement on liquefaction resistance of Solani sand. J. Geotech. Geoenviron. Eng. **138**(7), 831–840 (2012)
11. Padmanabhan, G., Shanmugam, G.K.: Liquefaction and reliquefaction resistance of saturated sand deposits treated with sand compaction piles. Bull. Earthq. Eng. **19**(11), 4235–4259 (2021). https://doi.org/10.1007/s10518-021-01143-8

Numerical Simulation of Caisson Supported Offshore Wind Turbines Involving Uniform Liquefiable Sand Layer

Alfonso Estepa Palacios, Manh Duy Nguyen, Vladimir Markovic, Sina Farahani, Amin Barari(✉), and Lars Bo Ibsen

Department of the Built Environment, Aalborg University, Thomas Manns Vej 23, 9220 Aalborg, Denmark
abar@build.aau.dk

Abstract. In this research, a series of the non-linear finite element (FE) analyses was conducted to analyze the influence of the contact pressure caused by the offshore wind turbine and motion characteristics on the settlement pattern and seismic demand of the structure. The procedure was validated against a database of well-documented centrifuge test. The FE results suggested that motion, and offshore wind turbine (OWT) characteristics greatly control liquefaction-induced OWT settlement.

Keywords: Offshore wind turbine · Suction caisson foundation · Nonlinear finite element · Arias Intensity

1 Introduction

Renewable energy has been emerging widely in developed countries in recent years and has been accepted as a more sustainable and ecological source of energy compared to the ones from fossil fuels. Even though a research and employment of renewable energy are steadily on the rise, it is still a long way to go to make it as a leading source of energy. Wind energy has shown optimistic results as it is seen as a cost-effective solution, thus a lot of research in that field is generated nowadays. As the deployment of offshore wind energy rises globally, there is an urgent need for exploring the rip coastal waters for the installation of wind farms where energy could be generated efficiently. Many of these sites lie in regions where seismic activity presents a significant threat to the stability and normal operation of the turbines [1–6]. Figure 1 shows the major offshore wind farm developments together with the global seismic hazard map in terms of peak ground acceleration (PGA). By far, most of the performed research has been exclusively related to either inertial interaction of ground founded OWTs or kinematic seismic response of foundations [2, 7, 8], in which studies have mostly focused on how the eccentric RNA to the tower top and rotary inertia due to blades can influence the structural response of the wind turbines or the forces induced to the piles connected to rigid pile caps while following the ground motions.

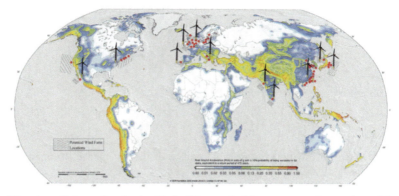

Fig. 1. Mapping of planned and submitted offshore wind farms proposals together with seismic hazard map according to the peak ground acceleration (PGA) [1]

A few studies have considered the soil-structure interaction problems of grounded offshore wind turbine (OWT) systems in liquefiable soils and many issues are still uncertain [4, 5, 9, 10]. Therefore, this study aims at investigating the seismic behavior of caisson-supported offshore wind turbine systems resting on liquefiable soils to provide insights into the key earthquake and state parameters governing the liquefaction-induced OWT settlement.

2 Numerical Predictions for Centrifuge Tests

2.1 Centrifuge Test Configuration and FEM Modeling Considerations

The centrifuge tests were conducted by Yu et al. [11] at the laboratory of Case Western Reserve University to evaluate the seismic behavior of the wind turbine with a suction caisson foundation resting on the liquefiable soil. The dimensions of the suction caisson model included a diameter and skirt length of 4 and 1.75 m (referred to as Test 1), respectively. Suction caisson weight was considered as 10.7 t. For the sake of simplicity, a concentrated lumped mass is assigned at the top of the tower of 10.6 t simulating the weight (W) of the rotor, blades, gearbox, and nacelle of the wind turbine, as shown in Fig. 2.

The rigid container was filled out by the well-graded Toyoura sand with $D_{50} = 0.17$ mm, which is poured from the level of 80 cm for ensuring the value of relative density equivalent to 68% consistently [11]. Moreover, the saturation process was conducted by using a de-aired water system and professional vacuum device in at least 24 h for getting the simulation as realistic as possible. The testing model was instrumented with accelerometers located 0.5 m below the lid and 0.5 m beneath the ground surface in the free field denoted by A1 and A2, respectively. As far as the soil domain is concerned, the constitutive model of pressure-dependent multi-yield surface (PDMY01) is utilized to simulate saturated sand.

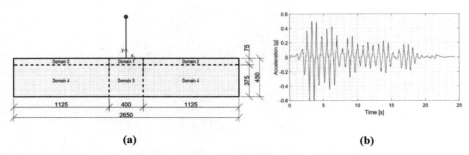

Fig. 2. (a) Domains distribution and mesh generation of the soil model. (b) Input earthquake applied to the model

In principle, this constitutive model is an extension of an original multi-surface plasticity concept, with flow and hardening rules [13] incorporated. The calibrated set of parameters of PDMY 01 material are reported in Table 1. The model has been implemented into OpenSees software [14], and its particular advantage is that it accounts for the accumulation of irreversible cycle-by-cycle shear strain, especially in clean medium-to-dense sands [15, 16]. In order to simulate properly the soil domain, QuadUP elements are chosen since they bring the possibility of simulating the generation of pore water pressure in saturated soil. QuadUP element is a four-node plane strain element suitable for 2D simulations.

Since the value of permeability is expected to differ depending on the depth and position of a specific point corresponding to the structure as previously discussed in [12], the soil model is divided into four different domains with the intention of having a higher control of the permeability, as indicated in Fig. 2.

2.2 Structure Model, Boundary Conditions and Input Excitation

In order to model the structural members of the wind turbine as well as the caisson bucket, the beam-column elements available in OpenSees [14] are used. Once having completed the definition of the soil and the structure domain, the interface between structure and soil is modeled by using the equalDOF command technique [16, 17]. As shown in Fig. 3, the nodes of the bucket surface (i.e., slave nodes) are restrained to follow the same displacements as the soil (i.e., master nodes). The input motions are applied to the model by the use of equivalent forces acting along both lateral and bottom boundaries according to method adopted in Joyner [18] and Ayala and Aranda [19]. In contrast to other approaches, the equivalent forces in this case are not applied only to the bottom of the model simulating the performance of a real earthquake. This fact is well-understood once one takes into account that the input motion received does not correspond to a time history recorded at the surface of the free field that can be deconvoluted and applied to a chosen depth. Thus, the above-mentioned equivalent forces are calculated applying the Eq. (1) in the case of the bottom boundaries $F_{e.s}$ or the Eq. (2) in the case of lateral boundaries $F_{e.p}$, as follows:

$$F_{e.s} = V_t \vartheta_s \rho A \tag{1}$$

$$F_{e.p} = V_t \vartheta_p \rho A \qquad (2)$$

where ϑ_s and ϑ_p denote shear wave velocity and compressive wave velocity for the rock medium in the same order, while ρ and A indicate density of the rock medium and tributary area of each node, respectively.

As observed in the expressions above, the input ground motion needs to be transformed from acceleration time-history to the equivalent velocity time-history (V_t).

2.3 Numerical Prediction Results

After analyzing the numerical model, spectral accelerations were computed for comparison purposes. A comparison of the OWT settlement measured at the centrifuge facility with that obtained by the numerical model is illustrated in Fig. 4 as well. One can immediately infer from Fig. 4 that the numerical model is able to predict the trend of all the recorders received from the centrifuge test.

Table 1. Calibrated set of parameter of PDMY 01 material

Parameters	Value	Unit
Density	2.5	$[ton/m^3]$
Reference shear modulus	140000	[kPa]
Reference bulk modulus	170000	[kPa]
Friction angle	24	[°]
Peak shear strain	0.1	[-]
Reference mean effective confining pressure	80	[kPa]
Pressure dependent coefficient	0.5	[-]
Phase transformation angle	20	[°]
Contraction parameter	0.08	[-]
Dilation coefficient 1	0	[-]
Dilation coefficient 2	0	[-]
Liquefaction coefficient 1	10	[-]
Liquefaction coefficient 2	0.02	[-]
Liquefaction coefficient 3	1	[-]
Number of yielding surfaces	20	[-]
Initial void ratio	0.55	[-]
Parameter straight critical state 1	0.9	[-]
Parameter straight critical state 2	0.02	[-]
Parameter straight critical state 3	0.7	[-]
Atmospheric pressure for normalization	101	kPa

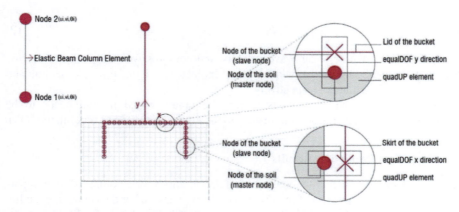

Fig. 3. Illustration of the structural elements and interface between soil and bucket foundation

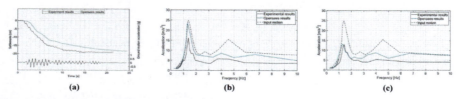

Fig. 4. Numerically computed and experimentally measured (a) average foundation settlement (b) and (c) comparison of FE-computed 5%-damped acceleration response to those measured in centrifuge experiment spectral acceleration at locations at locations A1 and A2 together with that of the base input motion

3 Effect of Contact Pressure and Seismic Demand on the Liquefaction-Induced Settlements

The benchmark numerical model was further evaluated under different input motions for five different values of masses at the top of the tower $m = 10$ t, 20 t, 29.3 t, 40 t and 50 t inducing the contact pressures 7.8 kPa, 15.61 kPa, 23.41 kPa, 31.22 kPa and 39.02 kPa to the soil underneath the caisson lid. It is worthwhile noting that the contact pressure is defined as pressure induced between the caisson foundation and the underneath soil. Table 2 provides details of selected earthquakes to perform nonlinear time-history analysis for evaluating the effects of input motion characteristics on the liquefaction-induced settlement history. All bedrock records were selected from the PEER NGA [20] database.

For the sake of clarity, the settlement time history recorded at the numerical model for five contact pressures as noted above due to the R10-San Fernando seismic scenario are plotted together with its corresponding Arias Intensity-time history, see Fig. 5.

It should be mentioned that Arias Intensity I_a is an index which represents the energy of the ground motion, and its value can be obtained through the Eq. (3):

$$I_a(T) = \frac{\pi}{2g} \int_0^\pi a^2(t)dt \qquad (3)$$

where g, a and T are gravity acceleration, acceleration value and time period, respectively.

Of particular interest here in this section, is to represent significant trends for liquefaction induced ground settlement against contact pressure P. Intuitively, the OWT settlement-time history pattern follows Arias Intensity-time history trend. Thus, it can be concluded that there is an obvious relationship between the settlement pattern and Arias intensity/time history of the corresponding motions. The results suggested that even though settlement increases proportionally with increasing contact pressure P, there is a clear adverse effect observed for higher values of P. The obtained results reveal that there is a threshold after a certain contact pressure (i.e., 31.22 kPa) beyond which any further contact pressure increase adversely influences the settlement trend.

In order to further investigate the effect of induced contact pressures on the liquefaction-induced settlement, the additional simulations with different masses of 60, 70, 80, 90, and 100 t which provide contact pressures of 46.83, 54.63, 62.43, 70.24, and 78.04 kPa were considered, respectively. Figure 6 illustrates representative trends for liquefaction-induced settlement versus contact pressure for a subset of the analyses. Also included is the Shaking Intensity Rate (SIR) parameter which is used to quantify the rate of earthquake energy as follows:

$$SIR = \frac{I_{a5-75}}{D_{5-75}} \quad (4)$$

where I_{a5-75} and D_{5-75} denote a change in Arias intensity from 5% to 75% of its total value and corresponding time duration, respectively.

It turns out that a threshold is not always present as the contact pressure increases; however, this controversial observation may be attributed to the Arias Intensity I_a or the corresponding SIR of motion. It is evident that for the higher values of I_a and SIR, the development of the settlements may be associated with a linear or a polynomial model, as illustrated in Figs. 6e and 6f. On the other hand, if the records with intermediate values of I_a or SIR are utilized to analyze the models, only a linearly increasing trend can be observed in the case of SIR, whereas linear and exponential trends can be found in the

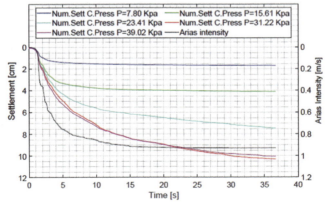

Fig. 5. Settlements time history and Arias intensity time history of the event R10 - San Fernando

Table 2. Characteristics of the input motions

ID	NGA NO.	Event	Year	M_w	Arias Intensity (m/s)	R(KM)	V_{S30} (m/s)
R0	–	Artificial	–	–	6.30	–	1000.00
R1	934	Big Bear-01	1992	6.46	0.1	35.41	659.09
R2	1245	Chi-Chi, Taiwan	1999	7.62	0.1	37.72	804.36
R3	285	Irpinia, Italy-01	1980	6.9	0.4	8.18	649.67
R4	1111	Kobe, Japan	1995	6.9	3.4	7.08	609.0
R5	1126	Kozani, Greece-01	1995	6.4	0.3	19.54	649.67
R6	801	Loma Prieta	1989	6.93	1.3	14.69	671.77
R7	454	Morgan Hill	1984	6.19	0.1	14.83	729.65
R8	497	Nahanni, Canada	1985	6.76	0.2	4.93	605.04
R9	1041	Northridge-01	1994	6.69	0.3	35.53	680.37
R10	71	San Fernando	1971	6.61	0.9	13.99	602.1
R11	8158	Christchurch, New Zealand	2011	6.2	5.7	2.52	649.67
R12	143	Tabas, Iran	1978	7.35	11.8	1.79	766.77
R13	73	San Fernando	1971	6.61	0.2	17.2	670.84
R14	6891	Darfield, New Zealand	2010	7.0	0.4	43.6	638.39
R15	5685	Iwate, Japan	2008	6.9	0.5	57.15	859.19
R16	897	Landers	1992	7.28	0.1	41.43	635.01
R17	369	Coalinga-01	1983	6.36	0.3	27.46	648.09
R18	501	Hollister-04	1986	5.45	0.1	12.32	608.37
R19	4872	Chuetsu-oki, Japan	2007	6.8	0.3	27.3	640.14
R20	1633	Manjil, Iran	1990	7.37	7.5	12.55	723.95

case of I_a, as shown in Figs. 6c and 6f. According to Figs. 6a and 6b, the less intense earthquakes with the lowest SIR values resulted in the exponential relation.

As can be seen in Fig. 7, the total value of settlements of the structure obtained for each record presented in Table 2 was compared with the corresponding Arias Intensity I_a and Shake Intensity Ratio SIR.

One can readily observe in Fig. 7a that increasing the value of Arias Intensity I_a leads to an increase in the settlements as well. Thus, the wind-turbine structures subjected to seismic input with $I_a > 1$ m/s underwent greater settlements in most of the studied models.

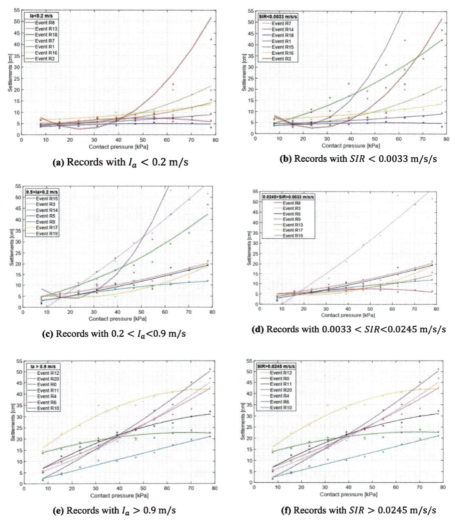

Fig. 6. Relation between settlements and contact pressure of the structure

This observation is less clear when the settlements are plotted against the SIR parameter, as illustrated in Fig. 7b. The caisson-supported wind turbines subjected to greater values of SIR, despite having different contract pressures settled maximum; the greater excitations with higher SIR are expected to intensify the seismic demand, and cyclic shear stresses leading to larger amount of excess pore pressure may be generated. The findings confirm that liquefaction induced settlement of offshore wind turbines is highly dependent on state and motion characteristics, whereas further research is warranted to better understand the interplay between different complex mechanisms.

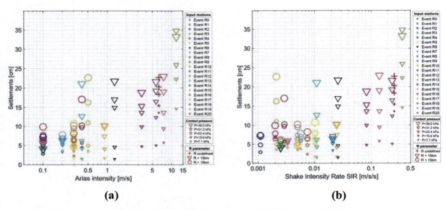

Fig. 7. Relation between settlements and **a)** Arias Intensity I_a **b)** Shake Intensity Rate SIR

4 Conclusions

The present study discusses the effects of seismic demand and state parameters (e.g., contact pressure) on the behavior of offshore wind turbines with suction caisson foundations considering the soil-structure interaction effect. Since the 2D modeling can provide less computational time as well as acceptable accuracy, a comprehensive numerical 2D model was developed and validated based on the available centrifuge test results to investigate the key parameters that can affect the liquefaction-induced settlement of offshore wind turbines with suction caisson foundation. The conclusions drawn for this study are as follows:

- The liquefaction-induced settlement increases with the increasing Arias Intensity I_a; therefore, the more intense the event is, the larger vertical displacements are resulted. This relation also exists when the settlements are compared with the SIR, with some clear exceptions for smaller values of this parameter.
- The results of the nonlinear dynamics analysis of twenty earthquake records showed that the vertical displacements do not increase linearly when the contact pressure is considerably greater than a certain magnitude that could be suppressed, indicating that a threshold may exist beyond which the linear trend is not though fulfilled.

References

1. Bhattacharya, S., et al.: Seismic design of offshore wind turbines: good bad and unknowns. Energies **14**, 3496 (2021)
2. Kaynia, A.: Effect of kinematic interaction on seismic response of offshore wind turbines on monopiles. Earthq. Eng. Struct. Dyn. **50**, 777–790 (2020)
3. Ahidashti, R.A., Barari, A., Haddad, A.: Insight into the post-liquefaction behaviour of skirted foundations. In: Proceedings of the 7th International Conference on Earthquake Geotechnical Engineering, Rome, pp. 1106–1113 (2019)

4. Amani, S., Prabhakaren, A., Bhattacharya, S.: Design of monopiles for offshore and nearshore wind turbines in seismically liquefiable soils: methodology and validation. Soil Dyn. Earthq. Eng **157**, 107252 (2022)
5. Haddad, A., Barari, A., Amini, R.: The remedial performance of suction caisson foundations for offshore wind turbines under seismically induced liquefaction in the seabed: shake table testing. Marine Struct. **83**, 103171 (2022). https://doi.org/10.1016/j.marstruc.2022.103171
6. Barari, A., Bayat, M., Saadati, M., Ibsen, L.B., Vabbersgaard, L.A.: Transient analysis of monopile foundations partially embedded in liquefied soil. Geomech. Eng. **8**(2), 257–282 (2015)
7. Ali, A., De Risi, R., Sextos, A.: Seismic assessment of wind turbines: how crucial is rotor-nacelle-assembly numerical modeling? Soil Dyn. Earthq. Eng. **141**, 106483 (2021)
8. Zhang, P., Xiong, P., Ding, H., Le, C.: Anti-liquefaction characteristics of composite bucket foundations for offshore wind turbines. J. Renew. Sustain. Energy **6**, 053102 (2014)
9. Wang, X., Zeng, X., Li, X., Li, J.: Liquefaction characteristics of offshore wind turbine with hybrid monopile foundation via centrifuge modelling. Renew. Energy **145**, 2358–2372 (2020)
10. Esfeh, P.K., Kaynia, A.: Modeling of liquefaction and its impact on anchor piles for floating offshore structures. Soil Dyn. Earthq. Eng. **127**, 105839 (2019)
11. Yu, H., Zeng, X., Lian, J.: Seismic behaviour of offshore wind turbine with suction caisson foundation. In: Geo-Congress 2014, Atlanta, Georgia, 23–26 February 2014, pp. 1206–1214 (2014)
12. Dashti, S., Bray, J.D.: Numerical simulation of Building response on liquefiable sand. J. Geotech. Geoenviron. Eng. **139**, 1235–1249 (2013)
13. Parra, E.: Numerical modeling of liquefaction and lateral ground deformation including cyclic mobility and dilative behaviour in soil systems. Ph.D. thesis, RPI (1996)
14. McKenna, F., Fenves, G.L., Scott, M.H.: Open system for earthquake engineering simulation. Pacific Earthquake Engineering Research Center, Berkeley, CA (2016). http://opensees.berkeley.edu
15. Yang, Z., Elgamal, A., Parra, E.: A computational model for cyclic mobility and associated shear deformation. J. Geotech. Geoenviron. Eng. **129**(12), 1119–1127 (2003)
16. Barari, A., Bagheri, M., Rouainia, M., Ibsen, L.B.: Deformation mechanisms of offshore monopile foundations accounting for cyclic mobility effects. Soil Dyn. Earthq. Eng. **97**, 439–453 (2017)
17. Barari, A., Zeng, X., Rezania, M., Ibsen, L.B.: Three-dimensional modeling of monopiles in sand subjected to lateral loading under static and cyclic conditions. Geomech. Eng. **26**, 175–190 (2021)
18. Joyner, W.B.: A method for calculating nonlinear seismic response in two dimensions. Bull. Seismol. Soc. Am. **65**(5), 1337–1357 (1975)
19. Ayala, G., Aranda, G.: Boundary conditions in soil amplification studies. In: Proceedings of the 6th World Conference on Earthquake Engineering, New Dehli, India (1977)
20. PEER Pacific Earthquake Engineering Research Ce Database (2013). http://www.peer.berkeley.edu

Pore-Pressure Generation of Sands Subjected to Cyclic Simple Shear Loading: An Energy Approach

Daniela Dominica Porcino[1(✉)], Giuseppe Tomasello[1], and Roohollah Farzalizadeh[2]

[1] University Mediterranea, Reggio Calabria, Italy
{daniela.porcino,giuseppe.tomasello}@unirc.it
[2] Urmia University, Urmia, Iran
st_r.farzalizadeh@urmia.ac.ir

Abstract. The energy-based approach to assess the residual pore water pressure build-up and the cyclic liquefaction resistance of sandy soils is receiving increasing attention in the last years since it has been found to better capture the effect of irregular loading cycles compared to the conventional equivalent stress-based methods. This paper reports the results of an experimental study carried out on two sands (Ticino and Emilia) under undrained cyclic simple shear loading with a focus on the development of specific correlations between residual pore water pressure (PWP) build-up and dissipated energy (W_S) up to the onset of liquefaction. Tests were carried out on specimens reconstituted at various relative densities (D_R) and subjected to various cyclic stress ratios (*CSR*). The results obtained were compared with two energy-based PWP models provided by other researchers leading to positive conclusions as for the general trend of W_S with the main influencing factors. A new simple energy-based PWP model was also proposed for sands and the good performance of the model for cyclic simple shear experiments was shown.

Keywords: Energy approach · Pore water pressure · Cyclic simple shear tests · Sand

1 Introduction

The energy-based approach (EBM) to assess the residual pore water pressure build-up and liquefaction potential of sandy soils is receiving increasing attention in the last years [1–3]. This approach seems to highlight some significant aspects of liquefaction potential evaluation from the viewpoint of energy demand versus energy supply and severity of liquefaction as well, which may not be sufficiently taken into account by the stress-based model (SBM) [4].

The amount of dissipated strain energy per unit volume of soils, which is also defined as energy density or unit energy, has been introduced, and for undrained cyclic simple shear test loadings, W_S can be computed numerically by the following equation:

$$W_S = \frac{1}{2} \cdot \sum_{i=1}^{n} (\tau_{i+1} + \tau_i) \cdot (\gamma_{i+1} - \gamma_i) \tag{1}$$

where n = number of load increments to liquefaction; τ_i and τ_{i+1} = applied shear stress at load increment i and i + 1, respectively; and γ_i and γ_{i+1} = shear strain at load increment i and i+1, respectively.

Several PWP-energy-based models were developed on the basis of the results of laboratory undrained cyclic tests, especially triaxial and torsional ones (e.g. [1, 3, 5]) whereas limited studies concerned cyclic simple shear tests (CSS) (e.g. [6]).

In the current study, 20 stress-controlled cyclic simple shear tests were conducted on Ticino and Emilia Romagna sands. The results were presented in terms of normalized dissipated energy (W_S/σ'_{v0}) to validate existing PWP energy-based models, and develop a new model for clean sands from CSS tests.

2 Experimental Program

The soils used in the present study were Ticino and Emilia Romagna silica sands. Ticino sand (TS) is a uniform coarse-to-medium natural sand taken from Ticino river in Italy; Emilia Romagna sand (ERS) was recovered from a site where liquefaction phenomena after the 2012 earthquake occurred in Italy. The soil properties are shown in Table 1.

Table 1. Material properties.

Material	G_S	e_{max}	e_{min}	D_{50} (mm)	C_U
Ticino sand	2.68	0.93	0.58	0.56	1.5
Emilia Romagna sand	2.70	0.93	0.54	0.32	2.8

Note: G_S = specific gravity; e_{max}, e_{min} = maximum and minimum void ratios; D_{50} = mean grain size; and C_U = uniformity coefficient.

Stress-controlled cyclic simple shear (CSS) tests were performed using a modified automated CSS apparatus (NGI type) [7]. The specimens were reconstituted by moist tamping (Ticino sand) and water sedimentation (Emilia sand).

The testing program comprised 20 CSS tests performed at (i) different relative density (D_R) ranging from 43% to 86%; (ii) different cyclic stress ratio (CSR) varying from 0.12 to 0.25; (iii) the same value of vertical effective stress (σ'_{v0}) equal to 100 kPa. For Emilia sand, the specimens were reconstituted at a prefixed relative density ($D_R = 60\%$) based on the interpretation of CPT results.

For the purpose of describing the triggering of liquefaction, a criterion of 3.75% shear strain (in single amplitude) was adopted.

3 Test Results

The dissipated energy was determined from CSS tests by plotting stress-strain hysteresis loops to compute the area bounded by the loops in each cycle (Eq. (1)).

Figure 1 shows that the application of the controlled cyclic shear stress enforces the corresponding cyclic shear strain to be increased; consequently, dissipated energy increases with cycles as shown in Fig. 1b. It is evident that just before reaching a liquefaction condition, W_S increases abruptly especially for loose sand specimens. The capacity energy W_{liq} is herein defined as the cumulative energy density required for liquefaction triggering for each test.

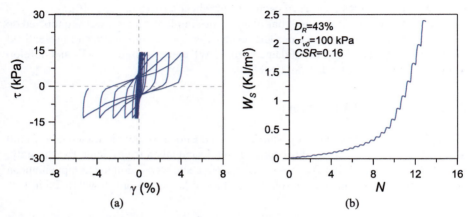

Fig. 1. Cyclic simple shear response of loose Ticino sand in terms of hysteresis loops and cumulative dissipated energy per unit volume.

3.1 Application of Energy-Based Pore Pressure Generation Models Under Cyclic Simple Shear Loading

Plots of measured pore water pressures against dissipated energy W_S for both loose and dense Ticino sand are reported in Fig. 2. Excess pore pressure ratio stands for the pore pressure normalized by initial vertical effective stress.

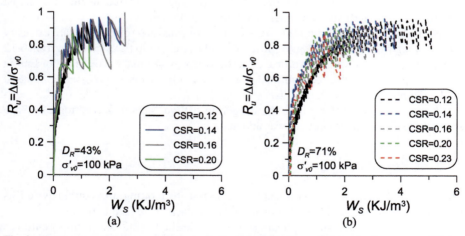

Fig. 2. Excess pore water pressure versus dissipated energy for cyclic simple shear tests on Ticino sand (loose and dense) at various cyclic stress ratios (*CSR*).

Although the results are scattered, Fig. 2 illustrates that in each cycle, the local minimum, maximum or residual values were used to represent pore water pressure history. In the present study, residual pore water pressure values at time corresponding to zero shear stresses have been considered for the analysis. A nonlinear increasing trend between R_u and W_S can be observed in both loose and dense TS specimens. Considering the residual points of the curves, it can be observed that cyclic stress ratio (CSR) has a minor effect on the relationship between R_u and dissipated energy, and a relatively narrow zone was obtained for any given D_R of the sand. This can be a considerable advantage when models for pore water pressure generation have to be developed based on laboratory stress-controlled test results.

A simple and widely used energy-based model implemented for use in earthquake site response analysis (e.g., DEEPSOIL [8]) was proposed by Berrill and Davis [9]. The authors proposed a power relation between R_u and normalized density energy which is expressed by the following equation:

$$R_u = \alpha \cdot \left(\frac{W_S}{\sigma'_{v0}}\right)^\beta \quad (2)$$

where σ'_{v0} is the initial vertical effective stress, α and β are positive calibration parameters and β is less than unity.

When Berril and Davis [9] model was applied to CSS test data on Ticino sand, it appears (Fig. 3) that all data points are well located within a narrow zone bounded by an upper and lower bound defined by the following parameters: $\alpha = 4.51 - 2.71$ and $\beta = 0.36 - 0.33$, respectively. Furthermore, the model suggested by Berrill and Davis [9] approximates residual excess pore pressure generation data for various samples of Ticino sands with a coefficient of determination $R^2 > 0.84$.

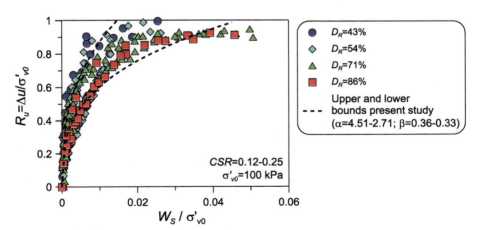

Fig. 3. Observed bounds of residual excess pore water pressure as a function of normalized dissipated energy using the model proposed by Berrill and Davis [9] for Ticino sand.

The model does not appear particularly sensitive to the value of β as suggested also by Berrill and Davis [9] and an optimal value for the tested sand is 0.32 (Fig. 4a) while a recommended value from Berrill and Davis [9] was 0.5. Since it is necessary to determine α values in Eq. (2), α is given in Fig. 4b as a function of relative density of sand for various CSR. As may be observed from this figure, α decreases with increasing D_R.

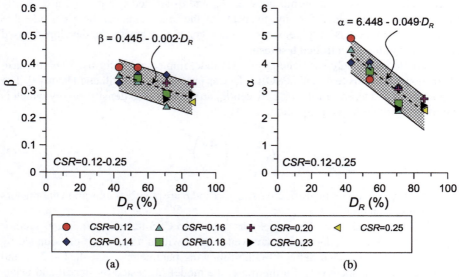

Fig. 4. Correlation for α and β of the Berrill and Davis [9] model as a function of D_R for Ticino sand in undrained cyclic simple shear tests.

Other energy-based models were proposed in the literature and among them, Green et al. [10] proposed a later empirical energy-based model (GMP) expressed by the following equation:

$$R_u = \sqrt{\frac{W_S/\sigma'_{v0}}{PEC}} \tag{3}$$

The model is defined by considering one calibration parameter termed *PEC* (the "pseudo energy capacity"). Setting the calibration parameters α and β in Eq. (2) equal to $1/PEC^{0.5}$ and 0.5, respectively, it can be seen that the GMP model is a special case of the general model initially proposed by Berrill and Davis [9]. The definition of *PEC* and the procedures for determining it were developed empirically from analyzing numerous cyclic tests [10, 11].

In the present paper, the correlation between *PEC* and relative density for Ticino sand from CSS tests (Fig. 5) evidences that *PEC* consistently increases with increasing D_R as observed by previous authors. It is noteworthy that *PEC* values predicted by Eq. (4) reported in Fig. 5 are lower than the corresponding ones obtained by Polito et al. [1] based on cyclic triaxial test data especially for higher densities of sand.

In order to validate the correlation for *PEC* with relative density inferred from cyclic simple shear tests, additional CSS test results performed on Emilia sand were interpreted for determining *PEC* following the procedure suggested by Green et al. [10] by plotting R_u versus the square root of W_S/σ'_{v0} as illustrated in Fig. 6. This figure evidences that the best-fit value of *PEC* for all the CSS tests at various *CSRs* was 0.0155 ($R^2 = 0.97$). Such a value is close to that one predicted by the correlation relating *PEC* to relative density (Eq. (4) in Fig. 5) demonstrating that it is a good approximation for deriving *PEC* values.

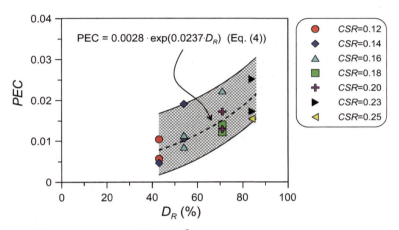

Fig. 5. Correlation for *PEC* ($1/\alpha^2$) with relative density for Ticino sand.

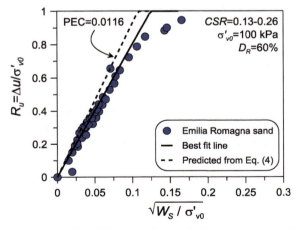

Fig. 6. Determination of *PEC* from cyclic simple shear test data on Emilia sand.

3.2 A New Energy-Based Pore Pressure Generation Model for Clean Sands Under Cyclic Simple Shear Loading

Development of improved functions would require an extensive laboratory testing program including not only cyclic triaxial tests or torsional tests but also cyclic simple shear tests. In the present study, a new relation between R_u and W_S/σ'_{v0} based on cyclic simple shear tests was introduced by the equation:

$$R_u = \frac{a \cdot \left(W_S/\sigma'_{v0}\right)}{b + \left(W_S/\sigma'_{v0}\right)} \quad (5)$$

where a and b are empirical calibration parameters. As can be seen in Fig. 7, the proposed model accurately captures residual excess pore pressure generation for various samples of both Ticino and Emilia sands subjected to cyclic simple shear loading with a coefficient of determination $R^2 > 0.95$.

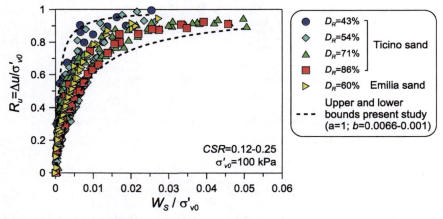

Fig. 7. Observed bounds of residual excess pore water pressure as a function of normalized dissipated energy for tested sands using the model proposed in the present study (Eq. (5)).

Correlations for estimating parameters required to calibrate the model are shown in Fig. 8. Since the variation of a is limited, it is reasonably assumed to be constant ($a = 1$); accordingly, b values vary from 0.001 to 0.0066 (upper and lower bounds).

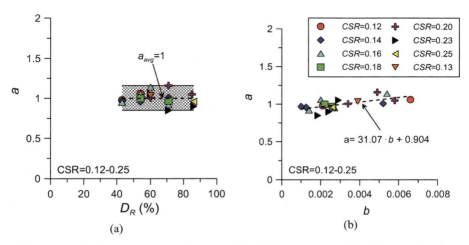

Fig. 8. Correlation for a and b of the proposed model as a function of D_R for tested sands.

4 Conclusions

In the present study, approximately 20 cyclic simple shear (CSS) tests were carried out on two clean sands to validate the commonly used energy-based PWP generation models, as well as develop new correlations applicable under cyclic simple shear loading. The main conclusions that can be drawn are the following:

- A nonlinear increasing trend between residual excess pore pressure ratio R_u and accumulated energy W_S was observed, and it was more dependent on the relative density of the sand while cyclic stress ratio (*CSR*) appeared to have a minor impact.
- Both models (i.e. Berril and Davis [9] and Green et al. [10]) used for predicting residual PWP generation in sands were found to be effective means of capturing the experimental trend but specific correlations for the calibration parameters of the models using data from CSS tests were found for tested soils.
- Several functional forms were examined for predicting residual PWP generation in sands and a new nonlinear simple but efficient model with two calibration parameters (i.e. a and b) was developed. The prediction performance of the model was shown to be very good ($R^2 > 0.95$). Since the variation of a was limited, it can reasonably be assumed to be constant ($a = 1$) while b values vary with the relative density of sand from 0.001 to 0.0066 (upper and lower bounds).

References

1. Polito, C.P., Green, R.A., Lee, J.: Pore pressure generation models for sands and silty soils subjected to cyclic loading. J. Geotech. Geoenviron. Eng. **134**(10), 1490–1500 (2008)
2. Baziar, M.H., Sharafi, H.: Assessment of silty sand liquefaction potential using torsional tests - an energy approach. Soil Dyn. Earthq. Eng. **31**(7), 857–865 (2011)

3. Yang, Z.X., Pan, K.: Energy-based approach to quantify cyclic resistance and pore pressure generation in anisotropically consolidated sand. J. Mater. Civ. Eng. **30**(9), 04018203 (2018)
4. Kokusho, T.: Liquefaction potential evaluation: energy-based method compared to stress-based method. In: Finn, W.D.L., Wu, G. (eds.) 7th International Conference on Case Histories in Geotechnical Engineering, Paper No. 4.01a, pp. 1–10 (2013)
5. Jafarian, Y., Towhata, I., Baziar, M.H., Noorzad, A., Bahmanpour, A.: Strain energy based evaluation of liquefaction and residual pore water pressure in sands using cyclic torsional shear experiments. Soil Dyn. Earthq. Eng. **35**, 13–28 (2012)
6. Law, K.T., Cao, Y.L., He, G.N.: An energy approach for assessing seismic liquefaction potential. Can. Geotech. J. **27**(3), 320–329 (1990)
7. Porcino, D., Caridi, G., Malara, M., Morabito, E. An automated control system for undrained monotonic and cyclic simple shear tests. In: Proceedings of Geotechnical Engineering in the Information Technology Age, pp. 1–6. ASCE, Reston (2006)
8. Hashash, Y.M.A., et al.: DEEPSOIL 7.0, User Manual. Urbana, IL, Board of Trustees of University of Illinois at Urbana-Champaign (2020)
9. Berrill, J.B., Davis, R.O.: Energy dissipation and seismic liquefaction of sands: revised model. Soils Found. **25**(2), 106–118 (1985)
10. Green, R.A., Mitchell, J.K., Polito, C.P.: An energy-based excess pore pressure generation model for cohesionless soils. In: Proceedings of the John Booker Memorial Symposium, pp. 1–9. A.A. Balkema Publishers, Rotterdam (2000)
11. Green, R.A.: Energy-based evaluation and remediation of liquefiable soils. Ph.D. thesis, Virginia Polytechnic Institute and State University (2001)

Constitutive Modeling of Undrained Cyclic Shearing of Sands Under Non-zero Mean Shear Stress

Andrés Reyes[1], Mahdi Taiebat[1](✉), and Yannis F. Dafalias[2,3]

[1] Department of Civil Engineering, University of British Columbia, Vancouver, BC, Canada
reyespa@mail.ubc.ca, mtaiebat@civil.ubc.ca
[2] Department of Civil and Environmental Engineering, University of California, Davis, CA, USA
jfdafalias@ucdavis.edu
[3] Department of Mechanics, Faculty of Applied Mathematical and Physical Sciences, National Technical University of Athens, Athens, Greece

Abstract. SANISAND-MSf is one the latest members of the SANISAND family of models within a critical state compatible bounding surface plasticity framework with kinematic hardening of the yield surface. In pursuance of enhancing the undrained cyclic response, the model incorporates a memory surface for controlling the stiffness affecting the deviatoric and volumetric plastic strains in the preliquefaction stage, and the concept of a semifluidized state for controlling stiffness and dilatancy in the post-liquefaction stage. The new model can precisely simulate the cyclic liquefaction of sands in undrained cyclic shearing in the absence of a mean shear stress under isotropic initial stress conditions. In this study we assess the performance of SANISAND-MSf in modeling the undrained cyclic shearing response of sands in the presence of non-zero mean shear stresses. Simulations have been compared with relevant experimental data, along with an assessment of the components of the model responsible for simulating such loading conditions.

Keywords: Sand · Plasticity model · Liquefaction · Mean shear stress

1 Introduction

It is widely recognized the undrained cyclic response of soils is dependent on the value of the mean shear stress imposed. With respect to isotropic stress conditions, extensive research has demonstrated that the presence of non-zero mean shear stresses can profoundly impact the cyclic resistance and deformation characteristics observed during undrained cyclic shearing. In cyclic triaxial testing [1–5], non-zero mean shear stress translates into an initial deviatoric stress q that is maintained during testing and symbolized by q_{mean}. In cyclic simple shear testing, a non-zero mean shear stress can refer to (a) having an initial lateral stress coefficient $K_0 \neq 1.0$ [6–8], which typically evolves towards 1.0 during cyclic shearing, to (b) the presence of an initial shear stresses on the horizontal plane, τ_{mean}, which is sustained throughout cyclic shearing [6, 7, 9–11], or (c) a combination of the latter two cases [12, 13]. In all the above scenarios, cyclic shearing with amplitude q_{cyc} or τ_{cyc} is superimposed to the corresponding non-zero mean values, producing circumstances where stress reversal, i.e., change of sign of shear stress q or τ, can take place during loading.

This paper evaluates the performance of a newly formulated constitutive model SANISAND-MSf in modeling cyclic triaxial shearing with non-zero mean shear stresses. This model introduces two new constitutive ingredients into the framework of a well-established reference model, DM04. These new components were designed to accurately control the rate of plastic deviatoric and volumetric strains in the pre-liquefaction stage, and to allow for an enhanced modeling of the ensuing cyclic shear strains in post-liquefaction. First, simulations of isotropic undrained cyclic triaxial tests on two types of sands showcase the performance constitutive model. Then, the model is evaluated for tests with non-zero mean shear stresses with varying levels of cyclic shear stresses.

2 SANISAND-MSf Model

Dafalias & Manzari [14], introduced a stress-ratio controlled, critical state compatible, bounding surface plasticity model, with a useful fabric-dilatancy quantity. This model is often referred to as DM04, which formed the basis of what was later on named SANISAND class of models [15]. The SANISAND class includes various extensions (e.g., Li and Dafalias [16]; Dafalias and Taiebat [17]; Petalas et al. [18, 19]). Most recently Yang et al. [20] introduced two major constitutive ingredients, to significantly improve the performance of this class of models. These two new constitutive ingredients suggested the name SANISAND-MSf for the model, because it is a member of the SANISAND family of models, with M standing for 'memory surface' and Sf for 'semifluidised state'. The memory surface was formulated to accurately control the plastic stiffness in pre-liquefaction. This memory surface (M), a modification from the propositions from Corti et al. [21] and Liu et al. [22], is a back-stress-ratio based bounding surface with kinematic and isotropic hardening. A normalized measure of the distance between memory surface and stress-ratio is used to affect the plastic modulus within DM04, which in turn allows for an improved control on the rate of plastic deviatoric and volumetric strains in pre-liquefaction. The impact of the memory surface on the plastic modulus is regulated with the introduction of model constants μ_0 and u. The concept of a semifluidized state (Sf) is employed to degrade the plastic stiffness and dilatancy of the model in order to simulate the progressive development of large cyclic shear strains in post-liquefaction. The Sf state was proposed by Barrero et al. [23], who postulated a new internal variable l that evolves at very low values of mean pressure p. The plastic modulus and dilatancy in DM04 is simultaneously degraded through l, granting the means to model increasing plastic shear strains in post-liquefaction without impacting the plastic volumetric strains. The degradation is controlled through model constants x and c_l.

The addition of these new ingredients provides much needed degrees of freedom to control the progressive generation of excess pore water pressures in pre-liquefaction, that is, before reaching a mean stress of zero or very close to zero, and the ensuing development of large cyclic shear deformation in post-liquefaction. The distinctive response observed at different amplitudes of cyclic shear stresses, both in terms of pore pressure generation and shear deformation, can also be captured by the new model, effectively overcoming major shortcomings of the reference model DM04. Figure 1 shows the simulation of an isotropic ($q_{mean} = 0$ kPa) undrained cyclic triaxial test on Karlsruhe sand [4] using both DM04 and SANISAND-MSf. The model constants, obtained from Yang et al. [20], are shown in Table 1. The DM04 simulations were performed by deactivating all the features presented in SANISAND-MSf. The simulations reveal that the MS allows for an improved modeling of the history of excess pore water pressure ratio r_u leading to initial liquefaction. Furthermore, the Sf state is the ingredient responsible for the precise modeling of the axial strains after initial liquefaction. Figures 2 and 3 further showcase the model enhanced response in capturing cyclic resistance at different levels of cyclic stress ratio $q_{cyc}/2p_0$ (CSR), initial p and two types of sand.

Table 1. SANISAND-MSf model constants for Karlsruhe fine sand and Toyoura sand.

Model constant	Symbol	Karlsruhe fine sand	Toyoura sand
Elasticity	G_0	100	125
	ν	0.05	0.05
Critical state	M	1.28	1.25
	c	0.75	0.712
	e_c^{ref}	1.038	0.934
	λ_c	0.056	0.019
	ξ	0.28	0.7
Yield surface	m	0.01	0.02
Dilatancy	n^d	1.2	2.1
	A_0'	0.56	0.704
	n_g	0.95	0.95
Kinematic hardening	n^b	1.0	1.25
	h_0'	7.6	7.05
	c_h	1.015	0.968
Fabric dilatancy	z_{max}	15	10
	c_z	1000	1000
Memory surface	μ_0	7.8	2.4
	u	0.87	2.6
Semifluidized state	x	3.3	4.0
	c_1	25	50

3 Non-zero Mean Shear Stresses in Cyclic Triaxial

Wichtmann and Triantafyllidis [4] and Pan and Yang [5] presented carefully curated sets of undrained cyclic triaxial tests on Karlsruhe fine sand and Toyoura sand, respectively, with non-zero mean shear stresses. The bulk of these experiments corresponding to medium dense consistencies were simulated using SANISAND-MSf with the model constants specified in Table 1 and calibrated based solely on isotropic tests.

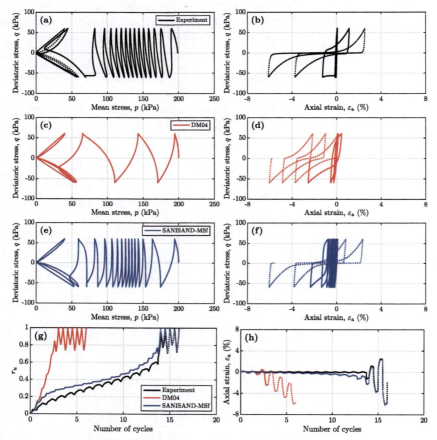

Fig. 1. Simulations compared with experiments in undrained cyclic triaxial tests on isotropically consolidated ($q_{\text{mean}} = 0$ kPa) samples of Karlsruhe fine sand with $D_r = 59\%$, $p_0 = 200$ kPa, and sheared with a CSR = 0.150: (a, b) experimental data from Wichtmann and Triantafyllidis [4]; and simulations of (c, d) DM04 and (e, f) SANISAND-MSf. Histories of (g) excess pore water pressure ratios r_u and (h) axial strains are presented in terms of number of cycles.

A first set of experiments on Karlsruhe fine sand consisted of samples cyclically sheared with the same CSR but with increasing the level of q_{mean} from 0 to 200 kPa. A selected test from this set, where stress reversal occurred, is shown in Fig. 4. The

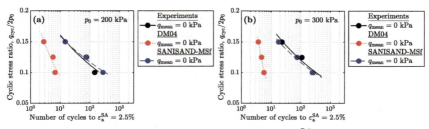

Fig. 2. Number of cycles to reach a single amplitude axial strain $\varepsilon_a^{SA} = 2.5\%$ in undrained cyclic triaxial tests on isotropically consolidated ($q_{mean} = 0$ kPa) samples of Karlsruhe fine sand with $D_r = 55$–65%, (a) $p_0 = 200$ kPa, and (b) $p_0 = 300$ kPa. Experimental data from Wichtmann and Triantafyllidis [4] and simulations of DM04 and SANISAND-MSf.

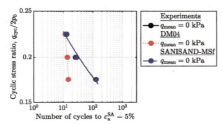

Fig. 3. Number of cycles to reach a single amplitude axial strain $\varepsilon_a^{SA} = 5.0\%$ in undrained cyclic triaxial tests on isotropically consolidated ($q_{mean} = 0$ kPa) samples of Toyoura sand with $D_r = 60\%$ and $p_0 = 100$ kPa. Experimental data from Pan and Yang [5] and simulations of DM04 and SANISAND-MSf.

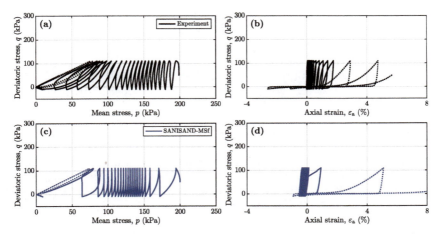

Fig. 4. Simulations compared with experiments in undrained cyclic triaxial tests on anisotropically consolidated ($q_{mean} = 50$ kPa) samples of Karlsruhe fine sand with $D_r = 56\%$, $p_0 = 200$ kPa, and sheared with a CSR = 0.150: (a, b) experimental data from Wichtmann and Triantafyllidis [4]; and (c, d) simulations of SANISAND-MSf.

simulation was found to reasonably capture the progressive decrease of mean pressure and the onset of liquefaction. However, upon reaching the Sf state, the model appears to overpredict the cyclic and accumulated axial strains in compression.

A second set of experiments on Karlsruhe fine sand was completed for samples with a level of q_{mean} of 150 kPa and sheared with three different CSR. Figure 5 presents a selected test from this last set, for which stress reversal was prevented. Note that results from the simulations are shown until attaining a single amplitude axial strain of 5%. The model can again reasonably simulate the rather slow decrease of mean pressure observed in the experiment. On the other hand, the rate of accumulation of axial strains in the simulation was overpredicted, because the number of cycles required to reach a specified level of undesired deformation was much less than what was shown in the experiment. Figure 6 presents a summary of the complete datasets of the number of cycles required to achieve a specified level of axial strain, where this latter observation can be extended for virtually all cases. Note that the only test where stress reversal was allowed is shown in Fig. 4 and summarized in Fig. 6a. Comparable experiments on Toyoura sand, where the magnitude of q_{mean} and CSR was varied, were correspondingly simulated and are summarized in Fig. 7. As was the case earlier, the number of cycles required to reach a specified level of deformation was largely underestimated. Unlike the set for Karlsruhe sand, stress reversal was prevented in all the experiments.

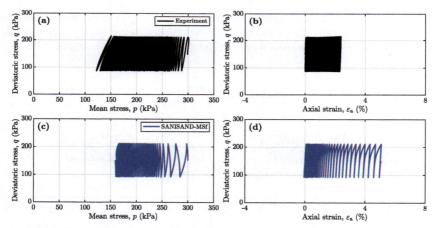

Fig. 5. Simulations compared with experiments in undrained cyclic triaxial tests on anisotropically consolidated ($q_{mean} = 150$ kPa) samples of Karlsruhe fine sand with $D_r = 63\%, p_0 = 300$ kPa, and sheared with a CSR = 0.100: (a, b) experimental data from Wichtmann and Triantafyllidis [4]; and (c, d) simulations of SANISAND-MSf.

The simulations presented in Figs. 4, 5, 6 and 7 suggest that SANISAND-MSf underestimates the plastic stiffness in undrained cyclic shearing when in the presence of non-zero mean shear stresses. This shortcoming becomes more noticeable the larger the magnitude of q_{mean}, particularly when stress reversal is prevented. It should be acknowledged that the introduction of the MS significantly improved the overall cyclic undrained response with respect to different amplitude of cyclic shear stress in isotropic symmetric shearing. This was possible primarily because the contribution of the MS into the model plastic stiffness is inversely proportional to the initial back-stress ratio α_{in}, where the smaller its value the larger the MS impact on the model stiffness. Such strategy proved effective for simulating cases with zero mean-shear stresses and varying cyclic shear amplitude, where higher pre-liquefaction plastic stiffness is expected from samples sheared at relatively low values of CSR. However, this is not generic, as at each stress reversal α_{in} is representative of the stress ratio associated with the sum of the non-zero mean and cyclic shear stresses. Ideally, a different class of internal variables needs to be introduced, so that the effect of the amplitude of the cyclic shear stress and the mean shear stress, q_{cyc} and q_{mean} in triaxial shearing, can be accounted for and controlled independently from each other into the plastic stiffness and dilatancy of the constitutive model. This task is under investigation at present.

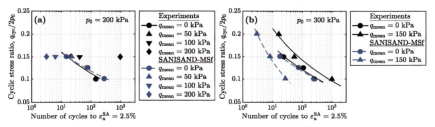

Fig. 6. Number of cycles to reach a single amplitude axial strain $\varepsilon_a^{SA} = 2.5\%$ in undrained cyclic triaxial tests on iso and anisotropically consolidated samples of Karlsruhe fine sand with $D_r = 55$–65%, (a) $p_0 = 200$ kPa, and (b) $p_0 = 300$ kPa. Experimental data from Wichtmann and Triantafyllidis [4] and simulations of SANISAND-MSf.

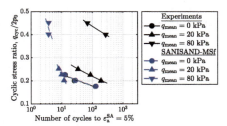

Fig. 7. Number of cycles to reach a single amplitude axial strain $\varepsilon_a^{SA} = 5.0\%$ in undrained cyclic triaxial tests on iso and anisotropically consolidated samples of Toyoura sand with $D_r = 60\%$ and $p_0 = 100$ kPa. Experimental data from Pan and Yang [5] and simulations of SANISAND-MSf.

4 Conclusions

The performance of SANISAND-MSf in simulating the undrained response in cyclic triaxial shearing in the absence and presence of mean shear stresses was evaluated. For medium dense samples of Karlsruhe and Toyoura sand, the model was shown to have an excellent fit to the experiments with zero mean shear stresses, that is, under isotropic initial conditions. The newly introduced constitutive ingredients, memory surface and semifluidized state, were shown to provide an accurate control of the response in the pre- and post-liquefaction stages, respectively, resulting in precise modeling of the cyclic resistance and large cyclic shear deformation observed in the absence of mean shear stresses. Despite this, the model was revealed to overestimate the cyclic and accumulated axial deformation when in the presence of non-zero mean shear stresses. This shortcoming is more evident the larger the magnitude of mean shear stress, essentially underpredicting the onset of the large deformations. The experimental evidence indicates it is necessary to account for the impact of non-zero mean shear stresses in the plastic stiffness of the model. The potential solution, however, must not be circumscribed to cyclic triaxial loading, but rather extended and validated for other shearing modes, e.g., simple shear or circular loading, where non-zero mean shear stresses vary in nature and can evolve during loading. This matter is the subject of ongoing research.

Acknowledgment. Support for this study was provided by the Natural Sciences and Engineering Research Council of Canada (NSERC) and BGC Engineering Inc.

References

1. Vaid, Y.P., Chern, J.C.: Effect of static shear on resistance to liquefaction. Soils Found. **23**(1), 47–60 (1983)
2. Hyodo, M., Murata, H., Yasufuku, N., Fujii, T.: Undrained cyclic shear strength and residual shear strain of saturated sand by cyclic triaxial tests. Soils Found. **31**(3), 60–76 (1991)
3. Yang, J., Sze, H.Y.: Cyclic behaviour and resistance of saturated sand under non-symmetrical loading conditions. Géotechnique **61**(1), 59–73 (2011)
4. Wichtmann, T., Triantafyllidis, T.: An experimental database for the development, calibration and verification of constitutive models for sand with focus to cyclic loading: part I—tests with monotonic loading and stress cycles. Acta Geotech. **11**(4), 739–761 (2015). https://doi.org/10.1007/s11440-015-0402-z
5. Pan, K., Yang, Z.X.: Effects of initial static shear on cyclic resistance and pore pressure generation of saturated sand. Acta Geotech. **13**(2), 473–487 (2017). https://doi.org/10.1007/s11440-017-0614-5
6. Tatsuoka, F., Muramatsu, M., Sasaki, T.: Cyclic undrained stress-strain behavior of dense sands by torsional simple shear test. Soils Found. **22**(2), 307–322 (1982)
7. Hosono, Y., Yoshimine, M.: Effects of anisotropic consolidation and initial shear load on liquefaction resistance of sand in simple shear condition. In: Liu, H., Deng, A., Chu. (eds.) Geotechnical Engineering for Disaster Mitigation and Rehabilitation, pp. 352–358. Springer, Heidelberg (2008). https://doi.org/10.1007/978-3-540-79846-0_37
8. Vargas, R., Ueda, K., Uemura, K.: Influence of the relative density and K0 effects in the cyclic response of Ottawa F65 sand - cyclic Torsional Hollow-Cylinder shear tests for LEAP-ASIA-2019. Soil Dyn. Earthq. Eng. **133**, 106111 (2020)
9. Sivathayalan, S., Ha, D.: Effect of static shear stress on the cyclic resistance of sands in simple shear loading. Can. Geotech. J. **48**, 1471–1484 (2011)
10. Chiaro, G., Koseki, J., Sato, T.: Effects of initial static shear on liquefaction and large deformation properties of loose saturated Toyoura sand in undrained cyclic torsional shear tests. Soils Found. **52**(3), 498–510 (2012)

11. Umar, M., Chiaro, G., Kiyota, T., Ullah, N.: Deformation and cyclic resistance of sand in large-strain undrained torsional shear tests with initial static shear stress. Soils Found. **61**(3), 765–781 (2021)
12. Vaid, Y.P., Finn, W.D.L.: Static shear and liquefaction potential. J. Geotech. Eng. Div. **105**(10), 1233–1246 (1979)
13. El Ghoraiby, M.A., Manzari, M.T.: Cyclic behavior of sand under non-uniform shear stress waves. Soil Dyn. Earthq. Eng. **143**, 1106590 (2021)
14. Dafalias, Y.F., Manzari, M.T.: Simple plasticity sand model accounting for fabric change effects. J. Eng. Mech. **130**(6), 622–634 (2004)
15. Taiebat, M., Dafalias, Y.F.: SANISAND: simple anisotropic sand plasticity model. Int. J. Numer. Anal. Meth. Geomech. **32**(8), 915–948 (2008)
16. Li, X.S., Dafalias, Y.F.: Anisotropic critical state theory: role of fabric. J. Eng. Mech. **138**(3), 263–275 (2012)
17. Dafalias, Y.F., Taiebat, M.: SANISAND-Z: zero elastic range sand plasticity model. Géotechnique **66**(12), 999–1013 (2016)
18. Petalas, A.L., Dafalias, Y.F., Papadimitriou, A.G.: SANISAND-FN: an evolving fabric-based sand model accounting for stress principal axes rotation. Int. J. Numer. Anal. Meth. Geomech. **43**, 97–123 (2018)
19. Petalas, A.L., Dafalias, Y.F., Papadimitriou, A.G.: SANISAND-F: sand constitutive model with evolving fabric anisotropy. Int. J. Solids Struct. **188–189**, 12–31 (2020)
20. Yang, M., Taiebat, M., Dafalias, Y.F.: SANISAND-MSf: a sand plasticity model with memory surface and semifluidised state. Géotechnique **72**(3), 227–246 (2022)
21. Corti, R., Diambra, A., Wood, D.M., Escribano, D.E., Nash, D.F.T.: Memory surface hardening model for granular soils under repeated loading conditions. J. Eng. Mech. **142**(12), 04016102 (2016)
22. Liu, H.Y., Abell, J.A., Diambra, A., Pisanó, F.: Modelling the cyclic ratcheting of sand through memory-enhanced bounding surface plasticity. Géotechnique **69**(9), 783–800 (2019)
23. Barrero, A.R., Taiebat, M., Dafalias, Y.F.: Modeling cyclic shearing of sand in the semifluidized state. Int. J. Numer. Anal. Meth. Geomech. **44**(3), 371–388 (2020)

Investigation of Lateral Displacement Mechanism in Layered and Uniform Soil Models Subjected to Liquefaction-Induced Lateral Spreading

Anurag Sahare(✉), Kyohei Ueda, and Ryosuke Uzuoka

Disaster Prevention Research Institute, Kyoto University, Uji, Kyoto 611-0011, Japan
sahare.anuragrahul.4m@kyoto-u.ac.jp

Abstract. The phenomenon of liquefaction-induced lateral spreading has been responsible for causing catastrophic damage during the past earthquakes. LEAP project was formed with an objective of developing a large centrifuge database using different centrifuge facilities in an effort to characterize the median response of a sloping ground during lateral spreading. Further these databases provide a unique opportunity to verify and validate numerical modeling and contribute to the further enhancement of constitutive modeling involving soil liquefaction. However, such centrifuge and numerical analysis did not consider the presence of a layered sloping ground, which is most likely to be encountered in an ideal scenario. At the same time, it is also important to assess the lateral displacement mechanism within different models having larger soil variability. This shortcoming hinders the development of a performance-based design involving liquefaction-induced lateral spreading considering larger variability in soil conditions. To overcome this, we developed a series of centrifuge experiments. In the centrifuge plan, tests were developed for a uniform loose sand model, a uniform dense sand model and a multi-layer soil model. The result portrays the significantly different deformation mechanism for the uniform denser soil model as compared to a multi-layer soil model and a uniform loose sand model. The lateral soil displacements were found to be largest near the center array for all the models. Due to the occurrence of liquefaction-induced lateral spreading within the loose sand layer, the above denser soil layer could be dragged alongside, resulting in significant mobilization of shear strains towards the ground surface.

Keywords: Centrifuge tests · Liquefaction · Lateral spreading · Earthquakes

1 Introduction and Background

Significant lateral stresses (monotonic in nature) might be imparted onto the structural systems due to liquefaction-induced lateral spreading of soil layers during a major seismic event. Recent earthquakes, e.g., 1989 Loma-Prieta, 1995 Kobe, 2010–11 Canterbury earthquake sequence, have shown the widespread damages induced to the soil-structure systems owing to liquefaction-induced lateral spreading (Ishihara 1997; Tokimatsu and

Asaka 1998; Ishihara and Cubrinovski 2004; Cubrinovski et al. 2014). This has motivated researchers to assess the structural demands due to lateral spreading or subgrade reaction arising due to the kinematic part of loading, particularly for pile foundations, which are used to carry large axial loads to a deeper stiff stratum passing through the liquefiable deposits (Abdoun et al. 2003; Brandenberg et al. 2005; Knappett and Madabhushi 2008; Motamed et al. 2013; Ebeido et al. 2019; Sahare et al. 2022). Liquefaction Experiment and Analysis Project (LEAP), an international joint research-based project, was formed with an objective of developing a high-quality large centrifuge database using different centrifuge facilities to study the seismic response of a saturated sloping ground. Centrifuge tests were developed as a part of LEAP-GWU 2015 at different facilities considering a uniform 5° sloped model made of Ottawa F-65 sand tested under a ramped sine waveform (Kutter et al. 2018). As a part of LEAP-UCD 2017 exercise, 24 centrifuge models were developed at nine centrifuge facilities with different sets of parameters in an effort to characterize the median response of the sloping ground during lateral spreading (Kutter et al. 2020). 3-D regression surfaces were developed to establish the correlations between the relative density, lateral displacement of soil and the earthquake characteristic in terms of PGA_{eff} from those 24 centrifuge experiments as shown in Fig. 1 (Kutter et al. 2020). Recently, during LEAP-ASIA 2019, additional 24 centrifuge tests were developed in an effort to validate the generalized scaling laws and to further refine the previously established relationship (Vargas et al. 2021). The developed centrifuge models served as a unique opportunity via numerical analysis to assess the effects of shear-induced dilatancy (He et al. 2020), influences of a rotating radius of geotechnical centrifuge (Sahare et al. 2020b). Further various soil constitutive models (Strain space multiple mechanism model, PDMY02, CycLiqCP, PM4Sand, SANISAND-SFR) were utilized to verify and validate the seismic behavior of a sloping ground following rigorous calibration of constitutive model parameters based on the

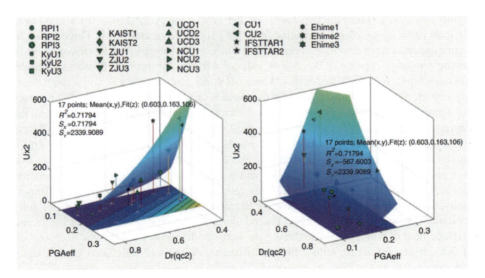

Fig. 1. Regression surface based on the six-parameter equation (Kutter et al. 2020)

series of cyclic triaxial and cyclic torsional tests (Sahare et al. 2020a; Qiu and Elgamal 2020; He et al. 2020; Chen et al. 2021; Reyes et al. 2021).

2 Problem Statement and Motivations for the Present Research

Although, these databases may serve or have served as a valuable input for the enhancement of existing constitutive soil models as well as for the verification and validation purposes concerning the numerical modeling. Yet, most of the centrifuge experiments and numerical analysis did not consider layered sloping ground deposits during an event of lateral spreading, which is more likely to be encountered in an ideal scenario (Cubrinovski and Ishihara 2004). A layered soil model would represent a more complex distribution of shear strains at the stiffness discontinuity and lateral demands on the structural systems. Moreover, due to different percentages of reduction in soil stiffness for different soil layers (having contrasting initial stiffness) under the excess pore pressure generation, a lateral displacement mechanism within a soil model needs to be assessed. It is important to study the spatial distribution of lateral spreading or lateral soil displacements during an earthquake; since some of the past studies (e.g., Motamed and Towhata 2010; Ebeido et al. 2019; Sahare et al. 2022) have shown structural systems to experience different magnitude of lateral stresses during liquefaction-induced lateral spreading.

3 Experimental Setup and Procedure

To overcome the limitations of the existing literature, a series of centrifuge experiments were developed for a sloping group under fully submerged conditions using a geotechnical beam-type centrifuge facility, located at the Disaster Prevention Research Institute, Kyoto University. The effective radius of the centrifuge is 2.50 m, with a maximum payload capacity of 24 g-ton reachable to 50G for the dynamic tests. Three series of centrifuge tests were developed considering a uniform loose sand model (toyoura sand), having a relative density (RD) of 40% (Model ARS-01), a uniform dense sand model (toyoura sand), having RD of 80% (Model ARS-02); shown in Fig. 2 and a multi-layer soil model; with toyoura sand used for the loose layer, while coulored silica S-7 to make the dense sand layer (Model ARS-03) as shown in Fig. 3 (with the positioning of all the instrumentations used for the tests). The multi-layer soil model comprised of different soil layers was used having distinct colors (as shown in Fig. 3) so as to effectively monitor the deformation mechanism. A rigid box was used for all the tests, which had one side as a polymethyl methacrylate window, which allowed monitoring of the deformation mechanism with the help of a high frame rate camera during a seismic event. Importantly, soba noodles, capable of flowing laterally with the liquefied soils (under fully saturated condition) were used and placed along the acrylate glass to trace the lateral displacement of soil layers. The three tests were conducted at a centrifugal acceleration of 50G. More details regarding the centrifuge modeling and a detailed experimental process (including the saturation procedure) could be found in Sahare et al. 2022.

4 Analysis of the Results

The centrifuge tests were conducted under a harmonic sinusoidal waveform having an excitation frequency of 1 Hz in the prototype scale.

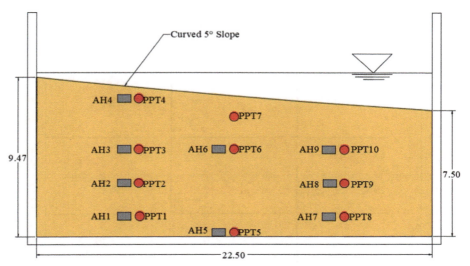

Fig. 2. Uniform model (Model ARS-01; having RD = 40% and Model ARS-02; having RD = 80%)

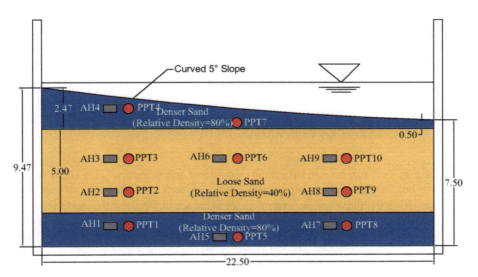

Fig. 3. Multilayer model (Model ARS-03)

A total of 10 shaking excitations were applied to the models to assess the re-liquefaction responses and the behavioral changes in soil fabric response to different levels of dynamic strains and build-up of excess pore pressure. Figure 4 shows the acceleration and excess pore pressure responses towards the downslope side for the three models during the third shaking excitation, having a targeted PGA value of 0.2 g. It is to be noted that prior two shakings were very small (Target 0.030 g and 0.06 g) and resulted in very minor (negligible) development of excess pore pressure.

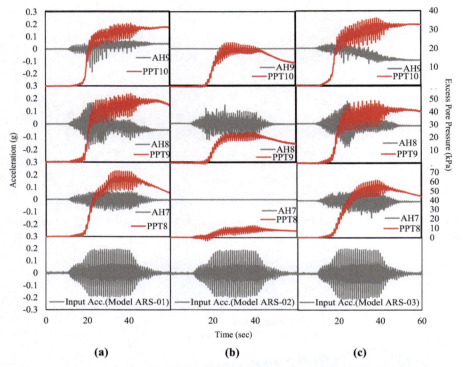

Fig. 4. Soil acceleration and excess pore pressure response near the downslope side **(a)** Model ARS-01 **(b)** Model ARS-02 **(c)** Model ARS-03

From Fig. 4, it can be said that due to significant softening and large shear strains, excessive build-up of excess pore pressure could be seen throughout the model depth for Model ARS-01; the result also signifies the presence of large dilative spikes towards the ground surface, having lesser confining stresses. On the other hand, limited excess pore pressures are generated for a very dense model, which shows considerably larger resistance to the propagating seismic waves. At the same time, the dissipation process is found to start early for a denser model, indicating soil to start regaining its lost shear strength during the shaking itself. Lesser peak excess pore pressure is found to occur at PPT8 for ARS-03 model, primarily due to its location within a dense sand layer.

However, similar responses are captured at PPT10 for ARS-01 and ARS-03, indicating the soil to undergo a similar amount of softening during lateral spreading. The acceleration response at AH9 shows significant drift, possibly due to large magnitude of lateral spreading, which may drag the denser soil layer alongside.

Fig. 5. Lateral soil displacement pattern; with soba noodles (**a**) Model ARS-01 (**b**) Model ARS-02 (**c**) Model ARS-03

Lateral soil displacement along the length of the model (along with soba noodles) before saturation and after the completion of tests (after ten shaking excitations) is shown in Fig. 5. The lateral soil displacements along the model are very minor (for the bottom half) for Model ARS-02, as seen from Fig. 5(b), which could be corroborated by the negligible development of excess pore pressure at PPT8 for Model ARS-02. However, substantial excess pore pressure at PPT10 (though below ru = 1.0 level) resulted in significant strains within the soil model, mobilizing larger displacements (particularly near the center array) for soba noodles towards the ground surface for Model ARS-02. The vertical displacement also seems to be smaller for the uniform denser soil model as compared to the other two cases. Focusing on the multi-layer model, one can observe larger lateral displacements for the soba noodles (compared to uniform loose model), signifying large lateral stresses to be mobilized due to the lateral spreading of dense colored sand. Importantly, the location of the maximum lateral displacement is found to occur at the interface of the two different layers, implying larger mobilization of shear strains at the stiffness discontinuity. The largest lateral stresses are found to occur at the ground surface for a multi-layer model, which are found to be slightly below the ground surface for a uniform loose sand model. Overall, the curved surface is found to change drastically for Model ARS-01 and Model ARS-03.

5 Conclusions

This study aims to investigate the lateral deformation mechanism for different soil models; in the loose and denser state as well as in a layered configuration. The centrifuge tests were conducted for the different soil models installed with soba noodles to trace the lateral deformation pattern along the length of the model. The result highlights limited disturbance/changes to the sloping configuration for a uniform denser soil model, primarily due to lesser mobilization of shear strains and limited excess pore pressure generation towards the bottom part. Maximum lateral displacement was found to occur at the interface of two distinct layers having different stiffness for a multi-layer soil model. Overall, maximum lateral displacements were recorded near the center of the soil model for all three cases, with considerable variability in the spatial distribution of lateral soil displacement, implying structural systems to be subjected to different magnitudes of soil pressure depending on their location within model during the course of lateral spreading.

References

Ishihara, K.: Terzaghi oration: geotechnical aspects of the 1995 Kobe earthquake. In: Proceedings of the International Conference on Soil Mechanics and Foundation Engineering-International Society for Soil Mechanics and Foundation Engineering. A.A. Balkema, Rotterdam (1997)

Tokimatsu, K., Asaka, Y.: Effects of liquefaction-induced ground displacements on pile performance in the 1995 Hyogoken-Nambu earthquake. Soils Found. **38**, 163–177 (1998)

Ishihara, K., Cubrinovski, M.: Case studies on pile foundations undergoing lateral spreading in liquefied deposits. In: Proceedings of 5th International Conference on Case Histories in Geotechnical Engineering. University of Missouri-Rolla, New York (2004)

Cubrinovski, M., et al.: Spreading-Induced damage to short-span bridges in Christchurch, New Zealand. Earthq. Spectra **30**, 57–83 (2014)

Abdoun, T., Dobry, R., O'Rourke, T., Goh, S.H.: Pile response to lateral spreads: centrifuge modeling. J. Geotech. Geoenviron. Eng. **129**(10), 869–878 (2003)

Brandenberg, S.J., Boulanger, R.W., Kutter, B.L., Chang, D.: Behavior of pile foundations in laterally spreading ground during centrifuge tests. J. Geotech. Geoenviron. Eng. **131**(11), 1378–1391 (2005)

Knappett, JA., Madabhushi, SPG.: Liquefaction-induced settlement of pile groups in liquefiable and laterally spreading soils. J. Geotech. Geoenviron. Eng. **134**(11), 1609–1618 (2008)

Motamed, R., Towhata, I., Honda, T., Tabata, K., Abe, A.: Pile group response to liquefaction-induced lateral spreading: E-Defense large shake table test. Soil Dyn. Earthq. Eng. **51**, 36–46 (2013)

Ebeido, A., Elgamal, A., Tokimatsu, K., Abe, A.: Pile and pile-group response to liquefaction-induced lateral spreading in four large-scale shake-table experiments. J. Geotech. Geoenviron. Eng. **145**(10), 04019080 (2019)

Sahare, A., Ueda, K., Uzuoka, R.: Influence of the sloping ground conditions and the subsequent shaking events on the pile group response subjected to kinematic interactions for a liquefiable sloping ground. Soil Dyn. Earthq. Eng. **152**, 107036 (2022)

Kutter, B., et al.: LEAP-GWU-2015 experiment specifications, results, and comparisons. Soil Dyn. Earthq. Eng. **113**, 616–628 (2018)

Kutter, B.L., et al.: Numerical sensitivity study compared to trend of experiments for LEAP-UCD-2017. In: Kutter, B.L., Manzari, M.T., Zeghal, M. (eds.) Model Tests and Numerical Simulations of Liquefaction and Lateral Spreading, pp. 219–236. Springer, Cham (2020). https://doi.org/10.1007/978-3-030-22818-7_11

Vargas, R., Ueda, K., Tobita, T.: Centrifuge modeling of the dynamic response of a sloping ground – LEAP-UCD-2017 and LEAP-ASIA-2019 tests at Kyoto University. Soil Dyn. Earthq. Eng. **140**, 106472 (2021)

He, B., Zhang, J.M., Li, W., Wang, R.: Numerical analysis of LEAP centrifuge tests on sloping liquefiable ground: influence of dilatancy and post-liquefaction shear deformation. Soil Dyn. Earthq. Eng. **137**, 106288 (2020)

Sahare, A., Tanaka, Y., Ueda, K.: Numerical study on the effect of rotation radius of geotechnical centrifuge on the dynamic behavior of liquefiable sloping ground. Soil Dyn. Earthq. Eng. **138**, 106339 (2020b)

Sahare, A., Ueda, K., Uzuoka R.: Sensitivity and numerical analysis using strain space multiple mechanism model for a liquefiable sloping ground. In: Geo-Congress 2020 ASCE: Geotechnical Earthquake Engineering and Special Topics, pp. 51–59 (2020a)

Qiu, Z., Elgamal, A.: Numerical simulations of LEAP centrifuge tests for seismic response of liquefiable sloping ground. Soil Dyn. Earthq. Eng. **139**, 106378 (2020)

Chen, L., Ghofrani, A., Arduino, P.: Remarks on numerical simulation of the LEAP-Asia-2019 centrifuge tests. Soil Dyn. Earthq. Eng. **142**, 106541 (2021)

Reyes, A., Yang, M., Barrero, A.R., Taiebat, M.: Numerical modeling of soil liquefaction and lateral spreading using the SANISAND-Sf model in the LEAP experiments. Soil Dyn. Earthq. Eng. **143**, 106613 (2021)

Cubrinovski, M., Ishihara, K.: Simplified method for analysis of piles undergoing lateral spreading in liquefied soils. Soils Found. **44**(5), 119–133 (2004)

Motamed, R., Towhata, I.: Shaking table model tests on pile groups behind quay walls subjected to lateral spreading. J. Geotech. Geoenviron. Eng. **136**(3), 477–489 (2010)

Probabilistic Calibration and Prediction of Seismic Soil Liquefaction Using quoFEM

Aakash Bangalore Satish[1(✉)], Sang-ri Yi[1], Adithya Salil Nair[2], and Pedro Arduino[3]

[1] University of California, Berkeley, Berkeley, CA 94804, USA
bsaakash@berkeley.edu
[2] The Ohio State University, Columbus, OH 43210, USA
[3] University of Washington, Seattle, WA 98195, USA

Abstract. Liquefaction under cyclic loads can be predicted through advanced (liquefaction-capable) material constitutive models. However, such constitutive models have several input parameters whose values are often unknown or imprecisely known, requiring calibration via lab/in-situ test data. This study proposes a Bayesian updating framework that integrates probabilistic calibration of the soil model and probabilistic prediction of lateral spreading due to seismic liquefaction. In particular, the framework consists of three main parts: (1) Parametric study based on global sensitivity analysis, (2) Bayesian calibration of the primary input parameters of the constitutive model, and (3) Forward uncertainty propagation through a computational model simulating the response of a soil column under earthquake loading. For demonstration, the PM4Sand model is adopted, and cyclic strength data of Ottawa F-65 sand from cyclic direct simple shear tests are utilized to calibrate the model. The three main uncertainty analyses are performed using quoFEM, a SimCenter open-source software application for uncertainty quantification and optimization in the field of natural hazard engineering. The results demonstrate the potential of the framework linked with quoFEM to perform calibration and uncertainty propagation using sophisticated simulation models that can be part of a performance-based design workflow.

Keywords: Liquefaction · PM4Sand · Bayesian calibration · Sensitivity analysis · Uncertainty propagation · quoFEM

1 Introduction

Earthquake-induced soil liquefaction can lead to unexpected casualties and property loss, and therefore, it is essential to predict its risk in advance. Soil behavior under cyclic loads can be simulated using liquefaction-capable constitutive models [1]. For example, PM4Sand is a sand plasticity model capable of simulating liquefaction under a broad mix of conditions in the field, including a wide range of density, shear stress, confining stress, and drainage/loading conditions [2]. This model has been implemented in numerical analysis software, including the open-source computational framework OpenSees [3, 4]. The flexibility of such a model is attained through a large number of parameters (although most of them are typically set to their recommended default values) as well as a wide search space of the possible parameter combinations [5]. While in-situ or lab test results provide information for calibrating unknown parameters [6], some challenges remain in using deterministic calibration methods. For example, the large search space of input parameters and potential multimodality of the calibration objective function make the optimizer susceptible to fall in local optima. Further, soil properties often can only be measured indirectly and reveal high spatial variability, thus limiting the credibility of parameter values estimated from limited test data. Consequently, after identifying a combination of optimal parameter values, a significant amount of uncertainty remains in the estimated parameter values as well as in the response predicted by the model [7]. Therefore, there is a need for adopting probabilistic calibration methods and performing forward uncertainty propagation. For example, by introducing the Bayesian calibration approach, correlations and interactions between parameters as well as multimodality can be captured in terms of multiple near-optimal parameter combinations, i.e., posterior distribution or samples, and such probabilistic representation allows prediction of the uncertainty propagating to the liquefaction-induced lateral spreading at a site of interest [8–10]. Additionally, probability-based sensitivity analysis of input parameters allows us to identify the importance of each parameter while taking inherent uncertainty into account [11, 12].

This study proposes a systemized approach for probabilistic liquefaction prediction updating that consists of three steps: parametric study, parameter calibration based on experimental data, and response prediction. Further, it shows that each probabilistic analysis can be greatly accelerated by using the research tool quoFEM developed by the NHERI SimCenter at UC Berkeley [13]; this is an open-source software application developed to assist researchers and practitioners in the field of natural hazard engineering. quoFEM allows users to link different simulation engines, including OpenSees and FEAP [14] and advanced uncertainty quantification (UQ) and optimization methods with a user interface. Each step of the framework introduces variance-based global sensitivity analysis, transitional Markov chain Monte Carlo-based Bayesian updating, and forward resampling algorithms, respectively, among other alternatives supported in quoFEM. Cyclic Direct Simple Shear (CyDSS) test data of Ottawa F-65 sand from Morales *et al.* (2021) is utilized to calibrate the PM4Sand model to match the prediction of the cyclic strength curve (i.e., the number of cycles to the onset of liquefaction given the cyclic shear stress ratio) from the model to that obtained from the tests [3, 15]. All the analyses are conducted on high-performance computers at DesignSafe-CI using the quoFEM user interface.

2 Probabilistic Framework for Calibration and Prediction of Lateral Spreading Induced by Seismic Soil Liquefaction

The systematic framework for updating predictions of lateral spreading due to soil liquefaction given experimental data consists of three main steps: (1) Variance-based global sensitivity analysis to identify the influence of the primary input parameters on the onset of liquefaction under cyclic loading, (2) Bayesian calibration of the primary input parameters of a soil constitutive model (PM4Sand in this case), and (3) forward uncertainty propagation to update the probability distribution of the lateral spreading under earthquake loading with the updated distribution of parameters.

Global Sensitivity Analysis. Variance-based sensitivity analysis is first introduced for a preliminary parametric study of the soil constitutive model. By identifying the proportion of response variance attributed to each parameter, the sensitivity measures, also known as Sobol indices, are assessed. The main Sobol index accounts for the contribution of a variable to a response, while the total Sobol index additionally accounts for the interaction effects [12, 16]. The main and total indices for i-th parameter, X_i, to a response $F(X)$ are defined respectively as $S_i = \text{Var}_{X_i}[\text{E}_{X_{\sim i}}[F(X)|X_i]]/Var(Y)$ and $S_i^T = \text{E}_{X_{\sim i}}[\text{Var}_{X_i}[F(X)|X_{\sim i}]]/Var(Y)$ where $\text{E}_x[\cdot]$ and $\text{Var}_x[\cdot]$ respectively represent mean and variance operator over a preset range of a variable or joint variables of x. The symbol $X_{\sim i}$ represents a set of all input parameters in the model except X_i, i.e. $X = \{X_i, X_{\sim i}\}$. For efficient estimation of the Sobol indices, a quasi-Monte Carlo-based estimation method is selected through the quoFEM Dakota UQ engine developed by Sandia National laboratory [11, 16].

Bayesian Calibration. Through Bayes' Theorem [8], the joint posterior distribution of all input parameters X as well as measurement noise levels Σ_ε are identified from the dataset D as $p(X, \Sigma_\varepsilon|D) \propto p(D|X, \Sigma_\varepsilon)p(X)p(\Sigma_\varepsilon)$, where $p(X)$ and $p(\Sigma_\varepsilon)$ are respectively prior distributions of parameter values and measurement variance levels, which are set to be uniform, and $p(D|X, \Sigma_\varepsilon)$ represents the likelihood of input parameters given the data D and measurement noise Σ_ε. The likelihood function is often defined by assuming that observation noises follow independent and identically distributed Gaussian distributions without loss of generality, i.e., if D are measurements corresponding to system outputs $F(X)$ with additive noise, $D = F(X) + \varepsilon$ and the noise term ε follows a Gaussian distribution, $N(0, \Sigma_\varepsilon)$, with Σ_ε being a diagonal matrix. The likelihood function of X is defined as $p(D|X, \Sigma_\varepsilon) = \text{N}(F(X), \Sigma_\varepsilon)$. The samples of the posterior distribution are obtained by transitional Markov chain Monte Carlo (MCMC) sampling through quoFEM's UCSD (University of California, San Diego) UQ engine [17, 18]. quoFEM also supports other probabilistic/deterministic calibration methods through the Dakota UQ engine.

Forward Propagation. Forward propagation allows researchers to see how the remaining uncertainty of the input parameters affects other estimations; in this case liquefaction-induced lateral spreading under an earthquake load. For this, the posterior samples obtained in the *Bayesian Calibration* process are directly imported back into quoFEM. Nonparametric naive resampling is conducted to generate samples from the posterior sample set, which are used as inputs to free-field analysis.

3 Problem Description

The framework described in Sect. 2 was applied to a problem of uncertainty analysis during seismic soil liquefaction simulations using the data and the models described in this section.

Table 1. Cyclic strength data from experiments

Cyclic shear stress ratio (CSR)*	Number of cycles to initial liquefaction
0.105	26
0.105	21
0.130	13
0.151	5
0.172	4
0.200	3

* CSR $= \tau_{cyc}/\sigma'_{vo}$, where τ_{cyc} is horizontal cyclic shear stress, σ'_{vo} is vertical consolidation stress

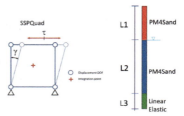

Fig. 1. [Left] – single element FE model used in sensitivity analysis and Bayesian calibration; [Right] – schematic of soil column used in free-field analysis

Experimental Data. Cyclic Direct Simple Shear (CyDSS) tests were conducted on Ottawa F-65 sand by Morales *et al.* (2021) for different Cyclic Stress Ratio (CSR) conditions and the data was shared on the DesignSafe-CI Data Depot [15]. From this data, the number of cycles to the onset of liquefaction was computed using a threshold of 3% on the amplitude of shear strain. The results of these computations are shown in Table 1, and this data is used to calibrate the parameters of the PM4Sand material model.

Material Model. PM4Sand is a constitutive model capable of simulating liquefaction response of sandy soils. In OpenSees, this model has 24 parameters [4], but we considered the three primary input parameters - apparent relative density D_r, shear modulus coefficient G_o, and contraction rate parameter h_{po}, for sensitivity analysis and calibration while the other parameter values were set to the default recommendations in Boulanger and Ziotopoulou (2017) [2].

Finite Element (FE) models. During sensitivity analysis and calibration, a single element, shown on the left-hand side of Fig. 1 was utilized to simulate the material response during a stress-controlled cyclic direct simple shear test. On the right-hand side of Fig. 1, a schematic representation of the synthetic layered soil profile used for 1D free-field analysis is shown. The soil column has a grade of 3%. The top layer, L1, represents a crust of thickness 2 m, L2 represents the liquefiable layer of 3 m thickness, and L3, the bottom 1 m of the soil, is linear elastic. The width of the soil column was 0.25 m, and the domain was discretized using undrained quadrilateral elements (SSPQuadUP) of size 0.25 m × 0.25 m. The response of this soil column was simulated by applying the Loma Prieta ground motion recorded at Gilroy Array #2.

4 Results and Discussion

Global Sensitivity Analysis. Values of the input parameters were randomly sampled from the probability distributions in Table 2 using the Latin Hypercube sampling method and the corresponding number of cycles to initial liquefaction was obtained from simulation with the single element FE model at five CSR values shown in Table 1. As results of variance-based global sensitivity analysis, quoFEM returns the main and total Sobol indices, sample values of the input parameters that were used to estimate the values of the Sobol indices, and corresponding model outputs.

Table 2. Probability distribution of the input parameters

Parameter	Distribution	Range
D_r	Uniform	0.1–0.9
h_{po}	Uniform	0.01–5
G_o	Uniform	200–2000

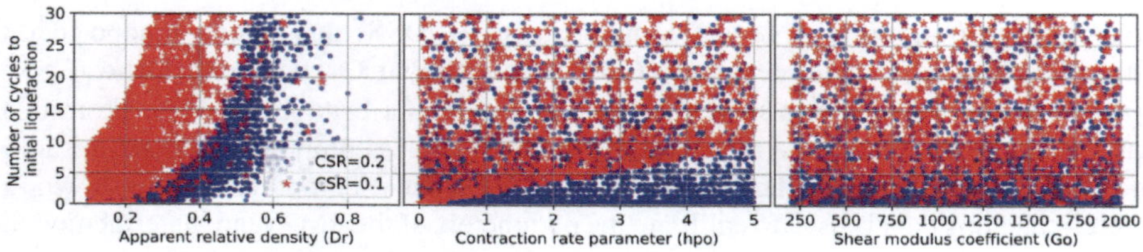

Fig. 2. Dependance of output on the primary input parameters of PM4Sand

Table 3. Main and total Sobol indices for the primary input parameters of PM4Sand to the number of cycles to initial liquefaction at different cyclic stress ratios

	CSR = 0.10		CSR = 0.13		CSR = 0.15		CSR = 0.17		CSR = 0.20	
	Main	Total	Main	Total	Main	Total	Main	Total	Main	Total
D_r	0.849	0.915	0.826	0.896	0.827	0.895	0.817	0.900	0.805	0.885
h_{po}	0.054	0.134	0.024	0.115	0.024	0.101	0.021	0.101	0.025	0.100
G_o	0.000	0.002	0.000	0.002	0.001	0.002	0.003	0.003	0.004	0.006

Figure 2 shows example scatter plots of the number of cycles to liquefaction, N_{cyc}, against each primary input parameter, X_i, given CSR of values 0.1 and 0.2. At any given value of one of the parameters X_i, there is a scatter in the output values which is due to the randomness in the sampled values of the other input parameters $X_{\sim i}$. At low apparent relative densities D_r, the sand is in a loose state and N_{cyc} is low, as expected. It can be observed from the figure that N_{cyc} exhibits a clear dependance on the value of D_r, while there is minor dependance on the value of the contraction rate parameter h_{po} at low CSR values and no clear dependance on the value of the shear modulus coefficient G_o. These qualitative observations of the influence of the input parameters on the outputs are quantified by the Sobol indices enumerated in Table 3. The estimated Sobol indices in the table correlate well with the observations from Fig. 2, with D_r having the highest main and total Sobol indices, followed by h_{po}, with G_o having near-zero Sobol index values at all CSR values studied. h_{po} has higher influence at lower CSR values than at higher CSR values. All the parameters, especially h_{po} have higher total Sobol indices compared to the main indices, which indicates that the inputs are interacting with each other when influencing N_{cyc} predicted by the model.

Bayesian Calibration. During Bayesian calibration, the data of cyclic strength obtained from lab tests at 5 different CSR values, shown in Table 1, were used to estimate the values of the three primary input parameters of the PM4Sand model.

The prior probability density chosen for the three primary input parameters was the same as that used for the sensitivity analysis, shown in Table 2. The prior probability density must be set in a way that reflects all the available information/knowledge about the parameter values before any data are available. In this example, uninformative (uniform) priors were used to reflect lack of information about the parameter values. With this uniform prior, the posterior probability distribution was determined by the information contained in the data; the prior only defining the possible range of values that the parameter estimates can take.

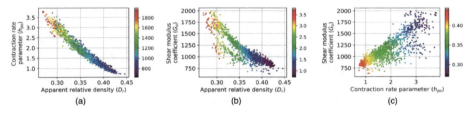

Fig. 3. Samples from posterior probability distribution, colored by (a) shear modulus coefficient, (b) contraction rate parameter, (c) apparent relative density

Figure 3 shows 2000 samples drawn using quoFEM from the joint posterior probability distribution of the three primary input parameters, which characterizes the parameter estimation uncertainty. Since the data used consisted of only the number of cycles to reach a threshold shear strain, there are multiple settings of the primary input parameters that could predict the same or very similar cyclic strength curve but with different dynamic response history. Hence, a unique set of parameter values could not be identified from this data. Samples shown in Fig. 3 capture the interdependence between the estimated parameter values that lead to predictions of similar cyclic strength curves.

quoFEM also returns predictions of the cyclic strength curve corresponding to the sample parameter values. The range of predicted cyclic strength shown in Fig. 4 (a) is due to the parameter estimation uncertainty. The mean of the predicted cyclic strength is also shown in Fig. 4 (a), along with the lab test data used for calibration, and a previous deterministic calibration to the same dataset by Ziotopoulou et al. (2018) [5]. From the figure, it is evident that the cyclic strength data from lab tests is of limited amount and noisy, introducing uncertainty about the true response of the soil underlying the measured values. Additionally, the computational model might not represent the true response exactly, which introduces another layer of uncertainty. Bayesian calibration accounts for these sources of uncertainty in parameter estimation and can lead to more robust predictions. For example, as seen in Fig. 4 (a), the range of predictions of cyclic strength from Bayesian estimation covers most of the data. The values of the root mean square error (RMSE) in the predictions of the number of cycles to initial liquefaction, shown in the legend in Fig. 4 (a), indicate that the mean of the predicted cyclic strength obtained by Bayesian calibration matches the data more closely on average than the prediction achieved by deterministic calibration.

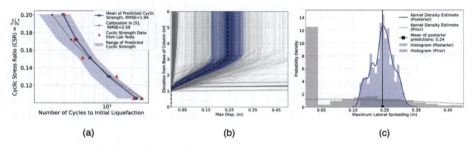

Fig. 4. Predictions from posterior sample values of the calibrated material model parameters

Uncertainty Propagation. After Bayesian calibration, the uncertainty in the estimated values of the primary input parameters was represented by the 2000 posterior samples shown in Fig. 3. These values were used as inputs to the PM4Sand model representing the behavior of the top two layers of soil, shown in Fig. 1, during simulation of the response of the soil column to a single earthquake ground motion record (Loma Prieta Gilroy Array #2) corresponding to a PGA of 0.37 g. The results of the uncertainty propagation analysis are shown in Fig. 4 (b) and (c). Figure 4 (b) shows the updated uncertainty in the predicted maximum lateral displacement profiles using samples from the posterior probability density of the parameters (blue lines) compared with prediction by samples from the prior (gray lines). The mean of the predictions from the posterior is shown with the black line. Figure 4 (c) shows the distribution of the maximum lateral spreading at the top of the soil column (mean: 0.24 m, standard deviation: 0.03 m).

5 Conclusions

This paper presented a framework for characterizing the uncertainty in the site response due to seismic soil liquefaction using the open-source software application quoFEM. Further directions can be to quantify the probability of damage to structures and economic loss because of seismic soil liquefaction, and regional scale analysis of risk to infrastructure induced by seismic liquefaction.

Acknowledgements. This material is based upon work supported by the National Science Foundation under Grant No. (1612843 & 2131111). Any opinions, findings, and conclusions or recommendations expressed in this material are those of the author(s) and do not necessarily reflect the views of the National Science Foundation.

References

1. Carey, T.J., Kutter, B.L.: Comparison of liquefaction constitutive models for a hypothetical sand. Geotech. Front. **2017**, 389–398 (2017)
2. Boulanger, R.W., Ziotopoulou, K.: PM4Sand (Version 3.1): A sand plasticity model for earthquake engineering applications. Department of Civil and Environmental Engineering, University of California, Davis, Davis, CA, Report UCD/CGM-17/01 (2017)
3. McKenna, F.: OpenSees: a framework for earthquake engineering simulation. Comput. Sci. Eng. **13**(4), 58–66 (2011)
4. Chen, L., Arduino, P.: Implementation, verification, and validation of the PM4Sand model in OpenSees. Pacific Earthquake Engineering Research (PEER) Center, University of California, Berkeley, Berkeley, USA, Report 2021/02 (2021)
5. Ziotopoulou, J., Montgomery, J., Bastidas, A.M.P., Morales, B.: Cyclic Strength of Ottawa F-65 sand: laboratory testing and constitutive model calibration. Geotech. Earthq. Eng. Soil Dyn. **293**, 180–189 (2018)
6. Ziotopoulou, K., Boulanger, R.W.: Calibration and implementation of a sand plasticity plane-strain model for earthquake engineering applications. Soil Dyn. Earthq. Eng. **53**, 268–280 (2013)
7. National Academics of Sciences, Engineering, and Medicine: State of the Art and Practice in the Assessment of Earthquake-Induced Soil Liquefaction and Its Consequences. The National Academies Press, Washington, DC, USA (2016)
8. Kruschke, J.: Doing Bayesian Data Analysis: A Tutorial with R, JAGS, and Stan, 2nd edn. Elsevier, Bloomington (2014)
9. Mercado, V., et al.: Uncertainty quantification and propagation in the modeling of liquefiable sands. Soil Dyn. Earthq. Eng. **123**, 217–229 (2019)
10. Chen, L.: Implementation, verification, validation, and application of two constitutive models for earthquake engineering applications. Dissertation, University of Washington (2020)
11. Adams, B.M., et al.: Dakota, A Multilevel Parallel Object-Oriented Framework for Design Optimization, Parameter Estimation, Uncertainty Quantification, and Sensitivity Analysis: Version 6.13 User's Manual (No. SAND2020-12495). Sandia National Laboratories (SNL-NM), Albuquerque, NM (United States) (2020)
12. Hu, Z., Mahadevan, S.: Probability models for data-driven global sensitivity analysis. Reliab. Eng. Syst. Saf. **187**, 40–57 (2019)
13. McKenna, F., Zsarnoczay, A., Gardner, M., Elhaddad, W., Yi, S., Aakash, B.S.: NHERI-SimCenter/quoFEM: Version 2.4.0 (v2.4.0). Zenodo (2021). https://doi.org/10.5281/zenodo.5558000
14. Taylor, R.L.: FEAP-A finite element analysis program (2014)
15. Morales, B., Humire, F., Ziotopoulou, K.: Data from: Cyclic Direct Simple Shear Testing of Ottawa F50 and F65 Sands (Feb. 1st, 2021). Distributed by Design Safe-CI Data Depot. https://doi.org/10.17603/ds2-eahz-9466. Accessed 28 June 2021
16. Weirs, V.G., et al.: Sensitivity analysis techniques applied to a system of hyperbolic conservation laws. Reliab. Eng. Syst. Saf. **107**, 157–170 (2012)
17. Ching, J., Chen, Y.C.: Transitional Markov chain Monte Carlo method for Bayesian model updating, model class selection, and model averaging. J. Eng. Mech. **133**(7), 816–832 (2007)
18. Minson, S.E., Simons, M., Beck, J.L.: Bayesian inversion for finite fault earthquake source models I—theory and algorithm. Geophys. J. Int. **194**(3), 1701–1726 (2013)

Fluid-Solid Fully Coupled Seismic Response Analysis of Layered Liquefiable Site with Consideration of Soil Dynamic Nonlinearity

Yiyao Shen, Zilan Zhong[✉], Liyun Li, and Xiuli Du

Beijing University of Technology, Beijing, China
zilanzhong@bjut.edu.cn

Abstract. The seismic response study of a layered liquefiable site is crucial in the seismic design of both aboveground and underground structures. This study introduces one-dimensional dynamic site response processes with advanced nonlinear soil constitutive models for non-liquefiable and liquefiable soils in the OpenSees computational platform. The solid-fluid fully coupled plane-strain u-p elements are used to simulate the soil elements. This study investigates the seismic response of a layered liquefiable site with specific focus on the development of excess pore water pressure, acceleration and post-earthquake ground surface settlement under two typical earthquake excitations. The numerical results show that the ground motion characteristics as well as the site profile have significant effects on the dynamic response of the layered liquefiable site. The loose sand layer with 35% relative density is more prone to liquefaction and contractive deformation under the same intensity of ground motion, resulting in irreversible residual deformation and vertical settlement. The saturated soil layer may efficiently filter the high-frequency components of ground motions while amplifying the low-frequency components. Meanwhile, during the post-earthquake excess pore pressure dissipation, the soil produce a large consolidation settlement.

Keywords: Liquefiable soil · Solid-fluid fully coupled analysis · Numerical simulation · Seismic analysis · Constitutive models

1 Introduction

Earthquake-induced ground failure, triggered by liquefaction of loose sand, silt, and gravelly soil deposits, has caused serious damage on engineering structures and infrastructure [1]. For instance, it was reported that settlement in the Tokyo Bay area was generally in the range between 30 to 50 cm during the 2011 Great East Japan Earthquake, and settlement in some extensively liquefied regions was as large as 70 to 100 cm. This macroscopic phenomenon indicates that the soil layers generates substantial consolidation settlement as a result of the generation and dissipation of excess pore water pressure (EPWP), which may result in full or partial instability of aboveground or underground structures. As a result, it is crucial to investigate the dynamic characteristics of the seismic liquefaction-prone sites.

Numerical simulation is a useful tool for estimating site response and evaluating the seismic safety of engineering sites. In recent years, many numerical studies on the seismic response of saturated soil sites have been conducted. Liyanathirana et al. [2] suggested a coupled effective stress method by combining the Hardin-Drnevich model with the pore pressure growth model proposed by Seed et al. [3]. The proposed method was then validated using seismic acceleration records from the Port Island downhole array obtained during the 1995 Hanshin earthquake in Japan. Ziotopoulou et al. (2012) [4] used three different constitutive models to numerically simulate the seismic liquefaction process at the two sites. According to the research results, dilatancy is an important factor in capturing appropriate ground motions in liquefiable sites. Wang et al. (2016) [5] established a three-dimensional finite element dynamic analysis model of piles in liquefiable ground using the OpenSees software platform and conducted numerical simulations of centrifuge shaking table tests of piles in liquefiable horizontal foundations. During the shaking, special attention was paid to the ground acceleration, EPWP, and the bending moment of the pile shaft. Limited attention has been paid to the ground settlement and pore water pressure dissipation process of the liquefiable soil strata following earthquakes.

In view of the above, the dynamic response of saturated soils under earthquake excitations is a complex mechanical process, which involves the soil's nonlinear hysteretic behavior and the coupling volumetric-distortional deformation. The effects of liquefaction on geotechnical structures are increasingly being evaluated in practice using nonlinear time history analyses, but previous studies often used a 'loosely coupled' approach that predicts pore pressure by using relationships in conjunction with total stress constitutive models [2, 6]. This approach fails to reasonably consider the influence of dilatancy and post-liquefaction shear deformation of the soil strata. In this study, concentrating on the effects of liquefaction, a 'fully coupled' approach is used to evaluate the nonlinear seismic response in liquefiable multi-layered soils under earthquake loading. To characterize the complicated nonlinear dynamic behavior of sand, the advanced soil constitutive model is used. The dynamic response characteristics associated with soil liquefaction are described in detail in this study based on numerical analyses.

2 Numerical Model of Liquefiable Site

2.1 Liquefiable Site Model

In this study, the numerical analyses are performed using the fully coupled dynamic effective stress method, which is available in the OpenSees platform [7]. Figure 1 depicts the schematic diagram of a generic site consisting of three soil layers based on previous studies [8] without the initial static shear stress. The soil domain was discretized using four-node quadrilateral elements with u-p formulation [9] such that water pressure and soil deformation could be estimated concurrently. The overall depth and width of the soil deposit are 21 m and 50 m, respectively. The soil deposit mainly consists of sand with relative density, $D_r = 35\%$ and 75% and crust clayed soil. The water table is assumed to be the same with the ground surface level. Therefore, the ground surface is supposed to be a free drainage surface, while the lateral and base boundaries are impermeable. Horizontal kinematic constraints are applied to the nodes on two side boundaries to ensure the identical horizontal movement of the two nodes at the same burial depth and effectively simulate the shear deformations of the soil layers under upward propagation of in-plane waves. The analyses are divided into two steps: gravity application and dynamic excitation. Before applying the seismic excitation, gravity is applied to the model with the nodes fixed at the base to obtain the distributions of initial effective confinement and pore pressure in the soil model for the site. Subsequently, the seismic input motion is applied to the base of the soil model, as a force-time history at the node which shares equal degrees of freedom with the Lysmer-Kuhlemeyer dashpot [10]. This dashpot demonstrates the finite stiffness of the underlying bedrock, with the soil profiles having a bedrock shear wave velocity of 760 m/s and a mass density of 2200 kg/m^3.

In the dynamic analyses, the Newmark integrator method with the parameters $\gamma = 0.6$ and $\beta = 0.3025$ is applied. Rayleigh damping with an initial damping ratio of 3% is used to stabilize the numerical results in the dynamic numerical analysis.

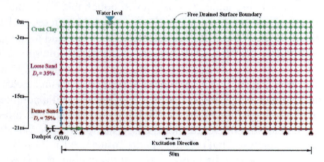

Fig. 1. Numerical configuration of soil column model for the simulation.

2.2 Constitutive Model and Input Parameters

The critical-state, bounding surface dynamic elastoplastic constitutive model [11], PM4Sand model, is used in this study to simulate the nonlinear response of the saturated sand layers, in which the model could capture the dilatancy, nonflow liquefaction, permanent shear strain accumulation, and continuously updating pressure dependent shear modulus. These characteristics are critical for simulating the undrained response of sands exposed to cyclic stress. Reasonable approximations of desired dynamic behavior of soil can be obtained from the numerical analyses including pore pressure generation and dissipation, limiting strains, and cyclic mobility by changing three primary input parameters, shear modulus coefficient, G_o, apparent relative density, D_r, contraction rate parameter, h_{po}. Default values are assigned to the rest secondary parameters. Table 1 shows the PM4Sand model parameters for loose sand and dense sand of two relative density levels, which have been rigorously calibrated to emphasize accurate simulation of liquefaction behavior [12]. Meanwhile, the nonlinear hysteretic behavior of clay soil under cyclic loading was represented using the Pressure Independent Multi-Yield kinematic plasticity model [13] with a von Mises multi-yield-surface, where plasticity arises exclusively in the deviatoric stress-strain response. In the linear-elastic response, the volumetric stress-strain response is independent of the deviatoric stress and insensitive to the confining pressure variation. Table 2 also listed the critical mechanical parameters of the crust clay.

Table 1. PM4Sand input parameters of the loose and dense sand layers.

Primary parameter	D_r	Relative density	35%	75%
	G_o	Shear modulus coefficient	476.0	890.0
	h_{po}	Contraction rate parameter	0.53	0.63
	ρ	Mass density (ton/m^3)	1.7	2.1
Secondary parameter	e_{max}	Maximum void ratio	0.8	0.8
	e_{min}	Minimum void ratio	0.5	0.5
	φ_{cv}	Critical state effective friction angle	33	40
	c_z	Controls the strain level at which fabric becomes important	200.0	200.0
	h_0	Variable that adjusts the ratio of plastic modulus to elastic modulus	0.4	0.4
	n_b	Bounding surface parameter	2	0.97
	n_d	Dilatancy surface parameter	0.1	0.1
	n_u	Poisson's ratio	0.3	0.3
	R	Critical state line parameter 1	1.5	1.5
	m	Yield surface constant	0.01	0.01
	Q	Critical state line parameter 2	10.0	10.0
Permeability coefficient/(m/s)			6.88×10^{-4}	1×10^{-5}

Table 2. Constitutive model parameters of the crust clay.

Parameters	Crust clay
Saturated soil mass density, ρ/(ton/m^3)	1.60
Reference shear modulus, Gr/kPa	50000
Reference bulk modulus, Br/kPa	250000
Cohesion, c/kPa	33
Octahedral peak shear strain, γ_{max}	0.1
Friction angle, ϕ/°	0.0
Reference mean effective confining pressure/kPa	100
Pressure coefficient	0.0
Number of yield surfaces	20
Permeability coefficient/(m/s)	1.2×10^{-8}

2.3 Input Ground Motions

Two acceleration time histories were employed as input motions in the horizontal direction of the elastic bedrock for the numerical simulations in this study. The two motions were recorded during the 1940 Imperial Valley earthquake at the El-Centro station (termed as El-Centro motion in this study) and the 1995 Kobe earthquake at the KJMA station (termed as Kobe motion in this study), respectively. The El-Centro motion, with an amplitude of 0.348 g and the main shaking duration is about 26 s, has the obvious characteristics of the middle- to far-field ground motions. The frequency content of the El-Centro motion concentrates in range of 0–10 Hz. The Kobe motion, with an amplitude of 0.83 g, has typical characteristics of near-fault ground motion with velocity pulse the concentration range of the excitation frequency in 0–5 Hz. Figure 2 presents the acceleration time histories, Arias intensity time histories and Fourier spectra of the two input motions after scaled to a peak bedrock acceleration of 0.30 g. It is worth noting that the energy flux for the scaled El-Centro motion is 2218 J/m^2/s, which is more than twice of the scaled Kobe motion (991 J/m^2/s).

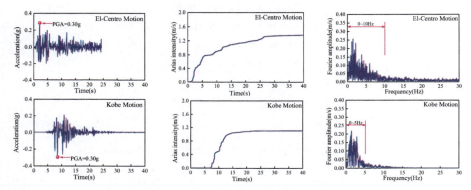

Fig. 2. Acceleration time histories, Arias intensity build-up and Fourier spectra of the input motions.

3 Numerical Analysis and Results

The computational domain is symmetrical in x = 25 m, as shown in Fig. 1. For brevity, in this study, the seismic dynamic analysis results at x = 25 m are used as the representatives to exhibit the dynamic responses of the layered liquefiable site.

3.1 Progressive Liquefaction of Site

EPWP is a critical indicator of liquefaction onset and stress state in saturated soils. Figure 3 depicts the EPWP along soil depth at different moments under two input motions. The initial effective vertical stress $\sigma'_{v0} = \gamma' h$ is also plotted as reference, where γ' is the submerged unit weight of the loose sand and h is the burial depth. When

EPWP reaches the initial effective vertical stress line at a specific depth, the soil loses its shear strength and acts like a liquid. It can be seen from Fig. 3 (a) that the liquefiable loose sand layer with an overlaying clay layer with exceptionally low permeability prevents pore water from dissipating upward, causing EPWP to progressively accumulate at the interface between the non-liquefied clay layer and the liquefied loose sand layer. Consequently, the upper layer of saturated loose sandy soil firstly achieves the liquefaction initiation state within 3 s, and then the soil liquefaction gradually spreads downward. As demonstrated in Fig. 3 (b), the majority of EPWP are generated during the first 6–20 s of the Kobe motion, which corresponds to most of its Arias intensity accumulation. Figure 4 depicts the shear strain distribution along soil depth at different moments under two input motions. Significant shear strain concentration can be observed in the loose sand.

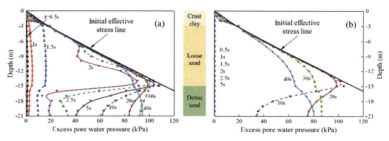

Fig. 3. Distribution of EPWP along soil depth at different moments (a) under the El-Centro motion; (b) under the Kobe motion.

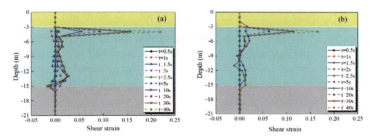

Fig. 4. Distribution of shear strain along soil depth at different moments (a) under the El-Centro motion; (b) under the Kobe motion.

3.2 Acceleration Response of Site

The acceleration time histories of free-field at different depths are shown in Fig. 5, and it can be shown that the acceleration response patterns at the depths of 15 m and 18 m are basically consistent with the input acceleration time histories. However, the acceleration amplitudes of the loose sand layer at 3 m and 9 m depths significantly reduce compared with the input acceleration, indicating that the liquefied soil layer dissipates seismic energy due to soil softening and increased damping on the seismic wave transmission. When the input acceleration is compared to the acceleration time histories of the loose sand layer and the clay layer, it can be found that the acceleration period is elongated owing to liquefaction of the loose sand layer. The acceleration responses at different depths under the Kobe motion are similar to those for the El-Centro case.

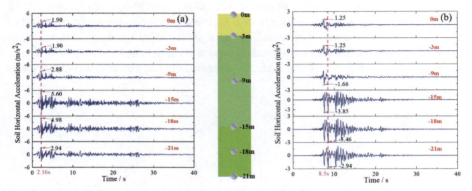

Fig. 5. Free-field acceleration time histories at different depths (a) under the El-Centro motion; (b) under the Kobe motion.

3.3 Settlement Deformation of the Ground Surface

The results of the layered liquefiable site obtained using OpenSees after the main shaking under the two ground motions are shown in Fig. 6. The ground surface settlement in the El-Centro case is 11.46 mm, with 5.3 mm generated during the post-earthquake pore pressure dissipation period, accounting for 31% of the overall settlement. In the Kobe case, a total settlement of 14.60 mm is achieved at the ground surface, and 9.77 mm occurs during the post-earthquake pore pressure dissipation period, accounting for 40% of the overall settlement.

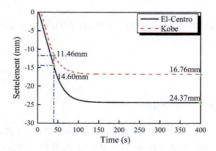

Fig. 6. Time history of settlement at the ground surface under the El-Centro and Kobe motions.

4 Conclusions

In this study, the fully-coupled nonlinear dynamic analyses of layered liquefiable site were conducted using the OpenSees program. The dynamic behavior of the liquefiable soils was modeled by an extensively validated soil constitutive model for sand soil. The dynamic behavior of the non-liquefiable soils was modeled by a von Mises multi-surface model, which can characterize the nonlinear hysteretic behavior of the clay soil under cyclic loading. The excess pore water pressure, acceleration, settlement deformation in liquefiable site were thoroughly investigated in detail during and after earthquakes, the following conclusions can be drawn from this study.

(1) The acceleration amplitude of the soil decreases dramatically after soil liquefaction, indicating that the liquefied soil significantly dissipates the input seismic energy. The peak accelerations of the ground surface under the El-Centro motion and the Kobe motion attenuate to 64% and 42% of the input ground motions, respectively.
(2) The low permeability clay prevents the upward dissipation of excess pore water pressure in the liquefiable sand, thus gradually accumulating excess pore water pressure at the interface of the clay and loose sand layers. The upper layer of loose sand layer experiences significant post-liquefaction flow deformation.
(3) The PM4Sand elastoplastic constitutive model is used to quantitatively simulate the liquefaction and reconsolidation settlement processes of a layered deposit. When the post-earthquake reconsolidation is considered, the overall settlement increases approximately by 40%.

Acknowledgements. This work presented in this paper was supported by the Scientific Research Fund of Institute of Engineering Mechanics, China Earthquake Administration (Grant No. 2018D09), the National Key Research and Development Program of China (2018YFC1504305), and the National Natural Science Foundation of China (51978020).

References

1. Sassa, S., Takagawa, T.: Liquefied gravity flow-induced tsunami: first evidence and comparison from the 2018 Indonesia Sulawesi earthquake and tsunami disasters. Landslides **16**(1), 195–200 (2018). https://doi.org/10.1007/s10346-018-1114-x
2. Liyanathirana, D.S., Poulos, H.G.: Numerical simulation of soil liquefaction due to earthquake loading. Soil Dyn. Earthq. Eng. **22**(7), 511–523 (2002)
3. Seed, H.B., Martin, P.P., Lysmer, J.: Pore-water pressure changes during soil liquefaction. J. Geotech. Eng. Div. **102**(4), 323–346 (1976)
4. Ziotopoulou, K., Boulanger, R.W., Kramer, S.L.: Site response analysis of liquefying sites. In: GeoCongress 2012: State of the Art and Practice in Geotechnical Engineering, pp. 1799–1808 (2012)
5. Wang, R., Fu, P., Zhang, J.: Finite element model for piles in liquefiable ground. Comput. Geotech. **72**, 1–14 (2016)
6. Cubrinovski, M., Rhodes, A., Ntritsos, N., Ballegooy, S.: System response of liquefiable deposits. Soil Dyn. Earthq. Eng. **124**, 212–229 (2019)

7. Mazzoni, S., McKenna, F., Fenves, G.: Open system for earthquake engineering simulation user manual. Pacific Earthquake Engineering Research Center, University of California, Berkeley (2006). http://opensees.berkeley.edu
8. Boulanger, R.W., Kamai, R., Ziotopoulou, K.: Liquefaction induced strength loss and deformation: simulation and design. Bull. Earthq. Eng. **12**(3), 1107–1128 (2013). https://doi.org/10.1007/s10518-013-9549-x
9. Zienkiewicz, O.C., Chan, A.H.C., Pastor, M., Schrefler, B.A., Shiomi. T. Computational Geomechanics with Special Reference to Earthquake Engineering. Wiley, New York (1999)
10. Lysmer, J., Kuhlemeyer, A.M.: Finite dynamic model for infinite media. J. Eng. Mech. Div. **95**, 859–877 (1969)
11. Boulanger, R.W., Ziotopoulou, K.: PM4Sand (Version 3): A sand plasticity model for earthquake engineering applications. Technical Report UCD/CGM-15/01, University of California, Davis (2015)
12. Idriss, I.M., Boulanger, R.W.: Soil liquefaction during earthquakes. Monograph MNO-12, Earthquake Engineering Research Institute, Oakland, CA, p. 261 (2008)
13. Yang, Z., Lu, J., Elgamal, A.: OpenSees Soil Models and Solid-Fluid Fully Coupled Elements User Manual. University of California, San Diego, California (2008)

Variation in Hydraulic Conductivity with Increase in Excess Pore Water Pressure Due to Undrained Cyclic Shear Focusing on Relative Density

Toshiyasu Unno[1](✉), Akiyoshi Kamura[2], and Yui Watanabe[1]

[1] Utsunomiya University, Utsunomiya, Japan
unno@cc.utsunomiya-u.ac.jp
[2] Tohoku University, Sendai, Japan

Abstract. This paper presents the variation in the hydraulic conductivity of a sandy soil with the increase in the excess pore water pressure by conducting falling head permeability tests under undrained cyclic shear while focusing on the relative density of the soil. A pressurized drainage tank with a double-pipe burette for water injection and a standpipe for drainage were newly installed in a hollow torsional shear apparatus. The testing results showed that the hydraulic conductivity remained practically unchanged after consolidation until the excess pore water pressure ratio increased to approximately 0.9 under the cyclic loading. However, when the excess pore water pressure ratio increased above 0.9 owing to the cyclic loading, the hydraulic conductivity rapidly increased regardless of the relative density.

Keywords: Hydraulic conductivity · Permeability test · Liquefaction · Cyclic shear test

1 Introduction

Postliquefaction behaviors, such as the differential settlement or lateral flow associated with buildup and the dissipation of the excess pore water pressure, are still important problems in geotechnical engineering, particularly in performance-based design. For example, Sento et al. [1] and Kazama et al. [2] indicated the necessity of evaluating the deformation behavior of a liquefied ground for performance-based design, and explained the importance of determining the permeability of the ground and the boundary conditions for it.

There have been some studies on the permeability during the liquefaction of soil; however, most of their findings are obtained by model experiments or numerical analysis (e.g., Ueng et al. [3]). In particular, few permeability tests have been conducted directly on specimens to measure the permeability during undrained cyclic shear tests (e.g., Watanabe et al. [4]). Hydraulic conductivity, which affects the behavior of the excess pore water pressure during an earthquake, is important for the verification of the deformation performance of a ground. Therefore, in this study, permeability tests under undrained cyclic shear were conducted to measure and evaluate the variation in the hydraulic conductivity of a sandy soil with the increase in the excess pore water pressure.

2 Test Series and Soil Properties

All tests were performed with two types of soils: Toyoura and Yahagi sand. Figure 1 shows the physical properties and grain size distribution curve of each sand. The hydraulic conductivity values measured by the standard permeability test (JIS A 1218 [5]) were 1.05×10^{-4} (m/s) for the Toyoura sand and 1.47×10^{-4} (m/s) for the Yahagi sand. In this study, the falling head permeability test was adopted to match the conditions of the permeability test using a hollow torsional shear apparatus, which is mentioned subsequently.

To evaluate the effects of the relative density of the soils, the test series consisted of two types of relative densities. Table 1 lists all test series and their soil properties. The dry densities of the specimens were adjusted to the same densities as after the cyclic shear tests.

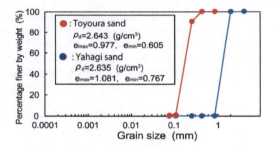

Fig. 1. Grain size distribution of each sand

Table 1. Test series and soil properties

Test series	Sample	Relative density (%)	Dry density (g/cm^3)
T50-1	Toyoura sand	50	1.48
T50-2	Toyoura sand	50	1.48
T90	Toyoura sand	90	1.61
Y50-1	Yahagi sand	50	1.37
Y50-2	Yahagi sand	50	1.39
Y90	Yahagi sand	90	1.49

3 Testing Apparatus and Methodology

Each soil specimen had an outer diameter of 7.0 cm, an inner diameter of 3.0 cm, a height of 10.0 cm, and a volume of 314.0 cm^3. A series of falling head permeability tests were conducted, and the method of the permeability tests and calculation of the conductivity were according to JIS A 1218 [5].

A schematic of the hollow torsional shear apparatus modified to perform the permeability tests is shown in Fig. 2. This apparatus enables the measurement of the flow rate and the hydraulic conductivity by pouring and draining water into a specimen subjected to undrained cyclic shear. A double burette pipe for water injection (upstream) and a pressurized drainage tank with a standpipe (downstream) were attached to the hollow torsional shear apparatus to allow water injection into the specimen under the backpressure condition. The water level in the upstream double burette pipe was varied from 120 cm to 90 cm above the bottom of the apparatus during the measurement of the hydraulic conductivity to conduct the falling head permeability test. The head of the standpipe in the drainage tank was set at a constant height of 28.5 cm above the bottom of the apparatus. The apparatus was designed with a tube having a uniform inner diameter (OD × ID: 6.4 × 4.6 mm) to reduce the hydraulic head loss. The thickness of the porous stone sandwiching the specimen was 1.0 cm each.

Excess pore water pressure was generated during undrained cyclic loading. Therefore, the falling head permeability tests were conducted by applying the sum of the backpressure and the excess pore water pressure to the double-bullet pipe and the pressurized drainage tank during the permeation of the specimens.

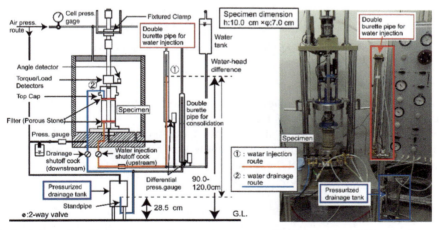

Fig. 2. Schematic of hollow torsional shear apparatus modified to conduct falling head permeability test

The cyclic shear loading applied to the specimens was set according to the same procedure used in general undrained cyclic shear tests, and only specimens with a B-value greater than 0.95 were used. The backpressure was 100 kPa, the initial effective

1720 T. Unno et al.

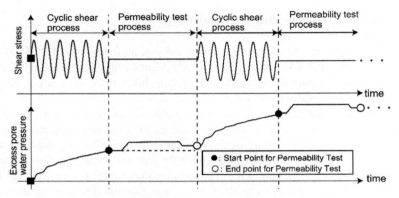

Fig. 3. Permeability test procedure during undrained cyclic shear process

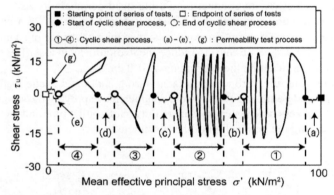

Fig. 4. Example of effective stress path in permeability test combined with undrained cyclic shear (Test series; T50-2)

confining pressure was 100 kPa, and the specimens were isotropically consolidated. The cyclic shear history was sinusoidal (f = 0.02 Hz).

Figure 3 shows a schematic of the permeability test procedure, and Fig. 4 presents an example effective stress path in the permeability tests combined with undrained cyclic shear. The actual procedure of the falling head permeability test of a specimen under cyclic shear loading is as follows:

1. After isotropic consolidation of the backpressurized specimen, cyclic shear loading is applied under undrained conditions. When the target value of the excess pore water pressure ratio is reached, the cyclic shear loading is stopped while maintaining the undrained condition. If the excess pore water pressure fluctuates, the specimen is allowed to stand until the value stabilizes (up to 10 min).
2. A pressure equivalent to the sum of the backpressure and the excess pore water pressure is applied to the double burette pipe (upstream) and the pressurized drainage tank (downstream). Specifically, the pressure values of the specimen, double burette

pipe, and pressurized drainage tank are equalized. The vertical displacement of the specimen due to the water injection is fixed using a clamp.
3. The piping routes of both the double burette pipe and pressurized drainage tank are opened to measure the values required for the hydraulic conductivity calculation. The injected water is poured into the bottom of the specimen from the upstream double burette pipe and drained into the standpipe of the downstream pressurized drainage tank. To limit the effect on the undrained cyclic shear process, the water shutoff cocks for the injection and drainage are closed simultaneously after the measurement is completed.
4. The above process is repeated until the excess pore pressure ratio of the specimen reaches 1.0.

Here, as shown in the time history of the excess pore water pressure ratio in Fig. 3, the excess pore water pressure ratio of a specimen increases when water is injected from the upstream double-burette pipe. This is due to the difference in the water head between the water levels in the upstream double-burette pipe and the position of the standpipe in the downstream pressurized drainage tank. Therefore, the excess pore water pressure ratio at the end of a cyclic shear loading does not coincide with that at the beginning of the subsequent cyclic shear loading, as shown in Fig. 4. However, because the difference between the excess pore water pressure ratios of the different sections is approximately 5% of the initial confining effective pressure, it is considered that it has no significant effect on the results described subsequently.

4 Hydraulic Conductivity of Specimens Subjected to Undrained Cyclic Shear Focusing on Relative Density

The relationships between the excess pore water pressure ratio and hydraulic conductivity of the Toyoura sand specimens are shown in Fig. 5. The white and red-filled plots show the results for Dr = 50% and 90%, respectively. Until the excess pore water pressure ratio increases to approximately 0.9, the hydraulic conductivity is similar to the value after consolidation (plots of water pressure ratio = 0). However, as the cyclic shear progresses and the excess pore water pressure ratio increases above 0.9, the hydraulic conductivity increases rapidly. These results are obtained as a common trend regardless of the relative density, and all plots seem to be univocally on a single curve. As pointed out by Ueng et al. [3], this may be a result of the soil particles gradually losing contact with each other with the increase in the excess pore water pressure ratio, allowing water to flow more easily through the void spaces.

The results for the Yahagi sand specimens shown in Fig. 6 also indicate that the hydraulic conductivity increases rapidly when the excess pore water pressure ratio exceeds 0.9, and the trend appears to be independent of the relative density. The results of all samples show an increase in the hydraulic conductivity of approximately 3.5–4.7 times compared to those of the specimens after consolidation. For all samples, the values of hydraulic conductivity increased with the generation of excess pore water pressure and were of similar orders of magnitudes, regardless of the density, and their results seemed to follow the same curve.

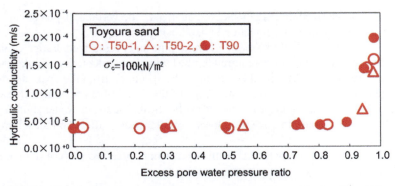

Fig. 5. Relationships between excess pore water pressure ratio and hydraulic conductivity of different Toyoura sand specimens (T50-1, T50-2, and T90)

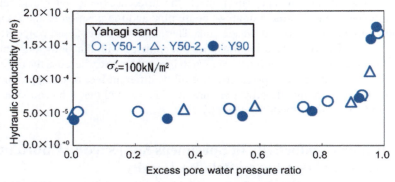

Fig. 6. Relationships between excess pore water pressure ratio and hydraulic conductivity of different Yahagi sand specimens (Y50-1, Y50-2, and Y90)

In contrast to the above, as shown in Fig. 7, when two different samples are subjected to different undrained cyclic shear histories, their behavior can be distinguished in terms of the accumulated shear strain versus the hydraulic conductivity. From the perspective of the cumulative shear strain, the hydraulic conductivity shows a sharp increase; however, the degree of increase cannot be distinguished by the relative density. Therefore, the accumulated shear strain is an effective index for representing the degree of shear strain history, and it is defined as $\gamma_{acm} = \int |\dot{\gamma}(t)| dt$, where $\dot{\gamma}(t)$ is the velocity of the shear strain at time t.

It should be noted that the permeability tests conducted using the hollow torsional shear apparatus may be affected by the hydraulic losses inherent to the apparatus itself. Consequently, the hydraulic conductivity measured by the modified hollow torsional shear apparatus (immediately after consolidation) does not match the value obtained by the general permeability test, as mentioned in Sect. 2. The apparatus that is used in this study has more pipelines and joints than in a general permeability testing machine. The influence of the hydraulic characteristics of the permeability testing apparatus itself was pointed out by Daniel et al. [6] and Black et al. [7], and it is considered necessary to

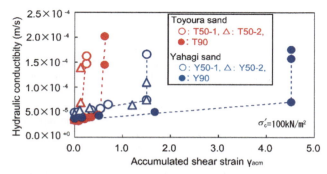

Fig. 7. Relationships between accumulated shear strain and hydraulic conductivity of all samples

evaluate these effects quantitatively for the apparatus used in this study. This point needs to be noted and is an issue for a future study.

5 Conclusions

The following conclusions were obtained from permeability tests conducted under undrained cyclic shear loading using a hollow torsional shear apparatus:

- A double-burette pipe for water injection (upstream) and a pressurized drainage tank with a standpipe (downstream) were newly attached to a hollow torsional shear apparatus to enable water injection into a specimen under the backpressure condition.
- Until the excess pore water pressure ratio increased to approximately 0.9, the hydraulic conductivity was similar to the value after consolidation (plots of water pressure ratio = 0). However, as the cyclic shear progressed and the excess pore water pressure ratio increased above 0.9, the hydraulic conductivity increased rapidly.
- These results were obtained as a common trend, regardless of the relative density, and all plots seemed to be univocally on a single curve for each soil. In addition, even in terms of the cumulative shear strain, the hydraulic conductivity shows a sharp increase, and the degree of increase cannot be distinguished by the relative density.
- The permeability tests using the hollow torsional shear apparatus may be affected by the hydraulic losses inherent to the apparatus itself. Therefore, the hydraulic conductivity measured by the modified hollow torsional shear apparatus (immediately after consolidation) did not match the value obtained using the general permeability test. This point needs to be noted and is an issue for a future study.

Acknowledgments. The authors express their sincere gratitude to the Japan Society for the Promotion of Science (JSPS, KAKENHI Grant Number 21H01423) for the financial support of this study.

References

1. Sento, N., Kazama, M., Uzuoka, R., Ohmura, H., Ishimaru, M.: Possibility of postliquefaction flow failure due to seepage. J. Geotech. Geoenviron. Eng. **130**(7), 707–716 (2004)
2. Kazama, M., Kamura, A., Kim, J., Han, J.: Liquefaction research on anti-catastrophic strategies for external forces exceeding the design seismic level. In: Proceedings of the 17th World Conference on Earthquake Engineering, 17WCEE, 2b-0185 (2020)
3. Ueng, T.S., Wang, Z., Chu, M., Ge, L.: Laboratory tests for permeability of sand during liquefaction. Soil Dyn. Earthq. Eng. **100**, 249–256 (2017)
4. Watanabe, Y., Unno, T., Kamura, A.: Variation of hydraulic conductivity in sandy soil due to undrained cyclic shear loading. J. Jpn. Soc. Civil Eng. **77**, 386–391 (2021). (in Japanese)
5. Japanese Industrial Standards: JIS A 1218, Test methods for permeability of saturated soils (2009). (in Japanese)
6. Daniel, D.E., Anderson, D.C., Boynton, S.S.: Fixed-wall versus flexible-wall permeameters. Hydraul. Barriers Soil Rock **874**, 107–126 (1985)
7. Black, D.K., Lee, K.L.: Saturating laboratory samples by back-pressure. ASCE J. Soil Mech. Found. Eng. Div. **99**, 75–93 (1973)

Framework and Demonstration of Constitutive Model Calibration for Liquefaction Simulation of Densified Sand

Hao Wang, Armin W. Stuedlein(✉) ⓘ, and Arijit Sinha ⓘ

Oregon State University, Corvallis, OR 97331, USA
`armin.stuedlein@oregonstate.edu`

Abstract. Two distinct mechanisms contribute to the increase in liquefaction resistance of densified granular soils: (1) the increased relative density, D_r, and the corresponding peak friction angle, ϕ'_p, and (2) the increased mean effective stress, p', resulting from vibro-compaction-induced lateral stresses. This paper presents a framework for simulating liquefaction phenomena in densified soil that accounts for the two mechanisms. The proposed framework is implemented in the *PDMY03* constitutive model calibrated to both laboratory tests and earthquake motions. The direct simple shear tests provide reference *CRR-N* curves at selected D_r of reconstituted, normally-consoliadated specimens, whereas prior experimental data is used to capture the effects of increased lateral stresses. The selected constitutive model is then calibrated to provide the correct cyclic resistance for the number of equivalent cycles inherent within selected ground motions, D_r, and coefficient of earth pressure at rest, K_0. The calibrated *PDMY03* constitutive model is then used within free-field nonlinear dynamic analyses to investigate the effect of densification on the free-field response of simple liquefiable soil profiles.

Keywords: Densification · Liquefaction · Cyclic laboratory tests · Equivalent number of cycles

1 Introduction

Whether using simplified methods or nonlinear dynamic analyses (NDAs), understanding the magnitude of cyclic shearing resistance mobilized, or exceeded, under cyclic shearing stresses produced by earthquake motions is essential for estimation of the consequences of liquefaction [1–4]. Such evaluations are particularly important when considering expensive ground improvements (e.g., densification and reinforcement). In general, the magnitude of at-rest lateral effective stress (expressed for example through the coefficient K_0) within a liquefiable soil deposit is not directly considered in simplified liquefaction triggering models, but is rather indirectly captured through penetration resistance. However, correlations to penetration test results, often based on calibration chamber test data [5], may not be able to adequately capture K_0 effects [6] when used to estimate post-densification relative densities. This may be of critical importance, given the recognition that densification-type ground improvements, such as vibro-compaction, can increase the lateral effective stress in contractive liquefiable soils [7–9], providing further improvement in cyclic resistance [10].

Densification of granular soils provides two distinct mechanisms for improving liquefaction resistance: (1) increasing the relative density, D_r, with the aim to change the state of liquefiable soils from contractive to dilative, with a corresponding increase in the peak friction angle, ϕ'_p, and (2) increasing the mean effective stress, p'. The former mechanism implies an increase in the failure envelope and thus a greater number of loading cycles is required to encounter the failure envelope in $q-p'$ space. The change in state however, provides the largest benefit by reducing shear strain potential and producing a cyclic mobility-type hysteretic response. The latter mechanism also contributes to a change in state, slightly offsetting the gains from the increased relative density, but will result in a greater requisite number of cycles to liquefaction relative to a normally-consolidated stress state. Although both mechanisms of improvement have been identified by the profession, the incorporation of the post-improvement stress state has received significantly less attention and is not typically accounted for in the design of ground improvement. Harada et al. [11] provides a detailed summary of the effects of increased K_0 on cyclic resistance for use with liquefaction triggering models; however, numerical simulations quantifying this effect on such aspects such as amplification and excess pore pressure generation have not been conducted.

This study investigates the effect of the densification-induced increases in relative density presumed to occur following vibro-compaction, through a laboratory cyclic testing program conducted on Hollywood beach sand, and K_0, through previously-reported experimental data, on liquefaction resistance and develops framework to quantify and calibrate constitutive models to capture the role of these effects during NDA-based liquefaction studies.

2 Laboratory Investigation on Liquefaction Potential

The main experimental basis for establishing the baseline cyclic resistance of unimproved, medium dense and dense sand from which to estimate the cyclic resistance of improved sand rests on cyclic direct simple shear (DSS) tests on reconstituted specimens of Hollywood beach (HB) sand. HB sand was retrieved from a well-described densification test site [8, 12, 13] in South Carolina and was used to reconstitute DSS test specimens. HB sand is characterized as subangular to subrounded uniformly-graded fine sand (SP) with a coefficient uniformity, $C_u = 1.51$, specific gravity, $G_s = 2.64$, minimum and maximum void ratios of $e_{min} = 0.66$ and $e_{max} = 1.06$, and a constant volume friction angle, $\phi'_{cv} = 33.4°$ [8, 12, 14, 15]. Dry specimens were produced by air-pluviation [4, 16–18] into a membrane-lined stack of thin shear rings to achieve a total height of 20.85 mm and an inner diameter of 72.47 mm; the use of different pluviation heights and rates allowed the production of consistent pre-consolidation relative densities, D_r, that achieved average $D_r = 48\%$, 60% and 74% after consolidation.

Prepared specimens were consolidated under the normal effective consolidation stress, σ'_{vc}, of 100 kPa prior to cyclic shearing. Thereafter, stress-controlled, constant-volume cyclic DSS tests were performed using uniform sinusoidal loading cycles, N, with a frequency of 0.1 Hz. Various cyclic stress ratios, $CSRs = \tau_{cyc}/\sigma'_{v0}$, were applied to specimens to produce the corresponding *CRR-N* relationship for a given D_r. Cyclic failure (i.e., liquefaction) was defined as $N_{\gamma=3\%}$ corresponding to the single amplitude shear strain, $\gamma = 3\%$ [4]. Figure 1 presents examples of the cyclic response for the case of $D_r = 74\%$, and *CSRs* of 0.175 and 0.230. In the first cycle of loading, the specimens experience 0.52% and 1.24% peak shear strain for *CSRs* of 0.175 and 0.230, respectively, as shown in Fig. 1a, corresponding to peak excess pore pressure ratios, r_u, of 0.15 and 0.42, respectively. Loading continued and resulted in cyclic failure and the number of cycles to liquefaction, $N_{\gamma=3\%}$, of 17.8 and 1.7 cycles for $CSR = 0.175$ and 0.230, respectively (Fig. 1d). Figure 1b presents the cyclic stress path and indicates a reduction in the vertical effective stress ratio from unity to approximately zero as a 3% single amplitude shear strain is reached and exceeded, as shown in Fig. 1c. A relatively uniform and more gradual increase in shear strain is observed for the specimen subjected to the smaller *CSR* of 0.175, resulting in a larger number of uniform cycles (i.e., $N_{\gamma=3\%} = 17.8$) for cyclic failure (Fig. 1c). Observed $r_u = 0.93$ and 0.98 corresponded $\gamma = 3\%$ for $CSR = 0.175$ and 0.230, respectively, slightly smaller than 1.0 often associated with "full" or instantaneous liquefaction.

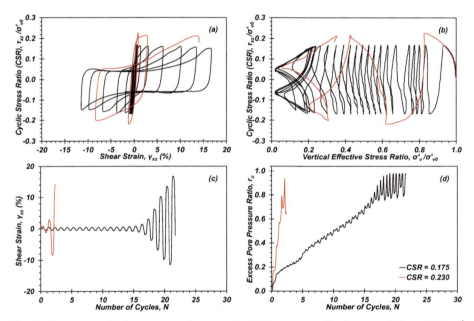

Fig. 1. Stress-controlled, constant-volume, cyclic DSS responses for the case of $D_r = 74\%$, $\sigma'_{vc} = 100$ kPa, and *CSRs* of 0.175 ($N_{\gamma=3\%} = 17.8$) and 0.230 ($N_{\gamma=3\%} = 1.7$), including: (a) shear stress-strain hysteresis, (b) effective stress path, (c) variation of γ with N, and (d) variation of r_u with N.

Figure 2a presents the CRR-$N_{\gamma=3\%}$ relationship for each target D_r considered. Naturally, specimens with a higher D_r required a greater number of cycles for liquefaction as compared to those with a lower D_r (e.g., $N_{\gamma=3\%} = 5.8$ for $D_r = 48\%$ compared to $N_{\gamma=3\%}$

$= 17.8$ for $D_r = 74\%$ for $CSR = 0.175$), which agrees with previously reported observations [2–4, 10, 19–22]. The CRR-N relationships are interpreted using a power-law in the form [4]:

$$CRR = aN^{-b} \qquad (1)$$

where, a and b are the fitted coefficient and exponent, respectively, determined using ordinary least squares (OLS) regression. A constant exponent of $b = 0.113$ was determined suitable to represent the curvature of the CRR-N relationship for normally-consolidated HB sand for all of the D_r considered, similar to that reported by Xiao et al. [23] for the cyclic responses of calcareous sand. However, the coefficient a, which controls the magnitude of CRR at $N = 1$, varies with D_r, as shown in Fig. 2b. The relationship is approximately linear, and using OLS regression, may be represented as:

$$a = 0.0013 D_r + 0.1533 \qquad (2)$$

Use of Eq. (2) implies that the sand is normally-consolidated (NC), with the corresponding NC K_0 which does not represent post-densification p', irrespective of D_r.

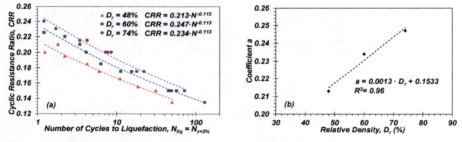

Fig. 2. Cyclic DSS results: (a) in terms of the cyclic resistance ratio, CRR, versus the number of cycles to liquefaction, $N_{\gamma=3\%}$, for specimens with average $D_r = 48\%$, 60% and 74%; and (b) variation of coefficient a with D_r.

3 Proposed Framework for Calibration of a 3D Nonlinear-Inelastic Constitutive Model for Densified Soils

The cyclic nonlinear characteristics of liquefiable sand are modeled using the pressure-dependent multi-yield surface PDMY03 constitutive model implemented in OpenSees [24]. An alternative calibration procedure is proposed for the calibration procedures provided by the PDMY03 developers, which were not developed or intended for use with densified sand. The new model calibration procedure is capable of capturing the effect of increases in D_r and K_0 on the combined cyclic resistance characteristics from the laboratory experiments and the number of uniform loading cycles associated with given earthquake motions.

3.1 Effect of Relative Density on Hollywood Beach Sand *CRR-N* Curve

The effect of relative density on the *CRR-N* curves is quantified by a coefficient a_{D_r}, which is linearly interpolated from correlation between coefficient a and D_r as expressed in Eq. (2). The corresponding increase in peak effective friction angle, ϕ'_p, due to sand densification is also accounted in the calibration of *PDMY03* model through use of the monotonic DSS test-based results for HB sand reported by Rauthause et al. [15]:

$$\phi'_p = 30.9(1 + 0.003D_r) - 3.2(1 + 0.003D_r)\ln(\sigma'_{vc}/P_a) \tag{3}$$

where P_a = atmospheric pressure (i.e., 101.3 kPa).

3.2 Effect of K_0 on *CRR-N* Curve for Silica Sands

The other mechanism contributing to the increased liquefaction resistance due to the effects of densification is attributed to the increase in p', as observed through changes in K_0 reported for example by Massarsch et al. [9]. Ishihara et al. [10] illustrated the effect of K_0 on *CRR-N* curves by conducting laterally-confined, strain-controlled, cyclic hollow, cylinder torsional shear tests on Fuji River sand ($C_u = 3.2$ and $D_r = 55\%$) with K_0 ranging from 0.5 to 1.5, and quantified the role of increased p' due to increases K_0 in the form [18]:

$$p' = (1 + 2K_0)/3 \cdot \sigma'_v \tag{4}$$

Fitting these experimental results to a power-law facilitated representation of the cyclic resistance as:

$$CRR = cN^{-d} \tag{5}$$

where, c and d are the coefficient and exponent, respectively, determined using ordinary least squares (OLS) regression. A constant exponent $d = 0.263$ is determined suitable to represent the curvature of the *CRR-N* relationship for each K_0 ranging from 0.5 to 1.5. The cyclic resistance for a given K_0 may therefore be obtained using linear interpolation of the coefficient c, which controls *CRR* magnitude at $N = 1$ to facilitate quantification of the effect of increased lateral stresses following densification, $K_{0,post}$, and the cyclic resistance. The effect of $K_{0,post}$ on liquefaction resistance may then be quantified in the form of a K_0 adjustment factor, a_{K_0}, to be applied to the representative unimproved *CRR-N* curve and given by:

$$a_{K_0} = \frac{CRR_{K_{0,post}}}{CRR_{K_{0,pre}}} \tag{6}$$

where, $CRR_{K_{0,post}}$ and $CRR_{K_{0,pre}}$ is the cyclic resistance for the K_0 conditions prior to and following sand densification, respectively. Use of Eq. (6) assumes that the relationship between K_0 and *CRR* for Fuji River sand is similar to that of HB sand; thus this approach may be considered tentative until further experimental work clarifies the effect of the governing variables on the K_0 effect.

3.3 Proposed *CRR* Modification

The increase in liquefaction resistance due to the increase in D_r, ϕ'_p, and p' has been established. The combined effect in liquefaction resistance, CRR_{post}, at a specific D_r of densified HB sand (i.e., $K_{0,post}$ condition) can then be expressed as:

$$CRR_{post} = a_{D_r} a_{K_0} \cdot N^{-0.11} \qquad (7)$$

Boulanger and Idriss [25] note that reconstituted and natural sands exhibit an increase in exponent *b* with D_r; Eq. (7) does not capture this effect for HB sand at this time. Future studies may be warranted to further improve the proposed framework for establishing the *CRR* for densified sands.

3.4 Role of Equivalent Cycles of Ground Motion Calibration

Use of ground motions, each characterized with their own equivalent number of cycles, N_{eq}, in NDAs considering liquefaction phenomena requires that the constitutive model for the liquefiable soil will provide the correct cyclic resistance at cyclic failure. In a larger overall study conducted by the authors, the *PDMY03* model was calibrated to 10 earthquake motion pairs selected from a PEER database [26] of 40 motion pairs representing crustal, strike-slip earthquakes with magnitude, $M_w = 7.0$, and rupture distance, $R_{rup} = 10$ km, at "rock site" conditions, by following Jayaram et al. [27] to capture the mean and dispersion of the target response spectrum of the 40 motions. Figure 3a presents the individual and median *RotD50* response spectra of the 10 motions, which exhibit a range in *PGA* of 0.084g to 0.437g.

The N_{eq} for each of the transient earthquake motion components is obtained by following the conversion procedure proposed by Seed et al. [2]. Due to the range in N_{eq} representing the 20 individual motions (Fig. 3a), three representative N_{eq} are selected to represent three bins of motions. Figure 3b presents the cumulative distribution of N_{eq} and the three bins of motions, represented using the average $\overline{N}_{eq} = 5.8$, 12.6, and 19.2. Due to space limitations, only the simulation results using motions M9(90), M4(N), and M2(180) and M2_02 are explored below (Table 1), which are characterized by $N_{eq} = 5.9$, 13.0, and 19.4, respectively.

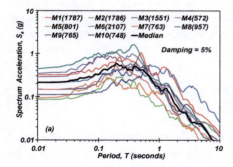 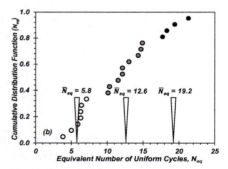

Fig. 3. Characteristics of the 10 selected earthquake motions from the PEER database [26] for: (a) the *RotD50* response spectra, (b) cumulative distribution function for the binned $\overline{N}_{eq}$. Note that the numbers in parenthesis correspond to the database motion [26].

3.5 Calibration of *PDMY03* for Numerical Simulation of Densified Sand

The baseline calibration of the *PDMY03* constitutive model is carried out by matching the output of numerically-simulated cyclic DSS tests to the results of the experimental cyclic DSS tests with $D_r = 48\%$ for the three "bins" or characteristic, average N_{eq} representing the 20 ground motions. Figure 4 compares the DSS test simulations generated using the calibrated *PDMY03* model to the experimental cyclic DSS test results with $D_r = 48\%$ for $\overline{N} = 5.8$. The calibration was performed by attempting to match the: (a) hysteretic shear stress-strain response, (b) effective stress path, (c) variation in the excess pore pressure, r_u, with shear strain and (d) r_u with N, and (e) the CRR-$N_{\gamma=3\%}$ at the average N_{eq} for the bin considered. The latter criterion is necessary owing to the differences in curvature between the power laws describing the cyclic DSS test specimens and the *PDMY03* model, which had originally been calibrated by Khosravifar et al. [24] to produce a power-law exponent of $b \approx 0.34$ based on the recommendations for clean sand (i.e., $FC \leq 5$; [4]). Good matches are observed between the numerical calibrations and the experimental cyclic DSS test results, indicating that the constitutive model is capable of reasonably capturing the shear stress-strain responses, especially at liquefaction triggering state with 3% of shear strain. The agreement between effective stress paths for the experiments and simulations indicates that the *PDMY03* model can accurately represent the contraction rate of HB sand at certain *CSR* levels. Note that the generation of r_u simulated by the *PDMY03* model appears to consistently produce $r_u = 80\%$ at a shear strain equal to

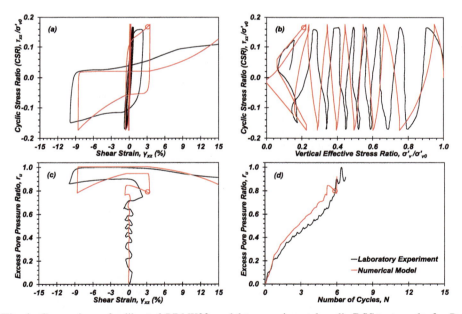

Fig. 4. Comparison of calibrated *PDMY03* model to experimental cyclic DSS test results for $D_r = 48\%$, $CSR = 0.175$, and $\overline{N} = 5.8$, in terms of: (a) hysteretic shear stress-strain response, (b) effective stress path, (c) excess pore pressure, r_u, versus shear strain, (d) excess pore pressure, r_u, versus the number of cycles, N.

3% regardless of the N, and continuing increases in r_u are observed as shear strains accumulate beyond 3% (e.g., Fig. 4c).

Figure 5a compares the calibrated *PDMY03* model results in terms of the *CRR-N* curves for each $\overline{N}$ and $D_r = 48\%$ and $K_0 = 0.42$ to that derived from the corresponding experimental cyclic DSS test results. The corresponding power-law model parameters from these calibrated *CRR-N* curves are $a = 0.308, 0.364$, and 0.418, and $b = 0.318$, 0.327, and 0.343 for $\overline{N} = 5.8, 12.6$, and 19.2, respectively. Although the experimental and *PDMY03*-derived *CRR-N* curves do not generally agree with one another, the calibrated model produces the correct *CRR* at the target $\overline{N}$; thus, when the calibrated model parameters are used in conjunction with ground motions sharing a similar N_{eq}, the simulated material will exhibit the appropriate cyclic response and liquefaction resistance.

Figure 5b presents the calibrated *PDMY03* model results in terms of the *CRR-N* curve at each $\overline{N}$ for densified sand with $D_r = 68\%$ and K_0 assumed equal to 0.73 following a review of the literature. The post-densification CRR_{post}-N curves, expressed in Eq. (7), are obtained by adjusting the reference laboratory NC *CRR-N* curve for $D_r = 48\%$ using a_{D_r} and a_{K_0}. Calibration of *PDMY03* additionally captures the effect increased D_r observed from the cyclic laboratory test program (e.g., contraction and dilation parameters) and the K_0 effect through the adjustment of the bulk modulus.

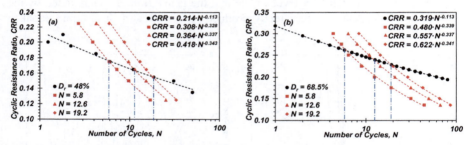

Fig. 5. Comparison of *CRR-N* curves for the calibrated *PDMY03* model with $\overline{N} = 5.8, 12.6$ and 19.6 cycles to: (a) the experimental cyclic DSS test-based *CRR-N* curve for $D_r = 48\%$, (b) densified sand *CRR-N* curve for $D_r = 68\%$.

4 Implementation of the Calibrated *PDMY03* Model

The implementation of the calibrated *PDMY03* model is carried out by the simulations of 3D free-field simulations of unimproved and vibro-compaction improved soil profiles in *OpenSees*. The simulations are conducted using a half unit cell (for comparison to results in the broader study, not described herein) and considered different liquefiable soil thicknesses, $T = 10$ and 22 m, and $D_r = 48, 58$ and 68% corresponding to $K_0 = 0.42, 0.59$, and 0.73, respectively, the latter of which are assumed to represent the vibro-compacted liquefiable layer. The liquefiable sand layer is overlain by a 1 m thick, and underlain by a 3 m thick, dense sand layer with NC sand and $D_r = 80\%$. The unimproved NC liquefiable sand (e.g., $D_r = 48\%$) serves as the baseline reference for the densified, and thus overconsolidated, liquefiable soil layer cases. The groundwater table is located at 1 m below the ground surface. The liquefiable layer was specified with depth- and relative density-dependent shear wave velocities and soil densities. The numerical domain is discretized using prismatic, 0.25 m × 0.125 m (vertical:horizontal), 8-node hexahedral BrickUP elements which facilitates simulation of excess pore pressure generation. For brevity, only the results obtained using three ground motions with widely varying characteristics, summarized in Table 1 each with N_{eq} closely matching the corresponding binned $\overline{N}_{eq}$, are described herein.

Table 1. Summary of characteristics of ground motions investigated.

Motion (component) (Fig. 3)	NGA record sequence number [26]	PGA (g)	Equivalent number of cycles, N	Arias intensity, I_A (m/s)
M2(180)	1786	0.080	19.4	0.14
M4(N)	572	0.142	13.0	0.38
M9(90)	765	0.473	5.9	1.69

Figure 6 presents examples of the simulations in terms of amplification (Figs. 6a and 6b: ratio of input to surface spectral acceleration) and excess pore pressure time histories (Figs. 6c–6g). In general, simulations for the densified case indicated that the PGA (relative to the unimproved, $D_r = 48\%$ case) increased by 2 to 40% for $T = 10$ and 22 m, respectively. Considering period-dependent changes in amplification, densification (and thus stiffening) of the liquefiable layer resulted in a reduction in the period where motions were amplified over short and medium periods, perhaps most readily-observable for Motion M2(90) in Figs. 6a and 6b, and large deamplification in the long periods. These results tentatively suggest that inertial structural loading may be anticipated to increase in stiffer structures, and decrease in flexible structures, following densification. On the other hand, peak ground displacements (not shown hereing) reduce by 9 to 38% and 4 to 32% for $T = 10$ and 22 m, respectively. Thus there is a tradeoff between the reduced seismic and post-seismic displacements provided by densification and the increased potential for inertial loading.

Figures 6c–6g show that: (1) the maximum r_u generated increases with Arias Intensity of the input motion, (2) the maximum r_u during a given motion reduces with depth, which is most apparent for the weakest motion (i.e., M2(90)), (3) the maximum r_u during any given motion reduces with increasing magnitudes of densification, (4) the duration of sustained maximum r_u decreases with increasing densification, and (5) the duration of sustained excess pore pressures increases with liquefiable soil thickness due to the longer drainage path. Furthermore, the benefit of densification, in terms of excess pore pressure generation, appears smallest for the weakest motion; in otherwords, the cyclic resistance available must be mobilized and phase transformation achieved, to reap substantial benefits of densification, similar field observations reported by Gianella and Stuedlein [12]. The examples presented here suggest that additional study of the effects of densification on ground response analysis and liquefaction triggering behavior will help to shed light on the seismic performance of improved ground.

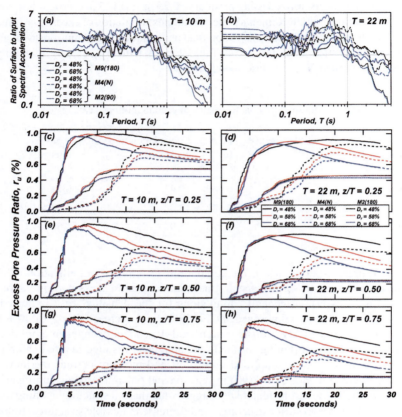

Fig. 6. Example simulation results from the application of the calibrated *PDMY03* model including the ratio of surface and input spectral acceleration for (a) T = 10 m, (b) T = 20 m, and r_u time histories for (c) depth, z, of 2.5, $T = 10$, (d) $z = 6$ m, $T = 22$ m, (e) $z = 5$ m, $T = 10$ m, (f) $z = 11$ m, $T = 22$ m, (g) $z = 7.5$ m, $T = 10$ m, and (h) $z = 16$ m, $T = 22$ m.

5 Concluding Remarks

This paper presents a framework to capture the effects of increased relative density and lateral stresses following densification of liquefiable sand on the seismic response of vibro-compaction improved soil profiles. A laboratory cyclic testing program conducted on Hollywood beach sand was used as the basis to capture the cyclic resistance of NC sand at varying relative densities, whereas the effect of increased lateral stresses was captured through previously-reported experimental data. A methodology was proposed to estimate the post-densification cyclic resistance ratio for a given number of loading cycles was developed, and the 3D nonlinear-inelastic PDMY03 constitutive model was calibrated to match the anticpated cyclic resistance for the number of equivalent cycles associated with selected earthquake motions. Example simulations of simple liquefiable soil profiles illustrated how amplification, peak ground displacement, and excess pore pressure generation varied with earthquake intensity, liquefiable soil layer thickness, and the magnitude of densification achieved. Although many aspects of the seismic performance of improved ground remain to be explored, the framework proposed herein may provide a fruitful avenue for capturing the effects of increased relative density and lateral stress following densification in nonlinear dynamic analyses of densified ground.

Acknowledgements. This research was funded in part by the USDA Agricultural Research Service in cooperation with the Tallwood Design Institute located at Oregon State University under grant #58-0204-6-002 and by the Deep Foundations Institute (DFI) under grant CPF-2018-GRIM-1. The authors would like to thank Amalesh Jana for laboratory assistance and Dr. Long Chen for helpful discussions on numerical model calibrations.

References

1. Seed, H.B., Idriss, I.M.: Simplified procedure for evaluating soil liquefaction potential. J. Geotech. Eng. Div. **97**(9), 1249–1273 (1971)
2. Seed, H.B., Idriss, I.M., Makdisi, F.I., Banerjee, N.: Representation of irregular stress time histories by equivalent uniform stress series in liquefaction analyses. EERC 75-29, Earthquake Engineering Research Center, University of California, Berkeley (1975)
3. Kramer, S.L.: Geotechnical Earthquake Engineering. Prentice Hall, Upper Saddle River (1996)
4. Idriss, I.M., Boulanger, R.W.: Soil liquefaction during earthquakes. Monograph MNO-12, Earthquake Engineering Research Institute (2008)
5. Salgado, R., Mitchell, J.K., Jamiolkowski, M.: Calibration chamber size effects on penetration resistance in sand. J. Geotech. Geoenviron. Eng. **124**(9), 878–888 (1998)
6. Nguyen, T.V., Shao, L.S., Gingery, J., Robertson, P.: Proposed modification to CPT-based liquefaction method for post-vibratory ground improvement. Geo-Congress **2014**, 1120–1132 (2014)
7. Boulanger, R.W., Hayden, R.F.: Aspects of compaction grouting of liquefiable soil. J. Geotech. Eng. **121**(12), 844–855 (1995)
8. Stuedlein, A.W., Gianella, T.N., Canivan, G.: Densification of granular soils using conventional and drained timber displacement piles. J. Geotech. Geoenviron. Eng. **142**, 04016075 (2016). https://doi.org/10.1061/(ASCE)GT.1943-5606
9. Massarsch, K.R., Wersäll, C., Fellenius, B.H.: Horizontal stress increase induced by deep vibratory compaction. In: Proceedings of the Institution of Civil Engineers-Geotechnical Engineering, pp. 1–26 (2019)
10. Ishihara, K., Iwamoto, S., Yasuda, S., Takatsu, H.: Liquefaction of anisotropically consolidated sand. In: Proceedings of Ninth International Conference on Soil Mechanics and Foundation Engineering, Japanese Society of Soil Mechanics and Foundation Engineering, Tokyo, Japan, vol. 2, pp. 261–264 (1977)
11. Harada, K., Orense, R.P., Ishihara, K., Mukai, J.: Lateral stress effects on liquefaction resistance correlations. Bull. N. Z. Soc. Earthq. Eng. **43**(1), 13–23 (2010)
12. Gianella, T.N., Stuedlein, A.W.: Performance of driven displacement pile–improved ground in controlled blasting field tests. J. Geotech. Geoenviron. Eng. **143**(9), 04017047 (2017)

13. Bong, T., Stuedlein, A.W.: Effect of cone penetration conditioning on random field model parameters and impact of spatial variability on liquefaction-induced differential settlements. J. Geotech. Geoenviron. Eng. **144**(5), 04018018 (2018)
14. Gianella, T.N.: Ground improvement and liquefaction mitigation using driven timber piles. Master Thesis, Oregon State University, Civil and Construction Department, Corvallis (2015)
15. Rauthause, M.P., Stuedlein, A.W., Olsen, M.J.: Quantification of surface roughness using laser scanning with application to the frictional resistance of sand-timber pile interfaces. Geotech. Test. J. **43**(4), 966–984 (2020)
16. Ladd, R.S.: Specimen preparation and liquefaction of sands. J. Geotechnical Eng. Div. **100**(10), 1180–1184 (1974)
17. Mulilis, J.P., Seed, H.B., Chan, C.K., Mitchell, J.K., Arulanandan, K.: Effect of sample preparation on sand liquefaction. J. Geotechn. Eng. Div. **103**(GT2), 91–108 (1977)
18. Ishihara, K., Yamazaki, A., Haga, K.: Liquefaction of K_0-consolidated sand under cyclic rotation of principal stress direction with lateral constraint. Soils Found. **25**(4), 63–74 (1985)
19. Seed, H.B., Lee, K.L.: Liquefaction of saturated sands during cyclic loading. J. Soil Mech. Found. Div. **92**, 25–58 (1966)
20. Seed, H.B., Idriss, I.M.: Analysis of soil liquefaction: Niigata earthquake. J. Soil Mech. Found. Div. **93**(SM3), 83–108 (1967)
21. Ishihara, K., Li, S.: Liquefaction of saturated sand in triaxial torsional shear test. Soils Found. **12**(2), 19–39 (1972)
22. Seed, H.B., Idriss, I.M., Arango, I.: Evaluation of liquefaction potential using field performance data. J. Geotech. Eng. Div. **3**(458), 458–482 (1983). https://doi.org/10.1061/(ASCE)0733-9410(1983)109
23. Xiao, P., Liu, H., Xiao, Y., Stuedlein, A.W., Evans, T.M., Jiang, X.: Liquefaction resistance of bio-cemented calcareous sand. Soil Dyn. Earthq. Eng. **107**, 9–19 (2018)
24. Khosravifar, A., Elgamal, A., Lu, J., Li, J.: A 3D model for earthquake-induced liquefaction triggering and post-liquefaction response. Soil Dyn. Earthq. Eng. **110**, 43–52 (2018)
25. Boulanger, R.W., Idriss, I.M.: Magnitude scaling factors in liquefaction triggering procedures. Soil Dyn. Earthq. Eng. **79**, 296–303 (2015)
26. Baker, J.W., Lin, T., Shahi, S.K., Jayaram, N.: New ground motion selection procedures and selected motions for the PEER transportation research program. PEER Technical Report 2011/03, p. 106 (2011)
27. Jayaram, N., Lin, T., Baker, J.W.: A computationally efficient ground-motion selection algorithm for matching a target response spectrum mean and variance. Earthq. Spectra **27**(3), 797–815 (2011)

Three-Dimensional Numerical Simulations of Granular Column Improved Layered Liquefiable Soil Deposit

Zhao Wang[✉], Rui Wang, and Jian-Min Zhang

Department of Hydraulic Engineering, Tsinghua University, Beijing, China
zhao-wan19@mails.tsinghua.edu.cn

Abstract. This paper presents numerical analysis on the mitigation effect of granular columns for layered liquefiable soil deposit. 3D simulations of the centrifuge shaking table tests conducted at the University of Colorado Boulder's 400 g-ton centrifuge facility are using a plasticity constitutive model for large post-liquefaction deformation of sand (CycLiq), via the FLAC^{3D} finite difference code. The simulations investigate the response of a layered liquefiable soil deposit improved by granular columns, under horizontal Kobe earthquake motion input. The results suggest that the constitutive model and FLAC^{3D} simulation method can be used to successfully capture the seismic response of granular columns improved liquefiable ground. Granular columns can significantly enhance drainage, which suppresses the build-up of excess pore pressure and increases the rate of excess pore pressure dissipation. Meanwhile, it also reinforces the ground by reducing the horizontal and vertical displacements.

Keywords: Granular columns · Soil liquefaction · Centrifuge tests simulation · FLAC^{3D}

1 Introduction

Soil liquefaction induced by earthquakes has caused many cases of foundation failure, leading to damage of geotechnical structures such as slopes, embankments and bridges. The installation of improvement measures offer an effective procedure for stabilizing an otherwise potentially liquefiable sand deposit. Hausler and Sitar [1] compiled over 90 case histories on the performance of improved sites from 14 earthquakes in Japan, China, Turkey, and the United States, of which the methods include from conventional densification methods like sand compaction piles to less common lateral restraint-based methods such as sheet pile walls or deep soil mixing grids. Seed and Booker [2] proposed the construction of granular columns as a mitigation strategy to ensure the stabilization of a medium dense sand layer.

Beyond the observations of case histories, several centrifuge tests have been conducted to investigate the mitigation mechanism of granular column for liquefiable soil deposit. Brennan et al. [3] carried out a series of centrifuge tests on level sand beds treated with a group of piles without surface structures, to investigate the effect of drainage. Adalier et al. [4] focused on the overall site stiffening effects rather than the drainage effects. Badanagki et al. [5] investigated how granular columns enhance drainage and shear reinforcement in both level and gentle slopes to achieve liquefaction mitigation and restrict lateral spreading. These results of centrifuge tests have provided excellent basis validation of numerical simulations for further study.

Li et al. [6] performed a two-dimensions simulation of the centrifuge tests conducted by Badanagki et al. [5], in OpenSees, using PDMY02 and SANISAND soil constitutive models. Dinesh et al. [7] simulated sites with various granular column area replacement ratio using PM4Sand constitutive model in FLAC program in plane strain.

This study performs 3D simulation of the centrifuge tests conducted by Badanagki et al. [5], using a plasticity constitutive model for large post-liquefaction deformation of sand (CycLiq) [8], via the FLAC^{3D} finite difference code [9]. The applicability of the constitutive model and the numerical platform for such simulations are assessed via comparisons between numerical and experimental results.

2 Modeling of a Gentle Slope Improved with Granular Columns

2.1 Model Description

The centrifuge test simulated in this study was conducted at the University of Colorado Boulder's (CU) 400 g-ton (5.5 m-radius) centrifuge facility to investigate the influence of granular column improvement on a gently sloping layered liquefiable soil deposit [5], as shown in Fig. 1. A grid of 1.75 m-diameter granular columns were installed in the left half of the ground, separated by 3.5 m (center-to-center) with an area replacement ratio of 20%, which will be taken as Model 1. The other half, taken as Model 2, was left untreated as a benchmark. The properties of sand, silt, and gravel are shown in Table 1.

The numerical simulation of the centrifuge test is performed using the Fast Lagrangian Analysis of Continua in 3 Dimensions (FLAC^{3D}). The nonlinear, elasto-plastic response of soil and columns was modeled with a plasticity constitutive model for large post-liquefaction deformation of sand proposed by Wang et al. [8–11], referred to as CycLiq [8]. Zou et al. [9] implemented the model in FLAC^{3D}, and validated it by simulating undrained cyclic torsional shear tests, triaxial compression tests, cyclic triaxial tests and centrifuge tests.

Figure 1 presents the schematic of the slopes, of which the length, width and height are 66.9 m, 26.3 m and 18 m. The thickness of the loose Ottawa sand is 8 m while that for the top two layers is 2 m [7]. The circular cross section of the column is modified to a 1.55 m-side length square with identical area for simplicity. As shown in Table 2, the model parameters of Ottawa sand were calibrated previously by He et al. [12] and Zou et al. [9], only minor modification was made in this study with respect to the dilatancy parameter $d_{re,1}$. The simulation of a typical clic torsional shear test by Vargas et al. [13] using this set of parameters in FLAC^{3D} is shown in Fig. 2, good agreement is achieved between test and simulation. Relevant element test results of the Monterey Sand and Silt used in the centrifuge test are not available, the model parameters are estimated using those of Toyoura sand, as reported by Zou et al. [9].

Table 1. Properties of sand, silt, and gravel.

Soil (Dr%)	Gs	emax	emin	Permeability (cm/s)
Monterey sand (90%)	2.64	0.84	0.54	0.0529
Silt	2.65	1.35	2.65	0.00003
Loose Ottawa sand (40%)	2.65	0.81	0.53	0.0141
Dense Ottawa sand (90%)	2.65	0.81	0.53	0.0119
Granular columns	1.732	0.92	0.62	2.9

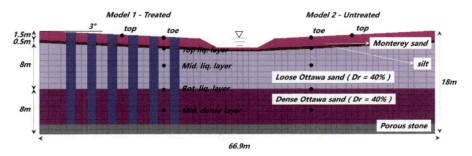

Fig. 1. A cross-sectional view of the model sloping ground treated with granular columns.

Table 2. Model parameters of Ottawa sand and gravel

	G_0	κ	h	$d_{re,1}$	$d_{re,2}$	d_{ir}	α	$\gamma_{d,r}$	n^p	n^d	M	λ_c	e_0	ξ
Ottawa sand	210	0.01	1.5	1.4	40	3.0	30	0.05	2.2	6.0	1.17	0.0112	0.78	0.715
Gravel	360	0.018	0.56	0.55	38	0.95	5.0	0.20	0.5	18.0	1.54	0.018	0.715	0.70

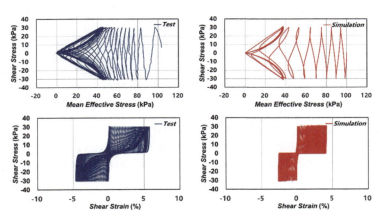

Fig. 2. Calibration of the cyclic torsional shear test with modified model parameters of Ottawa sand

2.2 Simulation Process

The simulation starts with the calculation of the initial stress field by elastoplastic calculation and seepage calculation. Subsequently, the earthquake motion is input during dynamic analysis with fluid-mechanical interaction. The mechanical and fluid boundary conditions are: 1) The base is fixed in the Z-axis direction, while the accelerations are input in the X-axis direction. 2) Considering the flexible-shear-beam (FSB) container was used to model free-field conditions, the nodes with the same height located at the left and right sides of the numerical model are tied, in the direction of the earthquake input. 3) The front and the back of the model are both fixed in the Y-axis direction. 4) All sides except the top surface are undrained, while the pore pressure of the top is 0 kPa.

The input motion at the base is shown in Fig. 3.

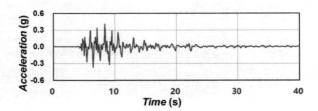

Fig. 3. The modified Kobe earthquake input motion.

3 Results and Discussion

3.1 Influence of Drainage in Granular Columns on Excess Pore Pressure Ratio

Figure 4 compares the computed excess pore pressure ratio (r_u) time histories at various depths in both treated and untreated sides for the test and numerical simulation. The test results show that liquefaction (defined asppened in both sides within the loose layer of Ottawa sand. Compared with the untreated side, the drainage of granular columns resulted in a notable speedup in the dissipation rate of excess pore pressure.

The numerical simulation results show good overall agreement with the recorded time histories of excess pore pressure ratio both during and after the shaking in most of the measurement points. At greater depths in the untreated side, the model slightly underestimates the rate of excess pore pressure dissipation, contributing to greater excess pore pressure ratio after shaking. This may be due to the constant soil permeability adopted in the numerical simulation. Soil permeability has been suggested to increase as excess pore pressure increases [6], which means the hydraulic conductivity (k) tends to increase during shaking. The simulation and test results at the two most shallow points in the treated side show some differences. However, the test results at these two locations does seem somewhat odd.

Figure 5 shows the numerical results of excess pore pressure contour immediately after shaking. Comparison between the two halves of the model clearly show that the granular columns can significantly reduce the excess pore pressure within the liquefiable ground.

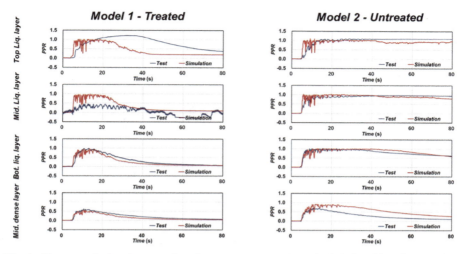

Fig. 4. The numerical and test results of excess pore pressure ratio (r_u) time histories at various depths in both treated and untreated sides.

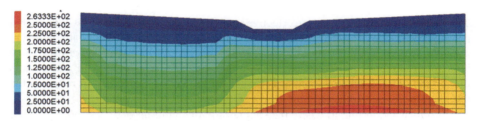

Fig. 5. Contour of excess pore pressure of the model ($t = 40$ s)

3.2 Influence of Granular Columns on Ground Displacement

Figure 6 compares the measured and numerically computed horizontal and vertical displacement time histories at the ground surface on both treated and untreated sides. The horizontal displacement are measured at the toes of two slopes, while the vertical displacements are measured at the top. Notable reduction of horizontal displacements in the treated slope in both test and simulation highlights the effectiveness of granular columns. This was due partly to the added shear resistance and partly to the faster dissipation of excess pore pressures [5]. The test results show a clear decrease in the settlement in the side treated with granular columns ($A_r = 20\%$) [6].

The numerical model underestimates the vertical settlement, especially for the untreated side. This deviation is likely due to the selection of model parameters, as the parameters were adopted from previous studies on Ottawa sand, which can be different from batch to batch [10]. However, the overall influence of granular columns is captured.

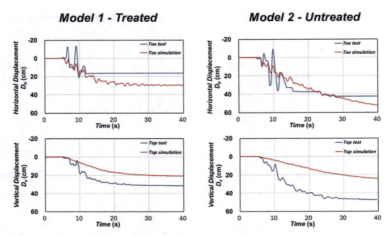

Fig. 6. The numerical and experimental comparisons of displacement time histories at tops and toes on both treated and untreated sides.

3.3 Stress-Strain Analysis

The shear stress-strain curves and effective stress paths in different layers on both treated and untreated sides are shown in Fig. 7 and Fig. 8. Figure 7 indicates that significant shear strain occurs at low mean effective stress values or zero mean effective stress state after initial liquefaction. The shear strain on the treated side is obviously greater than that on the untreated side. Granular columns are shown to be effective in reducing the shear strain in liquefiable soil profile.

As suggested in Fig. 4, where the suppression the build-up of excess pore pressure was indicated, Fig. 8 shows that the installation of granular columns can limit the decrease of effective stress during shaking.

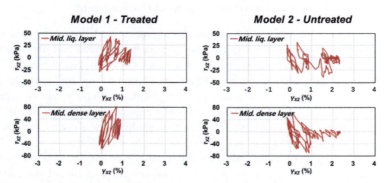

Fig. 7. Comparisons of calculated stress-strain relations ($\gamma_{xz} - \tau_{xz}$) in different layers on both treated and untreated sides.

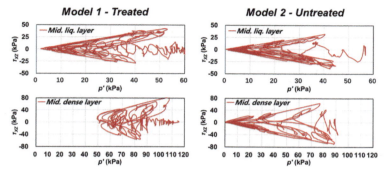

Fig. 8. Comparisons of calculated effective stress paths ($p' - \tau_{xz}$) in different layers on both treated and untreated sides.

4 Conclusions

The numerical simulation of a centrifuge test conducted at the University of Colorado Boulder's 400 g-ton centrifuge facility is reported in this study. A plasticity constitutive model for large post-liquefaction deformation of sand is used, within the FLAC^{3D} finite difference code.

The constitutive model can provide generally good simulation of the excess pore pressure ratio and horizontal and vertical displacements for both treated and untreated ground using a set of previously calibrated model parameters.

The increased excess pore pressure dissipation rate and decreased horizontal and vertical displacements of the treated side suggests the enhancement of drainage. Meanwhile, granular columns can significantly limit shear strain and effective stress decrease in the ground, highlighting the effectiveness of granular columns liquefaction mitigation effects.

References

1. Hausler, E.A., Sitar, N.: Performance of soil improvement techniques in earthquakes. In: 4th International Conference on Recent Advances in Geotechnical Earthquake Engineering and Soil Dynamics, San Diego, USA (2001)
2. Seed, H.B., Booker, J.R.: Stabilization of potentially liquefiable sand deposits using gravel drains. J. Geotech. Eng. Div. **107**(7), 757–768 (1977)
3. Brennan, A.J.: Vertical Drains as a countermeasure to earthquake-induced soil liquefaction. Ph.D. thesis, University of Cambridge, Cambridge (2004)
4. Adalier, K., Elgamal, A., Meneses, J., Baez, J.I.: Stone columns as liquefaction countermeasure in non-plastic silty soils. Soil Dyn. Earthq. Eng. **23**(7), 571–584 (2003)
5. Badanagki, M., Dashti, S., Kirkwood, P.: Influence of dense granular columns on the performance of level and gently sloping liquefiable sites. J. Geotech. Geoenviron. Eng. **144**(9), 04018065 (2018)
6. Li, P., Dashti, S., Badanagki, M., Kirkwood, P.: Evaluating 2D numerical simulations of granular columns in level and gently sloping liquefiable sites using centrifuge experiments. Soil Dyn. Earthq. Eng. **110**, 232–243 (2018)

7. Dinesh, N., Banerjee, S., Rajagopal, K.: Performance evaluation of PM4Sand model for simulation of the liquefaction remedial measures for embankment. Soil Dyn. Earthq. Eng. **152**, 107042 (2022)
8. Wang, R., Zhang, J.-M., Wang, G.: A unified plasticity model for large post-liquefaction shear deformation of sand. Comput. Geotech. **59**, 54–66 (2014)
9. Zou, Y.-X., Zhang, J.-M., Wang, R.: Seismic analysis of stone column improved liquefiable ground using a plasticity model for coarse-grained soil. Comput. Geotech. **125**, 103690 (2020)
10. Zhang, J.M.: Cyclic critical stress state theory of sand with its application to geotechnical problems. Ph.D. thesis, Tokyo Institute of Technology, Tokyo (1997)
11. Zhang, J.-M., Wang, G.: Large post-liquefaction deformation of sand, part I: physical mechanism, constitutive description and numerical algorithm. Acta Geotech. **7**(2), 69–113 (2012)
12. He, B., Zhang, J.-M., Li, W., Wang, R.: Numerical analysis of LEAP centrifuge tests on sloping liquefiable ground: influence of dilatancy and post-liquefaction shear deformation. Soil Dyn. Earthq. Eng. **137**, 106288 (2020)
13. Vargas, R.R., Ueda, K., Uemura, K.: Influence of the relative density and K0 effects in the cyclic response of Ottawa F-65 sand - cyclic Torsional Hollow-Cylinder shear tests for LEAP-ASIA-2019. Soil Dyn. Earthq. Eng. **133**, 106111 (2020)

Fundamental Study on Laboratory Test Method for Setting Parameters of Effective Stress Analysis

Masanori Yamamoto[1(✉)], Ryuichi Ibuki[1], Yasutomo Yamauchi[1], Taku Kanzawa[2], and Jun Izawa[1]

[1] Center for Railway Earthquake Engineer Research, Railway Technical Research Institute, 2-8-35, Hikari-cho, Kokubunji-shi, Tokyo 185-8540, Japan
yamamoto.masanori.21@rtri.or.jp

[2] Infrastructure Maintenance Department, Tokyo Metro Co., Ltd., 3-19-6, Higashi-ueno, Taito-ku, Tokyo 110-0015, Japan

Abstract. The authors of this paper discussed a practical method of setting parameters for an effective stress analysis using results of a laboratory test. As an example of effective stress models, we used GHE-Bowl model, which was composed of GHE model for a skeleton curve and Bowl model for a dilatancy model. The first step of the method is that the parameters of Bowl model are set so that the turning points of the dilatancy curve simulated by Bowl model coincide to those of the test results. The second step is that the parameters of GHE model are determine by fitting deformation properties of the model with those of the test results. In addition, an optimization method was applied for searching parameters, in which the error is minimized using the Shor's algorithm. Furthermore, to verify the validity of the method, we compared the ground response obtained from the effective stress analysis modeled by GHE-Bowl model with that of hybrid ground response analysis. As the results, we demonstrated that the effective stress analysis using the proposed method for setting parameters can give almost appropriate results.

Keywords: Effective stress analysis · Laboratory test method · GHE model · Bowl model · Optimization method

1 Introduction

Soil liquefaction of surface layer would greatly influence seismic stability and restorability of structures. It is, therefore, necessary to evaluate liquefaction potential and determine seismic actions in a seismic design of structures appropriately. F_L method has been used for an assessment of soil liquefaction over years in Japan. The F_L method determines the liquefaction resistant factor, F_L, which is ratio between applied shear stress and liquefaction strength of target layer. According to the F_L values, designers set seismic actions and the ground response in consideration of soil liquefaction. On

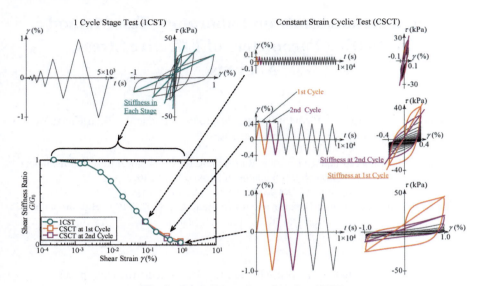

Fig. 1. An outline and results of an 1CST.

the contrary, in recent years, many design standards have shifted to the Performance Based Design, which permits designers to choose more advanced methods for evaluating behavior of the ground and structures. Therefore, effective stress analysis is often used to evaluate the behavior of surface ground liquefaction in seismic design recently. In general, designers determine parameters used in the effective stress analysis so that a liquefaction strength curve, that is, a R-N_C relation, fits the results of undrained cyclic shear tests. However, the detailed setting procedure has not been clearly indicated, and the accuracy of parameters and the results of analysis greatly depends on the judgment of designers. The authors discuss a practical procedure of setting parameters for an effective stress analysis in this paper.

2 Procedure of Setting Parameters for Effective Stress Analysis

In order to evaluate the seismic response of the non-liquefiable surface ground for seismic design, the seismic standards for railway structures in Japan [1] recommend conducting a time-domain nonlinear ground response analysis based on total stress analyses with GHE-S model [2] as nonlinearity of deformation properties of soil. Izawa et al. [3] have proposed a testing method to determine deformation properties of soil for the ground response analysis, which are originally composed of two types of laboratory tests: 1 cycle stage test (1CST) and constant strain cyclic test (CSCT). Figure 1 shows a loading history, τ-γ relationship, and G/G_0-γ relationship of an 1CST and a CSCT, where τ is shear stress; γ, shear strain; G, shear stiffness; and G_0, initial shear stiffness. An 1CST gives G/G_0-γ and h-γ relationship in a wide strain range eliminating the effect of excess pore water pressure, Δu, as much as possible. However, Δu is large at high strain level. To obtained more appropriate deformation properties at high strain level, a few CSCTs are conducted at a few strain levels. For example, 3 CSCTs were conducted in Fig. 1 ($\gamma = 0.1$, 0.4 and 1.0%). By replacing G and h of an 1CST with values of the first cycle of CSCTs at each strain level, appropriate deformation properties can be determined. Furthermore, this testing method gives appropriate deformation properties for earthquakes with long duration by replacing G and h with the values corresponding to the cyclic number obtained by considering the effect of Δu due to cyclic loading. Validity of the proposed method was verified by comparing the results of a dynamic analysis using deformation properties obtained from the proposed method with the results of hybrid ground response analysis (HGRA) [3], which can produce an appropriate ground seismic response of soil layers. Furthermore, the authors confirmed that an 1CST can give almost the same deformation properties of soil with a conventional method. In this paper, we attempt to use results of an 1CST for setting parameters of an effective stress analysis model. Incidentally, we regard the results of the test as pure shear deformation in one-direction because an 1CST is conducted with a torsion shear strain test apparatus under isotropic condition.

As an example of effective stress models, we use GHE-Bowl model, which is composed of GHE model [4] for a skeleton curve and Bowl model [5] for a dilatancy model. In GHE model, the skeleton curve, y-x relationship, is expressed as Eq. (1)

$$y = \frac{x}{\frac{1}{C_1(x)} + \frac{x}{C_2(x)}}, \qquad (1)$$

where, $y = \tau/\tau_f$ and $x = \gamma/\gamma_r$ are normalized stress and strain respectively; τ_f is shear strength of soil; and reference shear strain of soil, γ_r, is defined as Eq. (2)

$$\gamma_r = \frac{\tau_f}{G_0}. \qquad (2)$$

Corrective coefficients, $C_1(x)$ and $C_2(x)$ are obtained from Eqs. (3) and (4)

$$C_1(x) = \frac{C_1(0) + C_1(\infty)}{2} + \frac{C_1(0) - C_1(\infty)}{2} \cos \frac{\pi}{\alpha/x + 1}, \qquad (3)$$

$$C_2(x) = \frac{C_2(0) + C_2(\infty)}{2} + \frac{C_2(0) - C_2(\infty)}{2} \cos \frac{\pi}{\beta/x + 1}, \qquad (4)$$

where, $C_1(0)$, $C_1(\infty)$, α, $C_2(0)$, $C_2(\infty)$ and β are GHE model parameters. h-γ relationship obtained by Eq. (5)

$$h = h_{\max}\left(1 - G/G_0(\gamma)\right)^\kappa, \qquad (5)$$

where, $h_{\max}$ and κ are also GHE model parameters.

In Bowl model, volumetric strain due to shear loading, ε_v^s, is expressed as Eq. (6)

$$\varepsilon_v^s = \varepsilon_\Gamma + \varepsilon_G. \qquad (6)$$

The reversibly dilative component, ε_Γ, and the monotonously compressive component, ε_G, are defined as Eqs. (7) and (8)

$$\varepsilon_\Gamma = A\Gamma^B, \tag{7}$$

$$\varepsilon_G = \frac{G^*}{C + DG^*}, \tag{8}$$

where, A, B, C and D are Bowl parameters.

Assuming that an 1CST is pure shear deformation in one-direction, resultant shear strain, Γ, and cumulative shear strain, G^*, are obtained by Eqs. (9) and (10)

$$\Gamma = \sqrt{\gamma^2} = |\gamma|, \tag{9}$$

$$G^* = \sum \sqrt{\Delta\gamma^2} = \sum |\Delta\gamma|, \tag{10}$$

where, $\Delta\gamma$ is incremental value of γ. On the other hand, the increment of volumetric strain due to consolidation, $d\varepsilon_v^c$, is expressed as Eq. (11)

$$d\varepsilon_v^c = \frac{0.434\,C}{1+e_0} \frac{d\sigma'_m}{\sigma'_m} \quad \begin{cases} \text{for } d\sigma'_m < 0, C = C_s \\ \text{for } d\sigma'_m \geq 0, C = C_c \end{cases}, \tag{11}$$

where, σ'_m, the effective mean confining pressure; $d\sigma'_m$, the increment of σ'_m; C_s and C_c are the swelling and the compression index respectively; and e_0, void ratio of soil. Total volumetric strain, ε_v, can be obtained by Eq. (12).

$$\varepsilon_v = \varepsilon_v^s + \varepsilon_v^c. \tag{12}$$

Since ε_v is zero under undrained condition such as an 1CST, $d\varepsilon_v$ is also zero, and $d\sigma'_m$ is obtained by Eq. (13)

$$d\sigma'_m = \frac{-d\varepsilon_v^s \sigma'_m}{0.434\,C/1+e_0} \quad \begin{cases} \text{for } d\sigma'_m < 0, C = C_s \\ \text{for } d\sigma'_m \geq 0, C = C_c \end{cases}. \tag{13}$$

Excess pore water pressure ratio, $\Delta u/\sigma'_m$, can be obtained by calculating the summation of Eq. (13) at each step.

The procedure for setting parameters of GHE-Bowl model is summarized below using results of an 1CST with *Toyoura* sand with $D_r = 60\%$ as shown in Fig. 2.

1. The parameters of Bowl model are set so that turning points of the $\Delta u/\sigma'_m$-γ curve simulated by Bowl model coincide to those of the test result as shown in Fig. 2(a).
2. The parameters of GHE model are determined by fitting τ-γ and G/G_0-γ relationships of the model with those of the test results (Fig. 2(b)). Since the values of τ and G obtained from the test are affected by Δu in shearing, the effect of Δu is taken into account twice when Bowl model and GHE model are combined with the above parameter settings. Therefore, after τ and G are modified according to excess pore

water pressure, parameters are determined by fitting the modified τ-γ and G/G_0-γ relationships. Here, we assume that shear stiffness is proportional to the 0.5^{th} power of effective confining pressure as indicated in Eq. (14)

$$G/G_0 = \left(1 - \Delta u / \sigma'_{m0}\right)^{0.5}. \tag{14}$$

In the above 2 procedures, an optimization method was applied for searching parameters, in which the error is minimized using the Shor's algorithm [6]. Table 1 and 2 show parameters of GHE model and Bowl model for *Toyoura* sand with $D_r = 60\%$ obtained by optimization methods respectively.

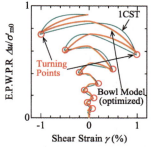

Setting the parameters of Bowl model so that the turning points of the $\Delta u/\sigma'_{m0}$-γ curve

(a) Setting parameters for $\Delta u/\sigma'_{m0}$-γ relationship

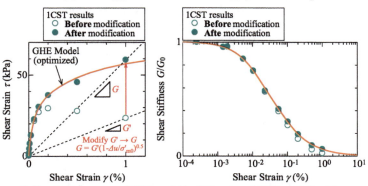

Setting the parameters of GHE model by fitting τ-γ and G/G_0-γ relationships
*At that time, degradation of τ and G due to $\Delta u/\sigma'_{m0}$ are modified.

(b) Setting parameters for τ-γ and G/G_0-γ relationships

Fig. 2. A concept of the proposed method.

An elemental simulation with a Finite Element Method (FEM) was conducted to obtain $\Delta u/\sigma'_m$-γ and τ-γ relationships from GHE-Bowl model. Figures 3(a) and (b) show comparison between results of the FEM analysis and the test results. The results clearly show that the proposed method can give appropriate parameters of GHE-Bowl model.

Table 1. Modified GHE model parameters.

$C_1(0)$	$C_1(\infty)$	$C_2(0)$	$C_2(\infty)$	α	β	h_{max}	κ
1.00	0.26	0.42	1.00	1.34	0.97	0.36	1.1

Table 2. Bowl model parameters.

A	B	C	D	$C_c/(1+e_0)$	$C_s/(1+e_0)$
−2.05	1.40	5.91	24.64	0.0092	0.0090

3 Verification of the Proposed Method

3.1 Model Ground Used for Verification

In order to verify the validity of the proposed method, we compare results of the effective stress analysis to the results of the HGRA [3]. Figure 4 shows soil profile of the model ground used in the analysis. A 4–6 m depth layer was modeled as a liquefiable layer by GHE-Bowl model in the effective stress analysis, and as a hybrid layer by the simple shear test element of *Toyoura* sand with $D_r = 60\%$. The other layers were modeled by GHE-S model based on the nonlinear total stress model and GHE-S standard parameters [2] were used. The level 2 spectrum II earthquake used for the seismic design of Japanese railway structures was applied to all the models. Rayleigh damping ($\alpha = 1.02$, $\beta = 0.00200$) was determined referring to Fukushima et al. [7]. A viscous boundary was applied at the basement ($\rho = 2.0$ g/cm^3, $V_S = 400$ m/s).

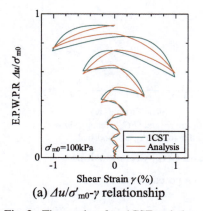
(a) $\Delta u/\sigma'_{m0}$-γ relationship

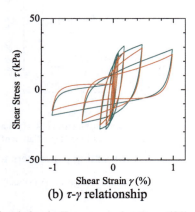
(b) τ-γ relationship

Fig. 3. The results of an 1CST and elemental simulation in *Toyoura* sand at $D_r = 60\%$.

3.2 Comparison of the Results

Figure 5 shows vertical distributions of the maximum response acceleration, displacement, shear stress and shear strain obtained from the HGRA and the effective stress analysis. Almost the same results can be seen in displacement, shear stress and shear strain distributions. On the other hand, the effective stress analysis gave the result that the maximum acceleration in the target layer and shallower is larger than that of HGRA.

Figure 6 and 7 show several responses in the target layer. The maximum acceleration can be attributed to large cyclic mobility that occurs at shear strain of approximately − 3.8%. The response at such a large strain level may not have been simulated accurately, because the parameter setting was implemented within a range of ±1% as shown in Fig. 3. In addition, though an 1CST contained cyclic loading in a small strain level which didn't influence the increase of Δu, $\Delta \gamma$ in a small strain level was counted in G^*. For more accurate parameter setting to reproduce cyclic mobility appropriately, further studies are required such as using the results of the cyclic shear in a larger strain level or a monotonic loading test from small strain to large strain.

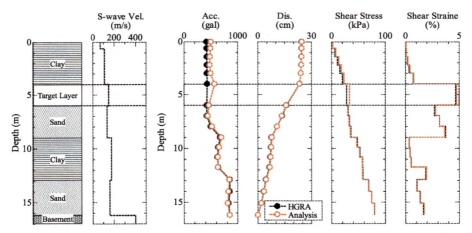

Fig. 4. Model ground for the HGRA and the effective stress analysis.

Fig. 5. Vertical distributions of maximum response obtained from the HGRA and the effective stress analysis.

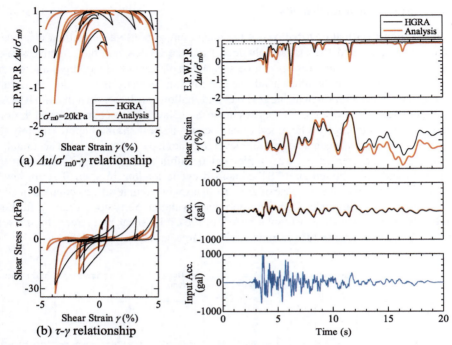

Fig. 6. Hysteresis loops obtained from the HGRA and the effective stress analysis.

Fig. 7. Time history responses obtained from the HGRA and the effective stress analysis.

4 Conclusion

The authors discuss a method of setting parameters for an effective stress analysis using result of 1 cycle stage test. As an example of effective stress models, we used GHE-Bowl model, which was composed of GHE model for a skeleton curve and Bowl model for a dilatancy model respectively. In addition, an optimization method was applied for searching parameters, in which the error is minimized using the Shor's algorithm. Furthermore, to verify the validity of the method, we compared the ground response obtained from the effective stress analysis modeled by GHE-Bowl model with that of hybrid ground response analysis. As the results, the effective stress analysis with GHE-Bowl model whose parameters are obtained by the method can give almost the same results in displacement, shear stress and shear strain in vertical distributions. However, further studies are necessary to produce more accurate simulation of cyclic mobility.

References

1. Railway Technical Research Institute: The Design Standards for Railway Structures and Commentary (Seismic Design), Supervised by Ministry of Land, Infrastructure and Transportation, Maruzen, Japan (2012). (in Japanese)

2. Nogami, Y., Murono, Y., Morikawa, H.: Nonlinear hysteresis model taking into account S-shape hysteresis loop and its standard parameters. In: Proceedings of 15th World Conference on Earthquake Engineering (2012)
3. Izawa, J., Suzuki, A., Toyooka, A., Kojima, K., Murono, Y.: Deformation properties of soils for a nonlinear dynamic response analysis. In: 7th International Conference on Geotechnical Earthquake Engineering (2019)
4. Tatsuoka, F., Shibuya, S.: Deformation characteristics of soils and rocks from field and laboratory tests. In: Proceedings of 9th Asian Regional Conference on Soil Mechanics and Foundation Engineering (1992)
5. Fukutake, K., Ohtsuki, A., Sato, M., Shamoto, Y.: Analysis of saturated dense sand-structure system and comparison with results from shaking table test. Earthq. Eng. Struct. Dyn. **19**(7), 977–992 (1990)
6. Kappel, F., Kuntsevich, A.V.: An implementation of Shor's r-algorithm. Comput. Optim. Appl. **15**(2), 193–205 (2000)
7. Fukushima, Y., Midorikawa, S.: Evaluation of site amplification factors based on average characteristics of frequency dependent Q^{-1} of sedimentary strata. J. Struct. Constr. Eng. (Trans. AIJ) **59**(460), 37–46 (1994). (in Japanese)

Experimental and Numerical Study of Rate Effect in Cone Penetration Tests

Jian-Hong Zhang[✉] and Hao Wang

Department of Hydraulic Engineering, State Key Laboratory of Hydroscience and Engineering, Tsinghua University, Beijing 100084, China
cezhangjh@tsinghua.edu.cn

Abstract. This paper investigates the rate effect of cone penetration tip resistance. The focus is on the relationship between the penetration resistance and the penetration rate under partially drained conditions. When the penetration rate is increased from a relatively low rate, soil resistance decreases. Beyond a certain value, any further increase in penetration rate causes an increase in the resistance. Numerical simulation of constant rate and variable rate cone penetration tests were carried out using the two-phase material point method (MPM). The penetration rate at which the calculated resistance started to increase was used to estimate the consolidation coefficient. Centrifuge modelling of variable rate cone penetration tests explained the rate effect in terms of consolidation and viscosity.

Keywords: Rate effect · Cone penetration test · Material point method

1 Introduction

Soil characterization using the pore pressure cone penetration test (CPT) is often difficult with respect to the penetration rate due to the variability of soil behavior caused by differing drainage and consolidation conditions [1]. The effect of the penetration rate is quantitatively expressed as a normalized rate related to consolidation. Drainage conditions during penetration depend on the penetration rate (v), the diameter of the penetrating object (D), and the coefficient of consolidation (c_v). The normalized velocity V, defined as $V = vD/c_v$, is proposed as a parameter for assessing the rate effect.

Dependency of the measured soil resistance on the penetration rate has been studied using the cone penetrometer since early pioneering work [2, 3]. The results show a two-stage trend. When the normalized penetration rate is high (V > 10), a decrease in the penetration rate results in a reduction of soil resistance. A further decrease in the penetration rate causes an increase in soil resistance [4–6].

This paper investigate the rate effect of cone penetration through numerical simulations. Centrifuge model test of CPTu were performed in sand at two penetration rate, i.e. 15 mm/s and 20 mm/s respectively. Two-phase Material Point Method (MPM) formulation were established for numerical simulation of the cone penetration process under axisymmetrical conditions. The results of numerical simulation are compared with centrifuge model tests.

2 Numerical Model

Cone penetration test in saturated soil is a complicated process involving solid–liquid coupling and large deformation. Numerically simulation of CPT process requires a robust numerical framework. The two-phase Material Point Method (MPM) is used in this study with NairnMPM [7], an open-source software dealing with two-dimensional (2D) and three-dimensional simulations. By combining the MPM process and stress update algorithm in specific steps, the USAVG (update stress average) method [8] used in the present work exhibits improved energy conservation. Figure 1 shows the algorithm with MPM.

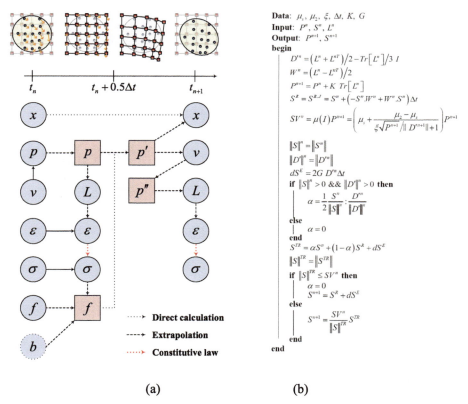

(a)   (b)

Fig. 1. MPM algorithm after [3]: (a) timesteps in USAVG (material point properties shown in circles; node properties in squares); (b) algorithm for updating stress.

As seen in Fig. 1a, at the beginning of a timestep, information (positions, momenta, stresses, strains and forces) is carried by material points. Then in the first half of Δt, the constitutive model calculates resultant forces. The momenta extrapolated to the mesh are updated initially using the forces. In the second half of Δt, particles are looped over to update their positions and velocities. Then, the new particle velocities are extrapolated to the grid to obtain a revised set of nodal momenta. The newly acquired momenta are used to determine the stresses and strains on the particle. Up until this point, all the information has been transferred to the particles. Therefore, at time tn + 1, the mesh is discarded and the model is ready to re-mesh.

Each MPM timestep contains two locations for updating stress (red line in Fig. 3a). The mean principal stress P, deviatoric stress tensor S and velocity gradient L are input at the beginning of the algorithm. An constitutive model proposed by Lin [9] is applied for an explicit time integration. The algorithm is concluded in Fig. 1b. Figure 2 illustrates an axisymmetrical model of the cone in a normally consolidated saturated soil. The diameter of the cone is 100 mm. The numerical model is discretized with four-point elements, consisting of 2752 nodes and 6672 material points. The geotechnical properties of the soil are listed in Table 1.

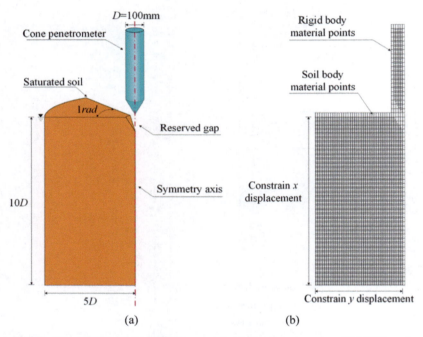

Fig. 2. Numerical model of CPT process simulation: (a) geometry; (b) MPM mesh

Frictional contact is applied at the soil–cone interface [10–12]. The contact surface is detected automatically. The dynamic and static friction coefficients were both empirically assumed to be 0.2. The water level was set at the ground surface of the soil body. Pore pressure was calculated by Biot's theory 13 to update the stress after each MPM step.

Table 1. Soil properties used in numerical simulation.

Parameter	Description	Value
G_s	Specific gravity of soil	2.70
e_0	Void ratio	0.52
E	Young's modulus (MPa)	150
ν	Poisson's ratio	0.30
k	Hydraulic conductivity (m/s)	1.0×10^{-5}
H	Biot's term (MPa)	41.67
Q	Biot's term (MPa)	50.57
μ_2	Maximum value of friction coefficient	0.84
μ_s	Minimum value of friction coefficient	0.70
I_0	Constant about inertial number	0.279

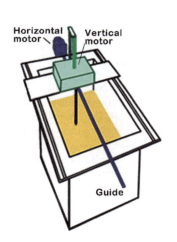

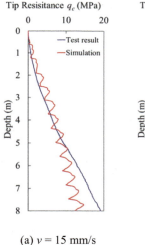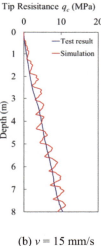

(a) v = 15 mm/s   (b) v = 15 mm/s

Fig. 3. Centrifuge model for CPTu.   **Fig. 4.** Comparison between tests and simulations

3 Results of Numerical Simulation

Two geotechnical centrifuge tests were performed of cone penetration in a saturated sand at a centrifugal acceleration of 50 g. The sand is a clean medium sand with a dry density of 1650 kg/m³. The effective internal friction angle was measured to be 38.5° through drained triaxial tests.

Figure 3 shows the strong box with a cone penetrometer probe fabricated by Thomas Broadbent & Sons Ltd., Huddersfield, UK. Two electric motors are mounted on the strong box, one to push the penetrometer vertically into the sand at set displacement rates of 15 mm/s and 20 mm/s adopted for this study, the other to move the penetrometer arrangement horizontally.

Figure 4 shows the results of the centrifuge tests and the numerical simulation in prototype scale. Within a penetration depth of 6 to 8 m, the numerical model gives quantitative predictions for tip resistance. However, the simulated tip resistance shows a certain fluctuation due to the fact that the grid size is not infinitely small. A slower penetration rate causes a greater measured soil resistance.

4 Conclusions

This paper investigate the rate effect of cone penetration using two-phase Material Point Method. Centrifuge model test of CPTu indicated that the measured soil resistance tended to be greater at a slower penetration rate. The numerical simulation of the cone penetration process under axisymmetrical conditions shows identical trend with that of centrifuge tests. The simulated soil resistance shows a certain fluctuation due to the relatively great grid size.

Acknowledgement. The authors would like to acknowledge financial support from the National Natural Science Foundation of China (No. 52090084 and 51879141).

References

1. Finnie, I.M.S., Randolph, M.F.: Punch-through and liquefaction induced failure of shallow foundations on calcareous sediments. In: Proceedings of the 17th International Conference on the Behaviour of Offshore Structures (BOSS 1994), pp. 217–230. Pergamon Press (1994)
2. Bemben, S.M., Myers, D.A.: The influence of rate of penetration on static cone resistance in Connecticut river valley Varved clay. In: Proceedings of the European Symposium on Penetration Testing (ESOPT), pp. 33–34. National Swedish Building Research (1974)
3. Leroueil, S., Marques, M.E.S.: Importance of strain rate and temperature effects in geotechnical engineering. In: Session on Measuring and Modeling Time Dependent Soil Behavior, ASCE Convention. Geotechnical Special Publication 61, Washington (1996)
4. House, A.R., Oliveira, J.R.M.S., Randolph, M.F.: Evaluating the coefficient of consolidation using penetration tests. Int. J. Phys. Model. Geotech. 1(3), 17–25 (2001). https://doi.org/10.1680/ijpmg.2001.010302
5. Lehane, B.M., O'Loughlin, C.D., Gaudin, C., Randolph, M.F.: Rate effects on penetrometer resistance in kaolin. Géotechnique 59(1), 41–52 (2009). https://doi.org/10.1680/geot.2007.00072
6. Schneider, J.A., Lehane, B.M., Schnaid, F.: Velocity effects on piezocone measurements in normally and over consolidated clays. Int. J. Phys. Model. Geotech. 7(2), 23–34 (2007). https://doi.org/10.1680/ijpmg.2007.070202
7. Oregon State University: The computational mechanics software in the research group of Prof. John A. Nairn (2019). http://osupdocs.forestry.oregonstate.edu
8. Nairn, J.A.: Material point method calculations with explicit cracks. Comput. Model. Eng. Sci. 4(6), 649–664 (2003). https://doi.org/10.3970/cmes.2003.004.649
9. Lin, H.L.: Failure mechanism of submarine slope and solid-fluid unified constitutive model. Ph.D. thesis, Tsinghua University, China (2019). (in Chinese)

10. Bardenhagen, S.G., Guilkey, J.E., Roessig, K.M., Brackbill, J.U., Witzel, W.M., Foster, J.C.: An improved contact algorithm for the material point method and application to stress propagation in granular material. Comput. Model. Eng. Sci. **2**, 509–522 (2001). https://doi.org/10.3970/cmes.2001.002.509
11. Nairn, J.A., Bardenhagen, S.G., Smith, G.D.: Generalized contact and improved frictional heating in the material point method. Comput. Particle Mech. **5**(3), 285–296 (2017). https://doi.org/10.1007/s40571-017-0168-1
12. Robertson, P.K.: Soil classification using the cone penetration test. Can. Geotech. J. **1990**(27), 151–158 (1990). https://doi.org/10.1139/t90-014
13. Biot, M.A.: General theory of three-dimensional consolidation. J. Appl. Phys. **12**, 155–164 (1941). https://doi.org/10.1063/1.1712886

Printed by Printforce, the Netherlands